Progress in Mathematics

Volume 201

Series Editors

H. Bass
J. Oesterlé
A. Weinstein

European Congress of Mathematics

Barcelona, July 10–14, 2000
Volume I

Carles Casacuberta
Rosa Maria Miró-Roig
Joan Verdera
Sebastià Xambó-Descamps
Editors

Birkhäuser Verlag
Basel · Boston · Berlin

Editors:

Carles Casacuberta
Departament de Matemàtiques
Universitat Autònoma de Barcelona
08193 Bellaterra
Spain
casac@mat.uab.es

Rosa Maria Miró-Roig
Departament d'Algebra i Geometria
Facultat de Matemàtiques
Universitat de Barcelona
08007 Barcelona
Spain
miro@mat.ub.es

Joan Verdera
Departament de Matemàtiques
Universitat Autònoma de Barcelona
08193 Bellaterra
Spain
verdera@mat.uab.es

Sebastià Xambó-Descamps
Departament de Matemàtica Aplicada II
Universitat Politècnica de Catalunya
08028 Barcelona
Spain
sebastia.xambo@upc.es

2000 Mathematics Subject Classification 00B25

A CIP catalogue record for this book is available from the Library of Congress, Washington D.C.,
USA

Deutsche Bibliothek Cataloging-in-Publication Data

ECM <3, 2000, Barcelona>:
European Congress of Mathematics : Barcelona, July 10-14, 2000 /
Carles Casacuberta ... ed. – Basel ; Boston ; Berlin : Birkhäuser
    ISBN 3-7643-6419-X

Vol. 1. - (2001)
    (Progress in mathematics ; Vol. 201)
    ISBN 3-7643-6417-3

ISBN 3-7643-6417-3 Birkhäuser Verlag, Basel – Boston – Berlin

© 2001 Birkhäuser Verlag, P.O. Box 133, CH-4010 Basel, Switzerland
Member of the BertelsmannSpringer Publishing Group
Printed on acid-free paper produced of chlorine-free pulp. TCF ∞
Printed in Germany
ISBN 3-7643-6417-3 (Vol. I/PM 201)
ISBN 3-7643-6418-1 (Vol. II/PM 202)
ISBN 3-7643-6419-X (Set)

9 8 7 6 5 4 3 2 1   www.birkhauser.ch

Table of Contents of Volume I

Preface.. xiii

    History and Significance.. xiv

    Committees.. xvi

    Congress Programme.. xvii

    Acknowledgements.. xxi

EMS Prize Winners... xxiii

Felix Klein Prize Winner... xxviii

Opening Ceremony

    Address by Sebastià Xambó... xxix

    Address by Ramón Marimón... xxxiii

    Address by Andreu Mas-Colell...................................... xxxiv

    Address by Joan Clos... xxxv

    Address by Rolf Jeltsch... xxxvii

    Address by Manuel Castellet.. xxxix

Closing Ceremony

    Address by Sebastià Xambó... xlv

    Address by Rolf Jeltsch... xlviii

Articles by Plenary Speakers

Robbert Dijkgraaf
The Mathematics of M-Theory 1

Hans Föllmer
Probabilistic Aspects of Financial Risk 21

Hendrik W. Lenstra, Jr.
Flags and Lattice Basis Reduction 37

vi

Yuri I. Manin
Moduli, Motives, Mirrors 53

Yves Meyer
The Role of Oscillations in Some Nonlinear Problems 75

Carles Simó
New Families of Solutions in N-Body Problems 101

Marie-France Vignéras
*Irreducible Modular Representations of a Reductive p-Adic Group
and Simple Modules for Hecke Algebras* 117

Oleg Viro
Dequantization of Real Algebraic Geometry on Logarithmic Paper 135

Articles by Parallel Speakers

Rudolf Ahlswede
Advances on Extremal Problems in Number Theory and Combinatorics 147

Volker Bach
Spectral Analysis of Nonrelativistic Quantum Electrodynamics 177

Viviane Baladi
Spectrum and Statistical Properties of Chaotic Dynamics 203

Joaquim Bruna
Sampling in Complex and Harmonic Analysis 225

Nicolas Burq
Lower Bounds for Shape Resonance Widths of Schrödinger Operators 247

Xavier Cabré
A Conjecture of De Giorgi on Symmetry for Elliptic Equations in $\mathbb{R}^n$ 259

Peter J. Cameron
The Random Graph Revisited 267

Zoé Chatzidakis
Difference Fields: Model Theory and Applications to Number Theory 275

Ciro Ciliberto
*Geometric Aspects of Polynomial Interpolation in More Variables and
of Waring's Problem* 289

Gianni Dal Maso
The Calibration Method for Free Discontinuity Problems 317

Jan Denef and François Loeser
Geometry on Arc Spaces of Algebraic Varieties 327

Barbara Fantechi
Stacks for Everybody 349

Alexander B. Goncharov
Multiple ζ-Values, Galois Groups, and Geometry of Modular Varieties 361

Alexander Grigor'yan
Heat Kernels on Manifolds, Graphs and Fractals 393

Michael Harris
*Local Langlands Correspondences and Vanishing Cycles on
Shimura Varieties* 407

Renato Iturriaga and Konstantin Khanin
Burgers Turbulence and Dynamical Systems 429

Kurt Johansson
Random Growth and Random Matrices 445

Pekka Koskela
Sobolev Spaces and Quasiconformal Mappings on Metric Spaces 457

Nicholas S. Manton and Bernard M. A. G. Piette
Understanding Skyrmions Using Rational Maps 469

Ieke Moerdijk
Models for the Leaf Space of a Foliation 481

Eric M. Opdam
Multivariable Hypergeometric Functions 491

Thomas Peternell
Contact Structures, Rational Curves and Mori Theory 509

Alexander Reznikov
Analytic Topology 519

Bernhard Schmidt
Towards Ryser's Conjecture 533

Klaus Schmidt
The Dynamics of Algebraic $\mathbb{Z}^d$-Actions 543

Bálint Tóth
Self-Interacting Random Motions 555

Erik van den Ban and Henrik Schlichtkrull
Harmonic Analysis on Reductive Symmetric Spaces 565

Table of Contents of Volume II

Articles by Prize Winners

Semyon Alesker
Classification Results on Valuations on Convex Sets 1

Raphaël Cerf
Towards a Microscopic Theory of Phase Coexistence 9

Dominic Joyce
Constructing Compact 8-Manifolds with Holonomy Spin(7)
from Calabi-Yau Orbifolds 21

Vincent Lafforgue
Banach KK-Theory and the Baum-Connes Conjecture 31

Michael McQuillan
An Introduction to Non-Commutative Mori Theory 47

Stefan Nemirovski
Geometric Methods in Complex Analysis 55

Paul Seidel
Vanishing Cycles and Mutation 65

Wendelin Werner
Critical Exponents, Conformal Invariance and Planar Brownian Motion 87

Mini-Symposium on Computer Algebra

Manuel Bronstein
*Computer Algebra Algorithms for Linear Ordinary
Differential and Difference Equations* 105

Wolfram Decker
Some Introductory Remarks on Computer Algebra 121

Gaston H. Gonnet
*A Study of Iteration Formulas for Root Finding,
Where Mathematics, Computer Algebra and Software Engineering Meet* 143

Laureano González-Vega and Tomás Recio
*Industrial Applications of Computer Algebra:
Climbing Up a Mountain, Going Down a Hill* 157

Gert-Martin Greuel
*Applications of Computer Algebra to Algebraic Geometry,
Singularity Theory and Symbolic-Numerical Solving* 169

x

Mini-Symposium on Curves over Finite Fields and Codes

Noam D. Elkies
Explicit Towers of Drinfeld Modular Curves 189

Arnaldo Garcia
Curves over Finite Fields Attaining the Hasse-Weil Upper Bound 199

Farshid Hajir and Christian Maire
Asymptotically Good Towers of Global Fields 207

Henning Stichtenoth
*Explicit Constructions of Towers of Function Fields
with Many Rational Places* 219

Gerard van der Geer
Curves over Finite Fields and Codes 225

Chaoping Xing
Authentication Codes and Algebraic Curves 239

Mini-Symposium on Free Boundary Problems

Giovanni Bellettini
*Some Aspects of Mean Curvature Flow in Presence of
Nonsmooth Anisotropies* 245

Klaus Deckelnick
A Phase-Field Model for Diffusion-Induced Grain Boundary Motion 255

Irina V. Denisova
Evolution of a Closed Interface between Two Liquids of Different Types 263

Gonzalo Galiano
*Applications of a Local Energy Method to Systems of
PDE's Involving Free Boundaries* 273

Harald Garcke
Phase Boundaries in Alloys with Elastic Misfit 281

Josephus Hulshof
Some Aspects of the Thin Film Equation 291

Regis Monneau
A Brief Overview on the Obstacle Problem 303

Henrik Shahgholian
The Impact of Monotonicity Formulas in Regularity of Free Boundaries 313

José Miguel Urbano
A Free Boundary Problem: Contributions from Modern Analysis 319

Mini-Symposium on Mathematical Finance: Theory and Practice

Tomasz R. Bielecki and Stanley R. Pliska
*Risk Sensitive Control with Applications to Fixed Income
Portfolio Management* 331

M. A. H. Dempster and A. Eswaran
Wavelet Based PDE Valuation of Derivatives 347

Ernst Eberlein and Sebastian Raible
Some Analytic Facts on the Generalized Hyperbolic Model 367

Hélyette Geman
Functionals of Brownian Motion in Path-Dependent Option Valuation 379

Ton Vorst
Optimal Portfolios under a Value at Risk Constraint 391

Mini-Symposium on Quantum Chaology

Eugene Bogomolny
Trace Formulas and Spectral Statistics of Diffractive Systems 399

Monique Combescure and Didier Robert
Semiclassical Results in the Linear Response Theory 413

Jens Marklof
The Berry-Tabor Conjecture 421

Zeév Rudnick
On Quantum Unique Ergodicity for Linear Maps of the Torus 429

Mini-Symposium on Quantum Computing

Nicolas Gisin, Renato Renner and Stefan Wolf
*Bound Information: The Classical Analog to Bound
Quantum Entanglement* 439

Mini-Symposium on String Theory and M-Theory

Michael R. Douglas
D-Branes on Calabi-Yau Manifolds 449

José M. F. Labastida
Knot Invariants and Chern-Simons Theory 467

Marcos Mariño
Topological Quantum Field Theory and Four-Manifolds 479

Christoph Schweigert and Jürgen Fuchs
D-Brane Conformal Field Theory and Bundles of Conformal Blocks 489

Angel M. Uranga
From Quiver Diagrams to Particle Physics 499

xii

Mini-Symposium on Symplectic and Contact Geometry and Hamiltonian Dynamics

Paul Biran
From Symplectic Packing to Algebraic Geometry and Back 507

Yuri Chekanov
New Invariants of Legendrian Knots 525

Hansjörg Geiges
Contact Topology in Dimension Greater than Three 535

Viktor L. Ginzburg
The Hamiltonian Seifert Conjecture: Examples and Open Problems 547

Àngel Jorba and Jordi Villanueva
The Fine Geometry of the Cantor Families of Invariant Tori in Hamiltonian Systems 557

Mikhail B. Sevryuk
Symplectic and Contact Geometry and Hamiltonian Dynamics 565

Vladimir Zakalyukin
Simple Coisotropic Projections and Caustics 575

Mini-Symposium on Wavelet Applications in Signal Processing

Eric Chassande-Mottin and Patrick Flandrin
Reassigned Scalograms and Singularities 583

Peter F. Craigmile, Donald B. Percival and Peter Guttorp
The Impact of Wavelet Coefficient Correlations on Fractionally Differenced Process Estimation 591

Emma J. McCoy
Wavelet-Based Modelling of Persistent Periodicities 601

Rudolf H. Riedi, Vinay J. Ribeiro, Matthew S. Crouse and Richard G. Baraniuk
Network Traffic Modeling Using a Multifractal Wavelet Model 609

Vasily Strela
Denoising Via Block Wiener Filtering in Wavelet Domain 619

Andrew T. Walden
Wavelet Analysis of Discrete Time Series 627

Preface

The Third European Congress of Mathematics (3ecm) took place from July 10th to July 14th, 2000 in Barcelona. It was organised by the Societat Catalana de Matemàtiques (Catalan Mathematical Society), under the auspices of the European Mathematical Society (EMS). As foreseen by the EMS and the International Mathematical Union, this Congress was a major event in World Mathematical Year 2000. In addition to reviewing outstanding research achievements, important aspects of the life of European mathematics were discussed.

Mathematics is undergoing a period of rapid changes. Effective computation and applications in science and technology go ever more hand in hand with conceptual developments. It was one of the aims of 3ecm to reflect this mutual enrichment, while steering present and future trends of mathematical sciences. In fact, the motto of the Congress, *Shaping the 21st Century*, was meant to capture these views.

Nearly 1400 people from 87 countries gathered together in the Palau de Congressos of Barcelona in order to take part in the activities of the 3ecm scientific programme: Nine plenary lectures, thirty invited lectures in parallel sessions, lectures given by EMS prize winners, ten mini-symposia on special topics, seven round tables, poster sessions, presentations of mathematical software and video exhibitions. Twenty events were satellites of 3ecm in various countries.

A special effort was made by 3ecm organisers to promote a broad participation of European mathematicians of all nationalities, in accordance with EMS principles. Thus, about one third of the Congress budget was devoted to grants, allowing funding for 240 participants.

As in previous EMS congresses, ten prizes were given to young researchers, no older than 32 years, who had been selected by a Prize Committee nominated by the EMS. In addition, the Felix Klein Prize was awarded for the first time, jointly by the EMS and the Institute of Industrial Mathematics of Kaiserslautern. The Ferran Sunyer i Balaguer Prize of the Institut d'Estudis Catalans was also awarded during the opening ceremony of the Congress.

Plenary talks and invited parallel talks were addressed to a non-specialised mathematical audience. All the speakers had been selected by a Scientific Committee nominated by the EMS. Mini-symposia were an innovative aspect of 3ecm. Ten emergent topics were chosen by the Scientific Committee. A coordinator was proposed for each of these topics, and entrusted with the responsibility to invite speakers and prepare a schedule. One century after the statement of Hilbert's problem list, it is no longer a one-man task to provide guidelines for mathematical research. Nevertheless, it is hoped that the impact of 3ecm mini-symposia will be influential on the mathematical community for the years to come.

The first volume of these proceedings contains articles written by plenary and parallel speakers. The second volume contains articles written by prize winners and participants at mini-symposia. A third volume, containing material from 3ecm round tables, will be published jointly by Societat Catalana de Matemàtiques and CIMNE (Centre Internacional de Mètodes Numèrics en Enginyeria, Barcelona).

Palau de Congressos de Barcelona

History and Significance

The series of European Congresses of Mathematics was conceived and set on course by Max Karoubi (France), shortly before the creation of the EMS in 1990. The objectives of European Congresses were agreed upon at that time and continued to hold at Paris 1992, Budapest 1996, and Barcelona 2000. The European Congresses intend to disseminate knowledge, stimulate new developments, and foster cooperation among mathematicians, particularly within Europe. They aim not only to present major scientific achievements, but also to be a forum for debate on topics of general interest to the mathematical community at round tables.

The Societat Catalana de Matemàtiques became a member of the EMS in 1992. After a successful bid, it was entrusted with the organisation of the Third European Congress in Mathematics by the EMS Council in 1996. This book is the last and best occasion to thank all those who supported the bid of the SCM, and those who continued to offer their collaboration during the years of preparation of the Congress.

The 3ecm Executive Committee was nominated by the Societat Catalana de Matemàtiques in November 1996. An Organising Committee was nominated in April 1997, consisting of members of Universitat de Barcelona, Universitat Autònoma de Barcelona, Universitat Politècnica de Catalunya and Universitat Pompeu Fabra. A Committee of Honour was also constituted, under the high Presidency of H. M. the King of Spain. Several other committees with specific tasks were later elected. In fact, the 3ecm was made possible by the generous dedication of many people from the Barcelona mathematical community, who devoted a great amount of time, energy and enthusiasm to this event.

3ecm Conference Room

Committees

Committee of Honour

President: S. M. el Rey de España
Molt Honorable President de la Generalitat de Catalunya
Excelentísima Ministra de Educación y Cultura
Excelentísima Ministra de Ciencia y Tecnología
Excel·lentíssim Alcalde de Barcelona
Honorable Conseller d'Universitats, Recerca i Societat de la Informació
Director General de la UNESCO
Excel·lentíssim President de l'Institut d'Estudis Catalans
Excel·lentíssim i Magnífic Rector de la Universitat de Barcelona
Excel·lentíssim i Magnífic Rector de la Universitat Autònoma de Barcelona
Excel·lentíssim i Magnífic Rector de la Universitat Politècnica de Catalunya
President de la Fundació Catalana per a la Recerca

Scientific Committee

Sir Michael Atiyah (Chair), Vladimir I. Arnold, Robert Azencott, Fabrizio Catanese, Ildefonso Díaz, Antti Kupiainen, Jack van Lint, Colette Moeglin, A.F.M. Smith, Johannes Sjöstrand, Domokos Szász, Stanislaw L. Woronowicz, Don Zagier.

Prize Committee

Jacques-Louis Lions (Chair), Noga Alon, Werner Ballmann, Jan Derezinski, Maxim Kontsevich, Eduard Looijenga, Angus Macintyre, David Nualart, A. N. Parshin, Ragni Piene, Itamar Procaccia, Mario Pulvirenti, Rolf Rannacher, Caroline Series, Vladimir Sverák, Dan Voiculescu.

Round Table Committee

Miguel de Guzmán (Chair), Andrey Bolibrukh, Heinz W. Engl, Juan José Manfredi, Carles Perelló, Tomás Recio, Zbigniew Semadeni, Vinicio Villani.

Executive Committee

Sebastià Xambó (Chair), Carles Casacuberta (*Information and Publications*), Julià Cufí (*Finances*), Rosa M. Miró-Roig (*Programming and Activities*), Marta Sanz-Solé (*Secretary of Organisation*), Marta València (*Infrastructure*).

Organising Committee

Lluís Alsedà, Jaume Amorós, Carles Broto, María J. Carro, Teresa Crespo, Josep M. Font, Gábor Lugosi, Jaume Moncasi, Antoni Montes, Joaquín M. Ortega, August Palanques (1957–1997), Antoni Ras, Jordi Saludes, Oriol Serra, Frederic Utzet, Joan Verdera, Santiago Zarzuela.

Congress Programme

Plenary Lectures

Robbert Dijkgraaf	*The mathematics of M-theory*
Hans Föllmer	*Probabilistic aspects of financial risk*
Hendrik W. Lenstra, Jr.	*Flags and lattice basis reduction*
Yuri I. Manin	*Moduli, motives, mirrors*
Yves Meyer	*The role of oscillations in non-linear problems*
Carles Simó	*New families of solutions in N-body problems*
Marie-France Vignéras	*Local Langlands correspondence for* $\mathrm{GL}(n, \mathbb{Q}_p)$ *modulo* $l \neq p$
Oleg Viro	*Dequantization of real algebraic geometry on a logarithmic paper*
Andrew J. Wiles	*Galois representations and automorphic forms*

Plenary Lecture by Andrew J. Wiles

Parallel Lectures

Rudolf Ahlswede	*Advances in extremal problems in number theory and combinatorics*
François Baccelli	*Flow control in stochastic networks*
Volker Bach	*Mathematical theory of matter and radiation*
Viviane Baladi	*Spectrum and statistical properties of chaotic dynamics*
Joaquim Bruna	*Sampling in complex and harmonic analysis*
Nicolas Burq	*Lower bounds for shape resonance width of Schrödinger operators*
Xavier Cabré	*A conjecture of De Giorgi on symmetry for elliptic equations in $\mathbb{R}^n$*
Peter J. Cameron	*The random graph and its relations*
Zoé Chatzidakis	*Difference fields: model theory and application to number theory*
Ciro Ciliberto	*Geometric aspects of polynomial interpolation in more variables and of Waring's problem*
Gianni Dal Maso	*The calibration method for free discontinuity problems*
Jan Denef	*Geometry or arc spaces of algebraic varieties*
Barbara Fantechi	*Stacks for everybody*
Alexander B. Givental	*Some aspects of symplectic field theory*
Alexander B. Goncharov	*Multiple ζ-values, Galois groups and geometry of modular varieties*
Alexander Grigor'yan	*Heat kernels on manifolds, graphs and fractals*
Michael Harris	*Local Langlands correspondence and vanishing cycles on Shimura varieties*
Kurt Johansson	*Random growth and random matrices*
Konstantin M. Khanin	*Burgers turbulence and dynamical systems*
Pekka Koskela	*Sobolev spaces and quasiconformal mappings on metric spaces*
Steffen L. Lauritzen	*Probabilistic networks and the mathematics of local computation*
Nicholas S. Manton	*Understanding skyrmions using rational maps*
Ieke Moerdijk	*Models for the leaf space of a foliation*
Eric M. Opdam	*Recent developments in the theory of hypergeometric functions*
Thomas Peternell	*Contact manifolds in algebraic geometry*

Alexander Reznikov	*Analytic topology*
Henrik Schlichtkrull	*Harmonic analysis on reductive symmetric spaces*
Bernhard Schmidt	*Sums of roots of unity and finite geometry*
Klaus Schmidt	*Algebraic $\mathbb{Z}^d$-actions*
Bálint Tóth	*Self-interacting random motions*

Mini-Symposia and Coordinators

Computer Algebra (Wolfram Decker); Curves over Finite Fields and Codes (Gerard van der Geer); Free Boundary Problems (José Francisco Rodrigues); Mathematical Finance: Theory and Practice (Hélyette Geman); Mathematics in Modern Genetics (Peter Donnelly); Quantum Chaology (Sir Michael Berry); Quantum Computing (Sandu Popescu); String Theory and M-Theory (Michael Douglas); Symplectic and Contact Geometry and Hamiltonian Dynamics (Mikhail Sevryuk); Wavelet Applications in Signal Processing (Andrew Walden).

Closing Round Table, *Shaping the 21st Century*

Round Tables and Moderators

Mathematics Teaching at the Tertiary Level (Vladimir Tikhomirov); The Impact of Mathematical Research on Industry and Viceversa (Irene Fonseca); How to Increase Public Awareness of Mathematics (Vagn Lundsgaard Hansen); What is Mathematics Today? (Zbigniew Semadeni); Building Networks of Cooperation in Mathematics (Friedrich Hirzebruch); The Impact of New Technologies on Mathematical Research (Rafael de la Llave); Shaping the 21st Century (Miguel de Guzmán).

Mathematical Software

CoCoA (Anna Maria Bigatti and Lorenzo Robbiano); LINSOL (Harmut Häfner); Internet Learning Math (Mario Essert); Newton (Jordi Guàrdia and Jesús Montes); FDEM (Torsten Adolph); MathType (Bob Mathews); SINGULAR (Gert-Martin Greuel); GRAMMARKOV (Inmaculada Fortes); CREP (Peter Dräxler and Rainer Nörenberg); Time-Delay System Toolbox (Arkadii Kim); Kenzo (Francis Sergeraert); Java View (Eike Preuss); MAGMA (Lancelot Pecquet); Polymake (Ewgenij Gawrilow).

Computer Facilities at the Congress Hall

Videos

Multicoloured fractals; Quadro, a compact soap bubble of genus 4; Geodesics and waves; Platonic solids; Knot energies; Interpolation of triangle hierarchies; Polytope sections; Surfaces, flows and holonomy; The optiverse; Distorted space; CMC, pictures of constant mean curvature tori; Numerical simulations of unstable detonations; On Gergonne's problem; Soap bubbles; The dynamics of the rabbit.

Exhibitors

Birkhäuser; Springer-Verlag; European Mathematical Society; London Mathematical Society; Cambridge University Press; Addlink Software; De Gruyter, Berlin, New York; Design Science; Academic Press; American Mathematical Society; Oxford University Press; Elsevier Science; France Edition Office, Société Mathématique de France; A. K. Peters, Ltd; Kluwer Academic Publishers; Gordon and Breach Publishing Group; Texas Instruments; CRC Press; Pearson Education; Timberlake Consulting; EDP Sciences; Facultat de Matemàtiques, UB; Publicacions Matemàtiques, UAB; CIMNE and Facultat de Matemàtiques i Estadística, UPC; John Wiley & Sons, Ltd; Princeton University Press; IEC, SCM and CRM.

Acknowledgements

It is a pleasant duty to acknowledge the support and encouragement of the European Mathematical Society, expressed by its two successive Presidents during the period of preparation of the Congress, Jean-Pierre Bourguignon and Rolf Jeltsch. The sponsoring granted by the International Mathematical Union is also gratefully appreciated.

Very special thanks are due to the Institut d'Estudis Catalans (IEC) and to its President, Manuel Castellet, for starting the whole initiative more than eight years ago and unfailingly supporting the mission of the Societat Catalana de Matemàtiques, not only with encouragement, advice and funding, but also by allowing to use the IEC infrastructure. The organisers of the Congress are also enormously indebted to the public universities of Barcelona for their investment of human resources in this venture and for their explicit financial support.

The Generalitat de Catalunya and the Spanish Government have strongly backed the Congress, through sponsorings by several of their bodies. The Barcelona City Council is also a major sponsor of the Congress. Fundación Retevisión provided the infrastructure for information and communications at the Palau de Congressos, while Compaq supplied the computer equipment. Many other institutions and private companies supported the 3ecm in various ways, including funds for the ten EMS prizes. The full list of sponsors is specified below. Our deepest recognition goes to all of them for their contributions.

Sponsors

Generalitat de Catalunya —Departament d'Universitats, Recerca i Societat de la
 Informació; Fundació Catalana per a la Recerca; Departament d'Ensenyament;
 Institut d'Estadística de Catalunya
Ministerio de Educación y Cultura
Fundación Retevisión
Ajuntament de Barcelona
Institut d'Estudis Catalans
Universitat de Barcelona
Universitat Autònoma de Barcelona
Universitat Politècnica de Catalunya
Compaq
European Commission
Port de Barcelona
Borsa de Barcelona
International Mathematical Union
Fundació Caixa Catalunya
Logic Control
Springer-Verlag
Codorníu
Diputació de Barcelona
Fundació Banc Sabadell
Fundació Caixa de Sabadell
COMSOL AB
Nokia

Collaborators

Fundació Caixa de Manresa
London Mathematical Society
Real Sociedad Matemática Española
Sociedad Española de Matemática Aplicada
Universitat Pompeu Fabra
Universitat Oberta de Catalunya
Consejo Superior de Investigaciones Científicas
Nestlé
Vichy Catalán
Panrico
Coca-Cola

Barcelona, February 2001

The 3ecm Executive Committee

EMS Prize Winners

EMS Prize Winners

Semyon Alesker   ISRAEL

Semyon Alesker contributed greatly both to the Asymptotic Theory of Convexity and to Classical Convexity Theory. His most significant work is on valuations (additive functionals) on convex bodies and it has remodeled a central part of convex geometry.

Group invariant valuations were studied since Dehn's solution of Hilbert's third problem, with later contributions by Blaschke and others, and culminating in Hadwiger's celebrated characterization theorem for the intrinsic volumes. The latter theorem was considered the top result in this area for almost fifty years. Alesker has now considerably extended this theory, obtaining very complete classification results under weaker invariance assumptions. He approximated (continuous) rotation invariant valuations by polynomial valuations and characterized the latter, making use of representations of the orthogonal group. This enabled him to extend Hadwiger's theorem to tensor valued valuations. In another direction, he solved a problem of McMullen, in a much stronger form, showing that translation invariant valuations are essentially (up to linear combinations and approximation) mixed volumes. The approach is via representation theory of the general linear group and involves a surprising application of D-modules. The new approach has also opened the way to finiteness results for valuations with other group invariances.

Raphaël Cerf

FRANCE

Raphaël Cerf became known through his results on Probability Theory. Using a large deviation principle in the proper topology, Raphaël Cerf has established a Wulff construction for the supercritical percolation model in three dimensions. This result is a very major advance in the subject, and provides the right formulation for the geometry of the problem. Raphaël Cerf has been able to carry out this program using a correct mixture of combinatorial arguments, geometric ideas and probabilistic tools.

In addition to this research, Raphaël Cerf has made original contributions in genetic algorithms. He has solved a central problem in bootstrap percolation and extended to three dimensions the metastable behavior of the stochastic Ising model in the limit of low temperatures.

Dennis Gaitsgory

U.S.A.

Dennis Gaitsgory is one of the leaders in the geometric Langlands correspondence and related areas. In the modern "geometric" representation theory one replaces functions by complexes of constructible sheaves on (infinite-dimensional) algebraic varieties. In this way many deep structures appear, and classical results in the theory of automorphic forms can be seen much more clearly.

In his thesis and in the subsequent work with Braverman, Gaitsgory established fundamental properties of Eisenstein series in the geometric setting. In a recent paper on nearby cycles, he proposed an extremely elegant construction of the convolution of equivariant perverse sheaves on so called affine Grassmannians. This implies that the center of the affine Hecke algebra coincides with the whole spherical Hecke algebra. Also, it gives the best conceptual explanation of the Satake equivalence.

Recent work of Gaitsgory relates finite quantum groups and chiral Hecke algebras. It is a very important step in the program of Beilinson and Drinfeld in the geometric Langlands theory.

Emmanuel Grenier   FRANCE

Emmanuel Grenier greatly contributed to the asymptotic analysis of Euler and Navier-Stokes equations with large Coriolis force. The simplest case (when the equations are set on the unit cube with periodic boundary conditions) has been solved by Grenier around 1995. Later, in collaboration with Desjardins, Dormy and Masmoudi, he gave rigorous derivations of several asymptotic models currently used in Ocean and Atmosphere modeling, or in Magnetohydrodynamics.

Grenier obtained both positive and negative important results on the problem of convergence of the Navier-Stokes equations to the Euler equations in a domain with solid boundary conditions. In particular, he showed that the positive results of Caflisch and Sammartino obtained for analytic initial data, cannot be extended to Sobolev data. He also justified the hydrostatic limit of the Euler equations in a two dimensional infinitesimally thin strip.

Grenier gave a very elegant proof of convergence for the semi-classical limit of the nonlinear Schrödinger equations (before appearance of shocks). He also obtained, simultaneously with Rykov and Sinai, a hydrodynamic limit for the Zeldovich adhesion particle model.

Dominic Joyce   U.K.

Dominic Joyce's work on the existence of metrics with special holonomy is among the best in Riemannian geometry in the last decade. The question of the existence of Riemannian metrics with special holonomy has a long history beginning with the work of Cartan. It includes some of the best work of such people as M. Berger, J. Simons, S. T. Yau and R. Bryant. Using a dazzling display of geometry and analysis, Joyce constructed compact examples in the exceptional cases where the holonomy is Spin_7 and G_2, the only remaining possibilities, since the others on Berger's list had been eliminated. Joyce also computed the dimension of the deformation spaces of such metrics and many other of their invariants. As a result he also discovered a totally unexpected version of mirror symmetry for such spaces. Dominic Joyce is one of the leading young differential geometers.

Vincent Lafforgue   FRANCE

Vincent Lafforgue's work is a major advance in the K-theory of operator algebras: the proof of the Baum-Connes conjecture for discrete co-compact subgroups of $SL_3(\mathbb{R})$, $SL_3(\mathbb{C})$, $SL_3(\mathbb{Q}_p)$ and some other locally compact groups. The Baum-Connes conjecture predicts the K-theory of the reduced C^*-algebra of a locally compact group (and of more general objects). The conjecture plays a central role in non-commutative geometry and has far-reaching connections to the Novikov conjecture on higher signatures in topology, to harmonic analysis on discrete groups and the theory of C^*-algebras. Lafforgue's result is the first passage of the barrier which property T of Kazhdan has posed for many years in the proof of the Baum-Connes conjecture. The proof involves several remarkable technical and conceptual developments, like a bivariant K-theory for Banach algebras (versus Kasparov's by now classical one for C^*-algebras) or establishing the conjecture for various completions of the L^1-algebras of the groups.

Professor Jacques-Louis Lions (1928-2001)
Chair of the Prize Committee

Michael McQuillan   U.K.

Michael McQuillan has created the method of dynamic diophantine approximation which has led to a series of remarkable results in complex geometry of algebraic varieties. Among these results one can mention a new proof of Bloch's conjecture on holomorphic curves in closed subvarieties of abelian varieties, the proof of the conjecture of Green and Griffiths that a holomorphic curve in a surface of general type cannot be Zariski-dense, and the hyperbolicity for generic hypersurfaces in a projective space P^3 of high enough degree (Kobayashi conjecture).

Stefan Yu. Nemirovski   RUSSIA

Stefan Yu. Nemirovski has obtained several strong results on topology and complex analysis. Using modern techniques like the famous Seiberg-Witten invariants he has solved some old classical problems about submanifolds in complex domains. First, he generalized the Thom inequality proved by P. Kronheimer and T. Mrowka. As a very particular case he proved that there are no nonconstant holomorphic functions in a neighbourhood of an embedded nontrivial 2-sphere in a complex projective plane. Another application of his main theorems is also very attractive. Suppose that an anlaytic disc is attached from the outside to a strictly pseudoconvex domain U in a complex 2-plane. Then there is no smooth disc inside of U with the same boundary. As a corollary one gets that it is impossible to attach an analytic disc from the outside to a strictly pseudoconvex domain that is diffeomorphic to a closed ball.

Paul Seidel   FRANCE

Paul Seidel became known through his work on symplectic topology. In his PhD thesis he studied the fundamental question whether symplectic diffeomorphisms which are diffeotopic to the identity are also symplectically diffeotopic to the identity. He showed that the answer is negative in many cases, already in dimension 4. His counterexamples are generalized Dehn twists; his proof involves Floer homology. In further work, Seidel constructed a natural representation of the fundamental group of the group of Hamiltonian symplectomorphisms into the quantum cohomology ring. This work was basic for later work of Lalonde, McDuff and Polterovich on the topology of the group of symplectomorphisms. There is more to say about other work. His latest work is related to mirror symmetry, showing his broad horizon.

Wendelin Werner   FRANCE

Wendelin Werner has obtained deep results on stochastic processes and, more precisely, he has proved a number of significant results on Brownian path properties, including the shape of Brownian islands and Brownian windings.

Wendelin Werner has made remarkable contributions to the study of self-avoiding random walks and the corresponding critical exponents. More specifically, he obtained the first non-trivial upper bound of the disconnection exponent, and he developed an elegant approach for studying the limiting behavior of the non-intersection exponents for a great number of independent Brownian motions. Among many other interesting works, he constructed, with a collaborator, the so-called true self-repelling motion using an ingenious method involving infinite systems of coalescing Brownian motions.

Felix Klein Prize Winner

David C. Dobson   U.S.A.

David C. Dobson started his work on the diffraction of electromagnetic waves from periodic structures when he was a postdoc at the famous Institute of Mathematics and its Applications of Professor Avner Friedman. The Honeywell Technology Center had posed the problem to model and analyse the diffraction and to develop appropriate numerical algorithms. In a next step an optimal shape design problem for phase lenses was solved. The fact that he used a "Fraunhofer approximation" was not(!) the reason to give him a prize endowed by the Fraunhofer Institute for Mathematics; what convinced the committee that he should be the first prize winner was that he used rigorous and sound mathematical methods in a quite tricky way for a problem, which Honeywell states to be of very high industrial importance.

Opening Ceremony

Address by SEBASTIÀ XAMBÓ
President of Societat Catalana de Matemàtiques and
Chair of the Organising Committee

Honourable authorities, guests, ladies and gentlemen,

Welcome to Barcelona and to the Third European Congress of Mathematics. Welcome on behalf of the Societat Catalana de Matemàtiques (Catalan Mathematical Society), and on behalf of the Organising Committee. We are grateful for your presence, and we hope that your stay will be pleasant and of much profit to you.

Let me briefly introduce the members of the presidential table:

- Excel·lentíssim Sr. Joan Clos, Alcalde de Barcelona (Mayor of Barcelona);
- Honorable Sr. Andreu Mas-Colell, Conseller d'Universitats, Recerca i Societat de la Informació de la Generalitat de Catalunya (Minister for Universities, Research and Information Society of the Catalan Government);
- Excelentísimo Sr. Ramón Marimón, Secretario de Estado de Política Científica y Tecnológica (Secretary of State for Scientific and Technological Policy of the Spanish Government);
- Excel·lentíssim Sr. Manuel Castellet, President de l'Institut d'Estudis Catalans (President of the Institute of Catalan Studies);
- Professor Rolf Jeltsch, President of the European Mathematical Society.

The composition of the presidential table reflects quite faithfully the public institutional support granted to the 3ecm: the Catalan Government, the Spanish Government, the Barcelona City Council, the Institute of Catalan Studies, the European Mathematical Society, and the three Universities in Barcelona which award degrees in mathematics, here represented by the minister.

It is a pleasant duty to thank you for accepting our invitation to be at the presidential table, and also to thank the institutions that you represent for the constant support granted to the 3ecm. At the same time, I am proud to say that participation in the Congress has perfectly matched this support, not only by the number of participants (over 1300 at this moment), but also through their very high professional qualifications.

The opening ceremony will have two parts. In the first part, which lasts about half an hour, we will listen to the speeches of the Secretary of State, of the Catalan Minister for Universities, Research and Information Society, and finally of the Mayor of Barcelona.

(The speeches are included on pages xxxiii–xxxvi below.)

(...)

Presidential Table at the Opening Ceremony

Now there will be a short musical interlude. The Toldrà String Quartet will play music of two Catalan composers: Eduard Toldrà and Xavier Turull. Meanwhile, I shall accompany the Mayor, the Minister and the Secretary of State to the door, for they have other duties in their schedules that do not allow them to remain with us any longer.

Let us now start the second part of the opening ceremony. Please allow me to say a few words before the addresses by the President of the European Mathematical Society and the President of the Institute of Catalan Studies (in this volume their speeches are on pages xxxvii–xliii).

First let us remember those involved with the 3ecm who passed away.

- August Palanques, a member of the Organising Committee: born in 1957, he was Associate Professor at the University of Barcelona, and died in 1997.

- Felipe Mellizo, who was to chair the round table "How to increase public awareness of mathematics": born in 1932, he was a civil engineer, a philosopher, a writer and a journalist. He died last Friday.

Ever since the European Mathematical Society Council meeting in Budapest entrusted the Catalan Mathematical Society with the organisation of 3ecm, we have been very busy in an involved and laborious collective process. Now it is a delight to crown it with this opening and all the activities that will follow. It is therefore appropriate to thank here all the people and institutions without whose involvement this event would not have been possible.

To organise the 3ecm, the Catalan Mathematical Society appointed an Executive Committee. Chaired by the President of the Society, it has five other members: Carles Casacuberta, in charge of information and publications; Julià Cufí, for finances; Rosa M. Miró-Roig, programming and activities; Marta Sanz, organisational secretary; and for the infrastructure, Ferran Puerta in the first year and Marta València since then. The organisation would not have been possible without their sustained dedication to the cause. On behalf of the Catalan Mathematical Society, our much-felt gratitude to them.

Our thanks must be extended to all the people who, proposed by the Executive Committee for diverse organisational tasks, were subsequently appointed by the Catalan Mathematical Society. An Organising Committee of 16 members was appointed soon after the Executive Committee, and their contributions have been numerous and indispensable in all areas. Later on, working groups were formed as the needs arose. Their accomplishments have also been excellent in number and kind. We are thankful to them all.

None of these endeavours would have been possible without the context of a strong and reliable local mathematical community, with a remarkably positive attitude and constant support. This includes numerous volunteer students whose help will be visible during the Congress. It also includes the members of the Catalan Mathematical Society, whose number has more than doubled in the last four years.

As for the institutions, our thanks to the Institute of Catalan Studies, for its decisive role in the sponsoring of the Congress, and for housing the 3ecm. Thanks are due to its President, Manuel Castellet, for his vision to submit a bid to organise the Congress and his subsequent constant support. Thanks also to all the Institute staff, who have been helping in one way or another.

We are much indebted to the public universities in the Barcelona area for their financial support. Moreover, the Universitat de Barcelona, the Universitat Autònoma de Barcelona and the Universitat Politècnica de Catalunya (the three universities that offer degrees in mathematics) have contributed substantial human resources, as they have encouraged the members of the committees and groups to work for the 3ecm.

It is a pleasant duty to acknowledge the continuous support and encouragement of the European Mathematical Society (EMS), particularly expressed by its successive Presidents during this period, namely Jean-Pierre Bourguignon and Rolf Jeltsch. We are also grateful for its sponsoring of 3ecm.

Many thanks also to the Scientific Committee (chaired by Professor Michael Atiyah), the Round Table Committee (chaired by Professor Miguel de Guzmán) and the Prize Committee (chaired by Professor Jacques-Louis Lions) for the enthusiasm and effectiveness with which they have undertaken their crucial tasks.

We much appreciate the sponsorship granted by the International Mathematical Union, and the fact that its President, Jacob Palis, informed us that the International

Mathematical Union regards the 3ecm as a main international mathematical event of World Mathematical Year.

The London Mathematical Society, the Real Sociedad Matemática Española and the Sociedad Española de Matemática Aplicada are sponsoring mathematical societies of the 3ecm and they are also to be duly thanked for their support.

On the governmental side, both the Catalan Government and the Spanish Government have strongly backed the Congress, through sponsorships by the Ministerio de Educación y Cultura, representing Spain, and of the Departament d'Universitats, Recerca i Societat de la Informació, the Fundació Catalana per a la Recerca, the Departament d'Ensenyament and the Institut d'Estadística de Catalunya, representing Catalonia.

The Barcelona City Council is also a major sponsor of the Congress, in part through related institutions such as the Institut de Cultura de Barcelona. Finally, the Diputació de Barcelona sponsors one of the prizes.

Many private institutions are sponsors of the 3ecm. Fundación Retevisión is providing, among other things, all the information and communications infrastructure. Compaq supplies the computing equipment that you can find in the main hall and which you are invited to use according to your needs, and Springer-Verlag has made possible the publication of the Barcelona Intelligencer. Our deepest recognition for these generous contributions.

We are indebted, for sponsoring prizes and round tables, to Borsa de Barcelona, Port de Barcelona, Fundació Caixa Catalunya, Fundació Banc Sabadell, Fundació Caixa de Sabadell, Fundació Caixa de Manresa, Logic Control, Comsol AB and Nokia. Codorníu, Nestlé, Panrico, Vichy and Coca-Cola have contributed to the catering, which all participants will surely notice and appreciate.

Our main task has been to provide an environment in which the scientific programme, and other important concomitant activities, could be decided upon and scheduled in detail, and an environment in which they could develop as planned. Our highest reward will be to have succeeded.

If this is the case, I would like that it be shared by our entire Catalan mathematical community, for after all this is the rich and strong soil out of which the organisation has grown. I hope that the Congress will definitely promote this community into playing a role proportional to its achievements on the international level.

As Allyn Jackson said in the article published in the May issue of the AMS Notices, "Barcelona is known for its great weather, arresting architechture, and excellent seafood". So, as well as wishing you an inspiring Congress, please have also some fun in this City which is the most wonderful in the world. Thank you very much again for your presence here today.

Palau de Congressos de Barcelona
July 10th, 2000

Address by RAMÓN MARIMÓN
Secretary of State for Science and Technology

Excel·lentíssim Alcalde, Honorable Conseller, all of you,

I want to welcome you as Secretary of State of the new Ministry of Science and Technology, and also to thank the organisers, the universities, the Societat Catalana de Matemàtiques, for bringing the Third European Congress of Mathematics to Spain, to Catalunya, to Barcelona.

Since just a few weeks ago I was on the other side of the table, I am not going to bore you with a political speech. Instead, let me share with you a small anecdote. When I was writing my thesis —at that time I was at Yale University—, I got interested in ergodic theory. This led me to talk quite a lot with Kakutani. He is a very friendly old man. One day, I met him while he was walking to his office. He looked terrible. I said, "What is wrong with you?" And his answer was, "It is not just me, it is mathematics". Seeing my puzzled face he said: "Look at art. You see a Picasso, a Van Gogh, and you recognise them immediately. You need not see their names. But I have been writing mathematics all my life and I still have to sign my articles. We all have to".

So I see your programme and I see that you still have your names on your articles, so according to Kakutani there is something wrong with mathematics. And probably you will agree with me in that he is a great mathematician, but less of an art critic. Because in art, what is important is to achieve the critical mass and creativity to form an art movement, and Picasso certainly contributed to that.

Last month, on June 15th, at the Council of Ministers for Research in Europe, we decided to move ahead with what is called the European Research Space Initiative, which is supposed to bring more mobility, more competition, to be at the forefront of the political agenda of the Union research. This is a very important initiative, although it would just be words without the work of scientists. So here you are, you are the mathematician artists of the 21st century in Europe and you should take up the initiative to make Europe competitive in research. And as far as I am concerned, you can certainly count with our Ministry's support. Thank you.

Palau de Congressos de Barcelona
July 10th, 2000

Address by ANDREU MAS-COLELL
Minister for Universities, Research and Information Society
of the Autonomous Government of Catalonia

Dear authorities and friends,

I would like to welcome you to the Third European Congress of Mathematics. I do this on behalf of the Generalitat de Catalunya (the Autonomous Government of Catalonia) and its President, who is currently on an official trip to the Far East. I should add that I am also holding the representation of the Rectors (the Presidents of the Catalan universities). Their representative at today's ceremony, Antoni Caparrós, Rector of the Universitat de Barcelona, has got suddenly ill and has asked me, just a few minutes ago, to put in the record of this ceremony the warm and enthusiastic welcome of the Catalan universities. Obviously I do so very gladly and at the same time I wish him a speedy recovery.

I wonder if you will allow me a somewhat unconventional approach to these opening remarks. Instead of addressing myself to the general audience, I will use the general audience as witness and I would like to direct my words to the Organising Committee and to the Catalan Mathematical Society.

What I would like to tell the organisers are essentially four things. The first is that we, the members of the Government, know the enormous effort that the organisation of this Congress has meant and we thank you warmly for having carried out this effort so well.

Second, that we know that you do this because it is professional good manners to do so, but also because you are proud, as we are, of the level of development that mathematics has reached in Catalonia and in Spain. You rightly see this European Congress as a kind of recognition of this fact from the larger mathematical community.

Third, that we, as you, see this Congress not as the end of a period, but as the beginning of a new one, as the occasion to mark a leap forward in the progress of mathematics and of mathematical research among us: in our universities, in our economy, in our society in general.

And finally, the fourth point is to tell you that, although this leap forward will not be a simple matter, you can count on the wholehearted support and the collaboration, the intense collaboration, of the Generalitat de Catalunya in the accomplishment of this task.

Let us therefore not rest and push ahead, all together, for the progress of mathematics in the 21st century, and for its progress among us. Hence —and now I am addressing the entire audience— there will be other congresses and other mathematical events that will tempt you to come back. So I conclude with an *au revoir*.

Palau de Congressos de Barcelona
July 10th, 2000

Address by JOAN CLOS
Mayor of Barcelona

Senyor Conseller, autoritats,

It is a pleasure for me also to welcome you here in Barcelona, a city that you will enjoy, I hope, during these days that you are staying with us, and a city where you will have the opportunity —if the organisers of the Congress allow you so— to see some applications of mathematics to the real city. I will focus on two of these applications.

First there is the *grid* of Barcelona, the square layout of our city which was planned in the middle of the 19th century, just after the demolishing of the medieval wall of the city. At that time (the middle of the last century), it was an empty space between the wall around the old city of Barcelona and the nearby villages. It was empty because of the royal ordinances concerned with "the shooting of a cannon distance". It was a distance away of the wall over which buildings were not allowed, and such distance was determined by one supposed shooting of a cannon placed on the wall.

This allowed the city to plan the biggest enlargement of its history, to decide what would happen on that big empty space. I imagine that for an architect, or an engineer, or a mathematician, this was a unique opportunity, and it really was. And the man who took the responsibility to plan for that space was the architect and engineer Mr. Cerdà, who planned this magnificent grid of l'Eixample de Barcelona, as a square layout, a huge square layout at that time, for it was to cover an area twenty times as big as the size of the old city. This twentyfold expansion of the city was neatly designed as a square system, which apart from its uniformity has another fantastic point: Namely, in the street crossings he created a cut of the façades in such a way that another square of much smaller size appears at each crossing just turned $45°$ of the big grid. This was a wonderful idea, for it meant more space in every corner, in every square of the grid, and hence it amounted to more room in the city.

This was the planning of Barcelona, and it was produced at a time in which Barcelona was healthy and going up economically with the industrial revolution. But such uniformity, which was certainly a consequence of the belief in science as a social organiser, science as provider of well-being, was then conflicting with the crisis of the industrialisation and the workers action. In fact, our city was to be also famous, apart from the grid, for giving birth to the anarchism movement, which in a way contradicts directly this grid.

But it was not only the anarchism movement that appeared in Barcelona in those times. Another thing appeared later on, about thirty years later, namely modernism: Gaudí, Domènech i Montaner, Puig i Cadafalch. And it was amazing to see how these architects —some of them, especially Gaudí, also proficient mathematicians— just created art and implanted these magnificent forms, these new biological forms, inside the square grid. And if you walk along Passeig de Gràcia, you will see different buildings in this grid that break the harmony, the uniform harmony, and which amount to the appearance of new forms. For example, a wavy stone façade, or chimneys in the roof that look like giants, or balconies which are the eyes of a big face, and so on. And you will see how these two tendencies mix together. You will also see such things in Sagrada Família or in Park Güell, where the application of mathematics was of course done according to the technology of that time.

I especially recommend you to visit the Gaudí Space in Passeig de Gràcia, in La Pedrera, where you will see how Gaudí calculated the arches and frames of a roof. He did so by taking a specular image with respect to a horizontal mirror of a net of hanging strings with weights (small sacks of sand) attached to them. This was the method by which Gaudí calculated the curvature, the optimal curvature, of its arches. He worked in a quite small space, and you can see a replica of his work in Gaudí Space.

This is just to tell you that, apart from attending the exciting sessions that you will have here, in the rooms of the Palau de Congressos, you must go outside. I request the organisers to allow people to go outside, to visit Barcelona, to visit these practical applications of mathematics in our city. And if you enter the shops, you will see more practical mathematics going on. If you wait a little bit, you will see marvellous machines that show you all what you can buy, and the price of everything. Apart from, of course, having a good Congress, go on, and discover Barcelona. It is an adventure!

It is a pleasure for me to declare open this Third European Congress of Mathematics. Thank you very much.

Palau de Congressos de Barcelona
July 10th, 2000

Address by ROLF JELTSCH
President of the European Mathematical Society

Dear President of the Catalan Mathematical Society and Chair of the Organising Committee, Professor Xambó;
Dear President of the Institute of Catalan Studies, Professor Castellet;
Dear colleagues, students, ladies and gentlemen,

It is a privilege and a pleasure for me to welcome you all to this Third European Congress of Mathematics. European Congresses are one of the central pieces of the activity of the EMS.

- *What is EMS?*

  I hope you allow me a few words to say concerning our Society, its aims and ambitions. The EMS is a society of societies covering all of Europe, from Iceland to Russia, from Norway to Israel. Mathematically we want to cover all of mathematics, pure and applied, up to and including applications in science and industry. It also has individual members.

- *What are the ambitions of EMS?*

  – We want to strengthen the feeling of a European identity among mathematicians.
  – We want to be *the* European partner in mathematics for the European Union and to the outside of Europe.
  – We want to raise the public awareness for mathematics.

Let me give an example. As you all know, there have been many activities during World Mathematical Year 2000 to raise public awareness. If you have come here today using the metro, you probably have seen mathematical posters. They came from a competition that the EMS had done. In order to continue this type of activity beyond the year 2000, the Executive Committee has created a new Committee to Raise Public Awareness with Vagn Lundsgaard Hansen in the chair. This committee will for example open a competition for papers for raising public awareness. The best results will be published in our Newsletter.

Let me come to a new and important development on the European level or, more precisely, EU level. The Secretary of State, Mr. Marimón, already touched upon this development in his speech. The new Commissioner in Brussels for Research, Mr. Busquin, has the vision to create a "European Research Area". In addition, the big change is that he listens to the scientists. He has made a meeting with Nobel Prize winners and Fields medalists to discuss his vision. Just two weeks ago, I received a preliminary summary of the answers to his questionnaire to business, industry and academic R & D communities. From these answers the Commissioner deduced seven priorities. I can only mention a few very briefly:

- Enhancing further the mobility of people geographically
    a) by establishing equivalences between academic diplomas;
    b) by harmonising social security, health care, pension plans.
- Improving even further electronic networks in all respects.
- Improving the reality and perception of careers in science and engineering in Europe to attract young Europeans.
- Improving science communication and organising public debates on science and innovation for the benefit of all European citizens.

Let me end by repeating that we consider it as very important that the Commissioner is willing to listen to the academic community. Unfortunately, due to this fact, I have to be in Brussels tomorrow.

When doing these discussions in Brussels, it is important that we have a large weight. Hence we have to increase the number of individual members. Therefore, if you are not already a member, come to our booth and join. It would be even better if you all come and pick up an extra application form and convince one of your friends to join.

Let me come back to the present event. This European Congress is completely in the spirit of the EMS. It is here to strengthen the feeling of a European identity among mathematicians. We shall learn about new developments from prominent speakers, we shall discuss in round tables developments in our profession and the influence of new technologies.

To end: let me thank, in the name of EMS, all those who helped create 3ecm. First and foremost, the success of the Congress is the result of our Catalan friends, the Catalan Mathematical Society and the Organising Committee, both headed by Professor Xambó. My thanks go to the Scientific Committee headed by Sir Michael Atiyah, who put together a very interesting programme. Maybe, for my taste, I could have lived with a little bit more *applied mathematics* in the main lectures. Let me also thank Professor de Guzmán and his Committee for finding excellent panelists for round tables. Finally, the EMS is thankful to the Prize Committees and their chairs, Professor Lions for EMS prizes and Professor Neunzert for the Felix Klein prize.

I wish to you all that you learn not only a lot of mathematics but that you make many new friends and, last but not least, have a few hours of leisure to learn about the local culture, Barcelona and its people. Thank you.

Palau de Congressos de Barcelona
July 10th, 2000

Address by MANUEL CASTELLET
President of the Institut d'Estudis Catalans

Ladies and gentlemen,

The opening of the Third European Congress of Mathematics in Barcelona gives me special satisfaction in my double capacity of mathematician and of president of the Institut d'Estudis Catalans. The Institut d'Estudis Catalans, Institute for Catalan Studies, is the National Academy of Catalonia and the parent body of the Catalan Mathematical Society, organiser of this Congress, which I hope will be to the satisfaction of all of you.

Catalonia is a small country in surface area and population, a country which never had a strong mathematical tradition. If you consult the Zentralblatt für Mathematik or the Mathematical Reviews for the 1970s you will notice that the probability of finding an article by a Catalan mathematician, even by a Spanish one, is effectively less than any epsilon. But now, 25 years later, there is an abundance of Catalan mathematical research published in prestigious journals, and there are outstanding Catalan mathematicians in all areas.

This reality is demonstrated by a recent study carried out by the Institut d'Estudis Catalans into mathematical research in Catalonia. The ratio between the number of articles published in Catalonia and the total population, scaled by the Gross National Product, is similar to the ratio in Norway, Great Britain or Germany; furthermore, 8.2% of all mathematical articles are published in the top quality journals. This situates Catalonia, according to these criteria, among the most scientifically significant countries of the world.

Doubtless this situation is fruit not only of the labours of recent generations of Catalan mathematicians, but also of the efforts of our whole society and of the recent entry of the country into the current democratic framework. But this alone would not be sufficient. It must also be because the famous phrase from Cicero's *Tusculanae* suits us perfectly: "Nature has placed in our spirit an insatiable desire to know the truth".

A truth which this country in its thousand year history has always sought, with more or less success, and almost always with limited resources. A truth which the Catalan countries, scientifically, have always sought looking towards Europe, a Europe of which we were an important nucleus in the Middle Ages, before a sharp decline, and of which we have been steadily forming a more integral part since the beginning of this 20th century. The Institut d'Estudis Catalans, founded in 1907, is an academic, scientific and cultural institution which has been an important factor in this cultural and scientific integration.

The aim of the Institut d'Estudis Catalans is to advance research, both in science and in all areas of Catalan culture. It contributes to the planning, co-ordination and realisation of research in various fields of science, technology and the humanities. It works to stimulate the progress and general development of Catalan society and, on occasion, serves as a consultancy for public bodies and institutions. The Institut d'Estudis Catalans is divided by broadly-defined subject areas into five sections (including the academy of the Catalan language) each of which has a maximum of 21 full fellows; associated to it there are also 25 scientific societies, one of which is the Catalan Mathematical Society, with a total of more than nine thousand members altogether.

Let me now give a brief review of our mathematical history in Catalonia.

Did you know that Gerbert d'Orlhac, a monk from the Catalan Pyrenees who in the year 1000 was better known as Pope Sylvester II, made an in-depth study of the *Quadrivium*, transformed calculation methods throughout Europe two centuries before Fibonacci, and built an original abacus?

Did you know that the Majorcan Ramon Llull, one of the most important writers in the Catalan language of the 13th century, already understood that the Earth was a sphere and wrote *Ars Combinatoria* and *Art de Navegar*, a book unsurpassed until the 16th century which describes the astrolabe and refers to the use of the magnetic compass? Alexander von Humboldt maintained over 100 years ago that these scientific advances were transmitted from Catalonia via other Mediterranean nations to the rest of the civilised world.

Did you know that in 1375 the Majorcan Jew Abraham Cresques drew the world map known as the *Catalan Atlas*, a representation of the known world which was essential for the navigators of the epoch?

Did you know that the second book of mathematics ever printed (after the anonymous arithmetic text of Treviso) was *Summa de l'art d'Aritmetica* by Francesc Santcliment, printed in Catalan in Barcelona in 1482? And that the first mathematics book printed in Spain was the Spanish translation of Santcliment's book?

Did you know that the Valencian Josep Chaix, author of works on differential and integral calculus, carried out with Pierre Mechain in 1793 the calculations to measure the meridian arc between the Pyrenees and Barcelona?

I do not wish to bore you with this history lesson, but I must mention two outstanding contemporary researchers from our country: Lluís Santaló, born in Girona and emigrated to Argentina, pioneer of integral geometry, stereology and geometric probability, who cannot attend this conference due to his delicate state of health. And Ferran Sunyer i Balaguer, possibly the greatest Catalan mathematician through the 1950s and 60s, born a tetraplegic, friend of Hadamard, Mandelbrot and Malliavin, among others, whose theorem on polynomials you can read on the stand of the Catalan Mathematical Society and in whose honour the Institut d'Estudis Catalans awards each year a prize which bears his name for a monograph which best expounds and summarises advances in an active area of mathematics.

But the greatest change, the integration of Catalan mathematicians to the ideas and scientific trends, both European and international, has happened in the last twenty or twenty-five years, when some of us made contact with André Lichnerowicz, Beno Eckmann or Paul Malliavin, to choose just three truly significant names of European mathematicians. The mathematics departments of the universities of Barcelona started to become active research centres and with great effort and dedication were able to gradually break through the scientific and cultural isolation to which we had been subjected by a dictatorial government and a closed society.

At the beginning of the 1980s the Catalan universities, that of Barcelona, the Autonomous University and the Polytechnical University, began to discover that they could compete in high quality mathematical research. The various mathematics departments have intensified the training of doctoral students and the production of research work. But to do so they had to keep in touch with the mathematical research being carried out in the developed world. Thus the Institut d'Estudis Catalans, in view of these changes, created in 1984 the Centre de Recerca Matemàtica, the only research institute in Spain, with the objective of facilitating contact between Catalan mathematicians and the European and international scientific elite; an institute for visiting professors and young researchers, whose aim is to stimulate growth in quality and quantity of Catalan mathematical research, organising specialised research semesters, awarding postdoctoral grants, inviting outstanding researchers, organising seminars, conferences, advanced courses, etc. The Centre de Recerca Matemàtica, which is a member of ERCOM (European Research Centres on Mathematics) has been and continues to be an infrastructure at the service of all Catalan mathematicians to facilitate access to the most advanced scientific trends and the incorporation of our local community in the international mathematical community.

It is not an isolated fact, then, that there are five mathematical events being held in Barcelona this year with the support of the European Commission: this Large Euroconference which we inaugurate today, two Euroconferences on Logic and on Algebraic Topology, a PhD Euroconference on Complex Dynamics, and a Euro Summer School on Quantum Groups; next year we celebrate the new century with six Euroevents organised by the Centre de Recerca Matemàtica: a Euroconference on Combinatorics and Graph Theory, a PhD Euroconference on Algebraic Topology, and four Euro Summer Schools on Modular Forms, Hamiltonian Systems, Riemannian Geometry and Algebraic Topology.

Neither is it an isolated fact that eight research groups in Catalonia (four at the Universitat Autònoma, one at the Politècnica and three at the Universitat de Barcelona) form part of various research training networks of the Improving Human Research Potential programme and that two of these groups, those of Operator Theory and of Algebraic Topology, both at the Universitat Autònoma, are the only mathematics groups in all of Spain which have been selected as Marie Curie Training Sites. The Centre de Recerca Matemàtica will host five Marie Curie postdoctoral fellows, during the next academic year.

For the first time in Catalonia, and in Spain, the words of Konrad Knopp, pronounced at the 1927 inaugural lecture of the University of Tübingen, are being understood: "Mathematics is the basis of all knowledge and contains all other culture". Certainly our world is more complex every day and the complexity of any system increases with the degree of interconnection. A more interconnected world, then, is a more complex system and at the same time more fragile and unstable. Mathematics has an increasingly decisive role to play in the management of complex systems, be they technological, financial or social, in the century we now enter. Mathematics will increasingly be an instrument of power which is sometimes dangerously underestimated.

I think all of you are aware that these words are not intended to convince mathematicians but rather the political and administrative authorities that we have with us here at the presidential table. Catalonia is prepared to face the complexity of the 21st century. In recent years a real effort has been made towards technological renovation and measures have been taken which have encouraged spectacular economic development and ensured good levels of social welfare. One must also take into account, however, that the intellectual resources of a country are at least as important, or even more important, than the material resources. Our administrators and governors should not forget that mathematics is the cheapest of all the sciences: it takes little more than brains, and networks for communication. And our country, rich in brains, must protect the second aspect, strengthening, for example, the two networks of mathematical communication that we have: the Centre de Recerca Matemàtica and the Catalan Mathematical Society.

I will close this parenthesis addressed to our authorities, and return to the desire of the Catalan mathematicians for a place in the European and international scientific community. It is true that there have always been internationally recognised Catalan researchers in some scientific areas and that in recent years this recognition has extended to areas where previously we had no presence, as is the case for mathematics; it is also true that our research groups are beginning to be visible in the literature. But it is one thing to be known and recognised individually and quite another for the country to be recognised in this way. Without the first we could not achieve the second, but this latter must always be our objective. We want Barcelona, and Catalonia, to be known internationally not just for Gaudí or Barça. We want to be recognised also as a scientific community, as a mathematical community, and this is the role of the Catalan Mathematical Society.

The Catalan Mathematical Society is a scientific society that, in a country with a population of six million, has 1000 members and acts on two different fronts: as a meeting ground for all Catalan mathematicians, organising conferences, courses, publishing periodically a *Butlletí* and a *Notícies* in Catalan; and on the other hand as a nucleus for direct connection with mathematical institutions of other countries and especially with the European Mathematical Society.

The participation of the Catalan Mathematical Society in the European Mathematical Society has been intense since the outset, collaborating on the Council and on the Executive Committee, and ultimately presenting its candidature four years ago now to host this Third European Congress of Mathematics, which was awarded to us at the Council meeting in Budapest.

We want not just to offer you good organisation, a welcoming city and country, a good climate and lots of sun; all this we do offer, but on top of all these ingredients we hope that you, the European mathematicians, and by extrapolation all the international mathematical community, take note that Catalonia is a country which is mathematically developed and prepared to face the complexities of which I spoke to you a moment ago.

I thank you for your attention.

Palau de Congressos de Barcelona
July 10th, 2000

Closing Ceremony

Address by SEBASTIÀ XAMBÓ
President of Societat Catalana de Matemàtiques and
Chair of the Organising Committee

Ladies and gentlemen,

First let me introduce the members of the presidential table: Rolf Jeltsch, President of the European Mathematical Society, and the members of the Executive Committee of the Congress: Marta Sanz, Secretary of Organisation; Rosa Maria Miró-Roig, vicepresident and in charge of Programming and Activities; Carles Casacuberta, for Publications and Communications; Julià Cufí, for Finances; and Marta València, for Infrastructure.

To begin, I would like to remind you of the main goals we envisioned for the Congress:

- To present new results to a wide audience, in the spirit of the unity of mathematics.
- To be a forum for the relations of mathematics with society.
- To foster cooperation among mathematicians.

On behalf of the Organising Committee, I am happy to say that we believe these objectives have been met.

It has been a truly international event, for there have been participants of 87 countries. The official participation is the following: there have been 1240 registered people and 130 volunteers.

As you have certainly seen, the lectures, mini-symposia, round tables and other activities have had a high quality for this kind of conference and you have seen the people talking and discussing in the halls and stands during all the week. Moreover, the computer area has been quite busy all the time.

As far as the image in the society at large, there has been some coverage in the press, interviews on TV and several radio programmes. I believe that these activities, together with all the activities developed in the frame of World Mathematical Year, have promoted, in our country, the image that mathematics is important for everybody, and not only for professional mathematicians. You possibly are witnesses of the posters and TV displays in the subway, and I would like to say that the Parliament of Catalonia organised, on June 19th, a special session to celebrate World Mathematical Year. In particular, there were two round tables, one on the present state of mathematics in Catalonia, including its teaching, and another on the theme of "Mathematics as a key for development".

I also hope that the Congress conveyed you a sharper image of our mathematical community, that we have managed to communicate you a realistic view of the quantity and quality of our mathematical community, including the very young students and volunteers that you have seen in action those days. If you report on the Congress, we would like to hear about it. For us, the 3ecm is not over, as we have to close the proceedings, the financial matters and all the thank-you correspondence to all people and institutions that have helped us. When we do this, it would be much valuable if we could add your points of view.

We want to express our deepest thanks to all the speakers, and also to all the chairmen of lectures, round tables and mini-symposia. In the case of mini-symposia, it is to be acknowledged that they had a particularly difficult task, for they had to invite the corresponding speakers, coordinate their contributions and chair all the sessions.

Their works will be recorded in the proceedings, which will be published by Birkhäuser as in the previous two European Congresses. I would like to remind you that the discount offered by Birkhäuser to the participants will be valid until the end of July.

I would also like to remind you that the discount offered by Springer-Verlag on the Video and Multimedia at 3ecm (in video and DVD formats) is valid until the end of August.

The efforts of the round table chairmen and panelists will also be recorded into one volume, published jointly by the Catalan Mathematical Society and the International Center for Numerical Methods in Engineering. To get a copy, you should get in touch with the Catalan Mathematical Society.

Finally, a few words of thanks.

I would like to stress the crucial role played by the public universities in the Barcelona area. We are much indebted to them for their explicit financial support. Moreover, the Universitat de Barcelona, the Universitat Autònoma de Barcelona and the Universitat Politècnica de Catalunya (the three universities that offer degrees in mathematics) have contributed substantial human resources, as they have permitted the members of the committees and groups to work for 3ecm.

To end, I would like to thank all the volunteers for the wonderful job they did. Here is a paragraph of an e-mail that I received this morning from a participant: "Considering the numerous participants, they were of much help to keep everything going and thus particularly long queues could be avoided. Also the high number of young Spanish students who were present impressed me". I think that they deserve a big applause.

And the same gratefulness goes to the Organising Committee and all persons that served on support committees, for whom I also would like to ask a big applause.

The last applause I would like to ask you, but certainly not the smallest, is for all the staff that has made possible running smoothly the 3ecm.

Shaping the 21st Century?

Palau de Congressos de Barcelona
July 14th, 2000

Presidential Table at the Closing Ceremony

Address by ROLF JELTSCH
President of the European Mathematical Society

Dear Professor Xambó;
Dear Members of the Organising Committee;
Dear colleagues;
Ladies and gentlemen,

You are apparently the persistent ones who stay until the end of this Congress, because maybe you want to have a glimpse at the future. I will come to this slightly later.

What are the goals of such a Congress? It has already been said, so I can only shortly repeat: You came here to listen to the lectures, the mini-symposia and the round tables. Some of you looked at the videos, and then of course I am sure that you enjoyed the Wednesday evening outside, the beautiful weather and the beautiful dinner at the Royal Palace.

Now I want to touch on something different. There was a lot of science, but such a Congress also does something for the European Mathematical Society. And what does it do? How did the Executive Committee and myself experience these eight days in Barcelona? Unfortunately, I have to confess that I had little time to listen to the lectures. But the ones that I attended have been really excellent. However, and this is more important for the development of the Society, the Executive Committee and myself listened to you; we have had many discussions with you; we talked with member societies which were present; we heard about their problems and their needs, but also about their successes; we learnt from them; we got ideas, and I got to hear what is good and what is not so good about our Society. I am very happy that I have received criticism, because this is helping me and the Executive Committee to improve things. Only like this a society can develop. We also received encouragement in what we are doing and this was of course very helpful.

During this week, at one point the future EMS Vicepresident Bodil Branner looked to me and said: "I think we proceed very fast. Look at all what has happened during this week. We usually need more than a year to do all these things." And this is really true! I think a lot has happened. It may not be visible for you at the moment, but it will become visible. And therefore we found that we should actually have large meetings, conferences, on an annual basis. Clearly these should not be full-blown European Mathematical Congresses —it would be nonsense to do them every year— but we could have specialised meetings, or other events. For the next two years this problem is already solved, so we do not have to worry about this. I have put a conference announcement for the next EMS conference. It is an EMS-SIAM conference, practically in a year, in Berlin, on September 2nd to 6th, 2001. You will find the details on the back side of the Friday 14th Daily News. This will be the first EMS-SIAM conference and the title is: "Applied Mathematics in our Changing World". I just mention some titles for preliminary topics, ordered by applications, not by mathematical subjects: Medicine, Biotechnology, Material Science, Environmental Science, Downscale Technology, Communications, Traffic, Market and Finance, Speech and Image Recognition, Engineering Design. I hope of course that many of you —even if you are not that applied— will attend and see where mathematics fits in these subjects.

The year after, there will be a Council meeting in Oslo, attached to the Abel meeting. We could again have some bigger event where many of us meet and we can develop not only science, but also the Society.

In addition to the idea that we should have more often meetings, it dawned on me that I, together with you and the Executive Committee, have to attack the following two major tasks in the immediate future, which I want to describe at the moment. One is the establishment of the EMS Publishing House. Without all the encouragement we received this week we would not have been where we are right now. With this encouragement we proceeded fast and we are convinced that we have to go this way. I hope to be able to report to you the formal establishment of this European Mathematical Publishing House. It will be a publishing house by mathematicians, through professionals, for mathematicians.

The other major topic and major task that I see is the following: to establish a structure, or procedure, to make EMS such that applied mathematicians feel that EMS is their home. They should feel that they are represented well by the EMS. We do not want a separation of pure and applied mathematics in Europe. This task is very urgent because it has to be established before I step down. It should not be that such a representation of applied mathematics is only depending on one person like the President. I and the Executive Committee need all your help to create the ideas to achieve this goal. And I am sure we will do it; I am sure that the EMS-SIAM meeting will be helpful to focus on the discussion.

I have talked enough now on the EMS. Let me come to the key question which is probably on your mind. You know that the idea for this European Mathematical Congress originated from Professor Karoubi from Paris. It is held every four years. It started in Paris in 1992, then it was in Budapest and now the third one is here in Barcelona. The question you are asking is: "Where is the next one?" Well, the call for bids was out last year and we have received one bid. Our site committee inspected the place and was convinced that it was a very good proposal. Unfortunately, when it came to the presentation at the Council meeting last Saturday, it became clear that the proposal had to be withdrawn. Hence, I can, unfortunately, not announce the location of 4ecm. However I can promise you, and the Executive Committee promises you, that we are determined to find an excellent location fairly soon —this means by the end of the year. So I will, by the end of this year, in my annual letter, hopefully announce this place.

To end, let me thank once again the Catalan Mathematical Society, its President, the Organising Committee, and all the people involved. Professor Xambó can name all

1

the different people much better than I can do, because he knows them all. I am happy that he already mentioned the enthusiastic students himself. I was impressed by how they did things and I think that they all created for us a wonderful and memorable experience. So thank you very much, and we will see you all next year in Berlin. And now I have to say formally that, on behalf of the Catalan Mathematical Society, I declare the Third European Congress of Mathematics closed.

Palau de Congressos de Barcelona
July 14th, 2000

The Mathematics of M-Theory

Robbert Dijkgraaf

Abstract. String theory, on its modern incarnation M-theory, gives a huge generalization of classical geometry. I indicate how it can be considered as a two-parameter deformation, where one parameter controls the generalization from points to loops, and the other parameter controls the sum over topologies of Riemann surfaces. The final mathematical formulation of M-theory will have to make contact with the theory of vector bundles, K-theory and non-commutative geometry.

1. Introduction

Over the years there have been many fruitful interactions between string theory [14] and various fields of mathematics. Subjects like algebraic geometry and representation theory have been stimulated by new concepts such as mirror symmetry [3], quantum cohomology [12] and conformal field theory [4]. But most of these developments have been based on the perturbative formulation of string theory, either in the Lagrangian formalism in terms of maps of Riemann surfaces into manifolds or the quantization of loop spaces. This perturbative approach is however only an approximate description that appears for small values of the quantization parameter.

Recently there has been much progress in understanding a more fundamental description of the theory that has become known as M-theory. M-theory seems to be the most complex and richest mathematical object so far in physics. It seems to unify three great ideas of twentieth century theoretical physics:

(1) General relativity – the idea that gravity can be described by the Riemannian geometry of space-time.
(2) Gauge theory – the description of forces between elementary particles using connections on vector bundles. In mathematics this involves K-theory and index theorems.
(3) Strings, or more generally extended objects, as a natural generalization of point particles. Mathematically this means that we study spaces primarily through their (quantized) loop spaces.

At present it seems that these three independent ideas are closely related, and perhaps essentially equivalent. To some extent physics is trying to build a dictionary between geometry, gauge theory and strings.

It must be said that in all developments there have been two further ingredients that are absolutely crucial. The first is quantum mechanics – the description of physical reality in terms of operator algebras acting on Hilbert spaces. In most attempts to understand string theory, quantum mechanics has been the foundation, and there is little indication that this is going to change.

The second ingredient is supersymmetry – the unification of matter and forces. In mathematical terms supersymmetry is closely related to de Rham complexes and algebraic topology. In some way many of the miraculous interconnections in string theory only work if supersymmetry is present. Since we are essentially working with a complex, it should not come as a surprise to mathematicians that there are various 'topological' indices that are stable under perturbation and can be computed exactly in an appropriate limit. From a physical perspective supersymmetry is perhaps the most robust prediction of string theory.

1.1. A two-parameter deformation of classical geometry

For pedagogical purposes in this lecture, M-theory will be considered as a two parameter family of deformations of "classical" Riemannian geometry. Let us introduce these two parameters heuristically. (We will give a more precise definition later.)

First, in perturbative string theory we study the loops in a space-time manifold. These loops can be thought to have an intrinsic length ℓ_s, the *string length*. At least at a heuristic level it is clear that in the limit $\ell_s \to 0$ the string degenerates to a point, a constant loop. The parameter ℓ_s controls the "stringyness" of the model. We will see how the quantity $\ell_s^2 = \alpha'$ plays the role of Planck's constant on the world-sheet of the string. That is, it controls the quantum correction of the two-dimensional field theory on the world-sheet of the string. An important example of a stringy deformation is quantum cohomology [12].

Secondly, strings can split and join, sweeping out a surface Σ of general topology in space-time. According to the general rules of quantum mechanics we have to include a sum over all topologies. Such a sum over topologies can be regulated if we can introduce a formal parameter $\lambda \in \mathbb{R}_+$, the *string coupling*, such that a surface of genus g gets weighted by a factor λ^{2g-2}. Higher genus topologies can be interpreted as virtual processes wherein strings split and join – a typical quantum phenomenon. Therefore the parameter λ controls the quantum corrections. In fact we can equate λ^2 with Planck's constant in space-time. Only for small values of λ can string theory be described in terms of loop spaces and sums over surfaces.

In fact, in the case of particles we know that for large values of λ it is better to think in terms of waves, or more precisely quantum fields. So we expect that for large λ and α' the right framework is string field theory [21]. This is partly true, but it is in general difficult to analyze this string field theory directly. In particular the occurrence of branes, higher-dimensional extended objects that will play an important role in the subsequent, is often obscure. (See however the recent work [17].)

Summarizing we can distinguish two kinds of deformations: stringy effects parametrized by α', and quantum effects parametrized by λ. This situation can be described with the following table

$\alpha' > 0$	conformal field theory *strings*	M-theory *string fields, branes*
$\alpha' = 0$	quantum mechanics *particles*	quantum field theory *fields*
	$\lambda = 0$	$\lambda > 0$

It is perhaps worthwhile to put some related mathematical fields in a similar table

$\alpha' > 0$	quantum cohomology *(Gromov, Witten)*	non-commutative geometry *(Connes)*
$\alpha' = 0$	combinatorical knot invariants *(Vassiliev, Kontsevich)*	4-manifold, 3-manifolds, knots *(Donaldson, Witten, Jones)*
	$\lambda = 0$	$\lambda > 0$

We will now briefly review these various generalizations. More background material can be found in [5].

2. Quantum Mechanics and Particles

In classical mechanics we describe point particles on a Riemannian manifold X that we think of as a (Euclidean) space-time. Pedantically speaking we look at X through maps

$$x \colon pt \to X$$

of an abstract point into X. Quantum mechanics associates to the classical configuration space X the Hilbert space $\mathcal{H} = L^2(X)$ of square-integrable wavefunctions. We want to think of this Hilbert space as associated to a point

$$\mathcal{H} = \mathcal{H}_{pt}\,.$$

For a supersymmetric point particle we instead work with the space of de Rham differential forms $\mathcal{H} = \Omega^*(X)$.

Classically a particle can go in a time t from point x to point y along some preferred path, typically a geodesic. Quantum mechanically we instead have a linear evolution operator

$$\Phi_t \colon \mathcal{H} \to \mathcal{H}$$

that describes the time evolution. Through the Feynman path-integral this operator is associated to maps of the line interval of length t into X. More precisely, the kernel $\Phi_t(x,y)$ of the operator Φ_t is given by the path-integral

$$\Phi_t(x,y) = \int_{x(\tau)} [dx] e^{-\int_0^t d\tau |\dot{x}|^2}$$

over all paths $x(\tau)$ with $x(0) = x$ and $x(t) = y$. Φ_t is the kernel of the heat equation

$$\frac{d}{dt}\Phi_t = \Delta\Phi_t, \qquad \Phi_0 = \delta(x-y).$$

These path-integrals have a natural gluing property: if we first evolve over a time t_1 and then over a time t_2 this should be equivalent to evolving over time $t_1 + t_2$.

$$\Phi_{t_1} \circ \Phi_{t_2} = \Phi_{t_1+t_2}. \tag{1}$$

This allows us to write

$$\Phi_t = e^{-tH}$$

with H the Hamiltonian. In the case of a particle on X the Hamiltonian is of course simply given by the Laplacian $H = -\Delta$.

The composition property (1) is a general property of quantum field theories. It leads us to Segal's functorial view of quantum field theory, as a functor between the categories of manifolds (with bordisms) to vector spaces (with linear maps) [15].

The Hamiltonian can be written as

$$H = -\Delta = -(dd^* + d^*d).$$

Here the differentials d, d^* play the role of the supercharges. Ground states satisfy $H\psi = 0$ and are therefore harmonic forms and in 1-to-1 correspondence with the de Rham cohomology group

$$\psi \in \mathrm{Harm}^*(X) \cong H^*(X).$$

We want to make two additional remarks. First we can consider also a closed 1-manifold, namely a circle S^1 of length t. Since a circle is obtained by identifying two ends of an interval we can write

$$\Phi_{S^1} = \mathrm{Tr}_{\mathcal{H}}\,\Phi_t.$$

Here the partition function Φ_{S^1} is a number associated to the circle S^1 that encodes the spectrum of Δ. We can also compute the supersymmetric partition function by using the fermion number F (defined as the degree of the corresponding differential form). It computes the Euler number

$$\mathrm{Tr}_{\mathcal{H}}(-1)^F \Phi_t = \chi(X).$$

Secondly, to make the step from the quantum mechanics to the propagation of a particle in quantum field theory we have to integrate over the metric on the

1-manifold. In case of an interval we so obtain the usual propagator, the Greens' function of the Laplacian

$$\int_0^\infty dt\, e^{t\Delta} = \frac{1}{\Delta}.$$

3. Conformal Field Theory and Strings

We will now introduce our first deformation parameter α' and generalize from point particles and quantum mechanics to strings and conformal field theory.

3.1. Sigma models

A string can be considered as a parametrized loop. So, in this case we study the manifold X through maps

$$x\colon S^1 \to X,$$

that is, through the free loop space $\mathcal{L}X$.

Quantization will associate a Hilbert space to this loop space. Roughly one can think of this Hilbert space as $L^2(\mathcal{L}X)$, but it is better to think of it as a quantization of an infinitesimal thickening of the locus of constant loops $X \subset \mathcal{L}X$. These constant loops are the fixed points under the obvious S^1 action on the loop space. The normal bundle to X in $\mathcal{L}X$ decomposes into eigenspaces under this S^1 action, and this gives a description (valid for large volume of X) of the Hilbert space $\mathcal{H}_{S^1}$ associated to the circle as the normalizable sections of an infinite Fock space bundle over X.

$$\mathcal{H}_{S^1} = L^2(X, \mathcal{F}_+ \otimes \mathcal{F}_-)$$

where the Fock bundle is defined as

$$\mathcal{F} = \bigotimes_{n\geq 1} S_{q^n}(TX) = \mathbb{C} \oplus qTX \oplus \cdots.$$

Here we use the formal variable q to indicate the $\mathbb{Z}$-grading of $\mathcal{F}$ and we use the standard notation

$$S_q V = \bigoplus_{N\geq 0} q^N S^N V$$

for the generating function of symmetric products of a vector space V.

When a string moves in time it sweeps out a surface Σ. For a free string Σ has the topology of $S^1 \times I$, but we can also consider at no extra cost interacting strings that join and split. In that case Σ will be an oriented surface of arbitrary topology. So in the Lagrangian formalism one is led to consider maps

$$x\colon \Sigma \to X.$$

There is a natural action for such a sigma model if we pick a Hodge star or conformal structure on Σ (together with of course a Riemannian metric g on X)

$$S(x) = \int_\Sigma g_{\mu\nu} dx^\mu \wedge * dx^\nu .$$

The critical points of $S(x)$ are the harmonic maps.

In the Lagrangian quantization formalism one considers the formal path-integral over all maps $x \colon \Sigma \to X$

$$\Phi_\Sigma = \int_{x \colon \Sigma \to X} e^{-S/\alpha'} .$$

Here the constant α' plays the role of Planck's constant on the string world-sheet Σ. It can be absorbed in the volume of the target X by rescaling the metric as $g \to \alpha' \cdot g$. The semi-classical limit $\alpha' \to 0$ is therefore equivalent to the limit $\mathrm{vol}(X) \to \infty$.

3.2. Functorial description

In the functorial description of conformal field theory the maps Φ_Σ are abstracted away from the sigma model definition.

One starting point is now an arbitrary (closed, oriented) Riemann surface Σ with boundary. This boundary consists of a collection of oriented circles. One declares these circles incoming or outgoing depending on whether then orientation matches that of the Σ. To a surface Σ with m incoming and n outgoing boundaries one associates a linear map

$$\Phi_\Sigma \colon \mathcal{H}_{S^1}^{\otimes n} \to \mathcal{H}_{S^1}^{\otimes n} .$$

These maps are not independent but satisfy gluing axioms that generalize the simple composition law (1)

$$\Phi_{\Sigma_1} \circ \Phi_{\Sigma_2} = \Phi_\Sigma$$

where Σ is obtained by gluing Σ_1 and Σ_2 on their outgoing and incoming boundaries respectively.

In this way we obtain what is known as a modular functor. It has a rich algebraic structure. For instance, the sphere, with three holes gives rise to a product

$$\Phi \colon \mathcal{H}_{S^1} \otimes \mathcal{H}_{S^1} \to \mathcal{H}_{S^1} .$$

Using the fact that a sphere with four holes can be glued together from two copies of the three-holed sphere one shows that this product is essentially commutative and associative. For more details see e.g. [12, 5].

3.3. *B*-fields and gerbes

There is a straightforward generalization of the sigma model action (3.1) that includes a 2-form B on X. This so-called B-field adds an extra phase

$$\exp i \int_\Sigma x^* B$$

in the path-integral. The two-form B should be considered as a two-form analogue of a connection. Its curvature H, which is locally given by $H = dB$, actually represents a class $[H/2\pi] \in H^3(X, \mathbb{Z})$. So more generally, if Σ is considered as the boundary of a 3-manifold W and the map x is extended over W, then the phase factor is defined as

$$\exp i \int_W x^* H.$$

By pulling back B to the loop space we see that it acts as a connection on a line bundle over $\mathcal{L}X$.

Such a B-field can be considered as a connection on a gerbe [9] over the manifold X. Gerbes are generalizations of line bundles that naturally support p-form connections. Extended objects such as strings and branes are closely connected to these p-form theories.

3.4. Strings and gravity

The way in which general relativity emerges from string theory is deep and I want to use some time to explain this here. We have already seen that at the classical level every two-dimensional sigma model is conformal invariant. In order to write down the classical action and the resulting equation of motion (the harmonic maps condition) for maps $\Sigma \to X$ we only need the Hodge star or a conformal structure on Σ. This is no longer true at the quantum level. Heuristically, we need a full metric γ on Σ to define the measure in the path-integral over all maps $\Sigma \to X$, and this metric dependence of the path-integral destroys the conformal invariance of the quantum theory.

So the partition function is actually a function of both the metric g on X and γ on Σ. We can now consider a (constant) rescaling

$$\gamma \to \mu \cdot \gamma$$

of the world-sheet metric and compute the μ-dependence of the partition function $\Phi(g, \gamma)$. In general one finds a non-zero result.

The amazing result of standard renormalization theory in quantum field theory is that a rescaling in γ can be absorbed into a redefinition of the metric g on X. This leads to the famous renormalization group flow equation [14]

$$\left[\mu \frac{\partial}{\partial \mu} + \int_X \beta_{\mu\nu} \frac{\partial}{\partial g_{\mu\nu}} \right] Z = 0.$$

Here the beta-function $\beta_{\mu\nu}$ is a vector field on the space $MET(X)$ of Riemannian metrics on X.

This effect can be observed most easily in a semi-classical expansion of the sigma model in α'. That is, one consider maps that are approximately constant

$$x = x_0 + \xi$$

8   R. Dijkgraaf

with $\xi\colon \Sigma \to T_{x_0}X$. In this approximation the action can be written to subleading order as (schematically)

$$\int g_{\mu\nu}(x_0)d\xi^\mu \wedge *d\xi^\nu + R_{\mu\nu\lambda\rho}(x_0)\xi^\mu\xi^\nu d\xi^\lambda \wedge *d\xi^\rho \,.$$

Here the Riemann curvature $R_{\mu\nu\lambda\rho}$ can be considered as a small perturbation in the large volume limit. To leading order one then finds that

$$\beta_{\mu\nu} = \alpha' R_{\mu\nu} + \mathcal{O}(\alpha'^2)\,.$$

So to this level of approximations the Ricci flat metrics with

$$R_{\mu\nu} = 0$$

correspond to the quantum CFTs.

An important property of the beta-function β is that it is a gradient vector field on $MET(X)$

$$\beta = \nabla\mathcal{S}, \qquad \mathcal{S} = \int \sqrt{g}R(g)$$

with $\mathcal{S}$ the Einstein-Hilbert action for the metric g on X. Conformal field theories are given by the zeroes of β and thus correspond to the critical points of $\mathcal{S}$. In this sense a quantum CFT corresponds to a semi-classical solution of gravity.

3.5. 'Stringy' geometry and T-duality

Two-dimensional sigma models give a natural one-parameter deformation of classical geometry. The deformation parameter is Planck's constant α'. In the limit $\alpha' \to 0$ we localize on constant loops and recover quantum mechanics or point particle theory. For non-zero α' the non-constant loops contribute.

This structure is most familiar now in the form of quantum cohomology for Kähler manifolds. Under suitable circumstances the path-integral localizes to holomorphic maps that get weighted by a phase factor q^d where d is the degree and $q = e^t$ with t the pull-back of the complexified Kähler form $\frac{\omega}{\alpha'} + iB$. So the CFT maps Φ_Σ have a typical expansion [3]

$$\Phi_\Sigma = \sum_d q^d N_\Sigma(d)$$

where $N_\Sigma(d)$ counts the number of holomorphic maps $\Sigma \to X$. Since essentially $q \sim e^{-1/\alpha'}$ we see that these corrections are non-perturbative from a world-sheet point of view. They are invisible in a Taylor expansion in α'.

In fact we can picture the moduli space of CFT's roughly as follows. It will have components that can be described in terms of a target space X. For these models the moduli parametrize Ricci-flat metrics plus a choice of B-field. These components have a boundary 'at infinity' which describes the large volume manifolds. We can use the parameter α' as local transverse coordinate on the collar around this boundary. If we move away from this boundary, stringy corrections set in.

In the middle of the moduli space exotic phenomena can take place. For example, the automorphism group of the CFT can jump, which gives rise to orbifold singularities at enhanced symmetry points.

The most striking phenomena is that the moduli space can have another boundary that allows again for a semi-classical interpretation in terms of a second classical geometry $\hat{X}$. These points look like quantum or small volumes in terms of the original variables on X but can also be interpreted as large volumes in terms of a set of dual variables on a dual or mirror manifold $\hat{X}$. In this case we speak of a T-duality. In this way two manifolds X and $\hat{X}$ are related since they give rise to the same CFT.

The simplest example of such a T-duality occurs for toroidal compactification. If $X = T$ is a torus, the CFT's on T and its dual T^* are isomorphic. We will explain this in more detail in §5.

4. M-Theory and Branes

We have seen how CFT gives rise to a rich structure in terms of the modular geometry as formulated in terms of the maps Φ_Σ. To go from CFT to string theory we have to make two more steps.

4.1. Summing over string topologies

First, we want to generalize to the situation where the maps Φ_Σ are not just functions on the moduli space $\mathcal{M}_{g,n}$ of Riemann surfaces but more general differential forms. In fact, we are particularly interested in the case where they are volume forms since then we can define the so-called string amplitudes as

$$A_g = \int_{\mathcal{M}_g} \Phi_\Sigma \, .$$

This is also the general definition of Gromov-Witten invariants [12].

Although we suppress the dependence on the CFT moduli, we should realize that the amplitudes A_g (now associated to a *topological* surface of genus g) still have (among others) α' dependence.

Secondly, it is not enough to consider a string amplitude of a given topology. Just as in field theory one sums over all possible Feynman graphs, in string theory we have to sum over all topologies of the string world-sheet. In fact, we have to assemble these amplitudes into a generating function

$$A(\lambda) \approx \sum_{g \geq 0} \lambda^{2g-2} A_g \, .$$

Here we introduce the so-called string coupling constant λ. Unfortunately this generating function can be at best an asymptotic series expansion of an analytical function $A(\lambda)$. A rough estimate of the volume of $\mathcal{M}_g$ shows that typically

$$A_g \sim 2g!$$

Indeed, general physics arguments tell us that the *non-perturbative* amplitudes $A(\lambda)$ have corrections of the form

$$A(\lambda) = \sum_{g \geq 0} \lambda^{2g-2} A_g + \mathcal{O}(e^{-1/\lambda}) \, .$$

Clearly to approach the proper definition of the string amplitudes these non-perturbative corrections have to be understood.

4.2. M-theory

The last five years have seen remarkable progress in this direction. It involves two remarkable new ideas.

1. String theory is not a theory of strings. It is not enough to consider loop spaces. We should also include other extended objects, collectively known as branes. One can try to think of these objects as associated to more general maps $Y \to X$ where Y is a higher-dimensional space. But the problem is that there is not a consistent quantization starting from 'small' branes along the lines of string theory, that is, an expansion where we control the size of Y (through α') and the topology (through λ). However, through the formalism of D-branes [13] these can be analyzed exactly in string perturbation theory.

2. As we stressed, the amplitudes A depend on many parameters or moduli. Apart from the string coupling λ, all other moduli have a geometric interpretation, in terms of the metric and B-field on X. The second new ingredient is the insight that string theory on X with string coupling λ can be given a fully geometric realization in terms of a new theory (called M-theory) on the manifold $X \times S^1$, where the length of the circle S^1 is λ [18].

5. Torus Compactifications

So we see that there is a natural hierarchy of generalized geometries, roughly associated to particles, strings and branes. According to this point of view a 'classical' manifold can be considered as an element of three different categories. Viewed as an object in such a category it can inherit different symmetries.

Although we are not yet in a position to give a completely rigorous definition of what M-theory is, we do know what kind of data we want to associate to a space X. These data contain at least the following:

1. A moduli space $\mathcal{M}$ of geometric structures on X. This can be a Ricci-flat metric, but also B-fields or their generalizations.
2. A charge lattice Γ that labels the various sectors of the theory.
3. A discrete symmetry group G (the duality group) that acts on the lattice Γ, which will typically form an irreducible representation.

4. A Hilbert space bundle

$$\mathcal{H} = \bigoplus_{\gamma \in \Gamma} \mathcal{H}(\gamma)$$

over the moduli space $\mathcal{M}$ that is graded by the charge lattice Γ.

This hierarchy of structures can be nicely illustrated with a very simple class of manifolds: n-tori, written as quotients

$$T = \mathbb{R}^n / L$$

with L a rank n lattice.

5.1. Particles on a torus

States of a quantum mechanical point particle on T are conveniently labeled by their momentum

$$p \in L^*.$$

The wave functions e^{ipx} form a basis of $\mathcal{H} = L^2(T)$ that diagonalizes the Hamiltonian $H = -\Delta = p^2$. So we can write

$$\mathcal{H} = \bigoplus_{p \in L^*}{}' \mathcal{H}(p),$$

where the graded pieces $\mathcal{H}(p)$ are one-dimensional. There is a natural action of the symmetry group

$$G = SL(n, \mathbb{Z}) = \operatorname{Aut} L$$

on the lattice $\Gamma = L$ and the Hilbert space $\mathcal{H}$. (These transformations will in general not leave the metric invariant, but instead give by pull-back another flat metric on T.)

5.2. Strings on a torus

In the case of a string moving on the torus T, states are also labeled by their winding number

$$w \in L.$$

The winding number simply distinguishes the various connected components of the loop space $\mathcal{L}T$, since

$$\pi_0 \mathcal{L}T = \pi_1 T \cong L.$$

We therefore see a natural occurrence of the so-called Narain lattice

$$\Gamma^{n,n} = L \oplus L^*.$$

This is an even self-dual lattice of signature (n, n) with inner product

$$p = (w, k), \qquad q^2 = 2w \cdot k.$$

It turns out that the symmetries of the lattice $\Gamma^{n,n}$ lift to symmetries of the full conformal field theory. The symmetry group

$$SO(n,n,\mathbb{Z}) = \operatorname{Aut} \Gamma^{n,n}$$

are called T-dualities. A particular example is the interchange $T \leftrightarrow T^*$. From this perspective the string can be considered to be moving on the space $T \times T^*$. If we consider chiral or BPS states in the superstring, then the graded Hilbert space $\mathcal{H}(q)$ is given by

$$\mathcal{H}(p) = \mathcal{F}(\tfrac{1}{2}p^2)\,,$$

in terms of the Fock space

$$\mathcal{F}_q = \bigotimes_{n=1}^{\infty} S_{q^n}(\mathbb{R}^8) \otimes \bigwedge\nolimits_{q^n}(\mathbb{R}^8) = \bigoplus_{N \geq 0} q^N \mathcal{F}(N)\,.$$

Note that the Dynkin diagram of the corresponding lie algebra $D_n \cong so(n,n)$ is obtained from $A_{n-1} \cong sl(n)$ by adding an extra root. Reflections in this root represent the T-duality that maps T to T^*.

5.3. Branes on a torus

If we move to the full M-theory the charge lattice becomes more complicated. For small values of n ($n \geq 4$) it can be written as

$$\Gamma^{n,n} \oplus H^{\mathrm{even/odd}}(T)\,.$$

Here we note that the lattice of branes (which are even or odd depending on the type of string theory that we consider)

$$H^{\mathrm{even/odd}}(T) \cong \bigwedge\nolimits^{\mathrm{even/odd}} L^*$$

transform as half-spinor representations under the T-duality group $SO(n,n,\mathbb{Z})$. The full duality group turns out to be the exceptional group over the integers

$$E_{n+1}(\mathbb{Z})\,.$$

So we see that our hierarchy is reflected in the symmetry groups

$$SL(n,\mathbb{Z}) \subset SO(n,n\mathbb{Z}) \subset E_{n+1}(\mathbb{Z})$$

of rank $n-1$, n and $n+1$ respectively. It is already a very deep (and generally unanswered) question what the 'right' mathematical structure is associated to an n-torus that gives rise to the group $E_{n+1}(\mathbb{Z})$.

In this case we also have indirect evidence how the graded Hilbert space $\mathcal{H}$ should behave. If one considers so-called BPS states, the graded pieces $\mathcal{H}(\gamma)$ should be finite-dimensional, and for large γ we can estimate their growth

$$\dim \mathcal{H}(\gamma) \sim \exp S(\gamma)$$

with $S(\gamma)$ the entropy. Arguments from black hole physics tell us that

$$S(\gamma) = \sqrt{Q(\gamma)}\,,$$

with $Q(\gamma)$ an algebraic invariant of the duality group G. For example, for $n = 5$ with $G = E_6$ the lattice Γ has rank 26 and Q is the famous cubic invariant. Similarly for $n = 6$ and $G = E_7$ we obtain the quartic invariant of the 56-dimensional representation.

6. D-Branes

The crucial ingredients to extend string theory beyond perturbation theory are D-branes [13]. From a mathematical point of view D-branes can be considered as a relative version of Gromov-Witten theory. The starting point is now a pair of relative manifolds (X, Y) with X a d-dimensional manifold and $Y \subset X$ closed. The string world-sheets are defined to be Riemann surfaces Σ with boundary $\partial\Sigma$, and the class of maps $x\colon \Sigma \to X$ should satisfy

$$x(\partial\Sigma) \subset Y.$$

That is, the boundary of the Riemann surfaces should be mapped to the subspace Y.

Note that in a functorial description there are now two kinds of boundaries to the surface. First there are the time-like boundaries that we just described. Here we choose a definite boundary condition, namely that the string lies on the D-brane Y. Second there are the space-like boundaries that we considered before. These are an essential ingredient in any Hamiltonian description. On these boundaries we choose initial value conditions that then propagate in time. In closed string theory these boundaries are closed and therefore are sums of circles. With D-branes there is a second kind of boundary: the open string with interval $I = [0, 1]$.

The occurrence of two kinds of space-like boundaries can be understood because there are various ways to choose a 'time' coordinate on a Riemann surface with boundary. Locally such a surface always looks like $S^1 \times \mathbb{R}$ or $I \times \mathbb{R}$. This ambiguity how to slice up the surface is a powerful new ingredient in open string theory.

To the CFT described by the pair (X, Y) we will associate an extended modular category. It has two kinds of objects or 1-manifolds: the circle S^1 (the closed string) and the interval $I = [0, 1]$ (the open string). The morphisms between two 1-manifolds are again bordisms or Riemann surfaces Σ now with possible boundaries. We now have two kinds of Hilbert spaces: closed strings $\mathcal{H}_{S^1}$ and open strings $\mathcal{H}_I$.

Semi-classically, the open string Hilbert space is given by

$$\mathcal{H}_I = L^2(Y, \mathcal{F})$$

with Fock space bundle

$$\mathcal{F} = \bigotimes_{n \geq 1} S_{q^n}(TX).$$

Note that we have only a single copy of the Fock space $\mathcal{F}$, the boundary conditions at the end of the interval relate the left-movers and the right-movers. Also the fields are sections of the Fock space bundle over the D-brane Y, not over the full space-time manifold X. In this sense the open string states are localized on the D-brane.

We have seen that an important new ingredient in the step from classical geometry to 'stringy' geometry was the 2-form B field, technically a connection on a gerbe. It coupled to the string world-sheet via

$$\int_\Sigma B\,.$$

The phase factor

$$\exp i \int_\Sigma B$$

that appears in the path-integral should be considered as the generalization of the holonomy of a connection on a line bundle associated to a loop. It satisfies a gauge invariance

$$B \to B + d\Lambda\,.$$

If we now work in the category of Riemann surfaces with boundary, we see that such maps are not gauge invariant but pick up a phase factor

$$\exp i \int_{\partial\Sigma} \Lambda\,.$$

However, on surfaces with boundary we can weight the path-integral with an extra phase factor. Let A be a connection (on a trivial line bundle) on Y. Then we can add the holonomy phase factor

$$\exp i \int_{\partial\Sigma} A\,.$$

Now we see that the system has a generalized gauge invariance where apart from the transformation of B we have

$$A \to A - \Lambda\,.$$

This leads to a generalized notion of gauge invariant curvature

$$F = dA + B\,.$$

This equation immediately implies that *when restricted to the D-brane Y* the curvature $H = dB$ should vanish.

6.1. Branes and matrices

One of the most remarkable facts is that D-branes can be given a multiplicity N which naturally leads to a non-abelian structure [19].

Given a modular category as described above there is a simple way in which this can be tensored over the $N \times N$ hermitean matrices. We simply replace the Hilbert space $\mathcal{H}_I$ associated to the interval I by

$$\mathcal{H}_I \otimes Mat_{N \times N}$$

with the hermiticity condition

$$(\psi \otimes M_{IJ})^* = \psi^* \otimes M_{JI}.$$

The maps Φ_Σ are generalized as follows. Consider for simplicity first a surface Σ with a single boundary C. Let C contain n 'incoming' open string Hilbert spaces with states $\psi_1 \otimes M_1, \ldots, \psi_n \otimes M_n$. These states are now matrix valued. Then the new morphism is defined as

$$\Phi_\Sigma(\psi_1 \otimes M_1, \ldots, \psi_n \otimes M_n) = \Phi_\Sigma(\psi_1, \ldots, \psi_n) \operatorname{Tr}(M_1 \cdots M_n).$$

In case of more than one boundary component, we simply have an additional trace for every component.

In particular we can consider the disk diagram with three open string insertions. By considering this as a map

$$\Phi_\Sigma \colon \mathcal{H}_I \otimes \mathcal{H}_I \to \mathcal{H}_I$$

we see that this open string interaction vertex is now given by

$$\Phi_\Sigma(\psi_1 \otimes M_1, \psi_2 \otimes M_2) = (\psi_1 * \psi_2) \otimes (M_1 M_2).$$

So we have tensored the associate string product with matrix multiplication.

If we consider the geometric limit where the CFT is thought of as the semi-classical sigma model on X, the string fields that correspond to the states in the open string Hilbert space $\mathcal{H}_I$ will become matrix valued fields on the D-brane Y, *i.e.* they can be considered as sections of $\operatorname{End}(E)$ with E a (trivial) vector bundle over Y.

This matrix structure naturally appears if we consider N different D-branes $Y_1, \ldots, Y_N$. In that case we have a matrix of open strings that stretch from brane Y_I to Y_J. In this case there is no obvious vector bundle description. But if all the D-branes coincide, $Y_1 = \cdots = Y_N$, a $U(N)$ symmetry appears.

6.2. D-branes and K-theory

The relation with vector bundles has proven to be extremely powerful. The next step is to consider D-branes with *non-trivial* vector bundles. It turns out that these configurations can be considered as a composite of branes of various dimensions [6]. There is a precise formula that relates the topology of the vector bundle E to the brane charge $\mu(E)$ that can be considered as a class in $H^*(X)$. (For convenience we consider first maximal branes $Y = X$.) It reads [8]

$$\mu(E) = ch(E)\widehat{A}^{1/2} \in H^*(X).\tag{2}$$

Here $ch(E)$ is the (generalized) Chern character $ch(E) = \operatorname{Tr}\exp(F/2\pi i)$ and $\widehat{A}$ is the genus that appears in the Atiyah-Singer index theorem. Note that the D-brane charge can be fractional.

Branes of lower dimension can be described by starting with two branes of top dimension, with vector bundles E_1 and E_2, of opposite charge. Physically two such branes will annihilate, leaving behind a lower-dimensional collection of branes. Mathematically the resulting object should be considered as a virtual bundle $E_1 \ominus E_2$ that represents a class in the K-theory group $K^0(X)$ of X [20]. In fact the map μ in (2) is a well-known correspondence

$$\mu \colon K^0(X) \to H^{\text{even}}(X)$$

which is an isomorphism when tensored with the reals. In this sense there is a one-to-one map between D-branes and K-theory classes [20]. This relation with K-theory has proven to be very useful.

6.3. Example: the index theorem

A good example of the power of translating between open and closed strings is the natural emergence of the index theorem. Consider the cylinder $\Sigma = S^1 \times I$ between two D-branes described by (virtual) vector bundles E_1 and E_2. This can be seen as a closed string diagram with in-state $|E_1\rangle$ and out-state $|E_2\rangle$

$$\Phi_\Sigma = \langle E_2, E_1\rangle .$$

Translating the D-brane boundary state into closed string ground states (given by cohomology classes) we have

$$|E\rangle = \mu(E) \in H^*(X)$$

so that

$$\Phi_\Sigma = \int_X ch(E_1)ch(E_2^*)\widehat{A}.$$

On the other hand we can see the cylinder also as a trace over the open string states, with boundary conditions labeled by E_1 and E_2. The ground states in $\mathcal{H}_I$ are sections of the Dirac spinor bundle twisted by $E_1 \otimes E_2^*$. This gives

$$\Phi_\Sigma = \operatorname{Tr}_{\mathcal{H}_I}(-1)^F = \operatorname{index}(D_{E_1 \otimes E_2^*}).$$

So the index theorem follows rather simply.

7. U-Duality

We indicated that in M-theory we do not want to include only strings but also D-branes (and even further objects that I will suppress in this discussion such as NS 5-branes and Kaluza-Klein monopoles). So in the limit of small string coupling λ the full (second quantized) string Hilbert space would look something like

$$\mathcal{H} = S^*(\mathcal{H}_{\text{string}}) \otimes S^*(\mathcal{H}_{\text{brane}}) \,.$$

Of course our discussion up to now has been very skew. In the full theory there will be symmetries, called U-dualities, that will exchange strings and branes.

There are at present very few formulations of M-theory that present such a manifest duality-invariant approach. Only for very special compactifications (such as low-dimensional tori) does matrix theory [1], or the famous AdS-CFT correspondence [11], give a precise non-perturbative definition.

We will give a rather simple example of such a symmetry that appears when we compactify the (Type IIA) superstring on a four-torus $T^4 = \mathbb{R}^4/L$. In this case the charge lattice has rank 16 and can be written as

$$\Gamma^{4,4} \oplus K^0(T^4) \,.$$

It forms an irreducible spinor representation under the U-duality group

$$G = SO(5,5,\mathbb{Z}) \,.$$

Notice that the T-duality subgroup $SO(4,4,\mathbb{Z})$ has three inequivalent 8-dimensional representations (related by triality). The strings with Narain lattice $\Gamma^{4,4}$ transform in the vector representation while the even-dimensional branes labeled by the K-group $K^0(T^4) \cong \wedge^{\text{even}} L^*$ transform in the spinor representation. (The odd-dimensional D-branes that are labeled by $K^1(T)$ and that appear in the Type IIB theory transform in the conjugate spinor representation.)

To compute the spectrum of superstrings we have to introduce the corresponding Fock space. It is given by

$$\mathcal{F}_q = \bigotimes_{n=1}^{\infty} S_{q^n}(\mathbb{R}^8) \otimes \wedge_{q^n}(\mathbb{R}^8) = \bigoplus_{N \geq 0} q^N \mathcal{F}(N) \,.$$

The Hilbert space of BPS strings with momenta $p \in \Gamma^{4,4}$ is then given by

$$\mathcal{H}_{\text{string}}(p) = \mathcal{F}(\tfrac{1}{2}p^2) \,.$$

For the D-branes we take a completely different approach. Since we only understand the system for small string coupling we have to use semi-classical methods. Consider a D-brane that corresponds to a K-theory class E with charge vector $\mu = ch(E) \in H^*(T)$. To such a vector bundle we can associate a moduli space $\mathcal{M}_\mu$ of self-dual connections. (If we work in the holomorphic context we could equally well consider the moduli space of holomorphic sheaves of this topological class.) Now luckily a lot is known about these moduli spaces. They are hyper-Kähler and (for primitive μ) smooth. In fact, they are topologically Hilbert schemes [10] which are deformations of symmetric products

$$\mathcal{M}_\mu \cong \mathrm{Hilb}^{\mu^2/2}(T^4) \sim S^{\mu^2/2}T^4 \, .$$

Computing the BPS states through geometric quantization we find that

$$\mathcal{H}_{\mathrm{brane}}(\mu) = H^*(\mathcal{M}_\mu) \, .$$

The cohomology of these moduli spaces have been computed [7] with the result that

$$\bigoplus_{N \geq 0} q^N H^*(\mathrm{Hilb}^N(T^4)) = \mathcal{F}_q \, .$$

This gives the final result

$$\mathcal{H}_{\mathrm{brane}}(\mu) = \mathcal{F}(\mu^2/2) \cong \mathcal{H}_{\mathrm{string}}(p)$$

where μ and p are related by an $SO(5,5,\mathbb{Z})$ U-duality transformation.

8. Non-Commutative Geometry

From the present point of view, D-branes and the corresponding open strings seem to indicate that in a more final formulation of M-theory a fundamental role will be played by non-commutative geometry [2, 16]. One of the indications is the occurrence of the B-field in string theory. Roughly in the presence of such a (constant) B-field the space-time coordinates no longer commute, but instead satisfy

$$[x_\mu, x_\nu] = B_{\mu\nu} \, .$$

One of the most striking results is that the D-brane gauge theory, in the presence of such a B-field indeed is invariant under the T-duality group through the concept of Morita equivalence. Further explorations of the links between string theory and noncommutative geometry can well give the key to a final understanding of M-theory.

References

[1] T. Banks, W. Fischler, S. H. Shenker and L. Susskind, *M Theory as a matrix model: a conjecture*, Phys. Rev. **D55** (1997) 5112–5128, hep-th/9610043.

[2] A. Connes, M. Douglas and A. Schwarz, *Noncommutative geometry and matrix theory: compactifications on tori*, JHEP **9802** (1998) 003, hep-th/9711162.

[3] D. A. Cox and S. Katz, *Mirror Symmetry and Algebraic Geometry* (Mathematical Surveys and Monographs, No. 68.), AMS, 1999.

[4] Ph. Di Francesco, P. Mathieu and D. Senechal, *Conformal Field Theory* (Graduate Texts in Contemporary Physics), Springer-Verlag, 1996.

[5] R. Dijkgraaf, *Les Houches Lectures on Fields, Strings and Duality*, in *Quantum Symmetries*, les Houches Session LXIV, Eds. A. Connes, K. Gawedzki, and J. Zinn-Justin, North-Holland, 1998, hep-th/9703136.

[6] M. Douglas, *Branes within branes*, hep-th/9512077.

[7] L. Göttsche, *The Betti numbers of the Hilbert scheme of points on a smooth projective surface*, Math. Ann. **286** (1990) 193–207; *Hilbert Schemes of Zero-dimensional Subschemes of Smooth Varieties*, Lecture Notes in Mathematics **1572**, Springer-Verlag, 1994.

[8] M. Green, J. Harvey and G. Moore, *I-brane inflow and anomalous couplings on D-branes*, Class. Quant. Grav. **14** (1997) 47–52, hep-th/9605033.

[9] N. Hitchin, *Lectures on special Lagrangian submanifolds*, math/9907034.

[10] D. Huybrechts, *Compact hyperkähler manifolds: basic results*, alg-geom/9705025.

[11] J. Maldacena, *The large N limit of superconformal field theories and supergravity*, Adv. Theor. Math. Phys. **2** (1998) 231–252, hep-th/9711200.

[12] Yu. I. Manin, *Frobenius Manifolds, Quantum Cohomology, and Moduli Spaces*, AMS, 1999.

[13] J. Polchinski, *Dirichlet-branes and Ramond-Ramond charges*, Phys. Rev. Lett. **75** (1995) 4724–4727, hep-th/9510017.

[14] J. Polchinski, *String Theory* (Cambridge Monographs on Mathematical Physics), Cambridge University Press, 1998.

[15] G. Segal, *The definition of conformal field theory*, preprint; *Two dimensional conformal field theories and modular functors*, in IXth International Conference on Mathematical Physics,. B. Simon, A. Truman and I. M. Davies Eds. (Adam Hilger, Bristol, 1989).

[16] N. Seiberg and E. Witten, *String theory and noncommutative geometry*, JHEP **9909** (1999) 032, hep-th/9908142.

[17] A. Sen and B. Zwiebach, *Tachyon condensation in string field theory*, JHEP **0003** (2000) 002, hep-th/9912249.

[18] E. Witten, *String theory in various dimensions*, Nucl. Phys. **B 443** (1995) 85, hep-th/9503124.

[19] E. Witten, *Bound states of strings and p-branes*, Nucl. Phys. **B460** (1996) 335, hep-th/9510135.

[20] E. Witten, *D-branes and K-theory*, JHEP **9812** (1998) 019, hep-th/9810188.

[21] B. Zwiebach, *Closed string field theory: quantum action and the B-V master equation*, Nucl. Phys. **B390** (1993) 33–152, hep-th/9206084.

Korteweg-de Vries Institute for Mathematics
University of Amsterdam
Plantage Muidergracht 24
1018 TV Amsterdam, The Netherlands
E-mail address: rhd@science.uva.nl

Probabilistic Aspects of Financial Risk

Hans Föllmer

Abstract. Problems arising in finance have become a significant source of new developments in stochastic analysis. We discuss some recent case studies, in particular some decomposition and representation theorems which are motivated by problems of hedging derivatives and of intertemporal consumption choice.

1. Introduction

In recent years, mathematics has become highly visible as a key technology in the area of Finance. This is reflected in new journals, new curricula, and in new job profiles. Increasingly, advanced methods from probability and statistics are being applied to the analysis of financial risk in its various forms. Their impact is not only felt on a computational level. To a surprising extent, concepts of stochastic analysis are shaping the discourse of the field, both on the academic level and in the financial industry. Conversely, finance has become a significant source of research problems which are of intrinsic mathematical interest, in particular in the area of stochastic analysis. The purpose of this lecture is to discuss some case studies which illustrate this converse impact.

We begin by explaining the key paradigm of a *perfect hedge*. It is concerned with financial *derivatives*, or *contingent claims*, which are defined in terms of some underlying financial asset. The mathematical analysis of such derivatives was started by Bachelier [1] who introduced Brownian motion as a model for the underlying price dynamics and computed prices for certain derivatives as expected values in this model. A nice diffusion model such as (geometric) Brownian is *complete* in the sense that any derivative can be replicated by a trading strategy in the underlying asset. From a mathematical point of view, this amounts to a version of Itô's representation theorem which represents functionals of Brownian motion as Itô integrals. From a financial point of view, the existence of a replicating strategy was the crucial insight in the work of Black and Scholes [6] and Merton [29] on arbitrage-free option pricing. Even though its standard version involves "high tech" methods from probability theory, the argument behind the pricing formula of Black and Scholes is essentially probability-free on a conceptual level; this is explained in Section 1.

However, the idea of a perfect hedge is limited in scope. Realistic models will be *incomplete*, i.e., derivatives typically will carry an intrinsic risk which cannot be hedged away completely. At this stage, probability comes in on a more fundamental level, and the hedging problem has become a source of new decomposition theorems in the theory of semimartingales. In this setting, we are going to focus on controlling the *shortfall*, i.e., that part of the contingent claim which will not be covered by the result of our hedging strategy. In Section 2.1 we insist on staying on the safe side by constructing a *superhedging* strategy which keeps the shortfall down to 0. In the general case, the construction of such a superhedging strategy involves a decomposition theorem of Kramkov [24] which may be viewed as a new version of the fundamental Doob–Meyer decomposition for supermartingales. In many situations the cost of superhedging a given claim will be too high from a practical point of view. Thus, we are led to the problem of constructing strategies which are *efficient* with respect to the cost and a suitably defined *shortfall risk*. In Section 2.2 we discuss the solution of this problem in Föllmer and Leukert [18]; in particular we will see how it is reduced to the problem of superhedging a suitably modified claim.

As a further case study, we consider in Section 3 a new variant of the problem of intertemporal consumption choice. Through the analysis of Bank and Riedel [3], it leads to a new stochastic representation problem which is closely related to the continuous-time theory of the Gittins index in El Karoui and Karatzas [10]. Its solution in Bank [2] involves a new stochastic analogue to the concept of a convex hull and is of interest from a purely mathematical point of view, quite independent of the initial motivation from finance.

2. The Idea of A Perfect Hedge

In this section we describe the idea of a *perfect hedge*, one of the key insights in mathematical finance. It is concerned with financial derivatives defined in terms of some underlying financial asset. A perfect hedge consists in replicating the derivative by a dynamical portfolio strategy which trades in the underlying asset. In a nice diffusion model for the stochastic fluctuation of asset prices such as geometric Brownian motion, this can indeed be done. In the usual presentation of this result a paradox arises: The argument uses "high tech" probabilistic methods, but the phenomenon itself does not really involve probability on a conceptual level.

In the standard formalism of mathematical finance, the evolution of the (properly discounted) price of some underlying asset over a given time interval $[0, T]$ is described as a semimartingale $X = (X_t)_{t \in [0,T]}$ on a probability space $(\Omega, \mathcal{F}, P)$ with filtration $(\mathcal{F}_t)_{t \in [0,T]}$. At this point, a reader not familiar with the notions of stochastic analysis may feel somewhat discouraged. But in this section we will drastically reduce the conceptual and technical difficulties by eliminating the probabilistic ingredients altogether. The basic uncertainty about the behaviour of asset prices will be reflected only in the fact that we specify a large set Ω of possible scenarios ω, and that we do not know in advance which scenario will actually take place.

A *derivative*, or a *contingent claim*, specifies a payoff $H(\omega)$ contingent on the scenario $\omega \in \Omega$ which will be realized. For example, a European call option with strike price c and exercise time T has payoff $H(\omega) = (X_T(\omega) - c)^+$. What is the *fair price* which should be payed by the buyer of such an option? In other words, what is the fair deterministic equivalent to the random variable H? This is a classical question, and the standard answer goes back to the founding fathers of probability theory, in particular to Jacob Bernoulli. It says that you should assign probabilities to the different scenarios ω and compute the expected value $E[H] = \int H dP$ of the random variable H with respect to the resulting probability measure P. Following Daniel Bernoulli, one might also want to add a risk premium in order to take account of risk aversion. But in our present financial context, the basic insight of Black and Scholes [6] and Merton [29] leads to a quite different result. In particular there will be no reason to add a risk premium, because the argument will show that there is no intrinsic risk. As we are going to see, the answer does not really involve the probabilities of the different scenarios; it is the structure of the space of scenarios which matters.

As *scenarios* we take the possible trajectories of the asset price starting from a fixed initial value x_0. Such a trajectory will be described as a positive continuous function $\omega \colon [0, T] \to R^+$. We use the notation $X_t(\omega) = \omega(t)$ for the corresponding price at time t; in particular we have $X_0(\omega) = x_0$. Let us introduce the space Ω of all such functions which admit a continuous quadratic variation

$$[X]_t(\omega) = \lim_n \sum_{t_i \leq t,\, t_i \in D_n} (X_{t_{i+1}}(\omega) - X_{t_i}(\omega))^2 \tag{1}$$

along the sequence of dyadic partitions D_n. For such functions we can use a strictly pathwise version of Itô's calculus; see Föllmer [12]. In fact, for any function $F(x, t) \in C^2$ and for any scenario $\omega \in \Omega$, Itô's formula holds in the form

$$F(X_t(\omega), t) = F(x_0, 0) + \int_0^t F_x(X_s(\omega), s) dX_s(\omega)$$

$$+ \int_0^t \frac{1}{2} F_{xx}(X_s(\omega), s) d[X]_s + \int_0^t F_t(X_s(\omega), s) ds. \tag{2}$$

The second integral is well defined as a Lebesgue–Stieltjes integral, while the first exists as an Itô integral, i.e., as limit of the non-anticipating Riemann sums

$$\sum_{t_i \leq t,\, t \in D_n} \xi_{t_i}(\omega)(X_{t_{i+1}}(\omega) - X_{t_i}(\omega)), \tag{3}$$

along the dyadic partitions D_n, where we put $\xi_t = F_x(X_t, t)$.

From a financial point of view where ω is seen as a possible scenario for the evolution of an asset price, the Itô integral can be interpreted as the *cumulative gain* from a self-financing trading strategy defined by ξ. In fact, if $\xi_t(\omega)$ is the number of shares held at time t, then the terms of the non-anticipating Riemann sums in (3) are the net gains generated by the price changes in successive periods. The non-anticipative construction of the Itô integral corresponds exactly to the natural financial requirement that the investment decision at a time does not anticipate the upcoming price change during the next time interval. With this interpretation in mind, it is quite crucial to insist on trajectories with a non-vanishing quadratic variation. Otherwise there would be arbitrage opportunities, i.e., trading strategies which would offer the chance of a strictly positive gain without any downside risk. In fact, suppose that we would admit only scenarios ω with quadratic variation $[X](\omega) = 0$. By Itô's formula, the strategy $\xi_t(\omega) = 2(X_t(\omega) - x_0)$ would then generate the gain $(X_T(\omega) - x_0)^2$, and this is clearly an arbitrage opportunity.

From a probabilistic point of view, the set Ω provides a natural framework because the law of a typical diffusion process will be carried by this set of paths. For a specific model like geometric Brownian motion we can be more precise and describe the typical trajectories in terms of a fixed volatility, given by a strictly positive (piecewise) continuous function σ on $[0, T]$. Without introducing any probabilities at this point, we will simply restrict the discussion to the set Ω_σ of all scenarios $\omega \in \Omega$ such that the quadratic variation is absolutely continuous with $d[X]_t = \sigma^2(t)X_t^2 dt$. In this context we can now explain the idea of a perfect hedge for a derivative of the form $H(\omega) = h(X_T(\omega))$ for some continuous function $h \geq 0$; a similar discussion can be found in Bick and Willinger [5]. Let $F(x, t)$ denote the solution of the parabolic equation

$$\mathcal{L}_F^\sigma(x, t) := (\frac{1}{2}\sigma(t)^2 x^2 \frac{\partial^2}{\partial x^2} + \frac{\partial}{\partial t})F(x, t) = 0 \qquad (4)$$

on the strip $R^+ \times [0, T]$ with boundary condition $F(x, T) = h(x)$. Consider the trading strategy $\xi_t = F_x(X_t, t)$. By Itô's formula, we obtain

$$H(\omega) = F(X_T(\omega), T) = F(x_0, 0) + \int_0^T \xi_s(\omega)dX_s(\omega) \qquad (5)$$

since the remaining part in Itô's formula vanishes due to (4). As we have seen, the Itô integral can be viewed as the net gain resulting from the self-financing trading strategy ξ. Thus, the contingent claim $H(\omega)$ is replicated as the final value of a portfolio generated by the initial amount $F(x_0, 0)$ and the strategy ξ. This replication of the derivative works simultaneously for all scenarios in our class Ω_σ. Thus, it does not involve any risk as long as we stick to the assumption that only scenarios with volatility profile σ will come up. Under this assumption, the initial cost $F(x_0, 0)$ of such a perfect replication is clearly the right price for the derivative. Otherwise there would be an arbitrage opportunity. If, for example, the price would be higher, then one could sell the derivative at that price and at the same time replicate the option at a lower cost. For any scenario one would then generate the required amount $H(\omega)$ at the final time T, and the difference between the price and the cost of replication would remain as a risk free gain.

So far, our discussion did not involve any probabilities on the space of possible scenarios. But we can now introduce probabilities as a "high tech" device which allows us to compute the correct price as an expected value rather than by solving a partial differential equation. By a theorem of Paul Lévy, there is exactly one probability measure P^* on the space Ω_σ such that the coordinate process $(X_t)_{t\in[0,T]}$ becomes a *martingale* under P^*, i.e.,

$$E^*[X_t|\mathcal{F}_s] = X_s \tag{6}$$

for times $s < t$. Under some growth condition on the function h, this will imply that the Itô integral in (5) has zero expectation under P^*. Thus, the price of the derivative can be computed as an expected value

$$F(X_0,0) = E^*[H] \tag{7}$$

with respect to the probability measure P^*. In this sense, probabilities do arise. But they do not appear as objective probabilities which govern the realization of the different scenarios. Instead, the probability measure P^* plays the role of a consistent pricing rule for derivatives of the underlying asset. Consistency means, in particular, that the price $X_s(\omega)$ of the asset at time s should be equal to the conditional expectation of any future price X_t under the measure P^* given the information up to time s, and this is exactly the martingale property required for P^*.

Let us emphasize that the preceding probabilistic argument is optional, not really essential. In principle, the argument of pricing a derivative by means of a perfect replication is probability-free. But note that the volatility profile σ has to be known in advance; this is needed in order to solve in advance the parabolic differential equation which determines both the initial cost and the self-financing strategy. If this is not the case, then we have an additional source of uncertainty. Does probability come in at this second stage? Not necessarily. Of course we could model volatility as a stochastic process, and this is an important chapter in mathematical finance. But we could also continue to insist on a probability free approach using the following argument by T. Lyons [28]. Suppose that we are ready to assume that the unknown volatility $\sigma(t)$ will in any case stay inside some moving interval $\Sigma(t)$. Under some regularity conditions, there is a solution $F \in C^2$ of the nonlinear Pucci equation

$$\mathcal{L}F(x,t) := (\sup_{\sigma(t)\in\Sigma(t)} \frac{1}{2}\sigma^2(t)x^2 \frac{\partial^2}{\partial x^2} + \frac{\partial}{\partial t})F(x,t) = 0 \tag{8}$$

with terminal value $F(x,T) = h(x)$. For any scenario ω with a volatility profile σ staying inside the given moving boundaries, our pathwise Itô formula together with equation (8) implies

$$F(X_t(\omega),t) = F(x_0,0) + \int_0^t F_x(X_s(\omega),s)dX_s(\omega) - B_t(\omega) \tag{9}$$

where

$$B_t(\omega) = \int_0^t (\mathcal{L} - \mathcal{L}^\sigma) F(X_s(\omega), s) ds \tag{10}$$

is increasing in t. From a financial point of view, this means that we can *superhedge* the contingent claim $H(\omega) = F(X_T(\omega), T)$ as follows. We start with the initial amount $F(x_0, 0)$ and use the trading strategy $\xi_t = F_x(X_t(\omega), t)$. At the same time we cumulatively withdraw the capital $B_t(\omega)$. The resulting portfolio value at time t is given by $F(X_t(\omega), t)$, and at the final time T we have generated exactly the required payoff $H(\omega)$ of our derivative. Allocating the initial capital $F(X_0, 0)$ to the purpose of hedging the derivative, we are thus staying on the safe side. As time goes on and uncertainty is revealed, it will typically become clear at a given time that the capital generated so far by the initial amount and by the strategy ξ is higher than what is actually needed to stay on the safe side from that time on, and this explains the refunding scheme described by the increasing process B. As we will see in the next section, the probabilistic analogue of this approach is provided in great generality by Kramkov's optional decomposition theorem.

3. Controlling The Short Fall

Now we return to the standard probabilistic setting where the (properly discounted) price process is given as an adapted stochastic process $X = (X_t)_{t \in [0,T]}$ on a probability space $(\Omega, \mathcal{F}, P)$ with filtration $(\mathcal{F}_t)_{t \in [0,T]}$. From an economic point of view, the model should be such that it excludes arbitrage opportunities. One of the fundamental theorems in mathematical finance states that this *absence of arbitrage* is essentially equivalent to the existence of an *equivalent martingale measure*, i.e., a measure $P^* \approx P$ such that X is a martingale under P^* in the sense of (6); for a precise formulation see Delbaen and Schachermayer [9] and the references given there. Let us therefore assume

$$\mathcal{P} \neq \emptyset \tag{11}$$

where $\mathcal{P}$ denotes the class of all martingale measures $P^* \approx P$.

This condition may be viewed as a minimal version of the "Efficient Market Hypothesis". In its strong form, the efficient market hypothesis would claim that (properly discounted) prices should be martingales under the objective measure P. This is based on a rough equilibrium argument: If the present price is below the market's average expectation, then the price should move immediately up, and vice versa. In fact, such an equilibrium argument was the starting point in the thesis of Bachelier [1] where Brownian motion was introduced as a model for price fluctuation on a speculative market. Condition (11) is of course much weaker since the martingale property is required only up to an equivalent change of measure. By Girsanov's theorem, it implies that X is a semimartingale under P, and so we are free to use the general tools of stochastic analysis.

Uniqueness of the equivalent martingale measure P^* is equivalent to *completeness*, i.e., any derivative H of the underlying price process admits (under some integrability conditions) a perfect replication in terms of a suitable trading strategy. As in the previous section, it follows that the price must be equal to the cost of replication, and that it can be computed as the expectation $E^*[H]$ under the unique equivalent martingale measure P^*.

There are, however, many reasons why the reality of financial markets should be seen as being *incomplete*: Not any contingent claim admits a perfect hedge. In other words, there are intrinsic risks which cannot be hedged away completely by a strategy which uses the available assets. From a mathematical point of view, this means that there is more than one equivalent martingale measure, and since $\mathcal{P}$ is convex we have

$$|\mathcal{P}| = \infty. \tag{12}$$

In this general context let us now look at the problem of hedging a derivative described by an $\mathcal{F}_T$-measurable random variable $H \geq 0$. An *admissible strategy* is determined by some initial amount V_0 and by a predictable integrand ξ for X such that the resulting portfolio process

$$V_t = V_0 + \int_0^t \xi_s ds \tag{13}$$

satisfies $V_t \geq 0$ for all $t \in [0, T]$. Whichever strategy we use, we will end up at time T with a *decomposition*

$$H = V_0 + \int_0^T \xi_s ds + C_T \tag{14}$$

of the contingent claim into a part which is perfectly hedged and hence priced by arbitrage, and into the remaining *hedging error* C_T. There are different approaches to the problem of choosing a hedging strategy, and hence a specific decomposition of the contingent claim. From an economic point of view, they involve different preferences towards risk. From a mathematical point of view, they have been a significant source of new decomposition theorems in the general theory of stochastic processes.

One approach consists in choosing a strategy which minimizes the hedging error C_T in $\mathcal{L}^2(P)$. This involves a projection of H on a space of stochastic integrals. Alternatively, we could insist on a decomposition where the process C is mean-self financing in the sense that it is a martingale under the objective measure P. In the special case where the efficient market hypothesis holds in its strong form $P \in \mathcal{P}$, both formulations of the problem are equivalent, and it is solved by the classical Kunita–Watanabe decomposition for square integrable martingales; see Föllmer and Sondermann [13]. In the case $P \notin \mathcal{P}$, the two versions of the problem are different but intimately related. Often, but not always, the problem can be reduced to the Kunita–Watanabe decomposition with respect to a suitable "minimal" martingale measure; see Föllmer and Schweizer [14]. In the general case, and in particular through the work of Martin Schweizer, the projection problem has become a source of new versions of the Kunita–Watanabe decomposition and of new closure results for spaces of stochastic integrals with respect to a semimartingale; see, e.g., Rheinländer and Schweizer [32]. For a survey of recent developments we refer to Schweizer [33].

Note, however, that there is a basic asymmetry in the financial interpretation of the hedging error which is not captured by the projection approach. It is really the *shortfall*, i.e., the positive part $C_T^+ = (H - V_T)^+$ of the hedging error, which should be kept under control. Here we will describe two approaches which focus on the shortfall. The first approach insists on staying on the safe side and keeping the shortfall down to 0. This is the idea of superhedging which has already appeared in the previous section. In our general probabilistic setting , it has led to a remarkable new version of the Doob–Meyer decomposition. The second approach consists in constructing partial hedges which are efficient in terms of the cost and of a suitably defined shortfall risk, and here the technique of superhedging comes in as a crucial tool.

3.1. Superhedging

In incomplete models it is not possible to replicate any given contingent claim H. But we could insist on keeping the shortfall C_T^+ down to 0. Thus, we would require

$$P[C_T \leq 0] = P[V_T \geq H] = 1 \,. \tag{15}$$

The program of constructing such a strategy with a minimal initial amount has been carried out on increasing levels of generality; see El Karoui and Quenez [11], Karatzas [21] and the references given there, Kramkov [24] and Föllmer and Kabanov [16]. More precisely, let us assume

$$U_0 := \sup_{P^* \in \mathcal{P}} E^*[H] < \infty \,, \tag{16}$$

and let us define (U_t) as a right-continuous version of the process defined by

$$U_t = \operatorname{ess.\,sup}_{P^* \in \mathcal{P}} E^*[H \mid \mathcal{F}_t] \,. \tag{17}$$

The process (U_t) is a $\mathcal{P}$-*supermartingale*, i.e., a supermartingale simultaneously for any $P^* \in \mathcal{P}$. In fact it is the smallest non-negative $\mathcal{P}$-supermartingale with terminal value $\geq H$.

The Doob–Meyer decomposition shows that a non-negative supermartingale with respect to a fixed probability measure can be represented as the difference of a local martingale and a predictable increasing process. Is there a reasonable analogue for $\mathcal{P}$-*supermartingales* which is valid simultaneously for all $P^* \in \mathcal{P}$? The study of this problem was initiated by El Karoui and Quenez [11] in a special context. As shown in full generality in Kramkov [24] and in Föllmer and Kabanov [16], any non-negative $\mathcal{P}$-supermartingale admits an *optional decomposition* of the form

$$U_t = U_0 + \int_0^t \xi_s dX_s - C_t \tag{18}$$

where C is an increasing optional process and ξ is an admissible strategy. Note that the stochastic integral is a local martingale simultaneously for all $P^* \in \mathcal{P}$. On the other hand, the increasing process is only optional but not necessarily predictable.

For the specific $\mathcal{P}$-supermartingale defined in (17), the optional decomposition can be interpreted as the following *superhedging* procedure: Put up the initial capital U_0, then follow the trading strategy ξ and withdraw the cumulative amount of capital C_t from the resulting portfolio as additional information comes in. As a corollary of the optional decomposition, the value U_t can be characterized as the least amount of capital needed at time t to cover the claim H by following an admissible strategy ξ from time t up to the final time T, i.e.,

$$U_t = \operatorname{ess.\,inf} V_t \tag{19}$$

where V_t runs through the class of $\mathcal{F}_t$-measurable random variables ≥ 0 such that

$$V_t + \int_t^T \xi_s dX_s \geq H \quad P\text{-a.s.} \tag{20}$$

for some admissible strategy ξ. In other words, U_t is an upper bound for any arbitrage-free price of the claim computed at time t. If additional constraints are imposed on the strategies ξ, then the dual description (19) of the process defined by (17) has a corresponding analogue in terms of a suitable extension of the class $\mathcal{P}$; see Karatzas [21] and Föllmer and Kramkov [15].

The minimal initial amount which is needed for staying on the safe side is given by U_0. From a practical point of view, this is usually too much. In fact, the whole idea of superhedging is more extreme, resp. conservative, than in any actuarial approach. It is also extreme from a mathematical point of view: For a convex derivative, the superhedging strategy can often be identified as a perfect replication strategy with respect to an associated complete model which is no longer equivalent to the initial model but sits on some kind of Choquet boundary of the initial space of scenarios.

In any case, superhedging provides the lowest upper bound U_0 for any arbitrage-free price of the given derivative. Moreover, as we shall see next, the optional decomposition of a suitably modified claim can be a crucial ingredient in the construction of hedging strategies which may be more realistic from a practical point of view.

3.2. Efficient hedging: cost versus shortfall risk

Together with the risk of a loss, superhedging will also take away the opportunity of making a profit. In fact, the cost of superhedging a contingent claim will typically exceed any actual price which could be obtained for that claim. Now suppose that the investor is unwilling to put up the initial amount U_0 required by a superhedging strategy and is therefore ready to accept some risk. What is the optimal *partial hedge* which can be achieved with a given smaller amount of capital V_0? In order to make this question precise we need a criterion expressing the investor's attitude towards the risk of a shortfall.

One possible criterion is the probability $P[C_T > 0]$ that some shortfall occurs, i.e., that the final amount V_T generated by the strategy falls short of the amount which is required to cover the contingent claim. We could fix an upper bound α for this shortfall probability and minimize the initial cost under this constraint. Alternatively, we could fix some initial cost $V_0 < U_0$ and look for a strategy which minimizes the probability of a shortfall. This problem is solved in Föllmer and Leukert [17], translating and extending an idea of Kulldorf [26] to the present financial context. The resulting strategy of *quantile hedging* arises as the superhedging strategy for a modified claim $\tilde{H}$. Typically, the modified claim takes the form of a knock-out option HI_A, and the set A is constructed as the optimal test in a Neyman–Pearson problem where the alternative is given by the objective measure P, and the compound hypothesis is defined in terms of the contingent claim and the class $\mathcal{P}$ of equivalent martingale measures. Quantile hedging may be viewed as a a dynamic version of the "value at risk" approach. It invites the same critique since it only takes into account the probability that a shortfall occurs, but not the size of the shortfall if it does occur. This motivates a modified approach where the investor's attitude towards the shortfall is specified in terms of a loss function l.

Let us assume that l is an increasing function on R^+ with $l(0) = 0$ and $E[l(H)] < \infty$; convexity of l would correspond to risk aversion. We introduce the notion of *shortfall risk*, defined as the expectation

$$E[l((H - V_T)^+)]$$

of the shortfall $C_T^+ = (H - V_T)^+$ weighted by the loss function l under the objective measure P. If we would be ready to allocate the initial amount U_0, then we could reduce the shortfall risk to 0 by using the superhedging strategy. For an initial capital $V_0 < U_0$, our problem is to find an admissible strategy which minimizes the shortfall risk under this cost constraint; alternatively, we could fix a bound on the shortfall risk and minimize the cost. In other words, we are looking for hedges which are *efficient* with respect to cost and shortfall risk. These efficient hedges interpolate in a systematic way between the extremes of a superhedge (no risk, no chance of making a profit) and no hedge (full risk of shortfall, full chance of profit), depending on the accepted level of shortfall risk.

This problem of minimizing the shortfall risk under a cost constraint is solved in Föllmer and Leukert [18]. The solution proceeds in two steps. In a first step, we consider the statistical decision problem

$$\min_{\varphi \in \mathcal{R}_0} E[l((1 - \varphi)H)] \tag{21}$$

where $\mathcal{R}_0$ denotes the class of all $\mathcal{F}_T$-measurable random values φ with values in $[0, 1]$ such that

$$\sup_{P^* \in \mathcal{P}} E^*[\varphi H] \leq V_0. \tag{22}$$

There exists a solution $\widetilde{\varphi}$, and we may assume $\widetilde{\varphi} = 1$ on $\{H = 0\}$. As pointed out in Delbaen and Schachermayer [9], existence results of this type follow very easily from a (simplified) version of Komlós's theorem that any sequence of random variables with bounded $\mathcal{L}^1$-norms admits a subsequence which satisfies the strong law of large numbers. Here we even have random variables bounded by 1, and in this case the theorem was already known to Rényi; see [23]. In fact, if (φ_n) is a sequence of functions in $\mathcal{R}_0$ whose shortfall risk approaches the infimum in (21), and if $\widetilde{\varphi}$ is the limit of an almost surely convergent sequence of averages along some subsequence, then it is easy to check that $\widetilde{\varphi}$ is indeed a solution of problem (21).

Let us introduce the modified claim $\tilde{H} = \widetilde{\varphi}H$, and let $\tilde{U}$ denote a right-continuous version of the process

$$\tilde{U}_t = \operatorname*{ess.\,sup}_{P^* \in \mathcal{P}} E^*[\tilde{H} \mid \mathcal{F}_t]. \tag{23}$$

Since $\tilde{U}$ is a $\mathcal{P}$-supermartingale, we can now apply the optional decomposition theorem. As shown in Föllmer and Leukert [18], this yields the solution of our optimization problem:

Theorem 3.1. *The strategy $\tilde{\xi}$ determined by the optional decomposition of the $\mathcal{P}$-supermartingale $\tilde{U}$ generated by the modified claim $\widetilde{H} = \widetilde{\varphi}H$ minimizes the shortfall risk under the constraint that the initial cost is bounded by V_0.*

In the *complete* case where the equivalent martingale measure P^* is uniquely determined, the structure of the optimal profile $\widetilde{H} = \widetilde{\varphi}H$ can be described explicitly, and the optimal strategy consists in replicating the contingent claim $\widetilde{\varphi}H$. For a smooth convex loss function let $I = (l')^{-1}$, the inverse of l', and denote by ρ^* the Radon–Nikodym derivative of P^* with respect to P. Then the solution $\widetilde{\varphi}$ of the optimization problem (21) is given by

$$\widetilde{\varphi} = 1 - \left(\frac{I(c\rho^*)}{H} \wedge 1 \right) \quad \text{on } \{H > 0\}, \tag{24}$$

where the constant c is determined by the condition $E^*[\widetilde{\varphi}H] = V_0$.

It is instructive to introduce a whole scale for the attitude towards risk and to consider the special case $l(x) = \frac{x^p}{p}$ where the risk is defined in terms of an upper partial moment of the hedging error. In the risk-averse case $p > 1$, the optimal hedge consists in replicating the modified claim

$$\varphi_p H = H - c_p(\rho^*)^{\frac{1}{p-1}} \wedge H \tag{25}$$

where the constant c_p is determined such that $E^*[\varphi_p H] = \tilde{V}_0$. In the limit $p \to \infty$ of ever increasing risk aversion, the modified claim $\varphi_p H$ converges to $(H - c)^+$, where c is the unique constant that satisfies $E^*[c \wedge H] = E^*[H] - \tilde{V}_0$. If H is a call at strike K, then the limit for $p \to \infty$ is again a call at the higher strike $\tilde{K} = K + c$ whose arbitrage-free price is given by V_0. We may also consider the case $p < 1$ where risk-averse behavior is replaced by risk-seeking behavior. As appetite for risk increases and p decreases from 1 to 0, the corresponding efficient hedges converge to the knock-out option which appears in the case of quantile hedging.

In the *incomplete* case, the problem of computing the optimal profile is much more subtle. A specific case study, where the problem consists in constructing efficient hedges for a volatility jump, can be found in Föllmer and Leukert [18]. The general case involves methods of convex duality, and in particular a deep result of Kramkov and Schachermayer [25] which describes the duality between the class of claims which can be generated by admissible strategies and a suitable set of supermartingale densities. We refer to Leukert [27] for details in the present context. Closely related results and extensions appear in Pham [31], Mnif and Pham [30], Cvitanić and Karatzas [7], and in Cvitanić, Schachermayer and Wang [8].

4. Stochastic Optimization under Intertemporal Preferences

So far, our discussion has involved preferences on the space of portfolio values V_T generated by a trading strategy at a fixed terminal time T. Initiated by von Neumann und Morgenstern, the structure of reasonable preferences on a space of random variables has been studied in depth and is now well understood; for a survey see Karni and Schmeidler [22]. Typically, they can be represented in terms of an expected utility functional. Our optimization problem above fits into this framework: For a convex loss function l, we have maximized the functional $E[u(V_T(\omega), \omega)]$ defined in terms of the increasing concave scenario-dependent utility function $u(x, \omega) = -l(H(\omega) - x)^+$.

In the microeconomic theory of intertemporal consumption choice, the natural commodity space is no longer a space of random variables but a space of positive *optional random measures* C defined on a given time interval $[0, T]$. Such a measure describes a consumption pattern contingent on the scenario and adapted to the incoming information. Identifying measures with their distribution functions, we take as our commodity space $\mathcal{C}$ the space of all optional increasing processes $C = (C_t)_{t \in [0,T]}$ on our probability space $(\Omega, \mathcal{F}, P)$ with filtration $(\mathcal{F}_t)_{t \in [0,T]}$.

As a corollary to the optional decomposition, the minimal cost of financing the consumption pattern C by a trading strategy on the underlying financial market can be identified as

$$\sup_{P^* \in \mathcal{P}} E^*\Big[\int_0^T \exp\Big(-\int_0^t r_s ds\Big) dC_t\Big]. \tag{26}$$

Here (r_t) is a predictable process describing the short term interest rate; in the previous sections it did not appear explicitly since all assets were already assumed to be discounted by a savings account growing at the short rate. For any preference functional U on the space of positive measures, we have a corresponding *optimization problem*

$$\max_{C \in \mathcal{C}_0} E[U(C(\omega))] \tag{27}$$

where $\mathcal{C}_0$ is the class of all optional measures $C \in \mathcal{C}$ such that the cost of financing C as defined in (26) is not greater than a given initial amount V_0.

Of course, the structure of the solution will depend both on the underlying price dynamics and on the choice of the preference functional U. A common choice consists in admitting only absolutely continuous measures $dC = c_t dt$ with a rate of consumption (c_t), and to use a functional of the form

$$U(C) = \int_0^T u(c_t, t) dt, \tag{28}$$

where u is some time-dependent utility function, typically of the form $u(x, t) = u(x) \exp(-\delta t)$ with some discounting factor δ and some classical utility function u. In this case, the corresponding optimization problem may be viewed as a space-time version of the optimization problem considered in Section 2, and it can be solved in an analogous way. It turns out, however, that the choice (28) of a preference functional on absolutely continuous measures is much less canonical than the expected utility functional on a space of random variables, both from a mathematical and from a microeconomic point of view. In particular, there is no reasonable extension to the space of all measures, no reasonable transition between discrete and continuous time, and no continuity with respect to the natural weak topology on the space of measures. Hindy, Huang, Kreps [19] provide a thorough analysis of why a functional of the form (28) is not satisfactory from a microeconomic point of view. Instead, they propose a utility functional of the form

$$U(C) = \int_0^T u(Y_t^C, t) dt \tag{29}$$

defined for all finite measures C on $[0, T]$, where Y_t^C is some index of past consumption up to time t, for example a weighted average

$$Y_t^C = \int_0^t \exp(-\beta(t - s)) \, dC_s. \tag{30}$$

The process Y^C will be called the *level of satisfaction* generated by C.

For a utility functional U of the form (29), the structure of the optimization problem (27) becomes much richer than in the case (28). Typically, the optimal optional measure $\widetilde{C}$ is no longer absolutely continuous; it may involve consumption in gulps and at a singular rate. If the underlying price dynamics is Markovian, then one can use a standard approach in terms of a Hamilton–Jacobi–Bellman equation; see Hindy and Huang [20]. Starting with the deterministic case, Bank and Riedel [3, 4] have initiated a quite different approach which is based on a new infinite-dimensional version of the Kuhn–Tucker theorem. In a general stochastic setting, and in particular in the thesis of Bank [2], this approach leads to some remarkable new problems, which are of independent interest from the point of view of general semimartingale theory.

Existence and uniqueness of a solution $\widetilde{C}$ can be shown in great generality. In the complete case, and for a functional of the form (30), Bank and Riedel [4] and

34 H. Föllmer

Bank [2] succeed in clarifying the structure of the optimal consumption plan $\widetilde{C}$ in terms of a reference process $I = (I_t)$, called the *minimal level of satisfaction*. They show that $\widetilde{C}$ can be explicitly constructed by "tracking" I. More precisely, $\widetilde{C}$ is reconstructed from its level of satisfaction $\widetilde{Y}$, and $\widetilde{Y}$ is computed explicitly as the minimal process in the class of all feasible index processes $Y^C \geq I$ with $C \in \mathcal{C}_0$. The crucial reference process I can be characterized in terms of the solution of a remarkable new *stochastic representation problem* of the following general form. Given an optional process $Z = (Z_t)$ (in our case defined in terms of the state-price densities) and a function $f(l,t)$ decreasing in l (in our case defined in terms of u), we want to construct an optional process $L = (L_t)$ such that

$$Z_t = E[\int_t^T f(\sup_{t \leq v \leq s} L_v, s) ds | \mathcal{F}_t]. \tag{31}$$

In discussions at an Oberwolfach meeting in May 2000, N. El Karoui and P. Bank clarified the connection to the theory of the Gittins index in continuous time; see El Karoui and Karatzas [10]. Using this connection, the complete solution of the representation problem (31) is carried out in Bank [2]. In the simple special case $f(l,t) = -l$, the process L appears as the solution of the optimal stopping problem

$$L_t = \text{ess.} \inf \frac{E[X_\tau - X_t | \mathcal{F}_t]}{E[\tau - t | \mathcal{F}_t]}, \tag{32}$$

where the essential infimum is taken over all stopping times τ with values in $(t, T]$. In the general case the construction of the solution is quite intricate; see Bank [2].

Thus, the initial problem of intertemporal consumption choice leads to new extensions of the theory of optimal stopping which are of intrinsic mathematical interest, quite independent of the financial motivation.

References

[1] L. Bachelier, *Théorie de la spéculation*, Ann. Sci. Ecole Norm. Sup. **17** (1900), 21–86.

[2] P. Bank, *Singular Control of optional random measures: Stochastic Optimization and Representation Problems arising in the Microeconomic Theory of Intertemporal Consumption Choice*, Dissertation, Humboldt-Universität zu Berlin (2000).

[3] P. Bank and F. Riedel, *Non-Time Additive Utility Optimization – the Case of Certainty*, J. Math. Econ. **33** (2000), 271–290.

[4] P. Bank and F. Riedel, *Optimal Consumption Choice with Intertemporal Substitution*, Preprint Humboldt-Universität zu Berlin (2000).

[5] A. Bick and W. Willinger, *Dynamic Spanning without Probabilities*, Stoch. Proc. Appl. **50** (1994), 349–374.

[6] F. Black and M. Scholes, *The Pricing of Options and Corporate Liabilities*, J. Political Econom. **72** (1973), 637–659.

[7] J. Cvitanić and I. Karatzas, *On dynamic measures of risk*, Finance and Stochastics **3** (1999), 451–482.

[8] J. Cvitanić, W. Schachermayer and H. Wang, *Utility Maximization in Incomplete Markets with Random Endowment*, Preprint (2000).

[9] F. Delbaen and W. Schachermayer, *A general version of the fundamental theorem of asset pricing*, Math. Annalen **300** (1994), 463–520.

[10] N. El Karoui and I. Karatzas, *Dynamic Allocation Problems in Continuous Time*, Ann. Appl. Probab. **4** (1994), 255–286.

[11] N. El Karoui and M. C. Quenez, *Dynamic programming and pricing of contingent claims in an incomplete market*, SIAM J. Control and Optimization **33 (1)** (1995), 29–66.

[12] H. Föllmer, *Calcul d'Itô sans Probabilités*, In: Sém. Probabilités Strasbourg XV, Lecture Notes in Mathematics 850, Springer (1981), 143–150.

[13] H. Föllmer and D. Sondermann, *Hedging of Non-redundant Contingent Claims*, In: Contributions to Mathematical Economics, In Honor of G. Debreu (Eds. W. Hildenbrand and A. Mas-Colell), Elsevier Science Publ., North-Holland (1986), 205–223.

[14] H. Föllmer and M. Schweizer, *Hedging of Contingent Claims under Incomplete Information*, In: Applied Stochastic Analysis, eds. M. H. A. Davis and R. J. Elliott, Gordon and Breach, London (1990), 389–414.

[15] H. Föllmer and D. Kramkov, *Optional decompositions under constraints*, Prob. Theory Relat. Fields, vol. 109 (1997), 1–25.

[16] H. Föllmer and Y. Kabanov, *Optional decomposition and Lagrange multipliers*, Finance and Stochastics, vol. 2, No. 1 (1998), 69–81.

[17] H. Föllmer and P. Leukert, *Quantile Hedging*, Finance and Stochastics, vol. 3, No. 3 (1999), 251–273.

[18] H. Föllmer and P. Leukert, *Efficient hedges: cost versus shortfall risk*, Finance and Stochastics **4** (2000), 117–146.

[19] A. Hindy, C.-F. Huang and D. Kreps, *On Intertemporal Preferences in Continuous Time - the Case of Certainty*, J. Math. Econ. **21** (1992), 401–440.

[20] A. Hindy and C.-F. Huang, *Intertemporal Preferences for Uncertain Consumption: a Continuous-Time Approach*, Econometrica **61** (1993), 85–121.

[21] I. Karatzas, *Lectures in Mathematical Finance*, Providence: American Mathematical Society (1997).

[22] E. Karni and D. Schmeidler, *Utility Theory with Uncertainty*, In: Handbook of Mathematical Economics, vol. IV, Ch. 33, eds. W. Hildenbrand and H. Sonnenschein. Elsevier Science Publ., North-Holland (1991), 1763–1831.

[23] J. Komlos, *A generalization of a problem of Steinhaus*, Acta Math. Acad. Sci. Hung. **18** (1967), 217–229.

[24] D. O. Kramkov, *Optional decomposition of supermartingales and hedging contingent claims in incomplete security markets*, Probability Theory and Related Fields **105** (1996), 459–479.

[25] O. Kramkov and W. Schachermayer, *The Asymptotic Elasticity of Utility Functions and Optimal Investment in Incomplete Markets*, Ann. Appl. Probab., Vol. 9, No. 3 (1999), 904–950.

[26] M. Kulldorff, *Optimal control of favorable games with a time-limit*, SIAM J. Control and Optimization **31** (1993), 52–69.

[27] P. Leukert, *Absicherungsstrategien zur Minimierung des Verlustrisikos*, Dissertation, Humboldt-Universität Berlin (1999).

[28] T. Lyons, *Uncertain volatility and the risk free synthesis of derivatives*, Appl. Math. Finance **2** (1995), 117–133.

[29] R. Merton, *Theory of Rational Option Pricing*, Bell J. Econom. Managem. Sci. **4** (1973), 141–183.

[30] M. Mnif and H. Pham, *Stochastic Optimization under Constraints*, Prepublication n° 599, Laboratoire de Probabilité, Université de Paris VI et VII (2000).

[31] H. Pham, *Minimizing Shortfall Risk and Applications to Finance and Insurance*, Preprint Université de Marne-la-Vallée (1999).

[32] T. Rheinländer and M. Schweizer, *On L^2-Projections on a Space of Semimartingales*, Ann. Probab. **25** (1997), 1810–1831.

[33] M. Schweizer, *A Guided Tour through Quadratic Hedging Approaches*, Preprint TU Berlin (1999).

Department of Mathematics
Humboldt University
Unter den Linden 6
D-10099 Berlin, Germany
E-mail address: foellmer@mathematik.hu-berlin.de

Flags and Lattice Basis Reduction

Hendrik W. Lenstra, Jr.

Abstract. In this lecture we give a self-contained introduction to the theory of lattices in Euclidean vector spaces. We reinterpret a large class of lattice basis reduction algorithms by using the concept of a "flag". In our reformulation, lattice basis reduction algorithms are more appropriately called "flag reduction" algorithms. We address a problem that arises when one attempts to find a particularly good flag for a given lattice.

1. Introduction

A *lattice* is a discrete subgroup of a Euclidean vector space. Every lattice has a *basis*, and a *lattice basis reduction algorithm* is an algorithm that transforms a given basis for a given lattice into a basis consisting of relatively short vectors.

The present lecture is devoted to a conceptual discussion of an important class of lattice basis reduction algorithms. This class includes an algorithm that I introduced in 1981 [5], its close relative that is known as the LLL or Lovász basis reduction algorithm from 1982 [4], and many variants of these algorithms that were proposed in subsequent years. The original applications of basis reduction algorithms to integer programming ([6, 1]) and to algorithmic number theory ([4, 3]) exemplify the scope of their importance in pure and applied mathematics.

The central notion in the present discussion is that of a *flag* for a lattice. A flag carries a little less information than a basis; the definition is given in Section 6. The basis reduction algorithms that I consider can be reinterpreted in terms of flags; one may say that they transform a given flag for a lattice into a 'reduced' flag for the same lattice. To this end they perform a series of successive steps, each step replacing a flag by a 'neighbouring' one that is closer to being reduced. We picture this procedure by means of a directed graph, of which the vertices represent all flags for a given lattice, and the arcs the steps that are permitted. For the algorithm to be efficient, it is necessary that not too many steps are performed in succession. This leads to the problem of giving an upper bound for the length of a directed path in the graph that starts from a given vertex. In Section 8, I present such an upper bound. It may be considered satisfactory if one considers only lattices of fixed rank. It is an interesting open problem to find an upper bound that has a better behaviour as a function of the rank.

2. Euclidean Vector Spaces

A *Euclidean vector space* is a finite dimensional vector space E over the field $\mathbf{R}$ of real numbers equipped with a map $\langle\,,\,\rangle\colon E \times E \to \mathbf{R}$ satisfying

$$\langle w + x, y\rangle = \langle w, y\rangle + \langle x, y\rangle, \quad \langle rx, y\rangle = r\langle x, y\rangle,$$
$$\langle x, y\rangle = \langle y, x\rangle, \qquad\qquad \langle z, z\rangle > 0$$

for all $r \in \mathbf{R}$ and $w, x, y, z \in E$, $z \neq 0$. We refer to the map $\langle\,,\,\rangle$ as the *inner product* on E. Any Euclidean vector space E is a *metric space* with distance function $d\colon E \times E \to \mathbf{R}$ defined by $d(x, y) = \langle x - y, x - y\rangle^{1/2}$. For each non-negative integer n, the vector space $\mathbf{R}^n$ is a Euclidean vector space with the *standard inner product* defined by $\langle (x_i)_{i=1}^n, (y_i)_{i=1}^n\rangle = \sum_{i=1}^n x_i y_i$.

Let E be a Euclidean vector space, and let $D \subset E$ be a subspace. Then the restriction of $\langle\,,\,\rangle$ to $D \times D$ makes D into a Euclidean vector space. Let $D^\dagger = \operatorname{Hom}(D, \mathbf{R})$ be the dual of D, and write $D^\perp$ for the kernel of the linear map $E \to D^\dagger$ sending $x \in E$ to the map $y \mapsto \langle x, y\rangle$. One has $D \cap D^\perp = 0$, so the natural maps $D \to E/D^\perp \to D^\dagger$ are injective, and by $\dim D = \dim D^\dagger$ they are isomorphisms. It follows that each $w \in E$ has a unique representation as $w = x + y$, with $x \in D$ and $y \in D^\perp$; this is the *orthogonal decomposition* of w with respect to D. The quotient space E/D becomes a Euclidean vector space as well, since it is canonically isomorphic to the subspace $D^\perp$ of E. One also concludes that E is canonically isomorphic to $E^\dagger$.

Applying the above to one-dimensional subspaces, one easily proves by induction on $\dim E$ that for every Euclidean vector space E there is a linear isomorphism from the standard Euclidean vector space $\mathbf{R}^{\dim E}$ to E that preserves inner products. For example, if one makes the field $\mathbf{C}$ of complex numbers into a Euclidean vector space by putting $\langle x, y\rangle = (x\bar{y} + y\bar{x})/2$ (= the real part of $x\bar{y}$), then the isomorphism $\mathbf{R}^2 \to \mathbf{C}$ sending (a, b) to $a + bi$ preserves inner products.

Proposition 2.1. *Let D be a finite dimensional vector space over $\mathbf{R}$, and let $\langle\,,\,\rangle\colon D \times D \to \mathbf{R}$ be a map satisfying $\langle w + x, y\rangle = \langle w, y\rangle + \langle x, y\rangle$, $\langle rx, y\rangle = r\langle x, y\rangle$, $\langle x, y\rangle = \langle y, x\rangle$, and $\langle x, x\rangle \geq 0$ for all $r \in \mathbf{R}$, $w, x, y \in D$. Write $\operatorname{rad} D = \{x \in D : \langle x, x\rangle = 0\}$. Then $\operatorname{rad} D$ is a subspace of D, and if we write $E = D/\operatorname{rad} D$ then $\langle\,,\,\rangle$ can be written as the composition of the natural map $D \times D \to E \times E$ and a uniquely determined map $E \times E \to \mathbf{R}$; moreover, the latter map makes E into a Euclidean vector space.*

Proof. Let $w \in \operatorname{rad} D$, $y \in D$. Then one has $2r\langle w, y\rangle + \langle y, y\rangle = \langle rw + y, rw + y\rangle \geq 0$ for every $r \in \mathbf{R}$, and therefore $\langle w, y\rangle = 0$. One readily deduces that $\operatorname{rad} D$ is a subspace and that for any $x, x', y, y' \in D$ with $x \equiv x' \bmod \operatorname{rad} D$ and $y \equiv y' \bmod \operatorname{rad} D$ one has $\langle x, y\rangle = \langle x', y'\rangle$. Hence $\langle\,,\,\rangle$ factors through $E \times E$. The last statement is now immediate. $\qquad\square$

3. Lattices in a Euclidean Vector Space

Let E be a Euclidean vector space. A *lattice in* E is an additive subgroup L of E for which there exists a positive real number l such that every $z \in L$, $z \neq 0$, satisfies $\langle z, z \rangle \geq l$; equivalently, it is an additive subgroup of E that is *discrete* in the topology induced by the metric on E.

Proposition 3.1. *Let L be a lattice in a Euclidean vector space E, and let $r, l \in \mathbf{R}$ be real numbers with $r \geq 0$, $l > 0$ such that every $z \in L$, $z \neq 0$, satisfies $\langle z, z \rangle \geq l$. Write n for the dimension of the subspace of E spanned by L. Then we have*

$$\#\{x \in L : \langle x, x \rangle \leq r\} \leq (1 + 2\sqrt{r/l})^n .$$

Proof. Replacing E by the subspace spanned by L we may assume $n = \dim E$. Since L is an additive subgroup, any two distinct elements of L differ by a non-zero element of L and therefore have distance at least $\sqrt{l}$. Hence the open n-dimensional balls with radius $\sqrt{l}/2$ centered at all $x \in L$ are pairwise disjoint. All of these balls whose center x satisfies $\langle x, x \rangle \leq r$ are contained in the open ball with radius $\sqrt{r} + \sqrt{l}/2$ centered at 0. Computing volumes we find

$$\#\{x \in L : \langle x, x \rangle \leq r\} \cdot (\sqrt{l}/2)^n \leq (\sqrt{r} + \sqrt{l}/2)^n .$$

(It is practical to rescale the volume on E so that the unit ball has volume 1.) $\square$

Let $\mathbf{Z}$ denote the ring of integers.

Proposition 3.2. *Let E be a Euclidean vector space, let b_1, b_2, $\ldots$, $b_n \in E$ be linearly independent, and for each i let $b_i = (b_i - b_i^*) + b_i^*$ be the orthogonal decomposition of b_i with respect to $\sum_{j<i} \mathbf{R}b_j$; so $\langle b_i^*, b_j \rangle = 0$ for $j < i$, and $b_i - b_i^* \in \sum_{j<i} \mathbf{R}b_j$. Then all b_i^* are non-zero, and for each $z \in \sum_i \mathbf{Z}b_i$, $z \neq 0$, one has $\langle z, z \rangle \geq \min_i \langle b_i^*, b_i^* \rangle$.*

Remark. If the b_i are not supposed to be linearly independent, then the dimension of the subspace they span equals the number of i for which $b_i^* \neq 0$.

Proof. We have $b_i^* \neq 0$ since $b_i \notin \sum_{j<i} \mathbf{R}b_j$. For $z = \sum_i n_i b_i \in \sum_i \mathbf{Z}b_i$, $z \neq 0$, choose i maximal with $n_i \neq 0$. Then $z = (z - n_i b_i^*) + n_i b_i^*$ is the orthogonal decomposition of z with respect to $\sum_{j<i} \mathbf{R}b_j$, so $\langle z, z \rangle \geq n_i^2 \langle b_i^*, b_i^* \rangle \geq \langle b_i^*, b_i^* \rangle$. $\square$

Proposition 3.3. *Let E be a Euclidean vector space and let L be a subset of E. Then L is a lattice in E if and only if there is a linearly independent subset $B \subset E$ with $L = \sum_{b \in B} \mathbf{Z}b$.*

If B and L are as in Proposition 3.3, then B is called a *basis* for L. Its cardinality $\#B$ equals the number n defined in Proposition 3.1, so it depends only on L; it is called the *rank* $\mathrm{rk}\, L$ of L. One has $\mathrm{rk}\, L \leq \dim E$.

Proof. The if-part follows from Proposition 3.2. For the only if-part, let $C \subset L$ be a basis for the subspace D of E spanned by L. Each $w \in L$ can be written as $w = x + y$ with $x \in M = \sum_{c \in C} \mathbf{Z}c$ and y in the intersection of L with the bounded set $\sum_{c \in C}[0, 1)c$. By Proposition 3.1, that intersection is finite, so M has finite index m (say) in L. Lagrange's theorem from group theory now implies that $mL \subset M$, so L is a subgroup of finite index of the free abelian group $m^{-1}M$. Therefore L has a basis B over $\mathbf{Z}$ with $\#B = \#C$, and B is linearly independent since its span contains C. $\qquad\qquad\qquad\qquad\qquad\qquad\qquad\qquad\qquad\qquad\qquad\qquad\qquad\quad\square$

4. Lattices

We next define lattices in an absolute sense, without reference to a Euclidean vector space. A *lattice* is a finitely generated abelian group L equipped with a map $q\colon L \to \mathbf{R}$ satisfying the following three conditions:

    (i) $q(x+y)+q(x-y) = 2q(x)+2q(y)$ for all $x, y \in L$ (the *parallelogram* law);

    (ii) $q(z) \neq 0$ for all $z \in L$, $z \neq 0$;

    (iii) for each real number r, the set $\{x \in L : q(x) \leq r\}$ is *finite*.

An *isomorphism* from a lattice L, q to a lattice L', q' is a group isomorphism $f\colon L \to L'$ such that for all $x \in L$ one has $q(x) = q'(f(x))$; if such a map exists then the lattices L and L' are called *isomorphic*.

By Proposition 3.1, any lattice in a Euclidean vector space becomes a lattice in the sense just defined if we put $q(x) = \langle x, x \rangle$. We prove that, up to isomorphism, any lattice can be obtained in this way.

Proposition 4.1. *Any lattice is isomorphic to a lattice in a Euclidean vector space.*

Proof. Let L, q be a lattice. For $x, y \in L$, define $\langle x, y \rangle = \big(q(x + y) - q(x - y)\big)/4$. The parallelogram law implies $q(x - y) = q(y - x)$, so we have $\langle x, y \rangle = \langle y, x \rangle$. Let $w, x, y \in L$. We have by the parallelogram law

$$q(w + x + y) + q(w - x + y) = 2q(w + y) + 2q(x),$$
$$q(w + x - y) + q(w - x - y) = 2q(w - y) + 2q(x),$$
$$q(w + x + y) + q(w - x - y) = 2q(x + y) + 2q(w),$$
$$q(w + x - y) + q(w - x + y) = 2q(x - y) + 2q(w).$$

Taking the alternating sum and dividing by 8 we find that $\langle w + x, y \rangle = \langle w, y \rangle + \langle x, y \rangle$. One readily checks that $q(0) = 0$ and $\langle x, x \rangle = q(2x)/4 = q(x)$, for $x \in L$. If $x \in L$ satisfies $q(x) < 0$, then one has $q(mx) = m^2 q(x) < 0$ for all non-zero $m \in \mathbf{Z}$, so x has infinite order, and one obtains a contradiction with (iii); hence $q(x) \geq 0$ for all $x \in L$. Write $D = \mathbf{R} \otimes_{\mathbf{Z}} L$, and let $\langle\ ,\ \rangle\colon D \times D \to \mathbf{R}$ be the $\mathbf{R}$-bilinear function induced by $\langle\ ,\ \rangle\colon L \times L \to \mathbf{R}$. For each positive integer m and each $x \in L$ the element $z = (1/m) \otimes x$ of D satisfies $\langle z, z \rangle = q(x)/m^2 \geq 0$, and since the set of all $z \in D$ of this form is dense in D one has $\langle z, z \rangle \geq 0$ for all $z \in D$. From Proposition 2.1 one now obtains a Euclidean vector space $E = D/\operatorname{rad} D$ such that the group homomorphism $f\colon L \to E$ sending x to the coset $(1 \otimes x) \bmod \operatorname{rad} D$ satisfies $q(x) = \langle f(x), f(x) \rangle$. By property (ii) the map is injective, and using (iii) one deduces that $f(L)$ is a lattice in E that is isomorphic to L. (Comparing ranks and dimensions one also finds $\operatorname{rad} D = 0$, so $D = E$.) $\square$

Remark. In the sequel, we shall write $\langle x, y \rangle = \big(q(x+y) - q(x-y)\big)/4$ for x, y in a lattice, and lattices may tacitly be assumed to be embedded in a Euclidean vector space. This is justified by Proposition 4.1 and its proof.

RANK AND DETERMINANT Two important numerical invariants attached to any lattice L are its *rank* $\operatorname{rk} L$ and its *determinant* $d(L)$. The rank is the unique non-negative integer n for which there is an isomorphism $L \cong \mathbf{Z}^n$ of abelian groups. The determinant is defined by

$$d(L) = \det\big(\langle b, b' \rangle\big)^{1/2}_{b,b' \in B},$$

where B is a basis of L; if L is a lattice in $\mathbf{R}^n$ with basis equal to the set of columns of a non-singular $n \times n$ matrix $\mathbf{B}$, then one has $d(L) = |\det \mathbf{B}|$. One way to prove that $d(L)$ is well defined is by showing the limit relation

$$\lim_{r \to \infty} \frac{\#\{x \in L : q(x) \leq r\}}{\omega_n r^{n/2}/d(L)} = 1,$$

which is valid for any lattice L of rank n. Here we write $\omega_n = \pi^{n/2}/\frac{n}{2}!$ for the standard volume of the unit ball in $\mathbf{R}^n$; the factor $\frac{n}{2}! = \Gamma(1+\frac{n}{2})$ may be computed from $0! = 1$, $\frac{1}{2}! = \sqrt{\pi}/2$, and $z! = z \cdot (z-1)!$. We have $d(L) = 1$ if $\operatorname{rk} L = 0$.

Proposition 4.2. *Let L, q be a lattice of positive rank n. Then there exists $x \in L$ with $x \neq 0$ and $q(x) \leq n \cdot d(L)^{2/n}$.*

Proof. Assume that L is a lattice in $\mathbf{R}^n$, and write vol for the standard n-dimensional volume. Let $B \subset \mathbf{R}^n$ be a basis for L, and write $F = \sum_{b \in B}[0,1)b$. Then $\operatorname{vol} F = d(L)$, and $\mathbf{R}^n$ is the disjoint union of the sets $x + F$, $x \in L$. Let $l = \min\{q(x) : x \in L, x \neq 0\}$, and write $t = \sqrt{l/n}$, so that the assertion of Proposition 4.2 is equivalent to $t^n \leq d(L)$. Let C be the cube $[0,t)^n$ in $\mathbf{R}^n$. Any two elements of C have distance smaller than $t\sqrt{n} = \sqrt{l}$, so their difference is not a non-zero element of L. Hence the sets $-x + C$, $x \in L$, are pairwise disjoint. Since C is the disjoint union of the sets $(x + F) \cap C$, $x \in L$, we conclude that

$$t^n = \operatorname{vol} C = \sum_{x \in L} \operatorname{vol}\big((x+F) \cap C\big) = \sum_{x \in L} \operatorname{vol}\big(F \cap (-x + C)\big)$$

$$= \operatorname{vol}\big(F \cap \bigcup_{x \in L}(-x + C)\big) \leq \operatorname{vol}(F) = d(L),$$

as required. $\square$

Remark. Replacing the cube in the proof by an open ball of radius $\sqrt{l}/2$ one finds the better inequality $q(x) \leq 4\omega_n^{-2/n} \cdot d(L)^{2/n}$, with ω_n as above, and further improvements are possible. One has $4\omega_n^{-2/n} = 2n/(\pi e + o(1))$ for $n \to \infty$.

SUBLATTICES AND QUOTIENT LATTICES   Let L, q be a lattice, and let K be a subgroup of L. Then the restriction of q to K makes K into a lattice, a *sublattice* of L. We next restrict to *pure* subgroups. In general, a subgroup K of an additively written abelian group L is called *pure* if for all positive integers m one has $mK = K \cap mL$. If L is a lattice, this property is equivalent to L/K being torsion-free; and if L is a lattice in a Euclidean vector space E, then it is equivalent to the existence of a subspace D of E such that $K = L \cap D$, and also to L having a basis that contains a basis for K. Now suppose that K is a pure sublattice of a lattice L. Then the map $q' \colon L/K \to \mathbf{R}$ defined by

$$q'(x + K) = \inf\{q(mx - y)/m^2 : m \in \mathbf{Z}, m \neq 0, y \in K\}$$

makes L/K into a lattice. To prove this, one embeds L as a lattice in a Euclidean vector space E, one defines D to be the subspace of E spanned by K, and one verifies that q' is induced by the inclusion of L/K in the Euclidean vector space E/D. One has

$$\mathrm{rk}\, K + \mathrm{rk}(L/K) = \mathrm{rk}\, L, \quad d(K) \cdot d(L/K) = d(L)\,.$$

Proposition 4.3. *Let L be a lattice and let r be a real number. Then the number of sublattices K of L with $d(K) \leq r$ is finite.*

Proof. For any subgroup $K \subset L$, with $\mathbf{R}$-linear span $\mathbf{R} \cdot K$, the subgroup $K' = L \cap (\mathbf{R} \cdot K)$ is pure, the number $m = \mathrm{index}[K' : K]$ is finite, and one has $d(K) = m \cdot d(K')$. Hence we may restrict to *pure* subgroups. We apply induction on $\mathrm{rk}\, L$. The set of non-zero b in L with $q(b) \leq \max\{i \cdot r^{2/i} : 1 \leq i \leq n\}$ is finite, and by Proposition 4.2 any non-zero subgroup $K \subset L$ with $d(K) \leq r$ contains at least one of them. If K is a pure subgroup containing a given such b, then it also contains the pure subgroup $L_b = L \cap \mathbf{R}b$, and K/L_b is a pure subgroup of L/L_b with $d(K/L_b) = d(K)/d(L_b)$. Now apply the induction hypothesis to each L/L_b. $\square$

Remark. An alternative proof of Proposition 4.3 makes use of exterior powers. For subgroups of rank 1, one uses defining property (iii) of lattices. Generally, if $K \subset L$ is a subgroup of rank i, then $\wedge^i K \subset \wedge^i L$ is a subgroup of rank 1, and $\wedge^i L$ has a natural lattice structure for which $d(\wedge^i K) = d(K)$; in addition, K is 'almost' determined by $\wedge^i K$ in the sense that another subgroup $J \subset L$ of rank i satisfies $\wedge^i J = \wedge^i K$ if and only if J is a subgroup of $L \cap (\mathbf{R} \cdot K)$ of the same index as K.

Remark. It follows from Proposition 4.3 that there is a positive lower bound for the determinants of the subgroups of a given lattice. Explicitly, any subgroup $K \subset L$ with $\mathrm{rk}\, K = i > 0$ satisfies $d(K) \geq \left(\min\{q(x) : x \in L, x \neq 0\}/i\right)^{i/2}$, by Proposition 4.2.

THE DUAL   Let L be a lattice in a Euclidean vector space E with $\dim E = \mathrm{rk}\, L$. Then $L^\dagger = \{x \in E : \langle x, L \rangle \subset \mathbf{Z}\}$ is also a lattice in E, the *dual* (or *polar*) of L. One has

$$\mathrm{rk}\, L^\dagger = \mathrm{rk}\, L, \quad d(L^\dagger) = d(L)^{-1}, \quad L^{\dagger\dagger} = L\,.$$

If L is a lattice in $\mathbf{R}^n$ with basis equal to the set of columns of a certain non-singular matrix, then the columns of the inverse transpose matrix form a basis for $L^\dagger$. If desired, one can also define the dual without reference to a Euclidean vector space, by taking $L^\dagger = \mathrm{Hom}(L, \mathbf{Z})$ and letting $q(f)$, for $f \in L^\dagger$, be the infimum of all non-negative real numbers r with the property that for all $x \in L$ one has $f(x)^2 \leq r \cdot q(x)$.

Let L be a lattice, with dual $L^\dagger$, and let $K \subset L$ be a pure sublattice. Then $K^\perp = \{x \in L^\dagger : \langle x, K \rangle = 0\}$ is a pure sublattice of $L^\dagger$ that may be identified with $(L/K)^\dagger$, and $K^\dagger$ may be identified with $L^\dagger/K^\perp$; in addition, one has $K^{\perp\perp} = K$.

5. Algorithmic Problems

In the present section we discuss a few fundamental and frequently encountered problems concerning lattices. The first is the *homogeneous approximation* problem: *given a non-zero lattice L, find a non-zero element $x \in L$ with $q(x)$ smallest possible*. The informal formulation allows many interpretations. For example, the lattice may be 'given' in some theoretical sense, and 'finding' x may be meant purely existentially, so that Proposition 4.2 goes some way towards solving the problem. We are mainly interested in an algorithmic interpretation, in which the lattice is 'given' in some numerical manner, and likewise its elements have a numerical representation; the problem of 'finding' x is then to be interpreted algorithmically, and one wants not just $q(x)$ but also the run time of the algorithm to be small. One will have to allow for a trade-off between the latter two quantities, and the requirement that $q(x)$ be 'smallest possible' may be taken to mean: smallest possible given the time that one is willing to spend.

One way of specifying a lattice L numerically is by means of a real $m \times n$ matrix $\mathbf{B}$ of rank n; then L is embedded in the Euclidean vector space $\mathbf{R}^m$, the columns of $\mathbf{B}$ forming a basis, and an element $x \in L$ is either represented as a real m-vector or as an integral n-vector consisting of the coefficients of x on that basis. In order to avoid rounding problems one may require the entries of $\mathbf{B}$ to be rational. A second way of specifying L is by means of a real positive definite symmetric $n \times n$ matrix $\mathbf{A}$; in this case L is the group $\mathbf{Z}^n$, its elements are represented as integral n-vectors, and $\langle x, y \rangle = x^T \mathbf{A} y$ for $x, y \in L$. Again one may require the entries of $\mathbf{A}$ to be rational. One easily transforms the first type of representation into the second by taking $\mathbf{A} = \mathbf{B}^T \cdot \mathbf{B}$, and this transformation preserves rationality. One can also transform the second representation into the first, but complications arise if one wishes to do this by means of a polynomial time algorithm that preserves rationality and keeps m low. There are other possibilities of representing lattices numerically, but the two that we just mentioned appear to be the most convenient ones for algorithmic purposes.

Of the many algorithmic situations giving rise to the homogeneous approximation problem we mention a single one; namely, the problem of factoring a given one-variable polynomial f with rational coefficients into irreducible factors, which was considered in [4]. In this case, one can take the lattice to consist of integer polynomials of a certain degree that assume a very small value in a suitably constructed p-adic zero of f, and one proves that any sufficiently short non-zero vector in that lattice must be an irreducible factor of f.

The homogeneous approximation problem has also appeared under the following guise: *given a lattice L in a Euclidean vector space E of dimension* $\mathrm{rk}\, L$, *find $x \in E$ with $L \subset (\mathbf{R}x)^{\perp} + \mathbf{Z}x$ and $\langle x, x \rangle$ largest possible.* Geometrically, this amounts to asking for a hyperplane H in E such that L is contained in the union of a collection of maximally widely spaced translates of H; namely, take $H = (\mathbf{R}x)^{\perp}$ and consider translates with successive distances equal to $\langle x, x \rangle^{1/2}$. Such a hyperplane is useful when one wishes to enumerate elements of L that lie in a certain bounded region, which occurs in the context of integer programming (see [6]). A given non-zero vector $x \in E$ satisfies $L \subset (\mathbf{R}x)^{\perp} + \mathbf{Z}x$ if and only if $x/\langle x, x \rangle$ belongs to the dual $L^{\dagger}$ of L, so the problem is equivalent to the homogeneous approximation problem for $L^{\dagger}$.

Finally, one frequently encounters the *inhomogeneous approximation* problem: *given a lattice L in a Euclidean vector space E, and $x \in E$, find $y \in E$ with $x - y \in L$ and $\langle y, y \rangle$ smallest possible.* In other words, one wishes to 'round' a given element x of E to an element w of L such that the 'error' $d(x, w)$ is minimal. It is a mistake to think that the special case $x = 0$ of the inhomogeneous approximation problem amounts to the homogeneous approximation problem (since one takes $w = y = 0$); but it is true that solving $2^{\mathrm{rk}\, L} - 1$ inhomogeneous approximation problems suffices to solve the homogeneous approximation problem; namely, let x range over coset representatives of all non-trivial elements of $\frac{1}{2}L/L$.

All problems that we mentioned can to a certain extent be solved if a *reduced basis* of the lattice is available. The notion of a 'reduced basis' has many different definitions, and one usually chooses the most convenient one for the purpose at hand. Different definitions are rarely logically equivalent, but typically bases that are reduced in different senses share many qualitative properties: they consist of 'fairly short' vectors that stand at 'almost right' angles, the product of their lengths is a 'fair' approximation to the determinant of the lattice, and, of course, they yield solutions to the three problems formulated above.

In the next section we shall consider *flags* of a lattice. The notion of a flag is a little weaker than the notion of a basis, but it still carries enough information to assist us in solving our three problems.

Finding a reduced basis for a given lattice is done by means of a *lattice basis reduction algorithm*, which replaces a given basis for a given lattice by a reduced basis for the same lattice. We shall not present any of these. Instead, we describe in very general terms a *flag reduction algorithm*, that is, a procedure that replaces a given flag of a given lattice by what might be called a 'reduced flag' of the same lattice; but we refrain from giving a rigorous definition of the latter term. Many existing lattice basis reduction algorithms, including those presented in [5] and [4], may be interpreted as flag reduction algorithms, and fit as such under our general description.

6. Flags

Let L be a lattice, and write $n = \operatorname{rk} L$. A *flag* of L is a sequence $\mathfrak{F} = (F_i)_{i=0}^n$ of pure sublattices F_i of L satisfying $\operatorname{rk} F_i = i$ (for $0 \le i \le n$) and $F_{i-1} \subset F_i$ (for $0 < i \le n$); clearly we must have $F_0 = \{0\}$ and $F_n = L$. Every basis $(b_i)_{i=1}^n$ of L gives rise to the flag $\left(\sum_{j \le i} \mathbf{Z} b_j\right)_{i=0}^n$, and one readily checks that every flag of L is of this form. In order to express when two bases $(b_i)_{i=1}^n$ and $(a_i)_{i=1}^n$ of L give rise to the same flag, let $(b_i^*)_{i=1}^n$ be defined as in Proposition 3.2, and $(a_i^*)_{i=1}^n$ analogously. Then the two bases give rise to the same flag of L if and only if for each i one has $b_i^* = \pm a_i^*$; or, equivalently, if and only if there are integers c_{ij}, for $1 \le j \le i \le n$, with $b_i = \sum_{j=1}^i c_{ij} a_j$ and $c_{ii} = \pm 1$ for all i.

In an algorithmic context one may wish to represent a flag numerically. Assuming L and its elements to be represented in one of the manners described in Section 5, one can do this by specifying a basis $(b_i)_{i=1}^n$ of L; then the flag is $\left(\sum_{j \le i} \mathbf{Z} b_j\right)_{i=0}^n$. As we just noted, certain changes in the basis do not change the flag. This freedom is often used in order to achieve that the real numbers μ_{ij} for which $b_i - b_i^* = \sum_{j=1}^{i-1} \mu_{ij} b_j^*$ satisfy $|\mu_{ij}| \le \frac{1}{2}$.

Let $\mathfrak{F} = (F_i)_{i=0}^n$ be a flag of L. The *size* $s(\mathfrak{F})$ of $\mathfrak{F}$ is defined by

$$s(\mathfrak{F}) = \prod_{i=0}^n d(F_i).$$

For $1 \le i \le n$, the *i-th successive distance* $l_i(\mathfrak{F})$ is defined by $l_i(\mathfrak{F}) = d(F_i/F_{i-1})$; if $\mathfrak{F}$ is obtained from a basis $(b_j)_{j=1}^n$, and $(b_j^*)_{j=1}^n$ is as above, then one has $l_i(\mathfrak{F}) = \langle b_i^*, b_i^* \rangle^{1/2}$. One has $d(F_i) = \prod_{j=1}^i l_j(\mathfrak{F})$ for $0 \le i \le n$, and $s(\mathfrak{F}) = \prod_{j=1}^n l_j(\mathfrak{F})^{n+1-j}$. It is an easy consequence of Proposition 4.3 that L has, for any real number r, only finitely many flags $\mathfrak{F}$ with $s(\mathfrak{F}) \le r$.

Let again $\mathfrak{F} = (F_i)_{i=0}^n$ be a flag of L. Then $\mathfrak{F}^\perp = (F_{n-i}^\perp)_{i=0}^n$ is a flag of $L^\dagger$, the flag *dual* to $\mathfrak{F}$. One has

$$l_i(\mathfrak{F}^\perp) = l_{n+1-i}(\mathfrak{F})^{-1}, \quad s(\mathfrak{F}^\perp) = s(\mathfrak{F})/d(L)^{n+1}, \quad \mathfrak{F}^{\perp\perp} = \mathfrak{F}$$

for $1 \le i \le n$.

We shall in particular be interested in flags $\mathfrak{F}$ with the property that $l_{i+1}(\mathfrak{F})$ is not much smaller than $l_i(\mathfrak{F})$, for each $i = 1, \ldots, n-1$. The following result and its proof show the relevance of such flags for the homogeneous approximation problem formulated in the previous section.

Proposition 6.1. *Let c be a real number with $c \ge 1$, let L be a non-zero lattice in a Euclidean vector space E of dimension $n = \operatorname{rk} L$, and let $\mathfrak{F}$ be a flag of L with the property $l_{i+1}(\mathfrak{F})^2 \ge c^{-1} \cdot l_i(\mathfrak{F})^2$ for $0 < i < n$. Then we have*

$$c^{1-n} \cdot l_1(\mathfrak{F})^2 \le \min\{q(x) : x \in L,\ x \ne 0\} \le l_1(\mathfrak{F})^2,$$

$$l_n(\mathfrak{F})^2 \le \max\{\langle x, x \rangle : x \in E,\ L \subset (\mathbf{R}x)^\perp + \mathbf{Z}x\} \le c^{n-1} \cdot l_n(\mathfrak{F})^2.$$

Proof. Let $(b_i)_{i=1}^n$ be a basis of L such that $\mathfrak{F} = \left(\sum_{j \le i} \mathbf{Z} b_j\right)_{i=0}^n$, and let $(b_i^*)_{i=1}^n$ be as in Proposition 3.2. By Proposition 3.2, we have $\min\{q(x) : x \in L,\ x \ne 0\} \ge$

$\min_i \langle b_i^*, b_i^* \rangle = \min_i l_i(\mathfrak{F})^2$. The hypotheses imply that $l_i(\mathfrak{F})^2 \geq c^{1-n} \cdot l_1(\mathfrak{F})^2$, and the first inequality follows. The second inequality follows from $l_1(\mathfrak{F})^2 = q(b_1)$. One proves the last two inequalities by applying the first two to the dual flag. Note that $x = b_n^*$ satisfies $L \subset (\mathbf{R} \cdot x)^\perp + \mathbf{Z}x$ and $l_n(\mathfrak{F})^2 = \langle x, x \rangle$. $\qquad\square$

Remark. The flags considered in Proposition 6.1 also yield a fairly good solution to the inhomogeneous approximation problem. Namely, if the notation and hypotheses are as in Proposition 6.1 and its proof, then for every $x \in E$ there is a unique element $y \in \sum_{i=1}^n (-\frac{1}{2}, \frac{1}{2}] \cdot b_i^*$ with the property $x - y \in L$, and this element y satisfies

$$\frac{c-1}{c^n - 1} \cdot \langle y, y \rangle \leq \min\{\langle z, z \rangle : z \in E, \, x - z \in L\} \leq \langle y, y \rangle \,,$$

where one should read $\frac{1}{n}$ for $\frac{c-1}{c^n-1}$ if $c = 1$.

Proposition 6.2. *Let c, L, n, $\mathfrak{F}$ satisfy the hypotheses of Proposition 6.1. Then we have*

$$\#\{x \in L : q(x) \leq l_1(\mathfrak{F})^2\} \leq 3^n \cdot c^{n(n-1)/4} \,.$$

Proof. Let E, $(b_i)_{i=1}^n$, and $(b_i^*)_{i=1}^n$ be as above. We write $r = q(b_1) = l_1(\mathfrak{F})^2$ and $l = c^{1-n} \cdot l_1(\mathfrak{F})^2$; the condition on l in Proposition 3.1 is then satisfied, by Proposition 6.1. From Proposition 3.1 we would now find $\#\{x \in L : q(x) \leq l_1(\mathfrak{F})^2\} \leq (1 + 2c^{(n-1)/2})^n$. To achieve the better bound $3^n \cdot c^{n(n-1)/4}$, we write $B_t = \{y \in E : \langle y, y \rangle < t^2\}$ for $t \in \mathbf{R}$, $t > 0$. The proof of 3.1 depended on the fact that the translates $x + B_{\sqrt{l}/2}$, for $x \in L$, $q(x) \leq r$, are pairwise disjoint subsets of $B_{\sqrt{r}+\sqrt{l}/2}$. Now let ϕ be the endomorphism of the real vector space E that maps b_i^* to $c^{(n-i)/2} \cdot b_i^*$, for $1 \leq i \leq n$. From $l_i(\mathfrak{F})^2 \geq c^{1-i} \cdot l_1(\mathfrak{F})^2$ one derives that $\phi B_{\sqrt{l}/2}$ is contained in the fundamental domain $\sum_{i=1}^n (-\frac{1}{2}, \frac{1}{2}] \cdot b_i^*$ for E modulo L, so the translates $x + \phi B_{\sqrt{l}/2}$ are still pairwise disjoint for $x \in L$. One has $\phi B_{\sqrt{l}/2} \subset B_{\sqrt{r}/2}$, so each translate $x + \phi B_{\sqrt{l}/2}$ with $q(x) \leq r$ is contained in $B_{3\sqrt{r}/2}$. Hence $\#\{x \in L : q(x) \leq r\}$ is at most the quotient of the volumes of $B_{3\sqrt{r}/2}$ and $\phi B_{\sqrt{l}/2}$, which equals $(3\sqrt{r/l})^n / \det \phi = 3^n \cdot c^{n(n-1)/4}$. $\qquad\square$

To obtain the best results in Propositions 6.1 and 6.2, one should take c smallest possible. In Section 8 we shall see that $c = \frac{4}{3}$ can be achieved; that is, every lattice L has a flag $\mathfrak{F}$ with the property $l_{i+1}(\mathfrak{F})^2 \geq \frac{3}{4} \cdot l_i(\mathfrak{F})^2$ for $0 < i < \operatorname{rk} L$. Also, $\frac{4}{3}$ is best possible in the sense that for any $n > 1$ there is a lattice L of rank n such that for every flag $\mathfrak{F}$ of L there exists i with $0 < i < n$ and $l_{i+1}(\mathfrak{F})^2 \leq \frac{3}{4} \cdot l_i(\mathfrak{F})^2$. Namely, one can take L to be the 'orthogonal sum' of the hexagonal lattice $\mathbf{Z}2 + \mathbf{Z}(1 + i\sqrt{3})$ in $\mathbf{C}$ with the lattice $N\mathbf{Z}^{n-2}$ in $\mathbf{R}^{n-2}$, for N large enough.

7. The Reduction Graph

Let L be a lattice, and let n be its rank. We write $\Gamma(L)$ for the set of flags of L. We make $\Gamma(L)$ into the set of vertices of a directed graph, the *reduction graph* of L, by drawing an arc from $\mathfrak{F} = (F_i)_{i=0}^n$ to $\mathfrak{F}' = (F_i')_{i=0}^n$ if and only if there exists j, $0 < j < n$, with the following properties:

(i) $F_i = F_i'$ for all $i \neq j$;

(ii) $F_j + F_j' = F_{j+1}$;

(iii) $s(\mathfrak{F}')$ is minimal, given (i) and (ii);

(iv) $s(\mathfrak{F}') < s(\mathfrak{F})$.

Condition (iii) means, more formally, that for all flags $\mathfrak{G} = (G_i)_{i=0}^n$ of L satisfying $F_i = G_i$ for all $i \neq j$ and $F_j + G_j = F_{j+1}$ one has $s(\mathfrak{F}') \leq s(\mathfrak{G})$. To reformulate this condition, suppose that (i) and (ii) are satisfied; then we can write $F_j/F_{j-1} = \mathbf{Z}w$, $F_j'/F_{j-1} = \mathbf{Z}x$ for a certain basis w, x of the rank 2 lattice F_{j+1}/F_{j-1}, and (iii) is now equivalent to the inequality $|\langle x, w\rangle| \leq \langle w, w\rangle/2$; also, one has $s(\mathfrak{F}')^2/s(\mathfrak{F})^2 = q(x)/q(w)$, so (iv) is equivalent to $q(x) < q(w)$.

We write $\mathfrak{F} \to \mathfrak{F}'$ to denote an arc from $\mathfrak{F}$ to $\mathfrak{F}'$, and refer to it as a *step* in $\Gamma(L)$. The *length* of such a step is defined to be $s(\mathfrak{F})^2/s(\mathfrak{F}')^2$, and the number j appearing above is called the *colour* of the step; by (i) and (ii) it is uniquely determined.

One readily checks that there are at most two steps in $\Gamma(L)$ that start from a given flag and have a given colour; and if there are two, then they have the same length.

Let K be a pure sublattice of L. The set of flags of L that comprise K may in an obvious manner be identified with $\Gamma(K) \times \Gamma(L/K)$. With this identification, one has $s((\mathfrak{E}, \mathfrak{F})) = s(\mathfrak{E}) \cdot s(\mathfrak{F}) \cdot d(K)^{(\mathrm{rk}\,K)\,\mathrm{rk}(L/K)}$, and there is a step $(\mathfrak{E}, \mathfrak{F}) \to (\mathfrak{E}', \mathfrak{F}')$ in $\Gamma(L)$ if and only if either $\mathfrak{F} = \mathfrak{F}'$ and there is a step $\mathfrak{E} \to \mathfrak{E}'$ in $\Gamma(K)$, or $\mathfrak{E} = \mathfrak{E}'$ and there is a step $\mathfrak{F} \to \mathfrak{F}'$ in $\Gamma(L/K)$; in the former case, $(\mathfrak{E}, \mathfrak{F}) \to (\mathfrak{E}', \mathfrak{F})$ has the same length and colour as $\mathfrak{E} \to \mathfrak{E}'$, and in the latter case $(\mathfrak{E}, \mathfrak{F}) \to (\mathfrak{E}, \mathfrak{F}')$ has the same length as $\mathfrak{F} \to \mathfrak{F}'$ but the colour is larger by $\mathrm{rk}\,K$.

Proposition 7.1. *Let L be a lattice. Then the map $\Gamma(L) \to \Gamma(L^\dagger)$ sending $\mathfrak{F}$ to $\mathfrak{F}^\perp$ is an isomorphism of directed graphs. Corresponding steps have the same length, and their colours add up to $\mathrm{rk}\,L$.*

Proof. This is entirely straightforward, and left to the reader. $\square$

The following result describes the effect of a step on the successive lengths.

Proposition 7.2. *Let L be a lattice, let $\mathfrak{F} \to \mathfrak{F}'$ be a step in $\Gamma(L)$, and let j be its colour. Then one has $l_i(\mathfrak{F}) = l_i(\mathfrak{F}')$ for all $i \neq j$, $j+1$, and*

$$l_{j+1}(\mathfrak{F}) \leq l_j(\mathfrak{F}') < l_j(\mathfrak{F}), \quad l_{j+1}(\mathfrak{F}) < l_{j+1}(\mathfrak{F}') \leq l_j(\mathfrak{F}).$$

Proof. The relation $l_i(\mathfrak{F}) = d(F_i)/d(F_{i-1})$ and (i) imply the first assertion. Write $F_j/F_{j-1} = \mathbf{Z}w$ and $F_j'/F_{j-1} = \mathbf{Z}x$, and let $\bar{x}$ be the component of x orthogonal to w. Then one has $l_{j+1}(\mathfrak{F})^2 = q(\bar{x}) \leq q(x) = l_j(\mathfrak{F}')^2$, and $l_j(\mathfrak{F}')^2 = q(x) < q(w) =$

$l_j(\mathfrak{F})^2$. This proves the first two inequalities. The last two follow from these and the equality $l_j(\mathfrak{F})l_{j+1}(\mathfrak{F}) = d(F_{j+1})/d(F_{j-1}) = l_j(\mathfrak{F}')l_{j+1}(\mathfrak{F}')$. $\qquad\square$

Note in particular that $l_1(\mathfrak{F}') \leq l_1(\mathfrak{F})$ in the situation of Proposition 7.2, and, dually, $l_n(\mathfrak{F}') \geq l_n(\mathfrak{F})$.

Proposition 7.3. *Let L be a lattice and let $\mathfrak{F}$ be a flag of L. Let j be an integer with $0 < j < \operatorname{rk} L$ and c a real number with $c \geq \frac{4}{3}$. Suppose that one has $l_{j+1}(\mathfrak{F})^2 < c^{-1} \cdot l_j(\mathfrak{F})^2$. Then there is a step $\mathfrak{F} \to \mathfrak{F}'$ in $\Gamma(L)$ with colour j and length greater than $\frac{4c}{c+4}$.*

Proof. Write $F_j/F_{j-1} = \mathbf{Z}w$, and choose $x \in F_{j+1}/F_{j-1}$ with $q(x)$ minimal subject to the condition $\mathbf{Z}x + \mathbf{Z}w = F_{j+1}/F_{j-1}$. With $\bar{x}$ as in the previous proof, we have $x = \bar{x} + rw$ with $|r| \leq \frac{1}{2}$, so $q(x) = q(\bar{x}) + r^2 q(w) = l_{j+1}(\mathfrak{F})^2 + r^2 l_j(\mathfrak{F})^2 < (\frac{1}{c} + \frac{1}{4})l_j(\mathfrak{F})^2 = \frac{c+4}{4c}q(w) \leq q(w)$. The proposition follows, with $\mathfrak{F}' = (F_i')_{i=0}^{\operatorname{rk} L}$ defined by (i) and $F_j'/F_{j-1} = \mathbf{Z}x$. $\qquad\square$

The expression $\frac{4c}{c+4}$ will reappear in the next section. It is increasing as a function of c, equal to 1 for $c = \frac{4}{3}$, and it tends to 4 for $c \to \infty$. The parameter y that appears in [4] may be viewed as the inverse of $\frac{4c}{c+4}$. A popular choice is $c = 2$, $\frac{4c}{c+4} = \frac{4}{3}$, $y = \frac{3}{4}$.

8. Paths in the Reduction Graph

Let L be a lattice, and put $n = \operatorname{rk} L$. A *path* in $\Gamma(L)$ is a finite sequence $\mathfrak{F}_1 \to \mathfrak{F}_2 \to \cdots \to \mathfrak{F}_t$ of steps $\mathfrak{F}_i \to \mathfrak{F}_{i+1}$ ($1 \leq i < t$) in $\Gamma(L)$; more properly, one could call this a 'directed path', but for 'undirected paths' – which would turn $\Gamma(L)$ into a *connected* graph – we have no use.

Let c be a real number with $c \geq \frac{4}{3}$. Proposition 7.3 leads to the following procedure for transforming a given flag $\mathfrak{F}$ of a lattice into a flag $\mathfrak{F}'$ satisfying the inequalities $l_{i+1}(\mathfrak{F}')^2 \geq c^{-1} \cdot l_i(\mathfrak{F}')^2$ ($0 < i < n$) from Proposition 6.1. If $\mathfrak{F}$ itself does not satisfy these inequalities, then by Proposition 7.3 one can take a step of length greater than $\frac{4c}{c+4}$ from $\mathfrak{F}$, and iterate. Since the number of flags of size smaller than $s(\mathfrak{F})$ is finite, this 'flag reduction algorithm' must terminate with a flag $\mathfrak{F}'$ with the required property. In particular, taking $c = \frac{4}{3}$, we see that we proved the statement made at the end of Section 6.

A good upper bound for the number of steps to be taken is of obvious interest for the analysis of actual algorithms that may be based on the procedure just described. Such a bound is easy to obtain in the case $c > \frac{4}{3}$. Namely, in that case we have $\frac{4c}{c+4} > 1$, and since the square of the size of the flag decreases by a factor greater than $\frac{4c}{c+4}$ in each step, the number of steps in the path $\mathfrak{F} \to \cdots \to \mathfrak{F}'$ is at most

$$2 \cdot \frac{\log\big(s(\mathfrak{F})/s(\mathfrak{F}')\big)}{\log\big(\frac{4c}{c+4}\big)}.$$

This is a satisfactory bound if a good lower bound for $s(\mathfrak{F}')$ is available, which is often the case; for example, if the lattice L is such that $\langle x, y \rangle \in \mathbf{Z}$ for all $x, y \in L$, then one has $d(K)^2 \in \mathbf{Z}$ for all sublattices K of L, so $s(\mathfrak{F}')^2$ is an integer, and $s(\mathfrak{F}') \geq 1$. In general, one has $s(\mathfrak{F}') \geq \prod_{i=1}^{n} (l/i)^{i/2}$ if l is as in Proposition 3.1, by the second remark after Proposition 4.3.

The argument just given fails in the case $c = \frac{4}{3}$, and more generally if we allow steps of length arbitrarily close to 1. It is, for fixed rank, nevertheless possible to prove a similar logarithmic upper bound for the length of any path $\mathfrak{F} \to \cdots \to \mathfrak{F}'$ in $\Gamma(L)$, as we shall see in Proposition 8.2. We first prove an auxiliary result on paths that consist of 'short' steps only.

Proposition 8.1. *For each integer $n \geq 0$ and each real number $c > \frac{4}{3}$ there is a positive integer $A = A(n, c)$ with the following property. Let L be a lattice of rank n, and let $\mathfrak{F}_1 \to \cdots \to \mathfrak{F}_t$ be a path in $\Gamma(L)$ such that each step $\mathfrak{F}_i \to \mathfrak{F}_{i+1}$ has length at most $\frac{4c}{c+4}$. Then one has $t \leq A$.*

Proof. The proof is by induction on n, the case $n \leq 1$ being trivial. Suppose that $n \geq 2$, and consider a path as in Proposition 8.1. We first show that there exists $m \in \{1, 2, \ldots, n\}$ with the following two properties:

   (i) $l_{i+1}(\mathfrak{F}_1)^2 \geq c^{-1} \cdot l_i(\mathfrak{F}_1)^2$ for $0 < i < m$;
   (ii) none of the steps $\mathfrak{F}_j \to \mathfrak{F}_{j+1}$ in the path has colour m.

If all $i = 1, \ldots, n-1$ satisfy the inequality in (i) then we can clearly take $m = n$. Now suppose that $i \in \{1, \ldots, n-1\}$ is such that $l_{i+1}(\mathfrak{F}_1)^2 < c^{-1} \cdot l_i(\mathfrak{F}_1)^2$. Then by Proposition 7.3, there is a step $\mathfrak{F}_1 \to \mathfrak{F}'$ of colour i and length greater than $\frac{4c}{c+4}$, so *any* step of colour i starting at $\mathfrak{F}_1$ has length greater than $\frac{4c}{c+4}$. By hypothesis, $\mathfrak{F}_1 \to \mathfrak{F}_2$ has length at most $\frac{4c}{c+4}$, so it does not have colour i. Therefore Proposition 7.2 implies $l_{i+1}(\mathfrak{F}_2) \leq l_{i+1}(\mathfrak{F}_1)$ and $l_i(\mathfrak{F}_2) \geq l_i(\mathfrak{F}_1)$. It follows that the inequality $l_{i+1}(\mathfrak{F})^2 < c^{-1} \cdot l_i(\mathfrak{F})^2$, which is satisfied for $\mathfrak{F} = \mathfrak{F}_1$, is likewise satisfied for $\mathfrak{F} = \mathfrak{F}_2$; by induction on j one now deduces that all steps $\mathfrak{F}_j \to \mathfrak{F}_{j+1}$ in the path have colour different from i and that all $\mathfrak{F} = \mathfrak{F}_j$ satisfy the inequality just stated. Hence $m = i$ satisfies (ii). If we take for m the *least* value of i violating the inequality in (i), then (i) is satisfied as well.

Write F_{j1} for the rank 1 lattice belonging to $\mathfrak{F}_j$. We claim that *among $F_{11}, \ldots, F_{t1}$ there are at most $(3^n \cdot c^{n(n-1)/4} - 1)/2$ different rank 1 lattices.* To prove this, let m be as above. By (ii), the rank m sublattice belonging to $\mathfrak{F}_j$ is the same for all j; let this lattice be called K. The lattices of rank at most m belonging to $\mathfrak{F}_1$ form a flag $\mathfrak{E}$ of K, and by (i) we have $l_{i+1}(\mathfrak{E}) \geq c^{-1} \cdot l_i(\mathfrak{E})^2$ for $0 < i < m$. Applying Proposition 6.2 to K and $\mathfrak{E}$ we see that the number of $x \in K$ with $q(x) \leq l_1(\mathfrak{F}_1)^2$ is at most $3^m \cdot c^{m(m-1)/4}$. Since x and $-x$ generate the same lattice, it follows that K has at most $(3^m \cdot c^{m(m-1)/4} - 1)/2$ sublattices M of rank 1 that satisfy $d(M) \leq l_1(\mathfrak{F}_1)$. Each F_{j1} is such an M, and $m \leq n$, so the claim follows.

In our path, the rank 1 sublattice changes only at steps of colour 1, and at each such step the determinant of that sublattice decreases. Hence our claim implies that we can write the path $\mathfrak{F}_1 \to \cdots \to \mathfrak{F}_t$ as the union of at most $(3^n \cdot c^{n(n-1)/4} - 1)/2$ subpaths connected by steps of colour 1, such that in each of the subpaths the rank 1 sublattice is held fixed. But when the rank 1 sublattice is held equal to M (say), one is really considering flags of the rank $n-1$ lattice L/M and paths in $\Gamma(L/M)$. Application of the induction hypothesis on n now leads in a straightforward way to the inequality in Proposition 8.1, with $A(n,c) = A(n-1,c) \cdot [(3^n \cdot c^{n(n-1)/4} - 1)/2]$. $\qquad\qquad\square$

We can now formulate and prove our main result.

Proposition 8.2. *For each non-negative integer n there exists a positive integer $B = B(n)$ with the following property. If L is a lattice of rank n, and $\mathfrak{F}$, $\mathfrak{F}'$ are flags of L, then every path from $\mathfrak{F}$ to $\mathfrak{F}'$ in $\Gamma(L)$ contains at most $B \cdot \bigl(1 + \log(s(\mathfrak{F})/s(\mathfrak{F}'))\bigr)$ flags.*

Proof. Fix a real number c with $c > \frac{4}{3}$, and call a step $\mathfrak{F}_1 \to \mathfrak{F}_2$ in $\Gamma(L)$ *long* if it has length greater than $\frac{4c}{4+c}$, and *short* otherwise.

Consider any path $\mathfrak{F} \to \cdots \to \mathfrak{F}'$. If k is the number of long steps, then one has $s(\mathfrak{F}')^2 \le \bigl(\frac{4+c}{4c}\bigr)^k \cdot s(\mathfrak{F})^2$, so $k \cdot \log \frac{4c}{4+c} \le 2\log\bigl(s(\mathfrak{F})/s(\mathfrak{F}')\bigr)$. Hence the path is the union of at most $1 + 2\bigl(\log(s(\mathfrak{F})/s(\mathfrak{F}'))\bigr)/\log \frac{4c}{4+c}$ subpaths connected by long steps, such that each of the subpaths consists of short steps only. By Proposition 8.1, the number of flags occurring in each of the subpaths is bounded by a function of the rank. The result follows. $\qquad\qquad\square$

The proposition just proved is useful in the analysis of algorithms that involve lattices of fixed rank. When the rank varies, it becomes important to express $B(n)$ as an explicit function of n; in particular, if one wishes such an algorithm to run in polynomial time, one may want to bound $B(n)$ by a polynomial function of n. I do not know whether this is possible. I do know the following much weaker result.

Proposition 8.3. *The numbers $B(n)$ in Proposition 8.2 can be chosen such that in addition one has $B(n) = (4/3)^{n^3/(24+o(1))}$ for $n \to \infty$.*

Proof. Making the proof of Proposition 8.2 explicit, one finds a value for $B(n)$ that is a function of c. One may choose c as a function of n that tends to $\frac{4}{3}$ for $n \to \infty$ sufficiently slowly for the factor $\log \frac{4c}{4+c}$ to be $\bigl(\frac{4}{3}\bigr)^{o(n)}$; for example, one may take $c = \frac{4}{3} + \frac{1}{n}$. This yields the result of Proposition 8.3, but with 12 instead of 24. To achieve 24, one improves the estimate obtained in the proof of Proposition 8.1. In that proof, we saw that the flags in a path $\mathfrak{F}_1 \to \cdots \to \mathfrak{F}_t$ consisting of short steps only comprise at most $(3^n \cdot c^{n(n-1)/4} - 1)/2$ different sublattices M of rank 1. One now notes that, by duality, they also comprise at most $(3^n \cdot c^{n(n-1)/4} - 1)/2$ different sublattices N of rank $n-1$. It follows that there are at most $3^n \cdot c^{n(n-1)/4} - 3$ steps of colour 1 or $n-1$, and that the path is the union of at most $3^n \cdot c^{n(n-1)/4} - 2$ subpaths, connected by steps of colours 1 and $n-1$, such that in each of the subpaths both N and M are fixed; it is then really a path in $\Gamma(N/M)$, where N/M has rank $n-2$. In this manner, one proves that one may take $A(n,c) = A(n-2,c) \cdot \left[3^n \cdot c^{n(n-1)/4} - 2\right]$ in Proposition 8.1. This improved bound leads to Proposition 8.3. $\qquad\square$

A. Akhavi obtained results similar to the propositions proved in this section [2].

Acknowledgements

The author is grateful to K. I. Aardal, A. Akhavi, and C. A. J. Hurkens, whose comments led to improvements in the paper.

References

[1] K. I. Aardal, *Lattice basis reduction and integer programming*, rapport UU-CS-1999–37, Informatica Instituut, Universiteit Utrecht, 1999.

[2] A. Akhavi, *Worst-case complexity of the optimal LLL algorithm*, pp. 355–366 in *Proceedings of Latin American theoretical informatics LATIN 2000*, Lecture Notes in Computer Science **1776**, Springer-Verlag, New York, 2000.

[3] H. Cohen, *A course in computational algebraic number theory*, Springer-Verlag, Berlin, 1993.

[4] A. K. Lenstra, H. W. Lenstra, Jr. and L. Lovász, *Factoring polynomials with rational coefficients*, Math. Ann. **261** (1982), 515–534.

[5] H. W. Lenstra, Jr., *Integer programming with a fixed number of variables*, report 81–03, Mathematisch Instituut, Universiteit van Amsterdam, April, 1981.

[6] H. W. Lenstra, Jr., *Integer programming with a fixed number of variables*, Math. Oper. Res. **8** (1983), 538–548.

Mathematisch Instituut
Universiteit Leiden
Postbus 9512
2300 RA Leiden, The Netherlands
E-mail address: hwl@math.leidenuniv.nl

Department of Mathematics # 3840
University of California
Berkeley, CA 94720–3840, U. S. A.

Moduli, Motives, Mirrors

Yuri I. Manin

Abstract. This talk is dedicated to various aspects of Mirror Symmetry. It summarizes some of the mathematical developments that took place since M. Kontsevich's report at the Zürich ICM and provides an extensive, although not exhaustive bibliography.

1. Introduction

This talk is dedicated to various aspects of Mirror Symmetry. It summarizes some of the developments that took place since M. Kontsevich's report [58] at the Zürich ICM and provides an extensive, although not complete bibliography.

1.1. Brief history

Mathematical history of Mirror Symmetry started in 1991, when an identity of a new type was discovered in the ground-breaking paper by four physicists [20] (it was reproduced in [85] where earlier works are also described and motivated).

The left-hand side (or A-side) of this identity was a generating series for the numbers $n(d)$ of rational curves of various degrees d lying on a smooth quintic hypersurface in $\mathbf{P}^4$. The right-hand side (B-side) was a certain hypergeometric function. The Mirror Identity states that the two functions become identical after an explicit change of variables which is defined as a quotient of two hypergeometric functions of the same type.

At the moment of discovery, not only the identity itself remained unproved, but even its A-side was not well defined: the correct way of counting rational curves was proposed by M. Kontsevich ([60]) only in 1994. In the same remarkable paper Kontsevich gave an explicit formula for $n(d)$ creatively using Bott's fixed point formula for torus actions at the target space. After the appearance of this paper one could hope that the Mirror Identity for quintics (and more general toric submanifolds) ought to be provable by algebraic manipulations with both sides. This turned out to be a difficult problem. A. Givental brought this program to a successful completion in 1996, by introducing a new torus action at the source space, stressing equivariant cohomology and inventing ingenious calculational strategy (see [40, 43, 45, 15, 90]). For subsequent important developments, see [66, 67, 14].

Plenary talk at the 3rd European Congress of Mathematicians, Barcelona, July 10–14, 2000.

This work however did not unveil the mystery of the Mirror Identity. The point is that the identity itself was discovered by the physicists as only one manifestation of a deeper principle. Physicists believe that with any Calabi-Yau manifold X one can associate two $N = (2,2)$ Superconformal Field Theories (SCFT) which are the respective A and B models (see e.g. [105]). The Mirror Correspondence between X and Y supposedly interchanges their A and B models. In particular, in the case of quintics the hypergeometric functions involved are actually periods of the mirror partner family of our quintics, and B-models generally reflect properties of variations of periods and Hodge structures.

Unfortunately, a precise and complete mathematical definition of what constitutes an $N = (2,2)$ ScFT is still lacking. Various components of this structure with varying degree of precision are described in the papers collected in [85] and [86]. In particular, a part of this structure is a modular functor in the sense of Segal, with possibly infinite dimensional Hilbert space. In turn, such theories are often constructed via representation theory of a vertex algebra. See [73] and [17, 18] for the most recent mathematical approach to this picture, achieving at least the construction of what seems to be the right vertex algebra.

The parts that are involved in the statement of Mirror Identity above refer correspondingly to the Quantum Cohomology (A-model, physicists' σ-model) and extended variations of Hodge structure. Both are now well understood mathematically: see [79] and [4] respectively. However, the Mirror partners are connected by many more ties than a mere Mirror Identity. These ties, in particular, relate Lagrangian and complex geometry in a remarkable way: see [96] and [58] for the basic conjectures to this effect.

Therefore now, more than a decade after it was discovered, the Mirror Symmetry mathematically looks like a complex puzzle, some of the pieces of which have found their respective places, some are still lying in disorder, and some, most probably, are missing.

1.2. Plan of the paper

This puzzle metaphor guided the organization of this report.

Section 2 is devoted to the binary relation of *mirror partnership* between families of Calabi-Yau manifolds endowed with additional structures which we call here *cusps*. This relation consists in the isomorphism of two Frobenius manifolds, constructed in two different ways for the respective families. In turn, Frobenius manifold isomorphisms generalize the Mirror Identity of [20].

Section 3 explains various versions of another mirror partnership relation, this time between certain symplectic, on the one hand, and complex, on the other hand, manifolds, endowed with additional structure which in this case is a choice of *a fibration by real tori*. Here I have taken as starting point a part of Kontsevich's package [58], with further details taken from [96, 93, 2], and other papers. I have chosen for these relations the word "partnership", or "duality", as opposed to "symmetry", because the definitions of both of them are explicitly un-symmetric.

The next part of Section 3 restores the idea of Mirror Symmetry: according to [96], the Mirror Symmetry relation connects Calabi-Yau manifolds endowed with Kähler structure, and thus simultaneously with compatible Lagrangian and complex structures, so that Kontsevich's duality can be imposed simultaneously upon two crossover Lagrangian/complex pairs.

Although the Frobenius manifold duality and the Lagrangian/complex duality some day are expected to become parts of a unified picture, at present contours of the latter are rather vague.

One common part of both dualities is the prediction of *mirror isomorphisms* connecting the cohomology spaces of mirror partners (X, Y). In particular, isomorphisms of the Frobenius manifolds restricted to their spaces of flat vector fields produces isomorphisms $\mu_{X,Y} \colon H^*(X, \mathbf{C}) \to H^*(Y, \wedge^*(\mathcal{T}_Y) \otimes V_Y^{-2}$ where $V_Y = H^0(Y, \Omega_Y^{\max})$.

Actually, the algebraic geometric model of the A-side, the theory of Gromov-Witten invariants, endows $H^*(X, \mathbf{C})$ with much stronger structure, which is motivic by its nature. It would be very interesting to understand the geometry of the mirror reflection of Calabi-Yau motives.

However, even when the intricate inner workings of Mirror Symmetry are understood, this will not be the end of the story.

1.3. Dualities in string theory

All this machinery emerged as an approximation in the quantum superstring theory whose aim is to provide a unified theory of matter and gravity (space-time). The first superstring revolution (1984–85) led to the belief that there are five consistent (perturbative, without ultraviolet divergencies) superstring theories, each on ten-dimensional space-time, or in other words, a 10d Poincaré-invariant vacuum. For all of them, the low-energy approximation is an effective 10d supergravity theory. The second superstring revolution (1994–??) started with Witten's suggestion that all five theories are limits of a single theory (see review in [94]). In other words, they are perturbarive expansions of a single underlying theory about distinct points in the moduli space of quantum vacua. Moreover, a sixth special point in this space is an 11d Poincaré-invariant vacuum.

C. Vafa suggests to look at the underlying *M-theory* as patched up from five/six local descriptions and their compactifications, in much the same way as a manifold is patched up from coordinate neighborhoods. The transition functions are called *dualities*, and according to [96], Mirror Duality is one of them.

If this is true, the Mirror Symmetry acquires an incredibly high epistemological status as one of the building blocks of the ambitious Unified Quantum Superstring Theory.

For mathematicians, this means that the puzzle we are trying to assemble is only a small piece of the still larger puzzle whose contours are yet barely visible.

2. Frobenius Manifolds and Mirror Partnership between Families of Calabi-Yau Manifolds

2.1. Calabi-Yau manifolds

In this section I will call a Calabi-Yau (CY) manifold *in the weak sense* any projective (or compact Kähler) complex manifold X with trivial canonical sheaf. Any such manifold admits a finite unramified covering $\widetilde{X}$ with the following property:

$\widetilde{X}$ *is a direct product of a complex torus A, of a simply connected CY Y with* $h^{2,0}(Y) = h^{2,0}(\widetilde{X})$, *and of a simply connected CY Z with $h^{2,0}(Z) = 0$.*

If the factors A and Y are absent in any finite unramified covering of X, then X is CY *in the strong sense.*

In dimension 1, the only CY manifolds are elliptic curves, in dimension 2, besides complex tori, there are $K3$-surfaces. In dimension 3, the first examples of CY in the strong sense appear. Quintics in $\mathbf{P}^4$ are the simplest of them.

More generally, anticanonical hypersurfaces in any compact toric manifold associated with a *reflexive polyhedron* are CY's: see [6]. This method produces 4319 families of $K3$-surfaces and 473 800 776 families of CY threefolds, among which at least 30 178 families can be distinguished by their Hodge numbers (see [63]). It is still unknown whether the number of maximal families of CYs in any dimension ≥ 3 is finite or not.

The toric construction (and its generalization to complete intersections in arbitrary Fano manifolds) remain the most important testing ground for basic conjectures about CYs.

A general approach to the complex moduli spaces of CY manifolds is furnished by the deformation theory. The Kodaira-Spencer local versal deformation of an n-dimensional CY manifold X is unobstructed and has dimension $h^{1,n-1}(X)$.

2.2. Mirror partner families of CYs: preliminaries

The notion which we will describe in this section is interesting mainly for CYs in the strong sense. It develops the discovery made in [20].

This notion is an asymmetric binary relation between versal local families $\{X_s \mid s \in S\}$, $\{Y_t \mid t \in T\}$ of CYs, satisfying the condition $h^{1,1}(X) = h^{1,n-1}(Y) = r$ and endowed with some additional structure.

On the A-side, this additional structure consists in a choice of a basis $(\beta_1, \ldots, \beta_r)$ of the group of numerically effective classes in $A_1(X_s)$. When $s \in S$ varies, elements of this basis must be horizontal with respect to the Gauss-Manin connection. Such a basis determines functions q_j^A on $H^2(X_s, \mathbf{C}/\mathbf{Z})$: $q_j^A(L) := e^{2\pi i(L, \beta_j)}$. We will refer to $H^2(X_s, \mathbf{C}/\mathbf{Z})$ together with q_j^A as a *Kähler cusp.*

On the B-side, this additional structure consists in the choice of a partial compactification $T \subset \overline{T}$ looking locally like the embedding of a product of pointed open unit discs in $\mathbf{C}$ into the product of non-pointed unit discs. The variation of Hodge structures of the family Y_t must have maximal unipotent monodromy on this compactification: see [88, 89, 27]. Geometrically, $\overline{T}$ contains a point of "maximal degeneration" of the family Y_t and T parametrizes Calabi-Yau manifolds "with large complex structure". The point of maximal degeneration is the transversal intersection of discriminantal divisors. Building upon [20] and [87, 88], Deligne has shown in [27] how to define a system of functions q_j^B on $\overline{T}$ in terms of the variation of Hodge structures determined by Y_t. We will refer to the germ of $\overline{T}$ at its point of maximal degeneration as *moduli cusp* of the relevant moduli space.

The relation of *mirror partnership* between such enhanced families X/S and Y/T, in particular, identifies q_j^A with q_j^B and thus establishes an isomorphism between a domain in $H^2(X, \mathbf{C}/\mathbf{Z})$ where q_j^A are sufficiently small and the respective domain in the moduli cusp. This isomorphism must identify two functions: *potential of the small quantum cohomology* at the A-side, and an integral involving a holomorphic volume form on the fibers Y_t at the B-side.

A fuller formulation of the mirror partnership relation consists in the identification of two *formal Frobenius manifolds (FM)*: quantum cohomology of any X_s at the A-side, and Barannikov-Kontsevich's FM on a formal *extended moduli space* at the B-side.

Most of the remaining part of this section will be devoted to the description of the relevant Frobenius manifolds.

However, the reader must be aware of more global aspects of this essentially local picture. In fact, moduli stacks of complex variations of Y may have many cusps; they are acted upon by the Teichmüller group $\mathrm{Diff}(Y)/\mathrm{Diff}_0(Y)$. One can speculate that mirror partnership is stable with respect to such moduli cusp changes. Then the question arises, what corresponds to them at the A-side?

Two partial answers were suggested. In [3] it was argued that different birational models of some X_s can produce canonically isomorphic cohomology groups, in particular H^2, in which however Kähler cones will form a non-trivial fan (in the sense of toric geometry). Maximal cones of this fan support Kähler cusps that might correspond to different moduli cusps of the same family at the B-side. In this picture, one does not see what should correspond to the Teichmüller group. M. Kontsevich suggested in the framework of his conjectured Lagrangian/complex duality that it must be the autoequivalence group of the derived category of coherent sheaves on X_s. Some evidence for this was furnished by comparison of the stabilizing subgroups of the cusps: see [52, 95].

2.3. Frobenius manifolds

Let M be an analytic or formal supermanifold. A structure of the Frobenius manifold on it is given by a flat metric g (symmetric non-degenerate form on the tangent sheaf) and a function (potential) Φ with the following property. Let (x_a) be a local g-flat coordinate system, $\partial_a = \partial/\partial x_a$, $\Phi_{abc} = \partial_a\partial_b\partial_c\Phi$. Raise one index of Φ_{abc} using g and define an $\mathcal{O}_M$-bilinear multiplication $\circ$ on $\mathcal{T}_M$ by $\partial_a \circ \partial_b := \sum_c \Phi_{ab}{}^c\partial_c$. Then this multiplication must be associative (it is obviously (super)commutative). Additional structures that are present in the mirror picture are the flat identity e for $\circ$ and an Euler vector field E satisfying the conditions $\mathrm{Lie}_E(g) = Dg$ for some constant D and $\mathrm{Lie}_E(\circ) = \circ$. It expresses homogeneity properties of Φ: we have $E\Phi = (D+1)\Phi + $ a polynomial in flat coordinates of degree ≤ 2.

At the A-side, the relevant Frobenius manifold is formal: M is the formal completion of the linear space $H^*(X_s, \mathbf{C})$, with its Poincaré pairing as g and the potential Φ constructed as formal series whose Taylor coefficients are Gromov-Witten invariants of X_s. At the B-side the relevant Frobenius manifold can be conceived as a certain formal neighborhood of the classical moduli space T near the relevant cusp: *extended moduli space* of the B-family. Both formal spaces can be refined to germs of analytic spaces.

Here are some details.

2.4. Quantum cohomology

Potential Φ of the quantum cohomology can be defined for any projective complex (or compact symplectic) manifold X. After taking into account the relevant homogeneity properties of the Gromov-Witten invariants, it can be written as a formal series in linear coordinates (x^a) on $\oplus_{k\neq 2} H^k(X, \mathbf{C})$ and their exponentials on $H^2(X)$:

$$\Phi(x) = \frac{1}{6}\left(\left(\sum x_a \Delta_a\right)^3\right)$$
$$+ \sum_{\beta \neq 0} e^{(\beta, \sum_{|\Delta_b|=2} x_b \Delta_b)} \sum_{n \geq 0, (a_i): |\Delta_{a_i}| \neq 2} \langle \Delta_{a_n} \dots \Delta_{a_1} \rangle_{0,n,\beta} \frac{x_{a_1} \dots x_{a_n}}{n!}. \quad (1)$$

Here (Δ_a) is the basis of $H^*(X, \mathbf{C})$ dual to (x_a), $\Delta_k \in H^{|\Delta_k|}(X)$, the first term in the rhs of (1) is the cubic self-intersection index, and β runs over numerically effective 1-classes in X. Finally, the Gromov-Witten invariant $\langle \Delta_{a_n} \dots \Delta_{a_1} \rangle_{0,n,\beta}$ counts the virtual number of stable maps of genus zero $(C; x_1, \dots, x_n; f: C \to X)$ such that $f_*([C]) = \beta$ and $f(x_i) \in D_{a_i}$ where D_{a_i} is a cycle representing the homology class dual to Δ_{a_i}. Physically, Δ_a are called *the primary fields* of the respective Conformal Field Theory, and the Gromov-Witten invariants are their *correlators*.

The *small quantum cohomology potential* is obtained by restricting $\Phi(x)$ to H^2, that is, putting $x_a = 0$ for $|\Delta_a| \neq 2$.

2.5. Barannikov-Kontsevich's construction

On the B-side, the relevant formal Frobenius potential is constructed on the completion at zero of the cohomology space $H^*(Y, \wedge^*(\mathcal{T}_Y))$ interpreted as a formal moduli space $\mathcal{M}_{A_\infty}$ of A_∞-deformations of Y. This construction was introduced in [4]; it refines the earlier proposal from [5]. Unlike the case of quantum cohomology, here it is essential to require Y to be a (weak) Calabi-Yau manifold. This condition will be used, in particular, through a choice of the global holomorphic volume form Ω on Y.

This geometric setup produces first of all an algebraic object $(\mathcal{A}, \delta, \Delta, \int)$, *special differential Batalin-Vilkovyski algebra (dBV)*, consisting of the following data which we will describe in axiomatized form.

(i) $\mathcal{A}$ *is a supercommutative* $\mathbf{C}$-*algebra.*

In the Calabi-Yau setup, $\mathcal{A} = \Gamma_{C^\infty}(Y, \wedge^*(\overline{T}_Y^*) \otimes \wedge^*(T_Y))$.

(ii) δ *is an odd* $\mathbf{C}$-*derivation of* $\mathcal{A}$, $\delta^2 = 0$.

In our case, $\delta = \overline{\partial}$, the operator defining the complex structure on Y and its tangent bundle, so that $\mathcal{A}$ is the Dolbeault resolution of the exterior algebra of the tangent bundle.

Therefore, the δ-cohomology space $H = H(\mathcal{A}, \delta) = \operatorname{Ker}\delta / \operatorname{Im}\delta$ in our case is identified with coherent cohomology $H^*(Y, \wedge^*(\mathcal{T}_Y))$. Generally, we assume it to be of finite dimension.

The space H plays the central role, because it will support the structure of the formal Frobenius manifolds.

We will denote by $K = \mathbf{C}[[x_a]]$ the ring of formal functions on H, (x_a) being coordinates on H dual to a basis (Δ_a).

(iii) Δ *is another odd differential,* $\Delta^2 = 0$, *which is a differential operator of order two with respect to the multiplication in* $\mathcal{A}$.

More precisely, we assume that for any $a \in \mathcal{A}$ the formula

$$\partial_a b = (-1)^{\tilde{a}}\Delta(ab) - (-1)^{\tilde{a}}(\Delta a)b - a\Delta b$$

defines a derivation ∂_a. Moreover, we assume that $\delta\Delta + \Delta\delta = 0$.

In our case, Δ is obtained from the ∂-operator on the complexified C^∞ de Rham complex of Y after the identification of this complex with $\mathcal{A}$ with the help of Ω: $\Delta(a) := (\vdash \Omega)^{-1} \circ \partial \circ (a \vdash \Omega)$.

From the $\partial\overline{\partial}$-Lemma in Kähler geometry, it follows that the two canonical embeddings of differential spaces

$$(\operatorname{Ker}\Delta, \delta) \to (\mathcal{A}, \delta), \quad (\operatorname{Ker}\delta, \Delta) \to (\mathcal{A}, \Delta) \tag{2}$$

are quasi-isomorphisms, and moreover, homology of all four differential spaces can be identified with $(\operatorname{Ker}\Delta \cap \operatorname{Ker}\delta)/\operatorname{Im}\delta\Delta$.

As a part of this package, one also obtains the following formality property: the natural map $\operatorname{Ker}\Delta \to H(\mathcal{A}, \Delta)$ induces a surjection of differential Lie algebras which is a quasi-isomorphism:

$$(\operatorname{Ker}\Delta, [\bullet], \delta) \to (H(\mathcal{A}, \Delta), 0, 0)\,.$$

In the axiomatized situation, we impose these conditions as an additional axiom. This condition can be weakened: it suffices to require only that cohomology of differentials $\delta + \Delta$ and δ have the same dimension.

(iv) $\int : \mathcal{A} \to \mathbf{C}$ *is a linear functional which must satisfy two integration by parts identities:*

$$\int (\delta a)b = (-1)^{\tilde{a}+1}\int a\delta b, \quad \int (\Delta a)b = (-1)^{\tilde{a}}\int a\Delta b\,. \tag{3}$$

The integral is given by the formula

$$\int a = \int_Y (a \vdash \Omega) \wedge \Omega \tag{4}$$

60   Y. I. Manin

where Ω means a holomorphic volume form on Y whose period over the unique monodromy invariant cycle at the chosen cusp is $(2\pi i)^d$, $d = \dim Y$.

(v) *Algebra grading* $\mathcal{A} = \oplus \mathcal{A}^n$, $\mathbf{C} \in \mathcal{A}^0$.

We assume that with respect to this grading, δ and Δ are of degree 1, and $\int$ has a definite degree. (This is at variance with [78, 79], but agrees with [4].)

Grading produces an Euler field on H, whereas the image of $1 \in \mathcal{A}$ serves as flat identity.

In the Calabi-Yau setup, we can grade $\wedge^p \overline{T}_Y^* \otimes \wedge^q T_Y$ by $q - p$.

2.5.1. FROBENIUS STRUCTURE   Having thus described the formal properties of a Batalin-Vilkovyski algebra $(\mathcal{A}, \delta, \Delta, \int)$, we can now explain the derivation of the Frobenius structure on H.

One starts with checking that the bilinear operation $[a \bullet b] = \partial_a b$, together with multiplication, endows $\mathcal{A}$ by the structure of *Gerstenhaber, or odd Poisson superalgebra*, in which the Lie bracket is a parity changing operation, and all the usual axioms are valid after inserting appropriate signs.

The basic ingredient of the construction from [4] is a certain exponential map Φ^W. In the Calabi-Yau setup it is an A_∞-analog $\mathcal{M}_{A_\infty} \to H^*(Y, \mathbf{C})[[\hbar^{-1}, \hbar]][d]$ of the classical period map. Roughly speaking the map Φ^W is described by the formula

$$\Phi^W(x_a, \hbar) = \left[\exp \frac{1}{\hbar} \widetilde{\Gamma} \right]$$

where $\widetilde{\Gamma} \in \mathcal{A} \widehat{\otimes} K[[\hbar]]$ is a W-normalized generic solution to the Maurer-Cartan equation $(\delta + \hbar\Delta)\widetilde{\Gamma} + \frac{1}{2}[\widetilde{\Gamma} \bullet \widetilde{\Gamma}] = 0$ and $[a]$ denotes the cohomology class with respect to the differential $\delta + \hbar\Delta$. Here δ and Δ are assumed to be extended to $\mathcal{A} \widehat{\otimes} K[[\hbar]]$ by linearity and $\widetilde{\Gamma}$ is supposed to be W-normalized generic in the following sense: firstly, $\left[\exp \frac{1}{\hbar} \widetilde{\Gamma} \right] \in 1 + L_W$, where L_W, $\hbar^{-1} L_W \subset L_W$ is a semi-infinite subspace associated with an increasing isotropic filtration on the cohomology of $\delta + \Delta$, and, secondly, the map $(\Phi^W - 1) \bmod (\hbar^{-1} L_W) : H \to L_W / \hbar^{-1} L_W$ is linear and is an isomorphism. In the Calabi-Yau setting W is the monodromy weight filtration associated with the relevant cusp. Existence of such a solution $\widetilde{\Gamma}$ for W satisfying certain transversality conditions can be proved by induction on the order of coefficients of Taylor expansion.

As a matter of fact, at this stage this construction exhibits certain common features with K. Saito's construction of FM structures on unfolding spaces of singularities. It seems that if one chooses for W a certain special filtration, then the primitive form from the K. Saito theory can be identified with an analog of $\Phi^W(x_a, \hbar)$. The existence of a primitive form in K. Saito's theory is a nontrivial fact which follows in general from the theory of mixed Hodge modules of M. Saito.

Let us put now $\Gamma = \widetilde{\Gamma}(x^a, \hbar = 0)$ and $\delta_\Gamma := id \otimes \delta + [\Gamma \bullet]$. The operator δ_Γ is a homological differential acting on $\mathcal{A}_K := K \widehat{\otimes} \mathcal{A}$. By continuity, one can canonically identify $H(\mathcal{A}_K, \delta_\Gamma)$ with $K \otimes H$. On the other hand, multiplication in $\mathcal{A}_K$ induces a multiplication on $H(\mathcal{A}_K, \delta_\Gamma)$. This is our $\circ$. The map $\Phi^W(x_a, \hbar)$ induces a pairing on the tangent sheaf to H:

$$\langle \partial_a, \partial_b \rangle^W := \int \partial_a \Phi^W(x_a, \hbar) \partial_b \Phi^W(x_a, -\hbar).$$

The properties of the map Φ^W imply that this pairing is constant: $\langle \partial_a, \partial_b \rangle^W = g_{ab}$. This is our flat metric.

2.5.2. MIRROR IDENTITIES FOR COMPLETE INTERSECTIONS IN PROJECTIVE SPA-CES   After these preparations, Barannikov's proof runs as follows. Barannikov invokes the famous Givental result ([40, 43, 66]) establishing the mirror identity on the level of "small quantum cohomology" (restriction to H^2) replacing the A-model, and a classical moduli space replacing the B-model. This furnishes identification of a part of Gromov-Witten invariants as coming from the relevant Picard-Fuchs equations. Now, Kontsevich-Manin's "First reconstruction theorem" from [61] shows that this part suffices for identification of the remaining invariants as soon as we know that Associativity Equations (= Frobenius structure) hold. In dimension 3 the latter supply no additional information, but the larger the dimension is, the more important Associativity Equations become.

2.5.3. EXTENDED MODULI SPACES   The context of Mirror Symmetry served to increase awareness of the importance of *extended moduli spaces* in many other contexts of algebraic geometry. Roughly speaking, any classical deformation problem is governed by a cohomology group H^k classifying infinitesimal extensions and the next cohomology group H^{k+1} classifying obstructions. In the stable and unobstructed case, H^k is the tangent space to the base of versal deformation. Extended moduli space in the unobstructed case has a total cohomology H^* as tangent space. Barannikov-Kontsevich's B-model is such an extended moduli space for Calabi-Yau manifolds.

See [62, 23, 74] for a discussion of this matter in general, and [84] for interesting constructions, related to the Frobenius structure.

2.6. Other mirror isomorphisms

There exist isomorphisms of auxiliary Frobenius manifolds connecting certain unfolding spaces of singularities (B-model) and moduli spaces of curves with spin structure (A-model) respectively, as was suggested by Witten [106] and mathematically developed in [55, 56]. See also [78] about possible relations to the Calabi-Yau mirror picture, developing the context in which the Mirror Symmetry was first discussed in [36, 37].

3. Lagrangian/Complex Duality and Mirror Symmetry

3.1. Classical phase spaces

Consider a C^∞ symplectic manifold (X, ω), endowed with a submersion $p_X : X \to U$ whose fibers are Lagrangian tori, and a Lagrangian section $0_X : U \to X$. This is the classical setup of action-angle variables in the theory of completely integrable systems.

The form ω identifies the bundle of Lie algebras of the tori $p_X^{-1}(u)$, $u \in U$, with the cotangent bundle T_U^*. Hence T_U^* can be seen as a fiberwise universal cover of X, and we have a canonical isomorphism $X = T_U^*/H$ where H is a Lagrangian sublattice in T_U^* with respect to the lift of ω which is the standard symplectic form on the cotangent bundle. There exists also a canonical flat symmetric connection on T_U^* for which H is horizontal.

Put $H^t = \mathrm{Hom}(H, \mathbf{Z})$. This local system is embedded as a sublattice into T_U, and we can define *the mirror partner* of $(p_X : X \to U, \omega, 0_X)$ as the toric fibration $Y := T_U/H^t$ endowed with the projection to the same base $p_Y : Y \to U$ and the zero section 0_Y.

3.2. Complex structure on Y

Passing from X to Y we have lost the symplectic form. To compensate for this loss, we have acquired a complex structure $J: T_Y \to T_Y$ which can be produced from $(p: X \to U, \omega, 0_X)$ in the following way. The flat connection on T_U obtained by the dualization from T_U^* produces a natural splitting $T_Y = p_Y^*(T_U) \oplus p_Y^*(T_U)$. With respect to this splitting, J acts as $(t_1, t_2) \mapsto (-t_2, t_1)$.

Conversely, suppose that we have a complex manifold Y endowed with a fibration by real tori $Y \to U$ with zero section, such that the operator of complex structure along the zero section identifies T_U with the bundle of Lie algebras of fibers. Then we can consecutively construct the lattice $H^t \subset T_U$, the dual fibration $X := T_U^*/H$ and the symplectic form on X coming from the cotangent bundle.

3.3. Fourier-Mukai transform and further relationships between Lagrangian and complex geometry

Consider first a pair of dual real tori $\mathbf{T} = H_{\mathbf{R}}/H$ and $\mathbf{T}^t = H_{\mathbf{R}}^t/H^t$ where H is a free abelian group of finite rank, H^t the dual group. Denote by $\langle , \rangle$ the scalar product $H^t \times H \to \mathbf{Z}$ and its real extensions. Each point $x^t \in \mathbf{T}^t$ can be interpreted as a local system of one dimensional complex vector spaces with monodromy $\pi_1(\mathbf{T}) = H \to S^1: h \mapsto e^{2\pi i \langle x^t, h \rangle}$. Hence $\mathbf{T}^t$ becomes the moduli space of such systems on $\mathbf{T}$, and similarly with roles of $\mathbf{T}$ and $\mathbf{T}^t$ reversed.

This can be conveniently expressed by introducing *the Poincaré bundle* $(\mathcal{P}, \nabla_{\mathcal{P}})$ on $\mathbf{T} \otimes \mathbf{T}^t$ which is a rank one complex bundle with connection. The connection is flat along both projections, but has curvature $2\pi i \langle \partial^t, \partial \rangle$ on $(\partial^t, \partial) \in H^t \times H$.

Using $(\mathcal{P}, \nabla_{\mathcal{P}})$, we can extend the correspondence between points of $\mathbf{T}$ and local systems on $\mathbf{T}^t$ in the following way. Call a skyscraper sheaf $\mathcal{F}$ on $\mathbf{T}$ a sheaf consisting of a finite number of vector spaces F_i supported by points x_i. We can define a functorial map

$$\mathcal{F} \mapsto p_{\mathbf{T}^t *}(p_{\mathbf{T}}^*(\mathcal{F}) \otimes \mathcal{P}) \tag{5}$$

whose image, if one takes in account the induced connection, is *a unitary local system on* $\mathbf{T}^t$, that is, a complex vector bundle with flat connection and semisimple monodromy with eigenvalues in S^1.

Let now X and Y be mirror partners in the sense of 3.1–3.2. The construction above shows first of all that points y of Y bijectively correspond to pairs consisting of a Lagrangian torus $L = p_X^{-1}(p_Y(y))$ and a unitary local system of rank one on it.

Moreover, $X \times_U Y$ carries the relative Poincaré bundle which we again will denote $(\mathcal{P}, \nabla_{\mathcal{P}})$: the connection is extended in an obvious way in the horizontal directions. An appropriate relative version of skyscraper sheaves is played by pairs $(L, \mathcal{L})$ consisting of a Lagrangian submanifold of X transversal to the tori and a unitary local system $\mathcal{L}$ on L. The Fourier transform (5) of such a system is defined by

$$(L, \mathcal{L}) \mapsto p_{Y*}(p_L^* \mathcal{L} \otimes (i \times \mathrm{id})^* \mathcal{P}) \tag{6}$$

where we denote by $i\colon L \to X$ the Lagrangian immersion, and $p_Y\colon L \times_U Y \to Y$, $i \times \mathrm{id}\colon L \times_U Y \to Y$, $p_L\colon L \times_U Y \to L$. The image of (6) also carries the induced connection. We can calculate the $\bar{\partial}$-component of it in the complex structure of Y and find out that it is flat. In other words, the rhs of (6) is canonically a holomorphic vector bundle on Y.

3.3.1. AN EXAMPLE: MIRROR DUALITY BETWEEN COMPLEX OR p-ADIC ABELIAN VARIETIES   In this subsection we propose a definition of mirror duality for abelian varieties which works uniformly well over arbitrary complete normed fields K. We will represent such a variety $\mathcal{A}$ as a quotient (in the analytic category) of an algebraic K-torus T by a discrete subgroup B of maximal rank. Such a "multiplicative uniformization" goes back to Jacobi. The passage to the algebraic-geometric picture is mediated by the classical or p-adic theta-functions which are defined as analytic functions on T with the usual automorphic properties with respect to shifts by elements of B, see e.g. [80] for details. The choice of multiplicative uniformization adequately models the choice of a cusp in the moduli space of abelian varieties.

To be precise, an algebraic torus T with the character group H over a field K is the spectrum of the group ring of H. The dual torus T^t, as above, has the character group H^t.

Consider now any diagram of the form

$$(j, j^t)\colon T(K) \leftarrow B \to T^t(K) \tag{7}$$

where B is a free abelian group of the same rank as H and j, resp. j^t, are its embeddings as discrete subgroups into $T(K)$, resp. $T^t(K)$.

We will say that pairs $(\mathcal{A} := T(K)/j(B), j^t)$ and $(\mathcal{B} := T^t(K)/j^t(B), j)$ are mirror dual to each other. The quotient spaces $\mathcal{A}$, $\mathcal{B}$ do not always have the structure of abelian varieties, but this is not important for the following.

In order to motivate this definition, we will show that for $K = \mathbf{C}$, we can produce from (7) a pair of dual real toric fibrations over a common base.

We have the Lie group isomorphism $\mathbf{C}^* \to S^1 \times \mathbf{R}\colon z \mapsto (z/|z|, \log|z|)$. This induces an isomorphism

$$(\alpha, \lambda)\colon T(\mathbf{C}) \to \operatorname{Hom}(H, S^1) \times \operatorname{Hom}(H, \mathbf{R}). \tag{8}$$

Since $j(B)$ is discrete of maximal rank, then $\lambda \circ j(B)$ is an additive lattice in the real space $\operatorname{Hom}(H, \mathbf{R})$. Thus (8) produces a real torus fibration of $T(\mathbf{C})$ over the base which is as well a real torus of the same dimension:

$$0 \to \operatorname{Hom}(H, S^1) \to T(\mathbf{C})/j(B) \to \operatorname{Hom}(H, \mathbf{R})/\lambda \circ j(B) \to 0. \tag{9}$$

Similarly, we have

$$0 \to \operatorname{Hom}(H^t, S^1) \to T^t(\mathbf{C})/j^t(B) \to \operatorname{Hom}(H^t, \mathbf{R})/\lambda^t \circ j^t(B) \to 0 \tag{10}$$

where λ^t is defined for T^t in the same way as λ for T. Let us identify linear real spaces $H_{\mathbf{R}}$ with $H_{\mathbf{R}}^t$ in such a way that lattice points $\lambda \circ j(b)$ and $\lambda^t \circ j^t(b)$ are identified for all $b \in B$. Then (2.5) and (2.6) become dual real torus fibrations over the common base.

The relevant complex structures in our context come from covering tori. They produce symplectic forms as was explained above.

3.4. Kontsevich's package

We now return to the general mirror dual toric fibrations. With some stretch of imagination, one can see the following pattern in the picture described above: *Lagrangian cycles with local systems on X, whose projection to U have real dimension k, must correspond to coherent sheaves on Y with support of complex dimension k.*

Kontsevich in [58] suggested a considerably more sophisticated conjecture. Namely, let X be a *compact* symplectic manifold with $c_1(X) = 0$, and Y some compact complex Calabi-Yau manifold.

Then the relation of mirror partnership between X and Y consists in an equivalence between the Fukaya triangulated category $D(\operatorname{Fuk}_X)$ concocted out of Lagrangian cycles with local systems on the one side, and (a subcategory of) $D^b(\operatorname{Coh}_X)$ on the other side.

Briefly, to construct $D(\operatorname{Fuk}_X)$ one proceeds in three steps: first, one constructs an A_∞-category Fuk_Y, then one produces from it another A_∞-category of twisted complexes, and finally, one passes to the homology category of the latter.

Objects $\Lambda = (L, \mathcal{L}, \lambda)$ of Fuk_Y are Lagrangian submanifolds L in X with unitary local systems $\mathcal{L}$, endowed with a lifting λ to the fiberwise universal cover of the Lagrangian Grassmannian of X.

A morphism space between a pair of such objects admits a transparent description in the case when their Lagrangian submanifolds L_1, L_2 intersect transversally. In this case it is simply $\operatorname{Hom}(\mathcal{L}_\infty, \mathcal{L}_2)$ in the category of sheaves on X. This space is $\mathbf{Z}$-graded with the help of a construction using λ and the Maslov index.

However, the composition of morphisms is not at all the composition of these morphisms of sheaves. In fact, a modification of Floer's construction using summation over pseudoholomorphic parametrized discs in X produces a series of polylinear maps

$$m_1 \colon \operatorname{Hom}(\Lambda_1, \Lambda_2) \to \operatorname{Hom}(\Lambda_1, \Lambda_2),$$

$$m_2 \colon \operatorname{Hom}(\Lambda_1, \Lambda_2) \otimes \operatorname{Hom}(\Lambda_2, \Lambda_3) \to \operatorname{Hom}(\Lambda_1, \Lambda_2),$$

and generally

$$m_r \colon \operatorname{Hom}(\Lambda_1, \Lambda_2) \otimes \cdots \otimes \operatorname{Hom}(\Lambda_{r-1}, \Lambda_r) \to \operatorname{Hom}(\Lambda_1, \Lambda_r).$$

If the respective sums converge, m_1 endows the graded Hom-spaces with the structure of a complex, m_2 becomes the morphism of complexes, and higher multiplications are interrelated by the A_∞-identities ensuring that the associativity constraints for the composition of morphisms are valid up to explicit homotopies.

For more detailed discussion, see [58, 93, 35], and the literature quoted therein. In particular, the case of elliptic curves is rather well understood thanks to Polishchuk and Zaslow, and Fukaya started treating abelian varieties and complex tori.

Both categories involved in Kontsevich's conjecture generally have non-trivial discrete symmetries, induced in the CY-context by monodromy at the Lagrangian side and by derived correspondences at the complex side. Thus some additional data have to be chosen in order to pinpoint the expected functor. The awareness of symmetries led Kontsevich to beautiful predictions about the correspondence between monodromy actions and automorphisms of derived categories: see [52, 95, 97]. We mentioned these predictions above, when we discussed the global properties of the Frobenius partnership relations.

Kontsevich was vague about both the origin of the equivalence functor and the exact geometric relation between X and Y. One can interpret the picture described in 3.1–3.3 which emerged later as a precise guess about the nature of several data left implicit in Kontsevich's presentation:

(i) *The character of additional data to be chosen: dual toric fibrations of X, Y over a common base.*

We will see below how this choice at the complex side is related to the notion of cusp of the relevant moduli space which we introduced in the context of Frobenius mirror partnership.

(ii) *The structure of the restriction of the equivalence functor acting on the simple objects: Fourier-Mukai transform corresponding to the choice (i).*

With exception of the case of complex tori, there is not much chance that X or Y would admit a global fibration by real tori: degenerate fibers are generally unavoidable, and their geometry and influence on the global geometry of the mirror picture are poorly understood. The case of $K3$-surfaces offers some testing ground, because $K3$-surfaces are hyperkähler, and Lagrangian tori can be transformed into a pencil of elliptic curves by an appropriate rotation of the complex structure.

Recently M. Kontsevich, Y. Soibelman and A. Todorov came up with a conjectural limiting metric picture of the maximally degenerating family of CY manifolds of dimension d (private communication). Namely, fix a cohomology class of Kähler forms and a moduli cusp. Deform the complex structure by moving to the maximal degeneration point, and the Calabi-Yau metric in the chosen class by multiplying it by a real number in such a way that the diameter of the space remains 1.

It is expected that the limit $\mathcal{X}$ in the Hausdorff-Gromov sense of this family of metric spaces will be a real d-dimensional manifold with a Riemannian metric which might have singularities in codimension two. Moreover, the remnants of the special real torus fibration consist in the following additional data: affine structure and a sublattice in the tangent bundle. In local affine coordinates, the metric must be the second derivative of a convex function H, and the volume form of the metric must be constant.

Conjecturally, a mirror dual family (endowed with appropriate cusps) produces the same limiting metric space $\mathcal{Y} = \mathcal{X}$, but with a different affine structure and sublattice in the lattice bundle.

A similar picture was envisioned and studied for $K3$-surfaces by M. Gross.

3.5. Mirror Symmetry between Calabi-Yau manifolds

Let now X, Y be two C^∞-manifolds each of which is endowed by a symplectic form, real toric fibration over a common base, and a complex structure, (ω_X, p_X, J_X) and (ω_Y, p_Y, J_Y) respectively. We will say that they are related by Mirror Symmetry, if (X, p_X, ω_X) is the mirror partner of (Y, p_Y, J_Y) and (X, p_X, J_X) is the mirror partner of (Y, p_Y, ω_Y) in the sense of Lagrangian/complex duality. An example of this setup is described in 3.3.1.

The structures J and ω at each side, of course, can be related. The most rigid connection between them is the presence of the Riemann metric g producing the Kähler package (J, ω, g). In the case of Calabi-Yau manifolds, the natural choice is Yau's Ricci-flat metric g.

The program of [96] develops this setup, in particular, supplying the topological and the metric characterization of the basic toric fibrations. Namely, the cohomology class of any toric fiber in X, resp. Y must be the generator of the cyclic group of invariant cycles in the middle cohomology with respect to the local monodromy action at the chosen cusp of moduli space. Moreover, non-degenerate toric fibers (and other relevant Lagrangian submanifolds) must be not simply Lagrangian, but *special* Lagrangian. This produces a version of Lagrangian geometry whose rigidity is comparable to that of the complex one, and makes it fit for comparison with the complex picture: see [48, 49, 98, 99] for many details.

It would be important to develop a version of Fukaya's category in this rigid context where the usual tools of homological algebra might work better.

3.6. Motives in the looking glass

One of the most basic expressions of the Mirror Symmetry of the Calabi-Yau manifolds is the existence of highly nontrivial isomorphisms between their cohomology spaces: the relation of mirror partnership between X and Y is expected to produce, roughly speaking, an isomorphism $H^*(X) \to H^*(Y)$.

More precisely, any isomorphism between the quantum cohomology of X and a Barannikov-Kontsevich formal Frobenius manifold of Y produces an identification of their spaces of flat vector fields, that is a mirror isomorphism of the cohomology spaces

$$\mu_{X,Y} \colon H^*(X, \mathbf{C}) \to H^*(Y, \wedge^*(\mathcal{T}_Y)) \otimes V_Y^{-2}, \quad V_Y := H^0(Y, \Omega_Y^{\max}). \tag{11}$$

Near a cusp in the moduli space of Y, V_Y can be trivialized by the choice of a volume form Ω having period $(2\pi i)^{\dim Y}$ along the invariant cycle. Then (11) becomes a ring isomorphism. Trace functionals and flat metrics on both sides are identified via (4). Comparing Euler fields, one sees that $H^{p,q}(X)$ is identified with $H^q(Y, \wedge^p(\mathcal{T}_Y)$. In particular, $H^{1,1}(X)$ becomes $H^2(Y, \mathcal{T}_Y)$, and the induced integral structure on the latter space (exponential coordinates near the cusp) are described in [27].

Notice now that the Frobenius structure at the left-hand side of (11) is essentially motivic, in the sense that numerical Gromov-Witten invariants of X come from algebraic correspondences between X^n and $\overline{M}_{0,n}$, $n \geq 3$. More generally, the theory of Gromov-Witten invariants can be conceived as a chapter of algebraic and/or non-commutative geometry *over the category of motives*, replacing the more common category of linear spaces. This geometry deals, for example, with affine groups whose function rings are Hopf algebras in the category of *Ind*-motives. P. Deligne developed basics of this geometry in [25, 26], in order to clarify the notion of motivic fundamental group. Further examples come from or are motivated by physics: besides Gromov-Witten invariants, one can mention Nakajima's theory of Heisenberg algebras related to Chow schemes of surfaces, and a recent paper [72].

It makes sense to ask then, what can be the mirror reflection of this motivic geometry. Since the mirror maps are highly transcendental, developing the adequate language presents an interesting challenge. Starting with the category of motives in the sense of [1] generated by Calabi-Yau manifolds, we can try to extend it by adding mirror isomorphisms as new motivated morphisms. In this context, Kontsevich's correspondence between CY Teichmüller groups and autoequivalences of derived categories might have an analog, saying that the mirror isomorphisms connect the motivic fundamental groups (see [26]) and motivic automorphism groups of CYs whose Lie algebras were studied in [71]. For abelian varieties, this phenomenon is stressed in [46].

Acknowledgement

I am grateful to S. Barannikov, M. Kontsevich and Y. Soibelman, who suggested revisions and corrections to the first version of this talk. Of course, I am fully responsible for the final text.

References

[1] Y. André. *Pour une théorie inconditionnelle des motives.* Publ. Math. IHES, 83 (1996), 5–49.

[2] D. Arinkin and A. Polishchuk. *Fukaya category and Fourier transform.* Preprint math. AG/9811023.

[3] P. Aspinwall, B. Greene and D. Morrison. *Calabi-Yau spaces, mirror manifolds and spacetime topology change in string theory.* In: [86, 213–279].

[4] S. Barannikov. *Extended moduli spaces and mirror symmetry in dimensions $n > 3$.* math. AG/9903124.

[5] S. Barannikov and M. Kontsevich. *Frobenius manifolds and formality of Lie algebras of polyvector fields.* Int. Math. Res. Notices, 4 (1998), 201–215.

[6] V. Batyrev. *Dual polyhedra and the mirror symmetry for Calabi-Yau hypersurfaces in toric varieties.* Journ. Alg. Geom., 3 (1994), 493–535.

[7] V. Batyrev. *Variation of the mixed Hodge structure of affine hypersurfaces in algebraic tori.* Duke Math. J., 69 (1993), 349–409.

[8] V. Batyrev. *Quantum cohomology ring of toric manifolds.* Astérisque, 218 (1993), 9–34.

[9] V. Batyrev and L. Borisov. *On Calabi-Yau complete intersections in toric varieties.* In: Proc. of Int. Conf. on Higher Dimensional Complex Varieties (Trento, June 1994), ed. by M. Andreatta, De Gruyter, 1996, 39–65.

[10] V. Batyrev and L. Borisov. *Dual cones and mirror symmetry for generalized Calabi-Yau manifolds.* In: Mirror Symmetry II, ed. by S. T. Yau, 1996, 65–80.

[11] V. Batyrev and L. Borisov. *Mirror duality and string-theoretic Hodge numbers.* Inv. Math., 126:1 (1996), 183–203.

[12] V. Batyrev and D. van Straten. *Generalized hypergeometric functions and rational curves on Calabi-Yau complete intersections in toric varieties.* Comm. Math. Phys., 168 (1995), 493–533.

[13] M. Bershadsky, S. Cecotti, H. Ooguri and C. Vafa. *Kodaira-Spencer theory of gravity and exact results for quantum string amplitudes.* Comm. Math. Phys., 165 (1994), 311–427.

[14] A. Bertram. *Another way to enumerate rational curves with torus actions.* math. AG/9905159.

[15] G. Bini, C. de Concini, M. Polito and C. Procesi. *On the work of Givental relative to Mirror Symmetry.* math. AG/9805097.

[16] L. Borisov. *On Betti numbers and Chern classes of varieties with trivial odd cohomology groups.* alg-geom/9703023.

[17] L. Borisov. *Vertex algebras and mirror symmetry.* math. AG/9809094.

[18] L. Borisov. *Introduction to the vertex algebra approach to mirror symmetry.* arXiv:math. AG/9912195.

[19] L. Borisov and A. Libgober. *Elliptic genera of toric varieties and applications to mirror symmetry.* math. AG/9904126 (to appear in Inv. Math.)

[20] P. Candelas, X. C. de la Ossa, P. S. Green and L. Parkes. *A pair of Calabi-Yau manifolds as an exactly soluble superconformal theory.* Nucl. Phys., B 359 (1991), 21–74.

[21] E. Cattani and A. Kaplan. *Degenerating variations of Hodge structures.* Astérisque, 179–180 (1989), 67–96.

[22] S. Cecotti. $N = 2$ *Landau-Ginzburg vs. Calabi-Yau σ-models: non-perturbative aspects.* Int. J. of Mod. Phys. A, 6:10 (1991), 1749–1813.

[23] I. Ciocan-Fontanine and M. Kapranov. *Derived Quot schemes.* Preprint math. AG/9905174.

[24] D. Cox and S. Katz. *Mirror symmetry and algebraic geometry.* AMS, Providence RI, 1999.

[25] P. Deligne. *Le groupe fondamental de la droite projective moins trois points.* In: Galois groups over **Q**, ed. By Y. Ihara, K. Ribet, J. P. Serre, Springer-Verlag, 1999, 79–297.

[26] P. Deligne. *Catégories tannakiennes.* In: The Grothendieck Festschrift, vol. II, Birkhäuser Boston, 1990, 111–195.

[27] P. Deligne. *Local behavior of Hodge structures at infinity.* In: Mirror Symmetry II, ed. by B. Greene and S. T. Yau, AMS-International Press, 1996, 683–699.

[28] P. Deligne and J. Milne. *Tannakian categories.* In: Springer Lecture Notes in Math., 900 (1982), 101–228.

[29] I. Dolgachev. *Mirror symmetry for lattice polarized K3 surfaces.* J. Math. Sci. 81 (1996), 2599–2630. alg-geom/9502005.

[30] R. Donagi and E. Markman. *Cubics, integrable systems, and Calabi-Yau threefolds.* In: Proc. of the Conf. in Alg. Geometry dedicated to F. Hirzebruch, Israel Math. Conf. Proc., 9 (1996).

[31] Ch. Doran. *Picard-Fuchs uniformization: modularity of the mirror map and mirror-moonshine.* math. AG/9812162.

[32] B. Dubrovin. *Geometry of 2D topological field theories.* In: Springer LNM, 1620 (1996), 120–348.

[33] B. Dubrovin. *Geometry and analytic theory of Frobenius manifolds.* Proc. ICM Berlin 1998, vol. II, 315–326. math. AG/9807034.

[34] K. Fukaya. *Morse homotopy, A^∞-categories, and Floer homologies.* In: Proc. of the 1993 GARC Workshop on Geometry and Topology, ed. by H. J. Kim, Seoul Nat. Univ.

[35] K. Fukaya. *Mirror symmetry of abelian variety and multi theta functions.* Preprint, 1998.

[36] D. Gepner. *On the spectrum of 2D conformal field theory.* Nucl. Phys., B287 (1987), 111–126.

[37] D. Gepner. *Exactly solvable string compactifications on manifolds of $SU(N)$ holonomy.* Phys. Lett., B199 (1987), 380–388.

[38] D. Gepner. *Fusion rings and geometry.* Comm. Math. Phys., 141 (1991), 381–411.

[39] A. Givental. *Homological geometry I: Projective hypersurfaces.* Selecta Math., new ser. 1:2 (1995), 325–345.

[40] A. Givental. *Equivariant Gromov-Witten invariants.* Int. Math. Res. Notes, 13 (1996), 613–663.

[41] A. Givental. *Stationary phase integrals, quantum Toda lattices, flag manifolds and the mirror conjecture.* In: Topics in Singularity Theory, ed. by A. Khovanski et al., AMS, Providence RI, 1997, 103–116.

[42] A. Givental. *Homological geometry and mirror symmetry.* In: Proc. of the ICM, Zürich 1994, Birkhäuser 1995, vol. 1, 472–80.

[43] A. Givental. *A mirror theorem for toric complete intersections.* In: Topological Field Theory, Primitive Forms and Related Topics, ed. by M. Kashiwara et al., Progress in Math., vol 60, Birkhäuser, 1998, 141–175. alg-geom/9702016.

[44] A. Givental. *Elliptic Gromov-Witten invariants and the generalized mirror conjecture.* math. AG/9803053.

[45] A. Givental. *The mirror formula for quintic threefolds.* math. AG/9807070.

[46] V. Golyshev, V. Lunts, D. Orlov. *Mirror symmetry for abelian varieties.* math. AG/9812003.

[47] B. Green. *Constructing mirror manifolds.* In: [86, 29–69].

[48] M. Gross. *Special Lagrangian fibrations I: Topology.* alg-geom/9710006.

[49] M. Gross. *Special Lagrangian fibrations II: Geometry.* math. AG/9809072.

[50] M. Gross and P. M. H. Wilson. *Mirror symmetry via 3-tori for a class of Calabi-Yau threefolds.* Math. Ann., 309 (1997), 505–531. alg-geom/9608004.

[51] N. Hitchin. *The moduli space of special Lagrangian submanifolds.* dg-ga/9711002.

[52] R. P. Horja. *Hypergeometric functions and mirror symmetry in toric varieties.* arXiv:math. AG/9912109.

[53] S. Hosono, B. H. Lian and S. T. Yau. *GKZ-generalized hypergeometric systems in mirror symmetry of Calabi-Yau hypersurfaces.* alg-geom/9511001.

[54] S. Hosono, B. H. Lian and S. T. Yau. *Maximal degeneracy points of GKZ systems.* alg-geom/9603014.

[55] T. Jarvis, T. Kimura and A. Vaintrob. *The moduli space of higher spin curves and integrable hierarchies.* math. AG/9905034.

[56] T. Jarvis, T. Kimura and A. Vaintrob. *Tensor products of Frobenius manifolds and moduli spaces of higher spin curves.* math. AG/9911029.

[57] M. Kontsevich. A_∞-*algebras in mirror symmetry.* Bonn MPI Arbeitstagung talk, 1993.

[58] M. Kontsevich. *Homological algebra of Mirror Symmetry.* Proceedings of the ICM (Zürich, 1994), vol. I, Birkhäuser, 1995, 120–139. alg-geom/9411018.

[59] M. Kontsevich. *Mirror symmetry in dimension 3.* Séminaire Bourbaki, n° 801, Juin 1995.

[60] M. Kontsevich. *Enumeration of rational curves via torus actions.* In: The Moduli Space of Curves, ed. by R. Dijkgraaf, C. Faber, G. van der Geer, Progress in Math. vol. 129, Birkhäuser, 1995, 335–368.

[61] M. Kontsevich and Yu. Manin. *Gromov-Witten classes, quantum cohomology, and enumerative geometry.* Comm. Math. Phys., 164:3 (1994), 525–562.

[62] M. Kontsevich and Y. Soibelman. *Deformation of algebras over operads and Deligne's conjecture.* math. QA/0001151.

[63] M. Kreuzer and H. Skarke. *Complete classification of reflexive polyhedra in four dimensions.* arXiv:hep-th/0002240.

[64] V. S. Kulikov. *Mixed Hodge structures and singularities.* Cambridge Univ. Press, 1998.

[65] N. C. Leung, Sh.-T. Yau and E. Zaslow. *From special Lagrangian to Hermitian-Yang-Mills via Fourier-Mukai transform.* Preprint math. DG/0005118.

[66] B. H. Lian, K. Liu and S.-T. Yau. *Mirror principle I.* Asian J. of Math., vol. 1, no. 4 (1997), 729–763.

[67] B. H. Lian, K. Liu and S.-T. Yau. *Mirror principle II.* Asian J. of Math., vol. 3, no. 1 (1999), 109–146.

[68] B. H. Lian, A. Todorov and S. T. Yau. *Maximal unipotent monodromy for complete intersection CY manifolds.* Preprint, 2000.

[69] A. Libgober. *Chern classes and the periods of mirrors.* math. AG/9803119.

[70] A. Libgober and J. Wood. *Uniqueness of the complex structure on Kähler manifolds of certain homology type.* J. Diff. Geom., 32 (1990), 139–154.

[71] E. Looienga and V. Lunts. *A Lie algebra attached to a projective variety.* Inv. Math., 129 (1997), 361–412.

[72] A. Losev and Yu. Manin. *New moduli spaces of pointed curves and pencils of flat connections.* To be published in Fulton's Festschrift, Michigan Journ. of Math., math. AG/0001003.

[73] F. Malikov, V. Schechtman and A. Vaintrob. *Chiral de Rham complex.* Comm. Math. Phys., 204:2 (1999), 439–473. math. AG/980341.

[74] M. Manetti. Extended deformation functors, I. math. AG/9910071.

[75] Yu. Manin. *Problems on rational points and rational curves on algebraic varieties.* In: Surveys of Diff. Geometry, vol. II, ed. by C. C. Hsiung, S.-T. Yau, Int. Press (1995), 214–245.

[76] Yu. Manin. *Generating functions in algebraic geometry and sums over trees.* In: The Moduli Space of Curves, ed. by R. Dijkgraaf, C. Faber, G. van der Geer, Progress in Math. vol. 129, Birkhäuser, 1995, 401–418.

[77] Yu. Manin. *Sixth Painlevé equation, universal elliptic curve, and mirror of* $\mathbf{P}^2$. AMS Transl. (2), vol. 186 (1998), 131–151. alg-geom/9605010.

[78] Yu. Manin. *Three constructions of Frobenius manifolds: a comparative study.* Asian J. Math., 3:1 (1999), 179–220 (Atiyah's Festschrift). math. QA/9801006.

[79] Yu. Manin. *Frobenius manifolds, quantum cohomology, and moduli spaces.* AMS Colloquium Publications, vol. 47, Providence, RI, 1999, xiii+303 pp.

[80] Yu. Manin. *Quantized theta-functions.* In: Common Trends in Mathematics and Quantum Field Theories (Kyoto, 1990), Progress of Theor. Phys. Supplement, 102 (1990), 219–228.

[81] R. C. McLean. *Deformations of calibrated submanifolds.* 1996.

[82] S. Merkulov. *Formality of canonical symplectic complexes and Frobenius manifolds.* Int. Math. Res. Notes, 14 (1998), 727–733.

[83] S. Merkulov. *Strong homotopy algebras of a Kähler manifold.* Preprint math. AG/9809172.

[84] S. Merkulov. *Frobenius$_\infty$ invariants of homotopy Gerstenhaber algebras.*

[85] S.-T. Yau, ed. *Essays on Mirror Manifolds.* International Press Co., Hong Cong, 1992.

[86] B. Greene, S. T. Yau, eds. *Mirror Symmetry II.* AMS-International Press, 1996.

[87] D. Morrison. *Mirror symmetry and rational curves on quintic threefolds: a guide for mathematicians.* J. AMS, 6 (1993), 223–247.

[88] D. Morrison. *Compactifications of moduli spaces inspired by mirror symmetry.* Astérisque, vol. 218 (1993), 243–271.

[89] D. Morrison. *The geometry underlying mirror symmetry.*

[90] R. Pandharipande. *Rational curves on hypersurfaces (after A. Givental).* Séminaire Bourbaki, Exp. 848, Astérisque, 252 (1998), 307–340. math. AG/9806133.

[91] A. Polishchuk. *Massey and Fukaya products on elliptic curves.* Preprint math. AG/9803017.

[92] A. Polishchuk. *Homological mirror symmetry with higher products.* Preprint math. AG/9901025.

[93] A. Polishchuk and E. Zaslow. *Categorical mirror symmetry: the elliptic curve.* Adv. Theor. Math. Phys., 2 (1998), 443–470. math. AG/980119.

[94] J. H. Schwarz. *Lectures on Superstring and M Theory Dualities.* hep-th/9607201.

[95] P. Seidel and R. Thomas. *Braid group actions on derived categories of coherent sheaves.* arXiv:math. AG/0001043.

[96] A. Strominger, S.-T. Yau and E. Zaslow. *Mirror symmetry is T-duality.* Nucl. Phys. B 479 (1996), 243–259.

[97] R. P. Thomas. *Mirror symmetry and actions of braid groups on derived categories.* arXiv:math. AG/0001044.

[98] A. Tyurin. *Special Lagrangian geometry and slightly deformed algebraic geometry (SPLAG and SDAG).* math. AG/9806006.

[99] A. Tyurin. *Geometric quantization and mirror symmetry.* Preprint math. AG/9902027.

[100] C. Vafa. *Topological mirrors and quantum rings.* In: Essays on Mirror Manifolds, ed. by Sh.-T. Yau, International Press, Hong-Kong 1992, 96–119.

[101] C. Vafa. *Extending mirror conjecture to Calabi-Yau with bundles.* hep-th/9804131.

[102] C. Vafa. *Geometric Physics.* Proc. ICM Berlin 1998, vol. I, 537–556.

[103] C. Voisin. *Symétrie miroir.* Panoramas et synthèses, 2 (1996), Soc. Math. de France.

[104] C. Voisin. *Variations of Hodge structure of Calabi-Yau threefolds.* Quaderni della Scuola Norm. Sup. di Pisa, 1998.

[105] E. Witten. *Mirror manifolds and topological field theory.* in: [85, 120–159].

[106] E. Witten. *Algebraic geometry associated with matrix models of two-dimensional gravity.* In: Topological Models in Modern Mathematics (Stony Brook, NY, 1991), Publish or Perish, Houston, TX (1993), 235–269.

[107] E. Zaslow. *Solitons and helices: the search for a Math-Physics bridge.* Comm. Math. Phys, 175 (1996), 337–375.

[108] I. Zharkov. *Torus fibrations of Calabi-Yau hypersurfaces in toric varieties and mirror symmetry.* Duke Math. J., 101:2 (2000), 237–258. alg-geom/9806091.

Director
Max-Planck-Institut für Mathematik
Vivatsgasse 7
53111 Bonn, Germany
E-mail address: manin@mpim-bonn.mpg.de

The Role of Oscillations in Some Nonlinear Problems

Yves Meyer

Abstract. The still image compression standard which is being developed under the name of JPEG-2000 (Section 2) is a technological challenge which relies on some advances in pure mathematics. This interaction between image processing and functional analysis also benefits partial differential equations.

Indeed new estimates on wavelet coefficients of functions with bounded variation (Theorems 6.6 and 7.5) imply new Gagliardo–Nirenberg inequalities (Section 7) and lead to a better understanding of blowup phenomena for solutions of some nonlinear evolution equations (Sections 10 to 13).

1. Introduction

Explaining the performances of JPEG-2000 requires a model for still images. Among several models, the one on which this discussion is based was proposed by Stan Osher and Leonid Rudin (Section 3). In this model, an image f is decomposed into a sum of two pieces u and v. The first piece is aimed to model the main features in f. The second one takes care of the textured components (Section 5), of the noise, and of what is unorganized.

In the Osher–Rudin model, the first component u is assumed to be a function with bounded variation (Section 4). Then the efficiency of wavelet based algorithms will be related to new estimates on wavelet coefficients of functions with bounded variations (BV). These estimates were discovered by Albert Cohen, Ronald DeVore, Ingrid Daubechies, Wofgang Dahmen, Pencho Petrushev and Hong Xu (Theorem 6.6 in Section 6 and Theorem 7.5 in Section 7). New Gagliardo–Nirenberg inequalities will then be obtained in Section 7.

A. Cohen, R. DeVore and Guergana Petrova went one step further (Section 9). They proved that wavelet coefficients of functions in $L^1(\mathbb{R}^n)$ had some remarkable properties. This was much against the general feeling that wavelet analysis would be inefficient if it was used for Banach spaces which did not admit an unconditional basis. Let me confess that it was my own belief. Moreover A. Cohen, R. DeVore and G. Petrova found a spectacular application of their theorem. This concerns the Boltzmann equation and more precisely the "averaging lemma" of P. L. Lions and Ronald DiPerna (Section 9).

Gagliardo–Nirenberg inequalities (Section 7) will manifest again in studying nonlinear evolution equations. Our first example will be the nonlinear heat equation for which blowup in finite time has been established for some smooth and compactly supported initial conditions. However there is no blowup when this initial condition is sufficiently oscillating and some Gagliardo–Nirenberg estimates will tell us why it is so and how these oscillations should be measured.

These successes led us to believe that the same was true for Navier–Stokes equations. We guessed that an oscillating initial condition should provide us with a solution which is global in time. In other words, the solution should not blow up and moreover the qualitative properties of the initial condition should be preserved under the evolution. This line of research started with some preliminary results by M. Cannone and F. Planchon and culminated with a beautiful theorem by Herbert Koch and Daniel Tataru. The Banach space which is used by Koch and Tataru for modeling the oscillations of the initial condition is exactly the same as the one we introduced for modeling textured components of images (Section 12).

2. Wavelets and Still Image Compression

Let us begin with some examples of technological applications of wavelets. The first example is extracted from the web page of the Pegasus company (`http://www.jpg.com`). It reads as follows:

> Pegasus Imaging Corporation has partnered with Fast Mathematical Algorithms & Hardware Corporation and Digital Diagnostic Corporation to develop new wavelet compression technologies designed for applications including medical imaging, fingerprint compression, video compression, radar imaging, satellite imaging and color imaging.
>
> Pegasus provides wavelet compression technology for both medical and non-medical application. Pegasus' wavelet implementation has received FDA market clearance for medical devices.
>
> This software is the only FDA-approved lossy compression software for image processing. Recent clinical studies have shown that the algorithm is comprehensively superior to other similar compression methods. It is licensed to multiple teleradiology developers and medical clinics including the Dutch software vendor Applicare Medical Imaging and the UK telecom giant British Telecom.

The second advertisement comes from a company named "Analog Devices". It reads:

> Wavelet compression technology is the choice for video capture and editing. The ADV601 video compression IC is based on a mathematical breakthrough known as wavelet theory ... This compression technology has many advantages over other schemes. Common discrete schemes, like JPEG and MPEG, must break an image into rectangular sub-blocks in order to compress it ... Natural images highly compressed with DCT schemes take on unnatural blocky artifacts ... Wavelet filtering yields a very robust digital representation of a picture, which maintains its natural look even under fairly extreme compression. In sum ADV601 provides breakthrough image compression technology in a single affordable integrated circuit.

A third success story tells us about the FBI and fingerprints. It says:

> The new mathematical field of wavelet transforms has achieved a major success,
> specifically, the Federal Bureau of Investigation's decision to adopt a wavelet-based
> image coding algorithm as the national standard for digitized finger-print records
> ... "

The interested reader is referred to a paper by Christopher Brislawn in the Notices of the AMS, November 1995, Vol. 42, Number 11, pp. 1278–1283 or to the remarkable web site of Christopher Brislawn [3].

Our next advertisement for wavelet-based image compression comes from the celebrated Sarnoff Research Center. It reads:

> A simple, yet remarkably effective, image compression algorithm has been devel-
> oped, which provides the capability to generate the bits in the bit stream in order
> of importance, yielding fully hierarchical image compression suitable for embedded
> coding or progressive transmission ...

Finally the last example will concern the upcoming JPEG-2000 still image compression standard. While the JPEG committee is still actively working, it is very likely that the JPEG-2000 standard will be based on a combination of wavelet expansion (the choice of the filter is not fixed, and could include biorthogonal filters such as the 9/7, as well as 2–10 integer filters) and trellis coding quantization. Applications range from Medical imagery, client/server application for the world wide web, to electronic photography and photo and art digital libraries.

These examples show that still image compression is a rapidly developing technology with far reaching applications.

A last remark concerns denoising by soft thresholding. This technique has been created and analyzed by David Donoho and his collaborators [16]. Donoho explains in IEEE spectrum (October 1996, pp. 26–35) what he is doing:

> Ridding signals and images of noise is often much easier in the wavelet domain than
> in the original domain ... The procedure works by taking the wavelet coefficients
> of the signal, setting to zero the coefficients below a certain threshold ... Wavelet
> noise removal has been shown to work well for geophysical signals, astronomical data,
> synthetic aperture radar, digital communications, acoustic data, infrared images and
> biomedical signals ...

3. Some $u + v$ Models for Still Images

Why do wavelet algorithms perform better than Fourier methods in image compression? One answer to this problem relies on an axiomatic model proposed by Osher and Rudin (among others). This model is named a $u + v$ model.

We start with the superficial approach that a black and white analog image on a domain Ω can be viewed as a function $f(x_1, x_2) = f(x)$ belonging to the Hilbert space $H = L^2(\Omega)$. The grey level of our image at a given pixel x is precisely $f(x)$. The energy of such an image is, by definition, $\int_\Omega |f(x)|^2 dx$. It is obvious that an arbitrary such function $f(x)$ in H is far from being a natural image or something looking similar but this hot issue will be clarified now. Indeed our main problem will be to try to understand how an image differs from an arbitrary L^2 function.

In a $u + v$ model, images $f(x) \in H$ are assumed to be a sum of two components $u(x)$ and $v(x)$. The first component $u(x)$ models the objects or features which are present in the given image while the $v(x)$ term is responsible for the texture and the noise. But the textures are often limited by the contours of the objects and $u(x)$ and $v(x)$ should be coupled by some geometrical constraints. These constraints are absent from most of the $u(x) + v(x)$ models.

In the Osher–Rudin model, the $u(x)$ component is assumed to be a function with bounded variation. We want to detect objects delimited by contours. Then these objects can be modeled by some planar domains $D_1, \ldots, D_n$ and the corresponding contours or edges will be modeled by their boundaries $\partial D_1, \ldots, \partial D_n$.

In this model, the function $u(x)$ is assumed to be smooth inside $D_1, \ldots, D_n$ with jump discontinuities across the boundaries $\partial D_1, \ldots, \partial D_n$. However we do not want to break an image into too many pieces and the penalty for a domain decomposition of a given image will be the sum of the lengths of these edges $\partial D_1, \ldots, \partial D_n$.

But this sum of lengths is indeed one of the two terms which appear in the BV norm of $u(x)$. The other one is the L^1 norm of the gradient of the restriction of u to the interior of the domains $D_1, \ldots, D_n$. The BV norm of a function $f(x)$ is defined as the total mass of the distributional gradient of $f(x)$ and we will return to this definition in the next section.

In the Osher–Rudin model, $v(x)$ will be measured by a simply minded energy criterion which says that $\|v\|_2$ is sufficiently small. In Section 5, a new model which takes care of the textured components will be proposed. In this model v can have a large energy but needs to be oscillating.

For the reader's convenience, some basic facts about functions with bounded variations are listed in the following section.

4. Functions with Bounded Variations

Assuming $n \geq 2$, we say that a function $f(x)$ defined on $\mathbb{R}^n$ belongs to BV if (a) $f(x)$ vanishes at infinity in a weak sense and (b) the distributional gradient of $f(x)$ is a bounded Radon measure. The BV norm of f is denoted by $\|f\|_{BV}$ and defined as the total mass of the distributional gradient of $f(x)$. The condition at infinity says that the convolution product $f \star \varphi$ should tend to 0 at infinity whenever φ is a function in the Schwartz class.

A second and equivalent definition reads as follows:

Definition 4.1. *A function $f(x)$ belongs to $BV(\mathbb{R}^n)$ if it vanishes at infinity in the weak sense and if there exists a constant C such that*

$$\int_{\mathbb{R}^n} |f(x + y) - f(x)|\,dx \leq C|y|, \qquad y \in \mathbb{R}^n . \tag{1}$$

We now return to functions defined on the plane. An example of a function in BV is given by the indicator (or characteristic) function χ_E of a domain E

delimited by a rectifiable Jordan curve ∂E. The BV-norm of χ_E is the length l_E of the Jordan curve ∂E.

The co-area identity tells us that any positive function in BV can be written as a convex combination of some normalized indicator functions. These normalized indicator functions should belong to the unit ball of BV and are therefore defined as $(l_E)^{-1}\chi_E$. They are called "atoms".

This remarkable "atomic decomposition" clarifies the relevance of BV in modeling geometrical features: the atoms are the objects to be detected.

5. Modeling Textures

The goal of this section is to address the issue of modeling textures by function spaces. We return to the Osher–Rudin model for representing images and we want to discuss the v component of our image. This v component contains both the textured components of our image and an additive noise. We will offer three choices for modeling these textured components.

Our first choice will be the Besov space $E = B_\infty^{-1,\infty}$ (see Definition 7.1, Section 7). If wavelet analysis [27] is being used, then this Besov space admits a trivial characterization which reads as follows.

Lemma 5.1. *Let $2^j\psi(2^j x - k)$, $j \in \mathbb{Z}$, $k \in \mathbb{Z}^2$, $\psi \in F$, be an orthonormal wavelet basis of $L^2(\mathbb{R}^2)$ where F is a finite set consisting of three analyzing wavelets belonging to C^2 and compactly supported. Then a generalized function f belongs to $B_\infty^{-1,\infty}$ if and only if its wavelet coefficients belong to $l^\infty(\mathbb{Z}^3 \times F)$.*

This lemma is extremely attractive since it nicely relates the functional norm in $B_\infty^{-1,\infty}$ to D. Donoho's wavelet shrinkage (see Section 2).

Wavelet shrinkage is a denoising algorithm which consists in putting to zero all wavelet coefficients which are less than a given threshold. Wavelet shrinkage will wipe out the v component of an image whenever its $B_\infty^{-1,\infty}$-norm is less than the threshold. Both the textured component and the noise meet this requirement. They will disappear in a wavelet shrinkage. In other words, Donoho's algorithm will treat the textured components of an image as being noise. We will return to this point at the end of this section.

Some slightly smaller Banach spaces F and G also provide some efficient modeling for textures or oscillating patterns in an image.

The space F consists of generalized functions f which can be written as

$$f = \partial_1 g_1 + \partial_2 g_2, \qquad g_j \in BMO, \quad j = 1, 2. \tag{2}$$

The norm of f in F is defined as the infimum of the sums of the BMO norms of g_1 and g_2 and this infimum is computed over all possible decompositions of f.

This Banach space F will be met again when Navier–Stokes equations are studied. Let us observe that BMO is a space of locally integrable functions, modulo constant functions. These floating constants will disappear in (2).

The Banach space G has a similar definition where BMO is replaced by L^∞.

Lemma 5.2. *We have*

$$BV \subset L^2(\mathbb{R}^2) \subset G \subset F \subset B_\infty^{-1,\infty}. \tag{3}$$

We are now coming to the heart of this section and we will study the relevance of our function spaces in texture modeling. It is given by the following remark.

Lemma 5.3. *Let f_n, $n \geq 1$, be a sequence of functions with the following three properties:*

  (a) *There exists a compact set K such that the supports of f_n, $n \geq 1$ are contained in K.*

  (b) *There exists an exponent $q > 2$ and a constant C such that $\|f_n\|_q \leq C$.*

  (c) *The sequence f_n tends to 0 in the distributional sense.*

Then $\|f_n\|_G$ tends to 0 as n tends to infinity.

This lemma says the following. If our sequence f_n is developing important oscillations, then $\|f_n\|_G$ tends to 0 (which obviously implies the same property for the two other norms).

Let us observe that Lemma 5.3 is wrong if $q = 2$. Indeed if $\varphi(x)$ is any smooth and compactly supported function, then $f_n(x) = n\varphi(nx)$ is an obvious counter-example.

Lemma 5.3 can be quantified. The following theorem describes a collection of functions $f(x)$ such that the F norm of $\exp(i\omega.x)f(x)$ decays as $|\omega|^{-1}$ when $|\omega|$ tends to infinity.

Theorem 5.4. *Let us assume that $f \in L^\infty$ and that a constant C exists such that the BV-norm of $f(x)$ on any ball B of radius R does not exceeds CR. Then*

$$\| \exp(i\omega.x)f(x)\|_F \leq C/|\omega|. \tag{4}$$

The second requirement on f means that the two measures $\mu_j = \partial f/\partial x_j$, $j = 1, 2$ should satisfy the celebrated Guy David condition saying that $|\mu|(B) \leq CR$ for any ball B with radius R. This space M of measures will be met again in Section 12.

If textures are modeled as above, then Donoho's denoising algorithm, named "wavelet shrinkage", will erase both the textures and the noise. This is exactly what Lemma 5.3 and Theorem 5.4 are saying. Let us now challenge Donoho's algorithm and define a "Fourier shrinkage" as the following nonlinear algorithm. One is given a small positive threshold η and one writes the Fourier expansion of a given function f. Then one retains only the terms for which $|c_k| \geq \eta$ in this expansion and this provides f_η.

If f represents an image which contains geometrical features, textured elements and some additive noise and if a "Fourier shrinkage" is applied to f, then this noise will be wiped out while most of the textured components will be kept. Extracting texture from noise is wanted in image processing. Does that mean that "Fourier shrinkage" performs better than "wavelet shrinkage"? It is not clear since a "Fourier shrinkage" would more seriously damage the BV component u than a "wavelet shrinkage" does.

6. Fourier Series vs. Wavelet Series: Expansions of BV Functions

For the sake of simplicity, let us first study periodic functions in BV. Let $f(x_1, x_2)$ be a function of two real variables which is 2π-periodic in each variable. We then abbreviate by saying that $f(x)$ is 2π-periodic. Let us write the Fourier series of $f(x)$ as $f(x) = \sum_{k\in\mathbb{Z}^2} c(k_1, k_2)\exp(ik.x)$ with $k = (k_1, k_2)$. Let us assume that $f(x)$ belongs to BV on $[0, 2\pi]^2$. Then we already know that $c(k)$ belongs to l^2. For such functions, Jean Bourgain proved the following

Theorem 6.1. *There exists a constant C such that for any 2π-periodic function $f(x)$ in $BV(\mathbb{R}^2)$, we have*

$$\sum_{k\in\mathbb{Z}^2} |c(k)|(|k| + 1)^{-1} \leq C\|f\|_{BV} . \tag{5}$$

This estimate complements $\sum |c(k)|^2 < \infty$ and these two results obviously follow from a sharper estimate given by

$$\sum_{j=0}^{\infty} s_j \leq C\|f\|_{BV} \tag{6}$$

where $s_j = (\sum_{2^j \leq |k| < 2^{j+1}} |c(k)|^2)^{1/2}$.

This is a mixed $l^1(l^2)$ estimate on Fourier coefficients of a BV function. It is optimal in the sense that there exists a function in BV for which $\sum |c(k)|^p = \infty$ for any $p < 2$. An example is given by $f(x) = |x|^{-1}(\log|x|)^{-2}\varphi(x)$ where $\varphi(x)$ is any smooth function which vanishes when $|x| > 1/2$ and is identically 1 around the origin. Then the Fourier coefficients $c(k)$ of $f(x)$ can be estimated by $|c(k)| \simeq |k|^{-1}(\log|k|)^{-2}$ which obviously implies $\sum |c(k)|^p = \infty$ as announced. The sorted Fourier coefficients of this function behave as $n^{-1/2}(\log n)^{-2}$. This counterexample shows that nothing better than l^2 can be expected inside the dyadic blocks of the Fourier series expansion of a function $f(x)$ in BV.

Now (6) can be rewritten as a Besov norm estimate. Indeed let $\Delta_j(f)$ denote the dyadic blocks of the Fourier series expansion of $f(x)$. For defining $\Delta_j(f)$ we retain only the frequencies $k \in \Gamma_j$ in the Fourier expansion of f where Γ_j is the dyadic annulus defined as $\{k \mid 2^j < |k| \leq 2^{j+1}\}$. We then obviously have $f(x) = c_0 + \sum_0^\infty \Delta_j(f)$ and our next theorem reads:

$$\sum_0^{\infty} \|\Delta_j(f)\|_2 \leq C\|f\|_{BV} . \tag{7}$$

This theorem will be further improved. This improved version no longer uses a Fourier series expansion and we can therefore give up the periodic setting and switch to the space $BV(\mathbb{R}^2)$ and to a Littlewood–Paley analysis.

Let us start with a compactly supported smooth function ψ with enough vanishing moments such that the Fourier transform Ψ of ψ satisfies

$$\sum_{0}^{\infty} \Psi(2^{-j}\xi) = 1, \qquad |\xi| \geq 1. \tag{8}$$

Next we write $\psi_j = 2^{2j}\psi(2^j x)$. Finally $\Delta_j(f)$ is the convolution product $f * \psi_j$.

With these notations (7) can be generalized to all exponents p in $(1, 2]$. Indeed the following theorem is an easy consequence of the co-area identity.

Theorem 6.2. *There exists a constant C such that, for every function f in $BV(\mathbb{R}^2)$, and for every exponent p with $1 < p \leq 2$, we have*

$$\sum_{-\infty}^{+\infty} 2^{js}\|\Delta_j(f)\|_p \leq C_p\|f\|_{BV} \tag{9}$$

with $s = -1 + (2/p)$ and $C_p \leq C/(p-1)$.

Corollary 6.3. *If $f(x)$ belongs to $BV(\mathbb{R}^2)$, and if $\psi_{j,k}(x) = 2^j\psi(2^j x - k)$, $j \in \mathbb{Z}$, $k \in \mathbb{Z}^2$, is an orthonormal wavelet basis of $L^2(\mathbb{R}^2)$, where the three wavelets ψ are smooth and localized as in Lemma 5.1, then the corresponding wavelet coefficients $c(j,k) = \langle f, \psi_{j,k}\rangle$ satisfy*

$$\sum_j \Big(\sum_k |c(j,k)|^p\Big)^{1/p} \leq C/(p-1)\|f\|_{BV}, \qquad 1 < p < 2. \tag{10}$$

Corollary 6.4. *With the same notations as above, we have*

$$\Big(\sum_j \sum_k |c(j,k)|^p\Big)^{1/p} \leq C/(p-1)\|f\|_{BV}. \tag{11}$$

Corollary 6.5. *With the same notations, let us assume $\|f\|_{BV} \leq 1$. For each integer m, let N_m be the cardinality of the set on indices (j,k) such that $|c(j,k)| > 2^{-m}$. Then*

$$N_0 + \cdots + 2^{-m}N_m \leq C(m+1). \tag{12}$$

It means that for most ms we have $N_m \leq C2^m$ since the average of $2^{-m}N_m$ is $O(1)$.

Indeed one has $N_m \leq C2^m$ for all m. Keeping the notation of Theorem 6.2, the sharp estimate $N_m \leq C2^m$ will be rephrased in the following theorem.

Theorem 6.6. *Let $\psi_\lambda, \lambda \in \Lambda$, be a two-dimensional orthonormal wavelet basis of class $\mathcal{C}^2$ with a rapid decay at infinity. Then for every f in $BV(\mathbb{R}^2)$, the wavelet coefficients $c_\lambda = \langle f, \psi_\lambda\rangle$, $\lambda \in \Lambda$ belong to weak $l^1(\Lambda)$.*

This theorem was proved by A. Cohen et al. [9] in the Haar system case. The general case was obtained by the author and the best reference is [36].

In other words, if $c_\lambda = \langle f, \psi_\lambda\rangle$ and if the $|c_\lambda|$, $\lambda \in \Lambda$, are sorted out by decreasing size, we obtain a non-increasing sequence c_n^* which satisfies $c_n^* \leq C/n$

for $1 \leq n$. This decay of the sorted wavelet coefficients was announced by S. Mallat in his book [25].

If f is the indicator function of any smooth domain, an easy calculation shows that $c_n^* \geq C/n$ which led me to believe that Theorem 6.6 was optimal. This issue will be addressed in the next section.

7. Improved Gagliardo–Nirenberg Inequalities

New Gagliardo–Nirenberg inequalities will now be proved using Theorem 6.6 and wavelet methods. This is the first outstanding application of wavelet techniques inside mathematics. This success story was so encouraging that we thought that better estimates might exist.

Then A. Cohen and his collaborators met our challenge and improved on Theorem 6.6. This will lead us to Theorem 7.5 and to more refined Gagliardo–Nirenberg inequalities.

But let us return to the Sobolev embedding of BV into $L^2(\mathbb{R}^2)$.

The estimate $\|f\|_2 \leq C\|f\|_{BV}$ is obviously consistent with translations and dilations. Indeed, for any positive a and $f_a(x) = af(ax)$, we have $\|f_a\|_2 = \|f\|_2$ and similarly $\|f_a\|_{BV} = \|f\|_{BV}$. But this estimate is not consistent with modulations: if M_ω denotes the pointwise multiplication operator with $\exp(i\omega x)$, then M_ω acts isometrically on L^2 while $\|M_\omega f\|_{BV}$ blows up as $|\omega|$ when $|\omega|$ tends to infinity.

For addressing this invariance through modulations, let us introduce an adapted Besov norm.

Definition 7.1. *Let B be the Banach space of all tempered distributions $f(x)$ for which a constant C exists such that*

$$|\langle f, g_{a,b}\rangle| < C \tag{13}$$

when $g(x) = \exp(-|x|^2)$, $g_{a,b} = ag(a(x-b))$, $a > 0$, $b \in \mathbb{R}^2$.

The infimum of these constants C is the norm of f in B and is denoted by $\|f\|_\epsilon$.

It is easily proved that this Banach space coincides with the space of second derivatives of functions in the Zygmund class. Therefore B is the homogeneous Besov space $B_\infty^{-1,\infty}$ of regularity index -1 which was already used for modeling textures.

We then have

Theorem 7.2. *There exists a constant C such that for any f in $BV(\mathbb{R}^2)$ we have*

$$\|f\|_2 \leq C[\|f\|_{BV}\|f\|_\epsilon]^{1/2} \tag{14}$$

and $\|f\|_\epsilon$ is the weakest norm obeying the same scaling laws as the L^2 or BV norm for which (14) is valid.

To better understand this theorem, let us stress that we always have $\|f\|_\epsilon \leq \|f\|_{BV}$ and the ratio $\|f\|_\epsilon / \|f\|_{BV}$ between these norms is denoted by β and is expected to be small in general. Then (14) reads

$$\|f\|_2 \leq C\beta^{1/2}\|f\|_{BV} \tag{15}$$

which yields a sharp estimate of the ratio between the L^2 norm and the BV norm of f. Moreover $\beta^{1/2}$ in (15) is sharp as the example of $f(x) = \exp(i\omega x)w(x)$ shows. Indeed if $|\omega|$ tends to infinity and $w(x)$ belongs to the Schwartz class, then $\|f\|_2$ is constant, $\|f\|_\epsilon \simeq |\omega|^{-1}\|f\|_\infty$, and finally $\|f\|_{BV} \simeq |\omega|\|w\|_1$. In this example β is of the order of magnitude of $|\omega|^{-2}$ which corresponds to $\beta^{1/2} \simeq |\omega|^{-1}$.

The proof of (14) is straightforward. One uses the following trivial estimate on sequences

$$\sum_{n=1}^{\infty} |c_n|^2 \leq 2\|c_n\|_\infty \|nc_n\|_\infty . \tag{16}$$

Then one applies Theorem 6.6 to an orthonormal wavelet basis of class $\mathcal{C}^2$. For concluding the proof, it suffices to make the following observation: if c_n^* denotes the non increasing rearrangement of the wavelet coefficients $|c(\lambda)|$, $\lambda \in \Lambda$, then $\|nc_n^*\|_\infty$ is precisely the norm of $c(\lambda)$ in the weak l^1 space.

Let us observe that (14) is an interesting improvement on the celebrated Gagliardo–Nirenberg estimates. These estimates read in the two-dimensional case

$$\|D^j f\|_p \leq C(\|D^m f\|_r)^\sigma (\|f\|_q)^{1-\sigma} \tag{17}$$

where $1 < p, q, r < \infty$, $j/m < \sigma < 1$ and $1/p - j/2 = \sigma(1/r - m/2) + (1-\sigma)/q$.

The notation $\|D^j f\|_p$ means $\sup\{\|\partial^\alpha f\|_p; \ |\alpha| = j\}$.

For comparing our new estimate to (17), we will assume $m = 2$, $j = 1$, $p = 2$ and $r = 1$. This either implies $s = 1$ or $q = \infty$. In the first case, (17) easily follows from the embedding of BV into L^2. In the second case (17) is weaker than (14). Indeed the L^∞ norm which is used in (17) is replaced in (14) by a much weaker one.

Theorem 7.2 generalizes to any dimension $n > 2$. It then reads

$$\|f\|_{n/n-1} \leq C(\|f\|_{BV})^{(1-1/n)}(\|f\|_\alpha)^{1/n} \tag{18}$$

where $\|f\|_\alpha$ is now defined as the optimal constant C for which one has $|\langle f, g_{a,b}\rangle| \leq C$ with $g_{a,b} = ag(a(x-b))$, $a > 0$, $b \in \mathbb{R}^n$, and $g(x) = \exp(-|x|^2)$. In other words $\|f\|_\alpha$ is the norm of f in the homogeneous Besov space $B_\infty^{-(n-1),\infty}$.

Every function with a bounded variation belongs to $L^{\frac{n}{n-1}}$ and it is natural to ask the following question: what would happen if a function f both belongs to BV and to L^q for some $q > n^*$ where $n^* = \frac{n}{n-1}$?

One guesses that this size estimate on a BV function should imply improved regularity. That is what the following theorem says.

The Lebesgue space L^q is contained inside the larger space $C^{-\beta}$ when $\beta = -n/q$. Then the theorem we have in mind reads as follows.

Theorem 7.3. *If $0 \leq s < 1/p$, $1 < p \leq 2$ and $\beta = \frac{1-sp}{p-1}$, then*

$$\|f\|_{L^{p,s}} \leq C\|f\|_{BV}^{1/p}\|f\|_{-\beta}^{1-1/p} \tag{19}$$

where $\|.\|_{-\beta}$ stands for the norm of f in the homogeneous Besov space $B_\infty^{-\beta,\infty}$.

The proof of Theorem 7.3 still relies on Theorem 6.6 and wavelet techniques. Returning to L^2 norms, we unsuccessfully tried to prove

Theorem 7.4. *In any dimension $n \geq 1$, let us assume that a function f both belongs to BV and to the homogeneous space $B_\infty^{-1,\infty}$. Then we have*

$$\|f\|_2 \leq C(\|f\|_{BV}\|f\|_\epsilon)^{1/2} \tag{20}$$

where $\|f\|_\epsilon$ is the norm of f in the Besov space $B_\infty^{-1,\infty}$.

The Besov norm of f can be defined as the optimal constant C for which one has $|\langle f, g_{a,b}\rangle| \leq Ca^{2-n}$, $a > 0$, $b \in \mathbb{R}^n$. Let us observe that BV is contained in L^2 if and only if $n = 2$. In other words when $n = 1$ or $n > 2$, the assumption $f \in B_\infty^{-1,\infty}$ complements $f \in BV$ and both are needed to get an L^2 estimate.

The proof of this theorem requires new estimates on wavelet coefficients of BV functions which sharpen Theorem 6.6. Indeed Albert Cohen, Wolfgang Dahmen, Ingrid Daubechies and Ron DeVore proved

Theorem 7.5. *In any dimension $n \geq 1$, let us assume $\gamma < n - 1$ where γ is a real exponent. Then for $f \in BV(\mathbb{R}^n)$ and $\lambda > 0$, one has*

$$\sum_{\{|c(j,k)|>\lambda 2^{-j\gamma}\}} 2^{-j\gamma} \leq C\lambda^{-1}\|f\|_{BV} \tag{21}$$

where $c(j,k) = \int_{\mathbb{R}^n} f(x)2^j\psi(2^j x - k)\,dx$.

It is easily seen that this estimate is false when $\gamma = n - 1$. If one returns to the two-dimensional case, it is clear that Theorem 7.5 implies Theorem 6.6. Indeed (21) is Theorem 6.6 when $\gamma = 0$. But Theorem 7.5 is saying more and it is not difficult to construct sequences $c(j,k)$ belonging to weak-l^1 for which (21) is not fulfilled. For instance pick $\alpha \in (0,1)$ and let F_j be a sequence of finite sets of integers of cardinality $N_j \approx 2^{\alpha j}$. Then the sequence defined by $c(j,k) = 2^{-\alpha j}$ for $k \in F_j$ will belong to weak-l^1 but does not fulfil (21) when $\gamma = \alpha$.

Knowing Theorem 7.5, the proof of Theorem 7.4 is an exercise.

8. Improved Poincaré Inequality

The standard Poincaré inequality reads as follows: If Ω is a connected bounded open set in the plane with a Lipschitz boundary $\partial\Omega$, then there exists a constant $C = C_\Omega$ such that for every f in $BV(\Omega)$ we have

$$\int_\Omega |f(x) - m_\Omega(f)|^2\,dx \leq C\|f\|_{BV}^2. \tag{22}$$

Here $m_\Omega(f)$ denotes the mean value of the function f over Ω.

Such an estimate cannot be true in $\mathbb{R}^n$ for $n \geq 3$ since BV is not locally embedded in L^2. However there exists an improvement on Poincaré's inequality which (a) is valid in any dimension and (b) sharpen the standard one in the plane.

We need a Banach space $C^{-1}(\Omega)$ for measuring the oscillations of a function f on Ω. Once more the Besov space $B_\infty^{-1,\infty}$ will be used. The first definition reads: $f \in C^{-1}(\Omega)$ iff f is the restriction to Ω of some distribution belonging to $B_\infty^{-1,\infty}$.

Here is an equivalent definition. We write $f \in C^{-1}(\Omega)$ if $f = \Delta F$ where F is the restriction to Ω of a function G defined on $\mathbb{R}^n$ and belonging to the Zygmund class. The Zygmund class is defined by the classical condition that a constant C should exist such that

$$|G(x + y) + G(x - y) - 2G(x)| \leq C|y| \qquad x, y \in \mathbb{R}^n \, . \tag{23}$$

The norm of f in $C^{-1}(\Omega)$ is denoted by $\|f\|_*$ and is defined as the infimum of these constants C. This infimum is computed over all extensions G of F such that $f = \Delta F$ on Ω.

The improvement we have in mind is valid in any dimension and reads as follows:

Theorem 8.1. *Let Ω be a connected bounded open set in $\mathbb{R}^n$. Let us assume that Ω has a smooth boundary $\partial\Omega$. With the preceding notation, there exists a constant $C = C_\Omega$ such that, for every function f both belonging to $BV(\Omega)$ and to $C^{-1}(\Omega)$, we have*

$$\int_\Omega |f(x) - m_\Omega(f)|^2 \, dx \leq C\|f\|_{BV}\|f\|_* \, . \tag{24}$$

Let us insist on the fact that $BV(\Omega)$ is not contained in $C^{-1}(\Omega)$. A counter-example is given by $f(x) = |x|^{-n+1}(\log|x|)^{-2}$ when Ω is the ball $|x| < 1/2$.

For proving Theorem 8.1, one introduces local coordinates on some annular neighbourhood R of $\partial\Omega$ defined by $x = y + t\nu$, $x \in R$, $y \in \partial\Omega$, $t \in (-\eta, \eta)$. We have denoted by ν the interior unit vector at to $\partial\Omega$ at y. Let us extend f into F as follows: $F(y + t\nu) = \pm f(y + t\nu)$ where $\pm$ is the sign of t.

Finally Theorem 7.4 is applied to this new function F, once it has been cut by a convenient cut-off function.

The key fact which enters in the proof of Theorem 8.1 is that this odd extension operator is both continuous with respect to the Besov norm and the BV norm. An even extension operator would also be fine for the BV norm but certainly not for our Besov norm.

9. Wavelet Coefficients of Integrable Functions

Albert Cohen et al. proved that wavelet coefficients of functions in $L^1(\mathbb{R}^n)$ have some interesting and important properties.

This work was motivated by a striking discovery by P. L. Lions and R. Di-Perna [15]. Lions and DiPerna observed that some velocity averages arising in the context of the Boltzmann equation are more regular than expected. More precisely they proved

Theorem 9.1. *Let Ω be a bounded open subset of $\mathbb{R}^n$ and $f(x, v)$, $x \in \mathbb{R}^n$, $v \in \Omega$ be any function satisfying the following two properties:*

(a) $f \in L^2(\mathbb{R}^n \times \Omega)$.
(b) $v.\nabla_x f \in L^2(\mathbb{R}^n \times \Omega)$.

Then the "velocity average"

$$\bar{f}(x) = \int_\Omega f(x, v)\,dv \tag{25}$$

belongs to the Sobolev space $H^{1/2}(\mathbb{R}^n)$.

This was later sharpened by M. Bézard [1]. Finally Ronald DeVore and Guerguana Petrova [14] proved

Theorem 9.2. *If $1 < p < \infty$ and if*

(a) $f \in L^p(\mathbb{R}^n \times \Omega)$,
(b) $v.\nabla_x f \in L^p(\mathbb{R}^n \times \Omega)$,

then the velocity average $\bar{f}(x)$ belongs to the homogeneous Besov space $B_p^{s,p}(\mathbb{R}^n)$ where $s = \inf(1/p, 1 - 1/p)$.

When $p = 2$ this is exactly the Lions–DiPerna theorem.

The proof of Theorem 9.2 has been simplified by A. Cohen and we will follow his presentation.

A. Cohen writes

$$g(x, v) = f(x, v) + v.\nabla_x f(x, v) \tag{26}$$

and denotes by $\bar{f}(x)$ the velocity average defined by (25). Finally A. Cohen studies the linear operator T which maps $g(x, v)$ on $\bar{f}(x)$.

A partial Fourier transformation in x gives

$$\hat{g}(\xi, v) = (1 + iv.\xi)\hat{f}(\xi, v). \tag{27}$$

Then Plancherel identity and standard calculations yield

$$\|\bar{f}(x)\|_{H^{1/2}} \leq C\|g\|_{L^2(\mathbb{R}^n \times \Omega)} \tag{28}$$

and T maps $L^2(\mathbb{R}^n \times \Omega)$ into $H^{1/2}(\mathbb{R}^n)$. This is the Lions–DiPerna theorem.

Moreover the uniform boundedness in v of $(1 + iv.\xi)^{-1}$ as a multiplier of $\mathcal{F}L^1(\mathbb{R}^n)$ implies that T maps $L^1(\mathbb{R}^n \times \Omega)$ to $L^1(\mathbb{R}^n)$.

In order to prove Theorem 9.2, it then suffices to use the real interpolation method of Lions and Peetre and show that $B_p^{s,p}$ is the interpolation space between $H^{1/2} = B_2^{1/2,2}$ and L^1. Since all Besov spaces admit trivial characterizations by size properties of wavelet coefficients, it remains to study the wavelet coefficients of functions in L^1. We now concentrate on that task.

The normalization which will be used is the following. We write ψ_λ for $\psi(2^j x - k)$ and the wavelet coefficients of f are now $c(\lambda) = \langle f, \psi_\lambda \rangle$. They are indexed by

$\Lambda = \mathbb{Z} \times \mathbb{Z}^n \times F$ where F is a finite set with cardinality $2^n - 1$. Next we denote by Q_λ the corresponding dyadic cube defined by $2^j x - k \in [0,1)^n$. The theorem on wavelet coefficients of $L^1(\mathbb{R}^n)$ functions says the following:

Theorem 9.3. *For any real exponent γ larger than 1, there exists a constant $C = C_{\gamma,n}$ such that for f in $L^1(\mathbb{R}^n)$ and for $\tau > 0$, one has*

$$\sum_{\{|c(\lambda)|>\tau|Q_\lambda|^\gamma\}} |Q_\lambda|^\gamma \leq C\tau^{-1}\|f\|_1 \,. \tag{29}$$

In other words the wavelet coefficients $c(\lambda)$, $\lambda \in \Lambda$, belong to to a weighted weak-l^1 space where the weighting factor is $|Q_\lambda|^\gamma$. Theorem 9.3 yields the required interpolation theorem we were looking for. Moreover Theorem 9.3 complements Theorem 7.5. Indeed Theorem 9.3 can be applied to the gradient of a BV function and the normalizations are adjusted in such a way that Theorem 9.3 corresponds to $\gamma > n$ in Theorem 7.5.

10. The Role of Oscillations in Some Nonlinear PDEs

For a number of nonlinear evolution equations, blowup may happen even if the initial condition is smooth and compactly supported. It is clear that such a blowup needs to be defined with respect to some functional norm and that some norms might become infinite when $t \to t_0$ while others would remain bounded. Proving that some strong norm does not blow up as long as a weaker norm remains under control is quite interesting but often rather hard. Such $0 - 1$ laws will be at the heart of our discussion and we will construct weak norms which do control stronger ones. Our favourite example is the nonlinear heat equation. For Navier–Stokes equations the occurence of an eventual blowup is still an open problem. The nonlinear Schrödinger equation will also be treated. In these first two examples the weaker norm which will be used is denoted by $\|.\|_*$ and $\|f\|_*$ is small when f is oscillating. If the nonlinear Schrödinger equation is excluded, the main message of this chapter is the following slogan: *blowup does not happen when the initial condition is oscillating.*

This assertion is easily proved for the nonlinear heat equation and was already known in the case of the Navier–Stokes equations, as both Peter Constantin and Roger Temam told me. We will later on return to their sharp comments.

What is completely new is that Besov spaces are manifesting themselves again (Theorems 12.1 and 12.7). Exactly as it happened when we were modeling textures, an oscillating pattern is defined as a function which has a small norm in a Besov space with negative regularity index. If the initial condition is such an oscillating pattern, the corresponding solution does not blow up.

As often in mathematics, a discovery raises new problems. Here we want to find the sharpest theorem in the direction given by these heuristic considerations. It means measuring oscillations with the weakest norm. In the case of Navier–Stokes equations, the best result was obtained by Herbert Koch and Daniel Tataru [24].

The space they used is no longer a Besov space and is defined as the collection of functions or vectors u_0 that can be written as $u_0 = \partial_1 f_1 + \partial_2 f_2 + \partial_3 f_3$ where f_j, $j = 1, 2, 3$ belong to the John and Nirenberg space BMO. This Koch and Tataru space is one of the spaces which were used for modeling textures.

11. A First Model Case: The Nonlinear Heat Equation

Our first model case will be the heat equation

$$\begin{cases} \frac{\partial u}{\partial t} = \Delta u + u^3, & (x,t) \in \mathbb{R}^3 \times (0, \infty) \\ u(x,0) = u_0(x) \end{cases} \tag{30}$$

where $u = u(x,t)$ is a real valued function of (x,t), $x \in \mathbb{R}^3$ and $t \geq 0$.

In the following calculation, u is assumed to be a classical solution to (30) with enough regularity and with appropriate size estimates. All L^p-norms will be finite by assumption and all integrations by parts will be legitimate.

Multiplying (30) by u and integrating over $\mathbb{R}^3$ yields $\frac{d}{dt}\|u\|_2^2 = -2\|\nabla u\|_2^2 + 2\|u\|_4^4$ which means that the evolution will depend on the competition between $\|u\|_4^4$ and $\|\nabla u\|_2^2$.

This remark paves the way to the following theorem.

Theorem 11.1. (J. Ball, H. A. Levine and L. Payne) *If u_0 is a smooth compactly supported function which does not vanish identically and if*

$$\|\nabla u_0\|_2 \leq \frac{1}{\sqrt{2}}\|u_0\|_4^2, \tag{31}$$

then the corresponding solution of (30) blows up in finite time: there exists a finite T_0 such that $\|u(.,t)\|_2$ is unbounded as t reaches T_0.

Theorem 11.1 raises the following problem: does there exist a function space norm $\|.\|_*$ and a positive η with the following properties:

(a) $\|f\|_4^2 \leq C\|\nabla f\|_2\|f\|_*$
   (then (31) would say that $\|f\|_*$ is large).
(b) If u_0 is smooth and compactly supported and if $\|u_0\|_* \leq \eta$, then there exists a global (in time) smooth solution $u(x,t)$ to (30) (no blowup).
(c) $\|f_\lambda\|_* = \|f\|_*$ if $f_\lambda(x) = \lambda f(\lambda x)$.
(d) $\|f\|_*$ is small if f is oscillating?

Let us comment (c). If $u(x,t)$ is a solution to (30), so are $u_\lambda(x,t) = \lambda u(\lambda x, \lambda^2 t)$ for every positive λ. If $p \neq 3$, the L^p-norm is not invariant under these rescalings and a condition like $\|u_0\|_p \leq \eta$ for some small η is not relevant for a global existence of the corresponding solution to (30). Indeed by a convenient rescaling of the initial condition u_0, this smallness requirement can be reached. This remark explains condition (c).

The simplest norm fulfilling (c) is the L^3 norm for which F. Weissler [40] proved

Theorem 11.2. *For a positive constant η, the condition $\|u_0\|_3 < \eta$ implies the existence of a global solution $u(x,t) \in C([0,\infty), L^3(\mathbb{R}^3))$ to (30).*

Uniqueness was proved by F. Weissler inside a Banach space Y which is smaller than the "natural space" $C([0,\infty), L^3(\mathbb{R}^3))$. This smaller space is defined by imposing on $u(x,t)$ the condition

$$t^{1/2}\|u(.,t)\|_\infty \leq \eta. \tag{32}$$

Let us denote the heat semigroup by $S(t)$. Then imposing this growth condition on the linear evolution $S(t)u_0$ is equivalent to saying that u_0 belongs to our old friend $B_\infty^{-1,\infty}$. This is not a restriction since this Besov space contains L^3.

In her thesis, Elide Terraneo constructed a striking counterexample showing that uniqueness of solutions $u(x,t)$ to (30) in $C([0,\infty), L^3(\mathbb{R}^3))$ could not be expected in general [39]. This explains the role of (32).

This situation sharply contrasts with what happens for Navier–Stokes equations. T. Kato proved the analogue of Theorem 11.2. Kato's proof is close to Weissler's approach and (32) plays a very important role in the construction. For quite a long time, uniqueness of Kato's solutions $v(x,t) \in C([0,\infty), L^3(\mathbb{R}^3))$ was an open problem. Finally uniqueness was proved by Giulia Furioli, Pierre-Gilles Lemarié-Rieusset and Elide Terraneo without assuming (32). The interested reader is referred to [20] or [28].

Theorem 11.2 says that $\|.\|_3$ fulfils (b). Moreover one has $\|f\|_4^2 \leq \|\nabla f\|_2\|f\|_3$ but these two answers to our program are far from being optimal ones. Indeed the L^3 norm can be replaced by a much weaker one for which (d) holds. This weaker norm is a Besov norm. The relevance of Besov norms in (a) is explained by the following Gagliardo–Nirenberg inequality.

Lemma 11.3. *For any function f belonging to the homogeneous Sobolev space H^1, we have*

$$\|f\|_4^2 \leq C\|\nabla f\|_2\|f\|_B \tag{33}$$

where B is the homogeneous Besov space $B_\infty^{-1,\infty}$.

Lemma 11.3 suggests that the weak norm $\|f\|_*$ fulfilling (a) to (d) might be the Besov norm $\|f\|_B$. This Besov norm is the weakest one since $B_\infty^{-1,\infty}$ is the largest function space whose norm is translation invariant and fulfils (c). We do not know if it is the case but the following theorem gives an example of a norm fulfilling (a) to (d).

Theorem 11.4. *Let $\|.\|_*$ denote the norm in the Besov space $B_6^{-1/2,\infty}$. Then there exists a positive constant η such that if the initial condition u_0 satisfies $u_0 \in L^3$ and $\|u_0\|_* \leq \eta$ then the corresponding solution of the nonlinear heat equation will be global in time and belongs to $C([0,\infty), L^3(\mathbb{R}^3))$. Moreover there exists a constant C such that*

$$\|u(t)\|_* \leq C\|u_0\|_*, \qquad t \geq 0. \tag{34}$$

The homogeneous Besov space $B_6^{-1/2,\infty}$ needs to be defined. We start with $\varphi(x) = (2\pi)^{-n/2}\exp(-|x|^2/2)$ and let S_j be the convolution operator with $2^{3j}\varphi(2^j x)$, $j \in \mathbb{Z}$. Then we have

Definition 11.5. *f belongs to $B_6^{-1/2,\infty}$ if and only if a constant C exists such that $\|S_j(f)\|_6 \le C 2^{j/2}$, $j \in \mathbb{Z}$. The optimal C being the norm of f in $B_6^{-1/2,\infty}$.*

Theorem 11.4 is not optimal but it improves on Theorem 11.2. On one hand,

$$L^3 \subset B_6^{-1/2,\infty} \subset B_\infty^{-1,\infty} \tag{35}$$

and these embeddings are provided by Bernstein's inequalities. On the other hand, Theorem 11.4 is consistent with the guess that the oscillating character of the initial condition implies the global (in time) existence of the corresponding solution. Indeed one can easily check that

$$\|\cos(\omega x)\varphi(x)\|_* \le C|\omega|^{-1/2}\|\varphi\|_6 \tag{36}$$

as $|\omega|$ tends to infinity. When $|\omega|$ is large enough, then the smallness requirement is met and the corresponding solution is global in time. A last observation concerns scale invariance. The norm in $B_6^{-1/2,\infty}$ has the same invariance as the L^3 norm does and this invariance is consistent with the one we found in the nonlinear heat equation.

The experience we gained on this specific example will now be used to attack the much more difficult Navier–Stokes equations.

12. The Navier–Stokes Equations

We now consider the Navier–Stokes equations describing the motion of some incompressible fluid. The fluid is assumed to be filling the space and there are no exterior forces. Then the Navier–Stokes equations read

$$\begin{cases} \frac{\partial v}{\partial t} = \Delta v - (v.\nabla)v - \nabla p \\ \nabla.v = 0 \\ v(x,0) = v_0 \end{cases} \tag{37}$$

Here $v = (v_1, v_2, v_3)$, $v_j = v_j(x,t)$, $x \in \mathbb{R}^3$, $t \ge 0$, the pressure is a scalar and the Navier–Stokes equations are a system of four equations with four unknown functions v_1, v_2, v_3 and p.

The notation ∇p means the gradient of the pressure, $\nabla.v$ means $\partial_1 v_1 + \partial_2 v_2 + \partial_3 v_3$. Moreover $(v.\nabla)v = v_1\partial_1 v + v_2\partial_2 v + v_3\partial_3 v$ which is a vector.

If the velocity $v(x,t)$ is not a smooth function of x, then multiplying some components of v with derivatives of some other components might be impossible. That is why $(v.\nabla)v$ should be rewritten as $\partial_1(v_1 v) + \partial_2(v_2 v) + \partial_3(v_3 v)$ which makes sense whenever v is locally square integrable.

Navier–Stokes equations have some remarkable scale invariance properties. First they commute with translations in x and $t \geq 0$. Moreover if the pair $v(x,t)$, $p(x,t)$ is a solution of (37) and if for every $\lambda > 0$ we dilate this solution into

$$\begin{cases} v_\lambda(x,t) = \lambda u(\lambda x, \lambda^2 t) \\ p_\lambda(x,t) = \lambda^2 p(\lambda x, \lambda^2 t), \end{cases} \tag{38}$$

then (v_λ, p_λ) is also a solution to the Navier–Stokes equations. The initial condition is replaced by

$$v_\lambda(x,0) = \lambda v_0(\lambda x). \tag{39}$$

We observe that this scale invariance is exactly the same as the one we met in the nonlinear heat equation.

The problem we want to address is the possible blowup of solutions in finite time. We are aiming to attack this problem by following the same heuristic approach as in the nonlinear heat equation setting. Our guess is the following: if the initial condition is (everywhere) oscillating, then the corresponding solution to Navier–Stokes equations should be global in time. Moreover this global solution will keep forever some additional qualitative regularity of the initial condition. For instance an initial condition which is $\mathcal{C}^\infty$ and is sufficiently oscillating will lead to a global solution which will also be $\mathcal{C}^\infty$ in the time-space variables. Such a result seems inconsistent. Indeed a function cannot at the same time be smooth and oscillating. However this objection disappears if the smoothness assumption is not given a quantitative form while the oscillations are defined by a specific threshold η and by imposing $\|v_0\|_* < \eta$. Here $\|f\|_*$ is a norm which is small whenever f is oscillating. This norm might be one of the norms which has been used in image processing in order to model textures. In other words, we are now assuming that our initial condition v_0 is a function which has a small norm in a function space containing generalized functions. The norm of a function f in such a function space takes advantage of the oscillating character of f. At the same time our v_0 may be extremely large in function spaces like the Hölder or Sobolev spaces. We will denote by B a Banach space of smooth functions. For instance B can be the Sobolev space H^m or the usual $\mathcal{C}^m$, $m \geq 1$.

Conjecture. *Let $\|.\|_*$ be one of the norms which has been used in image processing in order to model textures. Then there exists a positive η such that if v_0 is smooth and satisfies the following two conditions $\nabla.v_0 = 0$, and $\|v_0\|_* < \eta$, then the corresponding solution of the Navier–Stokes equations belongs to $\mathcal{C}([0,\infty), B)$.*

Notice that we are not requiring that $\|v_0\|_B$ be small. This conjecture will be our guideline in this chapter and the best result will be Theorem 12.7 which combines a deep theorem by Herbert Koch and Daniel Tataru [24] and a nice observation by Pierre-Gilles Lemarié-Rieusset and his team [21].

As Roger Temam pointed out, the first result along these lines has been proved by H. Fujita and T. Kato in 1964 [19]. It reads as follows:

Theorem 12.1. *Let $H^{1/2}(\mathbb{R}^3)$ denote the usual homogeneous Sobolev space. Then there exists a positive constant η such that if v_0 belongs to $H^1(\mathbb{R}^3)$ and fulfils*

$$\nabla.v_0 = 0 \tag{40}$$

$$\|v_0\|_{H^{1/2}(\mathbb{R}^3)} \leq \eta, \tag{41}$$

then there exists a unique global solution $v \in C([0,\infty), H^1(\mathbb{R}^3))$ to the Navier–Stokes equations.

As Roger Temam observed, this theorem is specially attractive if $\|v_0\|_{H^{1/2}}$ is much smaller than $\|v_0\|_{H^1}$. This is often the case since the first Sobolev norm is less demanding than the second one. Indeed in this situation, we do not pay too much for getting a global solution since the norm with which the initial condition is measured is weaker than the H^1 norm in which the global existence is proved. The following lemma tells us when $\|f\|_{H^{1/2}}$ is small while $\|f\|_{H^1}$ is large.

Lemma 12.2. *There exists a constant C such that*

$$\|f\|_{H^{1/2}} \leq C\sqrt{\|f\|_B \|f\|_{H^1}} \tag{42}$$

where B is the homogeneous Besov space $B_2^{-1/2,\infty}$.

The weak norm will be much smaller than the square root of the strong norm when our Besov norm is small. Does this Gagliardo–Nirenberg estimate mean that $B_2^{-1/2,\infty}$ is the space which needs to be used in our heuristic approach? It cannot be so since the Besov space B which is used does not enjoy the right scaling property. Indeed $f(x)$ and $f_\lambda(x) = \lambda f(\lambda x)$ do not have the same norm in B. However the homogeneous space $\|v_0\|_{H^{1/2}}$ enjoys this scale invariance. But an oscillating initial condition has a large norm in this Sobolev space. We can conclude in saying that the Fujita–Kato theorem does not meet our expectations.

The Fujita–Kato theorem should be compared to a second theorem proved by Y. Giga and T. Miyakawa [22]. These mathematicians are focusing on the vorticity field $\omega = \mathrm{curl}(v)$. Since u is divergence free, the mapping $v \mapsto \omega$ is an isomorphism for most of the function spaces which are being used in analysis. The inverse mapping is provided by The Biot–Savart law which reads

$$-4\pi v(x,t) = \int_{\mathbb{R}^3} |x-y|^{-3}(x-y) \times \omega(x,t)dy. \tag{43}$$

In other words, the mapping $\omega \mapsto v$ is smoothing of order 1 and any functional estimate on the vorticity field implies a corresponding estimate on the velocity field.

The motivation of Giga and Miyakawa was twofold. They wanted to model vorticity filaments in order to understand the evolution of such filaments. These vorticity filaments appear in numerical simulations of Navier–Stokes equations. At the same time Giga and Miyakawa wanted to construct some self-similar solutions to the Navier–Stokes equations. In order to achieve these goals, they modeled these vorticity filaments with the Guy David space of Radon measures μ which was already met in Section 5.

Definition 12.3. *A Radon measure satisfies the Guy David condition if and only if a constant C exists such that, for every ball B with radius R, we have*

$$|\mu|(B) \leq CR. \tag{44}$$

This Guy David space will be denoted by M. It is a dual space and will always be equipped with its weak* topology. In other words, $C([0,\infty), M)$ will always refer to the weak* topology.

Giga and Miyakawa proved the following theorem [22].

Theorem 12.4. *There exists a (small) positive number η such that whenever the initial condition $\omega_0(x)$ satisfies $\operatorname{div} \omega_0 = 0$ together with the size condition (44) with $C < \eta$, there exists a solution $\omega(x,t) \in C\big([0,\infty), M\big)$ to the Navier–Stokes equation which agrees with this initial condition. Moreover there exists a constant C_1 such that the corresponding velocity v satisfies*

$$t^{1/2}\|v(.,t)\|_\infty \leq C_1. \tag{45}$$

The Biot–Savart law enables us to lift this theorem from vorticities to velocities but we are not going to be more precise about the Banach space Γ describing these corresponding velocities. This space is a dual space, its norm is compatible with the scaling properties of the Navier–Stokes equations and it contains functions which are homogeneous of degree -1 which permitted Giga and Miyakawa to build self-similar solutions to the Navier–Stokes equations. Seven years later, Marco Cannone and Fabrice Planchon proposed another construction of self-similar solutions. We will later on explain their approach. However the Γ-norm of Giga and Miyakawa does not enjoy the crucial property that oscillating functions have small norms. That is why we still want to improve on their theorem.

Some progress was made by M. Cannone, F. Planchon. The norm in the Sobolev space $H^{1/2}$ which was used by Kato can be replaced by a weaker norm which is the Besov norm in the homogeneous space $B_q = B_q^{-(1-3/q),\infty}$ whenever $3 \leq q < \infty$. More precisely we have

Theorem 12.5. *There exists a positive constant η_q such that whenever the initial condition v_0 satisfies, for some $q \in [3,\infty)$,*

$$\nabla.v_0 = 0 \tag{46}$$

$$v_0 \in L^3(\mathbb{R}^3) \quad and \quad \|v_0\|_{B_q} < \eta_q, \tag{47}$$

then the corresponding solution to Navier–Stokes equations belongs to $C([0,\infty), L^3(\mathbb{R}^3))$ and is unique.

The homogeneous Besov space B_q is defined exactly the same way as in the special case $q = 2$ (see Theorem 11.4). We let φ be a function in the Schwartz class $\mathcal{S}(\mathbb{R}^3)$ such that $\int_{\mathbb{R}^3} \varphi(x)dx = 1$. Then $\varphi_j(x) = 2^{3j}\varphi(2^j x)$ and S_j denotes the convolution operator with $\varphi_j(x)$. Finally a function or a distribution f belongs to the homogeneous Besov space $B_q^{-\alpha,\infty}$ if and only if $\|S_j(f)\|_q \leq C2^{j\alpha}$, $j \in \mathbb{Z}$.

One should observe that this result does not contain the theorem obtained by Giga and Miyakawa. Indeed the function space used by these authors is not contained inside any space B_q, $3 \leq q < \infty$.

F. Planchon made the following remarks. The Banach spaces B_q are increasing with q in such a way that the conditions (45) seem to be less demanding as q grows. However the positive constant η_q which appears in (45) tends to 0 as q tends to infinity. Therefore comparing these distinct conditions is a delicate matter.

In this direction of large values of q, a main breakthrough was achieved by Herbert Koch and Daniel Tataru [24] who treated the limiting case $q = \infty$. As it is often the case, L^∞ should be replaced by the John and Nirenberg space BMO. Moreover the regularity exponent α which is $1 - 3/q$ tends to 1 as q tends to infinity and these remarks pave the way to

Definition 12.6. *We denote by $B = B_\infty$ the Banach space consisting of all generalized functions f which can be written as $f = \partial_1 g_1 + \cdots + \partial_3 g_3$ where g_j, $j = 1, 2$ and 3 belong to BMO.*

The norm in B_∞ is the infimum of the sum of the three BMO norms.

As usual H^m will denote the standard Sobolev space. Then a combination between the Koch-Tatar theorem [24] and a nice remark by Pierre-Gilles Lemarié-Rieusset [21] reads as follows.

Theorem 12.7. *There exists a positive constant η such that the conditions (a) $\|v_0\|_B \leq \eta$ together with (b) $v_0 \in H^m$ and (c) $\nabla.u_0 = 0$ imply the existence of a global solution v of the Navier–Stokes equations. This solution belongs to $\mathcal{C}([0, \infty), H^m(\mathbb{R}^3))$.*

As it might be guessed, the Koch and Tataru space contains all the previous Besov spaces which were used in Theorem 12.5. One also should observe that the Koch and Tataru space is exactly the one which was used for modeling the two-dimensional textures. Moreover the Koch and Tataru theorem implies the Giga–Miyakawa result. It is indeed a simple exercise to check that $\Lambda^{-1}(\mu)$ belongs to BMO whenever μ satisfies the Guy David condition. Here $\Lambda = (-\Delta)^{1/2}$.

Before ending this section, we would like to say a few words about the proof of the Koch and Tataru theorem. It follows the general organization which was pioneered by Kato and Weissler. That is to say that the Navier–Stokes equations are rewritten as an integral equation. This is achieved by solving the linear heat equation. Let $S(t)$ denote the heat semigroup. We have $S(t)[f] = f \star \Phi(t)$ where $\Phi(t) = t^{3/2}\Phi(x/\sqrt{t})$ and $\Phi(x)$ is the usual gaussian function. Then we obtain

$$v(t) = S(t)v_0 + \Pi \int_0^t S(t - \tau) \sum_1^3 \partial_j v_j v(\tau)d\tau. \tag{48}$$

Here Π denotes the Leray–Hopf projector on divergence free vector fields. In other words

$$\Pi(f_1, f_2, f_3) = (\sigma - R_1\sigma, \sigma - R_2\sigma, \sigma - R_3\sigma) \tag{49}$$

where $\sigma = R_1 f_1 + R_2 f_2 + R_3 f_3$. It implies that Π acts boundedly on all spaces which are preserved by the Riesz transforms R_1, R_2 and R_3.

Two points should be made. First the pressure $p(x,t)$ has disappeared from the Navier–Stokes equations and next the initial condition has been incorporated inside (48). Indeed the kernel of the Leray–Hopf operator is precisely the collection ∇p of curl-free vector fields.

We then rewrite (49) in a more condensed way as

$$v = g + B(v, v) \qquad (50)$$

where v is viewed as a vector inside some function space X and g is a given vector in X. All functions are defined on $\mathbb{R}^3 \times (0, \infty)$. The difficult part of the proof is the construction of this Banach space X to which a Picard fixed point theorem will be applied.

In the Koch and Tataru proof X is defined as follows.

Definition 12.8. *The Banach space X consists of all functions $f(x,t)$ which are locally square integrable on $\mathbb{R}^3 \times (0, \infty)$ and which satisfy the condition*

$$\|f\|_X = \sup \|t^{1/2} f(.,t)\|_\infty + \sup \left(|B(x,R)|^{-1} \int_{Q(x,R)} |f|^2 \, dydt \right)^{1/2} < \infty.$$

As usual $B(x,R)$ denotes the ball centered at x with radius R while $Q(x,R)$ is the Carleson box $B(x,R) \times [0, R^2]$. The supremum is computed over all such Carleson boxes and the right-hand side is the norm of f in X.

Two facts need to be proved. First the function g in (50) should belong to X. Next the bilinear operator $B(v,v)$ should act boundedly from $X \times X$ to X.

The first fact is an easy consequence of the characterization of BMO by Carleson measures. This characterization can also be interpreted as a characterization of BMO by size conditions on wavelet coefficients. The second part of the proof is much deeper and the reader is referred to the beautiful paper by Koch and Tataru [24].

13. The Nonlinear Schrödinger Equation

We now consider the nonlinear Schrödinger equation which obeys the same scaling laws as the two preceding nonlinear PDEs. There are indeed two such equations depending on a $\pm$ sign. The Schrödinger equations with critical nonlinearity are the evolution equations

$$\begin{cases} i \frac{\partial u}{\partial t} + \Delta u = \epsilon |u|^2 u \\ u(x,0) = u_0, \qquad x \in \mathbb{R}^3, \, t \in [0, \infty), \end{cases} \qquad (51)$$

where ϵ is either -1 or 1 and $u = u(x,t)$ is a complex valued function defined on $\mathbb{R}^3 \times (0, \infty)$. If λ is any positive scale factor, then, for every solution $u(x,t)$ of (51), $\lambda u(\lambda x, \lambda^2 t)$ is also a solution of (51) for which the initial condition is

$\lambda u_0(\lambda x)$. Therefore it is not unnatural to expect some similarities with both the nonlinear heat equation and the Navier–Stokes equations.

More precisely we might follow Kato and Fujita and expect (51) to be well posed for the critical Sobolev space $H^{1/2}(\mathbb{R}^3)$. Cazenave and Weissler [5] proved that it was the case under a smallness condition on the norm of the initial condition in $H^{1/2}(\mathbb{R}^3)$. Fabrice Planchon [38] extended this theorem and replaced the smallness condition $\|u_0\|_{H^{1/2}} \leq \eta$ by a weaker requirement which reads $\|u_0\|_* \leq \eta$ where the norm $\|.\|_*$ is the homogeneous Besov norm in $B_2^{1/2,\infty}$.

Theorem 13.1. *With the preceding notation, there exists a positive constant η such that for every initial condition u_0 in $H^{1/2}(\mathbb{R}^3)$ satisfying $\|u_0\|_* \leq \eta$, there exists a solution $u(.,t)$ to the Schrödinger equation (51) which belongs to $\mathcal{C}([0,\infty); H^{1/2}(\mathbb{R}^3))$.*

This theorem should be compared to Theorem 12.5. Indeed keeping the same notation as in Theorem 12.5, the Besov space which is used is B_2. Moreover this theorem implies the existence of many self-similar solutions to the nonlinear Schrödinger equation. Such solutions were previously proved to exist by Cazenave and Weissler [6] under much more restrictive regularity assumptions.

References

[1] M. Bézard. Régularité L^p précisée des moyennes dans les équations de transport. Bull. Soc. Math. France **22** (1994), 29–76.

[2] Aline Bonami and S. Poornima. Non-multipliers of Sobolev spaces. Journ. Funct. Anal. **71** (1987), 175–181.

[3] Christopher M. Brislawn. Fingerprints go digital. Notices of the AMS Vol. 42, no. 11, Nov. 1995, 1278–1283. Web site http://www.c3.lanl.gov/brislawn.

[4] Marco Cannnone. Ondelettes, paraproduits et Navier–Stokes. Diderot Editeur (1995).

[5] Thierry Cazenave and Frederic Weissler. The Cauchy problem for the critical nonlinear Schrödinger equation in H^s. Nonlinear Anal. T.M.A. **14** (1990), 807–836.

[6] Thierry Cazenave and Frederic Weissler. Asymptotically self-similar global solutions of the nonlinear Schrödinger and heat equations. Math. Zeit. **228** (1998), 83–120.

[7] Albert Cohen. Numerical analysis of wavelet methods. Handbook of numerical analysis. P. G. Ciarlet and J. L. Lions eds. (1999).

[8] Albert Cohen and Robert Ryan. Wavelets and multiscale signal processing. Chapman and Hall, London (1995).

[9] Albert Cohen, Ronald DeVore, Pencho Petrushev and Hong Xu. Nonlinear approximation and the space $BV(R^2)$. American Journal of Mathematics **121** (1999), 587–628.

[10] Ingrid Daubechies. Ten lectures on wavelets. SIAM Philadelphia (1992).

[11] Ronald DeVore. Nonlinear approximation. Acta Numerica (1998), 51–150.

[12] Ronald DeVore, Björn Jawerth and Vasil Popov. Compression of wavelet decompositions. American Journal of Mathematics **114** (1992), 737–785.

[13] Ronald DeVore and Bradley J. Lucier. Fast wavelet techniques for near optimal image compression. 1992 IEEE Military Communications Conference October 11–14 (1992).

[14] Ronald DeVore and Guergana Petrova. The Averaging Lemma. Journal of the American Mathematical Society (JAMS) Vol. 14 (2000), 279–296.

[15] Ronald DiPerna, Pierre-Louis Lions and Yves Meyer. L^p regularity of velocity averages. Ann. Inst. H. Poincaré (Analyse Non Linéaire) **8** (1991), 271–288.

[16] David Donoho. Nonlinear Solution of Linear Inverse Problems by Wavelet-Vaguelette Decomposition- Applied and Computational Harmonic Analysis **2** (1995), 101–126.

[17] David Donoho, Ron DeVore, Ingrid Daubechies and Martin Vetterli. Data compression and harmonic analysis. IEEE Trans. on Information Theory, Vol. 44, No. 6, October 1998, 2435–2476.

[18] David Donoho and Iain Johnstone. Wavelet shrinkage: Asymptopia? J. R. Statist. Soc. B **57**, no. 2 (1995), 301–369.

[19] Hiroshi Fujita and Tosio Kato. On the Navier–Stokes initial value problem I. Arch. Rational Mech. Anal. **16** (1964), 269–315.

[20] Giulia Furioli, Pierre-Gilles Lemarié-Rieusset et Elide Terraneo. Sur l'unicité dans $L^3(\mathbb{R}^3)$ des solutions "mild" des équations de Navier–Stokes. C. R. Acad. Sciences Paris, Série 1 (1997), 1253–1256.

[21] Giulia Furioli, Pierre-Gilles Lemarié-Rieusset, Ezzedine Zahrouni et Ali Zhioua. Un théorème de persistance de la régularité en norme d'espaces de Besov pour les solutions de Koch et Tataru des équations de Navier–Stokes dans $\mathbb{R}^3$. CRAS Paris, t. 330, Série I (2000), 339–342.

[22] Yoshikazu Giga and Tetsuro Miyakawa. Navier–Stokes flow in $\mathbb{R}^3$ with measures as initial vorticity and Morrey spaces. Comm. in PDE **14** (1989), 577–618.

[23] Jean-Pierre Kahane and Pierre-Gilles Lemarié-Rieusset. Fourier Series and Wavelets. Gordon and Breach Science Publishers (1996).

[24] Herbert Koch and Daniel Tataru. Well-posedness for the Navier–Stokes equations, Adv. Math. **157** (2001), no. 1, 22–35, CMP 1, 808–843.

[25] Stéphane Mallat. A Wavelet Tour of Signal Processing. Academic Press (1998).

[26] Yves Meyer. Wavelets, Algorithms & Applications. SIAM, Philadelphia, (1993).

[27] Yves Meyer. Wavelets and Operators. Cambridge studies in advanced mathematics **37** CUP (1992).

[28] Yves Meyer. Wavelets, paraproducts and Navier–Stokes equations, Current Developments in Mathematics, 1996, pp. 105–212, Ed. Raoul Bott, Arthur Jaffe, S. T. Yau, David Jerison, George Lusztig, Isadore Singer. International Press.

[29] Yves Meyer. Wavelets, vibrations and scalings. CRM Monograph Series, Vol. 9 (1998).

[30] Yves Meyer. Large time behavior and self-similar solutions of some semi-linear diffusion equations. Essays in Honor of Alberto Calderón, University of Chicago Press (1999).

[31] Sylvie Monniaux. Uniqueness of mild solutions to the Navier–Stokes equation and maximal L^p-regularity. CRAS. Paris, t. 328, Série I (1999), 663–668.

[32] Jean-Michel Morel. Filtres itératifs des images et équations aux dérivées partielles. Notes de cours du centre Emile Borel (1998).

[33] Jean-Michel Morel and Sergio Solimini. Variational methods in image segmentation. Birkhäuser, Boston (1995).

[34] David Mumford. Book reviews on [33]. Bulletin of the American Mathematical Society, Vol. 33, no. 2, April 1996.

[35] David Mumford and Jayant Shah. Boundary detection by minimizing functionals. Proc. IEEE Conf. Comp. Vis. Pattern Recognition, 1985.

[36] Frédéric Oru. Le rôle des oscillations dans quelques problèmes d'analyse non-linéaire. Thèse. CMLA. ENS-Cachan (9 Juin 1998).

[37] Stan Osher and Leonid Rudin. Total variation based image restoration with free local constraints. In Proc. IEEE ICIP, vol. I, p. 31–35, Austin (Texas) USA, Nov. 1994.

[38] Fabrice Planchon. On the Cauchy problem in Besov spaces for a non-linear Schrödinger equation. Communications in Contemporary Mathematics (2000), 243–254.

[39] Elide Terraneo. Sur la non-unicité des solutions faibles de l'équation de la chaleur non linéaire avec non-linéarité u^3. CRAS. Paris t. 328, Série I (1999), 759–762.

[40] Frederic Weissler. Local existence and nonexistence for semilinear parabolic equations in L^p. Indiana Univ. Math. J. **29** (1980), 79–102.

CMLA
Ecole normale supérieure de Cachan
94235 Cachan Cedex, France
E-mail address: Yves.Meyer@cmla.ens-cachan.fr

New Families of Solutions in N-Body Problems

Carles Simó

Abstract. The N-body problem is one of the outstanding classical problems in Mechanics and other sciences. In the Newtonian case few results are known for the 3-body problem and they are very rare for more than three bodies. Simple solutions, such as the so-called *relative equilibrium solutions*, in which all the bodies rotate around the center of mass keeping the mutual distances constant, are in themselves a major problem. Recently, the first example of a new class of solutions has been discovered by A. Chenciner and R. Montgomery. Three bodies of equal mass move periodically on the plane along the same curve. This work presents a generalization of this result to the case of N bodies. Different curves, to be denoted as *simple choreographies*, have been found by a combination of different numerical methods. Some of them are given here, grouped in several families. The proofs of existence of these solutions and the classification turn out to be a delicate problem for the Newtonian potential, but an easier one in strong force potentials.

1. Introduction

The classical N-body describes the motion of N punctual masses under the action of Newton's gravitation law of attraction. Let $z_j \in \mathbb{R}^d$, $j = 1, \ldots, N$ the positions of the bodies and $m_j > 0$, $j = 1, \ldots, N$ the respective masses. In this work we shall consider the *planar problem $d = 2$*. The equations of motion are

$$\ddot{z}_j = \Sigma_{i=1,\, i \neq j}^{N} m_i(z_i - z_j) r_{i,j}^{-3},$$

$$(1)$$

where $r_{i,j} = |z_i - z_j|$, $|\ \ |$ being the Euclidean norm and where the gravitational constant is taken equal to 1. The system (1) has the trivial integrals of the center of mass: $\Sigma_{i=1}^{N} m_i z_i$ moves on a straight line with constant velocity. It is not restrictive to assume that the center of mass is kept fixed at the origin and we shall assume this from now on: $\Sigma_{i=1}^{N} m_i z_i = 0$. Furthermore we have the angular momentum, $c = \Sigma m_i z_i \wedge \dot{z}_i$, and the energy, $H = K - U$, first integrals. No more first integrals exist in general. Here K and $-U$ denote the kinetic and potential energy

$$K = \frac{1}{2} \Sigma_{i=1}^{N} m_i |\dot{z}_i|^2, \qquad U = \Sigma_{1 \leq i < j \leq N} m_i m_j r_{i,j}^{-1}.$$

$$(2)$$

Very few solutions are known for the general N-body problem. Some results are available for the 3-body, for some *restricted* problems, where some masses are infinitesimal, or for some very special subproblems.

Simple solutions can be obtained from *central configurations*. A central configuration is defined as a configuration of the N bodies such that $\ddot{z}_j = \lambda z_j$, $j = 1, \ldots, N$ for some $\lambda < 0$ independent of j. Then, if the velocities are properly chosen, that is, if $|\dot{z}_j| = \gamma|z_j|$ with the same γ for all j and the angle between $\dot{z}_j$ and z_j is the same for all the bodies, the motion of the N bodies takes place on conics, all the conics being similar. In particular every body can move on a circle around the common center of masses. In these solutions the motion behaves as if the bodies form a *rigid body*. These solutions are also denoted as *relative equilibrium solutions*, being a fixed point of (1) if we use a rotating frame. They can also be obtained using a different approach. Let $I = \Sigma_{j=1}^N m_j|z_j|^2$ be the moment of inertia of the N bodies around the center of mass. The set of configurations with a fixed value $I > 0$ defines a sphere $\mathcal{S}$ in $\mathbb{R}^{2N-2}$. We can restrict U to $\mathcal{S}$. Due to the homogeneous character of I and U we can always assume the value $I = 1$. Then the central configurations correspond to the critical points of $U|_{\mathcal{S}}$. The problem of finding the number of central configurations for a given N and how it depends on the masses is still an open question. See [7] for general results and [8] for a numerical study for $N = 4$ and arbitrary masses.

Now we consider the special case of all equal masses, taking $m_j = 1$, $j = 1, \ldots, N$. The simplest relative equilibrium is the *regular N-gon*. It is obvious that all the bodies move then on the same circle with a periodic motion. This suggests the following question:

Are there other *periodic solutions* of the N-body problem such that all the bodies *travel along the same path* in the plane?

This is the main topic of the present work. It has been prompted by the recent discovery by A. Chenciner and R. Montgomery [4] of one such solution for $N = 3$, the bodies moving on a figure eight curve,[1] and by a similar solution found by J. Gerver with four bodies [5]. For historical details about these quite new solutions I refer to [3]. Looking for solutions with N equal mass bodies on the same curve poses several problems: a) Existence proofs; b) The admissible geometries of the supporting curves; c) Computation of the solutions; d) Study of the dynamical properties; e) Generalizations to other potentials.

All these topics are strongly related. In fact existence proofs are available only for some class of potentials which excludes the Newtonian case. The difficulties with the Newtonian and other potentials are related to the possible existence of *collisions*, i.e., values t^* of time such that there exist i and j, $i \neq j$ such that $\lim_{t \to t^{*-}} r_{i,j}(t) = 0$. This, in turn, is related to the *admissible curves* and their *time parameterization*. The basic method for the proof relies on a *variational approach*. This is also useful, but not enough, for the computation. The study of the *geometrical and dynamical properties* of the solutions we are looking for requires local information on orbits which are *far away* from any curve which could be described by analytical means. So the numerical approach seems to be the only possible way.

---

[1]Added in proof: see also C. Moore, Braids in classical gravity, Phys. Rev. Lett. **70** (1993), 3675–3679, where this solution, found numerically, was displayed.

It is instructive to look at the motion of the N bodies along the same path on the plane by means of an animation. The bodies are seen to *dance* in a somewhat complicated way. This suggests the name *choreographies* to denote this kind of motion. To be precise we should name them as *simple* choreographies, because all the bodies are on the same curve. One can imagine also *multiple* choreographies, the bodies travelling on $k > 1$ different curves.

In Section 2 we give some results about the *figure eight* choreography for $N = 3$. Section 3 is devoted to introducing the required notation about simple choreographies. The variational approach is presented in Section 4. To this end we generalize the problem to potentials of the form r^{-a}, $a > 0$, instead of using only $a = 1$. The existence proof is sketched for $a \geq 2$, the case known as *strong force*. Section 5 shows different kinds of choreographies found up to now. The changes in the behavior of the choreographies as a function of a are displayed in Section 6. This illustrates, also, the difficulties to be faced in the proofs for *weak forces*. Finally Section 7 gives a short description of the numerical methods used.

2. The Figure Eight Solution

The first choreography for $N = 3$ was found by Lagrange in 1772: the celebrated equilateral triangular solution. It was only in December 1999 that the next one was found.[2] The three bodies travel on a figure eight curve (see Fig 1.1). The period has been fixed equal to 2π for the periodic solution represented here and for all the solutions in this work. This fixes the size. Other periods are scaled to 2π by Kepler's third law $\ell^3 T^{-2} = $ constant, where ℓ is the length scale and T the period. It is also clear that $\mathbb{S}^1$ invariance allows the curve to be symmetrical with respect to the horizontal axis. Denote as P_j the body located at z_j. One can take initially the bodies on a collinear (Eulerian) configuration, as in the figure, with P_3 between P_1 (on the left) and P_2, with P_3 moving upwards. For $t = \pi/6$ they are in isosceles configuration and for $t = \pi/3$ they are again in collinear configuration with P_2 in the middle. In a full period the bodies pass twice for each one of the three collinear configurations.

The proof of the existence (see [4]) involves a variational argument. Let $q(t)$ be a parameterization of the solution such that if $z_1(t) = q(t)$ then $z_2(t) = q(t - 2\pi/3)$, $z_3(t) = q(t - 4\pi/3)$. The variational formulation of classical mechanics assures that any classical 2π-periodic solution is an extremal of the *action functional*

$$A = \int_0^{2\pi} L(t)dt, \quad L(t) = K(\dot{z}_1, \dot{z}_2, \dot{z}_3) + U(z_1, z_2, z_3), \tag{3}$$

the integral of the Lagrangian L. In fact the figure eight choreography is a minimum of A. The key point is to show that a minimization of A, inside the desired class of topological figure eight curves, does not lead to collisions. Using estimates in

---

[2]See the previous footnote.

[2] and a test path, the level curve of $U|_{\mathcal{S}}$ passing through the collinear points and traveling with constant speed with a suitable value of I, the exclusion of collisions is reduced to the evaluation of an integral.

The eight can be seen also as *a chain with two links*. Fig. 1 shows also the next simplest *chains* (3- and 4-chains) with four and five bodies. For completeness, initial data allowing the reader to reproduce the plots are given in Table 1. They have been rounded to 10^{-6} and, hence, should be refined for accurate purposes. Note that for the eight one has to rotate slightly Fig. 1.1 to agree with the initial conditions. The components of z_j are denoted as (x_j, y_j).

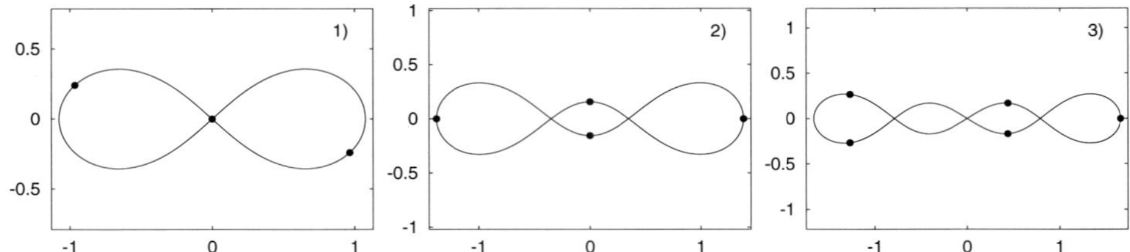

FIGURE 1. Chains with three, four and five bodies.

$N=3$	$x_2 = 0.995492$	$\dot{x}_3 = 0.695804$	$\dot{y}_3 = 1.067860$	$x_1 = -x_2,\ \dot{z}_1 = \dot{z}_2$
$N=4$	$x_1 = 1.382857$	$\dot{y}_1 = 0.584873$	$y_2 = 0.157030$	$\dot{x}_2 = 1.871935$
	$x_3 = -x_1$	$y_4 = -y_2$	$\dot{y}_3 = -\dot{y}_1$	$\dot{x}_4 = -\dot{x}_2$
$N=5$	$x_2 = 0.439775$	$y_2 = -0.169717$	$\dot{x}_2 = 1.822785$	$\dot{y}_2 = 0.128248$
	$x_3 = -1.268608$	$y_3 = -0.267651$	$\dot{x}_3 = 1.271564$	$\dot{y}_3 = 0.168645$
	$x_4 = x_3,\ y_4 = -y_3$	$\dot{x}_4 = -\dot{x}_3,\ \dot{y}_4 = \dot{y}_3$	$x_5 = x_2,\ y_5 = -y_2$	$\dot{x}_5 = -\dot{x}_2,\ \dot{y}_5 = \dot{y}_2$

TABLE 1. Initial conditions for Fig. 1. The data not given, and not following from the center of mass condition, are zero.

This solution has many remarkable properties. Beyond obvious symmetries,

- It seems that it is unique, leaving aside rotations. This is not proved in [4]. It is based on an extensive numerical search with zero angular momentum and $\dot{z}_1 = \dot{z}_2$.[3]
- The eight lives on $c = 0$, the zero angular momentum level. When $c \neq 0$ it is possible to use a rotating frame (with frequency ω) and to look for figure eight periodic solutions in this frame. This has been done by M. Hénon [6]. The loops become asymmetrical, but the general pattern is similar. These rotating solutions give rise to 2D tori if $\omega \notin \mathbb{Q}$, while they give new *satellite* choreographies if $\omega \in \mathbb{Q}$.

---

[3] Added in proof: The search is restricted to this simple topology. For other more complicated topologies see C. Simó, Dynamical properties of the figure eight solution of the three-body problem, in Proceed. of the Chicago Conf. on Cel. Mechanics dedicated to Don Saari, ed. A. Chenciner, R. Cushman, C. Robinson and Z. Xia, to appear in Contemp. Math. AMS.

- The orbit is *linearly stable*. It can be seen as a fixed point of a Poincaré map. The eigenvalues of the differential of that map are $\lambda = \exp(\pm 2\pi i \nu_j)$, with $\nu_1 = 0.00842272$, $\nu_2 = 0.29809253$. This comes as a surprise, because of the contrast with systems with two degrees of freedom, for which periodic orbits minimizing the action are unstable. Quoting Birkhoff ([1] p. 130) *"Doubtless analogous results hold for any number of degrees of freedom, and can be obtained by means of classical methods in the calculus of variations."*

- To obtain non-linear information a representation of the Poincaré map to high order is needed. Some methods are proposed in [9], by using higher order variationals. A different approach is given in Section 7. In particular this allows us to obtain the *Normal Form* around the periodic solution. The torsion matrix is indefinite and KAM theorem is applied to show the existence of invariant tori. Hence, most of the points close to the eight on $c = 0$ are stable. Furthermore, from the variation of the actions follows the existence of other classes of *satellite* choreographies, associated to periodic points, with suitable period, of the Poincaré map around the fixed point.

- It is possible to continue the periodic solution to other nearby masses, each moving then in a slightly different "eight". But stability is preserved only for relative variations of the order of 10^{-5}.

- Keeping equal masses we can explore different potentials of the form

$$U(z_1, \ldots, z_N) = \Sigma_{1 \le i < j \le N} f(r_{i,j}), \quad f(r) = r^{-a}, \quad a > 0. \tag{4}$$

The eight can be continued to all $a > 0$ and even to the limit case $f(r) = \log r$ and beyond. However it is found to be linearly stable only in a short range around $a = 1$.

3. Choreographies

We pass to $N > 3$. We look for 2π-periodic functions $q \colon \mathbb{S}^1 \mapsto \mathbb{R}^2$ such that if

$$z_j(t) = q(t - (j-1)2\pi/N), \; j = 1, \ldots, N, \tag{5}$$

we find a solution to (1). $\mathbb{Z}/N\mathbb{Z}$ acts on the set of bodies and in $\mathbb{S}^1$ by shifting to the next body (or to the next vertex of an N-gon). This can be used for theoretical and computational purposes. Note that $q(t)$ satisfies a differential equation with delays multiple of $2\pi/N$. This remark does not seem to reduce the difficulty.

A collision occurs if there exists a *double point* $q(t_1) = q(t_2)$ with $t_2 - t_1$ multiple of $2\pi/N$. We consider the class of collision-free functions. It has to be taken analytical (the potentials being analytical if $r \ne 0$), despite that for the variational approach it is enough to consider the Sobolev space $H^1(\mathbb{S}^1, \mathbb{R}^2)$ (or H^1 for brevity) of functions with square integrable first derivative. Let $\Delta \subset H^1$ be the functions associated to collisions. We would like to see that in each connected component of $H^1 \setminus \Delta$ there is a solution minimizing the action. Unfortunately this seems not to be true for the Newtonian potential.

To begin with we give in Fig. 2 most of the choreographies known to the present for five bodies, beyond the regular pentagon and the 4-chain of Fig. 1.3. They are numbered by increasing value of the action. The pentagon has an action less than all other choreographies and the 4-chain is located between cases 4 and 5 of Fig. 2. All of them have some symmetry.

Most of them can be seen as *linear chains* having loops of different size, with some of the loops eventually *folded*. Number 1 consists of a large loop and a small one. In the small loop there are either one or two bodies for all t.

case	action	momentum	energy	$\min r_{i,j}(t)$	$\max \lambda_L$
circle	58.308755	−6.186751	−3.093376	1.307660	0.939150
1	68.851604	−3.454467	−3.652691	0.364076	0.999344
2	71.331244	0.000000	−3.784240	0.690443	1.225034
3	75.184575	2.462827	−3.988666	0.213061	1.573730
4	77.158798	0.879793	−4.093401	0.423223	1.323903
4-chain	80.366525	0.000000	−4.263577	0.339434	1.727383
5	85.474715	2.631679	−4.534574	0.326152	1.639335
6	86.051192	−0.664730	−4.565158	0.399373	1.819673
7	88.439746	−2.562017	−4.691874	0.389835	1.757461
8	89.255582	−0.333911	−4.735156	0.236213	2.544799
9	90.108332	0.646869	−4.780395	0.381350	2.362498
10	93.864859	−1.839421	−4.979685	0.212068	2.189505
11	96.798960	−0.207731	−5.135344	0.280668	2.196531
12	102.751489	−0.484014	−5.451136	0.207233	3.381876
13	103.201740	0.000000	−5.475022	0.246947	3.001603
14	105.954031	−1.753301	−5.621036	0.266926	2.272570
15	108.787904	2.243734	−5.771378	0.073286	2.730382
16	109.636187	0.000000	−5.816380	0.355173	2.417178
17	109.882868	−4.476957	−5.829467	0.447649	2.628803
18	119.318405	2.002793	−6.330038	0.311516	3.177711

TABLE 2. Numerical data for 5-body choreographies. $\min r_{i,j}(t)$ is taken over all $i \neq j$, $t \in \mathbb{S}^1$. λ_L means Lyapunov exponent.

Definition 3.1. *Given a double point being the image of t_1 and t_2 by q, the images of the two arcs going from t_1 to t_2 in $\mathbb{S}^1$ are denoted as* loops associated to the point. *Assume $0 \leq t_1 < t_2 < 2\pi$. The lengths of these loops are $\ell = (t_2 - t_1)N/(2\pi)$ and the complement $\ell_c = N - \ell$. This extends in a simple way to multiple points.*

A key role is played by the *integer lengths* $[\ell]$ and $[\ell_c]$, where $[\ ]$ denotes the integer part. As for a collision-free function $[\ell] + [\ell_c] = N - 1$, we usually refer to the minimum between $[\ell]$ and $[\ell_c]$ as the *integer length* associated to the point. It is clear that if we deform q without passing through collisions, the integer length cannot change. But *small loops* of length less than 1 (integer length zero) can be created/destroyed without problem. Also two nearby double points on $q(\mathbb{S}^1)$ can collapse by deformation of q to the same point, if this one has $\ell \notin \mathbb{N}$, and the points can disappear. In a similar way new loops can be created.

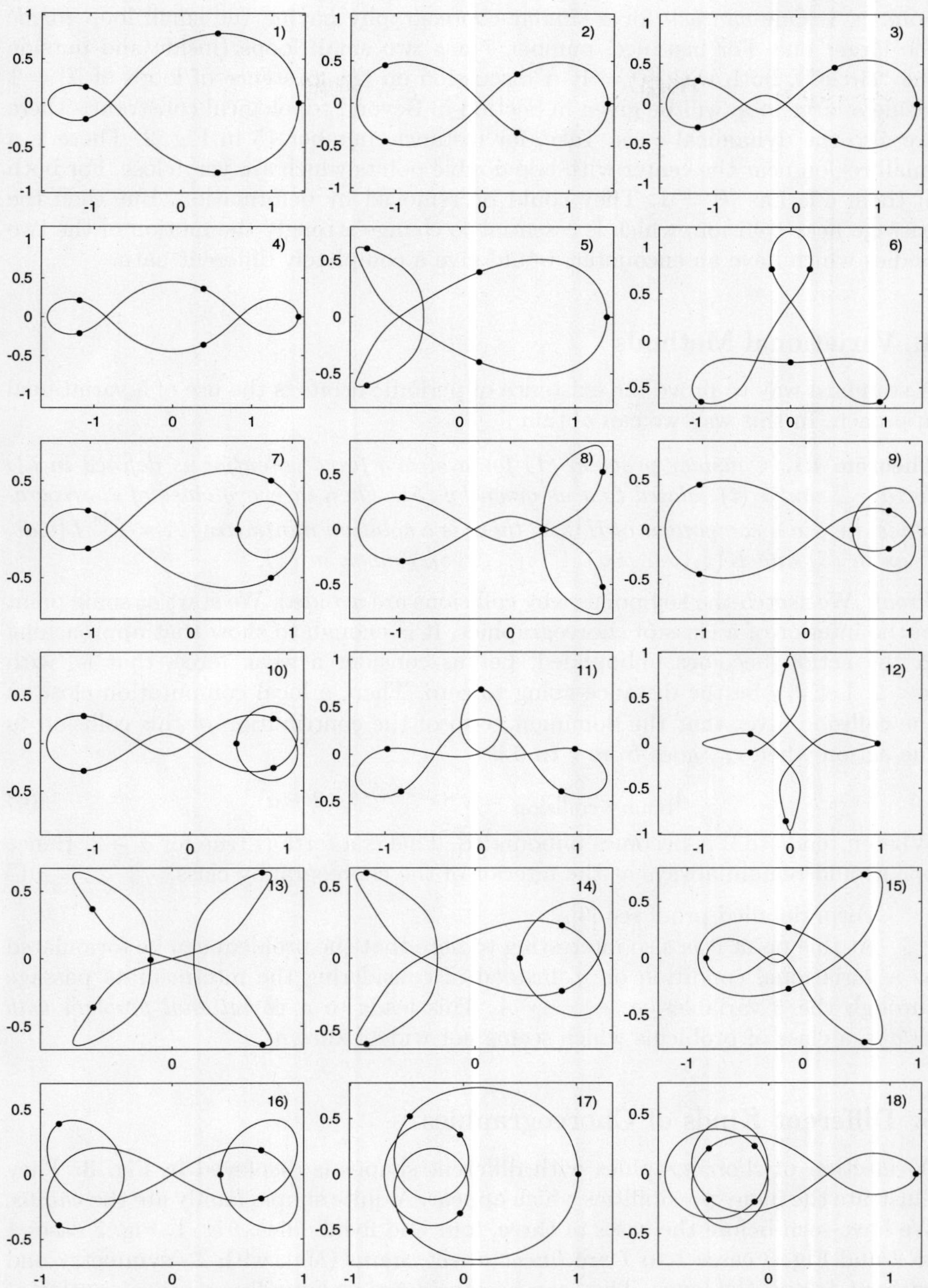

FIGURE 2. Choreographies found for five bodies. The dots denote initial conditions.

Going back to number 1 in Fig. 2, the integer length associated to the double point is 1. One can ask for a similar choreography having the small loop *inside* the larger one. For instance, number 7 has two small loops (inside and outside the "circle"), both with $[\ell] = 1$. A discussion on the existence of loops of $[\ell] = 1$ inside a large loop will be given in Section 6. Beyond topological constraints there are also the dynamical ones. Take, for instance, number 15 in Fig. 2. There is a small region near the center with two double points which are very close. For both of them one has $[\ell] = 1$. They could be removed by deformation. But then the passage near collision, which is essential to change strongly the motion of the two bodies which have an encounter, would give a completely different path.

4. Variational Methods

A standard way to prove the existence of periodic orbits is the use of a variational approach. In this way we can obtain

Theorem 4.1. *Consider problem (1) for a strong force potential as defined in (4) for $a \geq 2$ and $z_j(t)$ related to q as given by (5). Then in every class of choreographies, i.e., in a component of $H^1 \backslash \Delta$, there is a solution minimizing $A = \int_0^{2\pi} L(t)dt$, L as in (3) and $K(\dot{z}_1, \dots , \dot{z}_N)$, $U(z_1, \dots , z_N)$ given in (2).*

Proof. We sketch the key point: why collisions are avoided. We start at some point in the interior of a class of choreographies. It is enough to show that approaching Δ the action becomes unbounded. Let us consider a weak force, that is, with $a < 2$. Let $r_{i,j}$ be the distance going to zero. Then, a local computation close to the collision gives that the dominant term of the contribution of this collision to the action when $r_{i,j}$ goes from r to 0 is

$$A_{\text{binary collision}} = \sqrt{8}\, r^{(2-a)/2}/(2 - a). \tag{6}$$

When a tends to 2 it becomes unbounded. This is a fortiori true for $a \geq 2$. Hence one should remain always at the interior of the choreography class. $\quad\square$

For a detailed proof see [3].

At this point it is also interesting to note that the problem can be formulated as a variational condition on q, instead of considering the intermediate passage through the z variables: $q \rightarrow z \rightarrow A$. This leads to a *variational problem with delays*, a class of problems which seems not widely known.

5. Different Kinds of Choreographies

A selection of choreographies with different shapes is displayed in Fig. 3. They illustrate the many possibilities which appear. A quite simple family are the *chains*. We have seen before the cases of three, four and five bodies. Fig. 1, Fig. 2 cases 1 to 4 and Fig. 3 cases 1 to 7 are *linear direct chains* (*ldc*), with $\mathbb{Z}_2$ symmetry and without *folding* the loops. They can be considered as bounding a concatenation of k topological discs. Each piece of the chain between two consecutive double points

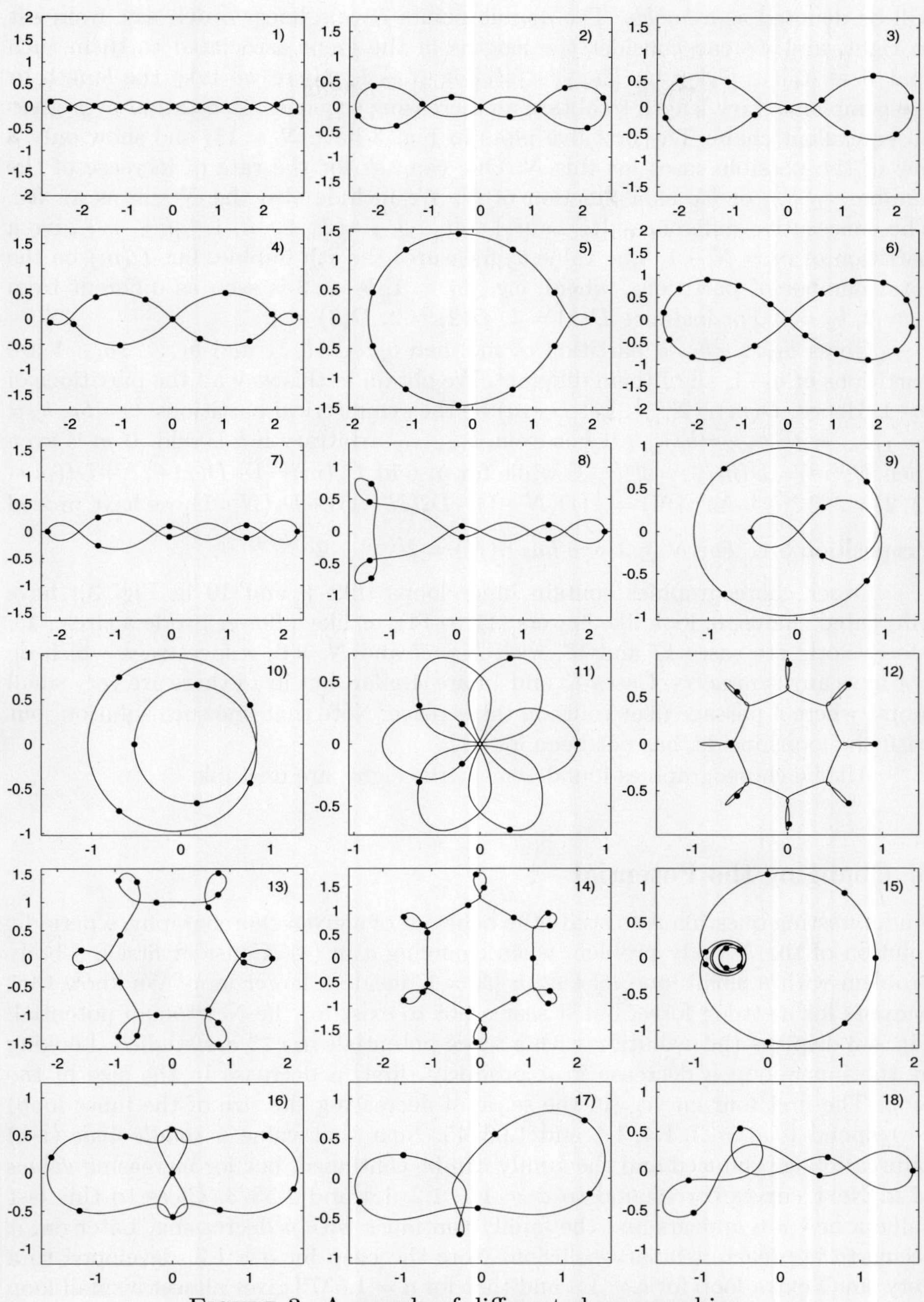

FIGURE 3. A sample of different choreographies.

will be denoted as a *bubble*. The double points have a linear order, say, from left to right, and we can consider the lengths of the loops associated to them such that $0 \equiv [\ell_0] < [\ell_1] < \ldots [\ell_{k-1}] < [\ell_k] \equiv N - 1$, where we take the length or the complementary length to obtain an increasing sequence. A rotation by π gives an equivalent chain. The first five plots in Fig. 3 have $N = 11$, and show only a few of the possible cases for this N. One can ask for the rate of increase of the number, $\gamma(N)$, of *ldc* as a function of N. We include also the N-gon as an *ldc*. The consecutive differences $\{b_j := [\ell_j] - [\ell_{j-1}], \, j = 1, \ldots, k, \, 1 \le k \le m\}$ give a partition of $m = N - 1$. The value b_j measures the jth bubble. Let $D(m)$ be the total number of partitions, where, e.g., $b_1 = 1$, $b_2 = 2$ is seen as different from $b_1 = 2$, $b_2 = 1$. For instance $D(1) = 1$, $D(2) = 2$, $D(3) = 4$.

Given $b_1, \ldots, b_k$, a partition of m, then $b_1, \ldots, b_k, 1$ and $b_1, \ldots, b_k + 1$ are partitions of $m+1$, all of them different. We obtain in this way all the partitions of $m+1$. Hence, $D(m) = 2^{m-1}$. Let $D_s(m)$ be the symmetrical partitions: $b_1 = b_k$, $b_2 = b_{k-1}, \ldots$ with $b_{k/2} = b_{k/2+1}$ if k is even, $b_{(k+1)/2}$ arbitrary if k is odd. If m is even then $D_s(m) = D(m/2) = 2^{m/2-1}$, while for m odd $D_s(m) = 1 + D(1) + \ldots + D((m-1)/2) = 2^{(m-1)/2}$. As $\gamma(N) = \frac{1}{2}(D(N-1) - D_s(N-1)) + D_s(N-1)$, we have proved

Proposition 5.1. *For $N \ge 3$ one has* $\gamma(N) = 2^{N-3} + 2^{[(N-3)/2]}$.

Other choreographies contain inner loops (like 9 and 10 in Fig. 3); have bifurcated chains, 8; look like flowers (11 to 14), or like a flower inside a circle, 15. More exotic are cases 17 and 18, with $N = 7$ and $N = 6$, respectively, which do not have any symmetry. Cases 13 and 14 are similar, but in 14 there are very small loops, where a passage near collision takes place. Note that one such solution, but with the loops inside, has not been found.

All the choreographies found, except the eight, are unstable.

6. Changing the Potential

An interesting question is to study the behavior of a given choreography, a periodic solution of the N-body problem when changing a in (4). Consider first a 4-body problem with a small loop, of length $[\ell] = 1$, inside a larger loop. We know that it exists for a strong force, but it seems not to exist for the Newtonian potential. Fig. 4.3 displays the evolution with a when potentials in r^{-a} are studied. Looking at the inner loop a decrease in a produces, first, a decrease in the size of the loop. The first four curves (in the sense of decreasing the size of the inner loop) correspond to $a = 2, 1.4, 1.1$ and 1.03445. Near that value a *saddle-node* (*s-n*) bifurcation is produced and the family can be continued, but for increasing values of a. Next curves correspond to $a = 1.1, 1.2, 1.4$ and 1.5373. Close to this last value a new *s-n* appears and the family continues with a decreasing. Later on, it seems to approach a binary collision. Note the cusp for $a \simeq 1.2$, developed to a very small extra loop for $a = 1.4$ and that for $a = 1.5373$ gives almost a small loop travelled twice. Hence, it seems that this solution is unable to reach $a = 1$.

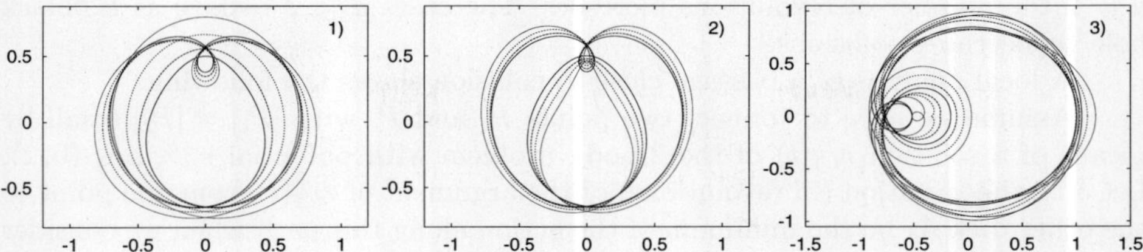

FIGURE 4. Several families for different potentials.

On the other hand, consider also $N = 4$ as displayed in Fig. 4.1. Looking at the small loop in the upper part, the values $a = 2, 1.4, 1.1$ and 1 have been used. Decreasing a the size of that loop decreases. The family can be continued by decreasing a again as shown in Fig. 4.2, but for $a \simeq 0.98267$ it has an s-n and the continuation requires an increase of a. The values shown are $a = 1, 0.98267, 1$ and 1.05. In particular there are two very close solutions for $a = 1$ and the solution for $a = 1.05$ is extremely close to a collision.

Therefore, any proof of existence or non-existence of a particular kind of solution, should be able to distinguish between very close values of a with different properties.

All the solutions displayed in Fig. 4 are minima of the action, but continuation has produced saddles for $N = 5$ (see [3]). An interesting question is to study what happens to a path giving a local minimum when we travel it several times. In particular we can consider a regular N-gon. The simplest example I have found appears for $N = 7$ travelled twice. Hence we consider the same path with period 4π instead of 2π. For $a = 1$ (the Newtonian case) it is no longer a local minimum. But a nearby orbit (Fig. 3.10), with two slightly different loops, is a local minimum. This subsists for a large range of values of a which contains $[1, 2]$ and values beyond $a = 2$. However, decreasing a there is again an s-n bifurcation close to $a = -0.3046$ and the continuation of the family requires to increase a.

All these difficulties are related to passage close to collisions. Assume that $r_{i,j}$ becomes small. Then it is better to study the relative motion, that is, the behavior of $r_{i,j}(t)$ (or $r(t)$ for short) and the contribution of that passage to the action. For simplicity we assume that all the other bodies are at finite distance. Going to a collision $r(t) = \alpha t^\beta \times (1 + o(1))$, where $\beta = 2/(2+a)$, α depending also on a, and the contribution to the action, going from $r = r^*$ to $r = 0$ is given in (6). In the Newtonian case, after passing close to a collision the two bodies (moving near a degenerate ellipse) essentially "bounce". This can be seen, using blow up, as a change by 2π in the argument of $r(t)$. For a general $0 < a < 2$ the variation of the argument is $1/(2 - a)$ revolutions. This shows why for $a = 2 - 1/n$, $n \in \mathbb{N}$ the collision can be regularized by bouncing and for $a = 2 - 2/(2n + 1)\, n \in \mathbb{N}$ by "crossing" through the collision. In the limit the bodies give an integer or half-integer number of revolutions around the common center of mass. Approaching

$a = 2$ the number of revolutions increases. The cases $a \geq 2$ behave as a "black hole" concerning collisions.

A local analysis of a passage close to collision shows the following:

Assume we have to connect two points P_1 and P_2 with $|P_1| = |P_2|$ small by means of a solution $r_{i,j}(t)$ of the 2-body problem with potential r^{-a}, $a \in (0, 2)$. Let δ be the variation (in revolutions) of the argument of $r_{i,j}(t)$ from one point to the other. Let A_δ be the minimum of the action along this path when we consider all values of the energy. We can also connect these points going from P_1 to collision and then to P_2 by ejection, matching two solutions of the problem. Then the action takes the value $A_c = 2A_{\text{binary collision}}$ as given in (6) with $r = |P_1|$ in this formula.

Proposition 6.1. $A_c < A_\delta$ *if and only if* $\delta > 1/(2 - a)$.

That is, the passage through collision is cheaper in terms of action only for large variations of the argument of r. Furthermore, if $\delta < 1/(2 - a)$ the value A_δ is achieved for $h = 0$.

This seems to explain why a small loop inside a large one cannot exist for the Newtonian potential, as shown in Fig. 4.3. The shape of the curve forces a variation of the argument larger than admissible without collisions. Of course, a complete proof requires estimates on the effect of all other bodies. It also suggests

Conjecture 6.2. *All linear direct chains exist for* $a = 1$.

There are some special chains. For N odd we can look for a symmetric eight shape curve having initially one body at the origin. An example is Fig. 3.8 with $N = 19$. Another special case is a chain with $\ell_j = j$, $j = 1, \ldots, N - 2$. In both examples it seems that the curves tend (after suitable scalings) to some limit.

7. Numerical Methods

We are faced with several problems: approximation of a choreography, refinement, continuation with varying exponent a and also the computation of good approximations of the Poincaré map around a periodic solution. In fact the last topic has been applied only to the eight with $N = 3$. The tools for continuation are standard. I refer to [9] for the description, applications and analysis of the bifurcations.

7.1. Implementing the variational method

The function q can be approximated by a function $\hat{q}$, either a trigonometrical polynomial, the values at a set of equi-spaced points or some other method. Let us denote as $\mathcal{P} \in \mathbb{R}^M$ the finite set of parameters needed for the approximation. Then the action A as in (3) is approximated by a *discretized* map $\hat{A} \colon \mathbb{R}^M \to \mathbb{R}$, where the integral is computed by a numerical quadrature formula. Due to the action of $\mathbb{Z}/N\mathbb{Z}$ (which must be preserved by the discretization), it is enough to do the integrals from 0 to $2\pi/N$. Taking into account that collision-free solutions are analytical, the trapezoid rule is highly convenient.

Then we proceed to minimize $\hat{A}$ using a combination of variants of the gradient method. The method has several

- Advantages: It is quite robust, and one can start from very rough approximations, like a few harmonics or a hand drawn curve. It is easy to program, to use any potential, etc. Furthermore the gradient can be obtained from the set $\mathcal{P}$ without need of numerical differentiation. It allows for checks, looking for the invariance of the energy and for the value of the *residual acceleration*: the difference between the acceleration of the masses computed from $\hat{q}$ and using (1).

- Inconveniences: Except in quite simple cases M has to be selected large, especially if there are passages not too far from collision. Typical values of M range in $[10^3, 10^4]$. Furthermore $\hat{A}$ is a very flat function, with lots of extrema. The number of iterations to achieve a good approximation is also in $[10^3, 10^4]$. This slows down the process.

The method detects clearly passages close to saddle points of $\hat{A}$. This can be used in the future to try to locate these solutions. An alternative approach is to implement numerically the *mountain pass lemma*, by following an arc between two local minima under the gradient flow.

The variational technique does not provide direct information on the stability properties of the periodic solutions.

7.2. Refinement of the solutions

Having some starting point, either provided by the variational method or from a solution for a nearby value of the exponent a, we can try to use a Newton method. Consider the values of z_j, $j = 1, \ldots, N$. The time $2\pi/N$ flow transports body j to body $j - 1$, where the indices have to be considered in $\mathbb{Z}/N\mathbb{Z}$. This can be converted immediately to the search for a solution of the equation $G(Z) := \Phi(2\pi/N, Z) - S(Z) = 0$, where Z denotes all the z_j and $\dot{z}_j$, Φ is the flow of (1) and S is the shift of indices.

To solve $G(Z) = 0$ by Newton's method one requires $D_Z G(Z)$. This is obtained by simultaneous integration of the system of ODE given by (1) and the first variational equations. This can be cumbersome if N is large. For instance, $N = 100$ leads, essentially, to a system of dimension $160,000$. The use of parallel computers allows a dramatic reduction in the computing time, because the first variational equations can be integrated by blocks.

One of the problems is that many solutions are quite unstable for N large and suffer from passages close to collision. Also the basin of convergence of the method can be small due to the existence of many nearby solutions. This can be solved by looking not only for the value of Z at time $t = 0$ but also at some intermediate values $0 = t_0 < t_1 < \ldots < t_{k-1} < t_k = 2\pi/N$. Let Z_m be the value at t_m. Then, instead of the equation $G = 0$, one has more equations, by requiring $\Phi(t_{m+1}; t_m, Z_m) - Z_{m+1} = 0$. This is the well-known *parallel shooting method* (see, e.g., [10]). The system has increased size, but the equations do not require additional effort and are better conditioned.

As a last comment, there is some freedom in the choice of initial time and the system is invariant by rotation. As presented here the method will fail because any solution has a related 6D family of solutions. So $D_Z G(Z)$ will be singular. This difficulty is avoided by selecting, at $t = 0$, some of the bodies to be on the x axis with $\dot{x} = 0$ and by using the center of mass reduction.

This method allows for very accurate solutions, can be implemented with arithmetics of large numbers of digits, in case of need, and, as the variational equations are solved simultaneously, allows us to obtain the stability properties as a by-product.

7.3. Computing the Poincaré map around a periodic solution

I want only to comment in passing that a routine providing an accurate computation of the Poincaré map (using suitable arithmetics and a high order Taylor recurrent integration scheme) allows us to obtain the coefficients of the polynomial expansion of the Poincaré map around the fixed point. It is enough to compute the coefficients by numerical differentiation, using a suitable number of points for the higher order derivatives. This procedure is also highly parallelizable. It is essential to select the optimal step size for the differentiation, which depends on the order.

Acknowledgements

This work has been possible thanks to the information that A. Chenciner, R. Montgomery and J. Gerver shared with me, and to the multiple discussions we are having on the topic. I am also indebted to M. Hénon, R. Martínez and R. Ortega for many useful suggestions. Most of the work was done during a sabbatical leave that I spent at the Institute de Mécanique Celeste (Observatoire de Paris), thanks to the support of CNRS. To all my colleagues at that institution my recognition for the hospitality and interest in the work. The parallel computing facilities of the UB Grup de Sistemes Dinàmics have been widely used. The support of grants DGICYT PB 94-0215 (Spain) and CIRIT 1998SGR-00042 is also acknowledged.

References

[1] G. D. Birkhoff, *Dynamical Systems*, Amer. Math. Soc., 1927.

[2] A. Chenciner, N. Desolneux, *Minima de l'intégrale d'action et équilibres relatifs de n corps*, C.R.A.S. Paris, **326**, Série I (1998), 1209–1212. Correction in C.R.A.S. Paris, **327**, Série I (1998), 193.

[3] A. Chenciner, J. Gerver, R. Montgomery, C. Simó, *Simple choreographic motions of N bodies with strong forces*, to appear in *Geometry, Mechanics, and Dynamics*, Springer-Verlag.

[4] A. Chenciner, R. Montgomery, *A remarkable periodic solution of the three body problem in the case of equal masses,* Annals of Mathematics, **152** (2000), 881–901.

[5] J. Gerver, private communication, (2000).

[6] M. Hénon, private communication, (2000).

[7] R. Moeckel, *On central configurations*, Math. Zeit., **205** (1990), 499–517.

[8] C. Simó, *Relative equilibrium solutions in the four-body problem*, Cel. Mechanics, **18** (1978), 165–184.

[9] C. Simó, *Analytical and numerical computation of invariant manifolds*. In D. Benest et C. Froeschlé, editors, *Modern methods in celestial mechanics*, 285–330, Editions Frontières, 1990.

[10] J. Stoer, R. Bulirsch, *Introduction to Numerical Analysis*, Springer-Verlag, 1983 (second printing).

Departament de Matemàtica Aplicada i Anàlisi, Universitat de Barcelona
Gran Via, 585
08007 Barcelona, Spain [4]
E-mail address: `carles@maia.ub.es`

[4]Temporary address: Astronomie et Systèmes Dynamiques, IMC-CNRS UMR 8028, 77 av Denfert-Rochereau, 75014 Paris.

Irreducible Modular Representations of a Reductive p-Adic Group and Simple Modules for Hecke Algebras

Marie-France Vignéras

Abstract. Let F be a local non-archimedean field of residual characteristic p, and let G be the group of rational points of a connected reductive group defined over F. The two main points in the search for a classification of the irreducible complex representations of G is to try to prove that any irreducible cuspidal representation is induced from an open compact subgroup and that the irreducible representations with a given inertial cuspidal support are classified by simple modules for the Hecke algebra of a type. Over a field R of characteristic $\neq p$ which is not the complex field, new serious difficulties arise and the purpose of this article is to indicate a way to avoid them. The mirabolic trick used when the group is $GL(n, F)$ does not generalize, but our new method is general and we can extend from the complex case to R the results of Morris and Moy–Prasad for level 0 representations.

1. Introduction

Let G be the group of rational points of a connected reductive group over a local non-archimedean field with finite residual field of characteristic p. Let R be a field, and let us first assume that the characteristic of R is 0 and that R is algebraically closed. The classification of smooth irreducible R-representations of G is achieved by solving the two following problems:

(a) Classification of cuspidal irreducible representations.

(b) Determination of the irreducible subquotients of the representations obtained by parabolic induction of the cuspidal representations from proper Levi subgroups.

Following the pioneer work of Roger Howe on the theory of types, the problems (a) and (b) are reduced to three other problems:

(c) Determination of a list of pairs (P, σ) of smooth irreducible representations σ of open compact subgroups P of G, called types. (Roughly speaking, each irreducible representation of G should be parametrized by a triple (P, σ, Φ) where (P, σ) is a type and Φ is a simple module of the endomorphism algebra of the compact induction of (P, σ) to G. A cuspidal irreducible representation of G

containing (P, σ) should be compactly induced from the normalizer $N_G(P)$ of P in G. These are the first important properties of the types [1].)

(d) Determination of the structure of the endomorphism algebra of a representation induced by compact induction of a type.

(e) Determination of the simple modules of these algebras.

These problems are completely solved only for the linear groups $G = GL(n, F)$ after the efforts of many mathematicians; (b) which implies (e) was solved by Zelevinski and later (c), (d) were solved by Bushnell and Kutzko.

For other groups, the problem (c) is solved only for irreducible representations of level 0. This results from the work of Morris, and also of Moy and Prasad. The structure of the endomorphism algebra of $\mathrm{ind}_P^G \sigma$ has been described by Morris; it is a product of algebras very close to affine Hecke algebras. In some cases the simple modules have been described by Morris and Lusztig. For representations of level > 0 of a general group, Moy and Prasad have described canonical unrefined types with very interesting properties. Partial results have recently been obtained by J. Adler, L. Blasco, C. Blondel, L. Corwin, J.-L. Kim, B. Lemaire, A. Moy, G. Prasad, A. Roche, A. Silberger, J.-K. Yu, T. Zink. The applications of the theory of types to the theory of characters and more generally to harmonic analysis are already numerous.

Here we are interested in the case where R is a field and the characteristic of R is 0 or a prime ℓ different from the residual characteristic p of F. Our philosophy is that the solution of (c) should be valid over R, when there is one. This is the case for the linear groups $GL(n, F)$ [10, 11]. Many new difficulties arise especially in the non-banal case, when ℓ divides the pro-order of an open compact subgroup of G. For instance the trivial representation of an Iwahori subgroup of G will be a type, but is not a type in the sense of Bushnell–Kutzko because the category of R-representations of G generated by their Iwahori fixed vectors is not abelian. The theory of types is particularly suitable to the study of R-representations, and some deep applications have already been given by Dat.

This article gives the solution of (c) for irreducible R-representations of level 0 of a general group G. The description of the endomorphism algebra of $\mathrm{ind}_P^G \sigma$ given by Morris can be extended to a general R which contains a square root of p, as in the case where G is a reductive finite group, P is a parabolic subgroup and σ is the inflation of an irreducible cuspidal representation of a Levi subgroup.

Let P be a parahoric subgroup of G. There is an exact sequence

$$1 \to U \to P \to M \to 1$$

---

[1]A pair (P, σ) is a type in the sense of Bushnell–Kutzko if the category $\mathcal{C}$ of representations generated by their (P, σ)-isotypic part satisfies any of the equivalent properties:

a) $\mathcal{C}$ is stable under subquotients,

b) $\mathcal{C}$ is a finite sum of blocs of the category of all R-representations of G,

c) $\mathcal{C}$ is equivalent to the category of right modules of the endomorphism algebra of the compact induction of (P, σ) to G.

where U is the pro-p-radical of P, M is the group of rational points of a connected reductive group over a finite field of characteristic p, and σ is a cuspidal irreducible R-representation of M inflated to P. The pairs (P, σ) constructed in this way are the types of level 0.

Let (P, σ) be an R-type of level 0. The normalizer $N_G(P)$ of P in G normalizes U and there is an exact sequence

$$1 \to U \to N_G(P) \to M^+ \to 1$$

for a finite mod center group M^+ which contains M. When P is maximal, let Λ be a smooth irreducible R-representation of $N_G(P)$ which contains σ. We will prove:

Theorem 1.1.

A) *The smooth irreducible R-representations π of G containing σ are in bijection with the simple modules of $\mathrm{End}_{RG}\mathrm{ind}_P^G\sigma$ by the functor $\pi \to \mathrm{Hom}_{RG}(\mathrm{ind}_P^G\sigma, \pi)$.*

B) $\mathrm{ind}_{N_G(P)}^G \Lambda$ *is irreducible when P is maximal and R is algebraically closed.*

Theorem 1.1 completed with the same methods as in the complex case gives:

Theorem 1.2. *Let π be any irreducible R-representation of G of level 0. Then there is a standard R type (P, σ) of level 0 contained in π unique modulo association, and a unique simple right module Φ for $\mathrm{End}_{RG}\,\mathrm{ind}_{RG}\,\sigma$ (modulo isomorphism), such that $\mathrm{Hom}_P(\sigma, \pi) \simeq \Phi$.*

The representation π is cuspidal, if and only if P is maximal. Then $\pi \simeq \mathrm{ind}_{N_G(P)}^G \Lambda$ as in Theorem 1.1 if R is algebraically closed.

Hence the solution of the problem (c) is the same over any field of characteristic $\neq p$.

Let us discuss the proof of Theorem 1.1.

A) is well known and easy when the characteristic ℓ of R does not divide the pro-order of P, because the representation $\mathrm{ind}_P^G \sigma$ is projective. Following the work of Dipper, it is noticed in [11] that the property A) remains true if $\mathrm{ind}_P^G \sigma$ is only almost projective, and this property was proved for a Bushnell–Kutzko R-type by a complicated argument.

B) is also well known and easy when ℓ does not divide the pro-order of the quotient of $N_G(P)$ by the center of G, and $\mathrm{End}_{RG}\,\mathrm{ind}_{N_G(P)}^G \sigma \simeq R$. The property B) was proved for a maximal Bushnell–Kutzko type [10] using a property of cuspidal irreducible representations specific to $GL(n, F)$.

There is in fact a general and simple argument to prove the theorem, valid also when (P, σ) is replaced by a simple Bushnell–Kutzko type. It relies on a basic simple result that was surprisingly not noticed yet. This is the following: for two standard parahoric subgroups P, Q of G with exact sequences

$$1 \to U \to P \to M \to 1, \quad 1 \to V \to Q \to L \to 1 \tag{1}$$

120   M.-F. Vignéras

the functor *"parahoric induction* (inflation from M to P then compact induction from P to G) *followed by parahoric restriction* (left adjoint of the parahoric induction)"

$$T^G_{L,M}\colon \operatorname{Mod} M \to \operatorname{Mod} P \to \operatorname{Mod} G \to \operatorname{Mod} Q \to \operatorname{Mod} L$$

is an (infinite) direct sum of functors "parabolic restriction followed by conjugation and by parahoric induction" (5.3.1); this is similar to the decomposition of the functor "parabolic induction followed by parabolic restriction" for representations of a finite reductive group. A similar decomposition for parabolics was noticed by Casselman and Bernstein–Zelevinski in the Grothendieck group of finite length representations of a reductive p-adic group. The proof of the decomposition is formal, and works simultaneously for the parabolics subgroups in a finite reductive group and for the parahorics subgroups in a p-adic reductive group.

A similar decomposition exists for the functor *"κ induction* (inflation then tensor product by κ then compact induction) *followed by κ restriction* (left adjoint of the κ induction)"

$$T^G_{\kappa,L,M}\colon \operatorname{Mod} M \to \operatorname{Mod} J \to \operatorname{Mod} G \to \operatorname{Mod} K \to \operatorname{Mod} L$$

for simple Bushnell–Kutzko subgroups J, K of $GL(n,F)$ corresponding to standard parahoric sequences (1) in the centralizer G_E of E in $GL(n,F)$, for an extension E/F contained in $M(n,F)$. We prove (8.3):

$$T^G_{\kappa,L,M} = T^{G_E}_{L,M}\,.$$

Probably similar decompositions will exist for all R-types of a general G, when they will be determined.

It is easy to deduce from this decomposition Theorem 1.1, thanks to the simple criterium for almost projectivity (3.1) and the simple criterium for irreduciblity (4.2).

It is a pleasure to thank Alberto Arabia for his work which led to the simple criterium of almost projectivity.

2. Almost-Projectivity

The notion of almost projectivity was introduced by Dipper for finite groups and is a particular case of the more general notion of quasi-projectivity.

Let R be a field and let A be an R-algebra. We consider the category $\operatorname{Mod}_{tf}(A)$ (resp. $\operatorname{Mod}^o_{tf}(A)$) of unital finite type left (resp. right) A-modules. A left module is called a module.

Definition 2.1. *A finite type unital A-module Q is called* **quasi-projective** *[11, A.3] when for any two morphisms*

$$Q \overset{\pi}{\underset{\alpha}{\rightrightarrows}} V$$

in $\operatorname{Mod}_{tf}(A)$ with π surjective, there exists $\beta \in \operatorname{End}_A Q$ such that $\alpha = \pi \circ \beta$.

It is called **almost projective** *when there exists a surjective morphism* $\pi\colon P \to Q$ *from a projective finite type unital A-module P such that for any morphism $\alpha \in \mathrm{Hom}_A(P, Q)$, there exists $\beta \in \mathrm{End}_A Q$ with $\alpha = \beta \circ \pi$.*

An almost projective module is quasi-projective [11, Proposition 7]. The fundamental property of quasi-projective modules is the following:

Main Property. (Arabia [11, Appendix th. 10]**)** *When Q is quasi-projective, the functor*

$$\mathrm{Hom}_A(Q, -)\colon \ \mathrm{Mod}_{tf}(A) \to \mathrm{Mod}^o_{ft}(\mathrm{End}_A Q)$$

induces a bijection between

a) *the isomorphism classes of simple A-modules V such that $\mathrm{Hom}_A(Q, V) \neq 0$ and*

b) *the isomorphism classes of simple right $\mathrm{End}_A Q$-modules.*

This functor is not in general an equivalence of category.

3. A Simple Criterium of Almost Projectivity

Let G be a locally profinite group, K an open compact subgroup of G, R a field and σ an irreducible smooth R-representation of K which admits a projective cover in $\mathrm{Mod}_R K$. We denote by:

$\mathrm{Mod}_R(G)$ the category of smooth R-representations of G.

V_σ the σ-isotypic part of $V \in \mathrm{Mod}_R(G)$, i.e., the biggest RK-submodule of V which is isomorphic to a direct sum $\oplus^I \sigma$.

The functor of compact induction $\mathrm{ind}_K^G\colon \mathrm{Mod}_R(K) \to \mathrm{Mod}_R(G)$ is exact when G has an R-Haar measure, has a right adjoint (the restriction from G to K denoted by $\pi \to \pi|_K$), respects projectivity and the property of being of finite type [10, I.5.10, I.5.7, I.5.9].

Lemma 3.1. *Suppose that the R-representation of G*

$$Q = \mathrm{ind}_K^G \sigma$$

admits a K-equivariant direct decomposition $Q = Q_\sigma \oplus Q^\sigma$ and no subquotient of Q^σ isomorphic to σ. Then Q is almost-projective.

This simple lemma which is basic for us was found by Arabia when G is a finite group.

Proof. Let $f\colon P_\sigma \to \sigma$ be a projective cover in $\mathrm{Mod}_R K$. We define

$$P := \mathrm{ind}_K^G P_\sigma, \quad \pi := \mathrm{ind}_K^G f.$$

The isomorphism of adjunction,

$$\mathrm{Hom}_{RG}(\mathrm{ind}_K^G P_\sigma, \mathrm{ind}_K^G \sigma) \simeq \mathrm{Hom}_{RK}(P_\sigma, \mathrm{ind}_K^G \sigma)$$

is given by restriction to the RK-submodule $\underline{P}_\sigma$ isomorphic to P_σ of functions with support in K in the canonical model of $\operatorname{ind}_K^G P_\sigma$. The image of $\operatorname{ind}_K^G f$ under the isomorphism of adjunction is $f\colon \underline{P}_\sigma \to \underline{\sigma}$ (where $\underline{\sigma} \simeq \sigma$ is the space of functions with support in K in the canonical model of $\operatorname{ind}_K^G \sigma$).

The hypothesis on $\operatorname{ind}_K^G \sigma$ and the definition of a projective cover imply that the map

$$\gamma \to \gamma \circ f\colon\ \operatorname{Hom}_{RK}(\sigma, \operatorname{ind}_K^G \sigma) \to \operatorname{Hom}_{RK}(P_\sigma, \operatorname{ind}_K^G \sigma)$$

is an isomorphism. With the isomorphisms of right adjunction of ind_K^G we obtain that the map

$$\beta \to \beta \circ \pi\colon\ \operatorname{End}_{RG} \operatorname{ind}_K^G \sigma \to \operatorname{Hom}_{RG}(\operatorname{ind}_K^G P_\sigma, \operatorname{ind}_K^G \sigma)$$

is an isomorphism. We deduce that $\operatorname{ind}_K^G \sigma$ is almost projective. $\square$

Exercise 3.2. *If $Q = \operatorname{ind}_K^G \sigma$ satisfies Lemma 3.1, then Q is quasi-projective.*

Let $\alpha, \pi\colon \operatorname{ind}_K^G \sigma \to V$ be two morphisms in $\operatorname{Mod}_R G$ with π surjective. We look for $\beta \in \operatorname{End}_{RG} \operatorname{ind}_K^G \sigma$ such that $\alpha = \pi \circ \beta$. The lemma implies that there exists a simple RK-submodule W' of Q_σ such that $\pi(W') = \alpha(\underline{\sigma})$. Let $\beta'\colon \underline{\sigma} \to \operatorname{ind}_K^G \sigma$ be the RK-equivariant morphism with image W' with $\pi \circ \beta' = \alpha|_{\underline{\sigma}}$, and $\beta \in \operatorname{End}_{RG} \operatorname{ind}_K^G \sigma$ the image of β' by adjunction. Then $\alpha = \pi \circ \beta$. $\square$

4. A Simple Criterium for Irreducibility

We replace the property "compact" for K by "compact mod center", and we denote by $\operatorname{Ind}_K^G\colon \operatorname{Mod}_R K \to \operatorname{Mod}_R G$ the induction without condition on the support. This functor has a left adjoint (the restriction $\pi \to \pi|_K$ from G to K) [10, I.5.7].

Lemma 4.1. *Let $\Lambda \in \operatorname{Irr}_R K$. When the space $\operatorname{End}_{RG}(\operatorname{ind}_K^G \Lambda)$ is finite dimensional, it is equal to $\operatorname{Hom}_{RG}(\operatorname{ind}_K^G \Lambda, \operatorname{Ind}_K^G \Lambda)$.*

Proof. Use the Mackey decomposition and the adjunction. $\square$

Lemma 4.2. *The R-representation $\operatorname{ind}_K^G \Lambda$ is irreducible when*

   a) $\operatorname{End}_{RG}(\operatorname{ind}_K^G \Lambda) = R$.
   b) *If Λ is contained in $\pi|_K$, then Λ is also a quotient of $\pi|_K$, for any $\pi \in \operatorname{Irr}_R G$.*

Proof. Suppose that a) and b) are true. Let $\pi \in \operatorname{Irr}_R G$ be a quotient of $\operatorname{ind}_K^G \Lambda$. By adjunction $\Lambda \subset \pi|_K$ and by b), Λ is a quotient of $\pi|_K$. By adjunction $\pi \subset \operatorname{Ind}_K^G \Lambda$. Hence there is a morphism $\operatorname{ind}_K^G \Lambda \to \operatorname{Ind}_K^G \Lambda$ with image π. By a) and (4.1) $\operatorname{ind}_K^G \Lambda = \pi$. Hence $\operatorname{ind}_K^G \Lambda$ is irreducible. $\square$

5. Restriction-Induction

It has been proved by Casselman, and by Bernstein–Zelevinski that the decomposition of the functor "parabolic induction followed by parabolic restriction" for representations of a finite reductive group, is also valid in the Grothendieck group of finite length representations of a reductive p-adic group. The main point of this paragraph is that the functor "parahoric induction followed by parahoric restriction" for a reductive p-adic group has a similar decomposition. The proof of the decomposition is formal, and works simultaneously for the parabolics subgroups in a finite reductive group and for the parahorics subgroups in a p-adic reductive group, as these subgroups have similar properties. The decomposition is given in (5.3), after a summary of well-known basic results in (5.1) and (5.2).

5.1. Basic functors

Let R be a commutative ring. We denote by Mod the category of R-modules and by $\mathrm{Mod}\,H$ the category of R-representations of a finite group H. Two representations $\pi, \pi' \in \mathrm{Mod}\,H$ are isomorphic if and only if the functors $\mathrm{Hom}(\pi, -)$, $\mathrm{Hom}(\pi', -)\colon \mathrm{Mod}\,H \to \mathrm{Mod}$ are isomorphic. We describe the relations between some basic functors. We restrict to those which will be used in the description of the functor parahoric induction-parahoric restriction, after generalisation to locally profinite groups.

5.1.1. CoINVARIANTS, INVARIANTS AND INFLATION   Let $1 \to U \to P \to M \to 1$ be an exact sequence of finite groups. The U-coinvariants $r_U = r_M^P\colon \mathrm{Mod}\,P \to \mathrm{Mod}\,M$ is left adjoint to the inflation $\mathrm{infl}_U = \mathrm{infl}_M^P\colon \mathrm{Mod}\,M \to \mathrm{Mod}\,P$ which identifies the representations $\rho \in \mathrm{Mod}\,M$ and the representations $\mathrm{infl}_U\,\rho \in \mathrm{Mod}\,P$ on which U acts trivially,

$$\mathrm{Hom}_M(r_U \pi, \rho) = \mathrm{Hom}_P(\pi, \mathrm{infl}_U\,\rho) \tag{2}$$

for $\pi \in \mathrm{Mod}\,P$, $\rho \in \mathrm{Mod}\,M$. The U-invariants $r^U\colon \mathrm{Mod}\,P \to \mathrm{Mod}\,M$ is the right adjoint of the inflation infl_U. When the order of U is invertible in R, the two functors r_U, r^U are canonically isomorphic.

5.1.2. INDUCTION AND RESTRICTION   Let Q be a subgroup of P. The induction $\mathrm{ind}_Q^P\colon \mathrm{Mod}\,Q \to \mathrm{Mod}\,P$ is left adjoint to the restriction $\mathrm{res}_Q^P\colon \mathrm{Mod}\,P \to \mathrm{Mod}\,Q$,

$$\mathrm{Hom}_P(\mathrm{ind}_Q^P\,\sigma, \pi) = \mathrm{Hom}_Q(\sigma, \mathrm{res}_Q^P\,\pi), \tag{3}$$

for $\pi \in \mathrm{Mod}\,P$, $\sigma \in \mathrm{Mod}\,Q$.

5.1.3. CONJUGATION   Let $p \in P$. We denote $\mathrm{Int}(p)(x) = pxp^{-1}$ for $x \in P$, and also $\mathrm{Int}(p)\colon \mathrm{Mod}\,Q \to \mathrm{Mod}\,pQp^{-1}$ the functor such that $\mathrm{Int}(p)\pi = \pi \circ \mathrm{Int}(p^{-1})$ for $\pi \in \mathrm{Mod}\,Q$.

124   M.-F. Vignéras

5.1.4. COMMUTATIVITY   By restriction to Q we have an exact sequence $1 \to U \cap Q \to Q \to M_Q \to 1$ and M_Q is a subgroup of M. The restriction and the inflation commute:

$$\mathrm{res}_Q^P \circ \mathrm{infl}_U = \mathrm{infl}_{U \cap Q} \circ \mathrm{res}_{M_Q}^M : \ \mathrm{Mod}\, M \to \mathrm{Mod}\, Q \tag{4}$$

(one may pass above through $\mathrm{Mod}\, P$ or below through $\mathrm{Mod}\, M_Q$). By adjunction, the coinvariants and the induction commute:

$$r_U \circ \mathrm{ind}_Q^P \sigma = \mathrm{ind}_{M_Q}^M \circ r_{U \cap Q} : \ \mathrm{Mod}\, Q \to \mathrm{Mod}\, M \tag{5}$$

(one may pass above through $\mathrm{Mod}\, P$ or below through $\mathrm{Mod}\, M_Q$). The image by $\mathrm{Int}(p)$ of the exact sequence defining M_Q is the exact sequence

$$1 \to pQp^{-1} \cap U \to pQp^{-1} \to M_{pQp^{-1}} \to 1.$$

Although M is not a subgroup of P, the conjugation by p defines an isomorphism $\mathrm{Int}(p) \colon M \to M_{pQp^{-1}}$. We denote also $\mathrm{Int}(p) \colon \mathrm{Mod}\, M \to \mathrm{Mod}\, pMp^{-1}$ the functor such that $\mathrm{Int}(p)\pi = \pi \circ \mathrm{Int}(p^{-1})$ for $\pi \in \mathrm{Mod}\, M$. The conjugation commutes with all the functors; for instance with the coinvariants

$$\mathrm{Int}(p) \circ r_{Q \cap U} = r_{p(Q \cap U)p^{-1}} \circ \mathrm{Int}(p) \colon \ \mathrm{Mod}\, Q \to \mathrm{Mod}\, M_{pQp^{-1}}. \tag{6}$$

5.1.5. MACKEY DECOMPOSITION   Let P_1, P_2 be two subgroups of a finite group G, and let $N_{2,1}$ be a set of representatives of the double classes $P_2 \backslash G / P_1$. We have $G = \cup_{n \in N_{2,1}} P_2 n P_1$ (disjoint union). The functor $t_{2,1}^G := \mathrm{res}_{P_2}^G \circ \mathrm{ind}_{P_1}^G \colon \mathrm{Mod}\, P_1 \to \mathrm{Mod}\, P_2$ (passing above through $\mathrm{Mod}\, G$) is a direct sum of functors

$$t_{2,1}^G = \sum_{n \in N_{2,1}} \mathrm{ind}_{P_2 \cap n P_1 n^{-1}}^{P_2} \circ \mathrm{Int}(n) \circ \mathrm{res}_{P_1 \cap n^{-1} P_2 n}^{P_1}, \tag{7}$$

(passing below through $\mathrm{Mod}\, P_1 \cap n P_2 n^{-1}$).

5.1.6. PARABOLIC RESTRICTION AND INDUCTION   Let G be a finite group and let $1 \to U_1 \to P_1 \to M_1 \to 1$ be an exact sequence of subgroups of G. We consider the "parabolic" or "Harish-Chandra" two functors,

$$r_{M_1}^G = r_{M_1}^{P_1} \circ \mathrm{res}_{P_1}^G : \ \mathrm{Mod}\, G \to \mathrm{Mod}\, M_1$$

$$i_{M_1}^G = \mathrm{ind}_{P_1}^G \circ \mathrm{infl}_{M_1}^{P_1} : \ \mathrm{Mod}\, M_1 \to \mathrm{Mod}\, G.$$

Let $1 \to U_2 \to P_2 \to M_2 \to 1$ be another exact sequence of subgroups of G. The Mackey decomposition of $\mathrm{res}_{P_2}^G \circ \mathrm{ind}_{P_1}^G$ gives a decomposition for the functor

$$T_{2,1}^G := r_{M_2}^G \circ i_{M_1}^G = r_{M_2}^{P_2} \circ \mathrm{res}_{P_2}^G \circ \mathrm{ind}_{P_1}^G \circ \mathrm{infl}_{M_1}^{P_1} : \ \mathrm{Mod}\, M_1 \to \mathrm{Mod}\, M_2.$$

The coinvariants and the induction commute (5), and we get for $n \in N$,

$$r_{M_2}^{P_2} \circ \mathrm{ind}_{P_2 \cap n P_1 n^{-1}}^{P_2} = \mathrm{ind}_{M_{P_2 \cap n P_1 n^{-1}}^2}^{M_2} \circ r_{U_2 \cap n P_1 n^{-1}}$$

where $M_{P_2 \cap n P_1 n^{-1}}^2$ is the image in M_2 of $P_2 \cap n P_1 n^{-1}$. The coinvariants and the conjugation commute (6), and we get

$$r_{U_2 \cap n P_1 n^{-1}} \circ \mathrm{Int}(n) = \mathrm{Int}(n) \circ r_{n^{-1} U_2 n \cap P_1} : \ \mathrm{Mod}\, n^{-1} P_2 n \cap P_1 \to \mathrm{Mod}\, M_{P_2 \cap n P_1 n^{-1}}^2.$$

We get

$$T_{2,1}^{G} = \oplus_{n \in N_{2,1}} \operatorname{ind}_{M_{P_2 \cap nP_1n^{-1}}^2}^{M_2} \circ \operatorname{Int}(n) \circ r_{n^{-1}U_2n \cap P_1} \circ \operatorname{res}_{n^{-1}P_2n \cap P_1}^{P_1} \circ \operatorname{infl}_{M_1}^{P_1} . \quad (8)$$

We see in the next paragraph that the right side of (8) can be simplified in important cases.

5.1.7. LOCALLY PROFINITE GROUPS   We may replace the finite groups by locally profinite groups such that U is closed in P, Q is open[2] in P, U_1, U_2 are closed in G, P_1, P_2 are open in G and the induction is the compact induction. All the relations between the functors remain true.

5.2. Intersection of parabolics or parahorics

Set:

$\underline{G}$ a reductive connected group over a field F which is either a finite field $\mathbf{F}_q$ of q elements and characteristic p (the finite case), or a local non-archimedean with residual field $\mathbf{F}_q$ (the p-adic case),

$\underline{T}$ a maximal split F-torus of $\underline{G}$,

$\underline{N}$ the $\underline{G}$-normalizer of $\underline{T}$,

$\underline{P}$ a minimal F-parabolic subgroup containing $\underline{T}$.

We denote by G, T, N, P the rational points of these groups, $W = N/T$ the Weyl group, isomorphic to the Weyl group of the roots of $(\underline{G}, \underline{T})$, Δ the simple roots determined by P.

5.2.1. PROPER PARABOLICS   For $J \subset \Delta$ and different from Δ, we denote by W_J the subgroup of W generated by the reflexions s_a associated to $a \in J$. The sets $P_J := PN_JP$, where N_J is the inverse image of W_J in N, are the standard proper parabolic subgroups of G. We have an exact sequence

$$1 \to U_J \to P_J \to M_J \to 1$$

where U_J is the unipotent radical of P_J, and M_J is the group of rational points of a reductive connected F-group. The images in M_J of the standard parabolic subgroups contained in P_J (the groups $P_H, H \subset J$) form a set of standard parabolic subgroups P_H^J of M_J.

Let $J, K \subset \Delta$ and different from Δ. The following basic properties are well known:

(P1) There is a unique w in any double coset of $W_J \backslash W / W_K$ characterized by any of the three following properties:

a) $w^{-1}J > 0$, $wK > 0$.
b) w has least length in $W_J w W_K$.
c) w has least length in $W_J w$ and also in $w W_K$.

We denote by $W_{J,K}$ the set of these elements w. We choose a set $N_{J,K}$ of representatives of $W_{J,K}$ in N. The image of $W_{J,K}$ by $w \mapsto w^{-1}$ is $W_{K,J}$, and we may suppose that $N_{K,J}$ is the image of $N_{J,K}$ by $n \to n^{-1}$.

(P2) For $n \in N_{J,K}$ with image $w \in W_{J,K}$, the intersection $n^{-1}P_J n \cap P_K$ of parabolic subgroups has the two following properties:

  i) $P_{w^{-1}J \cap K} = U_K(n^{-1}P_J n \cap P_K)$.

  ii) The image of $n^{-1}P_J n \cap P_K$ in M_K is the standard parabolic subgroup $P^K_{w^{-1}J \cap K}$ associated to $w^{-1}J \cap K$ with unipotent radical the image of $n^{-1}U_J n \cap P_K$.

We may replace n by $n^{-1} \in N_{K,J}$ and permute J, K. We get:

  i') $P_{wK \cap J} = U_J(nP_K n^{-1} \cap P_J)$.

  ii') The image of $nP_K n^{-1} \cap P_J$ in M_J is the standard parabolic subgroup $P^J_{wK \cap J}$ associated to $wK \cap J$ with unipotent radical the image of $nU_K n^{-1} \cap P_J$.

From these properties, we deduce that

$$\mathrm{Int}(n)M^K_{K \cap w^{-1}J} = M^J_{J \cap wK} \,;$$

in other terms the conjugation by n sends the first of these exact sequences to the second one

$$1 \to (n^{-1}P_J n \cap U_K)(n^{-1}U_J n \cap P_K) \to n^{-1}P_J n \cap P_K \to M^K_{K \cap w^{-1}J} \to 1$$

$$1 \to (P_J \cap nU_K n^{-1})(U_J \cap nP_K n^{-1}) \to P_J \cap nP_K n^{-1} \to M^J_{J \cap wK} \to 1\,.$$

(P3) By the Bruhat decomposition, $G = \cup_{n \in N_{J,K}} P_J n P_K$ (disjoint sum).

5.2.2. PARAHORICS [6] When F is local non-archimedean of residual field $\mathbf{F}_q$, the generalized affine Weyl group and the standard parahoric subgroups satisfy the same basic properties as the Weyl group and the standard parabolic subgroups. Set:

0 a special vertex of the apartment A defined by T in the semi-simple Bruhat–Tits building of G,

I the Iwahori subgroup determined by $(T, P, 0)$,

Δ^a the simple affine roots determined by $(T, P, 0)$,

$W^a = N/H$ where H is the kernel of the action of N on A.

For $J \subset \Delta^a$ and different from Δ^a, W^a_J is the subgroup of W^a generated by the reflexions s_α associated to $\alpha \in J$. The sets $P^a_J := IN^a_J I$, where N^a_J is the inverse image of W^a_J in N, are the standard parahoric subgroups of G. They are open and compact in G. We have an exact sequence

$$1 \to U^a_J \to P^a_J \to M^a_J \to 1$$

where U^a_J is the pro-p-nilpotent radical of P_J, and M^a_J is the group of rational points of a reductive connected $\mathbf{F}_q$-group. The images in M^a_J of the standard parahorics subgroups contained in P^a_J (the groups $P^a_H, H \subset J$) form a set of standard parabolics in M^a_J [2, 5.1.32]. We have the same basic properties when $(\Delta, W, P_J, \dots)$ is replaced by $(\Delta^a, W^a, P^a_J, \dots)$.

5.2.3. NORMALIZER OF A PARAHORIC SUBGROUP Let P_J^a be a standard parahoric subgroup of G. The normalizer $N_G(P_J^a)$ of P_J^a in G normalizes the pro-p-radical U_J^a of P_J^a. We have an exact sequence

$$1 \to U_J^a \to N_G(P_J^a) \to M_J^{+,a} \to 1$$

for a group $M_J^{+,a}$ which contains M_J^a as a normal subgroup. Let $W^{J,a}$ be the set of $w \in W^a$ with $wJ = J$. If J is a maximal proper subset of Δ, then $n \in N$ with image $w \in W^{J,a}$ belong to $N_G(P_J^a)$ ([6, Appendix]).

5.3. The functor $T_{K,J}$

From now on, we consider parabolic groups in the finite case, and parahoric groups in the p-adic case. With this convention, there is no confusion if we forget the upper indices a in the p-adic case; J is a proper subset of the simple, or simple affine roots Δ, P_J is a proper standard parabolic group of G in the finite case, and a standard parahoric group of G in the p-adic case.

The functor

$$r_{M_J}^G = r_{M_J}^{P_J} \circ \operatorname{res}_{P_J}^G : \operatorname{Mod} G \to \operatorname{Mod} M_J$$

and its left adjoint the functor

$$i_{M_J}^G = \operatorname{ind}_{P_J}^G \circ \operatorname{infl}_{M_J}^{P_J} : \operatorname{Mod} M_J \to \operatorname{Mod} G$$

(called the Harish-Chandra functor [5] in the finite case) can also be defined for the group M_H instead of G if $J \subset H \subset \Delta$, and denoted r_J^H, i_J^H to simplify. They can also be defined for the groups $N_G(P_J)$ instead of P_J in the p-adic case, and denoted $r_{M_J^+}^G$, $i_{M_J^+}^G$. For any proper subsets $J, K \subset \Delta$, we consider the functor

$$T_{K,J}^G = r_{M_K}^G \circ i_{M_J}^G : \operatorname{Mod} M_J \to \operatorname{Mod} M_K .$$

The formula (8) gives a decomposition of $T_{K,J}^G$. The property (P2) of (5.2) implies:

$$r_{n^{-1}U_K n \cap P_J} \circ \operatorname{res}_{n^{-1}P_K n \cap P_J}^{P_J} \circ \operatorname{infl}_{M_J}^{P_J} = r_{M_{w^{-1}K \cap J}^J}^{M_J} .$$

We get:

Basic Decomposition 5.1. $T_{K,J}^G \simeq \oplus_{w \in W_{K,J}} \; i_{K \cap wJ}^K \circ \operatorname{Int}(w) \circ r_{J \cap w^{-1}K}^J .$

In the decomposition, $\operatorname{Int}(w)$ denotes the functor

$$\operatorname{Int}(n) : \operatorname{Mod} M_{J \cap w^{-1}K}^J \to \operatorname{Mod} M_{K \cap wJ}^K$$

for any representative $n \in N_{K,J}$ of w, modulo isomorphism.

5.3.1. CUSPIDAL CASE AND $K = J$ We say that $\pi \in \operatorname{Mod} G$ is *cuspidal* when the coinvariants of π by the unipotent radicals of the proper parabolic subgroups of G are all 0. It suffices to consider only the standard proper parabolic subgroups. Cuspidality is respected by conjugation. We denote $\operatorname{Mod}_{cusp} G$ the set of cuspidal R-representations of G. When R is a field, we denote by $\operatorname{Cusp} G$ the irreducible cuspidal R-representations of G.

The basic decomposition for $T_J^G := T_{J,J}^G$ restricted to cuspidal representations is very simple. By definition of cuspidality, only the terms with $J \cap w^{-1}J = J$ do not vanish. As J is a finite set, the equality means $wJ = J$ and we get:

Corollary 5.2. *The functor $T_J^G = r_{M_J}^G \circ i_{M_J}^G$ restricted to $\operatorname{Mod}_{cusp} M_J$ is equivalent to the functor*

$$\oplus_{w \in W^J} \operatorname{Int}(w) \colon \operatorname{Mod}_{cusp} M_J \to \operatorname{Mod}_{cusp} M_J$$

where W^J is the set of $w \in W$ such that $wJ = J$.

We consider now the normalizer $N_G(P_J)$ in the p-adic case. The functor $r_{M_J^+}^G \circ i_{M_J^+}^G$ can be decomposed (8) as a sum on a system of representatives of the double cosets $N_G(P_J)\backslash G/N_G(P_J)$ which can be taken in $N_{J,J}$. Suppose that J is a proper maximal subset of Δ, and let $n \in N_{J,J}$. Let $\rho \in \operatorname{Mod} M_J^+$ with cuspidal restriction to M_J; if $wJ \neq J$, then

$$r_{n^{-1}U_Jn \cap N_G(P_J)} \circ \operatorname{res}_{n^{-1}N_G(P_J)n \cap N_G(P_J)}^{N_G(P_J)} \circ \operatorname{infl}_{M_J^+}^{N_G(P_J)} \rho = 0$$

because $r_{n^{-1}U_Jn \cap P_J} \circ \operatorname{res}_{n^{-1}P_Jn \cap P_J}^{P_J} \circ \operatorname{infl}_{M_J}^{P_J} \circ \operatorname{res}_{M_J}^{M_J^+} \rho = 0$. As $wJ = J$ implies $n \in N_G(P_J)$, we deduce:

Corollary 5.3. *Let P_J be a maximal standard parahoric subgroup and let $\rho \in \operatorname{Mod} M_J^+$ with cuspidal restriction to M_J. Then*

$$r_{M_J^+}^{N_G(P_J)} \circ i_{M_J^+}^G(\rho) \simeq \rho.$$

6. Almost Projectivity of Types of Level 0

We suppose that R is a field of characteristic $\ell \neq p$. Let G be as in Subsection 5.2. In the p-adic case, a *type of level* 0 is a conjugate in G of the inflation $\operatorname{infl}_{M_J}^{P_J} \sigma$ to a standard parahoric standard group P_J of a representation $\sigma \in \operatorname{Cusp}_R M_J$.

Proposition 6.1. *In the finite case for parabolic groups, the representation $\operatorname{ind}_{P_J}^G(\operatorname{infl}_{M_J}^{P_J} \sigma)$ compactly induced from a type of level 0 satisfies the simple criterium of almost projectivity (3.1).*

Proof. We have a P_J-equivariant decomposition

$$i_{M_J}^G \sigma = (i_{M_J}^G \sigma)^{U_J} \oplus W$$

where no subquotient of $W|_{U_J}$ is trivial. In particular, no subquotient of W has a subquotient isomorphic to $\lambda := \mathrm{infl}_{M_J}^{P_J} \sigma$. By (5.3.1), and the canonical equivalence between U_J-invariants and U_J-coinvariants,

$$(i_{M_J}^G \sigma)^{U_J} \simeq \oplus_{w \in W^J} \mathrm{Int}(w)\, \mathrm{infl}_{M_J}^{P_J} \sigma\,.$$

The representation $i_{M_J}^G \sigma$ satisfies the simple criterium of almost projectivity. $\quad\square$

7. Cuspidal Representations of Level 0

Let R be as in Section 6, and G as in Subsection 5.2 in the p-adic case.

Proposition 7.1. *In the p-adic case, when P_J is a maximal standard parahoric subgroup of G, R is algebraically closed and $\rho \in \mathrm{Irr}\, M_J^+$ with $\mathrm{res}_{M_J}^{M_J^+} \rho$ cuspidal, the representation $\mathrm{ind}_{N_G(P_J)}^G(\mathrm{infl}_{M_J^+}^{N_G(P_J)} \rho)$ satisfies the simple criterium of irreducibility.*

Proof. By adjunction, (5.3.3), and Schur's lemma, we have

$$\mathrm{End}_{RG}\, i_{M_J^+}^G \rho = \mathrm{Hom}_{RG}(\rho, r_{M_J^+}^G \circ i_{M_J^+}^G \rho) = \mathrm{End}_{RM_J^+} \rho \simeq R\,.$$

Let $\pi \in \mathrm{Irr}_R G$. If $\Lambda = \mathrm{infl}_{M_J^+}^{N_G(P_J)} \rho$ is contained in $\mathrm{res}_{N_G(P_J)}^G \pi$, then by adjunction ρ is contained in $r_{M_J^+}^G \pi$. The representations $r_{M_J^+}^G \pi$, π^{U_J} of M_J^+ are isomorphic, and π^{U_J} is a direct factor of $\mathrm{res}_{N_G(P_J)}^G \pi$. We deduce that Λ is a direct factor of $\mathrm{res}_{N_G(P_J)}^G \pi$. $\quad\square$

8. Bushnell–Kutzko Types

Let R be as in Section 6, and $G = GL(n, F)$.

We consider a "simple" type of Bushnell–Kutzko. To define it, we follow the notation of [3]. We fix $\beta \in G$ such that $E = F(\beta)$ is a field as in [3, 1.4.5, 1.4.13, 1.4.15, 2.4.1]. The degree $[E : F]$ divides n, and the centralizer G_E of E in G is isomorphic to $GL(d, E)$ where $n = d[E : F]$. As in Section 5, we consider a set of standard parahoric subgroups of G_E. For each standard parahoric subgroup P of G_E, there exists an exact sequence

$$1 \to U \to P \to M \to 1$$

where U is the pro-p-radical of P, and M a product of linear groups over the residual field of E. Let $\sigma \in \mathrm{Cusp}\, M$; then $(P, \mathrm{infl}_M^P \sigma)$ is a standard type of level 0 in G_E. We fix an endoclass Θ [3, 2.4.3, 5.1.8, 5.2.1 and 5.2.2]. Bushnell and Kutzko associate to (β, P, Θ) a pair (J, κ) with the following properties:

8.1.

a) There is an exact sequence of open compact subgroups of G

$$1 \to J^1 \to J \to M \to 1$$

where J^1 is the pro-p-radical of J, and $J^1 \cap G_E = U$, $J = J^1 P$, $J \cap G_E = P$.
b) The representation κ of J has an irreducible restriction η to J^1.
c) The pairs (J, κ), (J^1, η) are normalized by E^*.

The representation η depends only on (β, P, Θ), there is a choice for κ but only modulo torsion by a character of J normalized by E^*. In fact this choice can be made in a coherent way, and κ can be thought as determined by (β, P, Θ). This can be extracted with some pain from [3]. When $\sigma \in \operatorname{Cusp} M$, then $(J, \kappa \otimes \operatorname{infl}_M^P \sigma)$ is called a simple Bushnell–Kutzko type.

The following pair (HP, κ') satisfies (8.1), and will be more useful than the Bushnell–Kutzko pair (J, κ). The pair appears already in [3, 10, 11]. To define it, one considers first the maximal standard parahoric subgroup $P_{\max}$ of G_E (note that $U_{\max} \subset U \subset P \subset P_{\max} \simeq GL(d, O_E)$) and the Bushnell–Kutzko pair $(J_{\max}, \kappa_{\max})$ associated to $(\beta, P_{\max}, \Theta)$. Then

$$H = J^1_{max}, \quad \kappa' = \kappa_{\max}|HP \, .$$

The pro-p-radical of HP is HU, the restriction η' of κ' to HU is irreducible. We have the exact sequence

$$1 \to HU \to HP \to M \to 1 \, .$$

8.2. Double cosets

Let Q be another standard parahoric subgroup of G_E. The canonical exact sequence of Q is denoted

$$1 \to V \to Q \to L \to 1 \, .$$

(A1) *The map $QgP \to HQgPH$ for $g \in G_E$ is a bijection of double cosets*

$$Q \backslash G_E / P \to HQ \backslash HG_E H / HP \, .$$

Proof. This comes from the fact that the double coset HgH intersects G_E in a single double coset $U_{\max} g U_{\max}$. This is a consequence of [3, 1.6.1] and of the definition of $J^1_{\max}$. $\square$

(A2) *The groups HP, HQ verify the properties (P2) of (5.2), i.e., for $n \in N_{P,Q}$,*
   i) *$HR = HV(n^{-1}Pn \cap Q)$ for a standard parahoric subgroup R of G_E,*
   ii) *the image of $n^{-1}HPn \cap HQ$ in L is the standard parahoric subgroup with unipotent radical the image of $n^{-1}HUn \cap HQ$.*

Proof. This is a direct consequence of the same properties for P, Q. $\square$

Definition 8.1. *We consider the functors:*

a) *The κ inflation $\mathrm{infl}^J_{\kappa,M} = \kappa \otimes \mathrm{infl}^J_M \colon \mathrm{Mod}\,M \to \mathrm{Mod}\,J$ which identifies $\mathrm{Mod}\,M$ with the category of η isotypic representations of J; the κ coinvariant factor $\mathrm{res}^J_{\kappa,M}$ is the left adjoint and canonically identified to the right adjoint* (the functor of η coinvariants is canonically identified to the functor of η invariants because J^1 is a pro-p-group, and p is invertible in R).

b) *The κ induction $i^G_{\kappa,M} = \mathrm{ind}^G_J \circ \mathrm{infl}_\kappa \colon \mathrm{Mod}\,M \to \mathrm{Mod}\,G.$*

c) *The κ restriction $r^G_{\kappa,M} = \mathrm{res}^J_{\kappa,M} \circ \mathrm{res}^G_J \colon \mathrm{Mod}\,G \to \mathrm{Mod}\,M$ which is the left adjoint of the κ-induction and canonically identified to the right adjoint.*

d) *The κ restriction-κ induction*

$$T^G_{\kappa,L,M} = r^G_{\kappa,L} \circ i^G_{\kappa,M} \colon \mathrm{Mod}\,M \to \mathrm{Mod}\,L\,.$$

It is important to notice that the functors are more important than the pairs (J,κ), and we have the freedom to change (J,κ) as long as we do not change the functors. We can change (J,κ) by (HP,κ'). Indeed, we have:

Lemma 8.2. *The functors $i^G_{\kappa,M}$, $r^G_{\kappa,M}$ are equal to the functors $i^G_{\kappa',M}$, $r^G_{\kappa',M}$.*

Proof. [3, 5.2.5]. There is a pair (A,κ_A) of G which satisfies (8.1) and "contains" (J,κ) and (HP,κ'). This means that

$$J \subset A, \quad J^1 = J \cap A^1, \quad \kappa_A = \mathrm{ind}^A_J \kappa, \quad \eta_A = \mathrm{ind}^{A^1}_{J^1} \eta\,,$$

and the same relations for (HP,κ') instead of (J,κ). This implies the lemma. $\square$

Proposition 8.3. *The functor κ restriction-κ induction in G is equal to the functor parahoric restriction-parahoric induction in G_E:*

$$T^G_{\kappa,L,M} = T^{G_E}_{L,M}\,.$$

Proof. We choose a set N of representatives of the double cosets $HQ\backslash G/HQ$ in G. By (8.5) we have $T^G_{\kappa,L,M} = T^G_{\kappa',L,M} = r^{HQ}_{\kappa',L} \circ \mathrm{res}^G_{HQ} \circ \mathrm{ind}^G_{HP} \circ \mathrm{infl}^{HP}_{\kappa',L'}$ and the Mackey decomposition of $\mathrm{res}^G_{HQ} \circ \mathrm{ind}^G_{HP}$ gives a decomposition $T^G_{\kappa',L,M} = \oplus_{n\in N} F_{\kappa',n}$ where

$$F_{\kappa',n} := r^{HQ}_{\kappa',L} \circ \mathrm{ind}^{HQ}_{HQ\cap nHPn^{-1}} \circ \mathrm{Int}(n) \circ \mathrm{res}^{HP}_{HP\cap n^{-1}HQn} \circ \mathrm{infl}^{HP}_{\kappa',M}\,.$$

We check first that the $n \notin HG_E H$ give no contribution. If we forget the action of L, we have, for $\sigma \in \mathrm{Mod}\,M$,

$$F_{\kappa',n}(\sigma) = \mathrm{Hom}_{HQ}(\kappa', \mathrm{ind}^{HQ}_{HQ\cap nHPn^{-1}} \circ \mathrm{Int}(n) \circ \mathrm{res}^{HP}_{HP\cap n^{-1}HQn} \circ \mathrm{infl}^{HP}_{\kappa',M} \sigma)\,.$$

By adjunction, and using the commutation of the conjugation and the restriction we get

$$F_{\kappa',n}(\sigma) = \mathrm{Hom}_{HQ\cap nHPn^{-1}}(\kappa', \mathrm{Int}(n)(\kappa' \otimes \mathrm{infl}^{HP}_M \sigma))\,.$$

The restriction of κ' to H is η_{max}, hence $F_{\kappa',n}(\sigma) = 0$ if

$$\mathrm{Hom}_{H\cap nHg^{-1}}(\eta_{max}, \mathrm{Int}(n)\eta_{max}) = 0\,.$$

This happens when $n \notin HG_E H$ [3, 5.1.8 p. 160, 5.2.7 p. 170].

The restriction of $\kappa_{\max}$ to J' or K' is κ'. When $n \in G_E$ we have (loc.cit.)

$$\mathrm{Hom}_{H \cap nHn^{-1}}(\eta_{max}, \mathrm{Int}(n)\eta_{max}) = \mathrm{Hom}_{J_{max} \cap nJ_{max}n^{-1}}(\kappa_{max}, \mathrm{Int}(n)\kappa_{max}) \simeq R.$$

This implies that

$$\mathrm{Hom}_{HV \cap nHUn^{-1}}(\eta', \mathrm{Int}(n)\eta') = \mathrm{Hom}_{HQ \cap nHPn^{-1}}(\kappa', \mathrm{Int}(n)\kappa') \simeq R. \tag{9}$$

The end of the proof is now clear. We choose a system $N_{Q,P}$ of representatives of $Q \backslash G_E / P$ which satisfy the properties (P2) of (5.2). Let $n \in N_{Q,P}$. We have

$$F_{\kappa',n} = r_{\kappa',L}^{HQ} \circ \mathrm{ind}_{HQ \cap nHPn^{-1}}^{HQ} \circ \mathrm{res}_{HQ \cap nHPn^{-1}}^{nHPn^{-1}} \circ \mathrm{Int}(n) \circ (\kappa' \otimes \mathrm{infl}_M^{HP}).$$

The conjugation commute with the inflation and the tensor product, the inflation is transitive and we get

$$\mathrm{Int}(n)(\kappa' \otimes \mathrm{infl}_M^{HP}) = (\mathrm{Int}(n)\kappa') \otimes (\mathrm{infl}_{nPn^{-1}}^{nHPn^{-1}} \circ \mathrm{Int}(n) \circ \mathrm{infl}_M^P).$$

The induction and the restriction commute with the tensor product, and (9) implies that

$$r_{\kappa',L}^{HQ} \circ \mathrm{ind}_{HQ \cap nHPn^{-1}}^{HQ} \circ \mathrm{res}_{HQ \cap nHPn^{-1}}^{nHPn^{-1}} \circ \mathrm{Int}(n)(\kappa')$$

is the trivial representation of L. Hence $F_{\kappa',n} = F_n$ where

$$F_n = r_L^Q \circ \mathrm{ind}_{HQ \cap nHPn^{-1}}^{HQ} \circ \mathrm{res}_{HQ \cap nHPn^{-1}}^{nHPn^{-1}} \circ \mathrm{infl}_{nPn^{-1}}^{nHPn^{-1}} \circ \mathrm{Int}(n) \circ \mathrm{infl}_M^P.$$

By (8.2), the groups HQ, HP satisfy the properties (P2) of (5.2). We compare with (5.3.1) and we get the proposition. $\square$

Let $(J, \kappa \otimes \mathrm{infl}_M^J \sigma)$ be a Bushnell–Kutzko type. The normalizer of $J_{\max}$ is $J_{\max}E^*$. As in the level 0 case (5.3.2, 5.3.3, 5.3.4, 5.3.5) Proposition 8.3 implies:

Corollary 8.4.

1) $\mathrm{res}_{\kappa,M}^G \circ \mathrm{ind}_J^G(\kappa \otimes \mathrm{infl}_M^J \sigma) = \mathrm{res}_M^{G_E} \circ \mathrm{ind}_M^{G_E} \sigma$ *is a sum of conjuguates of* σ.

2) *Let* Λ *be a representation of* $J_{\max}E^*$ *which extends the maximal Bushnell–Kutzko type* $\kappa_{\max} \otimes \mathrm{infl}_M^{J_{\max}} \sigma$. *Then the* $\eta_{\max}$ *isotypic part of* $\mathrm{ind}_{J_{\max}E^*}^G \Lambda$ *is isomorphic to* Λ.

3) $\mathrm{ind}_J^G(\kappa \otimes \mathrm{infl}_M^J \sigma)$ *satisfies the simple criterium of almost projectivity.*

4) $\mathrm{ind}_{J_{\max}E^*}^G \Lambda$ *satisfies the simple criterium of irreduciblity, if* R *is algebraically closed.*

References

[1] F. Bruhat et J. Tits, *Groupes réductifs sur un corps local I.* Publications Mathématiques de l'I.H.E.S. n°41 (1997) 5–252.

[2] F. Bruhat et J. Tits, *Groupes réductifs sur un corps local II.* Publications Mathématiques de l'I.H.E.S. n°60 (1997) 5–184.

[3] J. Bushnell Colin and C. Kuzko Philip, *The admissible dual of GL(N) via compact open subgroups.* Annals of Mathematics Studies. Number 129. Princeton University Press 1993.

[4] M. Geck, G. Hiss and G. Malle, *Towards a classification of the irreducible representations in non-defining characteristic of a finite group of Lie type.* Math. Z. 221 (1996) 353–386.

[5] Harish-Chandra, *Eisenstein series over finite fields.* Collected papers IV pages 8–20. Springer-Verlag 1984.

[6] L. Morris, *Tamely ramified intertwining algebras.* Invent. Math. 114 (1993) 1–54.

[7] A. Moy and G. Prasad, *Jacquet functors and unrefined minimal K-types.* Comment. Math. Helvetici 71 (1996) 98–121.

[8] P. Schneider and U. Stuhler, *Representation theory and sheaves on the Bruhat–Tits building.* Publications Mathématiques de l'I.H.E.S. n°85 (1997) 97–191.

[9] J. Tits, *Reductive groups over local fields.* Proceedings of Symposia in Pure Mathematics Vol. 33 (1979) part 1 29–69.

[10] M.-F. Vignéras, *Représentations modulaires d'un groupe réductif p-adique avec $l \neq p$.* Progress in Math. 137 Birkhauser 1996.

[11] M.-F. Vignéras, *Induced R-representations of p-adic reductive groups.* with an appendix *Objets quasi-projectifs* by Alberto Arabia A. Sel. math. New ser. 4 (1998) 549–623.

Institut de Mathématiques de Jussieu
Université Paris 7, Denis Diderot
Case 7012, 2 place Jussieu
75251 Paris Cedex 05, France
E-mail address: vigneras@math.jussieu.fr

Dequantization of Real Algebraic Geometry on Logarithmic Paper

Oleg Viro

Abstract. On logarithmic paper some real algebraic curves look like smoothed broken lines. Moreover, the broken lines can be obtained as limits of those curves. The corresponding deformation can be viewed as a quantization, in which the broken line is a classical object and the curves are quantum. This generalizes to a new connection between algebraic geometry and the geometry of polyhedra, which is more straight-forward than the other known connections and gives a new insight into constructions used in the topology of real algebraic varieties.

1. Graphs of Polynomials on Logarithmic Paper

1.1. How to visualize a real polynomial?

If you ever tried to draw the graph for a polynomial of degree greater than, say, 4 and consisting of at least 4 monomials, you are aware of the natural difficulties. The graph is too steep. Whatever scale you choose, either some important details do not fit into the picture or are too small. The usual recipes from Calculus do not address the problem, but suggest, instead, to find roots of the first two derivatives, which does not seem to be much easier than the original problem.

1.2. Logarithmic paper

A physicist or engineer can give more practical advice, based on their experience: *use (double) logarithmic paper*. This is a graph paper, called also *log paper*, with a non-uniform net of coordinate lines and logarithmic scales on both axes. On a log paper a point with coordinates x, y is shown at the position with the usual, Cartesian coordinates equal to $\ln x$, $\ln y$. In other words, the transition to the log paper corresponds to the change of coordinates:

$$\begin{cases} u = \ln x \\ v = \ln y. \end{cases}$$

On a log paper the first quadrant is expanded homeomorphically to the whole plane, the line $x = 1$ occupies the position of the axis of ordinates, the line $y = 1$ occupies the position of the axis of abscissas, and the unit square bounded by these lines and the coordinate axes occupies the whole third quadrant.

1.3. A monomial on logarithmic paper

Let us try to follow the advice to use a log paper. Consider first the simplest special case: the graph of a monomial ax^k (i.e., the curve defined by $y = ax^k$). We are forced to consider only positive x, y and hence assume a to be positive as well. Then $v = \ln y = \ln(ax^k) = k \ln x + \ln a = ku + \ln a = ku + b$, where we denote $\ln a$ by b. Now the curve which we want to draw is defined in the coordinates u and v by the equation $v = ku + b$. Everybody knows that this is the straight line with slope k meeting the axis of ordinates (i.e., v-axis) at $(0, b)$.

1.4. A few slightly more complicated polynomials

First, consider a line $y = 1 + x$. Then

$$v = \ln y = \ln(1 + x) = \ln(1 + e^u).$$

See the left plot in Figure 1, where the graph of $v = \ln(1 + e^u)$ is shown together with lines $v = 0$ and $v = u$, which represent on the log paper the monomials 1 and x involved in our polynomial $1 + x$. The graph of $v = \ln(1 + e^u)$ looks like the broken line $v = \max\{0, u\}$ with a smoothed corner: it goes along and above this broken line getting very close to it as $|u|$ grows. For $|u| > 4$ the difference becomes beyond the resolution.

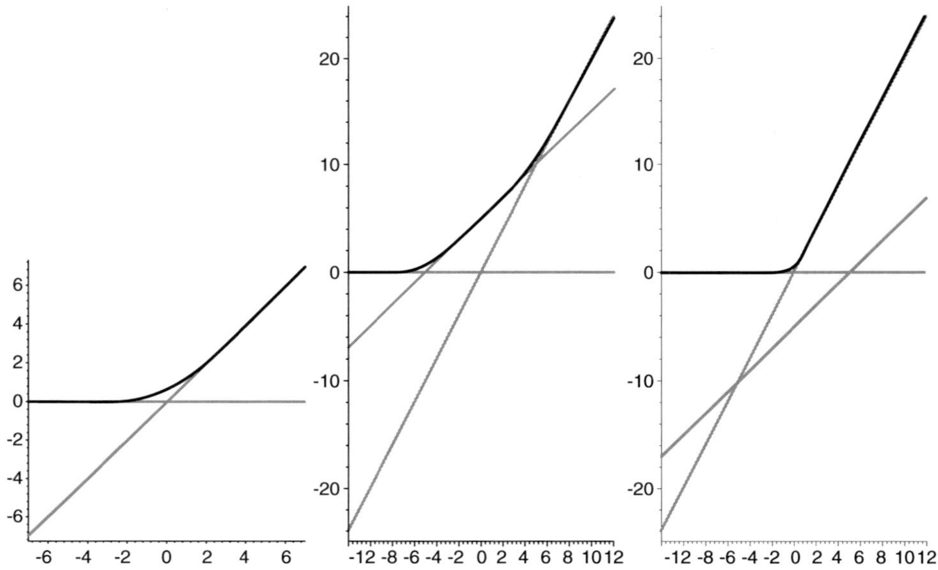

FIGURE 1. Log paper graphs of $1 + x$, $1 + e^5 x + x^2$ and $1 + e^{-5} x + x^2$.

Here are two further examples: the quadratic polynomials $1 + e^{\pm 5}x + x^2$. Then

$$v = \ln(1 + e^{\pm 5}x + x^2) = \ln(1 + e^{u \pm 5} + e^{2u}).$$

See the central and right plots in Figure 1. The graph of $v = \ln(1 + e^{u \pm 5} + e^{2u})$ looks like the broken line $v = \max\{0, u \pm 5, 2u\}$ with smoothed corners. It goes along and above this broken line getting very close to it far from its corners. Notice that the lines $v = 0$, $v = u \pm 5$ and $v = 2u$ represent on the logarithmic paper the monomials 1, $e^{\pm 5}x$ and x^2, respectively.

1.5. A polynomial versus the maximum of its monomials

This suggests, for a polynomial $p(x) = a_n x^n + a_{n-1}x^{n-1} + \cdots + a_0$ with positive real coefficients $a_i = e^{b_i}$, to compare the graphs on log paper for p and the maximum $M(p)(x) = \max\{a_n x^n, a_{n-1}x^{n-1}, \dots, a_0\}$ of its monomials. Denote the graph on log paper of a function f by Γ_f. With respect to the usual Cartesian coordinates, Γ_p is the graph of

$$L_p(u) = \ln\left(e^{nu+b_n} + e^{(n-1)u+b_{n-1}} + \cdots + e^{b_0}\right)$$

and $\Gamma_{M(p)}$ is the graph of a piecewise linear convex function

$$M_p(u) = \max\left\{nu + b_n, (n-1)u + b_{n-1}, \dots, b_0\right\}.$$

Obviously, $M_p(u) \leq L_p(u) \leq M_p(u) + \ln(n+1)$. Hence Γ_p is above $\Gamma_{M(p)}$, but below a copy of $\Gamma_{M(p)}$ shifted upwards by $\ln(n+1)$. The latter is in fact a rough estimate. It turns to equality only at u, where all linear functions, whose maximum is $M_p(u)$, are equal: $nu + b_n = (n-1)u + b_{n-1} = \cdots = b_0$.

For a generic value of u, only one of these functions is equal to $M_p(u)$. Say $M_p(u) = ku + b_k$, while $M_p(u) > d + lu + b_l$ for some positive d and each $l \neq k$. Then

$$L_p(u) < M_p(u) + \ln(1 + ne^{-d}) < M_p(u) + e^{-d}n.$$

If for some value of u the values of all of the functions $ku + b_k$ except two are smaller than $M_p(u) - d$, then

$$L_p(u) < M_p(u) + \ln(2 + (n-1)e^{-d}) < M_p(u) + \ln 2 + e^{-d}(n-1)/2.$$

Thus, on a logarithmic paper the graph of a generic polynomial with positive coefficients lies in a narrow strip along the broken line which is the graph of the maximum of its monomials. The width of the strip is estimated by characteristics of the mutual position of the lines which are the graphs of the monomials. The less congested the configuration of these lines, the narrower this strip.

1.6. Rescalings pushing the graph of a polynomial to a PL-graph

A natural way to make a configuration of lines less congested without changing its topology is to apply a dilation $(u, v) \mapsto (Cu, Cv)$ with a large $C > 0$. In what follows it is more convenient to use instead of C a parameter h related to C by $h = 1/C$. In terms of h the dilation acts by $(u, v) \mapsto (u/h, v/h)$. It maps the graph of $v = ku + b$ to the graph of $v = ku + b/h$. The parallel operation on monomials replaces ax^k by $a^{1/h}x^k$.

Consider the corresponding family of polynomials: $p_h(x) = \sum_k a_k^{1/h} x^k$. On log paper, the graphs of its monomials are obtained by dilation with ratio $1/h$ from the graphs of the corresponding monomials of p. Hence $\Gamma_{M(p_h)}$ is the image of $\Gamma_{M(p)}$ under the same dilation. However, Γ_{p_h} is not the image of Γ_p. It still lies in a strip along $\Gamma_{M(p_h)}$ and the strip is getting narrower as h decreases, but at the corners of $\Gamma_{M(p_h)}$ the width of the strip cannot become smaller than $\ln 2$.

To keep the picture of our expanding configuration of lines (the graphs of monomials) independent of h, let us make an additional calibration of coordinates: set $u_h = hu = h\ln x$, $v_h = hv = h\ln y$. Denote by Γ_f^h the graph of a function $y = f(x)$ in the plane with coordinates u_h, v_h.

Then $\Gamma_{M(p_h)}^h$ does not depend on h. The additional scaling reduces the width of the strip along $\Gamma_{M(p_h)}^h$, where $\Gamma_{p_h}^h$ lies, forcing the width to tend to 0 as $h \to 0$. Thus $\Gamma_{p_h}^h$ tends to $\Gamma_{M(p_h)}^h$ (in the C^0 sense) as $h \to 0$.

2. Quantization

2.1. Maslov dequantization of positive real numbers

The rescaling formulas $u_h = h\ln x$, $v_h = h\ln y$ bring to mind formulas related to the Maslov dequantization of real numbers, see e.g. [4, 5]. The core of the Maslov dequantization is a family of semirings $\{S_h\}_{h \in [0, \infty)}$ (recall that a semiring is a sort of ring, but without subtraction). As a set, each of S_h is $\mathbb{R}$. The semiring operations $\oplus_h$ and $\odot_h$ in S_h are defined as follows:

$$a \oplus_h b = \begin{cases} h\ln(e^{a/h} + e^{b/h}), & \text{if } h > 0 \\ \max\{a, b\}, & \text{if } h = 0 \end{cases} \tag{1}$$

$$a \odot_h b = a + b. \tag{2}$$

These operations depend continuously on h. For each $h > 0$ the map

$$D_h \colon \mathbb{R}_+ \smallsetminus \{0\} \to S_h \colon x \mapsto h\ln x$$

is a semiring isomorphism of $\{\mathbb{R}_+ \smallsetminus \{0\}, +, \cdot\}$ onto $\{S_h, \oplus_h, \odot_h\}$, that is

$$D_h(a + b) = D_h(a) \oplus_h D_h(b), \quad D_h(ab) = D_h(a) \odot_h D_h(b).$$

Thus S_h with $h > 0$ can be considered as a copy of $\mathbb{R}_+ \smallsetminus \{0\}$ with the usual operations of addition and multiplication. On the other hand, S_0 is a copy of $\mathbb{R}$ where the operation of taking maximum is considered as an addition, and the usual addition, as a multiplication.

Applying the terminology of quantization to this deformation, we must call S_0 a classical object, and S_h with $h \neq 0$, quantum ones. The analogy with Quantum Mechanics motivated the following *correspondence principle* formulated by Litvinov and Maslov [4] as follows:

"There exists a (heuristic) correspondence, in the spirit of the correspondence principle in Quantum Mechanics, between important, useful and interesting constructions and results over the field of real (or complex) numbers (or the semiring of all nonnegative numbers) and similar constructions and results over idempotent semirings."

This principle proved to be very fruitful in a number of situations, see [4, 5]. According to the correspondence principle, the idempotent counterpart of a polynomial $p(x) = a_n x^n + a_{n-1} x^{n-1} + \cdots + a_0$ is a convex PL-function $M_p(u) = \max\{nu + b_n, (n-1)u + b_{n-1}, \dots, b_0\}$. As we have seen above, p and M_p are related not only on an heuristic level. In Section 1.6 we connected the graph Γ_p of p on logarithmic paper and the graph $\Gamma_{M(p)}$ of M_p by a continuous family of graphs $\{\Gamma_{p_h}^h\}_{h \in (0,1)}$.

2.2. Logarithmic paper as a graphical device for the Maslov dequantization

As we saw in Section 1.5, the graph of a polynomial $p(x) = \sum_k a_k x^k$ with positive real coefficients $a_k = e^{b_k}$ on log paper is the graph of function $\mathbb{R} \to \mathbb{R}$ defined by $v = \ln\left(\sum_k e^{ku+b_k}\right)$. Observe that $\ln\left(\sum_k e^{ku+b_k}\right)$ is the value in S_1 of the polynomial $\sum_k b_k x^k$ at $x = u$. Therefore we can identify the graph Γ_p of $p(x) = \sum_k a_k x^k$ on log paper with the (Cartesian) graph of the polynomial $\sum_k b_k x^k$ on S_1^2.

Furthermore, $\Gamma_{p_h}^h$ is the graph of the function $\mathbb{R} \to \mathbb{R}$ defined by

$$v = h \ln\left(\sum_k a_k^{1/h} e^{(ku)/h}\right) = h \ln\left(\sum_k e^{(ku+b_k)/h}\right).$$

Observe, that the right-hand side is the value in S_h of the same polynomial $\sum_k b_k x^k$ at u. Therefore we can identify the graph $\Gamma_{p_h}^h$ of $p_h(x) = \sum_k a_k^{1/h} x^k$ on log paper with the (Cartesian) graph of $\sum_k b_k x^k$ on S_h^2.

At last, the graph of $\sum_k b_k x^k$ on S_0^2 is the the graph of M_p.

We see that the whole job of deforming Γ_p to the graph of a piecewise linear convex function can be done by the Maslov dequantization: the deformation consists of the graphs of the same polynomial $\sum_k b_k x^k$ in S_h^2 for $h \in [0,1]$. The coefficients b_k of this polynomial are logarithms of the coefficients of the original polynomial: $b_k = \ln a_k$. Since the map $x \mapsto \ln x \colon \mathbb{R}_+ \smallsetminus 0 \to S_1$ was denoted above by D_1, we denote by $D_1 F$ the polynomial obtained from a polynomial F with positive coefficients by replacing its coefficients with their logarithms. Thus $\sum_k b_k x^k = D_1 p(x)$. Since D_1 is a semiring homomorphism, the graph Γ_p of p on log paper is the graph of $D_1 p$ on S_1^2. The other graphs involved in the deformation are the graphs of the same polynomial $D_1 p$ on S_h^2. They coincide with the graphs on log paper of the preimages p_h of $D_1 p$ under D_h. Indeed, $p_h(x) = \sum_k a_k^{1/h} x^k$ and $D_h^{-1}(b_k) = D_h^{-1} D_1(a_k) = e^{D_1(a_k)/h} = e^{(\ln a_k)/h} = a_k^{1/h}$.

For a real polynomial $p(x) = \sum_k a_k x^k$ with positive coefficients, we shall call $p_h(x) = \sum_k a_k^{1/h} x^k$ with $h > 0$ the *dequantizing family* of polynomials.

2.3. Real algebraic geometry as quantized PL-geometry

The notion of polynomial is central in algebraic geometry. (I believe the subject of algebraic geometry would be better described by the name of *polynomial geometry*.) Since a polynomial over $\mathbb{R}$ is presented so explicitly as a quantization of a piecewise linear convex function, one may expect to find along this line explicit relations between other objects and phenomena of algebraic geometry over $\mathbb{R}$ and piecewise linear geometry. Indeed, in piecewise linear geometry the notion of piecewise linear convex function plays almost the same rôle as the notion of polynomial in algebraic geometry.

A representation of real algebraic geometry as a quantized PL-geometry may be rewarding in many ways. For example, in any quantization there are *classical objects*, i.e., objects which do not change much under the quantization. Objects of PL-geometry are easier to construct. If we knew conditions under which a PL object gives rise to a real algebraic object, which is classical with respect to the Maslov quantization, then we would have a simple way to construct real algebraic objects with controlled properties.

3. Put Algebraic Geometry Onto Logarithmic Paper

3.1. Speak of real algebraic geometry only for positive numbers

Above in Section 1 we discussed the drawing of graphs on a logarithmic paper only for a polynomial with positive coefficients. The graphs allowed us to see the behaviour of the polynomials only at positive values of the argument. This was for a good reason: we used logarithms of coordinates. The Maslov dequantization deals only with positive numbers. Therefore each fragment of algebraic geometry that we want to dequantize must be reformulated first only in terms of positive numbers. This seems to be possible for everything belonging to real algebraic geometry.

3.2. Visualizing roots of a polynomial on logarithmic paper

Above we could not meet roots of polynomials, for a polynomial with *positive* coefficients has no *positive* roots. However if we really want to do algebraic geometry on log paper, we must figure out how to use graphs on log paper for visualizing roots (well, only *positive* roots) of an arbitrary real polynomial.

Any real polynomial $p(x)$ is a difference $p^+(x) - p^-(x)$ of polynomials with positive coefficients. We can reformulate the problem of finding the positive roots of p as the problem of finding positive values of x at which $p^+(x) = p^-(x)$. The graphs of p^+ and p^- can be drawn on a log paper, where they are localised in the strips along broken lines, see Section 1.5 above. For some polynomials this picture gives decent information on the number and position of the positive roots.

The negative roots of $p(x)$ can be treated in the same way, since their absolute values are the positive roots of $p(-x)$.

3.3. Plane algebraic curves on logarithmic paper

Now consider a real polynomial $p(x, y) = \sum_{k,l} a_{k,l} x^k y^l$ in two variables. Similarly to the case of polynomials in one variable, in the *logarithmic space* the graph of a monomial $ax^k y^l$ with $a > 0$ is a plane $w = ku + lv + \ln a$, and the graph Γ_p of a polynomial $p(x, y)$ with positive coefficients lies in a neighborhood of a convex piecewise linear surface, which is the graph $\Gamma_{M(p)}$ of the maximum $M(p)$ of the monomials. Furthermore, p is included in a dequantizing family p_h defined as $\sum_{k,l} a_{k,l}^{1/h} x^k y^l$ for $h > 0$, cf. Section 2.2. The graph $\Gamma_{p_h}^h$ of p_h in the logarithmic space with scaled coordinates $u_h = h \ln x$, $v_h = h \ln y$, $w_h = h \ln z$ coincides with the graph of the polynomial $D_1 p(x, y) = \sum_{k,l} (\ln a_{k,l}) x^k y^l$ in S_h^3. These graphs with $h \in (0, 1]$ constitute a continuous deformation of $\Gamma_p = \Gamma_{p_1}^1$ to $\Gamma_{M(p)}$.

For a polynomial p in two variables with arbitrary real coefficients, denote by p^+ the sum of its monomials with positive coefficients, and put $p^- = p^+ - p$. Thus p is canonically presented as a difference $p^+ - p^-$ of two polynomials with positive coefficients. To obtain the curve defined on logarithmic paper by the equation $p(x, y) = 0$, one can construct the graphs Γ_{p^+} and Γ_{p^-} for p^+ and p^- in the logarithmic space, which are the surfaces defined in the usual Cartesian coordinates by $w = \ln(p^{\pm}(e^u, e^v))$, and project the intersection $\Gamma_{p^+} \cap \Gamma_{p^-}$ to the plane of arguments.

For the first approximation of this curve, one may take the broken line, which is the projection of the intersection of the piecewise linear surfaces $\Gamma_{M(p^+)}$ and $\Gamma_{M(p^-)}$ corresponding to p^+ and p^-.

Of course, it may well happen that the broken line does not even resemble the curve. This happens to first approximations. However, it is very appealing to figure out circumstances under which the broken line is a good approximation, for a broken line seems to be much easier to deal with than an algebraic curve.

3.4. Constructing classical algebraic curves from our quantum point of view

Recall that in the logarithmic space the graph of $ax^k y^l$ is a plane $w = ku + lv + \ln a$. It has a normal vector $(k, l, -1)$ and intersects the vertical axis at $(0, 0, \ln a)$. Thus if we want to construct a curve of a given degree m, we have to arrange planes whose normals are fixed: they are $(k, l, -1)$ with integers k, l, satisfying inequalities $0 \leq k, l, k + l \leq m$. The only freedom is in moving them up and down.

Consider the pieces of these planes which do not lie under the others. They form a convex piecewise linear surface U, the graph of the maximum of the linear forms defining our planes. The combinatorial structure of faces in U depends on the arrangement. Assume that at each vertex of U exactly three of the planes meet. This is a genericity condition, which can be satisfied by small shifts of the planes.

Divide now the faces of U arbitrarily into two classes. Denote the union of one of them by U^+, the union of the other by U^-. By genericity of the configuration, the common boundary of U^+ and U^- is the union of disjoint polygonal simple closed curves. It can be easily realized as the intersection of PL-surfaces $\Gamma_{M(p^+)}$

and $\Gamma_{(M(p^-)}$ as above: take for p^ε with $\varepsilon = \pm$ the sum of monomials corresponding to the planes of faces forming U^ε and put $p = p^+ - p^-$.

Consider now for $1 \geq h \geq 0$ the curve $C_h \subset S_h^3$ which is the intersection of the graphs in S_h^3 of the polynomials $D_1 p^+$ and $D_1 p^-$. At $h = 0$ this is the intersection of the convex PL-surfaces $\Gamma_{M(p^+)}$, $\Gamma_{M(p^-)}$. Due to the genericity condition above, this intersection is as transversal as one could wish: at all but a finite number of points the interior part of a face of one of them meets the interior part of a face of the other one, and at the rest of the points an edge of one of the surfaces intersects transversaly the interior of a face of the other surface.

When h becomes positive, the graphs become smooth, their corners are rounded off. The same happens to their intersection curve. While the graphs are transversal, the intersection curve is deformed isotopically.

Take the curve corresponding to a value of h such that the transversality is preserved between 0 and this value. The projection to the (u, v)-plane of C_h represents an algebraic curve of degree m on the scaled logarithmic paper and it can be obtained by a small isotopy of the projection of ∂U^+ to the (u, v)-plane.

3.5. Is this a patchworking?

A construction, which looks similar, has been known in the topology of real algebraic varieties for about 20 years as *patchworking*, or *Viro's method*. It has been used to construct real algebraic varieties with controlled topology and helped to solve a number of problems. For example, to classify up to isotopy non-singular real plane projective curves of degree 7 [7, 8] and disprove the Ragsdale conjecture [2] on the topology of plane curves formulated [6] as early as in 1906. To the best of my knowledge, the patchworking has never been related to the Maslov quantization.

4. Patchworking Real Algebraic Curves

4.1. The simplest patchworking

Here is a description of a simplified version of patchworking. The simplifications are of the following three kinds:

- we restrict to the case of nonsingular planar curves,
- we assume that all patches are trinomials, and
- we consider only the part of the curve contained in the first quadrant (what happens in other quadrants is described soon after).

4.1.1. INITIAL DATA   Let m be a positive integer (it will be the degree of the curve under construction) and Δ be the triangle in $\mathbb{R}^2$ with vertices $(0,0)$, $(m,0)$, $(0,m)$. Let τ be a *convex* triangulation of Δ with vertices having integer coordinates. The convexity of τ means that there exists a convex piecewise linear function $\nu \colon \Delta \longrightarrow \mathbb{R}_+$ which is linear on each triangle of τ and is not linear on the union of any two triangles of τ. Let the vertices of τ be equipped with signs. The sign (plus or minus) at the vertex with coordinates (k, l) is denoted by $\sigma_{k,l}$.

4.1.2. CONSTRUCTION OF THE PIECEWISE LINEAR CURVE   If a triangle of the triangulation τ has vertices of different signs, draw a midline separating pluses from minuses. Denote by L the union of these midlines. It is a collection of polygonal lines contained in Δ. The pair (Δ, L) is called the *result of combinatorial patchworking*.

4.1.3. CONSTRUCTION OF POLYNOMIALS   Given initial data m, Δ, τ and $\sigma_{k,l}$ as above and a positive convex function ν certifying, as above, that the triangulation τ is convex. Consider a one-parameter family of polynomials

$$b_t(x, y) = \sum_{\substack{(k, l) \text{ runs over} \\ \text{vertices of } \tau}} \sigma_{k,l} t^{\nu(k,l)} x^k y^l . \tag{3}$$

The polynomials b_t are called the results of *polynomial patchworking*.

Patchwork Theorem. *Let m, Δ, τ, $\sigma_{k,l}$ and ν be initial data as above. Denote by b_t the polynomials obtained by the polynomial patchworking of these initial data, and by L the PL-curve in Δ obtained from the same initial data by combinatorial patchworking.*

Then for all sufficiently small $t > 0$ the polynomial b_t defines in the first quadrant $\mathbb{R}^2_{++} = \{(x, y) \in \mathbb{R}^2 \mid x, y > 0\}$ a curve a_t such that the pair $(\mathbb{R}^2_{++}, a_t)$ is homeomorphic to the pair $(\operatorname{Int} \Delta,\ L \cap \operatorname{Int} \Delta)$.

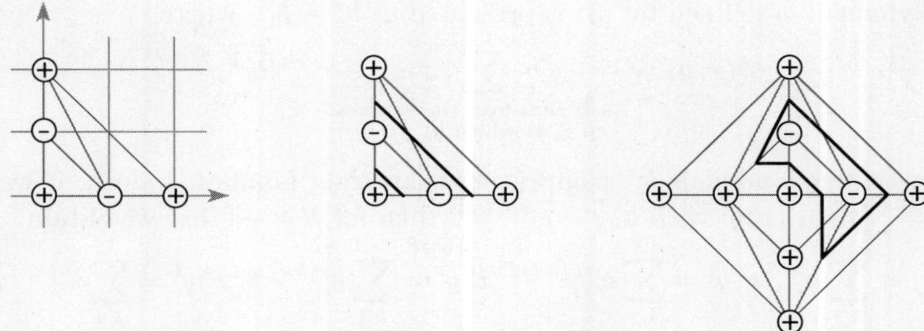

FIGURE 2. Patchworking: initial data, construction of the PL-curve in the first quadrant and on the whole plane. The corresponding algebraic curves are ellipses meeting the coordinate axes in their positive halves.

4.2. Patchwork in other quadrants

The Patchwork Theorem applied to $b_t(-x, y)$, $b_t(x, -y)$ and $b_t(-x, -y)$ gives a similar topological description of the curve defined in the other quadrants by b_t with sufficiently small $t > 0$. The results can be collected in the following natural combinatorial construction.

4.2.1. CONSTRUCTION OF THE PL-CURVE   Take copies $\Delta_x = s_x(\Delta)$, $\Delta_y = s_y(\Delta)$, $\Delta_{xy} = s(\Delta)$ of Δ, where s_x, s_y are reflections with respect to the coordinate axes and $s = s_x \circ s_y$. Denote by $A\Delta$ the square $\Delta \cup \Delta_x \cup \Delta_y \cup \Delta_{xy}$. Extend the triangulation τ to a symmetric triangulation of $A\Delta$, and the distribution of signs $\sigma_{i,j}$ to a distribution at the vertices of the extended triangulation by the following rule: $\sigma_{i,j}\sigma_{\varepsilon i,\delta j}\varepsilon^i\delta^j = 1$, where $\varepsilon, \delta = \pm 1$. In other words, passing from a vertex to its mirror image with respect to an axis we preserve its sign if the distance from the vertex to the axis is even, and change the sign if the distance is odd.

If a triangle of the triangulation of $A\Delta$ has vertices of different signs, select (as above) a midline separating pluses from minuses. Denote by AL the union of the selected midlines. It is a collection of polygonal lines contained in $A\Delta$. The pair $(A\Delta, AL)$ is called the *result of affine combinatorial patchworking*. Glue by s the sides of $A\Delta$. The resulting space $P\Delta$ is homeomorphic to the real projective plane $\mathbb{R}P^2$. Denote by PL the image of AL in $P\Delta$ and call the pair $(P\Delta, PL)$ the *result of projective combinatorial patchworking*.

Addendum to the Patchwork Theorem. *Under the assumptions of the Patchwork Theorem above, for all sufficiently small $t > 0$ there exist a homeomorphism $A\Delta \to \mathbb{R}^2$ mapping AL onto the the affine curve defined by b_t and a homeomorphism $P\Delta \to \mathbb{R}P^2$ mapping PL onto the projective closure of this affine curve.*

4.3. The simplest patchworking coincides with the construction of Section 3.4

The polynomial b_t defined by (3) is presented as $b_t^+ - b_t^-$, where

$$b_t^\varepsilon(x,y) = \sum_{\substack{(k,l)\ \text{runs over the vertices} \\ \text{of}\ \tau,\ \text{at which}\ \sigma_{k,l} = \varepsilon}} t^{\nu(k,l)}x^k y^l \,.$$

Observe that polynomials $b_t^\pm$ comprise dequantizing families. Indeed, if we take $p(x,y) = \sum_{k,l} a_{k,l}x^k y^l$ with $a_{k,l} = e^{-\nu(k,l)}$, then for $h = -1/\ln t$ we obtain

$$p_h(x,y) = \sum_{k,l} a_{k,l}^{1/h} x^k y^l = \sum_{k,l} e^{-\nu(k,l)/h} x^k y^l = \sum_{k,l} e^{\nu(k,l)\ln t} x^k y^l = \sum_{k,l} t^{\nu(k,l)} x^k y^l \,.$$

The Patchwork Theorem deals with sufficiently small positive t, while h in a dequantizing family of polynomials was a small positive number approaching 0. This is consistent with our setting $h = -1/\ln t$.

A monomial $a_{k,l}x^k y^l = e^{-\nu(k,l)}x^k y^l$ is presented in the logarithmic space by the graph of $w = ku + lv - \nu(k,l)$. Hence the graph of the maximum of linear forms corresponding to all monomials of p^+ and p^- is defined by

$$w = \max\{ku + lv - \nu(k,l) \mid (k,l) \text{ runs over vertices of } \tau\} \,. \tag{4}$$

In (4) we recognize the convex function conjugate to ν. The graph of (4) is a convex PL surface, whose natural stratification is dual to the triangulation τ of Δ: the face which lies on the plane $w = ku + lv - \nu(k,l)$ corresponds to the vertex (k,l) of τ, two such faces meet at an edge in the graph of (4) iff the corresponding

FIGURE 3. Patchworking of a counter-example to the Ragsdale Conjecture. A curve of degree 10 with 32 odd ovals, constructed by Itenberg [2].

vertices are connected with an edge of τ, three faces meet at a vertex iff the corresponding vertices of τ belong to a triangle of τ. In particular, we see that the configuration of planes satisfies the genericity condition of Section 3.4 and planes $w = ku + lv - \nu(k, l)$ corresponding to all monomials of $b_t^{\pm}$ show up in the graph of (4) as its faces.

Some of these faces correspond to monomials of b_t^{+}, the others to monomials of b_t^{-}. The edges which separate the faces of these two kinds constitute a broken line as in Section 3.4. These edges are dual to the edges of τ which intersect the result L of the combinatorial patchworking.

Therefore the topology of the projection to the (u, v)-plane of the broken line coincides with the topology of L in Δ.   □

5. Conclusion

We see that the quantum point of view (or its graphical log paper equivalent) gives a natural explanation to the simplest patchwork construction. The proofs become more conceptual and straight-forward. Of course, similar but slightly more involved quantum explanations can be given to all versions of patchwork. Let me shortly mention other problems which can be attacked using similar arguments.

First of all, this is the Fewnomial Problem. Although A. G. Khovansky [3] proved that basically all topological characteristics of a real algebraic variety can be estimated in terms of the number of monomials in the equations, the known estimates seem to be far weaker than conjectures. For varieties classical from the quantum point of view, strong estimates are obvious. It is very compelling to estimate how much the topology can be complicated by the quantizing deformations.

There seem to be deep relations between the dequantization of algebraic geometry considered above and the results of I. M. Gelfand, M. M. Kapranov and A. V. Zelevinsky on discriminants [1]. In particular, some monomials in a discriminant are related to intersections of hyperplanes in the dequantized polynomial.

Complex algebraic geometry also deserves a dequantization. Especially relevant may be amoebas introduced in [1].

References

[1] I. M. Gelfand, M. M. Kapranov and A. V. Zelevinsky, *Discriminants, resultants, and multidimensional determinants.* Mathematics: Theory & Applications. Birkhäuser Boston, Inc., Boston, MA, 1994. x+523 pp.

[2] I. Itenberg, Contre-exemples à la conjecture de Ragsdale, *C. R. Acad. Sci. Paris* **317**, Serie I (1993), 277–282.

[3] A. G. Khovansky, *Fewnomials.* Translated from the Russian by Smilka Zdravkovska. Translations of Mathematical Monographs, 88. American Mathematical Society, Providence, RI, 1991. viii+139 pp.

[4] G. L. Litvinov and V. P. Maslov, *The correspondence principle for Idempotent Calculus and some computer applications,* In book *Idempotency,* J. Gunawardena (Editor), Cambridge University Press, Cambridge, 1998, pp. 420–443.

[5] G. L. Litvinov, V. P. Maslov and A. N. Sobolevskiĭ, *Idempotent Mathematics and Interval Analisys,* Preprint math.SC/9911126, (1999).

[6] V. Ragsdale, *On the arrangement of the real branches of plane algebraic curves,* Amer. J. Math. **28** (1906), 377–404.

[7] O. Ya. Viro, Gluing algebraic hypersurfaces and constructions of curves, *Tezisy Leningradskoj Mezhdunarodnoj Topologicheskoj Konferentsii 1982,* Nauka, 1983, pp. 149–197 (Russian).

[8] O. Ya. Viro, Gluing of plane real algebraic curves and constructions of curves of degrees 6 and 7, *Lecture Notes in Math.,* vol. 1060, Springer-Verlag, 1984, pp. 185–200.

Department of Mathematics, Uppsala University, Box 480
S-751 06 Uppsala, Sweden
E-mail address: oleg@math.uu.se

Advances on Extremal Problems
in Number Theory and Combinatorics

Rudolf Ahlswede

1. Introduction

To keep an acceptable size, references not listed at the end are given by the Bibliography of the recent book [N] and/or the page number of [N].

1. Starting with solutions of extremal problems for *finite sets of numbers* under *divisibility constraints* [with L. H. Khachatrian, c.f. [PS]], then we describe the discovery of *correlation inequalities* implied by the AD-inequality [with L. H. Khachatrian, c.f. [N] P. C. Fishburn, L. Shepp, The Ahlswede-Daykin Theorem, 501–516] and conclude this section with results on extremal problems concerning *densities* of *primitive infinite sets of numbers* and related topics [with L. H. Khachatrian and A. Sárközy].

2. Divisibility properties for numbers naturally correspond to *intersection* properties for sets and thus there are also connections between methods. Especially, a new *pushing technique* for numbers led to the *discovery* of the method of *"generated sets"*, which made it possible to establish several Intersection Theorems in Combinatorics, which are highlighted by the *Complete Intersection Theorem* [with L. H. Khachatrian; P. Frankl wrote in [F1, p. 142] "At present this conjecture appears hopelessly difficult in general"], which, as a very special case, established the *4m-Conjecture* of Erdős/Ko/Rado from 1938 [[En]; [N] C. Bey, K. Engel, Old and New Results for the Weighted *t*-intersection Problem via AK-Methods, 45–74; [N] G. O. H. Katona, The Cycle Method and its Limits, 129–142; [CG]; B. Bollobás wrote in "Paul Erdős – Life and Work" the foreword of [GN]: "The third problem is from the 1961 paper of Erdős, Ko and Rado; it is, in fact, the last unsolved problem of that paper." "It is widely known that vast amounts of thought and ingenuity are required in order to earn \$ 500 on an Erdős problem; even so, this problem may be far harder than its price-tag suggests."].

3. We turn now to combinatorial work, which received its incentive from Information Theory and Computer Science. We demonstrate this for the area of information storage for rewritable memories, which led over Sperner type questions for "clouds" of antichains to *Higher Level Extremal Problems*. These problems are of one degree more complex than those usually considered: sets

take the role of elements, families of sets (clouds) take the role of sets, etc. [[N] P. L. Erdős, L. A. Székely, 117–124].

4. It also led to several kinds of *Vertex-Isodiametric Theorems in the Average* [with I. Althöfer, N. Cai] and *Edge Isoperimetric Theorems* [with N. Cai] which are rate-wise optimal and proved by novel *information theoretic* approaches.

5. We continue with several basic topics (partitions, monochromatic rectangles, shadows and isoperimetry under sequence-subsequence relations, antichains: splitting, AZ-identities, dimension constraints) from Sequence Spaces, which mostly were influenced by the analysis of Communication Complexity and Unconventional Coding. The most frequent coauthors are N. Cai, Z. Zhang, L. Bassalygo and M. Pinsker.

6. Finally, we conclude with counterexamples to known conjectures and with a list of seemingly basic open problems and new conjectures.

2. Contributions to Combinatorial Number Theory

2.1. Extremal problems under divisibility constraints for finite sets

Conjecture 2.1 (Erdős 1962). *The set $E(n, k)$ of integers not exceeding n, which are divisible by one of the first k primes, has maximal cardinality among the subsets of $[n] = \{1, 2, \ldots, n\}$ without $k + 1$ elements, which are pairwise relatively prime.*

He caught our interest in a lecture delivered in 1992. It stimulated us to make a systematic investigation of this and related number theoretical problems. When viewed combinatorially as an extremal problem for products of chains (by the prime decomposition) with a truncation condition (caused by the property "smaller than n") the issue is to understand whether the problem is just combinatorial in nature or does depend on the prime number distribution. The latter is the case, the conjecture was disproved, but proved for large n, in joint work [85], [101] with L. Khachatrian. Thus began a very fruitful cooperation. Immediate successes were proofs of other well-known conjectures of Erdős and Erdős/Graham (all in [E1] and [E2]).

More importantly we gained an understanding for the sensitivity towards the distribution of the primes and thus our program was very rewarding. The analysis led to the discovery of a new "pushing" method with wide applicability also in Combinatorics, where it led to the solution of several well-known problems like the 4m-conjecture, one of the oldest problems in combinatorial extremal theory (see Section 3) or the isodiametric problem in Hamming spaces (see Section 5). Complete proofs can be found in [En]. We describe now the results and for this adopt the following notation. $\mathbb{N}$ denotes the set of positive integers and $\mathbb{P} = \{p_1, p_2, \ldots\} = \{2, 3, 5, \ldots\}$ denotes the set of all primes.

For two numbers $u, v \in \mathbb{N}$ we write $u \mid v$ iff u divides v, (u, v) stands for the largest common divisor of u and v, and $[u, v]$ is the smallest common multiple of u

and v. The numbers u and v are called coprimes, if $(u, v) = 1$. We are particularly interested in the sets

$$\mathbb{N}_s = \left\{ u \in \mathbb{N} : \left(u, \prod_{i=1}^{s-1} p_i \right) = 1 \right\} \text{ and } \mathbb{N}_s(n) = \mathbb{N}_s \cap [n] . \tag{1}$$

Erdős introduced $f(n, k, s)$ as the largest integer r for which an $A \subset \mathbb{N}_s(n)$, $|A| = r$, exists with no $k + 1$ numbers in A being coprimes. Certainly there are no $k + 1$ coprimes in the set

$$\mathbb{E}(n, k, s) = \{ u \in \mathbb{N}_s(n) : u = p_{s+i} v \text{ for some } i = 0, 1, \ldots, k - 1 \} . \tag{2}$$

The case $s = 1$, in which we have $\mathbb{N}_1(n) = [n]$, gives Conjecture 2.1 above.

The papers [ESS2] and [ESS4] are centered around Conjecture 2.1. Whereas it is easy to show that it is true for $k = 1$ and $k = 2$, it was proved for $k = 3$ by C. Szabó and G. Tóth 1985 and for $k = 4$ by Motzkin 1988. The popularity of this problem is documented by R. Freud in [Fr].

General Conjecture (Erdős 1980). $f(n, k, s) = |\mathbb{E}(n, k, s)|$ *for all* $n, k, s \in \mathbb{N}$.

Theorem 2.2 (Ahlswede/Khachatrian [101]). *For every* $k, s \in \mathbb{N}$ *there exists an* $n(k, s)$ *such that for all* $n \geq n(k, s)$ $\mathbb{E}(n, k, s) = f(n, k, s)$ *and this optimal set is unique.*

Erdős mentions that he did not succeed in settling even the special case.

Conjecture 2.3. $f(n, 1, s) = |\mathbb{E}(n, 1, s)|$ *for all* $n, s \in \mathbb{N}$.

Whereas in [85] Conjecture 2.1 was disproved for $k = 212$, surprisingly Conjecture 2.3 is true.

Theorem 2.4 (Ahlswede/Khachatrian [104]). *For every* $s \in \mathbb{N}$ *and* n $f(n, 1, s) = |\mathbb{E}(n, 1, s)|$ *and the optimal configuration is unique.*

Notice that $\mathbb{E}(n, 1, s) = \{ u \in \mathbb{N}_1(n) : p_s \mid u; p_1, \ldots, p_{s-1} \nmid u \}$.

Studying square free numbers $\mathbb{N}^*$ one is naturally led to sets $\mathbb{E}^*(n, k, s) = \mathbb{E}(n, k, s) \cap N^*$ etc and to the function $f^*(n, k, s)$. Then

Theorem 2.4*. *For all* $s, n \in \mathbb{N}$ $f^*(n, 1, s) = |\mathbb{E}^*(n, 1, s)|$.

But even for squarefree numbers "Erdős sets" are not always optimal, that is, $f^*(n, k, 1) \neq |\mathbb{E}^*(n, k, 1)|$ can occur. Also $f^*(n, 2, s) \neq |\mathbb{E}^*(n, 2, s)|$ happens for $p_s = 101$ and $n \in [109 \cdot 113, 101 \cdot 127)$.

We generalize and analyse Conjecture 2.3 first for quasi-primes in order to understand how its validity depends on the distribution of the quasi-primes and primes. Our main result is a simply structured sufficient condition on this distribution. Using sharp estimates on the prime number distribution by Rosser and Schoenfeld we show that this condition holds for $\mathbb{Q} = \{ p_s, p_{s+1}, \ldots \}$, $s \geq 1$, as a set of quasi-primes and thus Theorem 2.4 follows.

Erdős/Graham asked for the maximal value $k = g(n)$ such that there are numbers $1 < a_1 < \cdots < a_k = n$, $(a_i, a_j) \neq 1$. Let $M(A) = \mathbb{N} \times A$ denote the set of multiples of A.

Conjecture 2.5 (After a little correction). *Let* $n = q_1^{\alpha_1} \ldots q_r^{\alpha_r}$, $\alpha_i \geq 1$, $\alpha_i \in \mathbb{N}$ *and* $Q = \{q_1 < \cdots < q_r\} \subset \mathbb{P}$, *then* $g(n) = \max_{1 \leq j \leq r} |M(2q_1, \ldots, 2q_j, q_1 \ldots q_j) \cap \mathbb{N}(n)| = f(n, Q)$, *say.*

We consider a more general and seemingly more natural problem by looking at sets of integers with pairwise common divisor and a factor from a specified set of primes with growth condition.

Let $Q = \{q_1, \ldots, q_r\} \subset \mathbb{P}$ and let

$$g(n, Q) = \max\{|A| : A \subset \mathbb{N}(n), (a, a') \neq 1, \left(a, \prod_{i=1}^{r} q_i\right) \neq 1 (\forall a, a' \in A)\}. \qquad (3)$$

Theorem 2.6. *For every* $Q = \{q_1, \ldots, q_r\} \subset \mathbb{P}$ *and* $n \geq \prod_{i=1}^{r} q_i$ $g(n, Q) = f(n, Q)$.

Clearly, this implies the **Conjecture of Erdős/Graham**. For $n < \prod_{q \in Q} q$ the conclusion of Theorem 2.6 does not hold.

Example 2.7. *Let* $Q = \{q_1, q_2, \ldots, q_{r-1}, q_r\} = \{5, 7, \ldots, p_{r+1}, q_r\}$ *that is* $q_i = p_{i+2}$ *for* $i = 1, \ldots, r - 1$. *Further* $q_{r-1} = p_{r+1} > 1000$, $n = 2 \cdot 3 \cdot 11 \cdot \prod_{i=1}^{r-1} q_i$ *and* $\frac{n}{1000} < q_r < \frac{n}{1000}$. *With Bertrand's postulate we establish the claim.*

2.2. Correlation inequalities from the AD-inequality

Next we report on new density inequalities for sets of multiples. For infinite sets $A, B \subset \mathbb{N}$ consider the set of least common multiples $[A, B] = \{[a, b] : a \in A, b \in B\}$, the set of largest common divisors $(A, B) = \{(a, b) : a \in A, b \in B\}$, the set of products $A \times B = \{a \cdot b : a \in A, b \in B\}$, and the sets of their multiples $M(A) = A \times \mathbb{N}, M(B), M[A, B], M(A, B)$, and $M(A \times B)$, resp. Our discoveries are the inequalities

$$\mathbf{d}M(A, B)\mathbf{d}M[A, B] \geq \mathbf{d}M(A) \cdot \mathbf{d}M(B) \geq \mathbf{d}M(A \times B), \qquad (4)$$

where $\mathbf{d}$ denotes the asymptotic density. The first inequality is by the factor $\mathbf{d}M(A, B)$ sharper than Behrend's well-known inequality. This in turn is a generalisation of an earlier inequality of Rohrbach and Heilbronn, which settled a conjecture of Hasse concerning an identity due to Dirichlet. Our second inequality does not seem to have predecessors.

Observing the similarity to the AD inequality led to our main discovery, the inequality (with L. H. Khachatrian [102])

$$\underline{\mathbf{D}}(A, B)\underline{\mathbf{D}}[A, B] \geq \underline{\mathbf{D}}A\underline{\mathbf{D}}B,$$

where A, B are arbitrary sets of positive integers, $(A, B) = \{(a, b) : a \in A, b \in B\}$ is the set of largest common divisors, $[A, B] = \{[a, b] : a \in A, b \in B\}$ is the set of least common multiples, and $\underline{\mathbf{D}}$ denotes the lower Dirichlet density. It is much more general than the previous inequality for multiples of sets.

This is more than an analogy: AD implies this number theoretical Correlation inequality. For reasons of scaling it is important to work with Dirichlet density!

Similarly, Behrend's inequality can be obtained as a *number theoretical form* of FKG, but it actually preceded it. The second inequality in (4) is the twin of the van der Berg/Kesten inequality etc.

Now AD has not only combinatorial and probabilistic correlation inequalities (see [N] Fishburn/Shepp, 501–516) as consequences, but also those above in Number Theory! We conclude with an important

Remark 2.8. *In the literature "AD-inequality" ("4-Function Theorem") refers to the inequality of [30], which holds for lattices. We emphasize that* **the much more general inequality** *of [33] makes no reference to lattices and* **should have a wider range of applications**.

2.3. Densities for primitive, prefix free, quotient, and squarefree sets

In this section we report on recent work with L. H. Khachatrian and A. Sárközy.

We begin with [150] "**On the counting function of primitive sets of integers**".

After **Besicovitch** (1934) gave an example of a set of multiples $M(B)$ *without* a density, interest arose in *primitive* sets $A \subset \mathbb{N}$: for $a, a' \in A$; $a \neq a'$ always $a \nmid a'$. Set $F(n) \triangleq$ greatest cardinality of primitive A in $[n]$ and notice that

$$F(n) = n - \left\lceil \frac{n}{2} \right\rceil \ \left(= \left(\frac{1}{2} + 0(1) \right) n \right) . \tag{5}$$

Besicovitch and Erdős 1935: $\forall \, \varepsilon > 0 \ \exists$ primitive $A \subset \mathbb{N}$ with $\overline{d}(A) > \frac{1}{2} - \varepsilon$.

Behrend 1935: For $A \subset [n]$ primitive $\sum_{a \in A} \frac{1}{a} < c_1 \frac{\log n}{(\log \log n)^{1/2}}$.

Erdős 1935: A primitive $A \subset \mathbb{N}$ (finite or infinite) satisfies $\sum_{a \in A} \frac{1}{a \log a} < c_2$.

Corollary 2.9. *If $A \subset \mathbb{N}$ is primitive, then for the counting function A*

$$A(x) < \frac{x}{\log \log x \log \log \log x} \quad \text{for } \infty \text{ many } x . \tag{6}$$

How far is this upper bound from the best possible? This question is closely related to one of the **favourite problems** of Erdős (mentioned in numerous papers). We use here the formulation of [ESS3]:

"The following problem seems difficult: Let $b_1 < b_2 < \ldots$ be an infinite sequence of integers. What is the necessary and sufficient condition that there should exist a primitive sequence $a_1 < a_2 < \ldots$ satisfying $a_n < b_n$ for every n? We must have $\sum_{i=1}^{\infty} \frac{1}{b_i \log b_i} < \infty \ldots$ It is not clear whether a simple necessary and sufficient condition exists." This is followed by a lengthy discussion of the problem how large one can make $\sum_{a \leq x} \frac{1}{a}$ uniformly in x for a primitive set $\{a_1 < a_2 < \ldots\}$ (see also [ESS1]).

It seems to be a *more natural* (although more difficult) problem to *replace* here the sum $\sum_{a \leq x} \frac{1}{a}$ by the counting function $A(x)$ i.e. to study the problem how large one can make $A(x)$ uniformly in x for a primitive set A. We provide a quite satisfactory answer by proving that (6) is best possible apart from a factor $(\log \log \log x)^{\varepsilon}$:

152 R. Ahlswede

Theorem 2.10. *For all $\varepsilon > 0 \; \exists$ infinite primitive set $A \subset \mathbb{N}$ such that for $x > x_0(\varepsilon)$ we have $A(x) > \frac{x}{\log\log x (\log\log\log x)^{1+\varepsilon}}$.*

Our recent interest in primitive sets arose while we investigated two related new concepts **"prefix-free sets"** and **"suffix-free sets"**, which are of information theoretical background.

For $a, b \in \mathbb{N}^*$ square free we write $a|_p b$ (a is prefix of b) if for primes $p_1 < \cdots < p_r < p_{r+1} < \cdots < p_t \quad a = p_1 \ldots p_r, \; b = p_1 \ldots p_{r+1} \ldots p_t$. Similarly, $c = p_{r+1} \ldots p_t$ is suffix of b and we write $c|_s b$.

If for $A \subset \mathbb{N}^*$ there are no $a, b \in A$ with $a|_p b$ (resp. $a|_s b$) then A is said to be prefix-free (resp. suffix-free). (Both notions could be extended to non-square-free cases.) There is a close connection between prefix-freeness and primitivity: if a set $A \subset \mathbb{N}$ is primitive, then it is prefix-free.

We present first the "prefix-free analog" of (5). Let $G(n)$ denote the cardinality of the greatest prefix-free set from $\mathbb{N}^*(n)$, and let $P^+(a)$ denote the smallest prime greater than $P(a)$.

Theorem 2.11 ([N, p. 1–16]).
  (i) *The set $B(n) = \{b : b \in \mathbb{N}^*(n), bP^+(b) > n\}$ is prefix-free and $G(n) = |B(n)|$.*
  (ii) $\lim_{n\to\infty} \frac{G(n)}{|\mathbb{N}^*(n)|} = 1$.
  (iii) *For all $\varepsilon > 0$ there is an infinite prefix-free set $A \subset \mathbb{N}^*$ with $\overline{d}^*(A) > 1 - \varepsilon$.*

The "prefix-free analog" to Behrend's result reflects an interesting difference between primitive sets and prefix-free sets. Indeed, consider now instead of $G(n)$
$$E(n) = \max_{\text{prefix-free } A \subset \mathbb{N}^*(n)} \sum_{a \in A} \frac{1}{a}.$$

Theorem 2.12 ([N, p. 1–16]). *For every $\varepsilon > 0$ and $n > n_2(y)$, suitable, $0,2689 - \varepsilon < \frac{E(n)}{\sum_{b \in \mathbb{N}^*(n)} \frac{1}{b}} < 0,7311 + \varepsilon$. Actually, we know for every $n \in \mathbb{N}$ the unique optimal prefix-free $A \subset \mathbb{N}^*(n)$ for which $E(n)$ is assumed, but the value is hard to estimate.*

Since Erdős 1935 above uses in the proof only the prefix property of a primitive sequence, obviously for prefix-free $A \subset \mathbb{N}$ $\quad \sum_{a \in A} \frac{1}{a \log a} < c_2$ and also (6) in Corollary 2.9 and Theorem 2.10 "primitive" can be replaced by "prefix-free".

While the behaviour of prefix-free and primitive sets is similar as far as the maximal growth of the counting function is concerned, the behaviour of the suffix-free sets is very different. Let $H(n)$ denote the cardinality of the largest suffix-free set selected from $\mathbb{N}^*(n)$.

Theorem 2.13 ([N, p. 1–16]).
  (i) *The set $C(n) = \{c \in \mathbb{N}^*(n) : 2 \mid c\} \cup \{\mathbb{N}^*(n) \cap \left(\frac{n}{2}, n\right]\}$ is suffix-free and $|C(n)| = H(n)$.*
  (ii) $\lim_{n\to\infty} \frac{H(n)}{|\mathbb{N}^*(n)|} = \frac{2}{3}$.
  (iii) *For every $\varepsilon > 0$ there exists a suffix-free set C with $\overline{d}^* C > \frac{2}{3} - \varepsilon$.*

Finally we consider logarithmic densities of suffix-free sets. Let $K(n) = \max_{\text{suffix-free } A \subset \mathbb{N}^*(n)} \sum_{a \in A} \frac{1}{a}$.

In contrast to the case of prefix-free sets here we have a simple description of the optimal set, which yields

Theorem 2.14 ([N, p. 1–16]). $\lim_{n\to\infty} \frac{K(n)}{\sum_{a\in\mathbb{N}^*(n)} \frac{1}{a}} = \frac{31}{72}$.

2.3.1. ON THE QUOTIENT SEQUENCE OF SEQUENCES OF INTEGERS   For $A \subset \mathbb{N}$, $a \in A$ let $Q_A^a \triangleq$ set of integers q such that

$$aq \in A \ a > 1 \ Q_A \triangleq \bigcup_{a\in A} Q_A^a. \tag{7}$$

So $Q_A =$ set of integers $q = \frac{a'}{a} > 1$ with $a, a' \in A$ called the quotient set of A.

By Behrend's and Erdős' Theorem the quotient set of a dense set A is *non-empty* (obvious).

The study of quotient sets of "dense" sets started with

Theorem 2.15 (Pomerance/Sárközy 1988). *There exist constants c_3, N_0 such that if $N \in \mathbb{N}$, $N > N_0$, $\mathcal{P}$ is a set of primes not exceeding N and if $A \subset \{1, 2, \ldots, N\}$ with*

$$\sum_{p\in\mathcal{P}} \frac{1}{p} > c_3 \ \text{and} \ \sum_{a\in A} \frac{1}{a} > 10 \log N \left(\sum_{p\in\mathcal{P}} \frac{1}{p}\right)^{-1/2} \tag{8}$$

then *there is a $q \in Q_A$ such that $q \mid \prod_{p\in\mathcal{P}} p$.*

Here we study *density related* properties of Q_A. Our first goal is to study connections between $\overline{\delta}(A)$ and $\overline{\delta}(Q_A)$. First we thought that for all $A \subset \mathbb{N}$

$$\overline{\delta}(Q_A) \geq \overline{\delta}(A). \tag{9}$$

Example 2.16. $A = \{2m, 3m \ or \ 5m \ with \ m \in \mathbb{N}, (m, 30) = 1\}$. *Then* $\overline{\delta}(A) = \delta(A) = d(A) = \frac{62}{225}$ *and* $\overline{\delta}(Q_A) = \delta(Q_A) = d(Q_A) = \frac{4}{15} = \frac{30}{31}\overline{\delta}(A)$.

Still, there is a connection between these densities, but they can be far apart.

Theorem 2.17.
  (i) *If for $A \subset \mathbb{N} \ \overline{\delta}(A) > 0$, then $\overline{\delta}(Q_A) > 0$.*
  (ii) *For all $\varepsilon > 0$, $\delta > 0$ there is a set $A \subset \mathbb{N}$ such that $\underline{\delta}(A) > 1 - \varepsilon$ however, $\overline{\delta}(Q_A) < \delta$.*

Proof. (i) Erdős/Davenport Theorem. (ii) Turan-Kubilius inequality.

Remark 2.18. *Erdős studied also* $Q_A^\infty = \bigcap_{n=1}^\infty \left(\bigcup_{a\geq n, a\in A} Q_A^a\right) = $ *integers $q > 1$ with ∞-many representations* $q = \frac{a'}{a}$, $a, a' \in A$.

We conclude with [159] "On primitive sets of squarefree integers". For $A \subset \mathbb{N}$ let $S(A) = \sum_{a\in A} \frac{1}{a}$. Erdős/Sárközy/Szemerédi proved in 1967:

$$\max_{A\subset[n] \text{ primitive}} S(A) = \big(1 + o(1)\big) \frac{\log n}{(2\pi \log \log n)^{1/2}} \ \text{as} \ n \to \infty. \tag{10}$$

154   R. Ahlswede

Theorem 2.19. *Let $Q = \{q_1, q_2, \ldots\} = \{p_1^{\alpha_1}, p_2^{\alpha_2}, \ldots\}$ (with $p_1 < p_2 < \ldots$) be a set of powers of distinct primes with $S(Q) < \infty$, then we have*

$$\max_{\substack{A \subset [n] \ primitive \\ q \nmid a \ for \ a \in A, q \in Q}} S(A) = \big(1 + o(1)\big) \prod_{q \in Q} \left(1 - \frac{1}{q}\right) \frac{\log n}{(2\pi \log \log n)^{1/2}} \quad as \ n \to \infty. \quad (11)$$

Note that here $Q = \varnothing$ is allowed and, indeed, in this special case we obtain the Theorem of Erdős/Sárközy/Szemerédi. For $Q = \{2^2, 3^2, 5^2, \ldots, p^2, \ldots\}$, the result was conjectured by Pomerance/Sárközy in 1988.

Another important special case is when Q consists of the primes not exceeding a fixed number K:

Corollary 2.20. *If $K \geq 2$, then we have*

$$\max_{\substack{A \subset [n] \ primitive \\ \left(a, \prod_{p \leq K} p\right) = 1 \ for \ all \ a \in A}} S(A) = \big(1 + o(1)\big) \prod_{p \leq K} \left(1 - \frac{1}{p}\right) \frac{\log n}{(2\pi \log \log n)^{1/2}} \quad as \ n \to \infty.$$

Moreover, we can prove that if Q is finite, then Q need not consist of prime powers, it suffices to assume coprimality:

Theorem 2.19'. *Let $Q = \{q_1, \ldots, q_t\}$ be a finite set of pairwise coprime positive integers: $(q_i, q_j) = 1$ for $1 \leq i < j \leq t$. Then (11) holds.*

It comes perhaps as a **surprise** that the heuristics that the density $\frac{6}{\pi^2}$ of the square free integers $\mathbb{N}^*$ should extend to the maximal primitive set, which would give

$$\max_{A \subset \mathbb{N}^*(n) \ primitive} |A| = \big(1 + o(1)\big) \frac{6}{\pi^2} \max_{A \in \mathbb{P}_n} |A| = \big(1 + o(1)\big) \frac{3}{\pi^2} n \ as \ n \to \infty, \ \textbf{fails}.$$

We will prove that this maximum is much greater, but can be estimated surprisingly well.

Theorem 2.21. *For N large we have $0.6362 \frac{6}{\pi^2} N < \max_{A \subset \mathbb{N}^*(N)} |A| < 0.6366 \frac{6}{\pi^2} N$.*

2.4. Cross-primitive sequences

With L. H. Khachatrian we introduced and analyzed the following concept. It is reported in [116] "Classical results on primitive and recent results on cross-primitive sequences", Paul Erdős 80, Graham/Nesetril edit. (A, B), $A, B \subset C \subset \mathbb{N}$ is *cross-primitive* in C, if $a \nmid b$; $b \nmid a$ for $a \in A, b \in B$.

Theorem 2.22. *For $M_n = \max_{(A,B) cross\text{-}primitive \ in \ [n]} |A||B|$*
  (i) $\frac{M_n}{n^2} \leq \frac{1}{4}$.
  (ii) $\lim_{n \to \infty} \frac{M_n}{n^2} = \frac{1}{4}$.

Theorem 2.23. *For $C = \mathbb{N}$*
  (i) $\max_{(A,B) cross\text{-}primitive} \overline{d}(A)\overline{d}(B) = \frac{1}{4}$.
  (ii) $\max_{(A,B) cross\text{-}primitive} \underline{d}(A) \cdot \underline{d}(B) = \frac{1}{16}$
     $= \max_{(A,B) cross\text{-}primitive, d(A), d(B) \ exist} d(A)d(B)$.

3. Combinatorial Intersection Theorems

3.1. The complete intersection theorem and some extensions

In the paper [122] together with L. H. Khachatrian we were concerned with one of the oldest problems in combinatorial extremal theory. A system of sets $\mathcal{A} \subset \binom{[n]}{k}$ is called t-intersecting, if $|A_1 \cap A_2| \geq t$ for all $A_1, A_2 \in \mathcal{A}$, and $I(n, k, t)$ denotes the set of all such systems. The investigation of the function $M(n, k, t) = \max_{\mathcal{A} \in I(n,k,t)} |\mathcal{A}|$, $1 \leq t \leq k \leq n$, and the structure of maximal systems was initiated by Erdős, Ko and Rado. Already in 1938 they proved and in 1961 they published

Theorem EKR. *For $1 \leq t \leq k$ and $n \geq n_0(k, t)$ (suitable) $M(n, k, t) = \binom{n-t}{k-t}$.*

This means that for large n the "naive" configuration $\mathcal{A} = \left\{ A \in \binom{[n]}{k} : [1, t] \subset A \right\}$ is optimal. The smallest $n_0(k, t)$ for which this is the case has been determined by Frankl 1978 for $t \geq 15$ in [F2] and Wilson 1984 for all t in [W]: $n_0(k, t) = (k - t + 1)(t + 1)$. All cases are settled in

Theorem 3.1 (Complete Intersection Theorem) (Ahlswede/Khachatrian [122]).
For $1 \leq t \leq k \leq n$ with $(k - t + 1)\big(2 + \frac{t-1}{r+1}\big) < n < (k - t + 1)\big(2 + \frac{t-1}{r}\big)$ for some $r \in \{0\} \cup \mathbb{N}$ we have

$$M(n, k, t) = |\mathcal{F}_r| = \left| \left\{ F \in \binom{[n]}{k} : |F \cap [1, t + 2r]| \geq t + r \right\} \right|$$

and $\mathcal{F}_r$ is —up to permutations— the unique optimum. (By convention $\frac{a}{0} = \infty$.)
For $(k - t + 1)\big(2 + \frac{t-1}{r+1}\big) = n$ for $r \in \mathbb{N}\{0\}$ we have $M(n, k, t) = |\mathcal{F}_r| = |\mathcal{F}_{r+1}|$ and an optimal system equals —up to permutations— either $\mathcal{F}_r$ or $\mathcal{F}_{r+1}$.

In particular, this theorem shows the validity of the **famous 4m-conjecture** of Erdős, Ko, Rado (1938), that is $M(4m, 2m, 2) = |\{F \in \binom{[4m]}{2m} : |F \cap [1, 2m]| \geq m + 1\}|$.

Remark 3.2. *The EKR Theorem is the most frequently quoted result of Erdős. In 1983 Deza and Frankl wrote a paper "The Erdős/Ko/Rado Theorem - 22 years later".*

The theorem presented and proved in [114] can be viewed as an extension or improvement of the Complete Intersection Theorem, just mentioned above. It goes considerably beyond the well-known Hilton/Milner Theorem and completely answers the question of determination of non-trivial t-intersecting families. (An $\mathcal{A} \in I(n, k, t)$ is called non-trivial if $\left| \bigcup_{A \in \mathcal{A}} A \right| < t$.)

In [137] together with H. Aydinian and L. H. Khachatrian the problem of maximal intersecting systems for direct products is considered. This problem was initiated by Frankl and arose in connection with a result of Sali. Let $n = n_1 + \cdots + n_m$, $k = k_1 + \cdots + k_m$, $[n] = [n_1] \cup [n_2] \cup \cdots \cup [n_m]$, $\mathcal{H} = \left\{ F \in \binom{[n]}{k} : |F \cap [n_i]| = k_i \right.$ for $i = 1, \ldots, m \}$. For given integers t_i, $1 \leq t \leq t_i \leq k_i$, $1 \leq i \leq m$, we may say that $\mathcal{A} \subset \mathcal{H}$ is $(t_1, \ldots, t_m)$-intersecting, if for every $A, B \in \mathcal{A}$ there exists an i, $1 \leq i \leq m$, such that $|A \cap B \cap \Omega_i| \geq t_i$ holds. Denote the set of such systems by $I(\mathcal{H}, t_1, \ldots, t_m)$. The problem is to determine $\max_{\mathcal{A} \in I(\mathcal{H}, t_1, \ldots, t_m)} |\mathcal{A}|$.

The case $t_1 = t_2 = \cdots = t_m = 1$ has been solved by Frankl. Here is the complete solution.

Theorem 3.3 (Ahlswede/Aydinian/Khachatrian [137]). *Let $n_i \geq k_i \geq t_i \geq 1$ for $i = 1, \ldots, m$, then $\max_{\mathcal{A} \in I(\mathcal{H}, t_1, \ldots, t_m)} = \max_i \frac{M(n_i, k_i, t_i)}{\binom{n_i}{k_i}} |\mathcal{H}|$.*

We emphasize that the combination of this theorem and Theorem 3.1 gives an explicit value. The proof is heavily (but not only!) based on ideas and methods from [112], in particular the method of "generated sets" (c.f. [N] Bey/Engel, "Old and New Results for the Weighted t-Intersection Problem via AK-Methods", 45–74;) it takes a central role in the recent book "Sperner Theory" by K. Engel.

3.2. The diametric theorem in Hamming spaces

For $\alpha^{[n]} = \{0, 1, \ldots, \alpha - 1\}^n$, endowed with the Hamming distance function d_H $\mathcal{A} \subset \alpha^{[n]}$ has a diameter d, if $\mathrm{diam}(\mathcal{A}) = \max_{a^n, b^n \in \mathcal{A}} d_H(a^n, b^n) = d$.

Given n, α, d find: $N_\alpha(n, d) = \max_{\mathcal{A} \subset \alpha^{[n]}, \mathrm{diam}(\mathcal{A}) = d} |\mathcal{A}|$ or equivalently "*find optimal anticodes*".

Previously known were the cases with pairs (d, α) of the form $(d, 2)$ [K], $(n - 1, \alpha)$ [B], $\{(d, \alpha) : n \leq d + \alpha - 1\}$ [FF], $\{(d, \alpha) : n \geq (\alpha - 1)^{d-1} + d\}$ [Ahlswede/Cai/Zhang [72]]. These authors conjectured the theorem below. It was conjectured in an equivalent form by Frankl/Füredi already in 1980. For $0 \leq i \leq \frac{d}{2}$ define $\mathcal{K}_i = \{a^n \in \alpha^{[n]} : (a_1, \ldots, a_n - d + 2i) \text{ has at least } n - d + i \text{ zeros}\}$. Clearly $\mathcal{K}_i$ has diameter d.

Theorem 3.4 (Ahlswede/Khachatrian [132]). *Let r be the largest integer s.t. $n - d + 2r < \min\left\{n + 1, n - d + 2 \cdot \frac{n-d+1}{\alpha-2}\right\}$, then $N_\alpha(n, d) = |\mathcal{K}_r|$.*

Moreover, up to permutation of $(12 \ldots n)$ and permutations of the alphabet in the components the **optimal configuration is unique**, *unless $n - d > 1$, $n - d + 2\frac{n-d-1}{\alpha-2} \leq n$ and $\frac{n-d-1}{\alpha-2}$ is integral, in which case we have* **two optimal configurations**: $\mathcal{K}_{\frac{n-d-1}{\alpha-2}}$ *and* $\mathcal{K}_{\frac{n-d-1}{\alpha-2}-1}$.

The result is derived from the Complete Intersection Theorem (which can also be viewed as a Diametric Theorem for d_H on the space $\binom{[n]}{k}$) via a Comparison Lemma.

3.3. A pushing-pulling method

It came as a surprise to us that at first we did not succeed to derive Katona's Intersection Theorem for the unrestricted case, that is in the space $2^{[n]}$, by the method of "generated sets". This led us to the discovery of another method, which yields Theorem 3.1 and Katona's Intersection Theorem. Subsequently we found a way to derive Katona's Theorem from Theorem 3.1 via another Comparison Lemma. This is the most complicated proof, where several simple proofs exist, but it teaches something about methods, which made progress possible on the t-Intersection Problem in the truncated Boolean Lattice covering the restricted and the unrestricted intersection problem as special cases (Ahlswede/Bey/Engel/Khachatrian [ABEK]). Whereas there are vertex- and edge-isoperimetric theorems (see Section 5) it went unsaid that diametric theorems are vertex-diametric theorems. We complete the story by introducing edge-diametric theorems into combinatorial extremal theory. Using again the pushing/pulling method we establish such a result for $\mathcal{V} = \{0,1\}^n$ and $\mathcal{E} = \{(a^n, b^n) : a^n, b^n \in \mathcal{V}\}$. Results and methods of this section are discussed in the survey [N, p. 45–74].

3.4. Other types of intersection problems

One type consists in looking at intersecting chains in the Boolean Lattice. It is mentioned in the context of Higher Level Extremal Problems in Section 4. Another one, with origin in and interest to Computer Science, was communicated to us by R. C. Mullin in 1990 in Oberwolfach.

For a finite alphabet $[\alpha] = \{0, 1, \ldots, \alpha - 1\}$ we consider the set $\alpha^{[n]}$ of words of length n and also its subsets W_α^n of words without repetition of letters, that is, $W_\alpha^n = \{x^n = (x_1, x_s, \ldots, x_n) \in \alpha^{[n]} : x_s \neq x_t \text{ for } s \neq t\}$. We write $x^n \diagup\!\diagdown y^n$ if for some $s \neq t$ $x_s = y_t$. The set $F \subset \alpha^{[n]}$ is "good", if for all $x^n, y^n \in F$ $x^n \diagup\!\diagdown y^n$. Denoting the family of all good sets in W_α^n by $\mathcal{F}_\alpha^n$ the quantity of interest is $f_\alpha^n = \max\{|F| : F \in \mathcal{F}_\alpha^n\}$. Its determination constitutes an extremal problem in a (growing) class of similar problems whose prototype or historically first candidate is the intersection problem of EKR.

Clearly, it is certainly also meaningful to study $\mathcal{G}_\alpha^n$, the family of all good sets in $\alpha^{[n]}$, and the quantity $g_\alpha^n = \max\{|E| : E \in \mathcal{G}_\alpha^n\}$.

The functions f_α^n and g_α^n are rather complex. We present here results for the first two non-trivial configurations of the parameters α and n, namely the cases $n = \alpha - 1$ and $n = 3$. Also, we have a limit theorem for α tending to infinity. Specifically, we have the following results.

Theorem 3.5 (Ahlswede/Cai [91]).
   (i) $f_\alpha^{\alpha-1} = \frac{1}{2}|W_\alpha^{\alpha-1}| = \frac{1}{2}\alpha!$ *Moreover, we determine all optimal configurations.*
   (ii) $f_\alpha^3 = f_\infty^3 = f_4^3 = 12$ *for* $\alpha \geq 4$.

Theorem 3.6 (Ahlswede/Cai [91]).
   (i) $g_\alpha^3 = 3\alpha + 7$ *for* $3 \leq \alpha < \infty$.
   (ii) $\lim_{\alpha \to \infty} \frac{g_\alpha^n}{\binom{\alpha-1}{n-2}} = \binom{n}{2}(n-2)!$ *or, equivalently,* $g_\alpha^n = \alpha^{n-2}(\binom{n}{2} + o(1))$ *as* $\alpha \to \infty$.

4. Higher Level Extremal Problems

4.1. Coding for write-efficient rewritable memories

Imagine a tape with n cells into which we can write letters from an alphabet $\mathcal{X}$. A word $x^n = (x_1, \ldots, x_n)$ stores some messages. When we want to update this record to a message represented by $y^n = (y_1, \ldots, y_n)$ the per letter costs $\varphi(x_t, y_t)$ add up to $\varphi_n(x^n, y^n) = \sum_{t=1}^{n} \varphi(x_t, y_t)$. In order to be able to update many messages under a cost constraint D we come to the diametric problem to characterize

$$M(\varphi_n, D) = \max\{|C| : \varphi_n(x^n, y^n) \leq D \text{ for all } x^n, y^n \in C\} \tag{12}$$

for the "sum-type" cost φ_n, which also can be a distance function like the Hamming, Taxi or Lee metric, etc. These problems are discussed in Section 5.

A simple, but basic, observation is that there is an advantage in having for every message i a set C_i (called "cloud") $\subset \mathcal{X}^n$ of possible representations such that for any representation $x^n \in C_i$ there exists a representation $y^n \in C_j$ with $\varphi_n(x^n, y^n) \leq D$.

Example 4.1. *For $\mathcal{X} = \{0,1\}$, $n = 3$, $\varphi_3 = d_H$ and $D = 1$ we have $M(d_H, 1) = 2$. On the other hand there are four clouds $C_1 = \{000, 111\}$, $C_2 = \{100, 011\}$, $C_3 = \{010, 101\}$, $C_4 = \{001, 110\}$, which can be used for updating four messages at cost 1.*

More generally we introduced together with Z. Zhang [62] *write-efficient memories* (WEM) as a new model for storing and updating information on a rewritable medium. There is a cost $\varphi \colon \mathcal{X} \times \mathcal{X} \to \mathbb{R}_\infty$ assigned to changes of letters. A collection of subsets $\mathcal{C} = \{C_i : 1 \leq i \leq M\}$ of $\mathcal{X}^n$ is an (n, M, D) WEM code, if $C_i \cap D_j = \varnothing$ for all $i \neq j$ and if $D_{\max} = \max_{1 \leq i,j \leq M} \max_{x^n \in C_i} \min_{y^n \in C_j} \sum_{t=1}^{n} \varphi(x_t, y_t) \leq D$. $D_{\max}$ is called the maximal correction cost with respect to the given cost function. The performance of a code $\mathcal{C}$ can also be measured by two parameters, namely, the maximal cost per letter $d_\mathcal{C} = n^{-1} D_{\max}$ and the rate of the size $r_\mathcal{C} = n^{-1} \log M$. The rate achievable with a maximal per letter cost d is thus $R(d) = \sup_{\mathcal{C}:d_\mathcal{C} \leq d} r_\mathcal{C}$.

This is the most basic quantity (the updating capacity for maximal per letter cost d) of a WEM $(\mathcal{X}^n, \varphi^n)_{n=1}^{\infty}$. We give a characterization of this quantity. For this we need some definitions. For a set $\mathcal{Z}$, $\mathcal{P}(\mathcal{Z})$ denotes the set of all probability distributions on $\mathcal{Z}$. Let (X, Y) be a pair of random variables with values in $\mathcal{X} \times \mathcal{X}$ and distribution P_{XY}. We denote the (marginal) distributions of X (resp. Y) by P_X (resp. P_Y) and $H(X \mid Y)$ is the conditional entropy. Finally we need distributions

$$\mathcal{P}_d = \{P_{XY} \in \mathcal{P}(\mathcal{X} \times \mathcal{X}) : P_X = P_Y, \mathbb{E}\varphi(X, Y) \leq d\}, \tag{13}$$

with equal marginals and an expectation of costs not greater than d and the quantity $\rho(d) = \max_{P_{XY} \in \mathcal{P}_d} H(Y \mid X)$.

Theorem 4.2. *For any $d \geq 0$ and $\varphi \colon \mathcal{X} \times \mathcal{X} \to \mathbb{R}^+$ $R(d) = \rho(d)$.*

4.1.1. THE STRUCTURE OF THEOREM 2.2 IN THE HAMMING CASE   Hypergraph $(\mathcal{V}, \mathcal{E}) = (\mathcal{X}^n, (S_D^n(x^n))_{x^n \in \mathcal{X}^n}$, $\kappa(\mathcal{V}, \mathcal{E}) =$ maximal number of colors assigned to vertices, such that every color occurs in every edge, that is ball $S_D^n(x^n)$ (General Ramsey Problem) $C_i = \{$vertices with color $i\}$. A key tool in the spirit of [35], [36] is

Color Carrying Lemma. *For every hypergraph* $\mathcal{H} = (\mathcal{V}, \mathcal{E})$

$$\kappa(\mathcal{V}, \mathcal{E}) \geq (\ell n |\mathcal{V}|)^{-1} \min_{E \in \mathcal{E}} |E|.$$

Since in our case $|\mathcal{V}| = |\mathcal{X}^n|$ and $|E| = |S_D^n(x^n)|$ grow exponentially in n, if $D = dn$, we get the optimal rate $R(d) = h(d)$, where h is the binary entropy function, – a very special case of Theorem 4.2. It is remarkable that in sequence spaces with cardinality of exponential growth these Ramsey type problems can be solved at least rate-wise.

There are still issues of code *constructions*. Further, instead of worst case costs one can consider also average costs, that is, diametric problems in the average. Several interesting questions arise, if *several persons or devices use the same tape* under various conditions: multi-user memories with constraints on privacy, hierarchy and technology [92].

Notice that the two models mentioned above can be described by the side information about the content of the tape available to the writer W, before he writes a new word. The reader R remembers no previous word.

Case (W^-, R^-) amounts to diametric problems. Case (W^+, R^-) gives our model with clouds. Formally, for all (i, j): $\forall x^n \in C_i \ \exists y^n \in C_j$ with $\varphi_n(x^n, y^n) \leq D$.

4.2. From here we pass to an independent combinatorial investigation

In [119] we studied with N. Cai and Z. Zhang families of clouds $(\mathcal{A}_i)_{i=1}^M$, $\mathcal{A}_i \subset 2^{[n]}$ with

(a) *relations* "$\subset$" (comparable), "$\bowtie$" (incomparable), "intersecting", "disjoint".
(b) *properties* $(\forall, \exists)$, $(\exists, \forall)$, $(\forall, \forall)$, $(\exists, \exists)$.
(c) "Disjoint" clouds, "distinct" clouds. (It is erroneously stated in [N] Erdős/Székely on page 118 that we always assume disjointness.)
(d) "restricted" case $|A| = k \ \forall A \in \bigcup_{i=1}^M \mathcal{A}_i$ and "unrestricted" case.
(e) For length M bound size $\max_{1 \leq i \leq M} |\mathcal{A}_i|$.

We explain now our symbolic notation in a particular case.

4.2.1. CLOUD ANTICHAINS
   Classical: $\{A_i\}_{i=1}^N$, $A_i \subset (2^{[n]})$ is an antichain, if $A_i \bowtie A_j \ \forall i \neq j$. Now sets are replaced by clouds, that is, families of sets. $(\mathcal{A}_i)_{i=1}^N$ is of
   type $(\forall, \forall)$, if for all $i \neq j$ $A_i \bowtie A_j \ \forall A_i \in \mathcal{A}_i, \ \forall A_j \in \mathcal{A}_j$
   type $(\exists, \forall)$, if for all $i \neq j$ $\exists A_i \in \mathcal{A}_i$ with $A_i \bowtie A_j \ \forall A_j \in \mathcal{A}_j$
   type $(\forall, \exists)$, if for all $i \neq j$ $\forall A_i \in \mathcal{A}_i \ \exists A_j \in \mathcal{A}_j$ with $A_i \bowtie A_j$
   type $(\exists, \exists)$, if for all $i \neq j$ $\exists A_i \in \mathcal{A}_i$, $\exists A_j \in \mathcal{A}_j$ with $A_i \bowtie A_j$.
   Maximal lengths $N_n(\forall, \forall)$, $N_n(\exists, \forall)$ for distinct and $M_n(\forall, \forall)$, $M_n(\exists, \forall), \ldots$ for *disjoint clouds.*

Convention: $f \sim g$ iff $\lim_{n \to \infty} f(n)g(n)^{-1} = 1$. From work with L. H. Khachatrian [86] we have

Theorem 4.3.

$$M_n(\exists, \forall) \sim 2^{n-1},$$

$$M_n(\forall, \exists) = \begin{cases} 2 & \text{if } n = 2 \\ 2^{n-1} - 1 & \text{if } n \geq 3 \end{cases}$$

$$M_n(\exists, \exists) = \binom{n}{\lfloor \frac{n}{2} \rfloor} + \left\lfloor \frac{2^n - 2 - \binom{n}{\lfloor \frac{n}{2} \rfloor}}{2} \right\rfloor.$$

Theorem 4.4 (Double exponential growth). $N_n(\exists, \forall) = \binom{k}{\lfloor \frac{k}{2} \rfloor}$ with $k = \binom{n}{\lfloor \frac{n}{2} \rfloor}$, $N_n(\forall, \exists) \sim 2^{2^n - 2}$, $N_n(\exists, \exists) \sim 2^{2^n}$.

The classical Sperner result takes the forms $M_n(\forall, \forall) = N_n(\forall, \forall) = \binom{n}{\lfloor \frac{n}{2} \rfloor}$.

For the maximal cardinality of families with the relations "comparable", "disjoint", and "intersecting" we choose the letters C, D, and I respectively. The types of problems such as $(\forall, \forall)$ etc. appear in the argument and n appears as index. In addition, in the restricted case a k appears in the argument.

4.3. Cloud-antichains (CAC) of length 2

Theorem 4.5 (Ahlswede/Zhang [68]). *A CAC $\{\mathcal{A}, \mathcal{B}\}$ in $2^{[n]}$ satisfies*
  (i) $|\mathcal{A}||\mathcal{B}| \leq 2^{2n-4}$. *Optimal configurations:*
       1. $\mathcal{A} = \{X \in 2^{[n]} : 1 \in X, 2 \notin X\}$,
          $\mathcal{B} = \{X \in 2^{[n]} : 1 \notin X, 2 \in X\}$.
       2. $\mathcal{A} = \{X \in 2^{[n]} : 1 \in X, |X| \leq \lfloor \frac{n-1}{2} \rfloor + 1\}$,
          $\mathcal{B} = \{X \in 2^{[n]} : 1 \notin X, |X| > \lceil \frac{n-1}{n} \rceil\}$ n odd.
  (ii) $\min\{|\mathcal{A}|, |\mathcal{B}|\} \leq 2^{n-2}$.

The proof uses AD-inequality. A sharper result is

Theorem 4.6 (Ahlswede/Khachatrian [107]). *For $0 \leq \alpha \leq 2^{n-1}$, if $f_n(\alpha) \triangleq \max\{|\mathcal{B}| : \exists \mathcal{A}$ such that $(\mathcal{A}, \mathcal{B})$ CAC in $2^{[n]}$ and $|\mathcal{A}| = \alpha$, then $f_n(\alpha) = 2^{n-1} + 2f_{n-2}^{(\alpha)} - \alpha$.*

For multisets of multiplicity k again AD and the arithmetic-geometric means inequality give $|\mathcal{A}|^{1/2} + |\mathcal{B}|^{1/2} \leq k^{n/2}$ for $(\mathcal{A}, \mathcal{B})$ CAC in $k^{[n]}$.

Using results of [DKW] we provide again with L. H. Khachatrian [107] partial results for

Question I. *For every k describe all CAC $(\mathcal{A}, \mathcal{B})$ with equality in 4.6 In the terminology of [100] this is an equality characterization problem.*

Question II. *How does $f_{n,k}(\alpha) = \max\{|\mathcal{B}| : \exists(\mathcal{A}, \mathcal{B})$ CAC in $k^{[n]}$, $|\mathcal{A}| = \alpha\}$ behave asymptotically in k, n, and α?*

In generalizing statements (i), (ii) in Theorem 2.11 we completely answer for every k.

Question III. *What is the growth of $g_{n,k} = \max_{(\mathcal{A},\mathcal{B})} \min_{CAC \ in \ k^{[n]}} (|\mathcal{A}|, |\mathcal{B}|)$?*

Question IV. *What is the growth of $S_{n,k} = \max_{(\mathcal{A},\mathcal{B})} \ _{CAC \ in \ k^{[n]}} |\mathcal{A}||\mathcal{B}|$?*

Question V. *What is the growth of $a_{n,k}(\Delta) = \max\{|\mathcal{B}| : \exists \mathcal{A} \ such \ that \ (\mathcal{A},\mathcal{B}) \ is$ $CAC \ in \ k^{[n]} \ and \ |\mathcal{A}| = |\mathcal{B}| + \Delta\} \ for \ -k^n \leq \Delta \leq k^n$?*

4.4. Intersection, clouds disjoint

Theorem 4.7 (Ahlswede/Cai/Zhang [95]). *In the* **restricted** *case $k = 2$*

$$I_n(\exists,\forall,2) = \begin{cases} n-1 & for \ n \in \mathbb{N} - \{3,5\} \\ n & for \ n = 3,5 \end{cases},$$

$$I_n(\forall,\exists,2) = \begin{cases} n & for \ n \in \mathbb{N} - \{1,2,4\} \\ n-1 & for \ n = 1,2,4 \end{cases},$$

$$I_n(\exists,\exists,2) \sim n^{3/2},$$

$$I_n(\forall,\forall,2) = \begin{cases} n-1 & for \ n \in \mathbb{N} - \{3\} \\ n & for \ n = 3 \end{cases} \quad \textit{(very special case of EKR).}$$

Theorem 4.8 (Ahlswede/Cai/Zhang [119]). *In the* **unrestricted** *case*

$$I_n(\forall,\forall) = I_n(\exists,\forall) = I_n(\forall,\exists) = 2^{n-1} \quad I_n(\exists,\exists) = 2^{n-1} + 2^{n-2} - 1.$$

4.5. Disjoint, clouds disjoint

Theorem 4.9 (Ahlswede/Cai/Zhang [95]). *In the* **restricted** *case $k = 2$*

$$\lim_{n\to\infty} D_n(\exists,\forall,2)n^{-2} = \frac{1}{6},$$

$$\lim_{n\to\infty} D_n(\forall,\exists,2)n^{-2} = \lim_{n\to\infty} D_n(\exists,\exists,2)n^{-2} = \frac{1}{4}.$$

Theorem 4.10 (Alon/Sudakov (see [N, p. 123])). *In the* **restricted** *case*

$$\lim_{n\to\infty} \frac{D_n(\exists,\forall,k)}{\binom{n}{k}} = \frac{1}{k+1},$$

$$\lim_{n\to\infty} \frac{D_n(\forall,\exists,k)}{\binom{n}{k}} = \lim_{n\to\infty} \frac{D_n(\exists,\exists,k)}{\binom{n}{k}} = \frac{1}{2}.$$

Conjectured by Ahlswede/Cai/Zhang [119], who settled the case $k = 2$.

4.6. Key tools are results on related graph coloring problems

This and the next paragraph give results from [119]. The study of cloud families of the $(\exists,\exists)$-type naturally leads to the following coloring concept. For any graph $\mathcal{G} = (\mathcal{V},\mathcal{E})$ a coloring of type $(\exists,\exists)$ is a map $f \colon \mathcal{V} \to M_f = \{1,2,\ldots,m_f\}$ such that for any two colors, say, $i,j \in M_f$, an edge $(a,b) \in \mathcal{E}$ exists with $f(a) = i$ and $f(b) = j$.

We are interested in the quantity $m(\mathcal{G}) = \max\{m_f : f \ is \ (\exists,\exists)\text{-coloring of } \mathcal{G}\}$.

Theorem 4.11. *For any graph $\mathcal{G} = (\mathcal{V}, \mathcal{E})$ we have with $N = 2|\mathcal{E}|$ $m(\mathcal{G}) \leq N^{1/2} + 1$.*
Moreover, if $D \triangleq \max_{V \in \mathcal{V}} \deg(x) \leq \left(\frac{N}{e^4 \log N}\right)^{1/2}$, then $m(\mathcal{G}) \geq \left(\frac{N}{e^4 \log N}\right)^{1/2}$.

Cloud families of $(\forall, \exists)$-type lead to a coloring of $(\forall, \exists)$-type, which is a map $g: \mathcal{V} \to M_g = \{1, 2, \ldots, m_g\}$ such that for any two colors, say, $i, j \in M_g$ and for any $a \in U$ with $g(a) = i$ there is an edge $\{a, b\} \in \mathcal{E}$ with $g(b) = j$. We are interested in $m^*(g) = \max\{m_g : g \text{ is } (\forall, \exists)\text{-coloring of } \mathcal{G}\}$

Theorem 4.12. *For any graph $\mathcal{G}$ we have*
 (i) *$(\log |\mathcal{V}|)^{-1}(d + 1) \leq m^*(\mathcal{G}) \leq d + 1$, where $d \triangleq \min_{x \in \mathcal{V}} \deg(x)$.*
 (ii) *$m^{*\prime}(\mathcal{G}) \triangleq \max|\{m^*(\mathcal{G}') : \mathcal{G}' \text{ is subgraph of } \mathcal{G}\}| \leq D + 1$.*

We mention that a coloring of type $(\exists, \forall)$ is a map $h: \mathcal{V} \to M_h = \{1, 2 \ldots, m_h\}$ such that for any two colors $i, j \in M_h$ an $a \in \mathcal{V}$ exists with $h(a) = i$ and $h(b) = j$ for all $b \in \mathcal{N}(a)$, $b \neq a$. The quantity $m^{**}(\mathcal{G}) = \max\{m_h : h \text{ is } (\exists, \forall)\text{-coloring of } \mathcal{G}\}$ is hard to analyse in general.

4.7. Asymptotic results via graph coloring
Theorem 4.13.
$$\lim_{n \to \infty} \frac{1}{n} \log C_n(\exists, \exists) = \lim_{n \to \infty} \frac{1}{n} \log D_n(\exists, \exists) = \frac{1}{2} \log 3.$$

Theorem 4.14.
$$\lim_{n \to \infty} \frac{1}{n} \log D_n(\exists, \exists, \varepsilon n) = \frac{1}{2} \left(h(\varepsilon) + (1 - \varepsilon) h\left(\frac{\varepsilon}{1 - \varepsilon}\right) \right),$$
$$\lim_{n \to \infty} \frac{1}{n} \log I_n(\exists, \exists, \varepsilon n) = h(\varepsilon).$$

Theorem 4.15.
$$\lim_{n \to \infty} \frac{1}{n} \log D_n(\forall, \exists, \varepsilon n) = (1 - \varepsilon) h\left(\frac{\varepsilon}{1 - \varepsilon}\right),$$
$$\lim_{n \to \infty} \frac{1}{n} \log I_n(\forall, \exists, \varepsilon n) = h(\varepsilon).$$

4.8. Hamming distance 1, clouds disjoint
An important relation is that of Hamming distance r for two words.

Theorem 4.16 (bf Ahlswede/Cai/Zhang [119]). *In the* **unrestricted** *case*
$$\lim_{n \to \infty} \frac{1}{n} \log H_n(\forall, \forall, \rho n) = \frac{1}{2}(1 + h(\rho)).$$

Notice that $H_n(\exists, \exists, 1)$ equals the maximal number of sets into which one can partition the n-cube, such that two different sets always have distance 1.

Theorem 4.17. (R. Ahlswede, S. L. Bezrukov, A. Blokhuis, K. Metsch and G. E. Moorhouse [80]) $\frac{\sqrt{2}}{2}\sqrt{n2^n} \leq H_n(\exists, \exists, 1) \leq \sqrt{n2^n} + 1$ *for all n.*

4.9. Another direction

Let $\mathcal{I}_n$ be the lattice of intervals in the Boolean lattice $\mathcal{L}_n$. For $\mathcal{A}, \mathcal{B} \subset \mathcal{I}_n$ the pair of clouds $(\mathcal{A}, \mathcal{B})$ is cross-disjoint, if $I \cap J = \phi$ for $I \in \mathcal{A}$, $J \in \mathcal{B}$. With N. Cai [109] we prove that for such pairs $|\mathcal{A}||\mathcal{B}| \leq 3^{2n-2}$ and that this bound is best possible.

Optimal pairs are up to obvious isomorphisms unique. The proof is based on a new bound on cross intersecting families in $\mathcal{L}_n$ with a weight distribution. It implies also an Intersection Theorem for multisets of Erdős and Schönheim from 1969.

Furthermore, in [115] in a canonical way we establish an AZ-identity and its consequences, the LYM-inequality and the Sperner-property. Further the Bollobás-inequality for the Boolean interval lattice turns out to be just the LYM-inequality for the Boolean lattice. We also present an Intersection Theorem for this lattice.

Perhaps more surprising is that by our approach the conjecture of P. L. Erdős, Seress, Székely [ErSS] and Füredi concerning an Erdős-Ko-Rado-type intersection property for the poset of Boolean chains could also be established. Actually we give two seemingly elegant proofs.

5. Diametric, Isoperimetric Theorems in Sequence Spaces

5.1. Introduction

Mankind has believed the Isoperimetric Theorem in Euclidean 3-space "For given surfaces the ball has maximal volumes" for more than 2000 years. The discovery of the 2-dimensional analog is often attributed to Dido, the daughter of a Phoenician King. Despite strong interest in *extremal problems* and *variational principles* in physics (and also philosophy: "Best of all worlds" for Leibnitz) in modern times after the invention of calculus a proof came only in the 19-th century by Schwarz —after an incomplete geometrical proof by Steiner, showing the uniqueness but not the existence of a solution.

Replacing surface by diameter leads to (iso)-diametric Theorems. A classic is Blaschke's "Kreis and Kugel". In sequence spaces $\alpha^{[n]}$ cardinalities take the role of volumes of subsets. For some distance function d (like Hamming, Lee or Taxi metrics) surface $\Gamma_d(A)$ is the set of points in the complement of A and with distances 1 to A.

Harper's solution of the isoperimetric problem in Hamming space $(2^{[n]}, d_H)$ is mentioned in Section 6. *The problem is open for $\alpha > 2$.* However, recently a **"rate-wise" optimal solution** was found with Z. Zhang for the r-th surface $\Gamma_{\Phi_n}^r(A) = \{b^n : b^n \notin A,\ \varphi_n(b^n, a^n) \leq r \text{ for some } a^n \in A\}$ with $r = \rho n$, where $\varphi_n(b^n, a^n) = \sum_{t=1}^n \varphi(b_t, a_t)$, $\varphi : [\alpha] \times [\alpha] \to \mathbb{R}$ is any symmetric "sum-type" function and not just the Hamming distance: $R(\lambda, \rho) = \lim_{n \to \infty} \frac{1}{n} \max_{|A| \leq \exp\{\lambda n\}} \log |\Gamma_{\varphi_n}^{\rho n}(A)|$.

Exact solutions are not even known for the *non-binary Hamming case*.

In Section 3 the Diametric Theorem in Hamming space is mentioned. For $\alpha = 2$ optimal are balls and for $\alpha > 2$ optimal are certain cartesian products of a ball and a suitable subcube (or cylinder set). Depending on the parameters this configuration can degenerate to a ball and up to isometries (with one exception of two solutions) there is only one solution.

Notice that the Complete Intersection Theorem for parameters (n, k, t) can be viewed as a Diametric Theorem on the *restricted* Hamming space $\left(\binom{[n]}{k}, d_H \right)$ for diameter $D = 2k - 2t$. Another kind of diametric theorem is for an *average diameter* constraint (in 5.2 below).

We have now gained by example an understanding of the following classification:

restricted case — unrestricted case,

vertex-isoperimetric — edge-isoperimetric,

exact solution — rate-wise optimal solution

worst case — average case

vertex-diametric — edge-diametric

Coauthors in this work are I. Althöfer, S. Bezrukov, N. Cai, L. H. Khachatrian, E. Yang, Z. Zhang.

5.2. Rate-wise optimal solutions for the average case (vertex)-diametric problem

Exact solutions for the worst case vertex-diametric problem have been discussed in connection with Intersection Theorems in Section 3 for the Hamming distance. An earlier result with Cai and Zhang [72] concerns the Taxi (or Manhattan) metric on $\alpha^{[n]}$ and gives solutions for almost all parameters. Bollobas/Leader noticed that the missing cases are covered by an earlier result of Kleitman/Fellow. A worst case diametric theorem for edges was mentioned in Section 3. The first diametric theorem for the *average* was obtained with I. Althöfer with a rate-wise optimal solution: $\mathcal{U} \subset \mathcal{X}^n$ has an *average* diameter not exceeding D, if $D_{\text{ave}} \triangleq \frac{1}{|\mathcal{U}|^2} \sum_{x^n \in \mathcal{U}} \sum_{y^n \in \mathcal{U}} \varphi_n(x^n, y^n) \leq D$. With Katona [31] already in 1978 the restricted case $k = 2$ was considered in the dual form, where the cardinality of $\mathcal{U}$ is specified and D_{ave} is minimized. An exact solution is given in the form that either $\mathcal{U}$ in lexicographic or in backwards lexicographic order is optimal. With I. Althöfer [87] we proved by entropy methods

Diametric Theorem in Average. *For the Hamming space $(\alpha^{[n]}, d_H)$ and rate $0 \leq R \leq \log \alpha$ the smallest* average *diameter per letter*

$$\frac{1}{n}\overline{d}_n(R) \triangleq \min_{A_n \subset \alpha^{[n]}, \frac{1}{n} \log |A_n| \geq R} D_{ave}(A_n), n \in \mathbb{N},$$

satisfies

$$\overline{d}(R) \triangleq \lim_{n \to \infty} \frac{1}{n}\overline{d}_n(R)$$
$$= \min \left[\lambda \sum_{x,y} d_H(x, y)P(x)P(y) + (1 - \lambda)d_H(x, y)P'(x)P'(y) \right],$$

where "min" is taken over $\lambda \in [0, 1]$, and probability distributions on $[\alpha]$ with $\lambda H(P) + (1 - \lambda)H(P') \geq R$. Here H stands for the entropy.

Writing $R = h(\beta)$ for $\alpha = 2$ we get $\overline{d}(R) = 2\beta(1 - \beta)$. For $\alpha = 3$ calculation shows that $P \neq P'$ occurs in the optimization.

For a general cost function the result holds with d_H replaced by φ_n as shown with Cai [127]. There are also extensions to several sets with some pairwise mutual average distance (or costs) and some internal average distances all simultaneously valid are treated. These generalizations are motivated by multiuser WEM (see Section 4). The proofs use a tool from Information Theory to bound the cardinality of ranges of auxiliary random variables:

Support Lemma. (Ahlswede/Körner [21]) *Let $\mathcal{P}(\mathcal{Z}) = $ set of all PD's on finite set $\mathcal{Z}$, let $f_i(j = 1, \ldots, k)\colon \mathcal{P}(\mathcal{Z}) \to \mathbb{R}$ be continuous functions, and let μ be a PD on $\mathcal{P}(\mathcal{Z})$ with Borel σ-algebra, then there exist elements $P_i \in \mathcal{P}(\mathcal{Z})$ and $\alpha_1, \ldots, \alpha_k \geq 0$, $\sum_{i=1}^{k} \alpha_i = 1$ such that $\int_{\mathcal{P}(\mathcal{Z})} f_j(P)\mu(dP) = \sum_{i=1}^{k} \alpha_i f_i(P_i)$ for $j = 1, 2, \ldots, k$.*

5.3. Edge-isoperimetric inequalities ratewise optimal

$\Psi_i\colon 2^{\mathcal{X}_i} \to \mathbb{R}(i = 1, 2)$, $\Psi_1 \times \Psi_2\colon 2^{\mathcal{X}_1 \times \mathcal{X}_2} \to \mathbb{R}$ defined by

$$\Psi_1 \times \Psi_2(A) = \sum_{x \in \mathcal{X}_2} \Psi_1(A_1(x)) + \sum_{x \in \mathcal{X}_1} \Psi_2(A_2(x)) \text{ for } A \subset \mathcal{X}_1 \times \mathcal{X}_2$$

$\Psi^n = (((\Psi \times \Psi) \times \Psi) \times \cdots \times \Psi)$ counts inner edges.

A. (Nestedness):
For $\mathcal{X} = \{0, 1, \ldots, \alpha - 1\}$, $k \in \mathcal{X}$, $[k] = \{0, 1, \ldots, k\}$, $A \subset \mathcal{X}$, $|A| = k + 1$
$\varphi(A) \leq \varphi([k])$ (Satisfied by $d_H, d_L, d_M, \ldots$).
B. (Submodularity):
For $A, B \subset \mathcal{X}$ $\varphi(A) + \varphi(B) \leq \varphi(A \cup B) + \varphi(A \cap B)$.
C. $\varphi(\varnothing) = 0$ and $\Delta\varphi(k) = \varphi([k]) - \varphi([k - 1])$.

A pair (R, δ) is achievable, if for all $\varepsilon_1, \varepsilon_2 > 0$ there exists an $n(\varepsilon_1, \varepsilon_2)$ such that for every $n \geq n(\varepsilon_1, \varepsilon_2)$ there is an $A_n \subset \mathcal{X}^n$ with $\left|\frac{1}{n} \log |A_n| - R\right| < \varepsilon_1$ and $\frac{1}{n|A_n|}\varphi^n(A) > \delta - \varepsilon_2$. $\mathcal{R}_\varphi$ is the set of all achievable pairs (R, δ).

Theorem 5.1. (Ahlswede/Cai [125], [126]) $\mathcal{R}_\varphi = \{(H(X|U), \mathbb{E}\Delta_\varphi(X)) : X, Y$ *satisfy* $(a), (b), (c)\}$
  (a) *Random variable X takes values in $\mathcal{X}$ and random variable U takes values in $\mathcal{U}$.*
  (b) $|\mathcal{U}| \leq |\mathcal{X}| + 1.$
  (c) $Pr(X = 0|U = u) \geq Pr(X = 1|U = u) \ldots \quad \geq Pr(X = \alpha - 1|U = u).$

Remark 5.2. *Bollobas/Leader solved the case $P_k^n = $ powers of k-paths and $C_k^n = $ powers of k-cycles, which are equivalent via Ψ! All trees on k vertices are equivalent.*

5.3.1. LOCAL-GLOBAL PRINCIPLE Let Ψ satisfy A, B, C. If the **Lexicographic order is exactly optimal** for edge-isoperimetry **for $n = 1, 2$**, then it is optimal *for every n.* There is related work with S. L. Bezrukov [98] and by him and his coauthors, surveyed in [N, p. 75–94].

6. Combinatorics on Sequence Spaces: Partitions, Monochromatic Rectangles, Shadows and Isoperimetry under Sequence-Subsequence Relation, Antichains Splitting, AZ-Identities, Dimension Constraints

6.1. Partitions

Consider $(\mathcal{V}, \mathcal{E})$, where $\mathcal{V}$ is a finite set and $\mathcal{E}$ is a system of subsets of $\mathcal{V}$. For the cartesian products $\mathcal{V}^n = \prod_1^n \mathcal{V}$ and $\mathcal{E}^n = \prod_1^n \mathcal{E}$, let $\pi(n)$ denote the minimal size of a partition of $\mathcal{V}^n$ into sets that are elements of $\mathcal{E}^n$, if a partition exists at all, otherwise $\pi(n)$ is not defined. This is obviously exactly the case if it is so for $n = 1$.

Whereas the packing number $p(n)$, that is the maximal size of a system of disjoint sets from $\mathcal{E}^n$, and the covering number $c(n)$, that is the minimal number of sets from $\mathcal{E}^n$ to cover $\mathcal{V}^n$, have been studied in the literature, this seems to be not the case for the partition number $\pi(n)$.

Obviously, $c(n) \leq \pi(n) \leq p(n)$, if $c(n)$ and $\pi(n)$ are well defined. The quantity $\lim_{n\to\infty} \frac{1}{n} \log p(n)$ is Shannon's zero error capacity. Although it is known only for very few cases, a nice formula exists for $\lim_{n\to\infty} (1/n) \log c(n)$ (see [73]).

The difficulties in analysing $\pi(n)$ are similar to those for $p(n)$. For the case of graphs with edge set $\mathcal{E}$ including all loops, we prove that $\pi(n) = \pi(1)^n$ ([83]). This result is derived from the corresponding result for complete graphs with the help of Gallai's Lemma in matching theory. Another interesting quantity is $\mu(n)$, the maximal size of a partition of $\mathcal{V}^n$ into sets that are elements of $\mathcal{E}^n$ (again only hypergraphs $(\mathcal{V}, \mathcal{E})$ with a partition are considered). We also call μ the maximal partition number. It behaves more like the packing number. Clearly, $\pi(n) \leq \mu(n) \leq p(n)$. It seems to us that an understanding of these partition problems would be a significant contribution to an understanding of the basic, and seemingly simple, notion of Cartesian products.

More generally, for hypergraphs $\mathcal{H}_i = (\mathcal{V}_i, \mathcal{E}_i)$ $(1 \leq i \leq n)$, we define the product hypergraph $\mathcal{H}^n = (\mathcal{V}^n, \mathcal{E}^n) = (\prod_{i=1}^n \mathcal{V}_i, \prod_{i=1}^n \mathcal{E}_i)$. Edges of cardinality 1 are called loops. Special hypergraphs are graphs $\mathcal{G} = (\mathcal{V}, \mathcal{E})$ defined by the property $|E| \in \{1, 2\}$ for all $E \in \mathcal{E}$ and, more generally, d-uniform hypergraphs (with or without loops) that satisfy $|E| \in \{1, d\}$ for all $E \in \mathcal{E}$.

In particular, there are d-uniform hypergraphs with all loops included, that is, $\{\{v\} : v \in \mathcal{V}\} \subset \mathcal{E}$.

When the set $\binom{\mathcal{V}}{d}$ of all vertex sets of cardinality d is contained in the edge set $\mathcal{E}$, we speak of a complete d-uniform hypergraph.

We introduced the partition number $\pi(\mathcal{H})$ as the minimal size of a partition of $\mathcal{V}$ into sets that are members of $\mathcal{E}$, if a partition exists, and ∞ otherwise. When $\mathcal{G}_i = (\mathcal{V}_i, \mathcal{E}_i)$ $(i = 1, 2, \ldots, n)$ are arbitrary finite graphs with all loops included, then we obviously have $\pi(\mathcal{G}_i) = |\mathcal{V}_i| - v(\mathcal{G}_i)$ for the partition number, where $v(\mathcal{G}_i)$ is the matching number of $\mathcal{G}_i$. A discovery of [83] is that for the hypergraph product $\mathcal{H}^n = \mathcal{G}_1 \times \cdots \times \mathcal{G}_n$ $\pi(\mathcal{H}^n) = \prod_{i=1}^n \pi(\mathcal{G}_i)$.

An important step in our proof is to show the above when all $\mathcal{G}_i$s are complete. Here we establish the following generalization.

Theorem 6.1. (Ahlswede/Cai [94]) *For complete d-uniform hypergraphs with all loops* $\mathcal{H}_i = (\mathcal{V}_i, \mathcal{E}_i)$, *that is,* $\mathcal{E}_i = \binom{\mathcal{V}_i}{d} \cup \{\{v\} : v \in \mathcal{V}_i\}$ $(i = 1, 2, \ldots, n)$, *write* $|\mathcal{V}_i| = dq_i + r_i, 0 \leq r_i < d$. *Then for* $\mathcal{H}^n = \prod_{i=1}^{n} \mathcal{H}_i$ *satisfying* $d > \prod_{i:r_i \neq 0} r_i$, *we have* $\pi(\mathcal{H}^n) = \prod_{i=1}^{n} \frac{|\mathcal{V}_i| + (d-1)r_i}{d} = \prod_{i=1}^{n}(q_i + r_i) = \prod_{i=1}^{n} \pi(\mathcal{H}_i)$. *(The result above is covered by the case $d = 2$.)*

Even in the case of non-identical factors $\mathcal{H}_i = (\mathcal{V}_i, \mathcal{E}_i)$, $i \in \mathbb{N}$, with $\max_i |\mathcal{E}_i| < \infty$, the asymptotics of $c(n)$ is known [73]:

$$\lim_{n\to\infty} \frac{1}{n} \left(\log c(n) - \sum_{t=1}^{n} \log \left(\max_{q \in Prob(\mathcal{E}_t)} \min_{v \in \mathcal{E}_t} \sum_{E \in \mathcal{E}_t} 1_E(v)q_E \right)^{-1} \right) = 0 \,,$$

where $Prob(\mathcal{E}_t)$ is the set of all probability distributions on $\mathcal{E}$, q_E is the indicator function of the set E.

6.2. Bounds on monochromatic rectangles

For a matrix consider the area $i \cdot j$ of an $i \times j$ minor with constant entries. This concept was introduced by Yao for estimating communication complexity. Some of our exact results and bounds, identities and inequalities, are reported by U. Tamm in [N, p. 589–602]. Interactive communication [120], a similar model, has striking phenomena and open problems. Methods from [35], [36] find application.

6.3. Shadows and isoperimetry under the sequence-subsequence relation

It has been suggested in 1988 on page 152 in project B_1 "Kombinatorik von Folgenräumen" of the SFB 343 "Diskrete Strukturen in der Mathematik" to study combinatorial extremal problems under the sequence-subsequence relation —in particular also shadow problems of the Kruskal-Katona type.

In September 1994 David Daykin wrote to us that he and Danh had a counterexample to the optimality of the B-G order. He also mentioned that they had a solution in the binary case with a very, very complicated proof (which we never have seen). Immediately thereafter a simple proof for the binary case (and also a simple counterexample to the B-G order in the general case) was given by Ahlswede/Cai [112]. Subsequently, in December 1994 the former author also gave a new shorter proof. (Also published in [112], see [N, p. 75–94].)

6.3.1. Shadows of arbitrary sets under deletion of any letter For $\mathcal{X}^n = \prod_1^n \mathcal{X}$, the sequences of length n over the alphabet $\mathcal{X}$, we consider for sets $A \subset \mathcal{X}^n$ their shadow $\nabla A = \{x^{n-1} \in \mathcal{X}^{n-1} : x^{n-1}$ is subsequence of some $a^n \in A\}$. The goal is to find for given cardinalities sets of minimal cardinality of the *shadow*.

Recall the H-order of Harper: For any integer $u \in [0, 2^n]$ the u-th initial segment consists of all $x^n \in \{0, 1\}^n$ with less than $n - k$ ones and all remaining elements with $n - k$ ones, whose complements are in the initial segment of the squashed order (used for instance in Kruskal-Katona).

As in the vertex isoperimetric problem in binary Hamming space it is optimal also for our shadows of sets in $\{0,1\}^n$. We use the unique binomial representation of an integer u

$$u = \binom{n}{n} + \cdots + \binom{n}{k+1} + \binom{\alpha_k}{k} + \cdots + \binom{\alpha_t}{t}; n > \alpha_k > \cdots > \alpha_t \geq 1,$$

and observe that for an initial, H-order segment S with $|S| = u$

$$|\triangledown S| = \binom{n-1}{n-1} + \binom{n-1}{n-2} + \cdots + \binom{n-1}{k} + \binom{\alpha_k-1}{k-1} + \cdots + \binom{\alpha_t-1}{t-1} = \overset{\triangledown}{G}(n,u),$$

say.

Theorem 6.2. *For every $A \subset \{0,1\}^n$ $|\triangledown A| \geq \overset{\triangledown}{G}(n,|A|)$ and the bound is achieved by the u-th initial segment in H-order.*

The proof is an immediate consequence of our main discovery, the

$\triangledown$-Inequality. (Ahlswede/Cai [112]) *If $w_1 \leq w_0 < \overset{\triangledown}{G}(n,w)$ and $w \leq w_0 + w_1$, then $\overset{\triangledown}{G}(n,w) \leq \overset{\triangledown}{G}(n-1,w_0) + \overset{\triangledown}{G}(n-1,w_1)$.*

In [128] Ahlswede/Cai established analogous results for

6.3.2. SHADOWS FOR FIXED LEVEL AND SPECIFIC LETTER

6.3.3. SHADOWS OF ARBITRARY SETS UNDER INSERTION OF ANY LETTER

6.3.4. TWO ISOPERIMETRIC INEQUALITIES   It has been emphasized in [27] that isoperimetric inequalities in discrete metric spaces are fundamental principles in combinatorics. The goal is to minimize the union of a specified number of balls of constant radius. We speak of an isoperimetric inequality, if this minimum is assumed for a set of ball-centers, which themselves form a ball (or quasi-ball, if numbers don't permit a ball).

For any $A \subset \{0,1\}^*$ and any distance d we define (the union of balls of radius r) $\Gamma_d^r(A) = \{x^{n\prime} \in \{0,1\}^* : d(x^{n\prime}, a^n) \leq r$ for some $a^n \in A\}$.

A prototype of a discrete isoperimetric inequality is the one discovered by Harper 1966, rediscovered by Ahlswede/Gács/Körner [23], and proved again by Katona for $d = d_H$.

The optimum is achieved by the $|A|$-th initial segment $I_{|A|}$ in H-order (this is a ball of radius k, if $|A| = \sum_{j=0}^{n} \binom{n}{j}$).

Ahlswede/Cai [128] define two distances, θ and δ, in $\{0,1\}^* = \bigcup_{n=0}^{\infty} \mathcal{X}^n$. For $x^m, y^{m\prime} \in \{0,1\}^*$ $\theta(x^m, x^{m\prime})$ counts the minimal number of insertions and deletions which transform one word into the other. $\Delta(x^m, x^{m\prime})$ counts the minimal number of operations, if also exchanges of letters are allowed. Thus $\delta(x^m, x^{m\prime}) \leq \theta(x^m, x^{m\prime})$.

Theorem 6.3. *For all $A \subset \mathcal{X}^n$ and $r \geq 0$*

   (i) $|\Gamma_\theta^r A| \geq |\Gamma_\theta^r I_A|$,

   (ii) $|\Gamma_\delta^r A| \geq |\Gamma_\delta^r I_A|$.

6.4. Antichains splitting

A novel type of result was found by Ahlswede/Erdős/Graham [103]: In any *dense* finite poset $\mathcal{P}$ (e.g. in the Boolean lattice) every maximal antichain S can be partitioned into disjoint subsets S_1 and S_2, such that the union of the downset of S_1 with the upset of S_2 yields the entire poset $\mathcal{D}(S_1) \cup \mathcal{U}(S_2) = \mathcal{P}$. Here $\mathcal{P}$ is called *dense* if every non-empty open interval $\{z \in \mathcal{P} : x < z < y\}$ contains at least two elements z', z''. It is called strongly dense if there are incomparable z', z''.

For finite posets the two concepts are equivalent, but for infinite posets they do not necessarily coincide. (For example the totally ordered chain of rational numbers is dense, but not strongly dense.)

A conjecture of [103] that every countable strongly dense poset has the splitting property was disproved by Ahlswede/Khachatrian with the poset of squarefree integers in [N, p. 29–44], where the reader finds also several open questions.

6.5. AZ-identities

Ahlswede and Zhang [64] found the following identity.

Theorem 6.4. (AZ$_1$-Identity) *For every family $\mathcal{A} \subset 2^\Omega$ of non-empty subsets of* $\Omega = \{1, 2, \ldots, n\}$ $\sum_{X \subset \Omega} \frac{W_\mathcal{A}(X)}{|X|\binom{n}{|X|}} = 1$, *where* $W_\mathcal{A}(X) = \left| \bigcup_{A \supset A \in \mathcal{A}} A \right|$.

We associate with every $\mathcal{E} \subset 2^\Omega$ the upset $\mathcal{U}(\mathcal{E}) = \{U \subset \Omega : U \supset E$ for some $E \in \mathcal{E}\}$ and the downset $\mathcal{D}(\mathcal{E}) = \{D \subset \Omega : D \subset E$ for some $E \in \mathcal{E}\}$. When $\mathcal{A}$ is an antichain in the poset $(2^\Omega, \supset)$, then the identity becomes $\sum_{X \in \mathcal{A}} \frac{1}{\binom{x}{|X|}} +$

$\sum_{X \in \mathcal{U}(\mathcal{A}) \setminus \mathcal{A}} \frac{W_\mathcal{A}(X)}{|X|\binom{n}{|X|}} = 1.$

The LYM inequality is obtained by omission of the second summand, which by definition of $W_\mathcal{A}$ can also be written in the form $\sum_{X \notin \mathcal{D}(\mathcal{A})} \frac{W_\mathcal{A}(X)}{|X|\binom{n}{|X|}}$. We call this the deficiency of the inequality. More generally, in [68] the Bollobas inequality was lifted to an identity.

Theorem 6.5. (AZ_2) *For two families $\mathcal{A} = \{A_1, \ldots, A_N\}$ and $\mathcal{B} = \{B_1, \ldots, B_N\}$ of subsets of Ω with the properties*
(a) $A_i \subset B_i$ *for* $i = 1, 2, \ldots, N$,
(b) $A_i \not\subset B_j$ *for* $i \neq j$ $\sum_{i=1}^N \frac{1}{\binom{n - |B_i \setminus A_i|}{|A_i|}} + \sum_{X \notin \mathcal{D}(\mathcal{B})} \frac{W_\mathcal{A}(X)}{|X|\binom{n}{|X|}} = 1.$

In [64] it was explained that Theorem AZ_1 gives immediately, what LYM does not, namely the uniqueness part in Sperner's Theorem. In [68] the uniqueness of an optimal configuration of unrelated chains of subsets due to Griggs, Stahl and Trotter 1984 [GST] was proved with the help of Theorem AZ_2.

Körner and Simonyi 1990 observed the LYM-type inequality:

For $\mathcal{A} = \{A_1, \ldots, A_N\}$, $\mathcal{B} = \{B_1, \ldots, B_N\} \subset 2^\Omega$ with $A_i \cap B_i = \varnothing, A_i \not\subset$ $A_j \cup B_j, B_i \not\subset A_j \cup B_j$ for $i \neq j$ $\sum_{i=1}^N \binom{n - |A_i|}{|B_i|}^{-1} + \binom{n - |B_i|}{|A_i|}^{-1} - \binom{n}{|A_i| + |B_i|}^{-1} \leq 1$ and they asked "Is this inequality ever tight?".

This rather modest question was a challenging test of the power of the identities above or, more precisely, of the procedure to produce new identities described in [64].

The outcome is an Ahlswede-Zhang type identity which goes considerably beyond Theorem AZ_2. From a special case of this identity we derive *a full characterization* of the cases with equality even for a generalized version of the inequality above. In other words we characterize the cases with deficiency zero.

Theorem 6.6. (Ahlswede/Cai [79]) *Suppose that for a family* $\mathcal{B} = \{B_1, \ldots, B_N\}$ *of subsets of* Ω *and a family* $\mathcal{A}^* = \{\mathcal{A}_1, \ldots, \mathcal{A}_N\}$ *of subsets of* 2^Ω, *where* $\mathcal{A}_i = \{A_i^t : t \in T_i\}$ *for a finite index set* T_i, *we have*
(a) $A_i^t \subset B_i$ *for* $t \in T_i$ *and* $i = 1, 2, \ldots, N,$
(b) $A_i^t \not\subset B_j$ *for* $t \in T_i$ *and* $i \neq j.$

Then $\sum_{i=1}^{N} \sum_{k=1}^{|T_i|} (-1)^{k-1} \sum_{S \subset T_i, |S|=k} \binom{n - |B_i - \bigcup_{t \in S} A_i^t|}{|\bigcup_{t \in S} A_i^t|}^{-1} + \sum_{X \notin \mathcal{D}(\mathcal{B})} \frac{W_{\mathcal{A}}(X)}{|X|\binom{n}{|X|}} = 1.$

The specialisation $|T_i| = 1$ for $i = 1, \ldots, N$ gives theorem AZ_1. Daykin and Thu 1994 presented a dual to the AZ-identity and for related identities see Thu, "Identities for Combinatorial Extremal Theory" in [112].

6.6. Extremal sets of vectors under linear dimension constraints

The paper [158] "Maximal Number of Constant Weight Vertices of the Unit n-Cube Contained in a k-Dimensional Subspace", together with H. Aydinian and L. H. Khachatrian, is the start of a *new direction* in Extremal Theory, which is indicated in the title.

We introduce and solve a seemingly basic geometrical extremal problem. For the set $E(n, w) = \{x^n \in \{0, 1\}^n : x^n \text{ has } w \text{ ones}\}$ of vertices of weight w in the unit cube of $\mathbb{R}^n$ we determine $M(n, k, w) \triangleq \max\{|U_k^n \cap E(n, w)| : U_k^n \text{ is a } k\text{-dimensional subspace of } \mathbb{R}^n\}$. We also present an extension to multi-sets and explain a connection to the (higher dimensional) Erdős-Moser problem.

The set $E(n, w)$ can also be viewed as the set in which constant weight codes are studied in Information Theory. Another interest there is in linear codes. This was a motivation for studying the interplay between two properties: constant weight and linearity. In particular we wanted to know $M(n, k, w)$.

Theorem 6.7. (Ahlswede/Aydinian/Khachatrian [158]) *For* $n, k, w \in \mathbb{N}$
(a) $M(n, k, w) = M(n, k, n - w).$
(b) *For* $w \leq \frac{n}{2}$ *we have*

$$M(n, k, w) = \begin{cases} \binom{k}{w}, & \text{if (i) } 2w \leq k \\ \binom{2(k-w)}{k-w} 2^{2w-k}, & \text{if (ii) } k < 2w < 2(k-1) \\ 2^{k-1}, & \text{if (iii) } k - 1 \leq w. \end{cases}$$

The sets giving the claimed values of $M(n, k, w)$ *in the three cases are*
(i) $S_1 = E(k, w) \times \{0\}^{n-k}.$
(ii) $S_2 = E(2(k - w), k - w) \times \{10, 01\}^{2w-k} \times \{0\}^{n-2w}.$
(iii) $S_3 = \{10, 01\}^{k-1} \times \{1\}^{w-k+1} \times \{0\}^{n-k-w+1}.$

A key tool in the proof is an extremal problem for families of w-element sets involving antichain properties for certain restrictions.

Lemma 6.8. *Let $X = X_1 \dot\cup \ldots \dot\cup X_s$ with $|X_i| = n_i$ for $i = 1, \ldots, s$ and let $\mathcal{A} \subset \binom{X}{w}$ be a family with the following property:*

(P) for any $A, B \in \mathcal{A}$ and $j = 1, \ldots, s$ $E \triangleq A \cap \left(\bigcup_{i=1}^{j} X_i \right) \neq B \cap \left(\bigcup_{i=1}^{j} X_i \right) \triangleq$
    F implies that E and F are incomparable (form an antichain). Then

$$g(n_1, \ldots, n_s, w) \triangleq \max \left\{ |\mathcal{A}| : \mathcal{A} \subset \binom{X}{w}, \mathcal{A} \text{ has property } (P) \right\}$$

$$= \max_{\sum_{i=1}^{s} w_i = w} \prod_{i=1}^{s} \binom{n_i}{w_i}.$$

For $s = 1$ we get Sperner's result. Evaluation of the max gives the formulas in the theorem.

7. Counterexamples, Conjectures and Problems

7.1. Counterexamples

7.1.1. FRANKL/PACH CONJECTURE FOR UNIFORM, DENSE FAMILIES   A family $\mathcal{F} \subset \binom{[n]}{\ell}$ is called ℓ-dense, if there exists an $F \in \mathcal{F}$, such that $|F \cap F_1 : F_1 \in \mathcal{F}| = 2^\ell$.

Frankl/Pach [FP] conjectured that every $\mathcal{F} \subset \binom{[n]}{\ell}$ with $|\mathcal{F}| > \binom{n-1}{\ell-1}$ is ℓ-dense.

Ahlswede/Khachatrian [129] provide a counterexample to this conjecture by the construction of a set $\mathcal{F} \subset \binom{[n]}{\ell}$, $|\mathcal{F}| = \binom{n-1}{\ell-1} + \binom{n-4}{\ell-3}$, which is not ℓ-dense.

7.1.2. KLEITMAN'S CONJECTURE   In [27] Ahlswede/Katona considered the following "Excess-problems": For $\mathcal{A} \subset \binom{[n]}{k}$ let $I(\mathcal{A}) = |\{(A_1, A_2) \in \mathcal{A}^2 : A_1 \cap A_2 \neq \phi\}|$, $G(\mathcal{A}) = |\{(A_1, A_2) \in \mathcal{A}^2 : |A_1 \cap A_2| \geq k - 1\}|$. Determine $f(M) = \max_{|\mathcal{A}|=M} I(\mathcal{A})$ or $g(M) = \max_{|\mathcal{A}|=M} G(\mathcal{A})$.

The problems are the same for $k = 2$ and here these authors described two configurations, quasi-ball and quasi-star, one of which is always optimal. For $k \geq 3$ none of the problems is solved. The second has also been called the Kleitman-West Problem. Ahlswede/Cai [147] disproved a conjecture of Kleitman for this problem.

7.1.3. CONJECTURE OF R. AHARONI FROM 1991 CONCERNING MATCHING THEORY FOR INFINITE FAMILIES OF FINITE SETS   The conjecture states that in any such family $\mathcal{A}$ there exists a subfamily $\mathcal{B}$ of disjoint sets (called strong maximal matching) such that no substitution of k of these sets by more than k sets from $\mathcal{A}$ results again in a subfamily of disjoint sets.

The counterexample of Ahlswede/Khachatrian [105] is the family $\mathcal{A}$ of those subsets of $\mathbb{N}$, whose cardinality equals its minimal element (in canonical order).

7.1.4. CONJECTURED SHARPENING OF EKR BY AHLSWEDE/CAI/ZHANG   The equation $I_n(\exists, \forall, k) = \binom{n-1}{k-1})$ for $n \geq 3k$ was shown by Ahlswede/Alon/P. L. Erdős/ Ruszinko/ Székely (see [N, p. 117–124]) to hold for $k = 2, 3$ and to be false for $k \geq 8$.

7.2. Open problems

Problem 7.1. *Prove or disprove Aharoni's Conjecture for families of sets of* bound- ed *cardinality k. For $k = 2$ it has been proved by Aharoni.*

Problem 7.2. *Exact solutions of isodiametric problems for general sumtype func- tions are hard to obtain. Solve the case of the Lee-metric already mentioned in [72].*

Problem 7.3. (Classification of D-perfect codes) *We learnt in Section 5 that op- timal anticodes need not be balls. This led to the new concept of D-perfect codes. Continue the classification by Ahlswede/Aydinian/Khachatrian [118].*

Problem 7.4. (Optimal rate for codes with localized errors) *The optimal rate R of codes over the alphabet $\{0, 1, \ldots, q-1\}$ with localized $t = \tau \cdot n$ errors does not exceed the Hamming bound $1 - h_q(\tau) - \tau \log_q(q-1)$: In 1987 Bassalygo, Gelfand, Pinsker showed that for $q = 2$ the bound is optimal. A series of investigations [75], [77], [90], [96], [97] gave in particular a bound for $q \geq 2$.*

Theorem 7.5. (Ahlswede/Bassalygo/Pinsker [151]) *Let $0 < \tau < 1/2 - \frac{q-2}{2q(2q-3)}$, then for any $\varepsilon > 0$ the Hamming bound can be achieved in rate up to ε with codes correcting τn localized errors.*

Prove or disprove: Not only for $q = 2$, but for all $q \geq 2$ this is true for $0 < \tau < 1/2$.

Problem 7.6. Equality characterisation *for AD constitutes by itself a rich area in combinatorial extremal theory. Less demanding are equality characterisation prob- lems for the consequences of AD. Aharoni/Holzman did this for Marica-Schönheim and Beck for another special case of AD (c.f. [100]). Already in 1979 Daykin/Kleit- man/West investigated Kleitman's inequality. Ahlswede/Khachatrian [100] com- pleted these investigations.*

Problem 7.7. *Determine $H_n(\exists, \exists, k, r)$. We believe that* $\lim_{n \to \infty} H_n(\exists, \exists, 3, 2)n^{-2} = \frac{1}{\sqrt{2}}$.

Problem 7.8. *Determine $H_n(\forall, \exists, k, r)$. We believe that* $\lim_{n \to \infty} H_n(\forall, \exists, 3, 2) = 2$.

There are **many problems** in Ahlswede/Zhang [60] and Ahlswede/Ye/ Zhang [63] on "Creating order". (See also P. Vanroose [N, p. 603–614] and in U. Tamm [N, p. 589–602].)

7.3. Conjectures

Conjecture 7.9. (Ahlswede/Khachatrian; also Erdös) *In Theorem 2.2 in Section 2 one can choose for every k $n(k) = cp_k^2$ for suitable constant c. Presently we have only $n(k) = \prod_{p \leq (p_{c_1 k})} p_{c_2 k}$.*

The theory of Communication Complexity led via Yao's monochromatic rectangles for special functions to the following

Conjecture 7.10. $\lim_{n \to \infty} I_n(\forall, \exists, 3)\binom{n}{2}^{-1} = \frac{5}{4}$, $\lim_{n \to \infty} I_n(\exists, \exists, k)\binom{n}{k-1}^{-1} = 1$ *[119]*.

Conjecture 7.11. *For $a^n = (a_1, \ldots, a_n)$, $b^n = (b_1, \ldots, b_n)$ define $m_n(a^n, b^n) = \sum_{t=1}^{n} a_t \wedge b_t$ and $M(\delta, n) = \max\{|A||B| : A, B \subset \{0,1\}^n, M_n(a^n, b^n) = \delta$, for all $a^n \in A, b^n \in B\}$; Ahlswede [57] guesses that $M(\delta, n) = \max_{0 \leq m \leq n - \delta} 2^m \binom{n-m}{\delta}$.*

Conjecture 7.12. *The pair (A, B) with $A, B \subset \{0, 1, \ldots, \alpha - 1\}^n$ is an (n, δ) **constant distance code pair**, if d_H $d(a, b) = \delta$ for all $a \in A, b \in B$. Let $S_\alpha(n, \delta)$ be the set of those code pairs, let $M_\alpha(n, \delta) = \max\{|A||B| : (A, B) \in S_\alpha(n, \delta)\}$, and let for $\alpha \geq 4$ $F_\alpha(n, \delta) = \max_{\delta_1 + \delta_2 = \delta} \left(\frac{\lceil \alpha \rceil}{2} \frac{\alpha}{\lfloor 2 \rfloor}\right)^{\delta_1} \binom{n-\delta}{\delta_2}(\alpha - 1)^{\delta_2}$.*

Theorem 7.13. (Ahlswede [57]) *For $n \in \mathbb{N}$, $0 \leq \delta \leq n$ $M_\alpha(n, \delta) = F_\alpha(n, \delta)$ for $\alpha = 4, 5$.*

Prove $M_\alpha(n, \delta) = F_\alpha(n, \delta)$ for all $\alpha \geq 6$ (300 US \$ are offered for first solver).

Conjecture 7.14. *($\frac{3}{4}$-Conjecture for fix-free codes by Ahlswede/Balkenhol/Khachatrian [118].) A (variable length) code is fix-free if no codeword is a prefix or a suffix of any other. A database construction by a fix-free code is instantaneously decodable from both sides. Prove or disprove: For numbers $\ell_1, \ldots, \ell_N$ satisfying $\sum_{i=1}^{n} 2^{-\ell_i} < \frac{3}{4}$ a fix-free code of word lengths $\ell_1, \ldots, \ell_N$ exists. If true, this bound is the best possible.*

Further Conjectures can be found in Ahlswede [56].

References

[N]   "Numbers, Information and Complexity, Special volume in honour of R. Ahlswede on the occasion of his 60th birthday"; I. Althöfer, N. Cai, G. Dueck, L. Khachatrian, M. S. Pinsker, A. Sárközy, I. Wegener, Z. Zhang editors. Kluwer Acad. Publ., Boston, Dordrecht, London, 2000.

[ABEK] R. Ahlswede, C. Bey, K. Engel and L. Khachatrian, "The t-intersection problem in the truncated boolean lattice", submitted to European Journal of Combinatorics.

[B]   C. Bérge, "Nombres de coloration de l'hypergraphe h-parti complet", Hypergraph Seminar (Proc. First Working Sem., Ohio State Univ., Columbus, Ohio, 1972; dedicated to Arnold Ross), pp. 13–20. Lecture Notes in Math., Vol. 411, Springer, Berlin, 1974.

[CG]	F. Chung and R. Graham, "Erdős on graphs. His legacy of unsolved problems", A K Peters, Ltd., Wellesley, MA, 1998.

[DKW]	D. E. Daykin, D. J. Kleitman and D. B. West, "On the number of meets between two subsets of a lattice", J. Comb. Theory, Ser. A 26, 135–155, 1979.

[E1]	P. Erdős, "Problems and results in combinatorial number theory", ch. 12 in a Survey of Combinatorial Theory, J. N. Srivastava et al. (eds) 1973.

[E2]	P. Erdős, "A survey of problems in combinatorial number theory", Ann. Discrete Math. 6, 89–115, 1980.

[En]	K. Engel, "Sperner Theory", Cambridge University Press, 1997.

[ESS1]	P. Erdős, A. Sárközy and E. Szemerédi, "On a theorem of Behrend", J. Austral. Math. Soc. 7, 9–16, 1967.

[ESS2]	P. Erdős, A. Sárközy and E. Szemerédi, "On some extremal properties of sequences of integers", Annales Univ. Sci. Budapest, Eőtvös 12, 131–135, 1969.

[ESS3]	P. Erdős, A. Sárközy and E. Szemerédi, "On divisibility properties of sequences of integers", Number Theory (Colloq., Jänos Bolyai Math. Soc., Debrecen, 1968) pp. 35–49, North-Holland, Amsterdam , 1970.

[ESS4]	P. Erdős, A. Sárközy and E. Szemerédi, "On some extremal properties of sequences of integers II", Publications Math. 27, 117–125, 1980.

[ErSS]	P. L. Erdős, A. Seress and L. A. Székely, "On intersecting chains in Boolean algebras". Combinatorics, geometry and probability (Cambridge, 1993), 299–304, Cambridge Univ. Press, Cambridge, 1997.

[F1]	P. Frankl, "Intersection theorems for finite sets and geometric applications", Proceedings of the International Congress of Mathematicians, Berkeley, USA, 1419–1430, 1986.

[F2]	P. Frankl, "The Erdős-Ko-Rado Theorem is true for $n = ckt$", Combinatorics (Proc. Fifth Hungarian Colloq. Keszthely, 1976), Vol. I, pp. 365–375, Colloq. Math. Soc. Janos Bolyai, 18, North-Holland, Amsterdam-New York, 1978.

[FF]	P. Frankl and Z. Füredi, "The Erdős-Ko-Rado theorem for integer sequences", SIAM J. Algebraic Discrete Methods 1, no. 4, 376–381, 1980.

[FP]	P. Frankl and J. Pach, "On disjointly representable sets", Combinat. 4, no. 1, 39–45, 1984.

[Fr]	R. Freud, "Paul Erdős, 80 — A Personal Account", Period. Math. Hungar. 26(2), 87–93, 1993.

[GN]	R. L. Graham and J. Nesetril Eds., "The Mathematics of Paul Erdős I", Springer-Verlag, 1997.

[GST]	J. R. Griggs, J. Stahl and W. T. Trotter, "A Sperner theorem on unrelated chains of subsets", J. Combin. Theory Ser. A 36, no. 1, 124–127, 1984.

[K]	D. J. Kleitman, "On a combinatorial conjecture of Erdős", J. Combinatorial Theory 1, 209–214, 1966.

[PS]   C. Pomerance and A. Sárközy, "Combinatorial Number Theory", pp. 967–1018, in Handbook of Combinatorics, edited by R. L. Graham, M. Grötschel and L. Lovász. Elsevier Science B.V., Amsterdam; MIT Press, Cambridge, MA, 1995.

[W]   R. M. Wilson, "The exact bound in the Erdős-Ko-Rado theorem", Combinatorica 4, no. 2-3, 247–257, 1984.

Department of Mathematics
University of Bielefeld
Postfach 100131
33501 Bielefeld, Germany
E-mail address: hollmann@mathematik.uni-bielefeld.de

Spectral Analysis of Nonrelativistic Quantum Electrodynamics

Volker Bach

Abstract. I review the research results on spectral properties of atoms and molecules coupled to the quantized electromagnetic field or on simplified models of such systems obtained during the past decade. My main focus is on the results I have obtained in collaboration with **Jürg Fröhlich** and **Israel Michael Sigal** [8, 9, 10, 11, 12, 13].

1. Introduction

In this lecture I review the progress achieved during the past decade on the mathematical description of quantum mechanical matter interacting with the quantized radiation field. My main focus will be on the results I have obtained in collaboration with **Jürg Fröhlich** and **Israel Michael Sigal** [8, 9, 10, 11, 12, 13].

1.1. Basic notions of quantum mechanics

I start by recalling some basic mathematical notions of quantum mechanics. The states of a quantum mechanical system to be described are vectors in a separable Hilbert space, $\mathcal{H}$. The dynamics on $\mathcal{H}$ is generated by the selfadjoint Hamiltonian operator, H. That is, given an initial state $\psi(0) = \psi_0 \in \mathcal{H}$ at time $t = 0$, the state at time $t > 0$ is given by $\psi(t) = \exp[-itH]\psi_0$. The corresponding differential equation fulfilled by $\psi(t)$ is Schrödinger's equation,

$$i\frac{d\psi(t)}{dt} = H\,\psi(t)\,. \tag{1}$$

Stone's theorem [43] states that the selfadjointness of H is equivalent for $t \mapsto \exp[-itH]$ to be a strongly continuous one-parameter unitary group. Thus selfadjointness of the Hamiltonian is the crucial property for the existence of quantum mechanical dynamics.

As a first example, I describe the above notions for a single, nonrelativistic electron moving in a potential $V\colon \mathbb{R}^3 \to \mathbb{R}$. The Hilbert space of states and the Hamiltonian are, in this case,

$$\mathcal{H}_{el} := L^2(\mathbb{R}^3 \times \mathbb{Z}_2)\,, \qquad H_{el} = -\Delta_x + V(x)\,, \tag{2}$$

Key words and phrases. Renormalization Group, Spectrum, Resonances, Fock space, QED.

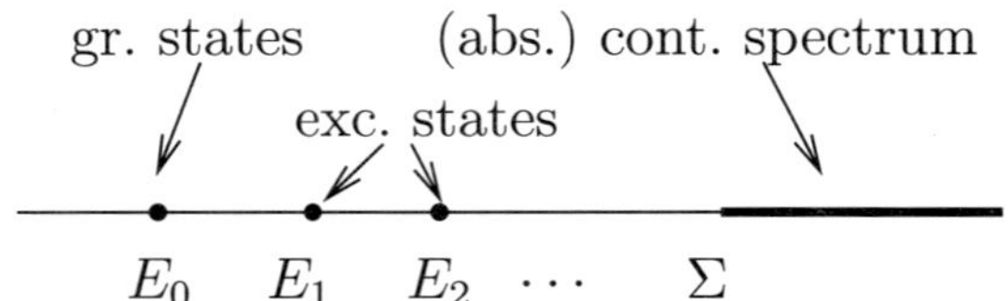

FIGURE 1. The spectrum of H_{el}

where Δ_x is the Laplacian on $\mathbb{R}^3$, and the potential $V(x)$ acts as a multiplication operator, $[V\psi](x,\sigma) := V(x)\psi(x,\sigma)$. Moreover, the $\mathbb{Z}_2$ factor in the definition of $\mathcal{H}_{el}$ accounts for the spin of the electron. Under the assumption that $V \in L^2 \cap L^\infty(\mathbb{R}^3; \mathbb{R})$, the Hamiltonian H_{el} is selfadjoint on the standard Sobolev space, $\mathrm{H}^2(\mathbb{R}^3 \times \mathbb{Z}_2) \subseteq L^2(\mathbb{R}^3 \times \mathbb{Z}_2)$, the domain $\mathrm{dom}(-\Delta_x)$ of selfadjointness of the Laplacian. If $\lim_{|x|\to\infty} V(x) = 0$, and if $\|(V)_-\|_{L^{3/2}}$ is not too small, then H_{el} has the following standard spectrum [42], see Fig. 1:

- Below 0, the spectrum is purely discrete, i.e., it consists only of isolated eigenvalues, $E_0 < E_1 < \cdots < 0$, each E_j being of finite multiplicity $n_j < \infty$. Thus there is an orthonormal basis of the corresponding spectral subspace of eigenvectors, $\{\varphi_{j,\alpha}\}_{\alpha=1,\ldots,n_j}$, i.e., $H_{el}\varphi_{j,\alpha} = E_j\varphi_{j,\alpha}$ and $\langle \varphi_{i,\alpha}|\varphi_{j,\beta}\rangle = \delta_{i,j}\,\delta_{\alpha,\beta}$. If there are infinitely many eigenvalues, they accumulate at 0.
- The positive half-axis supports the purely absolutely continuous spectrum.
- The singular continuous spectrum is empty.

$$\sigma(H_{el}) = \sigma_{\mathrm{disc}}(H_{el}) \cup \sigma_{\mathrm{ac}}(H_{el}), \tag{3}$$

$$\sigma_{\mathrm{disc}}(H_{el}) = \{E_0, E_1, E_2, \ldots\} \subseteq (-\infty, 0), \tag{4}$$

$$\sigma_{\mathrm{ac}}(H_{el}) = [0, \infty). \tag{5}$$

A typical potential to bear in mind is $V(x) := -|x|^{-1}$. Then $H_{el} = -\Delta_x - |x|^{-1}$ is the Hamiltonian of a *hydrogen atom*. It is actually not more difficult to include more than one electron in the model. The structure (3)–(5) of the spectrum of $\sigma(H_{el})$ would not change, qualitatively. I summarize the assumptions and definitions made in the following hypothesis.

Hypothesis 1.1. *Let $\mathcal{H}_{el} := L^2(\mathbb{R}^3 \times \mathbb{Z}_2)$, and assume that $V \in L^2 \cap L^\infty(\mathbb{R}^3; \mathbb{R})$ and $\lim_{|\vec{x}|\to\infty} V(x) = 0$. Let $H_{el} = -\Delta_x + V(x)$ be the corresponding selfadjoint, semibounded on the domain $\mathrm{H}^2(\mathbb{R}^3 \times \mathbb{Z}_2)$, and assume that H_{el} as (at least) one negative eigenvalue,*

$$E_0 = \inf \sigma(H_{el}) < 0 = \inf \sigma_{\mathrm{ess}}(H_{el}). \tag{6}$$

In my second example, for $N \in \mathbb{N}$, and real numbers $E_0 < E_1 < \cdots < E_N$, the Hilbert space of states is finite dimensional and the Hamiltonian is a diagonal $N \times N$ matrix,

$$\mathcal{H}_{el} := \mathbb{C}^N, \quad H_{el} = \mathrm{diag}[E_0, E_1, \ldots, E_N]. \tag{7}$$

While in itself this second example is trivial, it is of some importance as a model for the dynamics of the Schrödinger operator $-\Delta_x + V(x)$, restricted to the spectral subspace corresponding to (some part of) its discrete spectrum (implicitly assuming that the electron is spinless and that $n_j = 1$, for all j). Indeed, with this interpretation in mind and in the context of radiation theory, the 2×2 matrix $\mathrm{diag}[E_0, E_1]$ is also referred to in the physics literature as a *two-level atom*. I summarize the assumptions and definitions made in the following hypothesis.

Hypothesis 1.2. *Let* $\mathcal{H}_{el} := \mathbb{C}^N$, *for some* $N \in \mathbb{N}$, *and assume that* $\mathcal{H}_{el} = \mathrm{diag}[E_0, E_1, \ldots, E_N]$, *for some real numbers* $E_0 < E_1 < \ldots < E_N$.

My third example is the standard model for the quantized radiation field in quantum field theory. The Hilbert space carrying the field is a Fock space,

$$\mathcal{F} := \mathcal{F}_b[L^2(\mathbb{R}^3 \times \mathbb{Z}_2)] := \bigoplus_{n=0}^{\infty} \mathcal{F}^{(n)}, \tag{8}$$

where $\mathcal{F}^{(n)}$ is the state space of all n-photon states, the so-called n-photon sector. The space of no photons, $\mathcal{F}^{(0)}$, is one-dimensional, and the vacuum vector, Ω, is a unit vector in $\mathcal{F}^{(0)} := \mathbb{C}\,\Omega$. For $n \geq 1$, the n-photon sector is the subspace of the n-fold tensor product of $L^2(\mathbb{R}^3 \times \mathbb{Z}_2)$ which consists of all totally symmetric vectors (= wave functions),

$$\mathcal{F}^{(n)} := \left\{ \psi_n \in L^2\big[(\mathbb{R}^3 \times \mathbb{Z}_2)^n\big] \,\Big|\, \forall \pi \in \mathcal{S}_n : \right.$$

$$\left. \psi_n(k_{\pi(1)}, k_{\pi(2)}, \ldots, k_{\pi(n)}) = \psi_n(k_1, k_2, \ldots, k_n) \right\}$$

$$\subseteq \bigotimes_{j=1}^{n} L^2(\mathbb{R}^3 \times \mathbb{Z}_2), \tag{9}$$

where $k_j := (\vec{k}_j, \lambda_j) \in \mathbb{R}^3 \times \mathbb{Z}_2$ indicates that $\psi_n \in \mathcal{F}^{(n)}$ is given in momentum representation (Fourier transform). The symmetry of the wave functions accounts for the fact that photons are indistinguishable particles obeying Bose-Einstein statistics.

The Hamiltonian on $\mathcal{F}$ representing the energy of the free photon field is given by

$$H_f := \bigoplus_{n=0}^{\infty} H_f^{(n)}, \tag{10}$$

$$\big[H_f^{(n)} \psi_n\big](k_1, \ldots, k_n) := \big(\omega(k_1) + \ldots + \omega(k_n)\big)\, \psi_n(k_1, \ldots, k_n), \tag{11}$$

for suitable $\psi_n \in \mathcal{F}^{(n)}$, and $H_f \Omega := 0$. Here, $\omega(k) := |\vec{k}| = \sqrt{\vec{k}^2 + m^2}\,|_{m=0}$ is the *photon dispersion law*, in accordance with the principles of special relativity. From the explicit form of H_f it is clear that

$$\sigma(H_f) = [0, \infty), \quad \sigma_{\mathrm{pp}}(H_f) = \{0\}, \quad \sigma_{\mathrm{ac}}(H_f) = (0, \infty). \tag{12}$$

Note that $H_f^{(1)} = \sqrt{-\Delta_x}$. Further note that H_f leaves the n-photon sector invariant. The Hamiltonians from physics to be discussed do not have this invariance, however, and the representations (8)–(11) of $\mathcal{F}$ and H_f is rather cumbersome for those models. It is more convenient instead to express $\mathcal{F}$ and H_f in terms of creation and annihilation operators. Given $f \in L^2(\mathbb{R}^3 \times \mathbb{Z}_2)$, the creation operator $a^*(f)$ and the annihilation operator $a(f)$ are defined by $a(f)\,\Omega := 0$ and, for $n \geq 1$, by

$$a(f) \quad : \quad \mathcal{F}^{(n)} \to \mathcal{F}^{(n-1)}\,, \tag{13}$$

$$\big[a(f)\,\psi_n\big](k_1,\dots,k_{n-1}) \quad := \quad \sqrt{n} \int dk\,\overline{f(k)}\,\psi_n(k_1,\dots,k_{n-1},k)\,,$$

$$a^*(f) \quad : \quad \mathcal{F}^{(n-1)} \to \mathcal{F}^{(n)}\,, \tag{14}$$

$$\big[a^*(f)\,\psi_{n-1}\big](k_1,\dots,k_n) \quad := \quad \frac{\sqrt{n}}{n!} \sum_{\pi \in \mathcal{S}_n} f(k_{\pi(1)})\,\psi_{n-1}(k_{\pi(2)},\dots,k_{\pi(n)})\,,$$

and then extended to (a dense domain in) $\mathcal{F}$ by linearity and continuity. Note that $\big(a(f)\big)^* = a^*(f)$. The important feature of the creation and anihilation operators is that they represent the canonical commutation relations (CCR),

$$\forall f, g \in L^2(\mathbb{R}^3 \times \mathbb{Z}_2): \quad [a(f),\,a(g)] = [a^*(f),\,a^*(g)] = 0\,, \tag{15}$$

$$[a(f),\,a^*(g)] = \langle f|g\rangle\,\mathbf{1}_{\mathcal{F}}\,. \tag{16}$$

Here, $[A, B] := AB - BA$ on a suitable domain. Note that $f \mapsto a^*(f)$ is linear and $f \mapsto a(f)$ is antilinear in f. Hence, I may consider these maps as *operator-valued distributions* with formal distribution kernels $a^*(k)$ and $a(k)$, respectively. Bearing this interpretation in mind, one writes

$$a^*(f) \;=: \int dk\, f(k)\, a^*(k)\,, \quad a(f) \;=: \int dk\, \overline{f(k)}\, a(k)\,. \tag{17}$$

I remark that $a(k)$ is a densely defined operator, but not closable, while $a^*(k)$ is not even densely defined, because, e.g., $\Omega \notin \mathrm{dom}(a^*(k))$. In the sense of operator-valued distributions, i.e., with smearing by suitable test functions understood, I may rewrite the CCR as

$$\forall k, k' \in \mathbb{R}^3 \times \mathbb{Z}_2: \quad [a(k),\,a(k')] = [a^*(k),\,a^*(k')] = 0\,, \tag{18}$$

$$[a(k),\,a^*(k')] = \delta_{\lambda,\lambda'}\,\delta(\vec{k} - \vec{k}')\,\mathbf{1}_{\mathcal{F}}\,. \tag{19}$$

By means of creation and annihilation operators, I rewrite

$$\mathcal{F}^{(n)} = \overline{\mathrm{span}\Big\{a^*(f_1)\cdots a^*(f_n)\Omega \,\Big|\, f_1,\dots,f_n \in L^2(\mathbb{R}^3 \times \mathbb{Z}_2)\Big\}}\,, \tag{20}$$

$$H_f = \int dk\,\omega(k)\, a^*(k) a(k)\,. \tag{21}$$

As a fourth example, I describe a system consisting of an electron in an atom and the quantized radiation field. The appropriate Hilbert space for this

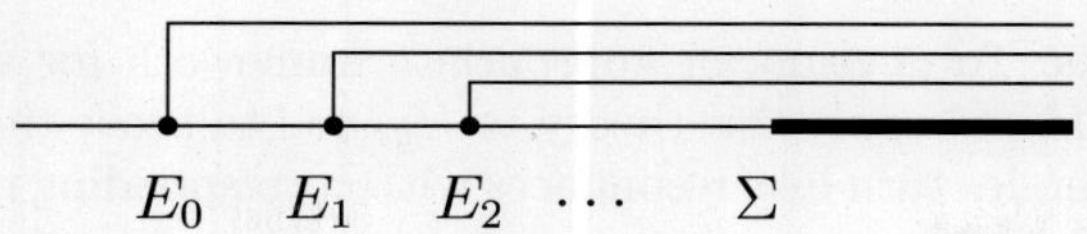

FIGURE 2. The Spectrum of $H_0 = H_{el} \otimes \mathbf{1} + \mathbf{1} \otimes H_f$

description is

$$\mathcal{H} := \mathcal{H}_{el} \otimes \mathcal{F}. \tag{22}$$

In the trivial case that the electron and the photon field do not interact, the Hamiltonian is given by

$$H_0 := H_{el} \otimes \mathbf{1}_f + \mathbf{1}_{el} \otimes H_f. \tag{23}$$

My ultimate goal is the study of an *interacting* electron-photon system. To develop sensible questions to be answered for such a system, however, it is instructive to first discuss the spectral properties of H_0. I note the general fact [42] that, for a sum of two selfadjoint operators as in (23), I have

$$\sigma(H_0) = \overline{\sigma(H_{el}) + \sigma(H_f)}, \tag{24}$$

and the spectral measure of H_0 is simply the product measure of the spectral measures of H_{el} and H_f,

$$\mu_{\varphi \otimes \psi}(H_0, \lambda + \mu) = \mu_\varphi(H_{el}, \lambda) \otimes \mu_\psi(H_f, \eta). \tag{25}$$

As a result, E_j is still an eigenvalue of H_0 with multiplicity n_j and corresponding eigenvectors $\{\varphi_{j,\alpha} \otimes \Omega\}_{\alpha=1,\dots,n_j}$. Note, however, that E_j are not isolated anymore. The lowest eigenvalue, the *ground state energy*, $\inf \sigma(H_0) = E_0$, is located at the bottom of $\sigma_{\mathrm{ac}}(H_0) = [E_0, \infty)$, and the higher eigenvalues, the *excited energies*, E_j, $j \geq 1$, are now embedded in a continuous spectrum, see Fig. 2.

   I now turn to the main object of study, the interacting electron-photon Hamiltonian,

$$H_g := H_0 + gW, \tag{26}$$

acting on $\mathcal{H} = \mathcal{H}_{el} \otimes \mathcal{F}$, as in (22). For the Hamiltonian H_g, I now formulate important tasks which have been addressed and/or even completed during the past decade.

(0.) **Models and Selfadjointness.** To give criteria for W ensuring that H_g defines a selfadjoint, semibounded Hamiltonian and general enough to include the most important applications for H_g in physics.
$\to$ See Hypothesis 2.1 and Corollary 2.3, below.

(1.) **Binding.** To specify conditions under which the Hamiltonian H_g has a ground state, i.e., under which $E_0(g) := \inf \sigma(H_g)$ is an eigenvalue.
$\to$ See Theorem 3.1, below.

182   V. Bach

(2.) **Resonances.** To develop an appropriate framework for a theory of resonances of H_g, to apply this theory to H_g, and to prove that the embedded excited energies turn into resonances with corresponding metastable states of finite life-time.
$\to$ See Theorems 4.3 and 4.5, below.

(3.) **Scattering Theory.** To derive continuous spectrum and scattering theory. To develop tools for the study of the asymptotic behaviour of e^{itH_g}, as $t \to \pm\infty$, like positive commutator estimates. Ultimately, to prove asymptotic completeness of scattering of such systems.
$\to$ See Theorem 5.3, below.

(4.) **Positive Temperatures.** To study the systems under consideration for nonzero temperature, given that the Hamiltonian and its spectral properties describe the dynamics of the system at zero temperature.
$\to$ See Theorem 6.1, below.

(5.) **Feshbach Renormalization Map.** To develop a renormalization group that allows for a direct analysis of the spectral properties of H_g and L_g.
$\to$ See Theorems 7.2, below.

In the remaining Sections 2–7, I discuss the topics (0.)–(5.) of the list above. Besides the papers mentioned or discussed below, there are many important contributions which cannot be discuss here but should, nevertheless, be mentioned: [2, 3, 4, 5, 1, 18, 19, 20, 21, 25, 26, 27, 38, 45, 46]

2. Models and Selfadjointness

2.1. Modelling the interaction

According to *first principles* in physics, the physically correct coupling of an electron to the electromagnetic field is the *minimal coupling*. Writing the Schrödinger operator $H_{el}=-\Delta_x+V(x)$ in Eq. (2) as $H_{el}=(\vec{\sigma}\cdot i\vec{\nabla}_x)^2+V(x)$ ($\vec{\sigma}=(\sigma^{(x)},\sigma^{(y)},\sigma^{(z)})$ being the three Pauli matrices), it amounts to replacing the momentum operator $-i\vec{\nabla}_x$ by $-i\vec{\nabla}_x - 2\pi^{1/2}\alpha^{3/2}\vec{A}(\alpha x)$ (to accommodate for gauge invariance),

$$H_\alpha \; := \; \left[\vec{\sigma}\cdot\left(-i\vec{\nabla}_{\vec{x}} - 2\pi^{1/2}\alpha^{3/2}\vec{A}(\alpha\vec{x}_j)\right)\right]^2 \; + \; V_c(x)\otimes\mathbf{1}_f \; + \; \mathbf{1}_{el}\otimes H_f, \quad (27)$$

where $\alpha \sim 1/137$ is the fine structure constant, and $\vec{A}(\vec{x})$ denotes the quantized vector potential of the transverse modes of the electromagnetic field in the Coulomb gauge, i.e.,

$$\vec{A}(\vec{x}) \; := \; \int dk\, \vec{G}_{\vec{x}}(k)\otimes a^*(k) \; + \; \overline{\vec{G}_{\vec{x}}(k)}\otimes a(k), \quad (28)$$

with *coupling function*

$$\vec{G}_{\vec{x}}(\vec{k},\lambda) \; := \; \frac{\sqrt{2}\,\kappa(|\vec{k}|/K)}{\sqrt{\pi\,K^3\,\omega(\vec{k})}}\, \exp[-i\vec{k}\cdot\vec{x}]\,\vec{\varepsilon}_\lambda(\vec{k}), \quad (29)$$

where $\vec{\varepsilon}_\lambda(\vec{k})$, $\lambda = 1, 2$, are photon polarization vectors satisfying

$$\vec{\varepsilon}_\lambda(\vec{k})^* \cdot \vec{\varepsilon}_\mu(\vec{k}) = \delta_{\lambda\mu}, \qquad \vec{k} \cdot \vec{\varepsilon}_\lambda(\vec{k}) = 0, \qquad \text{for } \lambda, \mu = 1, 2. \qquad (30)$$

Furthermore, κ is an entire function of rapid decrease on the real line, e.g., $\kappa(r) := \exp(-r^4)$. Hence, the factor $\kappa(|\vec{k}|/K)$ in (28) cuts off the vector potential in the ultraviolet domain, $|\vec{k}| \gg K$. It is artificial in the sense that physical principles actually imply that $\kappa \equiv 1$. With $\kappa \equiv 1$, however, $\|\vec{G}_x\|_{L^2}$ would diverge at $|\vec{k}| = \infty$, which, in turn, would imply that $\Omega \notin \mathrm{dom}(\vec{A}(x))$, for any $x \in \mathbb{R}^3 \times \mathbb{Z}_2$. In order to give a meaning to $\vec{A}(x)$ as a densely defined operator, I thus have to regularize $\vec{G}_x$ at (a preferably large) momentum scale $K \gg 1$. Indeed, by choosing κ to be a sufficiently rapidly decreasing, analytic function obeying $\kappa(0) = 1$, it is ensured that $\vec{G}_x \in L^2(\mathbb{R}^3 \times \mathbb{Z}_2)$, uniformly in $x \in \mathbb{R}^3 \times \mathbb{Z}_2$. The function $\kappa(|\cdot|/K)$ is called *ultraviolet cutoff*, and the construction of the limit $K \to \infty$, of this regularization is one of the open problems in nonrelativistic quantum electrodynamics.

I return to Eqn. (27), which I write as

$$H_g = H_0 + W_g, \qquad (31)$$

where H_0 is defined in (23), and I obtain

$$W_g + C_{no} = 4\pi^{1/2}\alpha^{3/2}\vec{A}(\alpha\vec{x}_j) \cdot (i\vec{\nabla}_{\vec{x}}) + 2\pi\alpha^3\vec{A}^2(\alpha\vec{x})$$
$$+ 2\pi^{1/2}\alpha^{5/2}\vec{\sigma} \cdot (\vec{\nabla} \wedge \vec{A})(\alpha\vec{x}) \qquad (32)$$

from expanding the square in (27). Note that W_g contains terms linear and quadratic in the creation and annihilation operators, $a^*(k)$, $a(k)$. Hence, I may write

$$W_g = gW_{1,0} + gW_{0,1} + gW_{2,0} + g^2W_{1,1} + g^2W_{0,2}, \qquad (33)$$

where $W_{1,0}$ and $W_{0,1}$ are linear in $a^*(k)$ and $a(k)$,

$$W_{1,0} := \int dk\, w_{1,0}(k) \otimes a^*(k), \qquad W_{0,1} := \int dk\, w_{0,1}(k) \otimes a(k), \qquad (34)$$

and $W_{2,0}$, $W_{1,1}$ and $W_{0,2}$ are quadratic in $a^*(k)$ and $a(k)$,

$$W_{2,0} := \int dk\, dk'\, w_{2,0}(k, k') \otimes a^*(k)a^*(k'), \qquad (35)$$

$$W_{1,1} := \int dk\, dk'\, w_{1,1}(k, k') \otimes a^*(k)a(k'), \qquad (36)$$

$$W_{0,2} := \int dk\, dk'\, w_{0,2}(k, k') \otimes a(k)a(k'). \qquad (37)$$

The tensor products in (34)–(37) indicate that I consider the *coupling functions* $w_{m,n}$ as functions on $(\mathbb{R}^3 \times \mathbb{Z}^2)^{m+n}$ with values in the operators on $\mathcal{H}_{el}$.

Comparing (34)–(37) to (32), I find that

$$w_{1,0}(k) = w_{0,1}(k)^* := 2i\,\vec{G}_{\vec{x}}(k) \cdot \vec{\nabla}_{\vec{x}} + \vec{\sigma} \cdot (\vec{B}_{\vec{x}}(k)), \qquad (38)$$

184 V. Bach

where the magnetic field $\vec{B}_{\vec{x}}(k)$ corresponds to the term $2\pi^{1/2}\alpha^{5/2}\vec{\sigma}\cdot(\vec{\nabla}\wedge\vec{A})(\alpha\vec{x})$ in W_g,

$$\vec{B}_{\vec{x}}(k) \ := \ \frac{\alpha\sqrt{2}\,\kappa(|\vec{k}|/K)}{i\sqrt{\pi\,K^3\,\omega(\vec{k})}}\,\exp[-i\alpha\vec{k}\cdot\vec{x}]\left(\vec{k}\wedge\vec{\varepsilon}_\lambda(\vec{k})\right). \tag{39}$$

Furthermore,

$$w_{2,0}(k_1,k_2) \ = \ w_{0,2}(k_1,k_2)^* \ := \ \vec{G}_{\vec{x}}(k_1)\cdot\vec{G}_{\vec{x}}(k_2)\,, \tag{40}$$

$$w_{1,1}(k_1,k_2) \ := \ \vec{G}_{\vec{x}}(k_1)^*\cdot\vec{G}_{\vec{x}}(k_2) \ + \ \vec{G}_{\vec{x}}(k_1)\cdot\vec{G}_{\vec{x}}(k_2)^*\,. \tag{41}$$

The constant C_{no} in (32) equals $\|\vec{G}_{\vec{x}}\|_{L^2}^2$, which is independent of $\vec{x}$. It results from *normal-ordering* one term contributing to $W_{1,1}$,

$$a(\vec{G}_{\vec{x}})\,a^*(\vec{G}_{\vec{x}}) \ = \ a^*(\vec{G}_{\vec{x}})\,a(\vec{G}_{\vec{x}}) \ + \ \|\vec{G}_{\vec{x}}\|_{L^2}^2\,\mathbf{1}\,. \tag{42}$$

Note that the finiteness of C_{no} is due to the introduction of the ultraviolet cutoff.

The next observation to be made is that the coupling functions $w_{m,n}$ obey the following bounds, pointwise in $k,k'\in\mathbb{R}^3\times\mathbb{Z}_2$,

$$\left\|w_{1,0}(k)\,(-\Delta_{\vec{x}}+\mathbf{1})^{-1/2}\right\|_{\mathcal{B}(\mathcal{H}_{el})} \ + \ \left\|w_{0,1}(k)\,(-\Delta_{\vec{x}}+\mathbf{1})^{-1/2}\right\|_{\mathcal{B}(\mathcal{H}_{el})} \ \leq \ J(k)\,, \tag{43}$$

$$\left\|w_{2,0}(k,k')\right\|_{\mathcal{B}(\mathcal{H}_{el})} \ + \ \left\|w_{1,1}(k,k')\right\|_{\mathcal{B}(\mathcal{H}_{el})} \ + \ \left\|w_{0,2}(k,k')\right\|_{\mathcal{B}(\mathcal{H}_{el})} \ \leq \ J(k)\,J(k')\,, \tag{44}$$

with

$$J(k) \ := \ \frac{4\,\kappa(|\vec{k}|/K)}{\omega(k)^{1/2}}\,, \tag{45}$$

and I note for later reference that, for any $0\leq\beta<2$,

$$\int\left(1+\omega(k)^{-\beta}\right)|J(k)|^2\,dk \ < \ \infty\,. \tag{46}$$

I use this example as a guideline for the following hypothesis on the form of the interaction W_g.

Hypothesis 2.1. *The interaction is of the form*

$$W_g \ = \ gW_{1,0} + gW_{0,1} + gW_{2,0} + g^2W_{1,1} + g^2W_{0,2}\,, \tag{47}$$

where

$$W_{1,0} \ := \ \int dk\,w_{1,0}(k)\otimes a^*(k)\,, \qquad W_{0,1} \ := \ \int dk\,w_{0,1}(k)\otimes a(k)\,, \tag{48}$$

$$W_{2,0} \;\; := \;\; \int dk\, dk'\, w_{2,0}(k,k') \otimes a^*(k)a^*(k')\,, \tag{49}$$

$$W_{1,1} \;\; := \;\; \int dk\, dk'\, w_{1,1}(k,k') \otimes a^*(k)a(k')\,, \tag{50}$$

$$W_{0,2} \;\; := \;\; \int dk\, dk'\, w_{0,2}(k,k') \otimes a(k)a(k')\,. \tag{51}$$

The coupling functions $w_{m,n}$ are functions on $(\mathbb{R}^3 \times \mathbb{Z}_2)^{m+n}$ with values in the operators on $\mathcal{H}_{el}$ obeying $w_{m,n} = w_{n,m}^*$. Moreover, there is a measurable function $J : \mathbb{R}^3 \times \mathbb{Z}_2 \to \mathbb{R}_0^+$ such that

$$\left\| w_{1,0}(k)\,(-\Delta_{\vec{x}} + 1)^{-1/2} \right\|_{\mathcal{B}(\mathcal{H}_{el})} + \left\| w_{0,1}(k)\,(-\Delta_{\vec{x}} + 1)^{-1/2} \right\|_{\mathcal{B}(\mathcal{H}_{el})} \;\le\; J(k)\,, \tag{52}$$

$$\left\| w_{2,0}(k,k') \right\|_{\mathcal{B}(\mathcal{H}_{el})} + \left\| w_{1,1}(k,k') \right\|_{\mathcal{B}(\mathcal{H}_{el})} + \left\| w_{0,2}(k,k') \right\|_{\mathcal{B}(\mathcal{H}_{el})} \;\le\; J(k)\,J(k')\,, \tag{53}$$

for all $k, k' \in \mathbb{R}^3 \times \mathbb{Z}_2$.

2.2. Relative bounds and selfadjointness

In this section I present the results on task (0.) in the list above, establishing the existence of the Hamiltonian H_g by deriving it from a semibounded quadratic form under minimal conditions. Furthermore, I give a criterion that ensures the stability of the domain of definition for H_g, i.e., $\mathrm{dom}(H_g) = \mathrm{dom}(H_0)$. The arguments are based on Kato perturbation theory and variations of the following simple estimate. Namely, given f such that $f/\sqrt{\omega} \in L^2(\mathbb{R}^3 \times \mathbb{Z}_2)$ and $\psi \in \mathrm{dom}(H_f)$, I observe that

$$
\begin{aligned}
\| a(f)\,\psi \| \;\; &\le \;\; \int |f(k)|\, \| a(k)\psi \|\, dk \\[2mm]
&\le \;\; \left(\int \frac{|f(k)|^2\, dk}{\omega(k)} \right)^{1/2} \left(\int \omega(k)\, \| a(k)\psi \|^2\, dk \right)^{1/2} \\[2mm]
&= \;\; \| \omega^{-1/2} f \|_{L^2}^2 \cdot \| H_f^{1/2} \psi \|\,,
\end{aligned}
\tag{54}
$$

hence, for any $\rho > 0$,

$$\left\| a(f)\,(H_f + \rho)^{-1/2} \right\| \;\le\; \| \omega^{-1/2} f \|_{L^2}^2\,. \tag{55}$$

The following lemma derives from (55).

Lemma 2.2. *Assume hypotheses 1.1 or 1.2 and 2.1.*

(i) *If $\omega^{-1} J^2 \in L^1(\mathbb{R}^3 \times \mathbb{Z}_2)$ then $W_{m,n}$ defines a quadratic form on $Q(H_0)$, and I have that*

$$\left\| (H_0 + i)^{-1/2}\, W_{m,n}\,(H_0 + i)^{-1/2} \right\| \;\le\; C(V)\, \| \omega^{-1} J^2 \|_{L^1}\,, \tag{56}$$

for all $1 \le m + n \le 2$, where $C(V) < \infty$ is a constant depending on the potential V.

(ii) *If $(1 + \omega^{-1})J^2 \in L^1(\mathbb{R}^3 \times \mathbb{Z}_2)$ then $W_{m,n}$ and $W^*_{m,n}$ define bounded operators on* $\mathrm{dom}(H_0)$, *and*

$$\left\| W^{\#}_{m,n}(H_0 + i)^{-1} \right\| \leq C(V) \left\| (1 + \omega^{-1})J^2 \right\|_{L^1},\tag{57}$$

where $W^{\#}_{m,n}$ is $W_{m,n}$ or $W^*_{m,n}$, and the constant $C(V) < \infty$ depends only on V.

Now, standard Kato perturbation theory implies that

Corollary 2.3. *Assume hypotheses 1.1 or 1.2 and 2.1.*

(i) *If $\omega^{-1}J^2 \in L^1(\mathbb{R}^3 \times \mathbb{Z}_2)$ and $|g| > 0$ is sufficiently small then H_g defines a symmetric, semibounded quadratic form on $Q(H_0)$, and hence the corresponding selfadjoint operator is essentially selfadjoint on* $\mathrm{dom}(H_0)$.

(ii) *If $(1 + \omega^{-1})J^2 \in L^1(\mathbb{R}^3 \times \mathbb{Z}_2)$ and $|g| > 0$ is sufficiently small then H_g is a semibounded, selfadjoint operator on $\mathrm{dom}(H_g) = \mathrm{dom}(H_0)$.*

The proofs for these basic statements can be found in many papers on this subject, e.g., [19, 10, 5]

3. Binding

In this section I focus on the bottom of the spectrum, $E_0(g) := \inf \sigma(H_g)$ of the interacting Hamiltonian H_g. Besides hypotheses 1.1 or 1.2 and 2.1, I will now assume that $(1 + \omega^{-1})J^2 \in L^1(\mathbb{R}^3 \times \mathbb{Z}_2)$. Then Corollary 2.3(ii) insures that $H_g = H_g^*$ on $\mathrm{dom}(H_0)$ and that $E_0(g) > -\infty$. Furthermore, from the discussion of the spectral properties of H_0 in Section 1, I know that $E_0(0) = E_0$ is an eigenvalue. Indeed, the corresponding eigenspace is spanned by $\{\varphi_{0,\alpha} \otimes \Omega\}_{\alpha=1,\dots,n_0}$.

The question of stability of this eigenvalue under perturbation now arises.

Theorem 3.1. *Assume hypotheses 1.1 or 1.2 and 2.1. Furthermore assume $W_{2,0} = W_{1,1} = W_{0,2} = 0$, $(1 + \omega^{-2})J^2 \in L^1(\mathbb{R}^3 \times \mathbb{Z}_2)$. There exists a constant $C(V) < \infty$ such that if $2\alpha := |E_0| - C(V)\|(1 + \omega^{-2})J^2\|_{L^1}g^2 > 0$ then $E_0(g)$ is an eigenvalue with corresponding eigenvector, $\Psi_0(g) \in \mathcal{H}$. Moreover,*

$$\left\| e^{\alpha|x|} \otimes N_f \Psi_0(g) \right\| < \infty.\tag{58}$$

Theorem 3.1 states that $\inf \sigma(H_g)$ is an eigenvalue and that the corresponding eigenfunction is exponentially localized about the origin. The physical interpretation of this statement is that the atom or molecule under consideration does not dissolve by switching on the interaction of the electron and the electromagnetic field. In fact, the spatial localization of the atom or molecule is continuous in $g \to 0$.

A first existence result for a ground state in the framework of Hypotheses 1.1 and 2.1, i.e., an eigenvalue at the bottom of the spectrum, was derived in [20], and another important result in the context of the Spin-Boson model was given in [45]. In the form stated above, Theorem 3.1 was proved under Hypotheses 1.1 and 2.1 in [10] and under Hypotheses 1.2 and 2.1 in [5]. The strategies of the proof in

[10] and in [5] are similar, and they both build on ideas given in [20]. The range of validity w.r.t. g was further enlarged in [46], and in [23] it was finally shown that no restriction on the magnitude of g is necessary, whatsoever.

Statements about uniqueness of the ground state, i.e., about the non-degeneracy of $E_0(g)$ as an eigenvalue, were given in [10, 27].

I outline the strategy of the proof of Theorem 3.1 as in [10].

- First, the coupling functions $w_{0,1}(k) = w_{1,0}(k)^*$, are replaced by $\chi[\omega(k) \geq m]\, w_{0,1}(k)$ and $\chi[\omega(k) \geq m]\, w_{1,0}(k)$, respectively, where $m > 0$ is interpreted to be a "photon mass". The resulting Hamiltonian is denoted $H_g^{(m)}$.
- By a suitable additional discretization, one shows that $E_0^{(m)}(g) + m = \inf \sigma_{\mathrm{ess}}(H_g^{(m)})$ where $E_0^{(m)}(g) := \inf \sigma(H_g^{(m)})$. Hence, $E_0^{(m)}(g)$ is an eigenvalue of finite multiplicity. Denote by $\Psi_0^{(m)}(g)$ a normalized eigenfunction, $H_g^{(m)}\, \Psi_0^{(m)}(g) = E_0^{(m)}(g)\, \Psi_0^{(m)}(g)$.
- From a simple norm bound follows the convergence $H_g^{(m)} \to H_g^{(0)} = H_g$ in norm-resolvent sense, as $m \to 0$. In particular, $\lim_{m\to 0} E_0^{(m)}(g) = E_0(g)$, and, possibly after passing to a subsequence, $w - \lim_{m\to 0} \Psi_0^{(m)}(g) =: \Psi$ is a ground state of H_g: $H_g\, \Psi = E_0(g)\, \Psi$.
- The key step in the proof is to show that $\Psi \neq 0$. At this point, Agmon estimates for the localization in the x-variable and *soft-photon bounds* insure that the sequence $\Psi_0^{(m)}(g)$ is compact and hence $\Psi \neq 0$.

4. Resonances

The notion of *resonances* discussed here is based on the analytic continuation of resolvent matrix elements by means of complex deformations (here: dilatations). More precisely, a resonance is a singularity of the function

$$F_{\varphi,\psi}(z) := \langle \varphi | \, (H - z)^{-1} \psi \rangle, \tag{59}$$

analytically continued from $z := \lambda + i\varepsilon \in \mathbb{C}^+$, $\lambda > E_0(g)$, across the real axis onto the second Riemann sheet in $\mathbb{C}^-$. Note that $\lambda \in \sigma_{\mathrm{ess}}(H_g)$, so such an analytic continuation cannot be expected to exist for all $\varphi, \psi \in \mathcal{H}$. Rather, the goal is to construct the analytic continuation of $F_{\varphi,\psi}$, for φ, ψ contained in a natural dense set $\mathcal{D}$.

The set $\mathcal{D}$ is not unique, but it is characterized by a maximality requirement: Denoting by $\mathcal{A}(\varphi, \psi)$ the domain of analyticity of $F_{\varphi,\psi}$, the intersection $\bigcap_{\varphi,\psi \in \mathcal{D}} \mathcal{A}(\varphi, \psi)$ should be the largest possible set under the requirement that $\mathcal{D} \subseteq \mathcal{H}$ be dense.

Our construction of the analytic continuation of $F_{\varphi,\psi}$ goes through complex dilatation [42, 15]. For $\theta \in \mathbb{R}$ and $\psi_n \in \mathcal{F}^{(n)}$, I define a unitary dilatation operator by

$$\left[U_\theta^{(n)}\, \psi_n\right](k_1, \ldots, k_n) := e^{-\frac{3\theta}{2}(|\vec{k}_1| + \cdots + |\vec{k}_n|)}\, \psi_n(e^{-\theta}k_1, \ldots, e^{-\theta}k_n), \tag{60}$$

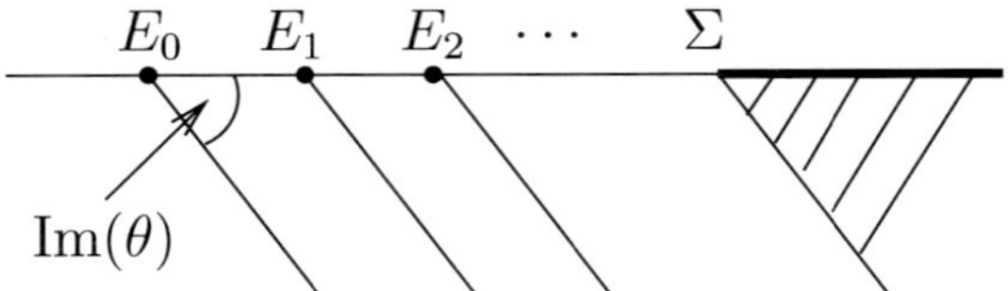

FIGURE 3. The spectrum of $H_0(i\vartheta)$, with $\vartheta > 0$.

where $e^{-\theta}k = e^{-\theta}(\vec{k}, \lambda) := (e^{-\theta}\vec{k}, \lambda)$. Furthermore, $U_\theta^{(0)}\Omega := \Omega$. Then, the unitary dilatation U_θ on $\mathcal{H}$ is defined by

$$U_\theta := \mathbf{1}_{el} \otimes \bigoplus_{n=0}^{\infty} U_\theta^{(n)}. \tag{61}$$

As in the introduction, it is instructive to discuss the action of U_θ on H_0 before applying it to H_g. I remark that $U_\theta \mathbf{1}_{el} \otimes H_f U_\theta^{-1} = e^{-\theta}\mathbf{1}_{el} \otimes H_f$ and hence

$$H_0(\theta) := U_\theta H_0 U_\theta^{-1} = H_{el} \otimes \mathbf{1}_f + e^{-\theta}\mathbf{1}_{el} \otimes H_f. \tag{62}$$

Observe that $H_0(\theta)$ extends from $\theta \in \mathbb{R}$ to an analytic family of type A [41] on the strip $\theta \in S_\pi := \{\theta| -\pi < \mathrm{Im}(\theta) < \pi\}$, i.e., the Banach space-valued map

$$S_\pi \ni \theta \mapsto H_0(\theta)\left(H_0 + i\right)^{-1} \in \mathcal{B}(\mathcal{H}) \tag{63}$$

is analytic. Note that, for $\theta \notin \mathbb{R}$, $H_0(\theta)$ is not selfadjoint. Yet, $H_0(\theta)$ is a normal operator, even for $\theta \notin \mathbb{R}$. Thus, the discussion of the spectral properties of $H_0(\theta)$ is as simple as the one for H_0. Namely,

$$\sigma[H_0(\theta)] = \overline{\sigma(H_{el}) + e^{-\theta}\sigma(H_f)} = \sigma(H_{el}) + e^{-\theta}\mathbb{R}_0^+. \tag{64}$$

For $j=0,1,\ldots$, the real numbers E_j are eigenvalues of $H_0(\theta)$ at the tips of branches of the continuous spectrum, the corresponding eigenvectors remain unchanged (see fig. 3).

I now construct the analytic continuation of $F_{\varphi,\psi}$, for $g = 0$. I define $\mathcal{D}$ to be the set of *dilatation analytic vectors*, i.e., those vectors $\psi \in \mathcal{H}$, for which the Hilbert space-valued map

$$S_\pi \ni \theta \mapsto \psi_\theta := U_\theta \psi \in \mathcal{H} \tag{65}$$

is analytic. Then, for any $z \in \mathbb{C}^+$ and $\theta = i\vartheta$, $0 < \vartheta < \pi/2$,

$$F_{\varphi,\psi}(z) = \langle\varphi| (H_0 - z)^{-1} \psi\rangle = \langle\varphi_{\bar{\theta}}| (H_0(\theta) - z)^{-1} \psi_\theta\rangle, \tag{66}$$

by analytic continuation in θ. (Note that φ continues to $\varphi_{\bar{\theta}}$ because of the anti-linearity of $\varphi \mapsto \langle\varphi|\psi\rangle$.) I now obtain the desired analytic continuation of $z \mapsto F_{\varphi,\psi}(z)$ into $\mathbb{C}^- \setminus \sigma[H_0(\theta)]$ by continuing the right side of (66) in z. From this point of view, the complex dilatation in θ defines a projection of the Riemann surface associated to $F_{\varphi,\psi}(z)$ onto the complex plane different from the one obtained for $\theta = 0$. The branches of continuous spectra appear as branch cuts associated to the chosen projection, and the eigenvalues coincide with the branch points of these cuts. Their position is independent of the chosen projection, i.e., invariant under (local) variations of the deformation parameter θ.

The construction of the analytic continuation of $F_{\varphi,\psi}(z)$ for $g > 0$ is similar to the one for $g = 0$, in principle. I recall from Hypothesis 2.1 that $W_g = \sum_{1 \leq m+n \leq 2} g^{m+n} W_{m,n}$ and that the coupling functions $w_{m,n}$ in $W_{m,n}$ are functions on $(\mathbb{R}^3 \times \mathbb{Z}_2)^{m+n}$ with values in the operators on $\mathcal{H}_{el}$. For the existence of resonances, I shall employ the following additional assumption.

Hypothesis 4.1. *There exists $0 < \theta_0 < \pi/2$ such that, for all $k \in \mathbb{R}^3 \times \mathbb{Z}_2$ and all $1 \leq m + n \leq 2$, the Banach space-valued maps*

$$D_{\theta_0} \ni \theta \;\mapsto\; w_{m,n}(e^{-\theta}k)\left(\Delta_x + 1\right)^{-1+\frac{m+n}{2}} \in \mathcal{B}(\mathcal{H}_{el}) \tag{67}$$

are analytic, where $D_{\theta_0} := \{|\theta| < \theta_0\} \subseteq \mathbb{C}^2$. Moreover, there is a measurable function $J \colon \mathbb{R}^3 \times \mathbb{Z}_2 \to \mathbb{R}_0^+$ such that

$$\sup_{|\theta|<\theta_0} \left\| w_{1,0}(e^{-\theta}k)\,(-\Delta_{\vec{x}} + 1)^{-1/2} \right\|_{\mathcal{B}(\mathcal{H}_{el})} \tag{68}$$

$$+ \sup_{|\theta|<\theta_0} \left\| w_{0,1}(e^{-\theta}k)\,(-\Delta_{\vec{x}} + 1)^{-1/2} \right\|_{\mathcal{B}(\mathcal{H}_{el})} \;\leq\; J(k),$$

$$\sup_{|\theta|<\theta_0} \left\| w_{2,0}(e^{-\theta}k, e^{-\theta}k') \right\|_{\mathcal{B}(\mathcal{H}_{el})} + \sup_{|\theta|<\theta_0} \left\| w_{1,1}(e^{-\theta}k, e^{-\theta}k') \right\|_{\mathcal{B}(\mathcal{H}_{el})} \tag{69}$$

$$+ \sup_{|\theta|<\theta_0} \left\| w_{0,2}(e^{-\theta}k, e^{-\theta}k') \right\|_{\mathcal{B}(\mathcal{H}_{el})} \;\leq\; J(k)\,J(k'),$$

for all $k, k' \in \mathbb{R}^3 \times \mathbb{Z}_2$.

For $j = 1, 2, \ldots$, let $\{\varphi_{j,\alpha}\}_{\alpha=1,\ldots,n_j} \subseteq \mathcal{H}_{el}$ be an orthonormal basis of eigenfunctions of H_{el} corresponding to the eigenvalue E_j. Then the following genericity assumption on the coupling function $w_{1,0}$ is assumed to hold, for all $j \geq 1$ and $1 \leq \alpha \leq n_j$,

$$\sum_{i=0}^{j} \sum_{\beta=1}^{n_i} \left| \mathrm{supp}\{ \langle \varphi_{i,\beta} |\, w_{1,0}(\,\cdot\,)\,\psi_{el,j,\alpha} \rangle \} \right| \;>\; 0\,. \tag{70}$$

I remark that the pointwise analyticity assumed in Hypothesis 4.1 is slightly stronger than what it necessary.

Furthermore, I remark that Hypothesis 4.1 does not hold for the coupling functions of the physical example in (38), (40), and (41) because the dilatation operator U_θ acts only on the photon Fock space and leaves the electron variable unchanged, and consequently, the factor $\exp[i\alpha\,\vec{k}\cdot\vec{x}]$ in $\vec{G}_{\vec{x}}(k)$, defined in (29), is turned into $\exp[\vartheta\,\alpha\,\vec{k}\cdot\vec{x}]$ which is exponentially growing, as $\vec{k}\cdot\vec{x}$ becomes large. This may be avoided by dilating both the electron and the photon variables, for then $\exp[i\alpha\,\vec{k}\cdot\vec{x}]$ is simply invariant under dilatation. The price to pay is that I have to require dilation analyticity of the potential V in H_{el} and that $H_{el}(\theta) := U_\theta H_{el} U_\theta^{-1}$ is not selfadoint anymore, if $\theta \notin \mathbb{R}$. The latter makes certain estimates on norms of

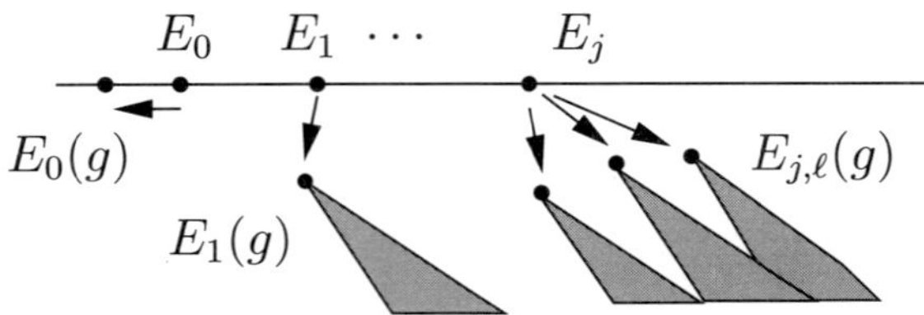

FIGURE 4. The spectrum of $H_g(i\vartheta)$, with $\vartheta > 0$, up to $\mathcal{O}(g^{2+\varepsilon})$-neighbourhoods, for some $\varepsilon > 0$.

resolvents of $H_{el}(\theta)$ slightly more complicated than for the selfadjoint case, $\theta = 0$. This has been carried out in [12].

Lemma 4.2. *Assume hypotheses 1.1 or 1.2 and 2.1 and 4.1. If $(1 + \omega^{-1})J^2 \in L^1(\mathbb{R}^3 \times \mathbb{Z}_2)$ then $H_g(\theta)$ defines an analytic family of type A, i.e., the Banach space-valued map*

$$D_{\theta_0} \ni \theta \;\mapsto\; H_g(\theta)\,(H_0 + i)^{-1} \;\in\; \mathcal{B}(\mathcal{H}) \tag{71}$$

is analytic.

Lemma 4.2 insures that, for all $\varphi, \psi \in \mathcal{D}$, for any $z \in \mathbb{C}^+$, and $\theta = i\vartheta$, with $0 < \vartheta < \theta_0$,

$$F_{\varphi,\psi}(z) \;=\; \langle \varphi | \, (H_g - z)^{-1}\, \psi \rangle \;=\; \langle \varphi_{\bar{\theta}} | \, (H_g(\theta) - z)^{-1}\, \psi_\theta \rangle, \tag{72}$$

by analytic continuation in θ. So, as for H_0, I can analytically continue in z from $\mathbb{C}^+$ to $\mathbb{C}^- \setminus \sigma[H_g(\theta)]$.

Theorem 4.3. *Assume hypotheses 1.1 or 1.2 and 2.1 and 4.1. Furthermore, assume that $\theta = i\vartheta$, where $\vartheta > 0$ is small but fixed, and that $(1 + \omega^{-\beta})J^2 \in L^1(\mathbb{R}^3 \times \mathbb{Z}_2)$, for some $\beta > 1$. Then, for each $j \geq 1$, there exist constants, $\Gamma_j > 0$ and $C_j < \infty$, such that, for $g > 0$ sufficiently small,*

$$\Big[E_{j-1} + C_j g \,,\, E_{j+1} - C_j g \Big] \;+\; i\Big(-g^2\,\Gamma_j \,,\, \infty \Big) \;\subseteq\; \rho[H_g(\theta)] \;:=\; \mathbb{C} \setminus \sigma[H_g(\theta)]. \tag{73}$$

Moreover, the spectrum is located in $\mathcal{O}(g^{2+\varepsilon})$-neighbourhoods of the comet-shaped regions depicted in Fig. 4, for some $\varepsilon > 0$.

Theorem 4.3 has the important consequence that the spectrum of H_g in the interval $[E_0(g) + Cg \,,\, \Sigma - Cg]$ is purely absolutely continuous (see, e.g., [15]).

Under more stringent conditions on the coupling functions it is possible to derive more precise information about the nature of the resonances than what is given in Theorem 4.3. The comet-shaped regions (see Fig. 4) are only a rough description of their location. The additional assumption that allows for a more precise statement is as follows.

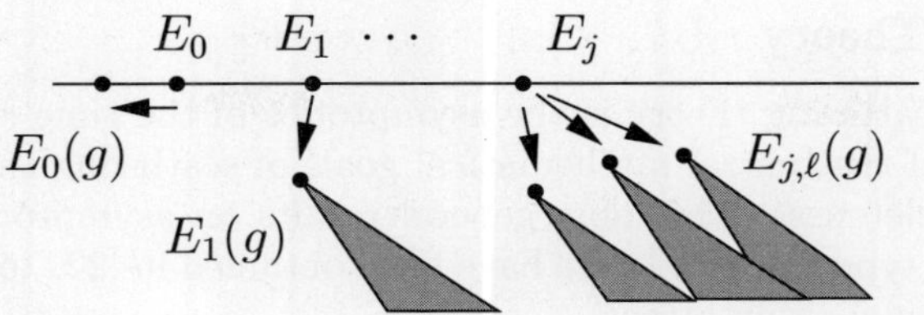

FIGURE 5. The spectrum of $H_g(i\vartheta)$, with $\vartheta > 0$.

Hypothesis 4.4. *Assume Hypothesis 4.1. For some $\mu > 0$, the function $J \colon \mathbb{R}^3 \times \mathbb{Z}_2 \to \mathbb{R}_0^+$ obeys the following additional bound,*

$$\sup_{k \in \mathbb{R}^3 \times \mathbb{Z}_2} \left\{ \omega(k)^{\frac{1-\mu}{2}} \, J(k) \right\} < \infty. \tag{74}$$

A renormalization group analysis, as described below in Section 7, reveals that the singularities of $F_{\varphi,\psi}$, are actually confined in cuspidal domains whose tip is an eigenvalue of $H_g(\theta)$, see Fig. 5.

Theorem 4.5. *Assume Hypotheses 1.1 or 1.2 and 2.1, 4.1, and 4.4. Furthermore, assume that $\theta = i\vartheta$, where $\vartheta > 0$ is small but fixed. Then, for each $j \geq 1$ and $g > 0$ sufficiently small, $H_g(\theta)$ possesses complex eigenvalues, $E_{j,\alpha}(g) = E_j + \mathcal{O}(g) \in \mathbb{C}^-$, with corresponding eigenvectors $\Psi_{j,\alpha}(g) = \varphi_{j,\alpha} \otimes \Omega + \mathcal{O}(g) \in \mathcal{H}$, where $\alpha = 1, \ldots, n_j$. The spectrum of H_g is locally (in a disk of radius $g^{2-\varepsilon}$ about E_j, where $\varepsilon > 0$) contained in the cuspidal domains*

$$E_j(g) + e^{-i\vartheta} \left\{ a + ib \,\big|\, a \geq 0, \, |b| \leq Ca^{1+\mu/4} \right\}, \tag{75}$$

see Fig. 5. Moreover, the eigenvalues $E_{j,\alpha}(g)$ and the corresponding eigenvectors $\Psi_{j,\alpha}(g)$ are obtained from a series expansion in (fractional) powers of g which is determined by the iterated application of the Feshbach renormalization map.

I remark that the same assertion holds for $j = 0$ and $\vartheta = 0$, in which case $E_0(g) = E_0 - \mathcal{O}(g) = \inf \sigma(H_g) \in \mathbb{R}$, is the perturbed ground state energy, and $\Psi_0(g)$ is the corresponding ground state. Under certain genericity assumptions similar to (70) the ground state will be unique, even if the multiplicity n_0 of the unperturbed ground state energy E_0 is 2 or even higher. The degeneracy of E_0, however, is not lifted in second but in higher order in g.

I further remark that, comparing (63) to the physical coupling function $\vec{G}_{\vec{x}}$ in (29), I observe that Hypothesis 4.4 is not fulfilled. Indeed, $\mu = 0$ in this case, and Theorem 4.5 does not apply. This should not cause any disappointment because, for $\mu = 0$, there are several counterexamples to the existence of a ground state $\Psi_0(g) \in \mathcal{H}$ known [19, 45, 1]. In these counterexamples it is shown that, if at all, the ground state of H_g is a density matrix in a different representation of Fock space, not unitarily equivalent to our original Fock space.

5. Scattering Theory

The subject of scattering theory is the asymptotics of the time evolution, e^{-itH}, as $t \to \infty$. One of the central mathematical goals of scattering theory is to prove asymptotic completeness. The most general results on asymptotic completeness for models of the type discussed here have been obtained in [22, 16, 17], essentially under two additional assumptions.

Hypothesis 5.1. *The photon field is massive, i.e., the photon dispersion $\omega(k) := |\vec{k}|$ has been replaced by*

$$\omega_m(k) := \sqrt{\vec{k}^2 + m^2}\,, \tag{76}$$

for some arbitrary but fixed $m > 0$, and

Hypothesis 5.2. *The particle system is confined, that is, either $\lim_{|x| \to \infty} V(x) = \infty$, or*

$$\left\| (|x| + 1)^{1+\mu}\, w_{1,0}(k)\, (-\Delta_{\vec{x}} + 1)^{-1/2} \right\|_{\mathcal{B}(\mathcal{H}_{el})} \leq J(k)\,, \tag{77}$$

for some $\mu > 0$.

It is additionally assumed that $w_{0,1} = w_{1,0}^*$, for H_g to be selfadjoint, and $w_{0,2} = w_{1,1} = w_{2,0} = 0$, for simplicity.

To formulate asymptotic completeness for the type of models discussed here, I first introduce the *asymptotic creation* and *annihilation operators*. For given $f \in L^2(\mathbb{R}^3 \times \mathbb{Z}_2)$, they are defined by

$$a_{\pm}^{\#}(f) := s - \lim_{t \to \pm\infty} \left\{ e^{-itH_g}\, e^{itH_0}\, a^{\#}(f)\, e^{-itH_0}\, e^{itH_g} \right\}, \tag{78}$$

where $a^{\#} = a$ or $= a^*$. Note that these operators act on the full space $\mathcal{H}$, rather than on $\mathcal{F}$. In [28, 29, 30] these operators were shown to exist, and this is the first time that the positivity of the mass $m > 0$ enters. The asymptotic creation and annihilation operators play the same role for H_g as the usual creation and annihilation operators do for H_0, namely

$$e^{-itH}\, a_{\pm}^{\#}(f)\, e^{itH} = a_{\pm}^{\#}\!\left(e^{it\omega} f\right). \tag{79}$$

It is easy to see that the asymptotic creation and annihilation operators yield another representation of the canonical commutation relations. It is, however, less clear, which vectors in $\mathcal{H}$ replace the vacuum vector, i.e., which vectors are contained in

$$\mathcal{K}_{\pm} := \left\{ \psi \in \mathcal{H} \,\middle|\, a_{\pm}(f)\,\psi = 0,\ \forall f \in L^2(\mathbb{R}^3 \times \mathbb{Z}_2) \right\}. \tag{80}$$

Observe that $\mathcal{K}_{\pm}$ contain all bound states of H_g, for if $H_g \psi = E\psi$ then

$$\left\| e^{-itH_g}\, e^{itH_0}\, a(f)\, e^{-itH_0}\, e^{itH_g}\, \psi \right\| = \left\| a\!\left(e^{-it\omega} f\right)\psi \right\| \to 0, \quad \text{as } t \to \infty, \tag{81}$$

because $e^{-it\omega} f \to 0$, weakly in L^2. Consequently,

$$\mathcal{H}_{pp}(H_g) \subseteq \mathcal{K}_+ \cap \mathcal{K}_-\,, \tag{82}$$

where $\mathcal{H}_{pp}(H_g)$ is the subspace corresponding to the pure point spectrum of H_g, i.e., onto all its eigenvectors. *Asymptotic completeness* is the statement that these three subspaces are all equal, in fact. The following result can be found in [17].

Theorem 5.3. *Assume Hypotheses 1.1 or 1.2 and 2.1, 5.1, and 5.2. Then*

$$\mathcal{H}_{pp}(H_g) \;=\; \mathcal{K}_+ \;=\; \mathcal{K}_- \,. \tag{83}$$

As a consequence of this theorem, there exists a unitary operator $J_\pm : \mathcal{H} \to \mathcal{H}_{pp}(H_g) \otimes \mathcal{F}$ such that $a_\pm^\#(f) = J\, a^\#(f)\, J^*$. By Theorem 3.1, I know that $\mathcal{H}_{pp}(H_g) \neq 0$, since $E_0(g)$ is an eigenvalue. The general belief is that $E_0(g)$ is simple, with corresponding eigenvector $\Psi_0(g)$, and that H_g has no other eigenvalues, i.e., $\mathcal{H}_{pp}(H_g) = \mathbb{C}\cdot\Psi_0(g)$. If that was the case then $J H_g J^* = \int dk\, \omega(k)\, a_\pm^*(k) a_\pm(k)$.

I remark that one of the basic inputs for the proof of asymptotic completeness is a positive commutator or Mourre estimate [39, 40], and such an estimate is indeed derived and applied in [22, 16, 17] to prove propagation estimates. The typical form of these estimates is

$$\chi_\Delta(H_g)\, i\big[H_g\,, A\big]\, \chi_\Delta(H_g) \;\geq\; \mu\,\chi_\Delta^2(H_g) + K\,, \tag{84}$$

where $\Delta \subseteq \mathbb{R}$ is a Borel set, A is a suitable observable, a customary choice being the dilatation generator, $\mu > 0$ is a strictly positive number, and K is a compact operator. Here is another point where positivity of the mass enters, as it guarantees that H_f is relatively bounded by N_f, the number operator on $\mathcal{F}$, and vice versa.

Positive commutator estimates, like (84), are interesting in their own right, for example, because they imply that in Δ, the spectrum of H_g is purely absolutely continuous. A variety of positive commutator estimates for the models discussed here were derived in [31, 33, 32, 44, 22, 14].

6. Positive Temperatures

The dynamics e^{-itH_g} generated by the Hamiltonian H_g on the state space $\mathcal{H}$ of the quantum mechanical system under consideration is the appropriate description for systems at zero temperature, $T = 0$. At positive temperature, $T > 0$, however, it is necessary to pass to a description in which the dynamics is generated by the Liouvillian, L_g, which acts on the tensor product $\mathcal{H} \otimes \mathcal{H}$ of two copies of $\mathcal{H}$. I briefly motivate this change and sketch the resulting mathematical objects, below.

It is well known that the Gibbs state of a finite quantum mechanical system, with Hamiltonian H and at inverse temperature $\beta := (kT)^{-1}$, is given by $\rho := \mathrm{Tr}\{\cdot\, e^{-\beta H}\}\big(\mathrm{Tr}\{e^{-\beta H}\}\big)^{-1}$, i.e., for a given observable $A = A^* \in \mathcal{B}(\mathcal{H})$, its expectation value is $\mathrm{Tr}\{A\, e^{-\beta H}\}\big(\mathrm{Tr}\{e^{-\beta H}\}\big)^{-1}$. The important point here is that I assumed the system to be *finite* or *confined*, meaning that $\mathrm{Tr}\{e^{-\beta H}\} < \infty$. Indeed, confining an infinite quantum system to a large but finite box $\Lambda \subseteq \mathbb{R}^3$ (with periodic b.c., say), I turn the continuous spectrum of the Hamiltonian into a discrete spectrum, and, for sufficiently large inverse temperature $\beta \gg 1$, the semigroup $e^{-\beta H}$ is trace class, and I call the corresponding state ρ_Λ.

For many questions concerning *static* thermodynamic properties (e.g., computation of the thermodynamic potentials or correlation functions), it usually suffices to work in finite boxes Λ, to prove estimates uniformly in $|\Lambda|$, and to pass finally to the thermodynamic limit, $\Lambda \nearrow \mathbb{R}^3$, by continuity. For example, the expectation value of a local observable A is then obtained in the thermodynamic limit as $\rho_\infty(A) := \lim_{\Lambda \nearrow \mathbb{R}^3} \rho_\Lambda(A)$.

For the study of *dynamical* questions, however, it may not be sufficient to work in finite boxes, but it might be necessary to formulate the dynamics in the thermodynamic limit right away. Indeed, the asymptotics of the time evolution, as $t \to \infty$, and the thermodynamic limit, $\Lambda \nearrow \mathbb{R}^3$, do not commute, in general. One example, for which this difference is crucial, is the property of *return to equilibrium*. If A_0 is an observable and $A_t := \alpha_t(A_0)$ its time evolution then the system under consideration is said to *return to equilibrium* iff, for all states ρ (with a certain trace-class property), we have

$$\text{(weak form)} \qquad \lim_{T \to \infty} \tfrac{1}{T} \int_0^T \rho(A_t)\, dt \;=\; \omega(A_0)\,, \tag{85}$$

$$\text{(strong form)} \qquad \lim_{t \to \infty} \rho(A_t) \;=\; \omega(A_0)\,. \tag{86}$$

Here, ω is a thermal equilibrium state, characterized by time-translation invariance and the KMS condition (see below). The existence and uniqueness of such a state is, in general, by no means trivial.

The framework for an infinite-volume theory at positive temperature was given in [24, 7, 6]. Two crucial properties that carry over from finite-volume Gibbs states, ρ_Λ, to the thermodynamic limit $\rho_\infty := \lim_{\Lambda \nearrow \mathbb{R}^3} \rho_\Lambda$ (provided it exists), are the time-translation invariance,

$$\rho_\infty\big(\alpha_t(A)\big) \;=\; \rho_\infty(A)\,, \tag{87}$$

for all t and A, and the KMS boundary condition,

$$\rho_\infty\big(A\,\alpha_t(B)\big) \;=\; \rho_\infty\big(\alpha_{-i\beta+t}(B)\,A\big)\,, \tag{88}$$

for A and B in a certain dense subalgebra $\overset{\circ}{\mathcal{A}}$ of the observable C^* algebra $\mathcal{A}$, invariant under α_t.

Using a GNS construction, the infinite-volume time evolution α_t of an observable $A \in \mathcal{A}$ can be unitarily implemented as

$$\rho_\infty\big(\alpha_t(A)\big) \;=\; \Big\langle \Omega_\beta \,\Big|\, e^{-itL_g}\, \ell[A]\, e^{itL_g}\, \Omega_\beta \Big\rangle\,, \tag{89}$$

where $\mathcal{H}_\beta := \mathcal{H} \otimes \mathcal{H}$, ℓ is a linear left-representation of $\mathcal{A}$ on $\mathcal{B}(\mathcal{H}_\beta)$, the KMS state ρ_∞ is identified with the projection onto a cyclic (vacuum) vector in $\mathcal{H}_\beta$, e.g., $\Omega_\beta = \varphi_0 \otimes \varphi_0 \otimes \Omega \otimes \Omega \in \mathcal{H} \otimes \mathcal{H}$ (tensor factors swapped), and the dynamics is generated by the selfadjoint operator L_g on $\mathcal{H}_\beta$, the *Liouvillian*.

The difference between finite and infinite systems is manifest in the form of the Liouvillian. For finite systems (i.e., for the models discussed here, discretized momentum space $|\Lambda|^{-1/3}\mathbb{Z}^3$ replacing $\mathbb{R}^3$), $\ell_\Lambda[A] = A \otimes \mathbf{1}$ and $L_g^\Lambda = H_g^\Lambda \otimes \mathbf{1} - \mathbf{1} \otimes$

H_g^Λ. For infinite systems, however, L_g is not of this form but rather

$$L_g = L_0 + \ell[W_g] - r[W_g] = H_0 \otimes \mathbf{1} - \mathbf{1} \otimes H_0 + \ell[W_g] - r[W_g], \qquad (90)$$

where $\ell[a(f)]$ is *not* simply $a(f) \otimes \mathbf{1}$ but, e.g.,

$$\ell[a(f)] = a\big(\sqrt{1+\rho_\beta}\, f\big) \otimes \mathbf{1} + \mathbf{1} \otimes a^*\big(\sqrt{\rho_\beta}\, \overline{f}\big), \qquad (91)$$

and $\rho_\beta(k) := (e^{\beta\omega(k)} - 1)^{-1}$.

The virtue of the GNS construction yielding the Liouvillian L_g is that it allows for tracing back the property of return to equilibrium to spectral properties of L_g. Namely, return to equilibrium follows if

- Zero is a simple eigenvalue of L_g, i.e., $\mathrm{Ker}\{L_g\} = \mathbb{C} \cdot \Omega_\beta(g)$, where $\Omega_\beta(g)$ is the unique KMS state of the system.
- Apart from zero, the spectrum is continuous, $\sigma_{\mathrm{cont}}(L_g)\backslash\{0\} = \sigma(L_g)\backslash\{0\}$. In this case, return to equilibrium holds at least in the weak form (85).
- If, apart from zero, the spectrum is even absolutely continuous, $\sigma_{\mathrm{ac}}(L_g) \backslash \{0\} = \sigma(L_g) \backslash \{0\}$, then return to equilibrium holds in the strong form (86).

This reformulation was proposed and applied to prove return to equilibrium for a system fulfilling Hypotheses 2.1–5.1 in [34, 35, 36, 37]. The spectral analysis of the Liouvillian then goes through a complex deformation, similar to the analysis of resonances in Section 3. The complex deformation used in [34, 35, 36, 37] is a special type of *complex translation*. This elegant method has the advantage that it yields fairly strong results already in second order perturbation theory, but the price to pay is the stringent analyticity assumptions on the coupling functions $w_{m,n}$ and the requirement of smallness of the coupling parameter g compared to the temperature $T > 0$.

The approach in [34, 35, 36, 37] has been generalized in [13] to allow for coupling functions that merely fulfill Hypotheses 2.1–5.1 and values of the coupling parameter uniform in the temperature $T \searrow 0$, by using complex dilatations. The trade-off here is that for the proof of return to equilibrium I need to use technically involved methods like the Feshbach renormalization map, described in Section 7, below.

To understand the results from [34, 35, 36, 37] and those in [13], it is again useful to discuss the trivial decoupled case, $g = 0$. Recall that, according to Hypothesis 1.2, I consider a simplified model of the particle system as a selfadjoint $N \times N$-matrix with non-degenerate eigenvalues, $H_{el} = \mathrm{diag}(E_0, E_1, \ldots, E_{N-1})$. Then the spectrum of $L_{el} := H_{el} \otimes \mathbf{1} - \mathbf{1} \otimes H_{el}$ is given by $\{E_{i,j} := E_i - E_j | 0 \leq i, j \leq N - 1\}$. Note that zero is an eigenvalue of multiplicity N. Next, the spectrum of $L_f := H_f \otimes \mathbf{1} - \mathbf{1} \otimes H_f$ covers the entire real axis, and according to $L_0 = L_{el} \otimes \mathbf{1} + \mathbf{1} \otimes L_f$, I have that $\sigma(L_0) = \overline{\sigma(L_{el}) + \sigma(L_0)} = \mathbb{R}$, and all $E_{i,j}$ become eigenvalues embedded in the continuum, see Fig. 6.

The complex translations used in [34, 35, 36, 37] now transform L_f into $L_f(\theta) = L_f - i\vartheta N_f$, where N_f is the number operator on $\mathcal{F} \otimes \mathcal{F}$ and $\theta = i\vartheta$,

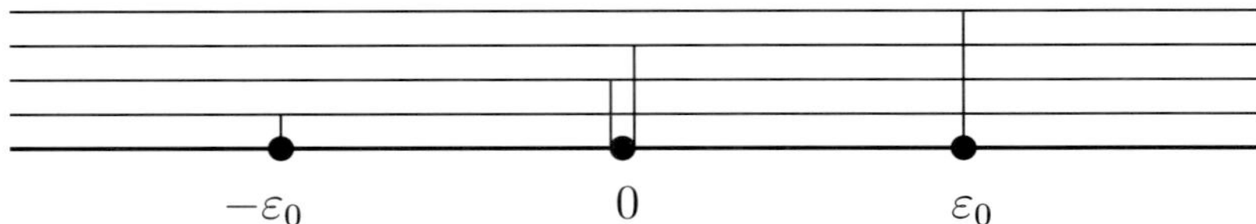

FIGURE 6. The spectrum of L_0 with $H_{el} = \mathrm{diag}[\varepsilon_0, -\varepsilon_0]$.

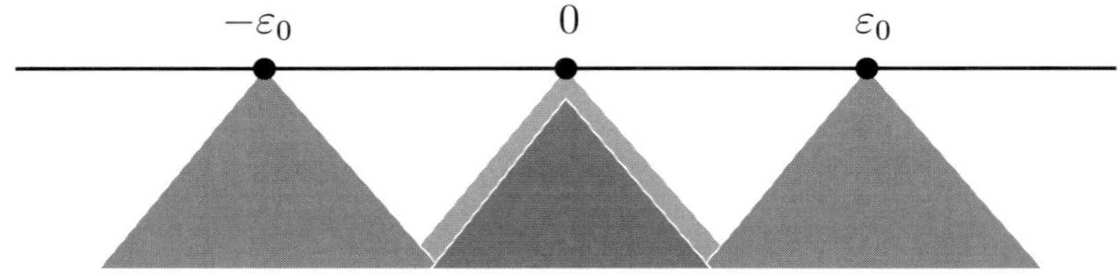

FIGURE 7. The spectrum of $L_0(\theta)$, for $\mathrm{Re}\,\theta = 0$, $\mathrm{Im}\,\theta = \vartheta > 0$.

$\vartheta > 0$. Therefore,

$$\sigma[L_0(\theta)] = \{E_{i,j} := E_i - E_j \,|\, 0 \le i,j \le N-1\} \cup \bigcup_{N \in \mathbb{N}} \{\theta N + \mathbb{R}\}. \tag{92}$$

I observe that the eigenvalues on the real axis are isolated. A simple application of second order perturbation theory now shows that, for $0 < g \ll |\vartheta|$, all non-zero eigenvalues are shifted into the lower half plane, $\mathrm{Im}E_{i,j}(g) < -\Gamma_{i,j}g^2$, $\Gamma_{i,j} > 0$. Furthermore, the N-fold degeneracy of the zero eigenvalue is lifted: all but one eigenvalues of $\mathrm{Ker}L_0(\theta)$ are also shifted into $\mathbb{C}_-$. The one vector remaining in $\mathrm{Ker}L_g$ in second order perturbation theory is, in fact, the approximate KMS state. Unfortunately, the domain of analyticity of the map $\theta \mapsto L_0(\theta)$ is the disk of radius T about 0, where $T > 0$ is the temperature. Thus, one has the restriction $|g| < T$.

Using a special form of complex dilatations, the unperturbed operator L_0 is mapped into $L_0(\theta) := L_{el} + \cos(\vartheta)L_f - i\sin(\vartheta)L_{\mathrm{aux}}$, where $L_{\mathrm{aux}} := H_f \otimes \mathbf{1} + \mathbf{1} \otimes H_f$ and $\theta = i\vartheta$, $\vartheta > 0$. Therefore, the spectrum of $L_0(\theta)$ is the union of sectors of opening angle $(\pi/2) - \vartheta$ in CC^-, with (real) eigenvalues $E_{i,j}$ as tips,

$$\sigma[L_0(\theta)] = \bigcup_{i,j=0}^{N-1} E_{i,j} + \{a - ib \,|\, b > 0,\ |a| \le \cot(\vartheta)b\}, \tag{93}$$

see Fig. 7.

The domain of analyticity of the map $\theta \mapsto L_0(\theta)$ now includes the open disk of radius $\pi/2$ about 0, uniformly in $T \to 0$. (I remark that this analytic continuation is more subtle than what is discussed in Section 3 because $L_g(\theta)$ is not an analytic family of type A.) Note, however, that the eigenvalues $E_{i,j}$ of $L_0(\theta)$ on the real axis are *not* isolated anymore, and their behaviour under switching on the coupling parameter $g > 0$ cannot be studied by standard perturbation theory, in general. Nevertheless, an argument adapted from second order perturbation theory now shows that, for $0 < g \ll 1$, all sectors attached to non-zero eigenvalues are (possibly slightly deformed and) shifted into the lower half-plane, $\mathrm{Im}E_{i,j}(g) < -\Gamma_{i,j}g^2$, $\Gamma_{i,j} > 0$. Furthermore, the N-fold degeneracy of the zero eigenvalue is lifted: $N - 1$ of the N overlapping sectors attached to the zero eigenvalues $E_{0,0} = \ldots = E_{N-1,N-1} = 0$ of $L_0(\theta)$ are also shifted into $\mathbb{C}_-$, and one sector at 0 remains there, for $g > 0$, in second order perturbation theory.

The most difficult part is now to show that the form of the spectrum of $L_g(\theta)$ described above is stable *beyond second order perturbation theory*, that is, to prove that higher order terms in a perturbation series do not change it qualitatively (although the sectors may become slightly deformed). This is established by applying the Feshbach renormalization map described in Section 7, below. As a result, the the following theorem is obtained in [13].

Theorem 6.1. *Assume Hypotheses 1.2 and 2.1, 5.1, 5.2. Then*

(i) *There exist $0 < \vartheta_0' < \vartheta_0$ such that, for $z \in \mathbb{C}^+$, the resolvent $(L_g(\theta) - z)^{-1}$ has an analytic continuation from $(z, 0)$ to $(z, i\vartheta)$, for any $\vartheta_0' < \vartheta < \vartheta_0$.*

(ii) *Zero is a simple eigenvalue of L_g and $L_g(i\vartheta)$ corresponding to a KMS state of the system, which, therefore, exists and is unique.*

(iii) *For $\vartheta_0' < \vartheta < \vartheta_0$, there exists $0 < \varepsilon$ such that, for $0 < g \ll 1$, the spectrum of $L_g(i\vartheta)$ is contained in*

$$\sigma[L_0(i\vartheta)] = \left\{ a - ib \,\middle|\, b > 0, \; |a| \leq \cot(\vartheta - g/\varepsilon)b \right\}$$

$$\cup \bigcup_{i,j=0,\,i\neq j}^{N-1} \left\{ E_{i,j}(g) + \left\{ a - ib \,\middle|\, b > 0, \; |a| \leq \cot(\vartheta)b \right\} + D(g^{2+\varepsilon}) \right\}. \tag{94}$$

Therefore, the spectrum of L_g away from zero is absolutely continuous, and return to equlibrium holds in the strong form (86).

7. Renormalization Map

In this final section I describe how the *Feshbach Renormalization Map* is used for spectral analysis, e.g., of resonances (see Section 3) or the zero eigenvalue of the Liouvillian (see Section 6), to prove the existence of eigenvectors and to derive an explicit, convergent series expansions for eigenvalues and the corresponding eigenvectors, for perturbation problems which are not of the standard type with isolated unperturbed eigenvalues.

To be concrete, I study the ground state energy of H_g, assuming hypotheses 1.2, 2.1, 5.1 and (without loss of generality) that $E_1 - E_0 = 2$.

The key ingredient is the Feshbach map which is well known in mathematics and physics, perhaps under a different name like "Grushin problem" or "Schur complement". I refer here to [11]. Given a closed operator H on a Hilbert space $\mathcal{H}$ and a bounded projection $P = P^2$, $\overline{P} := 1 - P$ denoting the complement.

Lemma 7.1. *Assume that $\overline{H} := \overline{P}H\overline{P}$ is bounded invertible on $\mathrm{Ran}\overline{P}$ and that PHP, $PH\overline{P}(\overline{H})^{-1}\overline{P}$, and $\overline{P}(\overline{H})^{-1}\overline{P}HP$ are bounded. Then*

(i) *The* Feshbach *map $H \mapsto \mathcal{F}_P(H)$,*

$$\mathcal{F}_P(H) := PHP - PH\overline{P}\,(\overline{H})^{-1}\overline{P}HP, \tag{95}$$

defines a bounded operator on $\mathrm{Ran}P$.

(ii) *H is invertible on $\mathcal{H}$ iff $\mathcal{F}_P(H)$ is invertible on $\mathrm{Ran}P$. In this case*

$$H^{-1} = \left(P - \overline{P}(\overline{H})^{-1}\overline{P}HP\right)\mathcal{F}_P(H)\left(P - PH\overline{P}(\overline{H})^{-1}\overline{P}\right) + \overline{P}(\overline{H})^{-1}\overline{P}. \tag{96}$$

(iii) $\dim \mathrm{Ker}(H) = \dim \mathrm{Ker}(\mathcal{F}_P(H))$.

I refer to (ii) and (iii) as *isospectrality* of the Feshbach map.

As a first application, I set $P := |\varphi_0\rangle\langle\varphi_0| \otimes \chi[H_f < 1]$ and apply the Feshbach map to $H := H_g - E_0 - z$, for $|z| < 1/2$. It is easy to see that the assumptions of lemma 7.1 are fulfilled, and thus I obtain a family of bounded operators, $H_g^{(0)}[z]$

$$|\varphi_0\rangle\langle\varphi_0| \otimes H_g^{(0)}[z] := \mathcal{F}_P(H_g - z), \quad \text{defined on } \mathcal{H}_{\mathrm{red}} := \mathrm{Ran}\chi[H_f < 1]. \tag{97}$$

To define the renormalization group map $\mathcal{R}_\rho$, I introduce a norm, $\|\cdot\|'$, on $\mathcal{B}(\mathcal{H}_{\mathrm{red}})$ that is stronger than the usual operator norm, $\|A\| \leq \|A\|'$ (Details can be found in [11]). In a small $\|\cdot\|'$-ball, $B \subseteq \mathcal{H}_{\mathrm{red}}$, about H_f, and for $0 < \rho < 1/32$, the renormalization map $\mathcal{R}_\rho$ is defined by

$$\mathcal{R}_\rho : B \to B, \quad H \mapsto \frac{1}{\rho}U_\rho\left(\mathcal{F}_{\chi[H_f<\rho]}(H) - \langle\mathcal{F}_{\chi[H_f<\rho]}(H)\rangle_\Omega\right)U_\rho^*, \tag{98}$$

where $\langle\cdot\rangle_\Omega$ denotes vacuum expectation value, U_ρ is the unitary dilatation that maps $k \mapsto \rho k$, thus $\rho^{-1}U_\rho H_f U_\rho^* = H_f$, $U_\rho\chi[H_f < \rho]U_\rho^* = \chi[H_f < 1]$, and hence $U_\rho\mathrm{Ran}\chi[H_f < \rho] = \mathcal{H}_{\mathrm{red}}$. The most important property of $\mathcal{R}_\rho$ is that it is a contraction on B, with fixed point H_f. This leads to the following theorem

Theorem 7.2. *For $0 < g \ll 1$, there is a unique number $E_0(g) \in B_{1/2}(E_0)$ such that*

$$H_g^{(n)} := \mathcal{R}_\rho^n\big(H_g^{(0)}[E_0(g)]\big) \to H_f, \quad in \ \|\cdot\|', \ as \ n \to \infty. \tag{99}$$

The number $E_0(g)$ can be interatively computed as $E_0(g) = \lim_{N\to\infty} E_0^{(N)}(g)$, where $E_0^{(N)}(g)$ is the unique solution of

$$E_0^{(N)}(g) = E_0 + \sum_{n=0}^{N} \rho^{-n}\big\langle\mathcal{F}_{\chi[H_f<\rho]}(H_g^{(n)})\big\rangle_\Omega. \tag{100}$$

The isospectrality of the Feshbach map, according to Lemma 7.1, and the fact that Ω is an eigenvector (corresponding to a zero eigenvalue) of the operator $H_f = \lim_{n\to\infty} H_g^{(n)}$ now yields the eigenvalue and eigenvector of H_g, sought for.

Corollary 7.3. *The number $E_0(g)$ defined in Theorem 7.2 is an eigenvalue of H_g. The corresponding eigenvector can be written as a limit of a sequence of approximate eigenvectors determined by an iterative equation similar to (100).*

References

[1] F. Hiroshima A. Arai, M. Hirokawa. On the absence of eigenvectors of hamiltonians in a class of massless quantum field models without infrared cutoff. *Preprint*, 1999.

[2] A. Arai. On a model of a harmonic oscillator coupled to a quantized, massless, scalar field. I. *J. Math. Phys.*, 22:2539–2548, 1981.

[3] A. Arai. On a model of a harmonic oscillator coupled to a quantized, massless, scalar field. II. *J. Math. Phys.*, 22:2549–2552, 1981.

[4] A. Arai. Spectral analysis of a quantum harmonic oscillator coupled to infinitely many scalar bosons. *J. Math. Anal. Appl.*, 140:270–288, 1989.

[5] A. Arai and M. Hirokawa. On the existence and uniqueness of ground states of the spin-boson Hamiltonian. *Preprint*, 1995.

[6] H. Araki. Relative Hamiltonian for faithful normal states of a von Neumann algebra. *Pub. R.I.M.S., Kyoto Univ.*, 9:165–209, 1973.

[7] H. Araki and E. Woods. Representations of the canonical commutation relations describing a non-relativistic infinite free bose gas. *J. Math. Phys.*, 4:637–662, 1963.

[8] V. Bach, J. Fröhlich, and I. M. Sigal. Mathematical theory of non-relativistic matter and radiation. *Lett. Math. Phys.*, 34:183–201, 1995.

[9] V. Bach, J. Fröhlich, and I. M. Sigal. Mathematical theory of radiation. *Found. Phys.*, 27(2):227–237, 1997.

[10] V. Bach, J. Fröhlich, and I. M. Sigal. Quantum electrodynamics of confined non-relativistic particles. *Adv. in Math.*, 137:299–395, 1998.

[11] V. Bach, J. Fröhlich, and I. M. Sigal. Renormalization group analysis of spectral problems in quantum field theory. *Adv. in Math.*, 137:205–298, 1998.

[12] V. Bach, J. Fröhlich, and I. M. Sigal. Spectral analysis for systems of atoms and molecules coupled to the quantized radiation field. *Commun. Math. Phys.*, 207(2):249–290, 1999.

[13] V. Bach, J. Fröhlich, and I. M. Sigal. Return to equilibrium. *J. Math. Phys.*, 2000.

[14] V. Bach, J. Fröhlich, I. M. Sigal, and A. Soffer. Positive commutators and spectrum of Pauli-Fierz Hamiltonian of atoms and molecules. *Commun. Math. Phys.*, 207(3):557–587, 1999.

[15] H. Cycon, R. Froese, W. Kirsch, and B. Simon. *Schrödinger Operators*. Springer-Verlag, Berlin, Heidelberg, New York, 1st edition, 1987.

[16] J. Derezinski and C. Gérard. *Scattering theory of classical and quantum N-particle systems*. Texts and Monographs in Physics. Springer-Verlag, 1997.

[17] J. Derezinski and C. Gérard. Asymptotic completeness in quantum field theory. massive Pauli-Fierz Hamiltonians. *Rev. Math. Phys.*, 11(4):383–450, 1999.

[18] J. Derezinski and V. Jaksic. Spectral theory of Pauli-Fierz Hamiltonians. I. *Preprint*, 1998.

[19] J. Fröhlich. On the infrared problem in a model of scalar electrons and massless scalar bosons. *Ann. Inst. H. Poincaré*, 19:1–103, 1973.

[20] J. Fröhlich. Existence of dressed one-electron states in a class of persistent models. *Fortschr. Phys.*, 22:159–198, 1974.

[21] J. Fröhlich and P. Pfeifer. Generalized time-energy uncertainty relations and bounds on lifetimes of resonances. *Rev. Mod. Phys.*, 67:795, 1995.

[22] Ch. Gérard. Asymptotic completeness for the spin-boson model with a particle number cutoff. *Rev. Math. Phys.*, 8:549–589, 1996.

[23] Ch. Gerard. On the existence of ground states for massless Pauli-Fierz Hamiltonians. *Preprint*, 1999.

[24] R. Haag, N. Hugenholz, and M. Winnink. On the equilibrium states in qauntum statistical mechanics. *Commun. Math. Phys.*, 5:215–236, 1967.

[25] M. Hirokawa. An expression for the ground state energy of the spin-boson model. *J. Func. Anal.*, 162:178–218, 1999.

[26] F. Hiroshima. Functional integral representation of a model in QED. *Rev. Math. Phys.*, 9(4):489–530, 1997.

[27] F. Hiroshima. Uniqueness of the ground state of a model in quantum electrodynamics: A functional integral approach. *Hokkaido U. Prepr. Series in Math.*, 429, 1998.

[28] R. Hoegh-Krohn. Asymptotic fields in some models of quantum field theory. I. *J. Math. Phys.*, 9(3):2075–2080, 1968.

[29] R. Hoegh-Krohn. Asymptotic fields in some models of quantum field theory. II. *J. Math. Phys.*, 10(1):639–643, 1969.

[30] R. Hoegh-Krohn. Asymptotic fields in some models of quantum field theory. III. *J. Math. Phys.*, 11(1):185–189, 1969.

[31] M. Hübner and H. Spohn. Atom interacting with photons: an N-body Schrödinger problem. *Preprint*, 1994.

[32] M. Hübner and H. Spohn. Radiative decay: nonperturbative approaches. *Rev. Math. Phys.*, 7:363–387, 1995.

[33] M. Hübner and H. Spohn. Spectral properties of the spin-boson Hamiltonian. *Ann. Inst. H. Poincare*, 62:289–323, 1995.

[34] V. Jakšić and C. A. Pillet. On a model for quantum friction. I: Fermi's golden rule and dynamics at zero temperature. *Ann. Inst. H. Poincaré*, 62:47–68, 1995.

[35] V. Jakšić and C. A. Pillet. On a model for quantum friction. II: Fermi's golden rule and dynamics at positive temperature. *Commun. Math. Phys.*, 176(3):619–643, 1996.

[36] V. Jakšić and C. A. Pillet. On a model for quantum friction. III: Ergodic properties of the spin-boson system. *Commun. Math. Phys.*, 178(3):627–651, 1996.

[37] V. Jakšić and C. A. Pillet. Spectral theory of thermal relaxation. *J. Math. Phys.*, 38(4):1757–1780, 1997.

[38] M. Merkli and I.M. Sigal. On time-dependent theory of quantum resonances. *Preprint*, 1999.

[39] E. Mourre. Absence of singular continuous spectrum for certain self-adjoint operators. *Comm. Math. Phys.*, 78:391–408, 1981.

[40] P. Perry, I. M. Sigal, and B. Simon. Spectral analysis of n-body Schrödinger operators. *Annals Math.*, 114:519–567, 1981.

[41] M. Reed and B. Simon. *Methods of Modern Mathematical Physics: Analysis of Operators*, volume 4. Academic Press, San Diego, 1st edition, 1978.

[42] M. Reed and B. Simon. *Methods of Modern Mathematical Physics I–IV*. Academic Press, San Diego, 2nd edition, 1980.

[43] M. Reed and B. Simon. *Methods of Modern Mathematical Physics: I. Functional Analysis*, volume 1. Academic Press, San Diego, 2nd edition, 1980.

[44] E. Skibsted. Spectral analysis of N-body systems coupled to a bosonic field. *Rev. Math. Phys.*, 10(7):989–1026, 1997.

[45] H. Spohn. Ground state(s) of the spin-boson Hamiltonian. *Commun. Math. Phys.*, 123:277–304, 1989.

[46] H. Spohn. Asymptotic completeness for Rayleigh scattering. *J. Math. Phys.*, 38:2281–2296, 1997.

FB Mathematik
Johannes Gutenberg-Universität
D-55099 Mainz; Germany
E-mail address: `vbach@mathematik.uni-mainz.de`

Spectrum and Statistical Properties
of Chaotic Dynamics

Viviane Baladi

Abstract. We present new developments on the statistical properties of chaotic dynamical systems. We concentrate on the existence of an ergodic physical (SRB) invariant measure and its mixing properties, in particular decay of its correlation functions for smooth observables. In many cases, there is a connection (via the spectrum of a Ruelle-Perron-Frobenius transfer operator) with the analytic properties of a weighted dynamical zeta function, weighted dynamical Lefschetz function, or dynamical Ruelle-Fredholm determinant, built using the periodic orbit structure of the map.

1. Introduction

We are concerned with the orbits of smooth iterated maps $f \colon M \to M$ on a (say, compact) Riemannian manifold, in situations where the dynamics is unpredictable because of sensitive dependence on initial conditions (related to positive Lyapunov exponents). We seek invariant ergodic Borel measures μ which describe the asymptotic behaviour of most or many points. Here, "most or many" is in the sense of the normalized Lebesgue measure.

To fix ideas, assume that f is a C^2 diffeomorphism. If μ is a *probability* measure, a desirable property is that for each $\varphi \in L^1(d\mu)$ there is a positive *Lebesgue* measure set of initial conditions x, with time averages $\frac{1}{n} \sum_{k=0}^{n-1} \varphi \circ f^k(x)$ converging to the space average $\int \varphi \, d\mu$. Even if μ is *ergodic*, this is more than what the Birkhoff theorem gives, and we call such an invariant (ergodic) probability measure a *physical measure*. Requiring absolute continuity with respect to Lebesgue is too strong, and a more appropriate sufficient condition to be a physical measure is often the *SRB* (Sinai-Ruelle-Bowen) property that the conditional measures along the unstable manifolds given by Pesin theory are absolutely continuous with respect to the measure induced by Lebesgue. (See, e.g., the efficient survey [104].)

If μ is a *mixing* physical or SRB probability measure, one would like to know how fast the *correlation function*

$$\rho_{\varphi\psi}(n) = \int \varphi \circ f^n \, \psi \, d\mu - \int \varphi \, d\mu \int \psi \, d\mu \tag{1}$$

1991 *Mathematics Subject Classification.* 37A25, 37C30, 37C40, 37DXX.

decays to zero for φ and ψ smooth (Hölder) *test functions* (or *observables*). Also, if μ is not volume itself, it is relevant to ask whether $(f^n)_*$(Lebesgue) converges to μ, and about the speed of this *convergence to equilibrium*. Such quantitative information on the *decay of correlations* may be used to prove a *central limit theorem* for smooth observables. (See [68, 97, 102] and references therein.)

Similar questions exist for continuous-time dynamics, but we shall completely leave aside Dolgopyat's [41]–[43] beautiful recent results on the speed of decay of correlations of Axiom A flows. (See [46] for an inspiring account. An alternative strategy is examined in [71].) Many of the techniques applied below to smooth attractors extend to repellors and saddles, or systems with holes, or other interesting (Gibbs or equilibrium) invariant measures, such as the measure of maximal entropy, but we do not delve into these topics. We do not venture into noncompact situations (including spatially extended systems such as chaotic weakly coupled map lattices, or symbolic situations with countably many states). Neither do we enter the very rich theory of hyperbolic and parabolic behaviour for holomorphic (polynomial, rational, etc.) dynamics. We completely neglect (Poincaré maps of) billiards, referring to [102, 33, 34] for recent results. We abstain from discussing the efficient numerical computation by Dellnitz et al. [39] of correlation spectrum via Ulam matrices. Finally, we do not take a "generic" approach to smooth dynamics on compact manifolds, referring to Palis [77] for a perhaps more holistic *Weltanschauung*.

Our aim is to give an overview of the constellation of recent and ongoing work on the decay of correlations associated to SRB invariant measures and the connection with *dynamical zeta functions or determinants*. In *Section 2* we recall "classical" results (a very subjective notion, appearing to coincide with "before I got my doctoral degree"), mentioning more modern proofs. In *Section 3*, we present some progress made during the last decade. The *final section* contains comments and questions. We refer, e.g., to the monographs [12, 97], or the surveys [104, 11, 10] for technical definitions and additional references. To the more physically inclined readers, we recommend Cvitanović's monograph [37] for many concrete applications (the Helium atom!) and bold insights on the periodic orbit structure of physically relevant chaotic dynamics.

2. Classical Results

2.1. Exponential mixing for uniformly hyperbolic systems

A diffeomorphism $f: M \to M$ is called *Axiom A* ([90], see also, e.g., [78]) if the periodic points are dense in the nonwandering set and the tangent bundle over the nonwandering set Ω splits as $T_\Omega M = E^u \oplus E^s$ with $\|Df^{-n}|_{E^u}\| \le C\lambda^{-n}$ and $\|Df^n|_{E^s}\| \le C\lambda^{-n}$, for some $\lambda > 1$, $C > 0$, and all $n \in \mathbf{Z}_+$. If $\Lambda_j \subset M$ is a (transitive) *attracting basic* set for a $C^{1+\epsilon}$ *Axiom A* diffeomorphism f, then by results of Sinai and Ruelle there is a unique ergodic physical probability invariant measure μ_j on Λ_j, which is also an *SRB measure*. If $f^{p_j}|_{\Lambda_{jk}}$ is topologically mixing, with $p_j \geq 1$ and $\Lambda_j = \sqcup_{k=1}^{p_j} \Lambda_{jk}$ from Smale's spectral decomposition, then work of Ruelle and Bowen shows that the *correlation function* for $(f^{p_j}|_{\Lambda_{jk}}, \mu_j)$ and Hölder observables *decays exponentially fast*. The first proofs used finite Markov partitions and a transfer operator associated to a one-sided (expanding) symbolic dynamics model. (This specific transfer operator is positive, and the relevant theorem deserves to be called a *Ruelle-Perron-Frobenius theorem*.) A more recent approach, exploiting Hilbert (projective) metrics to analyse the action of the transfer operator on Birkhoff cones of functions on M, was introduced by Liverani [67] and extended by Viana [97]. Yet another strategy, using a tower model (more precisely, estimating the asymptotics of return times to a suitable hyperbolic Cantor subset of Λ), was implemented by Young in two guises. She [102] first appealed to the spectral gap of a transfer operator on the tower, and then [103] adapted a probabilistic coupling method to the tower with recurrence bounds. The probabilistic coupling was recently carried out by Bressaud and Liverani [23] directly on the manifold (see also [44]).

Similar results (with simpler proofs) hold for $C^{1+\epsilon}$ (locally) *uniformly expanding* endomorphisms f. (I.e., assuming that $\|Df_x^n\| \geq C\lambda^n$ for some $C > 0$, $\lambda > 1$, all $x \in M$, and all $n \in \mathbf{Z}_+$.) In this case the physical probability measure is absolutely continuous with respect to Lebesgue, and the transfer operator $\mathcal{L}$ is defined on Hölder functions $\varphi \colon M \to \mathbf{C}$ by

$$\mathcal{L}\varphi(x) = \sum_{f(y)=x} \frac{\varphi(y)}{|\det Df(y)|}.$$

2.2. Exponential mixing for one-dimensional piecewise expanding maps

Here, $M = I \subset \mathbf{R}$ is just a compact interval, and f is assumed to be piecewise C^2 (or C^1, with inverse derivative of bounded variation). The easiest case is when there are finitely many intervals of monotonicity, i.e., finitely many singularities. Usually there does *not* exist a finite Markov partition: This is the main difference with previous settings. Assuming additionally that $|f'| > 1$ on the intervals of monotonicity, Lasota and Yorke showed that there are *finitely many absolutely continuous and ergodic* invariant probability measures μ_j (which are SRB and physical measures). Hofbauer's spectral decomposition gives $I = \bigsqcup_{\ell=1}^m W_\ell \sqcup \bigsqcup_{j=1}^k I_j$ with the W_ℓ wandering, $f|_{I_j}$ topologically transitive, and $I_j = \sqcup_{k=1}^{p_j} I_{jk}$ with each $(f^{p_j}|_{I_{jk}}, \mu_j)$ mixing. Using a spectral gap approach, Hofbauer and Keller [53] proved that the associated correlation functions *decay exponentially* fast for observables of bounded variation. More recently, the Birkhoff cone method was applied by Liverani to show the same result [69]. Finally, both the spectral gap and the probabilistic coupling methods on Young's tower [102, 103] also yield exponential decay of correlations. (See also the results from [16] inspired by the Milnor-Thurston theorem in Subsection 3.7.)

2.3. Dynamical zeta functions and dynamical Fredholm determinants

General references for dynamical zeta functions include the book [78] and the survey [10]. The *weighted dynamical zeta function* of a map $f\colon M \to M$ and a weight $g\colon M \to \mathbf{C}$ is formally defined by setting $g^{(n)}(x) = \prod_{k=0}^{n-1} g(f^k(x))$ and

$$\zeta_g(z) = \exp \sum_{n=1}^{\infty} \frac{z^n}{n} \sum_{x \in \mathrm{Fix}\, f^n} g^{(n)}(x). \tag{2}$$

Let f be a (transitive) $C^{1+\epsilon}$ Axiom A diffeomorphism, or expanding endomorphism, on a compact manifold M, and let $g\colon M \to \mathbf{C}$ be Hölder. Combining results of Pollicott [80], Ruelle [84], and Haydn [51], one gets that $\zeta_g(z)$ is analytic in the disc of radius $\exp(-P(\log|g|))$, where $P(\cdot)$ denotes topological pressure. It admits a meromorphic extension to the disc of radius $\theta^{-1/2} \exp(-P(\log|g|))$, where $0 < \theta < 1$ is related to the Hölder exponent of the invariant laminations and to the hyperbolicity factor $\lambda > 1$ of f. If g is positive, the poles of $\zeta_g(z)$ in this disc are in bijection with the poles of the Fourier transform of the correlation function (1) of the equilibrium measure μ of $\log g$ and Hölder φ, ψ in a strip.

In the case of interval dynamics, Keller and I [14] showed that the dynamical zeta function of a piecewise monotone interval map f with a generating partition, and a continuous weight g of bounded variation is analytic in the disc of radius $\exp(-P(\log|g|))$ and admits a meromorphic extension to a disc of inverse radius $\limsup_{n\to\infty} \sup |g^{(n)}|^{1/n}$. If g is positive, the poles of $\zeta_g(z)$ in this disc are related to the poles of the Fourier transform of the correlation function of the equilibrium measure of $\log g$ and observables of bounded variation.

The proof of both results is by studying the spectrum of a linear transfer operator $\mathcal{L}_g$ which (just as $\mathcal{L}$ in Subsection 2.1) is *not compact* but whose essential spectral radius can be estimated. For interval maps we simply have $\mathcal{L}_g \varphi(x) = \sum_{f(y)=x} g(y)\varphi(y)$. The poles of the zeta function are related to part of the discrete spectrum of the transfer operator and are sometimes called (Ruelle) *resonances*; I also like the phrase *correlation spectrum*.

The setting relevant to this note is $g = |\det Df|_{E^u}|$, respectively $g = 1/|f'|$. *Restricting f to an attracting basic set Λ, respectively to a *transitive* subinterval Λ*, the corresponding equilibrium measure is the physical (and SRB) ergodic measure. The associated zeta function $\zeta_g(z)$ (counting weighted periodic orbits in Λ) is meromorphic in a disc of radius > 1 with a simple pole at 1. If $f|_\Lambda$ is topologically mixing, then there are no other poles on the unit circle. The exponential decay of correlations for Hölder observables on a mixing attracting set Λ is thus reflected in the fact that the simple pole at 1 is the only singularity of the zeta function. More generally, the analytic properties of the full zeta function $\zeta_g(z)$ reflect the topological spectral decomposition: It is meromorphic in a disc of radius > 1, with a pole at $z = 1$ having the number of attracting transitive components (including the trivial sinks in the Axiom A case, on which $\det Df|_{E^u} = 1$) as multiplicity, the other poles of modulus 1 being roots of unity associated to the orders of mixing of the transitive components.

The *dynamical (Ruelle-Fredholm) determinant* of a differentiable map f: $M \to M$ and a weight $g \colon M \to \mathbf{C}$ is defined formally by

$$d_g(z) = \exp - \sum_{n=1}^{\infty} \frac{z^n}{n} \sum_{x \in \mathrm{Fix}\, f^n} \frac{g^{(n)}(x)}{|\det(1 - D_x f^{-n})|}. \tag{3}$$

In analytic expanding settings, the transfer operator $\mathcal{L}_g$ associated to the contracting inverse branches of f is compact. Applying Grothendieck's theory of Fredholm determinants for nuclear operators, Ruelle proved [83] that if g is real analytic and f is a real analytic expanding endomorphism, or if f a real analytic Anosov diffeomorphism *with real analytic stable and unstable foliations* (a very nongeneric assumption), then $d_g(z)$ is an entire function of finite order, whose zeroes are related to eigenvalues of a transfer operator. This result also holds in *piecewise* expanding, piecewise analytic situations, *when the partition is Markov*, e.g., if f is the Gauss map and $g = 1/|f'|$ [72]. (Dynamical Fredholm determinants and transfer operators are also useful in quantum chaos, at the triple intersection of number theory, geometry, and group theory. Mayer [73] wrote a very readable discussion on the Selberg zeta function and transfer operators applied to quantum chaos on the modular surface. See also [32] and references therein.)

Ruelle [86] later extended the Grothendieck-Fredholm theory to nonanalytic situations, showing in particular that if f is a C^r and λ-expanding endomorphism ($\lambda > 1$) on a compact manifold, and g is a C^r function (with $r \geq 1$), then, setting $g = 1/|\det Df|$ to fix ideas, $d_g(z)$ is an analytic function in the disc of radius λ^r, where its zeroes are inverse eigenvalues of the (noncompact) transfer operator $\mathcal{L}_g$.

3. Recent Developments

3.1. Low-dimensional nonuniformly hyperbolic dynamics: Unimodal maps, Hénon and Hénon-like maps; Multidimensional nonuniform hyperbolicity

In dimension 1, Keller-Nowicki [63] and Young [101] independently proved that a class of (nonrenormalizable) *good unimodal maps* containing the logistic map $x \mapsto a - x^2$ (for a positive measure-set of *good* parameters a) enjoys *exponential decay of correlations* for observables of bounded variation and their unique absolutely continuous and ergodic (thus, SRB and physical) invariant probability measure. Existence of this measure (and positive Lebesgue measure of the good parameter set) had been obtained by Jakobson. Exponential decay also holds for Hölder test functions [103].

Keller and Nowicki [63] also showed that the zeta function $\zeta_{1/|f'|}(z)$ (2) of such a good unimodal map is meromorphic in a disc of radius > 1 with a simple pole at $z = 1$ as its only singularity. Nowicki and Sands [76] list conditions (in particular the Collet-Eckmann condition $|(f^n)'(f(0))| \geq C\lambda^n$, for $n \in \mathbf{Z}_+$, with $\lambda > 1$) equivalent to exponential decay of correlations for the absolutely continuous invariant measure *(a.c.i.m)*.

Bruin proved that some unimodal maps with weaker (inverse-summable) growth of the postcritical derivative also possess an ergodic absolutely continuous invariant probability measure. Using Young's probabilistic coupling machinery [103], Bruin, Luzzatto, and Strien [24] estimated the rates of decay of correlations in such nonexponential situations.

Twenty years after Hénon studied numerically the surface diffeomorphism

$$(x, y) \mapsto \left(1 - ax^2 + y, bx\right), \tag{4}$$

for $a = 1.4$ and $b = 0.3$, Benedicks and Young [21] showed that the correlations functions of the SRB measure of the *Hénon maps with good parameters* (a, b) decayed exponentially (for Hölder observables). In between, a positive two-dimensional Lebesgue set of good parameters had been exhibited by Benedicks and Carleson, who aimed at a positive Lyapunov exponent, and the SRB measure had been constructed by Benedicks and Young [20] for good parameters. Existence of an SRB measure and exponential decay of its correlations has recently been obtained by Wang and Young [99] for a more general class of two-dimensional attractors *((generalized) solenoidal attractors)* "close" to a one-dimensional map with nonuniform hyperbolicity and arising from generic homoclinic bifurcations.

Viana [96] showed that an *open set of nonuniformly expanding endomorphisms* of the cylinder close to

$$(x, \theta) \mapsto (a - x^2 + \delta \sin \theta, D\theta(\mathrm{mod}\, 1)), \quad D \in \mathbf{Z}_+, \quad D \gg 16, \tag{5}$$

(with a a good parameter for the logistic map, and $\delta > 0$ small) possess *two* positive Lyapunov exponents. Alves [2] proved that they admit an absolutely continuous (physical, SRB) invariant probability measure which is mixing. (Viana [96] also studied the Lyapunov exponents of higher-dimensional invertible systems with multidimensional nonuniform hyperbolicity, replacing in particular $a - x^2$ by a Hénon map and $D\theta$ by a hyperbolic solenoid. To our knowledge the existence of an SRB measure has not been obtained yet for these systems.)

3.2. Piecewise expanding maps in arbitrary dimensions

Let f be an endomorphism of a compact manifold M partitioned into subsets on which the restriction of f is a $C^{1+\epsilon}$ uniformly expanding map. If the partition is not Markov (i.e., the image of each piece is not exactly a union of pieces), new problems appear in higher dimensions due to the accumulation boundary discontinuities under iterations. (See the introduction of [27] for references to piecewise C^r – arbitrary r – pathological surface transformations due to Tsujii, and piecewise affine pathological examples of Buzzi.) Another, more technical, difficulty consists in finding a convenient higher-dimensional version of bounded variation.

For higher-dimensional piecewise expanding maps, the existence of absolutely continuous invariant probability measure had only been proved in specific cases such as β-transformations, certain piecewise C^2 surface transformations (studied by Keller [60] over 20 years ago), or piecewise analytic and expanding surface endomorphisms (independently studied by Tsujii [94] and Buzzi [27], who also obtained exponential decay of correlations for Lipschitz observables). Around 1998, *three similar conditions* expressing that "the expansion must be stronger than the asymptotic growth of singularity accumulation" and *guaranteeing existence of an a.c.i. (probability) m.* were discovered independently (each of the three conditions is generic, i.e., open and dense in a natural class of piecewise expanding maps).

Saussol [91] introduced a weak sufficient (local) condition guaranteeing existence of the a.c.i. probability and its exponential mixing for Hölder observables. In dimension $N \geq 2$, when there are finitely many open domains of injectivity and their boundaries are piecewise C^1 codimension-one embedded compact submanifolds, the sufficient condition reduces to

$$s(f)^\epsilon + \frac{4s(f)}{1 - s(f)} Y(f) \frac{\gamma_{N-1}}{\gamma_N} < 1,$$

where $\gamma_N = \pi^{N/2}/(N/2)!$ is the volume of the N-dimensional ball in $\mathbf{R}^N$, $s(f) < 1$ is the largest contraction coefficient of the inverse branches, $0 < \epsilon \leq 1$ is the Hölder exponent of the Jacobian of the inverse branches, and $Y(f) \geq 1$ is the maximal number of boundaries of the partition having a common intersection point. Saussol's result holds more generally, for countable dynamical partitions, with possibly fractal boundary; we refer to [91] for the (technical) statement of the general sufficient condition (G).

Simultaneously, Buzzi [25], introduced another (local) condition sufficient to guarantee existence of a finite a.c.i.m.: Letting $\mathcal{Z}$ denote the finite partition into bounded open subsets of $\mathbf{R}^N$ on which $f \colon X \to X$ is uniformly expanding and $C^{1+\epsilon}$ (the approach works under weaker smoothness assumptions on the Jacobian), and writing $P(A, f)$ for the *topological pressure* of a subset $A \subset X$ and the weight $- \log |\det Df|$, the assumption is that the pressure of the boundary is smaller than the total pressure:

$$P(\partial \mathcal{Z}, f) < P(X, f). \tag{6}$$

Assuming additionally that the discontinuities are in general position, in the sense that for each $N + 1$-tuple $(Z_{k_i}, i = 1, \ldots, N + 1)$ with $Z_{k_i} \in \mathcal{Z}$ the intersection $\cap_{i=1}^{N+1} (f|_{Z_{k_{i-1}}} \circ \cdots \circ f|_{Z_{k_0}})^{-1} \partial Z_{k_i}$ is empty (this "transversality condition" in fact implies (6)), Buzzi and Maume [29] then proved exponential decay of the corresponding correlations when f is $C^{1+\epsilon}$ and slower rates of mixing when the Jacobian is allowed a (summable) modulus of continuity weaker than Hölder. The papers [25] and [29] also give results for other equilibrium states.

Cowieson [36] proved existence of an absolutely continuous invariant measure (and stochastic stability) for piecewise expanding, piecewise C^2 maps $f \colon M \to M$ satisfying another transversality condition. Restricting to a finite partition $\mathcal{Z}$ into C^2 submanifolds with corners his main assumption is (see [60] for surface maps)

$$\exists n \geq 1 \text{ with } \lambda(f^{n+1}) > F(\mathcal{Z}_n),$$

where $\lambda(f) > 1$ is the minimal local expansion coefficient, $\mathcal{Z}_n$ is the partition into domains of injectivity for f^n, and $F(\Pi)$ is the *cut index* of a partition Π, i.e., the maximum number of boundary pieces of the partition (which are not on the boundary of M) with a common intersection.

Buzzi and Keller [28] successfully carried through the analysis of the dynamical zeta function $\zeta_{1/|\det Df|}(z)$ (2) when f is a piecewise *affine and expanding surface* transformation (Jérôme Buzzi informed us that this can be extended to arbitrary dimensions by using Tsujii's [95] work.) They prove that the zeta function is analytic in the open unit disc and meromorphic in a disc of larger radius, where its poles are the inverse eigenvalues of the associated transfer operator acting on functions of higher-dimensional bounded variation. In particular, f is topologically mixing if and only if the zeta function has a simple pole at 1 as its only singularity in the larger disc. See also [75] and references therein.

3.3. Partially hyperbolic diffeomorphisms

Partially hyperbolic diffeomorphisms naturally generalize uniformly hyperbolic diffeomorphisms (see e.g. Sections 7–8 of [98] for an overview, including references to fundamental work by Brin-Pesin and Pesin-Sinai, as well as links with the theory of robust transitivity). The tangent bundle TM is assumed to split as either $E^c \oplus E^u$, with E^u uniformly expanding and dominating the expansion in E^c (the central bundle), or $E^s \oplus E^c$, with E^s uniformly contracting and dominating E^c. This includes time-one maps of hyperbolic flows, but also some maps without zero Lyapunov exponents. There does not seem to be a general description of the statistical properties of C^2 partially hyperbolic diffeomorphisms. If the central direction is *mostly contracting*, Bonatti-Viana [22] showed that there are finitely many SRB measures, and Castro [31] and Dolgopyat [44] independently showed, under mild additional technical assumptions, that correlation functions (for Hölder observables) decay *exponentially fast*. When the central direction is mostly expanding, Alves-Bonatti-Viana [3] proved that there exist SRB measures (and finitely many such measures if the Lyapunov exponents are bounded away from zero).

Skew-product constructions yielding partially hyperbolic diffeomorphisms are given by compact group extensions over a mixing uniformly hyperbolic system. The SRB measure is available for free, and in the volume preserving case mixing follows from work of Brin. Dolgopyat [45] combined key results of Brin and Katok on partially hyperbolic skew products with information gained from his previous study of hyperbolic flows [41, 42, 43] to prove that a diophantine condition on the Brin group implies rapid (i.e., faster than any polynomial) decay of correlations for C^∞ observables. The speed of mixing is not necessarily exponential for Hölder observables. These examples may not carry any periodic orbits.

3.4. Almost hyperbolic maps (neutral periodic points)

Consider the following "almost expanding" one-dimensional $C^{1+\epsilon}$ interval map: Assume that $I = [0,1]$ is partitioned into two subintervals which are mapped injectively by f onto I, that $f' \geq \lambda > 1$ on the right interval, and that $f' > 1$ on the left interval, except at $0 = f(0)$. Suppose also that f possesses an expansion

$$f(x) = x + x^{1+\gamma}(1 + u(x)),$$

with $\gamma > 0$ and $u(x) \to 0$ as $x \to 0$. (Additional technical assumptions are in fact useful.) A piecewise affine model for such maps was introduced by Gaspard-Wang [50] and has been much studied (see Isola [57, 58], also for references to work of Prellberg, Mori, Lopes, and others on intermittent maps).

Returning to f itself, it has been known for quite a while (see Pianigiani [79], Thaler [92], and for recent extensions to the non-Markov case, Zweimüller [106]) that if $\gamma \geq 1$ then f does *not have* an absolutely continuous invariant probability measure. However there always exists an *infinite (σ-finite) absolutely continuous measure*, about whose density much is known. This is perhaps the simplest (piecewise) smooth example where the SRB measure μ is not a probability. The physical measure is just the Dirac mass at 0, which is also an equilibrium measure for $-\log|f'|$. More generally, if μ is not a probability but only (conservative, invariant and) σ-finite, the Birkhoff ergodic theorem is not relevant. One finds inspiration in the Hopf ergodic theorem instead [1], seeking an *(anomalous) scaling rate*, i.e., a sequence $a_n \to \infty$, with $a_n = o(n)$, and $\frac{1}{a_n}\sum_{k=0}^{n-1}\varphi \circ f^k(x) \to \int \varphi \, d\mu$, for all $\varphi \in L^1(d\mu)$ and a positive Lebesgue measure set of x. Few rigorous results are available in general although the scaling rates a_n have been established for the one-dimensional maps discussed here [30, 35, 93].

If f is additionally assumed to be piecewise real analytic, Rugh [89] exploited a *regularization* idea to combine information from the Grothendieck-Fredholm theory and Fatou coordinates. (Leaving out the neutral periodic point somehow mirrors the algebraic identity for a dynamical determinant: $d(z) = (z - \mathcal{M} - \mathcal{N})^{-1} = (z-\mathcal{M})^{-1}(1-\mathcal{D}(z))^{-1}$, for $\mathcal{D}(z) = \mathcal{N}(z-\mathcal{M})^{-1}$ with $\mathcal{M}$ a bounded transfer operator associated to the neutral branch, and $\mathcal{N}$ a nuclear transfer operator associated to the expanding branch.) He unveiled the analytic structure of the dynamical determinant $d_{1/|f'|}(z)$, as well as the spectrum of the transfer operator $\mathcal{L}_{1/|f'|}$ acting on a suitable space of singular functions (there is no spectral gap).

For $0 < \gamma < 1$, there exists an absolutely continuous invariant probability μ, which is mixing. The speed of mixing is *polynomial*. Almost optimal upper bounds were obtained by Liverani-Saussol-Vaienti [70]. The exact upper bound

$$\left| \int \varphi \circ f^n \psi \, d\mu - \int \varphi \, d\mu \int \psi \, d\mu \right| \leq \frac{C_{\varphi\psi}}{n^{1/\gamma-1}},$$

(for Hölder continuous test functions φ, ψ) was obtained by Young [103] and Hu [54] who used different methods (Hu showed it is optimal).

Higher-dimensional "almost expanding" or "almost hyperbolic" models have been studied in recent years. Assuming the existence of two invariant (expanding and contracting) cone fields, except on a finite set of neutral periodic points, Hu [55, 56] found necessary conditions for the existence of finite or σ-finite SRB measures, as well as polynomial estimates for the rate of decay of correlations when the SRB measure is a probability.

One of the motivations to investigate such maps comes from hydrodynamics; we recommend the paper on intermittency by Pomeau-Manneville [81]. Expanding maps with neutral fixed points also appear naturally when studying the Selberg zeta function associated to arbitrary cofinite Fuchsian groups or simply investigating the asymptotic growth of closed geodesics on manifolds with cusps [7]. Almost expanding maps also occur in number theory when evaluating the quality of diophantine approximations of various continued fraction algorithms (from the Farey map to more exotic multidimensional continued fractions, see e.g. [105]).

3.5. Small random perturbations and stochastic stability

It is natural to ask whether the statistical properties of a deterministic dynamical system are stable under perturbation by small random noise. (More generally, one may investigate random compositions of dynamical systems within a given class, see Kifer's survey [64] and references therein.) We do not attempt to give general formal definitions here, and restrict to *independent identically distributed* (i.i.d.) noise where a Markov chain model χ^n is relevant: Assume we are given a probability space (Ω_0, ν), and consider the Bernoulli measure $\mathbf{P}_+(d\omega) = \prod_{i \in \mathbf{Z}_+} \nu(\omega_i)$ on $\Omega_+ = \Omega_0^{\mathbf{Z}_+}$, and for $\omega \in \Omega_+$ a "dynamical system valued random-variable" f_ω which depends only on ω_0. The transition probabilities are given by $\mathrm{Prob}(\chi^{n+1} \in E \mid \chi^n = \{x\}) = \int \mathbf{1}_E(f_\omega(x)) \, d\nu(\omega_0)$.

Letting σ_+ be the one-sided shift, we write $f_\omega^n = f_{\sigma_+^{n-1}\omega} \circ \cdots \circ f_\omega$. Mild assumptions suffice to ensure that the invariant measure for the Markov chain is the weak limit of the Birkhoff averages of dirac masses

$$\frac{1}{n} \sum_{i=0}^{n-1} \delta_{f_{\omega_{n-1}} \circ \cdots \circ f_{\omega_0}(x)}, \quad n \to \infty,$$

for Lebesgue almost all initial x and $\mathbf{P}_+$-almost all random itineraries ω.

Young [100] showed weak convergence of the unique invariant probability of the Markov chain associated to small i.i.d. perturbations of transitive Axiom A attractors towards the SRB measure.

For the *Markov chain* associated to small i.i.d. perturbations of *smooth uniformly expanding* or one-dimensional mixing piecewise uniformly expanding maps, very strong properties hold (Baladi-Young [18]): L^1 stability of the density of absolutely continuous invariant probability measures, and stability of the correlation spectrum, in particular the exponential rates of decay of the (averaged) correlation function of the Markov chain. Cowieson [36] extended the L^1 stability to higher-dimensional piecewise expanding dynamics. See also [62] for recent improvements of relevant spectral stability results by Keller and Liverani.

For *nonuniformly expanding dynamics*, such as good unimodal maps, results almost as strong, including L^1 stability of the density and some spectral stability, were obtained by Baladi-Viana [17]. (Weaker stability had been previously proved by Katok-Kifer, Collet, and Benedicks-Young in similar settings.) Benedicks and Viana [98] proved stochastic stability (weak convergence of the Markov chain invariant measure towards the SRB measure) of good Hénon maps.

Instead of considering the averaged behaviour given by a Markov chain, one may investigate *almost sure* random behaviour, i.e., consider the *two-sided*

Bernoulli space $(\mathbf{P}, \Omega)$ and study invariant measures of the form $\mu_\omega(dx)P(d\omega)$ of the random skew product $(x, \omega) \mapsto (f_\omega(x), \sigma\omega)$, where σ is the two-sided shift. In good situations one expects the disintegrations $\mu_\omega = (f_{\sigma^{-1}\omega})_*\mu_{\sigma^{-1}\omega}$ to be absolutely continuous or SRB quasi-invariant measures. It is also natural to conjecture

$$\lim_{n\to\infty} (f^n_{\sigma^{-n}\omega})_*(\text{Lebesgue}) = \mu_\omega\,, \tag{7}$$

for almost all ω. In uniformly expanding cases, Baladi-Kondah-Schmitt (see [12]) proved that the convergence in (7) takes place exponentially fast uniformly for all ω, that the μ_ω have Hölder densities which all converge to the invariant density as the noise level goes to zero and that the corresponding "random" correlations for Hölder observables decay exponentially (with uniform rate in the noise level).

In a Lasota-Yorke random context (non-Markov, expanding on the average, one-dimensional piecewise monotone and some particular multidimensional maps), Buzzi [26], who had proved existence of quasi-invariant absolutely continuous μ_ω, obtained almost sure exponential speed of convergence in (7), and exponential decay of the random correlation functions for Lipschitz observables, sometimes in the absence of exponential decay of averaged correlations. (His setting consists in random compositions of maps which are not necessarily close to a given map, and the result does not have the flavour of stochastic stability.)

Investigating the almost sure (as opposed to the averaged, Markov chain) behaviour of small random perturbations of good unimodal maps involves difficulties similar to those apearing when dealing with the deterministic skew product (5). With Benedicks and Maume-Deschamps [13], we consider a good unimodal map f (satisfying Benedicks-Carleson type assumptions), and for small $\epsilon > 0$ a probability ν_ϵ on $\Omega_0 = [-\epsilon, \epsilon]$, setting $f_\omega(x) = f(x) + \omega_0$. Assuming e.g. that ν_ϵ is absolutely continuous with a density bounded uniformly in ϵ, we show that the convergence to equilibrium (7) almost surely takes place at least stretched exponentially fast.

Theorem 3.1. (Almost sure rates of mixing) *For* $\mathbf{P} = \prod_{\mathbf{Z}} \nu_\epsilon$ *almost all* ω, *the quasi-invariant measure* $\mu_\omega(dx)$ *is absolutely continuous. For each small enough* $\epsilon > 0$ *there are* $C > 1$ *and a random variable* C_ω *with* $\mathbf{P}(\omega \mid C_\omega > K) \leq \frac{C}{K^u}$ *(for some* $u > 1$, *independent of* ϵ), *and such that for* $\mathbf{P}$-*almost all* ω

$$|(f^n_{\sigma^{-n}\omega})_*(\text{Lebesgue}) - \mu_\omega| \leq C_\omega \exp(-n^{1/16}/C), \forall n \in \mathbf{Z}_+\,. \tag{8}$$

No lower bounds are known, but analogous almost sure upper bounds hold for the *random correlations* of the μ_ω's and Lipschitz observables. The above theorem is *not* about stochastic stability: Our estimates on C_ω and C blow up as $\epsilon \to 0$.

See Section 2.7 in [12] for relations between the Lyapunov exponents of the random transfer operators, the decay of correlations, and random zeta functions.

3.6. Dynamical Fredholm determinants for hyperbolic dynamics

What makes transfer operators "nice" (compact, or at least quasi-compact) is that they are basically compositions by contractions, on smooth function spaces. Until the Ph.D. thesis of Rugh (see [88] and references therein) dynamical zeta functions for smooth hyperbolic dynamics were studied mainly through an expanding system (with contracting inverse branches) obtained by quotienting out along stable manifolds. This is the reason behind the very strong assumption of analyticity of the foliations required in Ruelle's [83] theorem from Subsection 2.3. In a two-dimensional analytic hyperbolic setting, Rugh introduced *pinning coordinates*, i.e., for each $n \in \mathbf{Z}_+$ two parametrized real analytic contractions φ_x^n and ψ_z^n, so that

$$f^n(x, \varphi_x^n(z)) = (\psi_z^n(x), z) . \tag{9}$$

Showing nuclearity of an associated transfer operator, he proved that if f is an *analytic Axiom A* diffeomorphism on a compact *surface*, the determinant

$$d_{SRB}(z) = \exp - \sum_{n=1}^{\infty} \frac{z^n}{n} \sum_{x \in \mathrm{Fix}\, f^n} \frac{1}{|\det(D_x f^n - 1)|} \tag{10}$$

is an entire function. (The dynamical *foliations* of f *need not be smoother* than Hölder.) Despite our optimistic notation, the relation of the zeroes of the determinant $d_{SRB}(z)$ with the correlation of the SRB measure has not been completely elucidated (see [87, 49]). Introducing more conceptual tools, Fried [48] proved that $d_{SRB}(z)$ is meromorphic in the whole complex plane in any dimensions. This applies to flows, proving a conjecture of Smale [90].

In a remarkable paper which deserves more attention, Kitaev [66] performed the breakthrough step from analyticity to *finite differentiability*, showing that if f is a C^r *Anosov* diffeomorphism on a compact manifold, with hyperbolicity factor $\lambda > 1$, then $d_{SRB}(z)$ defines an analytic function in the disc of radius $\lambda^{r/2}$. (The "loss" of half of the differentiability mirrors a similar phenomenon in symbolic dynamics, see [78].) Fried [49] has recently announced a spectral interpretation of the zeroes of $d_{SRB}(z)$ in this framework.

3.7. Sharp determinants and dynamical Lefschetz functions in dimension one

For continuous interval maps, assuming only existence of a finite partition into monotonicity intervals, Milnor and Thurston [74] defined a *negative zeta function* using the (finite) set $\mathrm{Fix}^- f^n$ of points with $f^n(x) = x$ such that f^n is decreasing in a neighbourhood of x, and putting $\zeta_-(z) = \exp \sum_{n \in \mathbf{Z}_+} \frac{z^n}{n} 2 \# \mathrm{Fix}^- f^n$. They showed that $\zeta_-(z)$ is the determinant of a finite matrix (with coefficients power series), the *kneading matrix* $D(z)$. (In fact, as was explained to us long ago by Jack Milnor, $\zeta_-(z)$ is essentially a Lefschetz zeta function.)

The desire, on the one hand to introduce weights of bounded variation (such as $g = 1/|f'|$) in the negative zeta function, and on the other to understand the link between the zeta function, the kneading matrix, and the transfer operator, led to a series of papers by Ruelle and/or myself ([16] and references therein). In the weighted case, a kneading operator $\mathcal{D}_g(z) = z\mathcal{N}_g(1 - z\mathcal{L}_g)^{-1}S$ (which has a bounded kernel) replaces the kneading matrix: The understanding emerged that the mechanism behind the Milnor-Thurston result is (again, see Subsection 3.4) *regularization*. A weighted *sharp zeta function* was introduced (where sums of integrals $\int \mathrm{sgn}(f_{\mathrm{br}}^{-1}x - x)/2 \, dg(x)$, with sgn the sign function, and f_{br}^{-1} an inverse branch, replace the usual sums over fixed points). The equality $\zeta_g^{\#}(z) = \det(1 + \mathcal{D}_g(z))$ was obtained and used to relate the poles of $\zeta_g^{\#}(z)$ and the discrete eigenvalues of the transfer operator $\mathcal{L}_g$. These results use technical assumptions, in particular the weight g should vanish on the boundary, but this can be weakened [82]. Just like the Milnor-Thurston theorem, they are true *also in the presence of uncountably many periodic orbits*. Integrating by parts, if the periodic points are isolated, the negative zeta function and the sharp zeta function can be viewed as counting all periodic orbits with *Lefschetz signs*. In the present not necessarily Markov context even the unweighted (Milnor-Thurston) Lefschetz zeta function is not always rational. Allowing weights is perhaps a way of revamping what Smale [90] called the "false" zeta function.

We refer to the survey [10] or the original papers for technical details. See also [15] for a partial extension of this analysis to one-dimensional complex dynamics and Baillif's article [8] for kneading determinants for tree maps. Alves and Sousa-Ramos [4, 5] have yet another, more functorial, approach to the unweighted Milnor-Thurston theorem for both interval and tree endomorphisms.

4. Closing Remarks

4.1. Low dimensions

Let us say that the *SRB zeta function of f has a gap* if there exist $\tau > 0$, and $r > 1$ so that for $t = 0$ and $t = \tau$ the zeta function

$$\zeta^{(t)}(z) = \exp \sum_{n \in \mathbf{Z}_+} \frac{z^n}{n} \sum_{x \in \mathrm{Fix}\, f^n} \frac{1}{|(f^n)'(x)|^{1-t}} \tag{11}$$

is meromorphic in the disc of radius r, with a simple pole at $q(t)$ as a unique singularity, and $q(0) \geq 1$. Let f be a nonrenormalizable unimodal interval map with negative Schwarzian derivative and at least two periodic orbits. Keller [61] proved last year that the SRB zeta function of f has a gap *if and only if* it admits an absolutely continuous invariant probability measure with exponential decay of correlations for Hölder observables. (The new implication is the "only if", this uses Nowicki-Sands [76] and Keller-Nowicki [63].)

The first question which comes to mind is whether one can weaken the SRB-gap property by taking $\tau = 0$ in Keller's theorem. Also, one wonders whether the topological spectral decomposition of multimodal maps with negative Schwarzian (assuming perhaps a Collet-Eckmann property to simplify) is readable in the multiplicity of the pole of the SRB (or unweighted) zeta function at 1 (or $\exp(h_{\mathrm{top}})$).

Recalling the results of Subsection 3.2 in particular [28], one can ask whether the SRB zeta function of a piecewise C^2 expanding map f (replacing $|f'|$ by $|\det Df|$ in (11) and assuming one of the three conditions in Subsection 3.2) has a gap (in the strong or weakened sense) if and only if f has an absolutely continuous invariant probability measure with correlations decaying exponentially fast (for Hölder observables). Is the presence of the gap generic?

Equivalence of the existence of a gap for the SRB zeta function and exponential mixing of an SRB measure is also open to our knowledge for the Hénon family (or more generally solenoidal attractors à la Wang-Young), or Viana's maps (5). (In view of the results by [13] mentioned above, one can expect the correlations of (5) to decay at least stretched exponentially.)

Away from the purely exponential setting, there are still few rigorous results relating the (nonpolar) singularity spectrum of dynamical zeta functions and the statistical properties of chaotic dynamics. For example, what can be said about the dynamical zeta function (11) associated to the multimodal maps studied by Bruin-Luzzatto-van Strien [24] which enjoy polynomial or stretched exponential mixing? (Véronique Maume suggested to investigate the abstract model from [103].) In the presence of neutral fixed points, numerical and heuristical work of Dahlqvist [38], and results of Isola [58] for countable Markov chains are inspiring rigorous enquiries in smooth settings such as the interval maps with neutral fixed points and the almost Anosov systems from Subsection 3.4. One would also like to know whether the analytic properties of dynamical zeta functions for analytic intermittent maps discovered by Rugh [89] may be linked to ergodic properties of an absolutely continuous invariant measure, also when it is only sigma-finite.

4.2. Higher dimensions: plethora and penury

The recent increase of our understanding of weighted dynamical zeta functions leaves me with the feeling that we have not pushed the theory to its limits yet. Can one generalize Keller's theorem and find a dynamical zeta function which describes statistical properties of (generic?) C^2 or piecewise C^2 dynamics on a compact manifold, going beyond hyperbolicity and assuming, e.g., existence of finitely many SRB measures without zero Lyapounov exponents (with sufficiently fast decay of correlations for Hölder observables)? If we seek this "mother of all" dynamical zeta functions, we must handle both overabundance and scantiness of periodic orbits.

We may be overwhelmed by a *plethora* of closed orbits even if we follow Artin-Mazur and count only *isolated* periodic points. Artin and Mazur [6] had showed that for a dense subset of C^r diffeomorphisms on a compact manifold the corresponding growth was at most exponential. However, recent work of Kaloshin (see [59] and references therein) reveals that superexponential growth of isolated periodic orbits occurs on a residual set of C^r diffeomorphisms, $r \geq 2$. Dima Dolgopyat recently pointed out to us that by combining Kaloshin's results with Castro's [31] one can construct a four-dimensional partially hyperbolic skew-product with superexponential growth of periodic orbits but having a single mixing SRB measure with exponential decay of correlations.

If zero Lyapunov exponents are present, we have to face *scarcity* even in a three-dimensional uniformly hyperbolic context: Just take the time-t map of a geodesic flow on a surface of constant negative curvature, with t not a common divisor of the orbit lengths, to find a volume preserving diffeomorphism with exponentially decaying correlations for Hölder observables, but not a single closed orbit. Other systems admitting an SRB measure with rapid mixing properties but not a single closed orbit are compact group extensions of hyperbolic systems as studied by Dolgopyat [45]. (Each of these discrete-time dynamical systems without closed orbits has a single zero Lyapunov exponent and is imbedded in a hyperbolic flow with the "right" periodic orbit structure, the zeta function of which has the expected analytic properties [41]–[43]. Fayad's [47] volume-preserving polynomially mixing flow on the two-torus has only a single fixed point, but it has zero entropy; it is not known if the decay is summable.)

Focusing mainly on the first of these two difficulties, let us briefly explore what tools are available, seeking insights from the low-dimensional results. In our opinion, a dynamical zeta function or determinant is interesting only if its poles or zeroes reflect topological or ergodic information: The topological entropy (or more generally the topological pressure), e.g., does not coincide in general with asymptotic (weighted) growth of periodic orbits. (See [19] and [52] for a characterization in terms of (n, α)-separated sets of ϵ-pseudo periodic orbits.) We have by now (re)learned sophisticated ways of counting the periodic orbits. This can mean introducing weights: The Jacobian along unstable directions, if one is concerned with SRB measures, but also analyticity-improving Jacobians in the denominators, as in the Ruelle-Fredholm determinants (3), useful in presence of strong smoothness. It can also signify rehabilitating Lefschetz-type signs: We saw in Subsection 3.7 that generalized Lefschetz signs attenuate superfluous growth for isolated periods and are a way to cleverly ignore the nonisolated orbits. Higher-dimensional version of the sharp traces of [16] could perhaps be devised (involving perhaps some kind of Leray-Schauder-Lefschetz index for compact continua of fixed points [40], as pointed out to us by André de Carvalho).

At this time we do not have candidates for a suitable class of smooth dynamics and its adapted dynamical zeta function. Let us finish by mentioning exciting ongoing work: In the case of a smooth diffeomorphism f satisfying a weak transversality assumption ensuring that periodic orbits are isolated, and a smooth weight g, Baillif [9] proves, using ideas of Kitaev [65], a higher-dimensional version of the Milnor-Thurston [74] theorem which should lead to a spectral interpretation of the zeroes of the following weighted Lefschetz zeta function ($L(\psi, x)$ denotes the usual Lefschetz index):

$$\exp - \sum_{n=1}^{\infty} \frac{z^n}{n} \sum_{x \in \mathrm{Fix}\, f^n} g^{(n)}(x) L(f_x^{-n}, x).$$

Acknowledgements

This text was prepared while I was enjoying the hospitality of the Courant Institute at New York University, I am grateful to Lai-Sang Young for inviting me there. I thank Mathieu Baillif, Jérôme Buzzi, Dima Dolgopyat, Bassam Fayad, Gerhard Keller, and Amie Wilkinson for useful comments.

References

[1] J. Aaronson, *An Introduction to Infinite Ergodic Theory,* Mathematical Surveys and Monographs, Amer. Math. Soc., (Providence, RI), **1997**.

[2] J. F. Alves, *SRB measures for nonhyperbolic systems with multidimensional expansion,* Annales scient. Ecole normale sup. (4), **33** (2000), 1–32.

[3] J. F. Alves, C. Bonatti and M. Viana, *SRB measures for partially hyperbolic systems whose central direction is mostly expanding,* Invent. Math., **140** (2000), 351–398.

[4] J. F. Alves and J. Sousa Ramos, *Kneading theory: a functorial approach,* Comm. Math. Phys., **204** (1999), 89–114.

[5] J. F. Alves and J. Sousa Ramos, *Zeta function of a tree map,* Preprint (1999).

[6] M. Artin and B. Mazur, *On periodic points,* Ann. of Math. (2), **21** (1965), 82–99.

[7] M. Babillot and M. Peigné, *Homologie des géodésiques fermées sur des variétés hyperboliques avec bouts cuspidaux,* Ann. scient. Ecole normale sup. (4), **33** (2000), 81–120.

[8] M. Baillif, *Dynamical zeta functions for tree maps,* Nonlinearity, **12** (1999), 1511–1529.

[9] M. Baillif, *Kneading operators, sharp determinants and weighted Lefschetz zeta functions in higher dimensions,* preprint (2001).

[10] V. Baladi, *Periodic orbits and dynamical spectra,* Ergodic Theory Dynam. Systems, **18** (1998), 255–292.

[11] V. Baladi, *Decay of correlations,* Preprint (1999), to appear Proceedings of the AMS Summer Institute on Smooth ergodic theory and applications, held in Seattle, 1999.

[12] V. Baladi, *Positive Transfer Operators and Decay of Correlations,* (2000) World Scientific (Advanced Series in Nonlinear Dynamics), River Edge, NJ.

[13] V. Baladi, M. Benedicks, and V. Maume, *Almost sure rates of mixing for i.i.d. unimodal maps,* Preprint (1999), to appear Ann. Ecole Normale Sup.

[14] V. Baladi and G. Keller, *Zeta functions and transfer operators for piecewise monotone transformations,* Comm. Math. Phys., **127** (1990), 459–479.

[15] V. Baladi, A. Kitaev, D. Ruelle, and S. Semmes, *Sharp determinants and kneading operators for holomorphic maps,* Proc. Steklov Inst. Math., **216** (1997) 186–228.

[16] V. Baladi and D. Ruelle, *Sharp determinants,* Invent. Math., **123** (1996), 553–574.

[17] V. Baladi and M. Viana, *Strong stochastic stability and rate of mixing for unimodal maps,* Annales scient. Ecole normale sup. (4), **29** (1996), 483–517.

[18] V. Baladi and L.-S. Young, *On the spectra of randomly perturbed expanding maps,* Comm. Math. Phys., **156** (1993), 355–385 (see also Erratum, Comm. Math. Phys., **166** (1994), 219–220).

[19] M. Barge and R. Swanson, *Pseudo-orbits and topological entropy,* Proc. Amer. Math. Soc., **109** (1990), 559–566.

[20] M. Benedicks and L.-S. Young, *Sinai-Bowen-Ruelle measures for certain Hénon maps,* Invent. Math., **112** (1993), 541–576.

[21] M. Benedicks and L.-S. Young, *Markov extensions and decay of correlations for certain Hénon maps,* Géométrie complexe et systèmes dynamiques, Proceedings of the Colloque en l'honneur d'Adrien Douady held in Orsay, June 1995, Astérisque, **261** (Paris), **2000**, 13–56.

[22] C. Bonatti and M. Viana, *SRB measures for partially hyperbolic systems whose central direction is mostly contracting,* Israel J. Math., **115** (2000), 157–193

[23] X. Bressaud and C. Liverani, *Anosov diffeomorphisms and coupling,* Preprint (2000).

[24] H. Bruin, S. Luzzatto and S. van Strien, *Decay of correlations in one-dimensional dynamics,* Preprint (1999).

[25] J. Buzzi, *A.c.i.m.s as equilibrium states for piecewise invertible dynamical systems,* Preprint (1998).

[26] J. Buzzi, *Exponential decay of correlations for random Lasota-Yorke maps,* Comm. Math. Phys., **208** (1999), 25–54.

[27] J. Buzzi, *Absolutely continuous invariant probability measures for arbitrary piecewise expanding and $\mathbf{R}$-analytic mappings of the plane,* Ergodic Theory Dynam. Systems, **20** (2000), 697–708.

[28] J. Buzzi and G. Keller, *Zeta functions and transfer operators for multidimensional piecewise affine and expanding maps,* Preprint, to appear Ergodic Theory Dynam. Systems (1999).

[29] J. Buzzi and V. Maume-Deschamps, *Decay of correlations for piecewise expanding maps in higher dimensions,* Preprint (1999).

[30] M. Campanino and S. Isola, *Infinite invariant measures for non-uniformly expanding transformations on $[0,1]$: weak law of large numbers with anomalous scaling,* Forum. Math., **8** (1996), 71–92.

[31] A. Castro, *Backward inducing and exponential decay of correlations for partially hyperbolic attractors with mostly contracting direction,* Ph.D. thesis, IMPA (1998).

[32] C.-H. Chang, *The transfer operator from statistical mechanics to quantum chaos,* Physics Bimonth., **21** (1999), 496–509.

[33] N. Chernov, *Statistical properties of piecewise smooth hyperbolic systems in high dimensions,* Discrete Contin. Dynam. Systems, **5** (1999), 425–448.

[34] N. Chernov, *Decay of correlations and dispersing billiards,* J. Statist. Phys., **94** (1999), 513–556.

[35] P. Collet and P. Ferrero, *Some limit ratio theorem related to a real endomorphism in case of a neutral fixed point,* Ann. Inst. H. Poincaré. Phys. Théor., **52** (1990), 283–301.

[36] W. Cowieson, *Piecewise smooth expanding maps in $\mathbf{R}^d$,* Ph.D. thesis UCLA (1999).

[37] P. Cvitanović, *Classical and Quantum Chaos: A Cyclist Treatise,* http://www.nbi.dk/ChaosBook.

[38] P. Dahlqvist, *The roles of singularities in chaotic spectroscopy,* Chaos Solitons Fractals, **8** (1997), 1011–1029.

[39] M. Dellnitz and O. Junge, *On the approximation of complicated dynamical behaviour*, SIAM J. Numer. Anal., **36** (1999) 491–515.

[40] A. Dold, *Fixed point index and fixed point theorem for Euclidean neighbourhood retracts*, Topology, **4** (1965), 1–8.

[41] D. Dolgopyat, *On decay of correlations in Anosov flows*, Ann. of Math. (2), **147** (1998), 325–355.

[42] D. Dolgopyat, *Prevalence of rapid mixing in hyperbolic flows*, Ergodic Theory Dynam. Systems, **5** (1998), 1097–1114.

[43] D. Dolgopyat, *Prevalence of rapid mixing-II: topological prevalence*, Ergodic Theory Dynam. Systems, **20** (2000), 1045–1059.

[44] D. Dolgopyat, *On dynamics of mostly contracting diffeomorphisms*, Comm. Math. Phys., **213** (2000), 181–201.

[45] D. Dolgopyat, *On mixing properties of compact group extensions of hyperbolic systems*, Preprint (2000).

[46] D. Dolgopyat and M. Pollicott, Addendum to *Periodic orbits and dynamical spectra*, Ergodic Theory Dynam. Systems, **18** (1998), 293–301.

[47] B. Fayad, *Polynomial decay of correlations for a class of smooth flows on the two torus*, Preprint (2000).

[48] D. Fried, *Meromorphic zeta functions for analytic flows*, Comm. Math. Phys., **174** (1995), 161–190.

[49] D. Fried, *The zeroes of dynamical zeta functions*, IHES talk (1999), in preparation.

[50] P. Gaspard and X.-J. Wang, *Sporadicity: between periodic and chaotic dynamical behaviors*, Proc. Nat. Acad. Sci. USA, **85** (1988), 4591–4595.

[51] N. T. A. Haydn, *Meromorphic extension of the zeta function for Axiom A flows*, Ergodic Theory Dynam. Systems, **10** (1990), 347–360.

[52] He Lianfa, *Pseudo-orbits and topological pressure*, Chinese J. Contemp. Math., **17** (1996), 405–414.

[53] F. Hofbauer and G. Keller, *Ergodic properties of invariant measures for piecewise monotonic transformations*, Math. Z., **180** (1982), 119–140.

[54] H. Hu, *Decay of correlations for piecewise smooth maps with indifferent fixed points*, Preprint (1998).

[55] H. Hu, *Conditions for the existence of SBR measures for "almost Anosov" diffeomorphisms*, Trans. Amer. Math. Soc., **352** (2000), 2331–2367

[56] H. Hu, *Equilibrium states with polynomial decay of correlations*, Preprint (1999).

[57] S. Isola, *On the rate of convergence to equilibrium for countable ergodic Markov chains*, Preprint (1997).

[58] S. Isola, *Renewal sequences and intermittency*, J. Stat. Phys., **97** (1999), 263–280.

[59] V. Kaloshin, *Generic diffeomorphisms with superexponential growth of number of periodic orbits*, Comm. Math. Phys., **211** (2000), 253–272.

[60] G. Keller, *Ergodicité et mesures invariantes pour les transformations dilatantes par morceaux d'une région bornée du plan*, C.R. Acad. Sci. Paris Sér. I Math., **289** (1979), 625–627.

[61] G. Keller, *A note on dynamical zeta functions for S-unimodal maps*, Colloq. Math., **84/85** (2000), 229–233.

[62] G. Keller and C. Liverani, *Stability of the spectrum for transfer operators*, Ann. Scuola Norm. Sup. Pisa. Cl. Sc., **28** (1999), 141–152.

[63] G. Keller and T. Nowicki, *Spectral theory, zeta functions and the distribution of periodic points for Collet-Eckmann maps*, Comm. Math. Phys., **149** (1992), 31–69.

[64] Yu. Kifer, *Random dynamics and its applications*, Proceedings of the International Congress of Mathematicians held in Berlin, August 1998, Doc. Math., Extra Vol. II (electronic), (Bielefeld), **1998**, 809–818.

[65] A. Kitaev, *Kneading functions for higher dimensions*, Personal communication (1995).

[66] A. Kitaev, *Fredholm determinants for hyperbolic diffeomorphisms of finite smoothness*, Nonlinearity, **12** (1999) 141–179 (see also Corrigendum, 1717–1719).

[67] C. Liverani, *Decay of correlations*, Ann. of Math. (2), **142** (1995), 239–301.

[68] C. Liverani, *Central limit theorem for deterministic systems*, Proceedings of the International Conference on Dynamical Systems held in Montevideo, 1995, Addison Wesley Longman, (Harlow), **1996**, 56–75.

[69] C. Liverani, *Decay of correlations for piecewise expanding maps*, J. Stat. Phys., **78** (1995), 1111–1129.

[70] C. Liverani, B. Saussol and S. Vaienti, *A probabilistic approach to intermittency*, Ergodic Theory Dynam. Systems, **19** (1999), 671–686.

[71] C. Liverani and D. Urbach, *A functional analytic approach to decay of correlations in Anosov flows*, private communication.

[72] D. H. Mayer, *Continued fractions and related transformations*, Proceedings of the conference on Ergodic Theory, Symbolic Dynamics and Hyperbolic Spaces held in Trieste, April 1989, Oxford University Press (New York), **1991**, 175–222.

[73] D. H. Mayer, *Thermodynamic formalism and quantum mechanics on the modular surface*, From Phase Transitions to Chaos, World Sci. Publishing (River Edge, NJ), **1992**, 521–529.

[74] J. Milnor and W. Thurston, *On iterated maps of the interval*, Proceedings of the conference on Dynamical Systems held at College Park, MD, 1986–87, Springer Lect. Notes in Math. **1342** (Berlin-New York), **1988**, 465–563.

[75] M. Mori, *Fredholm determinant for higher-dimensional piecewise linear transformations*, Japan. J. Math. (N.S.), **25** (1999), 317–342.

[76] T. Nowicki and D. Sands, *Non-uniform hyperbolicity and universal bounds for S-unimodal maps*, Invent. Math., **132** (1998), 633–680.

[77] J. Palis, *A global view of dynamics and a conjecture on the denseness of finitude of attractors*, Géométrie complexe et systèmes dynamiques, Proceedings of the Colloque en l'honneur d'Adrien Douady held in Orsay, June 1995, Astérisque, **261** (Paris) **2000** 335–348.

[78] W. Parry and M. Pollicott, *Zeta Functions and the Periodic Orbit Structure of Hyperbolic Dynamics*, Astérisque, **187–188**, Soc. Math. de France (Paris) **1990**.

[79] M. Pianigiani, *First return map and invariant measure*, Israel J. Math., **35** (1980), 32–48.

[80] M. Pollicott, *On the rate of mixing of Axiom A flows*, Invent. Math., **81** (1985), 413–426.

[81] Y. Pomeau and P. Manneville, *Intermittent transition to turbulence in dissipative dynamical systems*, Comm. Math. Phys., **74** (1980), 189–197.

[82] L. Rey-Bellet, unpublished notes on sharp determinants (1995).

[83] D. Ruelle, *Zeta functions for expanding maps and Anosov flows*, Invent. Math., **34** (1976), 231–242.

[84] D. Ruelle, *One-dimensional Gibbs states and Axiom A diffeomorphisms*, J. Diff. Geom., **25** (1987), 117–137.

[85] D. Ruelle, *The thermodynamic formalism for expanding maps*, Comm. Math. Phys., **125** (1989), 239–262.

[86] D. Ruelle, *An extension of the theory of Fredholm determinants*, Inst. Hautes Etudes Sci. Publ. Math., **72** (1990), 175–193.

[87] H. H. Rugh, *Fredholm determinants for real-analytic hyperbolic diffeomorphisms of surfaces*, Proceedings of the XIth International Congress of Mathematical Physics held in Paris, 1994, Internat. Press, (Cambridge, MA), **1995**, 297–303.

[88] H. H. Rugh, *Generalized Fredholm determinants and Selberg zeta functions for Axiom A dynamical systems*, Ergodic Theory Dynam. Systems, **16** (1996), 805–819.

[89] H. H. Rugh, *Intermittency and regularized Fredholm determinants*, Invent. Math., **135** (1999), 1–25.

[90] S. Smale, *Differentiable dynamical systems*, Bull. A.M.S., **73** (1967), 747–817.

[91] B. Saussol, *Absolutely continuous invariant measures for multidimensional expanding maps*, Israel J. Math., **116** (2000), 223–248.

[92] M. Thaler, *Transformations on [0, 1] with infinite invariant measures*, Israel J. Math., **46** (1983), 67–96.

[93] M. Thaler, *A limit theorem for the Perron-Frobenius operator of transformations on [0, 1] with indifferent fixed points*, Israel J. Math., **91** (1995), 111–127.

[94] M. Tsujii, *Absolutely continuous invariant measures for piecewise real-analytic expanding maps on the plane*, Comm. Math. Phys., **208** (2000), 605–622.

[95] M. Tsujii, *Absolutely continuous invariant measures for expanding piecewise linear maps*, Preprint (1999).

[96] M. Viana, *Multidimensional nonhyperbolic attractors*, Inst. Hautes Études Sci. Publ. Math., **85** (1997), 63–96.

[97] M. Viana, *Stochastic Dynamics of Deterministic Systems,* Col. Bras. de Matemática (Rio de Janeiro) **1997**.

[98] M. Viana, *Dynamics: a probabilistic and geometric perspective,* Proceedings of the International Congress of Mathematicians held in Berlin, August 18–27 1998, Doc. Math., Extra Vol. I (electronic), (Bielefeld), **1998**, 557–578.

[99] Q. Wang and L.-S. Young, *Analysis of a class of strange attractors,* Preprint (1999).

[100] L.-S. Young, *Stochastic stability of hyperbolic attractors,* Ergodic Theory Dynam. Systems, **6** (1986), 311–319.

[101] L.-S. Young, *Decay of correlations for certain quadratic maps,* Comm. Math. Phys., **146** (1992), 123–138.

[102] L.-S. Young, *Statistical properties of systems with some hyperbolicity including certain billiards,* Ann. of Math. (2), **147** (1998), 585–650.

[103] L.-S. Young, *Recurrence times and rates of mixing,* Israel. J. Math., **110** (1999), 153–188.

[104] L.-S. Young, *Ergodic theory of chaotic dynamical systems,* Proceedings of the XIIth Congress of Mathematical Physics held in Brisbane, 1997, Internat. Press (Cambridge, MA) **1999**, 131–143.

[105] M. Yuri, *Thermodynamic formalism for certain nonhyperbolic maps,* Ergodic Theory Dynam. Systems, **19** (1999), 1365–1378.

[106] R. Zweimüller, *Ergodic structure and invariant densities of non-Markovian interval maps with indifferent fixed points,* Nonlinearity, **11** (1998), 1263–1276.

Note added in proof: The master's thesis of S. Gouëzel (2001) extends the unpublished results of Rey-Bellet [82].

CNRS UMR 8628
Université de Paris-Sud
F-91405 Orsay, France
E-mail address: viviane.baladi@math.u-psud.fr

Sampling in Complex and Harmonic Analysis

Joaquim Bruna

Abstract. This is a survey article on uniqueness, sampling and interpolation problems in complex analysis. Most of these problems are motivated by applications of great practical importance in signal analysis and data transmission, but they also admit other mathematical formulations relating them to fundamental questions about existence of good bases in function spaces. This circle of ideas in complex analysis has experienced in recent years a notorious revitalization, mostly because of its connections with analogous problems in time-frequency and wavelet analysis, some of which will be discussed as well.

1. Sampling and Interpolation for Band-Limited Functions

In this paper we will be talking mostly about sampling, and the most well-known sampling result is the one dealing with band-limited functions, known as the Kotelnikov-Shannon-Whittaker theorem. A τ-band-limited function is one with finite energy, i.e. $\int |f(t)|^2 \, dt < +\infty$, whose Fourier transform $\hat{f}(\zeta)$

$$\hat{f}(\zeta) = \frac{1}{\sqrt{2\pi}} \int_{-\infty}^{+\infty} f(t) e^{-it\zeta} \, dt$$

vanishes for $|\zeta| > \tau$. Since $f(t) = \frac{1}{\sqrt{2\pi}} \int_{-\tau}^{+\tau} \hat{f}(\zeta) e^{it\zeta} \, d\zeta$, one may think of f as being an (infinite) linear combination of sine and cosine functions with frequencies $|\zeta| < \tau$. We denote by B_τ^2 the space of τ-band limited functions; note that the Fourier transform is an isometry between B_τ^2 and $L^2(-\tau, \tau)$.

The KSW sampling theorem states that the general form of such a function is given by the so-called *cardinal series*

$$f(t) = \sum_{k=-\infty}^{+\infty} a_k \frac{\sin(\tau t - \pi k)}{\tau t - \pi k}$$

with $\sum_k |a_k|^2 < +\infty$. Note that $a_k = f\left(\frac{k\pi}{\tau}\right)$ and, in fact,

$$\sum_k |a_k|^2 = \frac{\tau}{\pi} \int_{-\infty}^{+\infty} |f(t)|^2 \, dt \,.$$

There are two aspects in this statement which it is convenient to state separately. First, every $f \in B_\tau^2$ is completely recovered from its samples $f\left(\frac{k\pi}{\tau}\right)$ in a stable way, that is, for $f, g \in B_\tau^2$

$$\int_{-\infty}^{+\infty} |f(t) - g(t)|^2 \, dt = \frac{\pi}{\tau} \sum_k \left| f\left(\frac{k\pi}{\tau}\right) - g\left(\frac{k\pi}{\tau}\right)\right|^2 .$$

This means that a small error in the collection of samples $\left\{ f\left(\frac{k\pi}{\tau}\right)\right\}_{k \in Z}$ will produce a small error in the construction. Secondly, any square-summable sequence of numbers appears in this way, no other restriction appears.

Both aspects constitute the theoretical basis for the transition from analog signals $f(t)$ to discrete sequences $\{a_k\}_{k \in Z}$, and are of great practical importance in communications and data transmission in general.

The theorem is just a restatement of the fact that the exponential system $\left\{ \frac{1}{\sqrt{2\tau}} e^{i \frac{\pi}{\tau} k \zeta} \right\}_{k \in Z}$ is an orthonormal basis of $L^2(-\tau, \tau)$. In the expansion of $\hat{f}(\zeta)$ in this basis, the coefficients are

$$\left\langle \hat{f}, \frac{1}{\sqrt{2\tau}} e^{i \frac{\pi}{\tau} k \zeta} \right\rangle = \frac{1}{\sqrt{2\tau}} \int_{-\tau}^{+\tau} \hat{f}(\zeta) e^{-i \frac{\pi}{\tau} k \zeta} \, d\zeta = \sqrt{\frac{\pi}{\tau}} f\left(-\frac{k\pi}{\tau}\right)$$

so that

$$\hat{f}(\zeta) = \frac{\sqrt{\pi}}{\sqrt{2\tau}} \sum_{k=-\infty}^{+\infty} f\left(\frac{k\pi}{\tau}\right) e^{-i \frac{\pi}{\tau} k \zeta} .$$

Cotransforming this one gets the cardinal series expansion. Also,

$$\int_{-\infty}^{+\infty} |f(t)|^2 \, dt = \int_{-\tau}^{+\tau} |\hat{f}(\zeta)|^2 \, d\zeta = \frac{\pi}{\tau} \sum_k \left| f\left(\frac{k\pi}{\tau}\right)\right|^2 .$$

The rate of $\frac{\tau}{\pi}$ samples for a unit interval is called the *Nyquist rate*. It is well known that a slower rate of sampling is not possible in order to get exact reconstruction, while a higher rate (oversampling) may lead to reconstruction formulas with faster convergence. We will see below precise formulations of this fact.

It is quite natural to replace the equally spaced sequence $\{k\frac{\pi}{\tau}\}_{k \in Z}$ by a general sequence $\Lambda = \{t_k\}_{k=-\infty}^{+\infty}$, and ask to what extent the values $\{f(t_k)\}_{k=-\infty}^{+\infty}$ determine completely f, for every $f \in B_\tau^2$. If

$$f \in B_\tau^2, \ f(t_k) = 0 \quad \forall k \text{ implies } f \equiv 0$$

the sequence $\{t_k\}_{k=-\infty}^{+\infty}$ is called a *sequence of uniqueness* for B_τ^2. Note that whenever $f, g \in B_\tau^2$ and $f(t_k) = g(t_k) \ \forall k$ then $f = g$.

This notion alone is not sufficient to allow errors in the sampled values $\{f(t_k)\}_{k=-\infty}^{+\infty}$. For that purpose the notion of *stable sampling sequences* is introduced: these are the sequences $\Lambda = \{t_k\}_k$ for which there exist two constants A,

B such that

$$A \int_{-\infty}^{+\infty} |f(t)|^2 \, dt \leq \sum_k |f(t_k)|^2 \leq B \int_{-\infty}^{+\infty} |f(t)|^2 \, dt$$

for all $f \in B_\tau^2$. In particular, $\int |f(t) - g(t)|^2 \, dt$ is comparable to $\sum_k |f(t_k) - g(t_k)|^2$. Thus, f can be completely recovered, at least theoretically, from its samples $\{f(t_k)\}_k$ in a *stable way*, meaning that small errors in the samples will produce small errors in the reconstruction.

But, how does it perform this reconstruction? This is a matter of (infinite dimensional) linear algebra, that is, Hilbert spaces. To show that, it is convenient to work on the frequency side, that is in $L^2(-\tau, \tau)$; in terms of $g = \hat{f}$ the above inequality is written

$$A\|g\|_2^2 \leq \sum_k \left| \left\langle g, \frac{1}{\sqrt{2\tau}} e^{it_k \zeta} \right\rangle \right|^2 \leq B\|g\|_2^2 .$$

A family $\{e_k\}_{k \in Z}$ of vectors in a Hilbert space H satisfying

$$A\|u\|^2 \leq \sum_k |\langle u, e_k \rangle|^2 \leq B\|u\|^2, \quad u \in H$$

is called a *frame*. Frames were introduced in the article [8], which for this reason has been an influential paper in the last forty years. The frame condition is equivalent to the operator

$$\begin{aligned} T : H &\longrightarrow \quad \ell^2(Z) \\ u &\longmapsto \quad (\langle u, e_k \rangle)_k \end{aligned}$$

being one to one onto a closed subspace of $\ell^2(Z)$. This is in turn equivalent to the adjoint operator

$$T^* : \ell^2(Z) \longrightarrow H$$

being onto. A trivial computation shows that $T^*(\{c_k\}) = \sum_k c_k e_k$. Hence if $\{e_k\}_{k \in Z}$ is a frame, every $u \in H$ can be written

$$u = \sum_k c_k e_k .$$

In general this expression is not unique; there are relations $\sum_k c_k e_k = 0$, those $\{c_k\}$ in the kernel of T^*. Among all these expressions, the one minimizing $\sum_k |c_k|^2$ is of the form $c_k = \langle u, \tilde{e}_k \rangle$ for some family $\{\tilde{e}_k\}_k$ which turns out to be a frame as well, *the dual frame*. The dual frame of $\{\tilde{e}_k\}_k$ is $\{e_k\}_k$ so that the reconstruction formula for u is

$$u = \sum_k \langle u, \tilde{e}_k \rangle e_k = \sum_k \langle u, e_k \rangle \tilde{e}_k .$$

The redundancy of these expressions comes from linear relations between the vectors e_k, $\tilde{e}_k$, $k \in Z$, that is, from the kernel of T^*. Hence there is no redundancy as soon as $\ker T^* = \{0\}$; since $\ker T^*$ is the orthogonal of Range T, this leads in a natural way to the following notion: a family of vectors $(e_k)_{k \in Z}$ in a Hilbert space H is called a *Riesz-Fischer family* (or a *free family*) if

$$\langle u, e_k \rangle = a_k, \quad k \in Z$$

has a solution $u \in H$ for every $(a_k) \in \ell^2(Z)$. In our setting, when translated back to f, where $u = \hat{f}$, for $e_k(\zeta) = \frac{1}{\sqrt{2\tau}} e^{it_k \zeta}$, this corresponds to the sequence $\Lambda = \{t_k\}$ being an *interpolating* sequence in the sense that $f(t_k) = a_k$, $k \in Z$, has a solution $f \in B_\tau^2$ for every $(a_k) \in \ell^2(Z)$.

The frames with no redundancy are thus those for which T is an isomorphism; this amounts to T^* being an isomorphism, i.e. every u has a unique expression $u = \sum_k c_k e_k$ with $\|u\|_2^2$ comparable to $\sum_k |c_k|^2$. For this reason they are called as well *Riesz bases* or *exact frames*.

To summarize, Λ is a sequence of stable sampling for B_τ^2 iff the family of exponentials $\mathcal{E}(\Lambda) = \{e^{it_k \zeta}\}_{k \in Z}$ is a frame of $L^2(-\tau, \tau)$, and a sequence of free interpolation iff $\mathcal{E}(\Lambda)$ is a free family; $\mathcal{E}(\Lambda)$ is a Riesz basis of $L^2(-\tau, \tau)$ iff Λ is both of stable sampling and of interpolation (this is called too a *complete interpolating sequence*). Note that in this language, Λ is a set of uniqueness if and only if $\mathcal{E}(\Lambda)$ spans the whole of $L^2(-\tau, \tau)$.

Beurling ([5]) had considered sup-norm versions of these problems, that is, replacing functions in $L^2(-\tau, \tau)$ by measures or distributions μ supported in $[-\tau, \tau]$ giving rise to bounded functions

$$f(t) = \int_{-\tau}^{+\tau} e^{it\zeta} \, d\mu(\zeta)$$

with spectrum in $[-\tau, \tau]$. The corresponding space, larger than B_τ^2, is called B_τ^∞, *the Bernstein space*. In this setting, Λ is of stable sampling for B_τ^∞ if

$$\sup_{t \in \mathbb{R}} |f(t)| \leq A \sup_k |f(t_k)|$$

and interpolating if $f(t_k) = a_k$, $k \in Z$, has a solution for all bounded sequences $\{a_k\}_{k \in Z}$.

Complex analysis enters naturally into the picture because every function in B_τ^∞, B_τ^2 is the restriction to the real line of an entire function, the Fourier-Laplace transform

$$f(z) = \int_{-\tau}^{+\tau} e^{iz\zeta} \, d\mu(\zeta), \quad z \in \mathbb{C},$$

$(d\mu(\zeta) = g(\zeta) \, d\zeta, \ g \in L^2$ in the case of B_τ^2.)

By the Paley-Wiener theorem, B_τ^2 (resp. B_τ^∞) consists of those entire functions of exponential type at most τ, i.e., $|f(z)| = O(e^{\tau|z|})$ and such that $f_{|\mathbb{R}} \in L^2(\mathbb{R})$, (resp. $\in L^\infty(\mathbb{R})$).

From this point of view, the real sequence $\Lambda = (t_k)_{k \in Z}$ might be replaced by a general complex sequence $\Lambda \subset \mathbb{C}$; all considerations up to now remain the same. To simplify the exposition we will however limit ourselves to the case $\Lambda \subset \mathbb{R}$. This complex-analysis setting, as it is often the case, leads immediately to interesting new points of view. For instance, we mentioned that the KSW theorem is the theoretical basis to digitalize analog signals $f \in B_\tau^2$, through its samples $f(k\pi/\lambda)$; now, looking at f as an entire function gives immediately a discretization of f as well, because by Hadamard's factorization theorem, every entire function of exponential type is characterized up to a constant by its sequence of zeroes $\{z_k\}$

$$f(z) = ce^{az}z^m \prod \left(1 - \frac{z}{z_k}\right) e^{z/z_k}.$$

Vitushkin has studied the properties of this digitalization process.

The notions of frames, Riesz bases and so on in $L^2(-\tau, \tau)$ can of course be translated to B_τ^2 as well. Just note that

$$f(t) = \frac{1}{\sqrt{2\pi}} \int_{-\tau}^{+\tau} \hat{f}(\zeta)e^{it\zeta}\, d\zeta = \int_{-\infty}^{+\infty} f(x)K(t,x)\, dx$$

with

$$K(t,x) = \frac{1}{2\pi} \int_{-\tau}^{+\tau} e^{i\zeta(t-x)}\, d\zeta = \frac{1}{\pi} \frac{\sin \tau(t-x)}{\tau(t-x)}.$$

This last expression can be viewed as the inner product of f with the *Bergman kernel* $K(t, \cdot)$. Hence, statements about $\mathcal{E}(\Lambda)$ being a Riesz basis, a frame etc. of $L^2(-\tau, \tau)$ are equivalent to analogous statements about the family $K(\Lambda) = \{K(t_k, \cdot)\}_{k \in Z}$ in B_τ^2.

2. Sets of Uniqueness

In this paragraph, B_τ will denote any of the spaces B_τ^2, B_τ^∞.

Complexifying time, that is, viewing $f \in B_\tau$ as entire functions, it is evident that every Λ with a finite accumulation point is of uniqueness. These are the uninteresting uniqueness sets, whence we will assume Λ discrete. It is intuitively clear that such a Λ must have sufficiently many points; put in another way, if S is not a sequence of uniqueness, then there exists $f \in B_\tau$, $f \not\equiv 0$, such that $f_{|S} = 0$, that is, S is included in the zero set $S(f)$ of f and hence it cannot have too many points. Hence the question becomes one about distributions of (real) zeros of functions in B_τ. Since whenever $f \in B_\tau$ and $f(\alpha) = 0$ the function $g(z) = f(z)\frac{z-\beta}{z-\alpha}$ is again in B_τ and $g(\beta) = 0$, changing a finite number of points cannot have any effect. Altogether, this means that some asymptotic density must be "small" for sequences S and "big" for sequences Λ.

It is convenient to introduce the *characteristic function* of S

$$n_S(t) = \#\{S \cap (0,t]\},\ t > 0; \quad n_S(t) = -\#\{S \cap (t,0]\},\ t < 0$$

so that $n_S(b) - n_S(a) = \#\{S \cap (a,b]\}$, $n_S(0) = 0$.

A classical theorem of Levinson establishes that if $f \in B_\tau$ has real zeros $S(f)$, this sequence has a *finite asymptotic density*

$$D(S(f)) = \lim_{|t| \to \infty} \frac{n_{S(f)}(t)}{t} \leq \frac{\tau}{\pi}.$$

This leads in a natural way to define the *Polya maximal density* of a given discrete sequence $S \subset \mathbb{R}$ as

$$D^*(S) = \inf\{\rho : \text{there exists a sequence } T \supset S \text{ with density } \rho\}$$

which can be defined as well only in terms of S as

$$D^*(S) = \lim_{\substack{\zeta \to 1 \\ \zeta < 1}} \limsup_{r \to +\infty} \frac{n_S(r) - n_S(\zeta r)}{(1 - \zeta)r}.$$

Obviously, $D^*(S) \geq \overline{D}(S) = \limsup_{|t| \to +\infty} \frac{n_S(t)}{t}$, the real upper density, but $D^*(S)$ might be strictly bigger than $\overline{D}(S)$. For instance, if say S consists of the integers between 3^k and $3^k + 3^{k-1}$, $k \in Z$, then $D^*(S) = 1$, but $D(S) = \overline{D}(S) = 1/4$, so $D^*(S)$ takes care of holes.

Levinson's theorem gives then that Λ is a set of uniqueness for B_τ whenever $D^*(\Lambda) > \frac{\tau}{\pi}$. To obtain a sharp result one needs to introduce another density, and this is what Beurling and Malliavin did in the sixties in one of the finest works in the field. There are different equivalent definitions of the *Beurling-Malliavin effective density*; the most intuitive one uses the Rising-Sun construction of F. Riesz. Assuming for simplicity $\Lambda \subset \mathbb{R}^+$ we look at the graphic of the step function $n_\Lambda(t)$

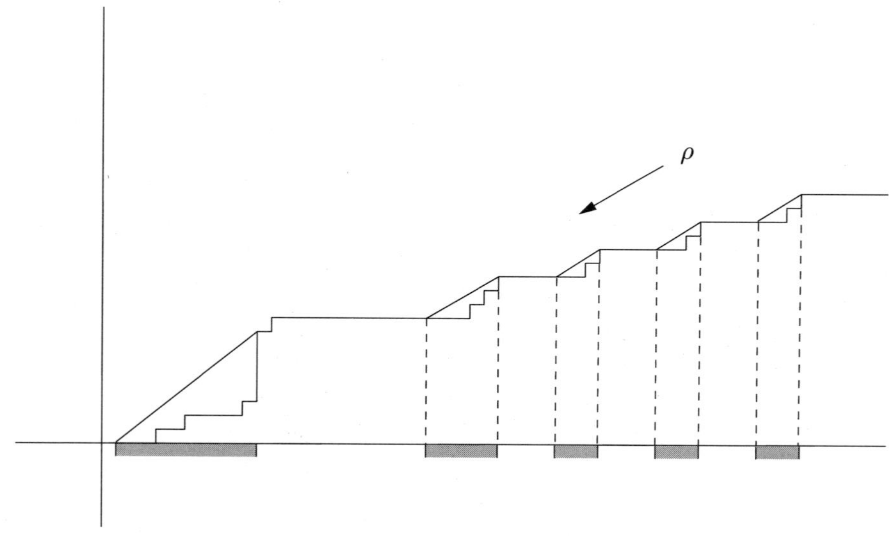

For a given ρ we look at the connected components (a_k, b_k) of the set $\{t > 0 : n_\Lambda(s) - n_\Lambda(t) > \rho(s - t) \text{ for some } s > t\}$. These are the intervals remaining in shadow when light rays with step ρ illuminate the stair. Note that this set is $(0, +\infty)$ if $\rho < \overline{D}(\Lambda)$; if $\rho > \overline{D}(\Lambda)$ all components have finite length, whereas if $\rho = \overline{D}(\Lambda)$ one unbounded component may appear. The BM-effective density is defined as

$$\tilde{D}(\Lambda) = \inf\left\{\rho : \sum_k \frac{(b_k - a_k)^2}{a_k^2} < +\infty\right\}.$$

For a general discrete $\Lambda \subset \mathbb{R}$, $\tilde{D}(\Lambda)$ is defined as the maximum of $\tilde{D}(\Lambda \cap \mathbb{R}^+)$ and $\tilde{D}(-(\Lambda \cap \mathbb{R}^-))$.

In their celebrated work, Beurling and Malliavin improved Levinson's result showing that $\tilde{D}(\Lambda) > \tau/\pi$ implies Λ being a sequence of uniqueness for B_τ. More than that, they showed that this is sharp in the following sense: if Λ is of uniqueness for B_τ, then $\tilde{D}(\Lambda) \geq \tau/\pi$. Taken together, both results imply that

$$\pi\tilde{D}(\Lambda) = \sup\{\tau : \Lambda \text{ is a sequence of uniqueness for } B_\tau^2\}$$

or which is the same

$$\pi\tilde{D}(\Lambda) = \sup\{\tau : \mathcal{E}(\Lambda) \text{ spans } L^2(-\tau, \tau)\}.$$

The right-hand side is called the *radius of completeness* of the family $\mathcal{E}(\Lambda)$ and it is easily seen to be independent of the metric chosen among functions in $[-\tau, \tau]$.

In spite of the Beurling-Malliavin theorem, the precise characterization of the uniqueness sequences for B_τ^2 remains, as far as I know, an open problem. A good reference for the Beurling-Malliavin results is [12].

3. Stable Sampling and Interpolation in Dimension One

A. Beurling characterized stable sampling and interpolating sequences in the sup-norm case, for B_τ^∞, using complex analysis methods ([5]); he introduced the *upper and lower uniform densities* of a *separated* sequence $\Lambda = \{t_k\}_{k=-\infty}^{k=+\infty}$ (meaning that $\inf_{k \neq \ell} |t_k - t_\ell| > 0$) as

$$D_u^+(\Lambda) = \lim_{r \to \infty} \frac{n^+(r)}{r}, \quad D_u^-(\Lambda) = \lim_{r \to \infty} \frac{n^-(r)}{r}$$

where $n^+(r)$ (resp. $n^-(r)$) denotes the maximum (resp. minimum) number of points of Λ to be found in an interval of length r. Note that $\tilde{D}(\Lambda) \geq D_u^-(\Lambda)$. Beurling proved that $\Lambda \subset \mathbb{R}$ is of stable sampling for B_τ^∞ if and only if it contains a separated sequence Λ_0 with $D_u^-(\Lambda_0) > \frac{\tau}{\pi}$ and $\Lambda \subset \mathbb{R}$ is interpolating for B_τ^∞ if and only if it is separated and $D_u^+(\Lambda) < \frac{\tau}{\pi}$. In particular, no stable sampling interpolating sequences exist in the sup-norm case. Beurling results had been recently extended to arbitrary sequences $\Lambda \subset \mathbb{C}$ in [24].

In the L^2-case B_τ^2, it is not hard to see that an interpolating sequence must be separated, and, on the other hand every stable sampling sequence contains a separated stable sampling subsequence. Thus one can restrict attention to separated sequences. For separated sequences Beurling results for B_τ^∞ imply in a trivial way that

$$D_u^-(\Lambda) > \tau/\pi \Rightarrow \Lambda \text{ stable sampling} \quad \Rightarrow D_u^-(\Lambda) \geq \tau/\pi$$

$$D_u^+(\Lambda) < \tau/\pi \Rightarrow \Lambda \text{ free interpolation} \Rightarrow D_u^+(\Lambda) \leq \tau/\pi$$

and hence it is clear that the sequences of stable sampling with no redundancy ($\mathcal{E}(\Lambda)$ Riesz basis) must have uniform density

$$D(\Lambda) = \lim_{r \to \infty} \frac{\#\{\Lambda \cap [x, x+r]\}}{r} = \frac{\tau}{\pi}$$

meaning that for large r, every integral of length r must have $\frac{\tau}{\pi}r + o(r)$ points of Λ. Of course, the prototype is the sequence $Z\frac{\pi}{\tau}$.

The description of the stable sampling no redundant sequences of B_π^2 (that is the Riesz basis of $L^2(-\pi, \pi)$ of type $\mathcal{E}(\Lambda)$) was for a while a very central problem, and was solved by the Russian school ([26, 22, 21]). On the way a number of perturbative results appeared, among which Kadec's 1/4-theorem is the most well-known: if $|t_k - k| \leq r < 1/4$, $\Lambda = \{t_k\}$ is such a sequence.

Every such sequence Λ gives rise to a series analogous to the cardinal series. Indeed, (assuming $t_0 = 0$ in case $0 \in \Lambda$)

$$S(z) = (z - \lambda_0) \lim_{R \to \infty} \prod_{\substack{|t_k| < R \\ k \neq 0}} \left(1 - \frac{z}{t_k}\right)$$

defines the so-called *generating function* of Λ and the reconstruction from $a_k = f(t_k)$ is given by

$$f(t) = \sum_{k \in Z} a_k \frac{S(t)}{S'(t_k)(t - t_k)}.$$

There are different formulations of Pavlov's theorem none of which is easy, as they are all related to BMO functions or A_2 weights. We record one of them for completeness (see [11]): $\mathcal{E}(\Lambda)$ is a Riesz basis for $L^2(-\pi, \pi)$ if and only if S has an exponential type π and $\frac{|S(x)|^2}{d(x,\Lambda)^2}$ satisfies the Muckenhoupt A_2-condition:
"There exists a constant C such that

$$\left(\int_I \frac{|S(x)|^2}{d(x, \Lambda)^2} \, dx\right) \left(\int_I \frac{d(x, \Lambda)^2}{|S(x)|^2} \, dx\right) \leq C \quad \forall \text{ interval } I \text{.''}$$

In spite of all this progress, the description of stable sampling (separated) sequences for B_τ^2 has been an open and subtle question till very recently. Indeed, it has been known since 1995 that the stable sampling sequences Λ with $D_u^-(\Lambda) > \frac{\tau}{\pi}$ contain a stable sampling and non-redundant (interpolating) subsequence, but also examples of stable sampling sequences with no such subsequences are known ([32]).

A very important breakthrough has been recently achieved in [25]. Ortega-Cerdà and Seip succeed, by bringing into the picture the de Branges's theory of Hilbert spaces of entire functions, to give a complete characterization of the stable sampling sequences for the Paley-Wiener space B_π^2, or, which is the same thing, the Fourier frames $\mathcal{E}(\Lambda)$. This is one more example of the remarkable influence of de Branges work in modern analysis.

A de Branges space is a Hilbert space H of entire functions fulfilling:

(H1) If $f \in H$, $\zeta \notin \mathbb{R}$ and $f(\zeta) = 0$, the function $g(z) = f(z)\frac{z-\bar{\zeta}}{z-\zeta}$ is in H with the same norm.

(H2) For every $\zeta \notin \mathbb{R}$, $f \mapsto f(\zeta)$ is a continuous functional.

(H3) If $f \in H$, the function $f^*(z) = \overline{f(\bar{z})}$ is also in H with the same norm.

A fundamental theorem of de Branges ([6]) states that such a space H is of the following form $H(E)$: there exists a function E in the Hermite-Biehler class HB (entire functions with no zeros in the upper half plane and such that $|E(z)| \geq |E(\bar{z})|$ for $\operatorname{Im} z > 0$) and H consists of the entire functions f such that

$$\sup_y \int_{-\infty}^{+\infty} \frac{|f(x+iy)|^2}{|E(x+i|y|)|^2}\, dx < +\infty$$

the norm of f being $\|f\|_E^2 = \int_{-\infty}^{+\infty} \frac{|f(x)|^2}{|E(x)|^2}\, dx$. The Paley-Wiener space B_π^2 corresponds to $E(z) = e^{-i\pi z}$.

With these definitions, the result by Ortega-Cerdà and Seip is stated as follows: a separated sequence Λ of real numbers is stable sampling for B_π^2 if and only if there exist two entire functions $E, F \in HB$ such that $H(E) = B_\pi^2$ and Λ is the zero-sequence of $EF + E^*F^*$.

The functions $E \in HB$ such that $H(E) = B_\pi^2$ have been described in [18]. An obvious comment about this theorem is that the description does not involve Λ itself alone; such a description is probably not possible with a simple formulation. In spite of this, Ortega-Cerdà and Seip are able to draw some interesting applications of the theorem. For instance, they show that there exists a complete interpolating sequence $\Gamma = \{\gamma_k\}_{k \in Z}$ such that for every $k \in Z$ there is at least one $t \in \Lambda$ with $\gamma_k \leq t < \gamma_{k+1}$, that is, a stable sampling sequence is always denser than some complete interpolating sequence. Another interesting feature that they show is the following: if Λ is a stable sampling sequence for B_π^2 there is another Hilbert space H of entire functions, bigger than B_π^2, such that Λ becomes a complete interpolating sequence for H in the appropriate sense.

As far as I know, no analogous progress has been made in the problem of describing the interpolating sequences for B_π^2.

In connection with signal analysis applications to band-limited functions the following problem might be interesting: for which sequences of measures $\{\mu_k\}_{k \in Z}$, say with disjoint supports, does it hold that

$$A \int_{-\infty}^{+\infty} |f(t)|^2 \, dt \leq \sum_k \left| \int_{-\infty}^{+\infty} f \, d\mu_k \right|^2 \leq B \int_{-\infty}^{+\infty} |f(t)|^2 \, dt, \quad f \in B_\pi^2$$

for some constants A, B? The case in which the μ_k are the Lebesgue measures of a sequence of disjoint intervals should of course be studied first.

To close this section we briefly consider now *multiband signals*. This means functions in $L^2(\mathbb{R})$ whose spectrum is contained in a set S consisting of a finite number of disjoint intervals $I_1, \ldots, I_m$, which we call B_S^2. Here, no easy description of B_S^2 as a space of entire functions is available, and no usual complex analysis methods work. Using operator-theory methods, Landau ([15]) was able to prove that whenever Λ is of stable sampling for B_S^2,

$$n^-(r) \geq \frac{m(S)}{2\pi} r - A \log^+ r - B$$

for some constants A, B, whereas

$$n^+(r) \leq \frac{m(S)}{2\pi} r + A \log^+ r + B$$

in case Λ is of free interpolation. In particular, one has that $D_u^-(\Lambda) \geq \frac{m(S)}{2\pi}$, $D_u^+(\Lambda) \leq \frac{m(S)}{2\pi}$ are necessary conditions for sampling and interpolation, respectively, in B_S^2. It must be pointed out that in the multiband case the Beurling-Landau conditions cannot provide a complete solution to these problems; arithmetic relations among the points of Λ then play an important role and no density condition seems appropriate. Indeed, for instance Landau [14] constructed a symmetric sequence Λ arbitrarily close to the integers for which $\mathcal{E}(\Lambda)$ is complete in $L^2(S)$, where S is any finite union of the intervals $|x - 2\pi n| < \pi - \delta$, with arbitrarily large measure. A recent improvement of Landau's results is due to Ulanovskii [33]. As far as I know the following basic question is still unanswered.

Question. *Does there at all exist, for every finite union $S = I_1 \cup \cdots \cup I_m$ of finite intervals, a real sequence Λ such that $\mathcal{E}(\Lambda)$ is a Riesz basis in $L^2(S)$?*

It is known that there exist complex sequences Λ lying in horizontal strips such that $\mathcal{E}(\Lambda)$ is a Riesz basis. The answer to the question is yes if the lengths of the intervals I_i are commensurable, and also for two intervals ([17]). The question is essentially a trivial one in case S is a convenient explosion of an interval of the same length $|S|$, as pointed out in general in the next section.

4. Uniqueness and Stable Sampling for Multidimensional Signals

Let us replace the interval $[-\tau, \tau]$ by a general measurable set S in $\mathbb{R}^n$, and let B_S^2 denote the closed subspace of $L^2(\mathbb{R}^n)$ consisting of functions whose Fourier transform $\hat{f}(\zeta) = (2\pi)^{-n/2} \int_{\mathbb{R}^n} f(t) e^{-it \cdot \zeta} \, dt$ is supported on S. The problem is: what can be said about uniqueness sequences, stable sampling sequences, interpolating sequences, complete interpolating sequences $\Lambda = \{t_k\}_{k\in Z}$ for B_S^2? As before, these notions correspond to the family of exponentials $\mathcal{E}(\Lambda) = \{e^{it_k\cdot\zeta}\}_{k\in Z}$, where now $t_k \cdot \zeta$ denotes the inner product in $\mathbb{R}^n$, being complete, a frame, free or a Riesz basis, respectively, of $L^2(S)$.

There are of course certain trivial cases that can be treated just making "direct product" of the one-variable results: if S is a rectangle $|\zeta_i| \leq \tau_i$, $i = 1,\ldots,n$, and $\Lambda_i \subset \mathbb{R}$ is a uniqueness sequence for $B_{\tau_i}^2$ (that is $\mathcal{E}(\Lambda_i)$ complete in $L^2(-\tau_i,\tau_i)$) it is immediate that $\Lambda_1 \times \cdots \times \Lambda_n$ is a uniqueness sequence for B_S^2, and an analogous fact occurs with frames or Riesz basis. It is also easy to produce examples for some special sets S. We show this for the case of orthonormal basis; from the proof of the KSW theorem indicated at the beginning, it is clear that one will have a "cardinal series" development for functions in B_S^2 as soon as the exponentials $\{e^{ik\cdot\zeta}\}_{k\in Z^n}$ constitute an orthonormal basis of $L^2(S)$. This is of course the case for $S_0 = [-\pi,\pi]^n$ but also for $L^2(S)$ if S is what results if we break S_0 into a finite number of pieces and move each piece by an integer vector keeping them disjoint. Moreover, exploiting the nice behaviour of the Fourier transform under the action of the linear group, we may replace S_0 by US_0 and Z^n by $(U^T)^{-1}Z^n = \Lambda$. In this way one can obtain an orthonormal basis of exponentials for $L^2(S)$ and a cardinal series development for functions in B_S^2 whenever S is a fundamental region of a translation group.

In this connection there is an interesting conjecture by Fuglede [9]: $L^2(S)$ admits an orthonormal basis of exponentials $\mathcal{E}(\Lambda)$ if and only if S *tiles* $\mathbb{R}^n$ *by translations*, in the sense that there exists a discrete set $T \subset \mathbb{R}^n$ such that up to sets of measure 0, the sets $S + t$, $t \in T$ are disjoint and fill $\mathbb{R}^n$. Fuglede himself proves the conjecture if either Λ or T is a lattice. The convex sets S tiling $\mathbb{R}^n$ by translations have been completely characterized ([20]); in dimension $n = 2$, they are exactly the symmetric polygons of four or six sides. Moreover, in this case, T may be chosen to be a lattice, and hence, by Fuglede's theorem, $L^2(S)$ has an orthonormal basis of exponentials. Significant progress has been made on Fuglede's conjecture in some other special cases (see [13] and the references there).

As in dimension one, functions in B_S^2 are restrictions to $\mathbb{R}^n$ of certain entire functions in $\mathbb{C}^n$. As a space of entire functions, B_S^2 can be described in neat terms only if S is convex: in this case $f \in B_S^2$ if and only if $f \in L^2(\mathbb{R}^n)$ and $|f(x + iy)| = O(e^{\varphi_S(y)})$, $x, y \in \mathbb{R}^n$ where $\varphi_S(y) = \sup_{\zeta\in S} \zeta \cdot y$ is the support function of S. For instance, when S is the rectangle $|\zeta_i| \leq \tau_i$, $i = 1,\ldots,n$ we get the condition $\log|f(z)| \leq \tau_1|y_1| + \cdots + \tau_n|y_n| + c$.

Certain generalizations of Levinson's theorem are known ([28, 1]). For instance, if Λ is a separated sequence with separation constant $h = \inf_{k\neq\ell}\|t_k - t_\ell\|_\infty$, (where $\|x\|_\infty = \max_j |x_j|$) and $\limsup_{r\to\infty} \frac{\#\{\Lambda\cap T_r\}}{(2r)^n} = d > 0$ (here $T_r = \{\zeta : |\zeta_i| \leq r\}$), then Λ is a uniqueness sequence for B_S^2, for every rectangle $S = \{|\zeta_i| \leq \tau_i\}$ with $\tau_1 + \cdots + \tau_n < \frac{\pi^n}{(n-1)!2^{n-1}}h^{n-1}d$. Note however that this result does not cover the trivial case $\Lambda = Z^n$ for $S = T_\pi$. Another result of this type can be found in [5, page 310].

In general dimension n, the most important contribution remains that of Landau ([15]). He was able to generalize Beurling's necessary conditions for stable sampling and interpolation, for general S, using methods from operator theory. The uniform upper and lower densities $D_u^\pm(\Lambda)$ of a separated sequence are defined in an analogous way: if $n^+(r)$, $n^-(r)$ denote respectively the maximum and minimum number of points of Λ to be found in a translate of rU, U being the unit cube,

$$D_u^+(\Lambda) = \limsup_{r \to \infty} \frac{n^+(r)}{r^n}, \quad D_u^-(\Lambda) = \liminf_{v \to \infty} \frac{n^-(r)}{r^n}.$$

It may be seen that the result does not change replacing U by another set of measure 1. Landau's results are:

if Λ is a sequence of stable sampling for B_S^2, then $D_u^-(\Lambda) \geq \dfrac{m(S)}{(2\pi)^n}$;

if Λ is a sequence of free interpolation, then $D_u^+(\Lambda) \leq \dfrac{m(S)}{(2\pi)^n}$.

The method of proof of Landau is to give a firm-ground justification of the following, which a fortiori follows from the existence of stable sampling sequences. Suppose a rectangle R fixed, it has a finite number of points of Λ: if one thinks of functions well concentrated on R it is intuitively clear from the stability of the sampling that the sampled values of R must have little affect and may be thought as 0. In this way they are essentially characterized by a finite number of samples, whence one concludes that the space of functions in B_S^2 which "live essentially in a rectangle R" has a "finite dimension". In a certain way, Landau reverses this line of reasoning. Landau's results imply that the Nyquist rate of sampling cannot be improved by any means.

Let us assume that $n = 2$, $S = [-\tau_1, \tau_1] \times [-\tau_2, \tau_2]$, and let Λ_1, Λ_2 be two separated sequences in $\mathbb{R}$. Then it is not hard to see that $\Lambda = \Lambda_1 \times \Lambda_2$ is of stable sampling for B_S^2 if and only if each Λ_i is for $B_{\tau_i}^2$. On the other hand, $D_u^-(\Lambda)$ is at least $D_u^-(\Lambda_1)D_u^-(\Lambda_2)$, whence it may be made $\geq \frac{\tau_1 \tau_2}{\pi^2}$ taking $D_u^-(\Lambda_1)$ big enough and $D_u^-(\Lambda_2)$ small, say $< \frac{\tau_2}{\pi}$. Combining both facts we see that $D_u^-(\Lambda) > \frac{m(S)}{(2\pi)^n}$ cannot be a sufficient condition if $n \geq 2$. The same considerations suggest that some "directional uniform density conditions" should come into the picture in order to obtain sharp results.

Concerning Riesz basis in $L^2(S)$, the following natural question seems unanswered.

Question. *Does there at all exist a sequence $\Lambda \subset \mathbb{R}^n$ such that $\mathcal{E}(\Lambda)$ is a Riesz basis for $L^2(S)$?*

Besides the "direct product" situations and the trivial ones described before the only known result, as far as I know, is a recent one in [19]: the answer is yes if S is a convex symmetric polygon. This question is of interest for instance in computerized tomography (CT), where it is needed to sample the Radon transform Rf of the unknown function f. The case when S is a ball seems particularly appealing. For this case, Beurling gave a sufficient condition for stable sampling in B_S^∞: if S has radius r, and $\sup \operatorname{dist}(\zeta, \Lambda) < \pi/2r$, Λ is of stable sampling for B_S^∞ ([5, page 30]).

Let us close this section with some basic comments about entire functions in several variables. Dealing with these kind of questions, and leaving aside that only for convex S can we understand B_S^2 as a space of entire functions in $\mathbb{C}^n$, the common situation one encounters is: $f \in B_S^2$ vanishes on $\Lambda \subset \mathbb{R}^n$. Either f is given and we want to understand how small Λ must be or else Λ is given and we must *construct* f. In dimension 1, complete interpolating sequences and also stable sampling sequences appear, as we have mentioned, as zero sequences of *generating functions*, and one can use infinite products to construct entire functions. Now, in dimension $n > 1$, the zeros of analytic functions have complex dimension $n - 1$, and one needs $n - 1$ analytic functions to define, generically speaking, a discrete set. In one dimension, division of analytic functions is an easy task, not at all so in dimension ≥ 2. All this helps explain the difficulties in dealing with the multidimensional problem.

In several complex variables, varieties of zeros of analytic functions and interpolation problems from these varieties have of course been considered, and very intensively, in the last decades. But the motivation comes from other sources. Among them, one of the most important is of course Partial Differential Equations, and we show by an example what the connexion is. Let $P(D)$ be a partial differential operator with constant coefficients $D = \left(-i\frac{\partial}{\partial x_1}, \ldots, -i\frac{\partial}{\partial x_n}\right)$: if $z_0 \in \mathbb{C}^n$ satisfies $P(z_0) = 0$ then a solution of $P(D)u = 0$ is $u(x) = \exp(ix \cdot z_0)$, and also one has solutions $Q(x)e^{ix \cdot z_0}$ with a polynomial Q depending on the multiplicity. These are called *exponential-polynomial solutions*. Euler proved that in dimension 1, the finite linear combinations of exponential-polynomial solutions give all solutions. An n-dimensional version of this is the Fundamental Principle of Ehrenpreis-Palamodov: $f \in C^\infty(\mathbb{R}^n)$ is a solution of $P(D)f = 0$ if and only if it can be represented as

$$f(\tau) = \sum_{j=1}^{J} \int_{V_j} \partial_j(e^{ix-z}) \, du_j(z)$$

where the V_j are algebraic varieties contained in $V = \{z \in \mathbb{C}^n : P(z) = 0\}$, the ∂_j are partial differential operators with constant coefficients and the u_j are measures supported on V_j. $P(D)f = 0$ is equivalent to $0 = \langle f, {}^t P(D)u \rangle$ for every distribution u with compact support, whence to f defining a well-defined functional on the quotient space $\mathcal{E}'(\mathbb{R}^n)/{}^t P(D)\mathcal{E}'(\mathbb{R}^n)$: by Fourier transform this is $\widehat{\mathcal{E}}'(\mathbb{R}^n)/p \cdot \widehat{\mathcal{E}}'$ which is a space of analytic functions on V. Proving the theorem amounts to understanding this quotient space, and this is about functions vanishing on V and also about extending functions defined on V.

Incidentally, Beurling applied his condition about sampling for B_S^2, S a ball, to deal with the equation $P(D)u = f$, say with $f \in L^2(S)$. By Fourier transform the solution u must satisfy $P(\zeta)\hat{u} = \hat{f}$, whence $\hat{u} = \hat{f}/P$ but this may lead to problems because of the zeros of P. Beurling's idea is quite nice and simple: if Λ is of stable sampling for B_S^2, without losing information, the above is equivalent to $P(\zeta)\hat{u}(\zeta) = \hat{f}(\zeta)$ for $\zeta \in \Lambda$, and choosing Λ so that $|P(\zeta)|$ is bounded below on Λ (which is always possible) one can invert $P(D)$ in a stable way. In terms of the dual frame $\tilde{e}_k$ of $e^{i\lambda_k \zeta}$ in $L^2(S)$, if

$$f(\zeta) = \sum_k \langle f, \tilde{e}_k \rangle e^{i\lambda_k \zeta}$$

the solution is simply

$$u(z) = \sum_k \frac{\langle f, \tilde{e}_k \rangle}{P(\lambda_k)} e^{i\lambda_k \zeta} \,.$$

This method can be used as well to show the existence of a fundamental solution.

Note that the fundamental principle has something in common, in fact, with the questions we are discussing, namely, the issue is to represent certain functions as superpositions of complex exponentials. A general setting would be as follows: we are given N, M in $\mathbb{C}^n$ and ask which functions on N can be represented as Fourier-Laplace transforms of functions or measures on M

$$f(\omega) = \int_M e^{iz\omega} g(z)\, d\sigma(z), \quad \omega \in N \,.$$

A fortiori, these functions must be restrictions to N of entire functions but if N is totally real this would imply no restriction. Usually g is obtained from f by some kind of Fourier-Laplace transform too, so this is, vaguely, a question about inverting Fourier-Laplace transforms. A general scheme leading to such formulas, which includes most of the known cases, has been shown in [3]. Stable sampling sequences Λ for B_S^2 correspond to $N = S$, $M = \Lambda$. In dimension $n > 1$, there is more choice for the dimension of M, e.g. it could be of dimension k, $0 \leq k \leq n - 1$. For instance, for Paley-Wiener spaces B_S^2 with $S \subset \mathbb{R}^n$ it makes sense, mathematically speaking, (maybe not so from the signal analysis point of view) to ask for "sampling manifolds" M of dimension $k \geq 1$, which would correspond to "continuous frames". The point we want to emphasize is that the techniques developed in several complex variables so far are better adapted to deal with the case $k = n - 1$.

5. Time-Frequency Analysis and Gabor Wavelets

In time-frequency analysis, the exponentials $e^{i\zeta \cdot t}$ are replaced by localized versions $G_{b,\zeta}(t) = G(t - b)e^{i\zeta t}$. Here G is a *window function*, that is, an L^2-function with $\|G\|_2 = 1$, which we may think of as being centered at 0, and localized in time. The following analogue of Fourier's formula holds for every $f \in L^2(\mathbb{R})$

$$f(t) = \frac{1}{2\pi} \int_{-\infty}^{+\infty} \int_{-\infty}^{+\infty} \langle f, G_{b,\varsigma} \rangle G_{b,\varsigma}(t) \, db \, d\varsigma$$

$$\int_{-\infty}^{+\infty} |f(t)|^2 \, dt = \frac{1}{2\pi} \int_{-\infty}^{+\infty} \int_{-\infty}^{+\infty} |\langle f, G_{b,\varsigma} \rangle|^2 \, db \, d\varsigma \, .$$

The function $\tilde{f}(b, \varsigma) = \langle f, G_{b,\varsigma} \rangle$ is called the *windowed Fourier transform* of f; in view of the above $\frac{1}{2\pi}|\tilde{f}|^2$ is a density energy in the time-frequency plane. In time, $G_{b,\varsigma}$ is thought of as centered at b and with most of its energy in $[b-\sigma, b+\sigma]$, where $\sigma = \sigma(G) = \int t^2 |G(t)|^2 \, dt$; a computation shows that $(G_{b,\varsigma})^\wedge(w) = e^{ib(\varsigma-w)} \hat{G}(w-\varsigma)$; assuming $\hat{G}$ centered at 0 as well, in loose terms one can thus think of the atom $G_{b,\varsigma}$ as occupying in the time-frequency plane a box centered at (b, ς) with sides $\sigma(G)$, $\sigma(\hat{G})$ respectively. The above formulas say that $f \mapsto \tilde{f}$ is an isometry, up to a constant, of $L^2(\mathbb{R})$ onto a closed subspace H_G of $L^2(\mathbb{R}^2)$ and that the family $\{G_{b,\varsigma}\}_{b,\varsigma \in \mathbb{R}}$ is a sort of "doubling continuous" orthonormal basis in $L^2(\mathbb{R})$ because

$$f = \frac{1}{2\pi} \iint \langle f, G_{b,\varsigma} \rangle G_{b,\varsigma} \, db \, d\varsigma \, .$$

The windowed Fourier transform $\tilde{f}$ is of interest in signal analysis because it provides local information about f simultaneously in time and frequency. The "precision" of this information is measured by σ in time and $\hat{\sigma} = \sigma(\hat{G})$ in frequency (because $\tilde{f}(b, \varsigma) = \langle f, G_{b,\varsigma} \rangle = \langle \hat{f}, \hat{G}_{b,\varsigma} \rangle$). The *uncertainty principle* in Harmonic Analysis states that $\sigma\hat{\sigma}$ is bounded below, meaning that it is impossible to make precise both pieces of information simultaneously. It is well known that $\sigma(G)\sigma(\hat{G})$ achieves its minimum value for Gaussian windows. The $G_{b,\varsigma}$ are called Gabor atoms or *Gabor wavelets*.

It is intuitively clear that $\tilde{f}(b, \varsigma)$ being a function of two variables, there exists redundancy in the above representation. It is therefore quite natural to try to discretize this representation and to extract redundancy. In this way we are led to the following questions: for which discrete sequences $\Lambda = \{(b_k, \varsigma_k)\}_{k \in Z}$ of points in $\mathbb{R}^2$ is the family $G(\Lambda)$ of time-frequency atoms $G_k = G_{b_k, \varsigma_k}$ complete, a frame, or a Riesz basis in $L^2(\mathbb{R})$? This is equivalent to the sequence Λ being of uniqueness, of stable sampling or complete interpolating for the space H_G. For some special choices of the window function G the space H_G is isometric to a Hilbert space of entire functions. For instance, for $G(x) = \pi^{-\frac{1}{4}} e^{-\frac{x^2}{2}}$, one gets, with $z = b - i\varsigma$

$$\tilde{f}(b, \varsigma) = e^{-\frac{i}{2}b\varsigma} e^{-\frac{1}{4}|z|^2} Bf(z)$$

where $Bf(z)$ is the entire function (Bargmann transform)

$$Bf(z) = \pi^{-\frac{1}{4}} e^{-\frac{z^2}{4}} \int_{-\infty}^{+\infty} f(t) e^{-\frac{t^2}{2}} e^{tz} \, dt \, .$$

The entire functions $F = Bf$ arising in this way satisfy thus

$$\int |F(z)|^2 e^{-\frac{|z|^2}{2}} \, dA(z) < +\infty$$

and constitute the so called *Bargmann-Fock space*.

For separated sequences $\Lambda = \{(b_k, \zeta_k)\}_{k \in Z}$, the densities to be taken under consideration are again the Beurling-Landau uniform densities

$$D_u^-(\Lambda) = \liminf_{r \to \infty} \frac{n^-(r)}{\pi r^2}, \quad D_u^+(\Lambda) = \limsup_{r \to \infty} \frac{n^+(r)}{\pi r^2}$$

where $n^+(r)$, $n^-(r)$ respectively denote the maximum and minimum number of points of Λ to be found in a disc of radius r. The Landau necessity conditions hold as well in this case for general windows ([27]), that is, a frame $G(\Lambda)$ must satisfy $D_u^-(\Lambda) \geq \frac{1}{2\pi}$ and a free system $G(\Lambda)$ must satisfy $D_u^+(\Lambda) \leq \frac{1}{2\pi}$. In particular, Riesz basis $G(\Lambda)$ can occur only if Λ has a uniform density $\frac{1}{2\pi}$. For a regular lattice Λ of the form $(nb_0, m\zeta_0)_{n,m \in Z}$, this amounts to $b_0\zeta_0 = 2\pi$. But for regular lattices with $b_0\zeta_0 = 2\pi$, the *Balian-Low theorem* establishes that if $G(\Lambda)$ is a frame then either $\sigma(G) = +\infty$ or $\sigma(\hat{G}) = +\infty$. That is, at the critical density, one cannot have a Riesz basis with good localization properties both in time and frequency (see [7]). The situation is thus quite different from the one with band-limited functions.

I do not know whether there exists a version of the Balian-Low theorem for general (separated) sequences. It is known that for windows satisfying

$$\int |x|^{2+\varepsilon} |g(x)|^2 \, dx < +\infty, \quad \int |\zeta|^{2+\varepsilon} |\hat{g}(\zeta)|^2 \, d\zeta < +\infty$$

for some $\varepsilon > 0$, no orthonormal basis of $G_{b,\zeta}$'s exists. One might guess for instance that whenever $\sigma(G) + \sigma(\hat{G})$ is finite (or some other localization property of G, $\hat{G}$) then Landau's necessary conditions can be strengthened to strict inequalities $D_u^-(\Lambda) > \frac{1}{2\pi}$, $D_u^+(\Lambda) < \frac{1}{2\pi}$.

For gaussian windows this is indeed the case, and even the converse is true: $G(\Lambda)$ is a frame if and only if $D_u^-(\Lambda) > \frac{1}{2\pi}$ and a free system if and only if $D_u^+(\Lambda) < \frac{1}{2\pi}$. This has been proved by complex analysis methods, using the Bargmann-Fock model space, in [29] and [30]. Incidentally, $G(\Lambda)$ for a regular lattice with $b_0\zeta_0 = 2\pi$ is an example of a complete system which is not a frame (unstable sampling). Under mild conditions on the window G, it is not hard to see that if $\text{dist}(p, \Lambda) \leq \varepsilon = \varepsilon(G)$ for every $p \in \mathbb{R}^2$, then $G(\Lambda)$ is a frame for $L^2(\mathbb{R})$. In particular a regular lattice $\Lambda = (nb_0, m\zeta_0)_{n,m \in Z}$ works if $b_0\zeta_0$ is small enough (an obviously expected fact in view of the doubly continuous representation above).

In particular, no complete interpolating sequences exist in the Bargmann-Fock space, no Riesz basis of Gabor wavelets exists in $L^2(\mathbb{R})$. This is in strong contrast with the Paley-Wiener space B_π^2 and leads naturally to the following question.

Question. *Which is the property of a Hilbert space of entire functions upon which depends the existence of complete interpolating sequences, that is, the existence of a Riesz basis of Bergman kernels?*

The boundedness properties of the Hilbert transform with respect the space norm seems to be related with the above question.

It must be pointed out that if one replaces the exponentials by suitable cosines and sines then it is possible to obtain an orthonormal basis of $L^2(\mathbb{R})$ (Wilson and Malvar bases) with good time-frequency localization properties (see [7, Section 4.2.2] and the references there).

The results on the Bargmann-Fock space have been successfully generalized to more general spaces F_ϕ^2 of the following type

$$\|F\|_\phi^2 = \|Fe^{-\phi}\|_2^2 = \int_{\mathbb{C}} |F(z)|^2 e^{-2\phi(z)}\, dA(z) < +\infty$$

where ϕ is a subharmonic function in $\mathbb{C}$ satisfying $0 < m \leq \Delta\phi(z) \leq M$. In this case the densities are defined as follows:

$$D_\phi^-(\Lambda) = \liminf_{r \to +\infty} \inf_{z \in \mathbb{C}} \frac{\#(\Lambda \cap D(z,r))}{\int_{D(z,r)} \Delta\phi}$$

$$D_\phi^+(\Lambda) = \limsup_{r \to +\infty} \sup_{z \in \mathbb{C}} \frac{\#(\Lambda \cap D(z,r))}{\int_{D(z,r)} \Delta\phi} .$$

In this setting, a sequence $\Lambda = \{\gamma_k\}_{k \in Z}$ is called of stable sampling if $\|F\|_\phi^2$ is comparable to $\sum_k |F(\gamma_k)|^2 e^{-2\phi(\gamma_k)}$ and interpolating if the interpolation problem $F(\gamma_k) = a_k$ has a solution F with $\|F\|_\phi < +\infty$ for all sequences (a_k) such that $\sum_k |a_k|^2 e^{-2\phi(\gamma_k)} < +\infty$. In the papers [2, 23] it has been proved that a (separated) sequence Λ is interpolating if and only if $D_\phi^+(\Lambda) < \frac{1}{2\pi}$ and stable sampling if and only if $D_\phi^-(\Lambda) > \frac{1}{2\pi}$ (the Bargmann-Fock space corresponds to $\phi(z) = \frac{|z|^2}{4}$ for which $\Delta\phi = 1$).

For certain window functions $G = e^{-\Psi}$ there is a variant of the Bargmann transform allowing us to identify H_G with F_ϕ^2 for some ϕ depending on Ψ; for these cases one has thus a way to obtain strict inequalities.

The results for the spaces F_ϕ^2 have been recently generalized to several variables in [16]. In this case ϕ is a 2-homogeneous, plurisubharmonic function, and the densities $D_\phi^-(\Lambda)$, $D_\phi^+(\Lambda)$ are defined analogously replacing the discs $D(z,v)$ by balls $B(z,r)$ and $\Delta\phi$ by $4^n(i\partial\overline{\partial}\phi)^n$; the result is that a (separated) stable sampling sequence satisfies $D_\phi^-(\Lambda) \geq \frac{1}{(2\pi)^n n!}$, and an interpolating sequence must be separated and satisfy $D_\phi^+(\Lambda) \leq \frac{1}{(2\pi)^n n!}$ (the strict inequalities should hold too). It is worthwhile explaining the reason why $(i\partial\overline{\partial}\phi)^n$ appears instead of $\Delta\phi$. The simple example $\phi(z_1, z_2) = \alpha_1|z_1|^2 + \alpha_2|z_2|^2$ in $\mathbb{C}^2$ and Λ a lattice of the type $\Lambda = \Lambda_1 \times \Lambda_2$, $\Lambda_i = a_i(Z \times iZ)$ will do; if Λ is to be of stable sampling, then both Λ_1 and Λ_2

must be so in one variable, and hence $a_i^{-2} > 2\alpha_i/\pi$. The asymptotic number of points from Λ in a big ball $B(z,r)$ is $\mathrm{vol}(B(z,r))/a_1^2 a_2^2$, which thus exceeds $\mathrm{vol}(B(z,r))\frac{4\alpha_1\alpha_2}{\pi^2}$, an expression which involves $(i\partial\bar{\partial}\phi)^n$ rather than $\Delta\phi$.

The situation is similar to the one discussed before for the Paley-Wiener spaces in several variables and Landau's necessary conditions. Similar considerations show as well that these necessary density conditions can not be sufficient if $n > 1$. It is interesting to note a (formal) connection between the two problems. The Paley-Wiener space B_S^2 is not a space of type F_ϕ^2, but it is "almost it" with $\phi(z) = \varphi_S(\mathrm{Im}\,z)$. For instance, for $S = \prod_{i=1}^{n}[-\tau_i, \tau_i]$, $\phi(z) = \sum_i \tau_i|\mathrm{Im}\,z_i|$; in the distribution sense, $4^n(i\partial\bar{\partial}\phi)^n$ equals $n!2^n\tau_1\tau_2\ldots\tau_n\,dm(x)$ where dm denotes the Lebesgue measure in $\mathbb{R}^n$. For a real sequence $\Lambda \subset \mathbb{R}^n$, one has thus $D_u^+(\Lambda) = 2^n n!\tau_1\tau_2\ldots\tau_n D_\phi^+(\Lambda)$, and we see that the critical value $\frac{1}{(2\pi)^n n!}$ becomes $\frac{2^n\tau_1\ldots\tau_n}{(2\pi)^n} = \frac{m(S)}{(2\pi)^n}$ as in Landau's result. This computation can be seen to hold for a general convex S. As with the Paley-Wiener spaces, one might guess that some condition involving "directional densities" should be necessary.

6. Time-Scale Analysis

In time-scale analysis, the atoms $G_{b,\zeta}$ are replaced by wavelets $\Psi_{b,a}(t) = a^{-1/2}\Psi\left(\frac{t-b}{a}\right)$ obtained by translation and dilation of a (real) fixed function $\Psi \in L^2(\mathbb{R})$ with $\|\Psi\|_2 = 1$ (here $b \in \mathbb{R}$, $a > 0$). If $2\pi\int_0^\infty|\hat{\Psi}(\zeta)|^2|\zeta|^{-1}\,d\zeta = c(\Psi) < +\infty$, then one has again a reconstruction formula

$$f = \frac{1}{c(\Psi)}\int_{-\infty}^{+\infty}\int_0^\infty \langle f, \Psi_{b,a}\rangle\Psi_{b,a}\frac{da}{a^2}\,db$$

with

$$\int_{-\infty}^{+\infty}|f(t)|^2\,dt = \frac{1}{c(\Psi)}\int_{-\infty}^{+\infty}\int_0^\infty |\langle f, \Psi_{b,a}\rangle|^2\frac{da}{a^2}\,db.$$

The function $W_\Psi f(b,a) = \langle f, \Psi_{b,a}\rangle = \int_{-\infty}^{+\infty} f(t)\Psi_{b,a}(t)\,dt$ is called the (continuous) *wavelet transform of f*, and establishes an isometry from $L^2(\mathbb{R})$ to a closed subspace W_Ψ of $L^2\left(\mathbb{R}_+^2, \frac{da\,db}{a^2}\right)$. Analogously as before, it is quite natural to try to discretize this representation and to extract redundancy: for which sequences $\Lambda = \{z_k\}_{k\in Z}$ of points in the upper half space is the family $W\Psi(\Lambda)$ of wavelets Ψ_{t_k} a frame, a Riesz (or orthonormal) basis of $L^2(\mathbb{R})$? This amounts to requiring Λ to be of stable sampling or a complete interpolating sequence for the space W_Ψ (it is precisely in this context that frames came back after their introduction in [8]). The first historical example is the Haar basis, for which $\Psi(x)$ equals 1 in $(0, 1/2)$, -1 in $(1/2, 1)$ and is zero otherwise, and Λ is the grid $(n2^m, 2^m)_{n,m\in Z}$. Notice that a grid of type $(nb_0 a_0^m, a_0^m)_{n,m\in Z}$ with $a_0 > 1$, $b_0 > 0$ is *hyperbolically regular*, that is, the dyadic squares $Q_{n,m} = \{(b,a) : a_0^m < a < a_0^{m+1}, nb_0 a_0^m < b < (n+1)b_0 a_0^m\}$ have constant measure with respect to the hyperbolically invariant measure $\frac{da\,db}{a^2}$ of $\mathbb{R}_+^2$ appearing above; these are the regular lattices $\Lambda(a_0, b_0)$ to be considered here.

Under very general conditions on Ψ (for instance Ψ in the Schwarz class and $\int \Psi(t)\,dt = 0$) it is not hard to see that if a_0, b_0 are small enough then $W\Psi(\Lambda)$ is a frame for the regular lattice $\Lambda = \Lambda(a_0, b_0)$ (see [7]). This holds as well for a general Λ which is hyperbolically separated and such that $\mathrm{dist}_H(p, \Lambda) \leq \varepsilon = \varepsilon(\Psi)$ (hyperbolic distance) is small enough.

If $W\Psi(\Lambda)$ is an orthonormal basis for a regular lattice $\Lambda = \Lambda(a_0, b_0)$ then $c(\Psi) = b_0 \ln a_0$ (see [7, p. 63]). Normalizing $c(\Psi)$, this suggests, in analogy with the Gabor wavelet case, that $b_0 \ln a_0$ could maybe play the same role as $b_0 \zeta_0$ and that some critical density might exist for orthonormal or Riesz bases. Pushing the analogy further one might think as well of a version of the Balian-Low theorem for wavelets, establishing for instance that no orthonormal basis $W\Psi(\Lambda)$ exists for regular Λ and Ψ well localized in time and frequency.

All this turns out to be (fortunately) false. Stromberg and Meyer constructed nice Ψ's for which the $\Psi_{n,m}(t) = 2^{-m/2}\Psi(2^{-m}t - n)$, $n, m \in Z$ constitute an orthonormal basis of $L^2(\mathbb{R})$. Later, the theory of multiresolution analysis (MRA) developed by Mallat and Meyer provided a natural framework and led to a procedure to construct plenty of such bases, even with a regular and compactly supported Ψ. Concerning the existence of a critical density, the Meyer wavelet is a counterexample too, for it can be proved that $W\Psi(\Lambda(2, b))$ is a Riesz basis for all b close enough to 1 (see [7] for all these facts).

But Riesz bases $W\Psi(\Lambda)$ do not always exist. For some mother wavelets Ψ the situation is in fact similar to the Gabor case and a critical density exists. This is so, for instance, for the Poisson wavelet $\Psi(t)$ for which $\hat{\Psi}(\zeta) = c|\zeta|e^{-|\zeta|}$; in this case the space W_Ψ is isometric to the space of all holomorphic functions $F(z)$ of $z = b + ia$ in $\mathbb{R}_+^2$ such that

$$\|F\|_2^2 = \int_{-\infty}^{+\infty} \int_0^{+\infty} a|F(z)|^2\,db\,da < +\infty.$$

This is called a (weighted) Bergman space.

In this isometry, the frames $W\Psi(\Lambda)$ correspond as usual with the stable sampling sequences, which here are those $\Lambda = \{z_k = b_k + ia_k\}$ for which $\|F\|_2^2$ is comparable to $\sum_k |F(z_k)|^2 a_k^3$, and the interpolating sequences (corresponding to free systems $W\Psi(\Lambda)$) are those for which $F(z_k) = \lambda_k$ has a solution for all sequences $(\lambda_k)_k$ such that $\sum_k |\lambda_k|^2 a_k^3 < +\infty$. Seip ([31]) characterized both type of sequences in terms of Beurling-type densities. The (pseudo)hyperbolic distance between $z, w \in \mathbb{R}_+^2$ being $d_H(z, w) = \left|\frac{z-w}{z-\bar{w}}\right|$, for a pseudohyperbolically separated sequence Λ, let $n(z, r)$ be the number of points of Λ in the d_H-disc centered at z of radius r, and let $a(r)$ denote the hyperbolic area of that disc. Writing

$$N(z, r) = \int_0^r n(z, t)\,dt, \qquad A(r) = \int_0^r a(t)\,dt$$

the densities $D_u^+(\Lambda)$, $D_u^-(\Lambda)$ are now defined

$$D_u^+(\Lambda) = \limsup_{r \to 1} \sup_z \frac{N(z,r)}{A(r)}$$

$$D_u^-(\Lambda) = \liminf_{r \to 1} \inf_z \frac{N(z,r)}{A(r)} \,.$$

For a regular lattice $\Lambda = \{nb_0 a_0^m + ia_0^m, n, m \in Z\}$ it turns out that $D_u^+(\Lambda) = D_u^-(\Lambda) = 2\pi/b_0 \ln a_0$; Seip proved that Λ is a stable sampling sequence if and only if $D_u^-(\Lambda) > 1$, and interpolating if and only if $D_u^+(\Lambda) < 1$. In particular, no wavelet bases $W\Psi(\Lambda)$ exist for the Poisson wavelet.

We will finish this section with some comments about wavelet bases in $L^2(\mathbb{R}^n)$ with $n > 1$. One of the basic constructions is the "tensor product technique" leading to separable wavelet bases of $L^2(\mathbb{R}^n)$ (see [7]), but other constructions which do not single out the n axis directions are possible. In the more general context, for a mother wavelet $\Psi \in L^2(\mathbb{R}^n)$, $\|\Psi\|_2 = 1$, a family of wavelets $\{\Psi_{b,M}\}$ is obtained from Ψ replacing the scale parameter a in one dimension by a matrix M in the linear group $GL(\mathbb{R}^n)$:

$$\Psi_{b,M}(x) = |\det M|^{-1/2} \Psi(M^{-1}(x-b)), \quad M \in GL(\mathbb{R}^n), \, b \in \mathbb{R}^n \,.$$

The continuous wavelet transform $f \in L^2(\mathbb{R}^n)$ is then the function $W_\Psi f(b, M) = \langle f, \Psi_{b,M} \rangle = \int_{\mathbb{R}^n} f(x) \Psi_{b,M}(x) \, dx$.

Let now H be a (connected) closed subgroup of $GL(\mathbb{R}^n)$ and let $d\sigma(M)$ denote the left-invariant Haar measure on H. It is not hard to see that if Ψ is H-admissible in the sense that

$$\int_H |\hat{\Psi}(\zeta M)|^2 \, d\sigma(M) = c(\Psi) < +\infty \text{ independently of } \zeta$$

then one has again a reconstruction formula

$$f = \frac{1}{c(\Psi)} \int_H \int_{\mathbb{R}^n} \langle f, \Psi_{b,M} \rangle \Psi_{b,M} \, db \frac{d\sigma(M)}{|\det M|} \,.$$

The subgroup H is called *admissible* if some H-admissible Ψ exists. For instance, the whole $H = GL(\mathbb{R}^n)$ is not if $n > 1$; in general, if H is transitive acting on $\mathbb{R}^n \backslash \{0\}$, H is admissible if and only if the isotropy group $S_H(x) = \{M : Mx = x\}$ is compact for all $x \neq 0$. More generally, if H has an open orbit U such that $S_H(x)$ is compact for $x \in U$, H is admissible (see [4] and [10]). Of course, other examples of admissible groups, with no open orbits, are known. For instance for $H = \mathbb{R} \, \text{Id}$, H-admissible functions exist (necessarily radial).

The question of discretizing the continuous representation, that is, sampling it on a discrete $\Lambda \subset \mathbb{R}^n \times H$ is very little studied, as far as I know. When H has an open orbit U with $S_H(x)$ trivial, then H must have dimension n, as a Lie group. For $n = 2$ there are, up to conjugation, four connected subgroups of $GL(\mathbb{R}^2)$ of dimension 2: the diagonal matrices, the triangular ones, and

$$H = \left\{ \begin{pmatrix} a & 0 \\ b & a^\varepsilon \end{pmatrix}, a, b \in \mathbb{R},\ a \neq 0 \right\}$$

$$H = \left\{ \begin{pmatrix} a & b \\ -b & a \end{pmatrix}, a, b \in \mathbb{R},\ ab \neq 0 \right\}.$$

For all of them it is possible to choose an admissible function Ψ (with Fourier transform $\hat{\Psi}$ being an indicator function) and a discrete subgroup H_0 of H such that the $\Psi_{b,M}$ with $M \in H_0$ and $b \in M(Z^n)$ form an orthonormal basis of $L^2(\mathbb{R}^n)$.

Acknowledgements

The author thanks Joaquim Ortega-Cerdà for his help in the preparation of this survey.

References

[1] B. Berndtsson, *Zeros of analytic functions in several variables*, Ark. Mat. **16** (1978), 251–262.

[2] B. Berndtsson and J. Ortega-Cerdà, *On interpolation and sampling in Hilbert spaces of analytic functions*, J. Reine Angew Math. **464** (1995), 109–128.

[3] B. Berndtsson, *An inequality for Fourier-Laplace transforms of entire functions, and the existence of exponential frames in Fock space*, J. Funct. Anal. **149** (1997), 83–101.

[4] D. Bernier and K. F. Taylor, *Wavelets from square-integrable representations*, Siam J. Math. Anal. **27(2)** (1996), 594–608.

[5] L. Carleson, P. Malliavin, J. Neuberger and J. Wermer, editors, "The Collected Works of Arne Beurling", vol. 2, Birkhäuser, 1989, 341–365.

[6] L. de Branges, "Hilbert spaces of entire functions", Prentice Hall Inc. Englewood Cliffs, N.J., 1968.

[7] I. Daubechies, "Ten Lectures on Wavelets", SIAM, Philadelhia, Pennsylvania, 1992.

[8] R. J. Duffin and A. C. Schaeffer, *A class of nonharmonic Fourier series*, Trans. Amer. Math. Soc. **72** (1952), 341–366.

[9] B. Fuglede, *Commuting self-adjoint partial differential operators and a group theoretic problem*, J. Funct. Anal. **16** (1974), 101–121.

[10] H. Führ, *Discrete and semidiscrete wavelet transforms in higher dimensions*, preprint 1999.

[11] S. V. Khrushchev, N. K. Nikol'skii and B. S. Pavlov, *Unconditional bases of exponentials and reproducing kernels*, in: "Complex Analysis and spectral theory", Lecture Notes in Math. vol. **864**, Springer-Verlag, Berlin-Heidelberg, 1981, 214–335.

[12] P. Koosis, *Leçons sur le théorème de Beurling et Malliavin*, Les Publications CRM, Montreal, 1996.

[13] I. Laba, *Fuglede's conjecture for a union of two intervals*, preprint Princeton Univ.

[14] H. J. Landau, *A sparse regular sequence of exponentials closed on large sets*, Bulletin A.M.S. **70** (1964), 566–569.

[15] H. J. Landau, *Necessary density conditions for sampling and interpolation of certain entire functions*, Acta Math. **117** (1967), 37–52.

[16] N. Lindholm, *Sampling in weighted L^p spaces of entire functions in $\mathbb{C}^n$ and estimates of the Bergman kernel*, preprint Göteborg University.

[17] Y. Lyubarskii and K. Seip, *Sampling and interpolating sequences for multiband-limited functions and exponential bases on disconnected sets*, J. of Fourier Analysis and Applications **3(5)** (1997), 599–615.

[18] Y. Lyubarskii and K. Seip, *Weighted Paley-Wiener spaces*, preprint 1999.

[19] Y. Lyubarskii and A. Rashdovskii, *Complete interpolating sequences for Fourier transforms supported by convex symmetric polygons*, to appear in Arkiv f. Math.

[20] P. McMullen, *Convex bodies which tile space by translation*, Mathematika **27** (1980), 113–121.

[21] A. M. Minkin, *Reflection of exponents, and unconditional bases of exponentials*, St. Petersburg Math. J. **3** (1992), 1043–1068.

[22] N. K. Nikol'skii, *Bases of exponentials and the values of reproducing kernels*, Dokl. Akad. Nauk SSSR **252** (1980), 1316–1320; english transl. in Sov. Math. Dokl. **21** (1980).

[23] J. Ortega-Cerdà and K. Seip, *Beurling-type density theorems for weighted L^p spaces of entire functions*, J. Analyse Mathematique **75** (1998), 247–266.

[24] J. Ortega-Cerdà and K. Seip, *Multipliers for entire functions and an interpolation problem of Beurling*, J. Funct. Anal. **162** (1999), 400–415.

[25] J. Ortega-Cerdà and K. Seip, *On Fourier Frames*, preprint 2000.

[26] B. S. Pavlov, *Basicity of an exponential system and Muckenhoupt's condition*, Dokl. Akad. Nauk SSSR **247** (1979), 37–40; english transl. in Sov. Math. Dokl. **20** (1979).

[27] J. Ramanathan and T. Steger, *Incompleteness of sparse coherent states*, Applied and Computational Harmonic Analysis **2** (1995), 148–153.

[28] L. I. Ronkin, *On discrete uniqueness sets for entire functions of exponential type in several variables*, Sib. Mat. Zh. **19(1)** (1971), 142–152; english transl. in Sib. Math. J. **19** (1978), 101–108.

[29] K. Seip, *Density theorems for sampling and interpolation in the Bargmann-Fock space I*, J. Reine Angew Math. **429** (1992), 91–106.

[30] K. Seip and R. Wallstén, *Density theorems for sampling and interpolation in the Bargmann-Fock space II*, J. Reine Angew Math. **429** (1992), 107–113.

[31] K. Seip, *Beurling type density theorems in the unit disk*, Invent. Math. **113** (1993), 21–39.

[32] K. Seip, *On the connection between exponential bases and certain related sequences in $L^2(-\pi, \pi)$*, J. Funct. Anal. **130** (1995), 131–160.

[33] A. Ulanovskii, *Sparse systems of functions closed on large sets in $\mathbb{R}^n$*, to appear in Proc. London Math. Soc.

Departament de Matemàtiques
Universitat Autònoma de Barcelona
08193 Bellaterra (Barcelona), Spain
E-mail address: bruna@mat.uab.es

Lower Bounds for Shape Resonances Widths of Schrödinger Operators

Nicolas Burq

Abstract. In this lecture we present some results giving general lower bounds of shape resonance widths near positive energy levels in the semi-classical limit for Schrödinger operators in the exterior of smooth compact obstacles with Dirichlet or Neuman boundary conditions and with long range dilation analytic potentials. These lower bounds are exponentially small with respect to the Planck constant. We also give some consequences of these lower bounds on the asymptotic behaviour in large time of solutions of wave equations.

1. Introduction

The purpose of this talk is the study of the localization of resonances for the semi-classical Schrödinger operator near positive energy levels. The shape resonances are the metastable states of a system whose evolution is described by a Hamiltonian, H, depending upon the Planck constant h. An important feature of the resonances is their lifetime (the reverse of their width) which describes the time needed by a wave packet associated to the resonance and initially localized in a compact region of the space, to develop a nonzero probability density in the outside region. In this paper we shall study the semi-classical approximation when the planck constant tends to 0. Most results about the shape resonances are concerned about the *existence* of resonances and *upper* bounds for their width (see the works by J. M. Combes, P. Duclos, M. Klein and R. Seiler [4], P. D. Hislop and I. M. Sigal [9] and by B. Helffer and J. Sjöstrand [8]). However, in the case of trapped energy levels, there are very few results about *nonexistence* of resonances or more generally about *lower* bounds for the width of shape resonances. The main result we want to present in this talk is the following:

Theorem 1.1. *Consider a long range $(\mathcal{O}\left(1/|x|^{\beta}\right), \beta > 0)$ self-adjoint analytic dilation perturbation of the semi-classical Laplace operator $-h^2\Delta$ (satisfying the assumptions [1 ... 4] of section 2). Then for any $0 < E_m < E_M < +\infty$, there exists $C, d, h_0 > 0$ such that for any $0 < h < h_0$, the operator $P(h)$ has no resonance in the set*

$$\{z \in \mathbb{C};\ \operatorname{dist}(z, [E_m, E_M]) \leq Ce^{-d/h}\}. \tag{1}$$

The text is organized as follows: in Section 2 we give the assumptions and define the resonance in the complex scaling framework. The material in this part is due to J. Sjöstrand and M. Zworski. In Section 3 we give the geometric properties which allow us to distinguish between the trapping and non-trapping cases and present the main results about pole-free regions in the non-trapping and trapping cases. For the trapping case, most results are due in successive generality to B. Vainberg [19], P. Lax and R. Phillips [10], C. Morawetz [15, 16] (for Helmoltz equation) and R. Melrose and J. Sjöstrand [14] (see also S. H. Tang and M. Zworski [18] for an abstract result and N. Burq [3] for an elementary and natural proof).

To our knowledge, the only known results in the trapping case (the general case) were particular cases: E. M. Harrel [7] gave lower bounds for a Schrödinger operator in space dimension 1 equal to $-\partial_x^2 + V(x)$ with a compactly supported potential. In space dimension $d \geq 1$, B. Helffer and J. Sjöstrand [8] gave a complete asymptotic expansion of the resonances for analytic dilation potential with a unique non-degenerate local minimum and energy near the bottom of the well. Later on C. Fernandez and R. Lavine [5] extended the results by E. Harrel to the case of space dimension $d \geq 1$ and potential with compact support. Finally the author [2] has given such lower bounds for the Helmoltz operator (null potential) outside any obstacle (whose complement is connected), with Dirichlet conditions on the boundary. In fact, the methods developed in [2] apply to any compactly supported smooth self-adjoint perturbations of the Laplace operator (in the semi-classical framework). These results have recently been extended by G. Vodev, using the same kind of methods, to super-exponentially decaying perturbations of the Laplace operator. In Section 4 we give some by-products of our work about the decay of the local energy of solutions of the wave equation in an asymptotically flat manifold. Finally in Section 5 we give an outline of the proof of our results. The strategy of the proof relies on three main steps: Carleman estimates where the perturbation is large, some dissipation estimates related to analytic dilation close to infinity and some Carleman–Mourre inequalities in between. Carleman estimates are due to L. Hörmander for semi-classical second order boundary value problems in the interior. For boundary value problems these results are, in the context of the Laplace operator, due to G. Lebeau and L. Robbiano [13, 12]. Carleman inequalities are robust enough to be performed where our operator is not close to the standard Laplace operator (in any bounded set). The idea in [2] was to use the outgoing behaviour close to infinity to perform explicit computations on the standard Laplace operator. Here, since the operator we consider is no longer equal to the Laplace operator close to the infinite, these computations are much more difficult and the idea developed is to switch from Carleman inequalities (subelliptic inequalities with loss of 1/2 derivative) to Mourre type estimates (positive commutators with loss of 1 derivative) close to the infinite; and then to perform analytic dilation (see Section 2) and conclude by means of elliptic estimates (see Section 4).

2. Assumptions, Resonances

2.1. Assumptions

Let $\Theta \subset \mathbb{R}^d$ be a smooth obstacle, such that $\Omega = \Theta^c$ is connected. Consider $P(h)$ a *self-adjoint* operator on $L^2(\Omega)$ depending upon the constant h, $0 < h < 1$ (with $(D_{x_i} = \frac{1}{i}\partial_{x_i})$):

$$P = -h^2 \sum_{i,j=1}^{d} a_{i,j}(x)\partial_{x_i}\partial_{x_j} + h \sum_{i=1}^{d} (b_i(x,h)) D_{x_i} + V(x,h) \qquad (2)$$

with boundary conditions $(\partial\Omega = \Gamma_D + \Gamma_N, \overline{\Gamma_D} \cap \overline{\Gamma_N} = \emptyset)$

$$u\mid_{\Gamma_D} = 0, \qquad \frac{\partial u}{\partial n}\mid_{\Gamma_N} + b(x)\, u\mid_{\Gamma_N} = 0$$

where $b \in C^\infty(\Gamma_N)$ and $\frac{\partial}{\partial n}$ is the outgoing normal vector field to the boundary.

The assumptions we make about the coefficients of the operator P are the following:

1. **(Smoothness.)** The functions $a_{i,j}$, b_i, V and their derivatives are uniformly with respect to $0 < h < 1$ bounded for the $C^\infty(\mathbb{R}^d)$ topology (in any compact set).

2. **(Analytic Dilation.)** There exists $\theta_0, \varepsilon > 0$ and $R_0 > 0$ such that the functions $a_{i,j}$, a_i, b_i, V and W are analytic in x in the domain

$$\Lambda_{\theta_0,R_0,d_0} = \left\{ r\omega; \; \omega \in \mathbb{C},\, \mathrm{dist}\left(\omega, \mathbb{S}^{d-1}\right) < d_0, \right.$$
$$\left. r \in \mathbb{C},\, |r| > R_0,\, \arg(r) \in [-\theta_0, \theta_0] \right\}. \qquad (3)$$

3. **(Ellipticity.)** The operator P is supposed to be elliptic

$$\exists C > 0;\; \forall x \in \Omega,\, \forall \xi \in \mathbb{R}^d \sum_{i,j=1}^{d} a_{i,j}(x)\,\xi_i\xi_j \geq C|\xi|^2. \qquad (4)$$

4. **(Asymptotic flatness.)** The operator P is close to $-\Delta$ at infinity:

$$\exists \beta > 0,\, M > 0;\; \forall x \in \Lambda_{\theta_0,R_0,d_0}$$
$$|a_{i,j}(x) - \delta_{i,j}| + |b_i(x,h)| + |V(x,h)| \leq \frac{M}{|x|^\beta}. \qquad (5)$$

Remark 2.1. *The assumptions above are satisfied if the operator is a* smooth *and* compactly supported *perturbation of* $h^2 \times$ *the Laplacean outside an obstacle.*

Remark 2.2. *According to the Cauchy formula, we obtain from* (5),

$$\exists \beta > 0,\, \forall \alpha \in \mathbb{N}^d \setminus \{0\},\, \exists M_\alpha > 0;\; \forall x \in \Lambda_{\theta_0/2,R_0,d_0/2}$$
$$|\partial_x^\alpha(a_{i,j})(x)| + |\partial_x^\alpha(b_i)(x,h)| + |\partial_x^\alpha(V)(x,h)| \leq \frac{M_\alpha}{|x|^{\beta+|\alpha|}}. \qquad (6)$$

2.2. Resonances

Note for $e \in \mathbb{C}^- = \{z \in \mathbb{C}; \operatorname{Im} z < 0\}$

$$R(e, h) = (P - e)^{-1}$$

the resolvent such that

$$(P(h) - e)R(e, h)f = f \in L^2(\Omega)$$
$$R(e, h)f \mid_{\Gamma_D} = 0, \qquad \partial_N R(e, h)f \mid_N = 0.$$

Since the operator is self-adjoint, the family

$$e \in \mathbb{C}^- \mapsto R(e, h)$$

is an analytic function with values in $\mathcal{L}(L^2(\Omega))$.

Proposition 2.3. *The family $e \mapsto R(e, h)$ has a meromorphic extension to*

$$\mathbb{C}^-_{\theta_0} = \{z \in \mathbb{C}; z = re^{i\theta}; \theta \in] - \pi - 2\theta_0, 2\theta_0[\}$$

(as operators from L^2_{comp} to L^2_{loc}):
For any $\chi \in C_0^\infty(\mathbb{R}^d)$, the family

$$e \in \mathbb{C}^-_{\theta_0} \mapsto \chi(x)R(e, h)\chi(x)$$

is meromorphic and the poles of this family are independent with respect to χ (provided that $\chi = 1$ on a sufficiently large set).

Definition 2.4. *The resonances are the poles of the meromorphic extension of $e \mapsto \chi(x)R(e, h)\chi(x)$. We denote by $\mathrm{Res}(P, h)$ the set of resonances.*

Remark 2.5. *It is possible to show that if P satisfies the assumptions above, then there is no resonance of real positive energy (i.e., there is no resonance in $\mathbb{R}^{*,+}$).*

2.3. Analytic dilation

The analytic dilation framework gives an alternative way to define the resonances. Consider $f_\theta(t) : [0, \theta_0] \times [0, +\infty[\to \mathbb{C}$, injective for any θ with the following properties:

(i) $f_\theta(t) = t$ for $0 \le t \le R_1$.
(ii) $f_\theta(t) = e^{i\theta}t$ for $t \ge T_0$.
(iii) $\arg(f_\theta(t))$ is an increasing function of θ and t.

Consider the map

$$\kappa_{\theta_0} : \mathbb{R}^d \ni x = t\omega \mapsto f_{\theta_0}(t)\,\omega \in \mathbb{C}^d, \, t = |x|. \tag{7}$$

Then the image of κ_{θ_0}, Γ_{θ_0} is a totally real manifold which coincides with $\mathbb{R}^d$ along $B(0, R_1)$.

Define

$$L_{\theta_0}^2 = L^2(\Gamma_{\theta_0} \setminus \Theta) \tag{8}$$

$$\mathcal{H}_{0,\theta_0} = \{u \in H^1(\Gamma_{\theta_0})\,;\, u\,|_{\Gamma_D} = 0\} \tag{9}$$

$$\mathcal{D}_{\theta_0} = \left\{u \in H^2(\Gamma_{\theta_0})\,;\, u\,|_{\Gamma_D} = 0,\ \left(\frac{\partial u}{\partial n} + b(x)u\right)|_{\Gamma_N} = 0\right\}. \tag{10}$$

Let us consider the operator $P_{\theta_0}(h)$ acting on $L^2(\Gamma_{\theta_0} \setminus \Theta)$ with domain $\mathcal{D}_{\theta_0}$ which inside $B(0, R_1)$ coincides with $P(h)$ and outside is (in the polar coordinates coming from κ_{θ_0}) equal to

$$P_{\theta_0}(h) = P\left(f_{\theta_0}(t)\,\omega,\ f_{\theta_0}'(t)\,D_t,\ D_\omega\right). \tag{11}$$

Proposition 2.6. (Sjöstrand) *The resonances are the eigenvalues of the operator P_{θ_0}.*

Remark 2.7. *It was also shown by J. Sjöstrand that if $\chi \in C_0^\infty(\mathbb{R}^d)$ is supported in the set $\{x;\, f_\theta(|x|) = |x|\}$, then*

$$\chi R(e,h)\chi = \chi(P_\theta - e)^{-1}\chi. \tag{12}$$

3. Geometry, Results

3.1. Geometry

Denote by

$$p_e(x,\xi) = \sum_{i,j} a_{i,j}(x)\xi_i\xi_j + \sum_i b_i(x,h=0)\xi_i + V(x,h=0) - e \tag{13}$$

the semi-classical principal symbol of the operator P.

Definition 3.1. *The geometry is called* non-trapping *at energy e if the x-projection of any integral curve on the surface $p_e = 0$ of*

$$H_{p_e} = \partial_\xi p_e \partial_x - \partial_x p_e \partial_\xi,$$

the hamiltonian vector field of p, reflecting on the boundary of Ω according to the law of geometric optics leaves any compact set in finite time. The geometry is called trapping *at energy e otherwise (which by a compactness argument is equivalent to the existence of a periodic integral curve of p_e).*

3.2. The results

3.2.1. THE NON-TRAPPING CASE   The following result is in its most general setting a consequence of the results on propagation of singularities.

Theorem 3.2. (Vainberg, Morawetz, Lax-Phillips, ... , Melrose-Sjöstrand) *For any compact set $K = [a,b] \subset\,]0, +\infty[$ and any $N \in \mathbb{N}$ there exists $h_0 > 0$ such that for any $0 < h < h_0$ there is no resonance in the set*

$$[a, b] + i[0, +(Nh\log(1/h))]$$

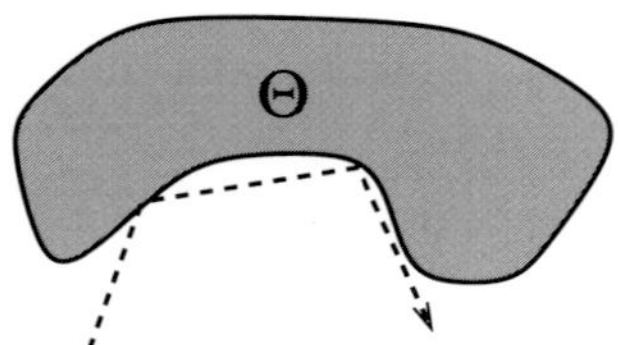

FIGURE 1. Non-trapping geometry $(a_{i,j} = \delta_{i,j} \; b = V = 0)$

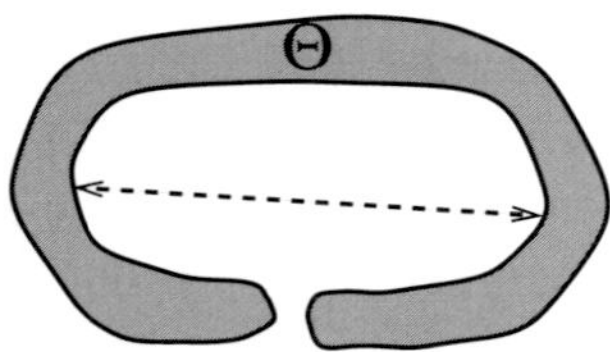

FIGURE 2. Elliptic trapping geometry $(a_{i,j} = \delta_{i,j} \; b = V = 0)$

and the truncated resolvent $R(e)$ satisfies for any e in the set above the estimate

$$\|\chi(x)R(e,h)\chi(x)\|_{\mathcal{L}(L^2(\Omega))} \leq C_N h^{-1} e^{C \operatorname{Im} e/h} . \tag{14}$$

Remark 3.3. *In some special cases (analyticity or convexity assumptions), it is possible to get a better result than above (see Lebeau [11] and C. Bardos, G. Lebeau, J. Rauch [1] for analyticity assumptions and T. Hargé–G. Lebeau [6], J. Sjöstrand–M. Zworski [17] for convexity conditions).*

3.2.2. THE GENERAL CASE   In this case it is known (see the works by Stefanov–Vodev, Tang–Zworski and Stefanov) that in general there exists resonances that are *exponentially close to the real axis*: $\exists C > 0$ and $0 < h_n$ such that $\lim_{n \to +\infty} h_n = 0$ and

$$\operatorname{Res}(P, h_n) \cap [a, b] + i]0, e^{-C/h_n}] \neq \emptyset$$

(this result gives an upper bound for the width of some resonances). The following result gives a similar general lower bound:

Theorem 3.4. (Burq, 2000) *Consider a long range $(\mathcal{O}\left(1/|x|^{\beta}\right), \beta > 0)$ self-adjoint analytic dilation perturbation of the semi-classical Laplace operator $-h^2 \Delta$ (satisfying the assumptions [1 ... 4] of Section 2). For any $K = [a,b] \subset]0, +\infty[$ there exists $h_0 > 0$ and $C_0 > 0$ such that for any $0 < h < h_0$ there is no resonance in the set*

$$[a, b] + i[0, e^{-C_0/h}]$$

and the truncated resolvent $R(e, h)$ satisfies for any e in the set above the estimate

$$\|\chi(x)R(e,h)\chi(x)\|_{\mathcal{L}(L^2(\Omega))} \leq C_\chi e^{C_0/h} . \tag{15}$$

(Even if there are resonances which are exponentially close to the real axis, they cannot be closer than exponentially.)

3.2.3. Comments Such results as in Theorem 3.4 were previously known in particular cases:

1. For a Schrödinger operator in space dimension 1 equal to $-\partial_x^2 + V(x)$ with a compactly supported potential, E. M. Harrel [7] gave lower bounds.
2. In space dimension $d > 1$, for analytic dilation potential with a unique non-degenerate local minimum and energy near the bottom of the well, B. Helffer and J. Sjöstrand [8] gave a complete asymptotic expansion of the resonances.
3. In space dimension $d > 1$, C. Fernandez and R. Lavine [5] gave lower bounds for potentials with compact supports.
4. In space dimension $d > 1$, N. Burq [2] gave lower bounds for a potential equal to 0, and the Laplacean with Dirichlet boundary conditions outside any obstacle. In fact, the methods developed in [2] apply for any *compactly supported perturbation* of the semi-classical Laplace operator and the result has been recently extended by G. Vodev [20] to super exponentially decaying perturbations

$$|V(x, h)| \leq C_\chi e^{-(2+\varepsilon)|x^2|},$$

using the same kind of methods.

4. Some Remarks and Consequences of The Lower Bound

4.1. Estimates for the resolvent truncated in annulus

A by-product of the proof of Theorem 3.4 is that, in the particular case of a cut-off function $\chi \in C_0^\infty(\mathbb{R}^d)$ equal to 0 on a sufficiently large ball (i.e., χ is supported in a ring with large internal diameter), we have for any $e \in [a, b] + i]0, e^{-C_0/h}]$, the estimate

$$\|\chi(x)R(e, h)\chi(x)\|_{\mathcal{L}(L^2(\Omega))} \leq C_\chi h^{-1} \tag{16}$$

($e^{C/h}$ is replaced by h^{-1} in the estimate).

Remark 4.1. *The estimate (16) for the resolvent truncated in* rings *in the trapping case is the same as the estimate (14) for the resolvent truncated in* balls *in the non-trapping case.*

The physical meaning of this result is roughly speaking that the essential part of the energy of the resonant states is distributed in a compact set (the trapped set).

4.2. Local energy decay for solutions of wave equations

Consider a smooth (positive definite) metric $g = (g_{i,j})_{i,j=1,\ldots,d}$ in an unbounded connected smooth domain $\mathbb{R}^d \ni \Omega = \Theta^c$ satisfying the *smoothness, analyticity at infinity, ellipticity and asymptotic flatness* assumptions above. Consider for $(u_0, u_1) \in C_0^\infty(\Omega)$, $u(t)$ the solution of the wave equation:

$$\left(\partial_t^2 - \sum_{i,j=1}^d \frac{1}{\sqrt{\det g}} \partial_{x_i} \sqrt{\det g}\, g^{i,j}(x)\, \partial_{x_j} \right) u = 0$$

$$u \mid_{\Gamma_D} = 0$$

$$\frac{\partial u}{\partial n} \mid_{\Gamma_N} + b(x) u \mid_{\Gamma_N} = 0 \tag{17}$$

$$u \mid_{t=0} = u_0$$

$$\partial_t u \mid_{t=0} = u_1$$

with ∂_n the normal derivative associated to the metric g and $(g^{i,j}) = (g_{i,j})^{-1}$.

A consequence of estimates (15) and (16) on the asymptotic behaviour in large time of solutions of wave equations is the following:

Theorem 4.2. *For any $k \in \mathbb{R}^+$ and any $R > 0$ there exists a constant $C > 0$ such that for any initial data, (u_0, u_1) with support in the ball $\Omega \cap B(0, R)$ and any $t > 1$ we have*

$$\|u(t)\|_{H^1(\Omega \cap B(0,R))} + \|\partial_t u(t)\|_{L^2(\Omega \cap B(0,R))}$$

$$\leq \frac{C}{(\log t)^k} \left[\|u_0\|_{H^{1+k}} + \|u_1\|_{H^k} \right]. \tag{18}$$

There exists $R_0 > 0$ such that for any $R > R_0$ there exists a constant $C > 0$ such that for any initial data, (u_0, u_1) with support in the ring $\Omega \cap (B(0, R) \setminus B(0, R_0))$ we have

$$\int_0^{+\infty} \|u(t)\|_{H^1(\Omega \cap B(0,R) \setminus B(0,R_0))}^2 + \|\partial_t u(t)\|_{L^2(\Omega \cap B(0,R) \setminus B(0,R_0))}^2$$

$$\leq C \left[\|u_0\|_{H^1}^2 + \|u_1\|_{L^2}^2 \right]. \tag{19}$$

5. Ideas for The Proof of Theorem 3.4

The strategy of the proof of Theorem 3.4 is to prove the estimates directly on the global resolvent of the operator P_{θ_0}. We prove the following:

Theorem 5.1. *For any $K = [a, b] \subset]0, +\infty[$ there exists $h_0 > 0$ and $C_0 > 0$ such that for any $0 < h < h_0$ and any $e \in K$ the operator $P_{\theta_0} - e$ is invertible and we have*

$$\|(P_{\theta_0} - e)^{-1}\|_{\mathcal{L}(L^2(\Omega))} \leq C e^{C_0/h}. \tag{20}$$

Remark 5.2. *Remark that estimate (20) for $e \in K$ imply by a perturbation argument a similar estimate (with different C) for $e \in K + i[0, \varepsilon e^{-C_0/h}]$.*

Remark 5.3. *Remark also that (12) (and the remark above) show that Theorem 5.1 implies Theorem 3.4.*

To deal with what happens in a large ball (containing the principal part of the perturbation), we use Carleman Estimates. Then we show that it is possible to make a transition to positive commutator estimates (Mourre type estimates) in a larger ball. Finally, we conclude using the fact that the operator is (after the analytic dilation) elliptic near infinity and we can use elliptic estimates.

5.1. Carleman estimates

Consider $\varphi \in C^\infty(\mathbb{R}^d)$. The operator P_φ^e is defined by

$$P_\varphi^e = e^{\varphi/h} P^e e^{-\varphi/h} .$$

His principal symbol is $p_\varphi^e = p^e(x, \xi + i\varphi_x')$. Suppose that φ satisfies the hypoellipticity assumptions of Hörmander:

$$\exists c > 0; \ \forall x \in \overline{\Omega} \ \forall \xi \in \mathbb{R}^d, \ p_\varphi^e(x, \xi) = 0 \Rightarrow \{\mathcal{R}e\, p_\varphi^e, \operatorname{Im} p_\varphi^e\} \geq c \qquad (21)$$

where $\{f, g\} = \sum_i \partial_{\xi_i} f \partial_{x_i} g - \partial_{x_i} f \partial_{\xi_i} g$ is the Poisson bracket of f and g. Suppose also that

$$\frac{\partial \varphi}{\partial n} \big|_{\partial\Omega} \neq 0 \text{ and } \nabla\varphi \neq 0 \text{ in } \overline{\Omega} . \qquad (22)$$

Proposition 5.4. (Lebeau–Robbiano) *There exists $C > 0$, $h_1 > 0$, such that for any $0 < h < h_1$ and any $g \in C^\infty(\overline{\Omega})$*

$$\int_\Omega |P_\varphi^e(g)|^2 + h \int_{\partial\Omega} \left(|g\,|_{\partial\Omega}\,|^2 + |h\partial_{x'} g\,|_{\partial\Omega}\,|^2 + |h\partial_{x_n} g\,|_{\partial\Omega}\,|^2\right)$$
$$\geq Ch \int_\Omega \left(|g|^2 + |h\nabla g|^2\right) . \quad (23)$$

Proposition 5.5. (Lebeau–Robbiano) *There exists $C > 0$, $h_1 > 0$, such that for any $0 < h < h_1$ and any $g \in C^\infty(\overline{\Omega})$, $\Gamma = \Gamma_D \cup \Gamma_N \subset \partial\Omega$ a union of connected components of the boundary $\partial\Omega$. Suppose that*

$$\frac{\partial \varphi}{\partial n} < 0 \ on \ \Gamma$$

($\frac{\partial}{\partial n}$ is the outgoing normal unit on the boundary), then there exists $c > 0$, $h_1 > 0$, such that for any $0 < h < h_1$ and any $g \in C^\infty(\overline{\Omega})$ such that $g\,|_{\Gamma_D} = 0$, $(\partial_n g + b(x)g)\,|_{\Gamma_N} = 0$

$$\int_\Omega |P_\varphi^e(g)|^2 + h \int_{\partial\Omega\backslash\Gamma} \left(|g\,|_{\partial\Omega}\,|^2 + |h\partial_{x'} g\,|_{\partial\Omega}\,|^2 + |h\partial_{x_n} g\,|_{\partial\Omega}\,|^2\right)$$
$$\geq Ch \int_\Omega \left(|g|^2 + |h\nabla g|^2\right) . \quad (24)$$

5.2. Carleman–Mourre estimates

Proposition 5.6. *There exists $0 \ll R_1 < R_2$ and a phase function $\varphi(r, h)$ such that $\partial_r \varphi(r, h) \geq c$ for r close to R_1, $\partial_r \varphi \leq Ch$ close to R_2, $\partial_r \varphi(r, h) \geq ch$ for $r \geq R_1$;*

and $h_0 > 0$ such that for any $0 < h < h_0$, $e \in [1,2]$ and any $u \in \mathcal{D}_\theta$, if $v = 0$ for $|x| \notin [R_1, R_2]$,

$$2\|(P_\varphi - e)v\|_{L^2}^2 \geq \|\varphi' h \partial_r v\|_{L^2}^2 + h \left(\frac{\varphi'}{r} \frac{h\nabla_\sigma}{r} v, \frac{h\nabla_\sigma}{r} v \right)_{L^2} + \frac{h}{2} \left(\frac{\varphi'}{r} v, v \right)_{L^2}. \quad (25)$$

Remark 5.7. *The estimate* (25) *is a subelliptic estimate with loss of* $1/2$ *derivative close to* R_1 *and loss of* 1 *derivative close to* R_2.

5.3. Elliptic estimates

Suppose that the analytic dilation has been performed for $| x | > B = R_2 - 1$.

Proposition 5.8. *For any* $\delta > 0$ *there exists* h_0 *such that any* $0 < h < h_0$ *and* $u \in \mathcal{D}_{\theta_0}$

$$- \operatorname{Im} \left(e^{i\kappa_{\theta_0}(r)} (P_{\theta_0} - e) u, u \right)_{L^2} \geq \frac{1}{2} \int (f_{\theta_0}(r) + r f_{\theta_0}(r)') |h\partial_r u|^2$$
$$+ \frac{1}{2} \int f_{\theta_0}(r) (|\frac{h\nabla_\sigma}{r} u|^2 + |u|^2) dr dL - h^2 \varepsilon(\delta) \int_{B \leq r \leq B+\delta} |u|^2 \quad (26)$$

with

$$\lim_{\delta \to 0} \varepsilon(\delta) = 0. \quad (27)$$

The proof of Proposition 5.8 is a simple integration by parts. The proof of Theorem 1.1 is obtained by glueing together Carleman estimates for $|x| \leq R_1$ Carleman–Mourre estimates for $|x| \in [R_1, R_2]$ and elliptic estimates for $|x| \geq R_2 + 1$.

References

[1] C. Bardos, G. Lebeau, and J. Rauch. Scattering frequencies and Gevrey 3 singularities. *Inventiones Mathematicae*, 90:77–114, 1987.

[2] N. Burq. Décroissance de l'énergie locale de l'équation des ondes pour le problème extérieur et absence de résonance au voisinage du réel. *Acta Mathematica*, 180:1–29, 1998.

[3] N. Burq. Semi-classical estimates for the resolvent in non-trapping geometries. *Préprint*, 2000.

[4] J. M. Combes, P. Duclos, M. Klein, and R. Seiler. The shape resonance. *Communications in Mathematical Physics*, 110(2):215–236, 1987.

[5] C. Fernandez and R. Lavine. Lower bounds for resonance width in potential and obstacle scattering. *Communications in Mathematical Physics*, 128:263–284, 1990.

[6] T. Hargé and G. Lebeau. Diffraction par un convexe. *Inventiones Mathematicae*, 118:161–196, 1994.

[7] E. Harrel. General lower bounds for resonances in one dimension. *Communications in Mathematical Physics*, 86:221–225, 1982.

[8] B. Helffer and J. Sjöstrand. Resonances en limite semi-classique. *Mémoire de la S.M.F.*, 114(24–25), 1986.

[9] P. D. Hislop and I. M. Sigal. Semiclassical theory of shape resonances in quantum mechanics. *Memoirs of the American Mathematical Society*, 78(399):123, 1989.

[10] P. D. Lax and R. S. Phillips. *Scattering theory*. Number 26 in Pure and Applied Mathematics. Academic Press, 2 edition, 1989.

[11] G. Lebeau. Régularité gevrey 3 pour la diffraction. *Communications in Partial Differential Equation*, 9:1437–1494, 1984.

[12] G. Lebeau and L. Robbiano. Contrôle exact de l'équation de la chaleur. *Communications in Partial Differential Equation*, 20:335–356, 1995.

[13] G. Lebeau and L. Robbiano. Stabilisation de l'équation des ondes par le bord. *Prépublications de l'université de Paris-Sud*, 95–40, 1995.

[14] R. B. Melrose and J. Sjöstrand. Singularities of boundary value problems I. *Communications in Pure Applied Mathematics*, 31:593–617, 1978.

[15] C. S. Morawetz. The decay of solutions of the exterior initial-boundary value problem for the wave equation. *Communications on Pure and Applied Mathematics*, 14:561–568, 1961.

[16] C. S. Morawetz. Decay of solutions of the exterior problem for the wave equation. *Communications on Pure and Applied Mathematics*, 28:229–264, 1975.

[17] J. Sjöstrand and M. Zworski. Asymptotic distribution of resonances for convex obstacles. *Prépublication du Centre de Math. de l'Ecole Polytechnique*, 98–6, 1998.

[18] S. H. Tang and M. Zworski. Resonance expansions of scattered waves. *Comm. Pure Appl. Math.*, 53 (2000) 1305–1334.

[19] B. R. Vainberg. *Asymptotic methods in equations of matheatical physics*. Gordon and Breach, New York, 1988.

[20] G. Vodev. Exponential bounds of the resolvent for a class of noncompactly supported perturbations of the laplacian. *Math. Res. Letters*, 7:287–298, 2000.

Mathématiques
Bât. 425
Université de Paris-Sud, Orsay
91405 Orsay Cedex, France
E-mail address: Nicolas.burq@math.u-psud.fr

A Conjecture of De Giorgi on Symmetry for Elliptic Equations in $\mathbb{R}^n$

Xavier Cabré

Abstract. In 1978 De Giorgi formulated the following conjecture. *Let u be a solution of $\Delta u = u^3 - u$ in all of $\mathbb{R}^n$ such that $|u| \leq 1$ and $\partial_{x_n} u > 0$ in $\mathbb{R}^n$. Is it true that all level sets $\{u = \lambda\}$ of u are hyperplanes, at least if $n \leq 8$?* Equivalently, does u depend only on one variable? When $n = 2$, this conjecture was proved in 1997 by N. Ghoussoub and C. Gui. More recently, L. Ambrosio and the author have proved it for $n = 3$. The question, however, remains open for $n \geq 4$. A connection with the Bernstein problem for minimal hypersurfaces suggests that the conjecture may be true at least if $n \leq 8$. The results for $n = 2$ and 3 apply also to the equation $\Delta u = F'(u)$ for every nonlinearity $F \in C^2$.

1. Introduction

In 1978 De Giorgi [7] stated the following conjecture:

Conjecture. ([7]) *Let $u \in C^2(\mathbb{R}^n)$ be a solution of*

$$\Delta u = u^3 - u \quad in \ \mathbb{R}^n$$

such that

$$|u| \leq 1 \quad and \quad \partial_{x_n} u > 0$$

in the whole $\mathbb{R}^n$. Is it true that all level sets $\{u = \lambda\}$ of u are hyperplanes, at least if $n \leq 8$?

When $n = 2$, this conjecture was proved by Ghoussoub and Gui [9] in 1997. More recently, Ambrosio and the author [2] have proved it in dimension $n = 3$. The conjecture, however, remains open for all $n \geq 4$.

Note that the level sets of u are hyperplanes if and only if u depends only on one variable. Thus, the question of De Giorgi is concerned with the one-dimensional character or symmetry of bounded solutions u of semilinear elliptic equations in the whole space $\mathbb{R}^n$, under the assumption that u is monotone in one direction,

1991 *Mathematics Subject Classification.* 35J60, 35B05, 35B40, 35B45.
Key words and phrases. Nonlinear elliptic PDE, Symmetry and monotonicity properties, Energy estimates, Liouville theorems.

say, $\partial_{x_n} u > 0$ in $\mathbb{R}^n$. The proofs for $n = 2$ and 3 use some techniques developed by Berestycki, Caffarelli and Nirenberg in [4] for the study of symmetry properties of solutions of semilinear equations in half spaces.

We will see in Section 3 that the conjecture of De Giorgi has some relations with the theories of minimal hypersurfaces and phase transitions. In particular, a connection with the Bernstein problem for minimal graphs is probably the reason why De Giorgi [7] includes "at least if $n \leq 8$" in his statement of the question.

The positive answers to the conjecture in dimensions 2 and 3 apply not only to the scalar Ginzburg-Landau equation $\Delta u + u - u^3 = 0$, but to general nonlinearities. Indeed, we have the following:

Theorem 1.1. ([9, 2, 1]) *Assume that $F \in C^2(\mathbb{R})$. Let u be a bounded solution of*

$$\Delta u - F'(u) = 0 \quad in \ \mathbb{R}^n \tag{1}$$

satisfying

$$\partial_{x_n} u > 0 \ in \ \mathbb{R}^n . \tag{2}$$

If $n = 2$ or $n = 3$, then all level sets of u are hyperplanes, i.e., there exist $a \in \mathbb{R}^n$ and $g \in C^2(\mathbb{R})$ such that

$$u(x) = g(a \cdot x) \quad for \ all \ x \in \mathbb{R}^n .$$

For the model case $F'(u) = u^3 - u$, the function $\tanh(s/\sqrt{2})$ is the unique (up to a translation of the independent variable s) one-dimensional solution of the equation. Hence, in this case the conclusion of Theorem 1.1 is that

$$u(x) = \tanh \left(\frac{a \cdot x - c}{\sqrt{2}} \right) \quad \text{in } \mathbb{R}^n ,$$

for some $c \in \mathbb{R}$ and $a \in \mathbb{R}^n$, with $|a| = 1$ and $a_n > 0$. It is also easy to verify that if $F \in C^2(\mathbb{R})$ satisfies $F > F(-1) = F(1)$ in $(-1, 1)$ and $F'(-1) = F'(1) = 0$, then $h'' - F'(h) = 0$ has an increasing solution $h(s)$ satisfying $\lim_{s \to \pm\infty} h(s) = \pm 1$, which is unique up to a translation in s.

Several articles have also considered the question of De Giorgi in a slightly simpler version. It consists of assuming the hypothesis of the conjecture and, in addition, that

$$\lim_{x_n \to \pm\infty} u(x', x_n) = \pm 1 \quad \text{for all } x' \in \mathbb{R}^{n-1} . \tag{3}$$

Here, the limits are *not* assumed to be uniform in $x' \in \mathbb{R}^{n-1}$. Even in this simpler form, the conjecture was first proved in [9] for $n = 2$, in [2] for $n = 3$, and it remains open for $n \geq 4$.

In Theorem 1.1 the direction a of the variable on which u depends is not known a priori. Indeed, if u is a one-dimensional solution satisfying (2), we can "slightly" rotate coordinates to obtain a new one-dimensional solution still satisfying (2). The same remark applies to assumption (3). Instead, if one further assumes that the limits in (3) are uniform in $x' \in \mathbb{R}^{n-1}$, then an a priori choice of the direction a is imposed, namely $a \cdot x = x_n$. Furthermore, with this additional assumption one knows a priori that every level set of u is contained between two parallel hyperplanes. In this respect, it has been established in [3, 5, 8] (independently and using different techniques) that, for every dimension n, if the limits in (3) are assumed to be uniform in $x' \in \mathbb{R}^{n-1}$ then u only depends on the variable x_n, that is, $u = u(x_n)$. This result applies to equation (1) for various classes of nonlinearities F which always include the Ginzburg-Landau model $\Delta u + u - u^3 = 0$.

The first partial result towards the question of De Giorgi was proved by Modica and Mortola [14] in 1980. They gave a positive answer to the conjecture when $n = 2$ under the additional assumption that the level sets of u are the graphs of an equi-Lipschitz family of functions. Note that, since $\partial_{x_n} u > 0$, each level set of u is the graph of a function of x'. In 1985 Modica [11] proved that if $F \geq 0$ in $\mathbb{R}$ then every bounded solution u of $\Delta u - F'(u) = 0$ in $\mathbb{R}^n$ satisfies the gradient bound

$$\frac{1}{2}|\nabla u|^2 \leq F(u) \quad \text{in } \mathbb{R}^n . \tag{4}$$

In 1994 Caffarelli, Garofalo and Segala [6] generalized this bound to more general equations. They also showed that, if equality occurs in (4) at some point of $\mathbb{R}^n$, then the conclusion of the conjecture is true.

2. Sketch of the Proofs in Dimensions 2 and 3

The proof of Theorem 1.1 relies on the following method, already used by Berestycki, Caffarelli and Nirenberg in [4] to establish (also for low dimensions) a very general result on the symmetry of positive solutions in half spaces. The idea is to consider the functions

$$\varphi := \partial_{x_n} u > 0 \quad \text{and} \quad \sigma_i := \frac{\partial_{x_i} u}{\partial_{x_n} u}$$

for each $i \in \{1, \ldots, n-1\}$. Since

$$\varphi^2 \nabla \sigma_i = \partial_{x_n} u \, \nabla \partial_{x_i} u - \partial_{x_i} u \, \nabla \partial_{x_n} u$$

and $\Delta \partial_{x_j} u = F''(u) \partial_{x_j} u$, the function σ_i satisfies the equation

$$\operatorname{div}(\varphi^2 \nabla \sigma_i) = 0 \quad \text{in } \mathbb{R}^n . \tag{5}$$

The goal is to prove that σ_i is constant, since then the theorem follows. Indeed, if for every $i \in \{1, \ldots, n-1\}$ we have

$$\partial_{x_i} u = c_i \, \partial_{x_n} u$$

for some constant c_i, then u is constant along the $n-1$ directions $\partial_{x_i} - c_i \partial_{x_n}$. We then conclude that u is a function of the variable $a \cdot x$ alone, where $a = (c_1, \ldots, c_{n-1}, 1)$.

The proof that σ_i is necessarily constant in dimensions 2 and 3 uses the following Liouville theorem for equation (5). Its proof is already contained in the paper [4] by Berestycki, Caffarelli and Nirenberg; see also [2].

Proposition 2.1. *Let $\varphi \in L^\infty_{\mathrm{loc}}(\mathbb{R}^n)$ be positive in $\mathbb{R}^n$. Suppose that $\sigma \in H^1_{\mathrm{loc}}(\mathbb{R}^n)$ satisfies*

$$\sigma \operatorname{div}(\varphi^2 \nabla \sigma) \geq 0 \quad in \ \mathbb{R}^n \tag{6}$$

in the distributional sense. For every $R > 1$, let $B_R = \{|x| < R\}$ and assume that

$$\int_{B_R} (\varphi \sigma)^2 \leq C R^2 \,, \tag{7}$$

for some constant C independent of R. Then σ is constant.

The proof of this proposition is based in a simple Cacciopoli type estimate for the function σ.

Returning to the conjecture of De Giorgi, we note that

$$\varphi \sigma_i = \partial_{x_i} u \,.$$

Hence, assumption (7) in the Liouville theorem will hold if one shows that, for each $R > 1$,

$$\int_{B_R} |\nabla u|^2 \leq C R^2 \tag{8}$$

for some constant C independent of R. Using standard local elliptic estimates and that $u \in L^\infty(\mathbb{R}^n)$ is a solution of $\Delta u - F'(u) = 0$, we deduce that ∇u also belongs to $L^\infty(\mathbb{R}^n)$. Therefore, estimate (8) is obviously true if $n = 2$. This finishes the proof of Theorem 1.1, and hence of the conjecture of De Giorgi, when $n = 2$.

In order to prove the theorem in dimension 3, it suffices to establish (8). This bound is a consequence (when $n = 3$) of new energy estimates established in [2] and [1]. A first estimate which holds in all dimensions is given by the following:

Theorem 2.2. *([2]) Let $F \in C^2(\mathbb{R})$ and u be a bounded solution of $\Delta u - F'(u) = 0$ in $\mathbb{R}^n$. Assume that*

$$\partial_{x_n} u > 0 \ in \ \mathbb{R}^n \quad and \quad \lim_{x_n \to +\infty} u(x', x_n) = 1 \quad for \ all \ x' \in \mathbb{R}^{n-1} \,.$$

Then, for every $R > 1$, we have

$$\int_{B_R} \left\{ \frac{1}{2} |\nabla u|^2 + F(u) - F(1) \right\} dx \leq C R^{n-1} \tag{9}$$

for some constant C independent of R.

Note that the previous estimate is clearly true (and optimal) for one-dimensional solutions. The energy functional in B_R,

$$E_R(u) = \int_{B_R} \left\{ \frac{1}{2} |\nabla u|^2 + F(u) - F(1) \right\} dx \,,$$

has $\Delta u - F'(u) = 0$ as Euler-Lagrange equation. In 1989 Modica [12] proved that if $F - F(1) \geq 0$ in $\mathbb{R}$ and u is a bounded solution of $\Delta u - F'(u) = 0$ in $\mathbb{R}^n$, then the quantity

$$\frac{E_R(u)}{R^{n-1}}$$

is a nondecreasing function of R. Theorem 2.2 establishes that this quotient is, in addition, bounded from above. Moreover, the upper bound of Theorem 2.2 is optimal. Indeed, if $E_R(u)/R^{n-1} \to 0$ as $R \to \infty$ then the monotonicity formula of Modica would give that $E_R(u) = 0$ for every $R > 0$, and hence that u is constant in $\mathbb{R}^n$.

The proof of the energy estimate of Theorem 2.2 relies on the consideration of the path of functions $u^t(x) = u(x', x_n + t)$ connecting u (for $t = 0$) and 1 (for $t = +\infty$). Note that the functions u^t have different boundary values on $\partial B_R(0)$, but they are all solutions of the same Euler-Lagrange equation. The desired energy estimate follows from a bound for the variation of energy with respect to t, in a fixed ball $B_R(0)$, which is obtained using the key hypothesis $\partial_{x_n} u > 0$; see [2].

For the model case $F'(u) = u^3 - u$, we have $F(u) = (1 - u^2)^2/4$ and hence $F(u) - F(1) \geq 0$. It follows then that (8) is a consequence of (9) when $n = 3$. Moreover, the assumption $\lim_{x_n \to +\infty} u(x', x_n) = 1$ in Theorem 2.2 may be removed when $n = 3$ (see [2]). In this way, the proof of the conjecture of De Giorgi in dimension 3 is completed. The argument applies not only to $F(u) = (1 - u^2)^2/4$ but also to a large class of nonlinearities F.

The proof of the conjecture in dimension 3 for *every* nonlinearity $F \in C^2$ (see Theorem 1.1) has been obtained by Alberti, Ambrosio and the author in [1]. It relies on the following local minimality property of u.

Theorem 2.3. ([1]) *Let $F \in C^2(\mathbb{R})$ and let u be a bounded solution of $\Delta u - F'(u) = 0$ in $\mathbb{R}^n$ satisfying $\partial_{x_n} u > 0$ in $\mathbb{R}^n$. Consider the functions*

$$\underline{u}(x') = \lim_{x_n \to -\infty} u(x', x_n) \quad and \quad \overline{u}(x') = \lim_{x_n \to +\infty} u(x', x_n).$$

Then

$$\int_{B_R} \left\{ \frac{1}{2}|\nabla u|^2 + F(u) \right\} dx \leq \int_{B_R} \left\{ \frac{1}{2}|\nabla v|^2 + F(v) \right\} dx$$

for every function $v \in C^1(\mathbb{R}^n)$ such that $\{v \neq u\}$ has compact closure contained in B_R and

$$\underline{u}(x') < v(x', x_n) < \overline{u}(x') \quad for \ all \ x = (x', x_n) \in B_R. \tag{10}$$

The theorem states that every solution u which is monotone in one direction is a minimizer of the energy in every ball with respect to the class of functions v with the same boundary values as u and satisfying condition (10). Before we discovered Theorem 2.3, a hint for its validity had been given in [2], where we pointed out that the condition $\partial_{x_n} u > 0$ implies (by the maximum principle) that the second variation of energy at u is nonnegative definite with respect to compact perturbations – a necessary condition for minimality.

The proof of Theorem 2.3 is based on the construction of an appropriate calibration or null-lagrangian (a classical tool in the Calculus of Variations) associated to the extremal field $u^t(x) = u(x', x_n + t)$ for $t \in \mathbb{R}$. Note that the graphs of the functions u^t are disjoint due to the monotonicity condition $\partial_{x_n} u > 0$.

In [1] we use Theorem 2.3 to obtain an improved version of the energy estimate of Theorem 2.2, in which $F(1)$ is replaced by the infimum of F on the range of u. The improved estimate leads to the gradient bound (8), and hence to the one-dimensional symmetry result (Theorem 1.1 in dimension 3) for all nonlinearities $F \in C^2$.

3. Relation with the Bernstein Problem for Minimal Graphs

In this section we present the heuristic argument that relates the conjecture of De Giorgi with the Bernstein problem for minimal graphs. We suppose that $F(u) = (1 - u^2)^2/4$, for simplicity. With u as in the conjecture, one considers the blown-down sequence

$$u_R(y) = u(Ry) \quad \text{for } y \in B_1 \subset \mathbb{R}^n \,,$$

and the penalized energy of u_R in B_1

$$H_R(u_R) = \int_{B_1} \left\{ \frac{1}{2R} |\nabla u_R|^2 + RF(u_R) \right\} dy \,.$$

Note that $H_R(u_R)$ is a bounded sequence, by Theorem 2.2. By a result of Modica and Mortola [13], as $R \to \infty$ the functionals H_R Γ-converge to a functional which is finite only for characteristic functions with values in $\{-1, 1\}$ and equal (up to the multiplicative constant $2\sqrt{2}/3$) to the area of the hypersurface of discontinuity.

The sequence u_R is therefore expected to converge to a characteristic function whose hypersurface of discontinuity S has minimal area, by the local minimality property of Theorem 2.3. Moreover, S is expected to be the graph of a function defined on $\mathbb{R}^{n-1}$ (since the level sets of u are graphs due to hypothesis $\partial_{x_n} u > 0$), possibly with vertical parts. The set S describes the behavior at infinity of the level sets of u. The conjecture of De Giorgi states that the level sets of u are hyperplanes. The connection with the Bernstein problem (see Chapter 17 of [10] for a complete survey on this topic) is due to the fact that every minimal graph of a function defined on $\mathbb{R}^m = \mathbb{R}^{n-1}$ is known to be a hyperplane whenever $m \leq 7$, i.e., $n \leq 8$. Hence, S is expected to be a hyperplane when $n \leq 8$. Using the local minimality property of u (Theorem 2.3), we have given in [1] a rigorous proof of this convergence result, up to subsequences of radii $R_k \to \infty$.

References

[1] G. Alberti, L. Ambrosio and X. Cabré, *On a long-standing conjecture of E. De Giorgi: old and recent results*, to appear in Acta Applicandae Mathematicae.

[2] L. Ambrosio and X. Cabré, *Entire Solutions of Semilinear Elliptic Equations in $\mathbb{R}^3$ and a Conjecture of De Giorgi*, Journal Amer. Math. Soc. **13** (2000), 725–739.

[3] M. T. Barlow, R. F. Bass and C. Gui, *The Liouville property and a conjecture of De Giorgi*, Preprint.

[4] H. Berestycki, L. Caffarelli and L. Nirenberg, *Further qualitative properties for elliptic equations in unbounded domains*, Ann. Scuola Norm. Sup. Pisa Cl. Sci. **25** (1997), 69–94.

[5] H. Berestycki, F. Hamel and R. Monneau, *One-dimensional symmetry of bounded entire solutions of some elliptic equations*, Preprint.

[6] L. Caffarelli, N. Garofalo and F. Segala, *A gradient bound for entire solutions of quasi-linear equations and its consequences*, Comm. Pure Appl. Math. **47** (1994), 1457–1473.

[7] E. De Giorgi, *Convergence problems for functionals and operators*, Proc. Int. Meeting on Recent Methods in Nonlinear Analysis (Rome, 1978), 131–188, Pitagora, Bologna (1979).

[8] A. Farina, *Symmetry for solutions of semilinear elliptic equations in $\mathbb{R}^N$ and related conjectures*, Ricerche di Matematica **XLVIII** (1999), 129–154.

[9] N. Ghoussoub and C. Gui, *On a conjecture of De Giorgi and some related problems*, Math. Ann. **311** (1998), 481–491.

[10] E. Giusti, *Minimal Surfaces and Functions of Bounded Variation*, Birkhäuser, Basel-Boston, 1984.

[11] L. Modica, *A gradient bound and a Liouville theorem for nonlinear Poisson equations*, Comm. Pure Appl. Math. **38** (1985), 679–684.

[12] L. Modica, *Monotonicity of the energy for entire solutions of semilinear elliptic equations*, Partial differential equations and the calculus of variations, Vol. II., Progr. Nonlinear Differential Equations Appl. **2**, 843–850, Birkhäuser, Boston (1989).

[13] L. Modica and S. Mortola, *Un esempio di Γ^--convergenza*, Boll. Un. Mat. Ital. B **14** (1977), 285–299.

[14] L. Modica and S. Mortola, *Some entire solutions in the plane of nonlinear Poisson equations*, Boll. Un. Mat. Ital. B **17** (1980), 614–622.

Departament de Matemàtica Aplicada 1
Universitat Politècnica de Catalunya
Diagonal, 647
08028 Barcelona, Spain
E-mail address: cabre@ma1.upc.es

The Random Graph Revisited

Peter J. Cameron

Abstract. The random graph, or Rado's graph (the unique countable homogeneous graph) has made an appearance in many parts of mathematics since its first occurrences in the early 1960s. In this paper I will discuss some old and new results on this remarkable structure.

1. The Random Graph

Consider the following two countably infinite graphs.

For the first graph, take any countable model of the Zermelo-Fraenkel axioms for set theory. (The existence of such models, the so-called *Skolem paradox*, follows from the downward Löwenheim-Skolem theorem of first-order logic.) Abstractly, this is a countable set with an asymmetric membership relation. Now form an undirected graph by symmetrising the relation; that is, x and y are adjacent (written $x \sim y$) if either $x \in y$ or $y \in x$.

For the second graph, the vertices are the primes congruent to 1 mod 4. Let $p \sim q$ hold if p is a quadratic residue mod q. By the law of quadratic reciprocity, this relation is already symmetric.

Remarkably, these two graphs are isomorphic.

To see this, we first note that each of them has the following property: if U and V are finite disjoint sets of vertices, then there exists a vertex z such that $u \sim z$ for all $u \in U$, while $v \not\sim z$ for all $v \in V$. (We write $v \not\sim z$ to indicate that v and z are not adjacent.) In the first graph, this property is demonstrated using the Pairing, Union, and Foundation axioms. In the second graph, the proof uses the Chinese Remainder Theorem and Dirichlet's Theorem on primes in arithmetic progression.

This property of a graph G can be stated another way. If A and B are finite graphs with A an induced subgraph of B, then every embedding of A into G extends to an embedding of B into G. The stated property asserts this when $|B| = |A| + 1$, and the general case follows by induction. We call this the *I-property*, because it is a form of injectivity.

Now any two countable graphs having the I-property are isomorphic. This is a standard application of the *back-and-forth* technique from model theory. If G_1 and G_2 are countable graphs with the I-property, we build successively larger finite partial isomorphisms between them by going alternately from G_1 to G_2 and from G_2 to G_1; the result after countably many steps is an isomorphism.

The argument shows more. If we take $G_1 = G_2 = G$, and begin with a given finite partial isomorphism, we see that *every isomorphism between finite subgraphs of G extends to an automorphism of G.* We call a graph with this property *homogeneous*; we have shown that a graph with the I-property is homogeneous. The converse also holds, provided that we formulate the I-property correctly: we require that the two finite graphs A and B must belong to the *age* of G, the class of all finite graphs embeddable in G.

So the two graphs with which we started this section, despite appearances, are isomorphic, and are homogeneous, although no symmetries are visible to the casual observer.

There are several more features of this story. First, Fraïssé [11] showed that two countable homogeneous graphs with the same age are isomorphic (essentially by the argument given above); he also characterized the ages of homogeneous graphs. They are the classes of finite graphs which are closed under isomorphism, closed under taking induced subgraphs, and have the *amalgamation property.* (This asserts that, if two members of the age have isomorphic induced subgraphs, then they can be embedded in another graph in the class in such a way that the isomorphism becomes equality.) If these conditions are satisfied by the class $\mathcal{C}$ of finite graphs, then the unique countable homogeneous graph G whose age is $\mathcal{C}$ is called the *Fraïssé limit* of $\mathcal{C}$.

Subsequently, all the countable homogeneous graphs were determined by Lachlan and Woodrow [18]:

Theorem 1.1. *The countably infinite homogeneous graphs are the following:*

(a) *the disjoint union of m complete graphs of size n, where m and n are finite or countable (and at least one is infinite);*

(b) *the complements of the graphs under (a);*

(c) *the Fraïssé limit of the class of graphs containing no complete subgraph of size r, for given finite $r \geq 3$;*

(d) *the complements of the graphs under (c);*

(e) *the Fraïssé limit of the class of all finite graphs.*

The graphs in case (c) were first constructed and studied by Henson [12]. The unique graph under (e) is the one with which we began this section. It is called the *random graph* (for reasons which will shortly appear) or *Rado's graph* (since the first explicit construction of it was given by Rado [19]).

We digress to give Rado's construction. Let us first define an asymmetric relation $\rightarrow$ on the set of non-negative integers by the rule that $x \rightarrow y$ if the xth digit in the base-2 expression of y is 1 (that is, when y is written as a sum of distinct powers of 2, then 2^x is one of these powers). Now the symmetrised form of $\rightarrow$ defines a graph on $\mathbb{N}$ which is the graph R. For the relation $\rightarrow$ just defined is a model of the Zermelo-Fraenkel axioms (including the Axiom of Choice) except for the Axiom of Infinity: every finite set X of natural numbers is represented by the single natural number $y = \sum_{x \in X} 2^x$, and $x \to y$ if and only if $x \in X$. Now the Axiom of Infinity was not used in our proof that the symmetrised membership relation defines the graph R.

Fraïssé also observed that there is nothing special about graphs: the relation between homogeneity, the I-property, and the amalgamation property holds for arbitrary relational structures. (We need to include one extra condition in the statement of Fraïssé's Theorem in general, namely that there are only countably many structures in the class $\mathcal{C}$, up to isomorphism. This is to rule out the possibility of, for example, infinitely many unary relations, any subset of which might hold on a particular element of the domain; by Cantor's Theorem, a countable set does not contain enough elements for all possibilities to be realized.)

The homogeneous relational structures of various types have been determined by various authors: partially ordered sets by Schmerl [20]; tournaments by Lachlan [17], simplified by Cherlin [7]; directed graphs by Cherlin [8]. Remarkably, there are only three countable homogeneous tournaments: the linear order $\mathbb{Q}$; the *circular tournament*, otherwise known as the *local order* or *locally transitive tournament*, and has arisen in diverse areas such as permutation groups [4] and computer science [16]; and the analogue of R, the *random tournament*, which admits a construction similar to our second construction of the random graph, but using primes congruent to -1 mod 4 instead. On the other hand, there are uncountably many countable homogeneous directed graphs (this had been pointed out earlier by Henson [13]).

There is no need to stick to relational structures. The argument essentially extends to all locally finite first-order structures, and even beyond this class with care. For a recent treatment of these ideas in group theory, see Higman and Scott [14].

Finally, we come to the reason for the term *random graph*. We use the simplest possible model for the random graph on a given finite or countable vertex set: we decide, with probability $1/2$, whether a given pair of vertices is an edge or a non-edge, independently of all other choices. If the vertex set is finite, then every possible graph occurs, with probability inversely proportional to the order of its automorphism group; indeed, graphs with no symmetry at all predominate. On the other hand, Erdős and Rényi [9] showed the following:

Theorem 1.2. *There is a countable graph R such that, with probability 1, the random graph on a countable vertex set is isomorphic to R.*

Of course, R is the graph we met above, the Fraïssé limit of the class of all finite graphs. The proof is simple: we have to show that the I-property holds with probability 1. Since the union of countably many null sets is null, and there are only countably many pairs of finite disjoint sets of vertices, it suffices to show that, for given finite disjoint sets U and V, the probability that no vertex z with the correct joins exists is 0.

But the probability that a given vertex z is not correctly joined is $(1 - 1/2^n)$, where $n = |U \cup V|$; by independence, the probability that no vertex is correctly joined is

$$\lim_{k \to \infty} (1 - 1/2^n)^k = 0\,.$$

The result of Erdős and Rényi was described, in their account of it by Erdős and Spencer [10], as demolishing the theory of countable random graphs. In its place, we have the theory of *the* countable random graph! A survey of some of the remarkable properties of R and their generalisations to other classes of structures is given in [5]. In the remainder of this paper, some recent results will be mentioned.

2. The Pigeonhole Property

If a countably infinite set is partitioned into two parts, then at least one of the two is countable, and hence is isomorphic to the original set. We say that a structure has the *pigeonhole property* if, whenever it is partitioned into two parts, one of the parts is isomorphic to the original structure.

Theorem 2.1. *The only countable graphs with the pigeonhole property are the complete and null graphs and the random graph.*

For the proof, see [5]. The proof that the random graph has the pigeonhole property is very short, and goes as follows. Suppose that it has a partition such that neither of the two parts X_1 and X_2 is isomorphic to R. Then there exist finite disjoint subsets U_i and V_i of X_i such that no vertex of X_i is joined to all vertices in U_i and none in V_i, for $i = 1, 2$. But then no vertex of R is joined to all vertices in $U_1 \cup U_2$ and to none in $V_1 \cup V_2$, a contradiction.

This remarkable characterisation of R suggests either weakening the definition of the pigeonhole property or considering other classes of structures. The first approach has not yielded much yet. I merely pose a question:

> Which countable graphs G have the property that, whenever G is partitioned into two parts, one of the parts has an induced subgraph isomorphic to G?

More is known about the second case. Bonato and Delić [3, 2] determined the tournaments with the pigeonhole property. Note that there are no complete or null tournaments.

Theorem 2.2. *Let T be a countable tournament with the pigeonhole property. Then either*

(a) *T or its converse is an ordinal power of ω; or*
(b) *T is the random tournament.*

Some results are known about more general binary relations such as directed graphs, but there is not yet a complete classification of these.

3. Automorphism Groups

The random graph R is homogeneous. Hence, its automorphism group $\text{Aut}(R)$ is transitive on finite subgraphs of any given isomorphism type; in particular, it is transitive on vertices, edges, and non-edges, and so is a primitive permutation group of rank 3 on vertices.

A number of properties of this group are known:

(a) It has cardinality $2^{\aleph_0}$.
(b) It is simple [21].
(c) It has the strong small index property (see below) [15].

A permutation group on a countable set X is said to have the *small index property* if every subgroup of index less than $2^{\aleph_0}$ contains the pointwise stabiliser of a finite set; it has the *strong small index property* if every such subgroup lies between the pointwise and setwise stabilisers of a finite set.

Truss [21] found all the cycle structures of automorphisms of R. In particular, R has cyclic automorphisms. The easiest way to see this is as follows. Choose a set S of positive integers independently at random. Construct a graph $\Gamma(S)$ with vertex set $\mathbb{Z}$ by joining x and y if and only if $|x - y| \in S$. The resulting graph admits the cyclic shift $x \mapsto x + 1$ as an automorphism; and, with probability 1, it is isomorphic to R. Similarly it follows from Truss's classification that R admits an automorphism with one fixed vertex v and two infinite cycles (necessarily the neighbours and the non-neighbours of v).

Bhattacharjee and Macpherson [1] settled a question of Peter Neumann by the following remarkable combination of the two types of automorphisms just described.

Theorem 3.1. *There exist automorphisms f, g of R such that*

(a) *f, g generate a free subgroup of $\text{Aut}(R)$,*
(b) *f has a single cycle on R, which is infinite,*
(c) *g fixes a vertex v and has two cycles on the remaining vertices (namely, the neighbours and non-neighbours of v),*
(d) *the group $\langle f, g \rangle$ is oligomorphic, and transitive on vertices, edges, and non-edges of R, and each of its non-identity elements has only finitely many cycles on R.*

The existence of cyclic auutomorphisms of R shows that in fact it is a Cayley graph for the infinite cyclic group. Cameron and Johnson [6] generalised this as follows.

Theorem 3.2. *Let G be a countable group which cannot be expressed as the union of finitely many translates of square-root sets of non-identity elements. Then, with probability 1, a random Cayley graph for G is isomorphic to R.*

Here, the *square-root set* of an element g is $\{x : x^2 = g\}$.

Wielandt called a group G a *B-group* if every primitive permutation group containing the regular representation of R is doubly transitive. The B stands for Burnside, who gave the first examples: the finite cyclic groups of prime-power (non-prime) order. In fact, it follows from the Classification of Finite Simple Groups that, for almost all n (that is, all but a set of density 0), every group of order n is a B-group. (However, the finite non-B-groups have not been determined.) In the infinite case, things are very different: *no countable B-group is known*, although many groups are known not to be B-groups.

It follows from the above theorem of Cameron and Johnson that, if a countable group G is not the union of finitely many translates of square-root sets of non-identity elements, then G is not a B-group; for it is a regular subgroup of the primitive but not doubly transitive group $\mathrm{Aut}(R)$. Other countable homogeneous structures have been examined in the hope of finding further non-B-groups, but this has not been successful. (Note however that the above result does not exhaust the list of groups which are not B-groups. For example, any group which is the direct product of two countable subgroups is a non-B-group, since it is contained in the primitive group $S_\omega \wr S_2$ (in its product action).) The two results together show that no countable abelian group is a B-group. For let G be a countable abelian group. If the subgroup A_2 of involutions has infinite index in A, then A satisfies the hypotheses of the Cameron-Johnson theorem. If not, it has finite exponent, and so is a direct product of two infinite factors.

I mention an unsolved problem arising from the search for regular groups of automorphisms of countable structures. Recall Henson's construction [12] of countable homogeneous graphs H_r, where H_r is the Fraïssé limit of the class of finite graphs containing no complete subgraph of size r. Henson showed in that paper that H_3 has cyclic automorphisms but H_r does not for $r \geq 4$. His argument in fact shows more:

Theorem 3.3. *For $r \geq 4$, H_r is not a normal Cayley graph of any countable group.*

A Cayley graph for G is *normal* if it admits both left and right translation by G, in other words, if the set of neighbours of the identity is a normal subset. Now we can pose the open problem: *Is the Henson graph H_r for $r \geq 4$ a Cayley graph?*

References

[1] M. Bhattacharjee and H. D. Macpherson, Strange permutation representations of free groups, *J. Austral. Math. Soc.*, to appear.

[2] A. Bonato, P. Cameron and D. Delić, Tournaments and orders with the pigeonhole property, *Canad. Math. Bull.*, **43** (2000), 397–405.

[3] A. Bonato and D. Delić, A pigeonhole property for relational structures, *Mathematical Logic Quarterly* **45** (1999), 409–413.

[4] P. J. Cameron, Orbits of permutation groups on unordered sets, II, *J. London Math. Soc.* (2) **23** (1981), 249–265.

[5] P. J. Cameron, The random graph, pp. 331–351 in *The Mathematics of Paul Erdős*, (J. Nešetřil and R. L. Graham, eds.), Springer-Verlag, Berlin, 1996.

[6] P. J. Cameron and K. W. Johnson, An essay on countable B-groups, *Math. Proc. Cambridge Philos. Soc.* **102** (1987), 223–232.

[7] G. Cherlin, Homogeneous tournaments revisited, *Geometriae Dedicata* **26** (1988), 231–239.

[8] G. Cherlin, The classification of countable homogeneous directed graphs and countable homogeneous n-tournaments, *Memoirs Amer. Math. Soc.* **621**, American Mathematical Society, Providence, RI, 1998.

[9] P. Erdős and A. Rényi, Asymmetric graphs, *Acta Math. Acad. Sci. Hungar.* **14** (1963), 295–315.

[10] P. Erdős and J. Spencer, *Probabilistic Methods in Combinatorics*, Academic Press, New York/Akad. Kiadó, Budapest, 1974.

[11] R. Fraïssé, Sur certains relations qui généralisent l'ordre des nombres rationnels, *C. R. Acad. Sci. Paris* **237** (1953), 540–542.

[12] C. W. Henson, A family of countable homogeneous graphs, *Pacific J. Math.* **38** (1971), 69–83.

[13] C. W. Henson, Countable homogeneous relational structures and $\aleph_0$-categorical theories, *J. Symbolic Logic* **37** (1972), 494–500.

[14] G. Higman and E. A. Scott, *Existentially Closed Groups*, Oxford University Press, Oxford, 1988.

[15] W. A. Hodges, I. M. Hodkinson, D. Lascar and S. Shelah, The small index property for ω-stable, ω-categorical structures and for the random graph, *J. London Math. Soc.* (2) **48** (1993), 204–218.

[16] D. E. Knuth, *Axioms and Hulls*, Lecture Notes in Computer Science **606**, Springer-Verlag, Berlin, 1992.

[17] A. H. Lachlan, Countable homogeneous tournaments, *Trans. Amer. Math. Soc.* **284** (1984), 431–461.

[18] A. H. Lachlan and R. E. Woodrow, Countable ultrahomogeneous undirected graphs, *Trans. Amer. Math. Soc.* **262** (1980), 51–94.

[19] R. Rado, Universal graphs and universal functions, *Acta Arith.* **9** (1964), 331–340.

[20] J. H. Schmerl, Countable homogeneous partially ordered sets, *Algebra Universalis* **9** (1979), 317–321.

[21] J. K. Truss, The group of the countable universal graph, *Math. Proc. Cambridge Philos. Soc.* **98** (1985), 213–245.

Added in proof: Theorem 2.1 has been extended by Bonato, Delić, Thomassé and the author, who determined all countable graphs with the property that, for any partition of the vertex set into three parts, the induced subgraph on the union of some two of the parts is isomorphic to the original graph.

School of Mathematical Sciences
Queen Mary, University of London
London E1 4NS, U.K.
E-mail address: p.j.cameron@qmw.ac.uk

Difference Fields: Model Theory and Applications to Number Theory

Zoé Chatzidakis

Abstract. This is an overview of the main model-theoretic results obtained on difference fields and of some applications.

1. Introduction

A difference field is a field with a distinguished endomorphism σ. Algebraists have extensively studied difference fields in parallel with differential fields, and we refer to Cohn's book [6] for the basic algebraic results. In this talk we will focus on inversive difference fields, i.e., we impose that σ be an automorphism, and on what we will call generic difference fields. The model theory of these fields has been studied by various people, first by Macintyre, Van den Dries and Wood (see [15]), then by Hrushovski, Peterzil and myself (see [4, 5], and also [8, 9]). In this paper, I will not explain how model theory led to the various results obtained, but instead focus on their use. The paper is organized as follows: the first section gives the preliminaries in model theory and in algebra. The second section states the main model-theoretic results on generic difference fields, and how they can be used in applications.

2. Preliminaries on Difference Fields and Their First-Order Theory

First some conventions: all difference fields will be **inversive** (i.e., the endomorphism σ is onto). We let $\mathcal{L}$ be the language $\{+, -, \cdot, 0, 1, \sigma\}$, where $+$, $-$, $\cdot$ are binary operations and σ is a unary function symbol. Any difference field K is naturally an $\mathcal{L}$-structure, with $+$, $-$, $\cdot$ interpreted as addition, subtraction and multiplication respectively, σ as the distinguished automorphism of K, and 0, 1 as 0, 1 of the field K. The algebraic closure of a field K is denoted by K^{alg}. All results in this section can be found in [6] and in [4].

2.1. Difference polynomial rings, σ-ideals, σ-closed sets

Let (K, σ) be a difference field. We define the *difference polynomial ring* $K[X_1, \ldots, X_n]_\sigma$ by taking the ring $K[X_1, \ldots, X_n]_\sigma$ to be the ordinary polynomial ring $K[\sigma^j(X_i) \mid i = 1, \ldots, n, j \in \mathbb{N}]$, and extending σ to $K[X_1, \ldots, X_n]_\sigma$

in the way suggested by the name of the generating elements. Note that σ is not onto. The order of a difference polynomial f is the largest m such that some indeterminate $\sigma^m(X_i)$ appears in f.

Ideals I of $K[X_1, \ldots, X_n]_\sigma$ satisfying $\sigma(I) \subseteq I$ are called σ-*ideals*. A *perfect* σ-ideal of $K[X_1, \ldots, X_n]_\sigma$ is a σ-ideal I satisfying moreover that $a\sigma(a^m) \in I$ implies $a \in I$ for all $m \in \mathbb{N}$. Thus a perfect σ-ideal is radical. A *prime σ-ideal* is a σ-ideal which is prime and perfect. Quotients of $K[X_1, \ldots, X_n]_\sigma$ by prime σ-ideals are domains, on which σ defines an embedding. Thus they embed uniquely in a smallest difference field.

The perfect σ-ideals of $K[X_1, \ldots, X_n]_\sigma$ satisfy the ascending chain condition. Consider the topology on K^n, whose basic closed sets are the σ-closed sets, i.e., subsets of K^n which are zero-sets of some finite subset A of $K[X_1, \ldots, X_n]_\sigma$. The ascending chain condition on perfect σ-ideals implies that this topology is Noetherian, and that every closed set is σ-closed. This topology is finer than the Zariski topology on K^n.

2.2. Formulas and definable sets

Let K be a difference field, let E be a difference subfield of K. A σ-*equation* (over E) is a formula of the form $f(x_1, \ldots, x_n) = 0$, where $f(X_1, \ldots, X_n) \in E[X_1, \ldots, X_n]_\sigma$. If the coefficients of $f(X_1, \ldots, X_n)$ are in $\mathbb{Z}$, then we say that the σ-equation is over $\emptyset$, or *has no parameters*. The zero-set in K^n of a σ-equation $f(x_1, \ldots, x_n) = 0$ is called the *set defined by the formula* $f(x_1, \ldots, x_n) = 0$.

The quantifier-free formulas over E (or with parameters in E) are obtained by closing the set of σ-equations over E under the logical connectives $\wedge$, $\vee$, and $\neg$ ("and", "or", and "not"). A quantifier-free formula $\varphi(x_1, \ldots, x_n)$ then defines in K a Boolean combination of σ-closed subsets of K^n ($\wedge$ corresponds to intersection, $\vee$ to union and $\neg$ to complement).

The set of formulas with parameters in E is the smallest set of formulas containing the quantifier-free formulas over E, and closed under prefixing by quantifiers $\exists x$, $\forall x$; so, if $\varphi(x_1, \ldots, x_n)$ is a formula, so are $\exists x_1\, \varphi(x_1, \ldots, x_n)$ and $\forall x_1\, \varphi(x_1, \ldots, x_n)$. The subset of K^n defined by a formula $\varphi(x_1, \ldots, x_n)$ is the obvious one, corresponding to the interpretation of $\exists x\, \psi(x, \bar{y})$ as "there exists x such that $(x, \bar{y})$ is in the set defined by $\psi(x, \bar{y})$", and so on. It is then easy to check that the set of definable subsets of K^n is closed under Boolean combinations. A formula φ is called a *sentence* if φ is over $\emptyset$ and all variables occuring in φ are bounded, (see any logic book for the correct definition), which roughly speaking means that φ is logically equivalent to a formula $Q_1 x_1, \ldots, Q_n x_n \psi(x_1, \ldots, x_n)$, where the Q_i's are quantifiers, and $\psi(x_1, \ldots, x_n)$ is a quantifier-free formula with variables among $x_1, \ldots, x_n$. We also have *sentences over E*, i.e., formulas over E in which all variables are bounded. Another way of viewing formulas or sentences over E, is as formulas over $\emptyset$ in which one has plugged parameters from E, i.e., formulas of the form $\varphi(\bar{x}, \bar{e})$, where $\bar{e}$ is a tuple of elements of E and $\varphi(\bar{x}, \bar{y})$ a formula over $\emptyset$.

If a tuple $\bar{a}$ is in the set defined by $\varphi(\bar{x})$ in K, then we also say that $\bar{a}$ *satisfies* $\varphi(\bar{x})$ (in K), or that K *satisfies* $\varphi(\bar{a})$. Be aware that the notion of satisfaction will in general depend on the difference field we are working in. When we speak about a definable subset D of K^n, we mean that there is a formula $\varphi(\bar{x})$ such that D is the set defined by φ, and we write $\bar{a} \in D$ to mean that $\bar{a}$ satisfies $\varphi(\bar{x})$.

2.3. Generic difference fields

A difference field K is *generic* if whenever L is a difference field containing K, and $f_1(\bar{X}), \ldots, f_m(\bar{X}) \in K[\bar{X}]_\sigma$ ($\bar{X}$ a tuple of variables) have a common zero in L^n, then they already have a common zero in K.

Generic difference fields can be characterized by the following properties: (K, σ) is generic if K is an algebraically closed field, σ is an automorphism of K, and whenever U and V are (absolutely irreducible algebraic) varieties defined over K, such that $V \subseteq U \times U^\sigma$ and the projection maps $V \to U$ and $V \to U^\sigma$ are generically onto, then there is a tuple $\bar{a}$ in K such that $(\bar{a}, \sigma(\bar{a})) \in V$.

Here U^σ denotes the variety whose defining equations are obtained by applying σ to the coefficients of a set of defining equations of U. These properties can be expressed by a collection of sentences of our language.

2.4. Useful properties of generic difference fields and their definable sets

(1) Every difference field embeds in a generic difference field.

(2) Let K be a generic difference field, and $D \subseteq K^n$ a definable set. Then there is a σ-closed subset U of K^{n+m} for some m, such that the projection map $\pi \colon K^{n+m} \to K^n$ on the first n coordinates restricts to a finite-to-one onto map $: U \to D$.

   So, this says, that if D is definable, using arbitrarily many quantifiers, then in fact D can be defined existentially, i.e., by a formula of the form $\exists \bar{y}\, \varphi(\bar{x}, \bar{y})$, where $\varphi(\bar{x}, \bar{y})$ is a conjunction of σ-equations and $\bar{x}$, $\bar{y}$ are tuples of variables.

(3) If $K \subseteq L$ are generic difference fields, then every formula with parameters in K is satisfied in L if and only if it is satisfied in K. Thus, in order to study definable subsets of K, we may in fact go to a generic L containing K which has many automorphisms (as a difference field, i.e., field-automorphisms which commute with σ). This type of technique is standard in model-theory.

(4) Let K and L be two generic difference fields. Then K and L satisfy the same sentences if and only if there is a difference field isomorphism between the algebraic closure of the prime subfield of K and the algebraic closure of the prime subfield of L. If that is the case, then there will be a generic difference field in which K and L embed. So, to each generic difference field K, we associate an *invariant*, which is the isomorphism type of the difference subfield of K consisting of elements algebraic over the prime subfield of K. These invariants play the same role for generic difference fields that the characteristic does for algebraically closed fields.

(5) If E is an algebraically closed difference subfield of the generic difference field K, then the isomorphism type of the difference field E determines the set of sentences with parameters in E satisfied by K. This means, that if $F \subseteq K$ and $f\colon E \to F$ is an isomorphism of difference fields, then there is a generic difference field L containing K, such that f extends to an automorphism of L. This also means that if K' is another generic difference field containing E, then K and K' both E-embed in some L.

(6) Let K be a generic difference field, $D \subseteq K^n$ a definable set, and R a definable equivalence relation on D. Then D/R is in definable bijection with a definable subset S of K^m, for some m.

(7) Let $D \subseteq \mathrm{Fix}(\sigma)^n$ be definable in K. Then D is definable in the pseudo-finite field $\mathrm{Fix}(\sigma)$. Thus the results of [12] and [13] can be used.

2.5. Notation, transformal transcendence and algebraicity

Let K be an algebraically closed difference field, and let $A \subseteq K$. We denote by $\mathrm{acl}_\sigma(A)$ the smallest algebraically closed difference subfield of K containing A, i.e., $\mathrm{acl}_\sigma(A)$ is the algebraic closure of the field generated by $\bigcup_{n\in\mathbb{Z}} \sigma^n(A)$. If E is a difference subfield of K, and $A \subseteq K$, we denote by $E(A)_\sigma$ the difference subfield $E(\sigma^i(A) \mid i \in \mathbb{Z})$ of K.

We say that an element $a \in K$ is *transformally transcendental* over E if $E(a)_\sigma$ has infinite transcendence degree over E, and that a is *transformally algebraic over E* otherwise. We say that A is *transformally algebraic over E* if all elements of A are transformally algebraic over E. There are things like transformal transcendence bases, degree of transformal transcendence, which behave the way they should, but we will not go into them, see [6] for details.

2.6. Independence and rank

Let K be a generic difference field, and let A, B, C be subsets of K. We say that A and B are *independent over* C, if $\mathrm{acl}_\sigma(CA)$ and $\mathrm{acl}_\sigma(CB)$ are linearly disjoint over $\mathrm{acl}_\sigma(C)$. This is a natural notion of independence, and satisfies all properties that ordinary algebraic freeness does. Observe that if $C = \mathrm{acl}_\sigma(C) \subseteq B = \mathrm{acl}_\sigma(B)$ and $\bar{a}$ is a finite tuple, then $\bar{a}$ and B are independent over C if and only if the ideal $\{f(\bar{X}) \in B[\bar{X}]_\sigma \mid f(\bar{a}) = 0\}$ is generated by its intersection with $C[\bar{X}]_\sigma$.

For each n one can define a rank on definable subsets of K^n, which takes its values in the set of ordinal numbers $\leq \omega n$. Several notions of rank have been introduced in model theory; they are meant to define a notion similar to that of dimension of algebraic varieties, but usually are more complex. One of these ranks (the U-rank), turns out to be particularly simple and well adapted to our structures. We will not go into the details (see [4]), what we are really interested in is the dichotomy finite rank/infinite rank, and we will only mention how it translates in our particular case.

We say that a definable subset D of K^n has *finite rank* if there is a positive integer m such that, if E is the smallest difference subfield of K over which there is a formula $\varphi(\bar{x})$ defining D, whenever $\bar{a} \in D$ then $\mathrm{tr.\,deg}(E(\bar{a})_\sigma/E) \leq m$. Otherwise we say that D has *infinite rank*. (For example, in our case, K^n, or any definable subset of K^n which is dense in K^n for the σ-topology, has rank ωn.)

It turns out that definable sets of finite rank can be analysed in terms of definable sets of rank 1, see the definition below. The structure of definable sets of infinite rank is much more complicated.

Definition. *Let $E = \mathrm{acl}_\sigma(E) \subset K$, and assume that $D \subseteq K^n$ is definable over E. We say that D has rank 1 if $D \nsubseteq E$, and whenever $\bar{a} \in D$ and $F = \mathrm{acl}_\sigma(F)$ contains E, then either $\bar{a} \in F$, or $\bar{a}$ and F are independent over E.*

Let $D \subseteq K^n$ be a definable set of rank 1. Then any definable subset of D is either finite or has rank 1. It may happen that D can be partitioned into infinitely many definable sets, but the definable sets appearing in this partition will be defined using quantifiers, and will not be defined uniformly. For quantifier-free definable subsets of D one has the following result: let X_i, $i \in I$, be a set of quantifier-free definable subsets of K^n, and assume that the sets $X_i \cap D$, $i \in I$, are non-empty infinite and form a partition of D. Then I is finite.

2.7. Properties of the rank

Let K be a generic difference field, and $G \subseteq K^n$ a group definable in K, and H a definable subgroup of G, both defined over $E = \mathrm{acl}_\sigma(E)$.

(1) The rank is preserved under definable bijection.
(2) If $D \subseteq \mathrm{Fix}(\sigma)^n$ is definable in K, then the rank of D equals the dimension of its Zariski closure.
(3) G and H have the same rank if and only if $[G : H] < \infty$.
(4) If G has finite rank, we can then define a notion of *generic* of G, namely $\bar{a} \in G$ is a generic of G over F containing E, if and only if $\mathrm{tr.\,deg}(F(\bar{a})_\sigma/F)$ is maximal. This notion extends the notion of generic of algebraic groups.
(5) Assume that G is a subgroup of some algebraic group defined over K. If $\bar{G}$ is the σ-closure of G, then $[\bar{G} : G] < \infty$.

2.8. Examples of sets of rank 1

We start with some obvious examples. If $D \subseteq K^n$ is infinite and defined over the algebraically closed difference field E and is such that $\mathrm{tr.\,deg}(E(\bar{a})_\sigma/E) \leq 1$ for every $\bar{a} \in D$, then certainly D has rank 1. For instance the sets defined by the equations $\sigma(x) = x^2 + 1$, $\sigma(x) = x^2$, or $\sigma(x) = x$ have rank 1.

One can also show that the sets defined by the equations $\sigma^2(x) = x^2 + 1$ and $\sigma^2(x) = x^2$ have rank 1. However the set defined by $\sigma^2(x) = x$ has rank > 1: let a satisfy $\sigma^2(x) = x$, and such that $\mathrm{tr.\,deg}(\mathbb{Q}(a)_\sigma/\mathbb{Q}) = 2$; then a and $a\sigma(a)$ are not independent over $\mathbb{Q}$, but $a \notin \mathrm{acl}_\sigma(\mathbb{Q}, a\sigma(a)) = \mathbb{Q}(a\sigma(a))^{\mathrm{alg}}$.

With some work, one can show that in characteristic 0 the set defined by the equation $\sigma^2(x) = (x^2 + 1)^2 + 1$ has rank 1 (one shows that the only infinite proper σ-closed subset of this set is, up to a finite set, defined by the equation $\sigma(x) = x^2 + 1$).

3. Some of the Main Results on Difference Fields and How to Use Them

In this section, I will state some of the main model-theoretic results on generic difference fields, and discuss some of their applications. I will start with a striking result, proved independently by Hrushovski [9] and Macintyre [16]. I chose to state this result first, because of its prettiness and also because one of the motivations in the study of generic difference fields was the hope that such a result would be true. I then continue with the dichotomy results ([4, 5]) which are at the heart of many of the applications, and with the results of Hrushovski on definable subgroups of algebraic groups ([8]). I then state some applications, to the Manin-Mumford conjecture [8], to the Jacobi conjecture for difference fields [10], to the Denis conjecture [18], and to Galois groups.

Let me mention a very regrettable fact: the results of Hrushovski and Macintyre cited here have not yet appeared in print!! The existing preliminary preprints are however available, have been carefully checked, and except for minor details, are correct. A survey of [8] can be found in [1] and in [3]. Special cases of the proof are given in [17], see also [2] for the results on definable groups.

3.1.

For each prime p and $q = p^n$, consider the difference field $F_q = (\mathbb{F}_p^{\mathrm{alg}}, \sigma_q)$, where $\sigma_q\colon x \mapsto x^q$. Let $\mathcal{U}$ be a non-principal ultrafilter on the set Q of prime powers, and consider $(K, \sigma) = \prod_{q \in Q}(\mathbb{F}_p^{\mathrm{alg}}, \sigma_q)/\mathcal{U}$.

Theorem 1. (Hrushovski [9], Macintyre [16]) *K is a generic difference field. Every generic difference field has the same invariant (see (2.4)(4)) as some non-principal ultraproduct of the F_q's.*

Remarks.

(1) *Note the following logical content of this theorem. Let φ be a sentence. All generic difference fields satisfy φ if and only if all but finitely many F_q satisfy φ. Equivalently (replacing φ by its negation): there is a generic difference field satisfying φ if and only if the set of difference fields F_q satisfying φ is infinite.*

(2) *If K is a difference field, then there is an index set I, and an ultrafilter $\mathcal{U}$ on $Q \times I$, such that K embeds into $\prod_{(q,i)} F_q/\mathcal{U}$.*

(3) *For each $q \in Q$, let K_q be an algebraically closed field of characteristic p with the distinguished automorphism $\sigma_q\colon x \mapsto x^q$. If $\mathcal{U}$ is a non-principal ultrafilter on Q then $\prod_{q \in Q} K_q/\mathcal{U}$ is a generic difference field.*

3.2.

Hrushovski's proof of theorem 1 gives estimates on the number of points of σ-closed subsets of finite rank of F_q^n for q sufficiently large, which are similar to the Lang-Weil estimates for the number of $\mathbb{F}_q$-rational points of varieties defined over $\mathbb{F}_q$. His bounds are of the form $||D| - aq^d| < Cq^{d-1/2}$, where a, C and d depend only on the algebraic family of the defining equations of the σ-closed set D. Let me mention an easy corollary of this theorem and of Hrushovski's bound:

Corollary. ([9]) *Let K be a generic difference field.*

(1) *Let G be an algebraic group defined over K, and B a definable subgroup of $G(K)$ of finite rank. Assume that the tuple of difference polynomials $f(\bar{X})$ restricts to an endomorphism of B. Then $[B : f(B)] = |\operatorname{Ker}(f) \cap B|$.*

(2) *Similarly, if U is a σ-closed subset of K^n and the tuple $f(\bar{X})$ defines an injective map $U \to U$, then $f(U) = U$.*

Proof. Both proofs are very much inspired by Ax's proof of (2) in case U is an algebraic set and f is a tuple of rational functions. Let me sketch a proof of (1). The proof of (2) is similar in spirit, but uses the fact that if U is a definable subset of F_q^n, then $U = \bigcup_m (U \cap \mathbb{F}_{p^m}^n)$.

Let $\bar{c}$ be a tuple over which the algebraic group G and its group operation, the polynomials f and the subgroup B are defined. Let $\varphi(\bar{t}, \bar{x})$, $\psi(\bar{t}, \bar{x}, \bar{y}, \bar{z})$ be the formulas of the language of rings such that $\varphi(\bar{c}, \bar{x})$ defines $G(K)$ in K, and $\psi(\bar{c}, \bar{x}, \bar{y}, \bar{z})$ defines the graph of multiplication. Let $\theta(\bar{c}, \bar{x})$ and $\eta(\bar{c}, \bar{x}, \bar{y})$ be the formulas which define B and the graph of f respectively within $G(K)$. The following properties of the tuple $\bar{c}$ can then be expressed by formulas in the variables $\bar{t}$:

(A1) The formula $\psi(\bar{t}, \bar{x}, \bar{y}, \bar{z})$ is the graph of a group operation on the set S defined by $\varphi(\bar{t}, \bar{x})$.

(A2) The formula $\theta(\bar{t}, \bar{x})$ defines a subgroup H of S, and the formula $\eta(\bar{t}, \bar{x}, \bar{y}) \wedge \theta(\bar{t}, \bar{x})$ defines the graph of a group homomorphism g which takes its values in H. All the elements of H satisfy a difference equation over $\bar{t}$ of order $\leq n$ (some fixed n).

(A3m) If $\operatorname{Ker}(g)$ has size m, then $[H : g(H)] = m$.

(A4m) If $\operatorname{Ker}(g)$ has size $> m$, then $[H : g(H)] > m$.

It suffices to show that every tuple $\bar{d}$ in F_q which satisfies (A1) and (A2) also satisfies (A3m) and (A4m) for all $m \geq 1$. But this is clear: by Hrushovski's estimates on the number of points, we get that if (A2) is satisfied by $\bar{d}$ then the set defined by $\theta(\bar{d}, \bar{x})$ is finite; hence $\bar{d}$ also satisfies (A3m) and (A4m).

3.3. Modularity, and the dichotomy theorem

I am not going to give the model-theoretic definitions of some of the terms used below (modularity, full stability): model theorists know them, and the definition will not bring any illuminating insight to the other mathematicians. Let me simply say that stability and modularity are abstract properties, which were introduced in a very general model-theoretic context, and which turn out, in concrete examples, to have a totally algebraic translation with very useful consequences. Full stability is a generalization of stability to an unstable context like ours. That a field is never modular, and that a modular set has little more structure than a vector space. Maybe the example given below and the applications to algebraic groups will help to see the strength of the modularity property. I first give the translation of these concepts in our particular case ((2) is not simply a translation, but needs some argumentation):

Definitions. *Let K be a generic difference field, $D \subseteq K^n$ a definable subset, defined over some $E = \mathrm{acl}_\sigma(E) \subset K$.*

(1) *D is modular if whenever $\bar{a}_1, \dots, \bar{a}_m \in D$ and $\bar{b}$ is a tuple of elements of K, then $(\bar{a}_1, \dots, \bar{a}_m)$ and $\bar{b}$ are independent over $C = \mathrm{acl}_\sigma(E, \bar{a}_1, \dots, \bar{a}_m) \cap \mathrm{acl}_\sigma(E, \bar{b})$.*

(2) *D is fully stable if for any $\bar{a} \in D$ and $F = \mathrm{acl}_\sigma(F)$ containing E, any F-isomorphism $F\,\mathrm{acl}_\sigma(E, \bar{a})_\sigma \to K$ extends to $\mathrm{acl}_\sigma(F, \bar{a})$.*

3.4. Examples

We said above that a field k is never modular. Let K be a generic difference field, and let p equal either 1 if $\mathrm{char}(K) = 0$, or $\mathrm{char}(K)$ otherwise. We will call a field defined by a formula of the form $\sigma^n(x) = x^{p^m}$, where $n \geq 1$, $m \in \mathbb{Z}$, a *fixed field*. A particular example is of course $\mathrm{Fix}(\sigma)$, and results which one proves for $\mathrm{Fix}(\sigma)$ are valid for the other fixed fields, because one can show that if K is a generic difference field, and $\tau\colon x \mapsto \sigma^n(x^{p^m})$ for some $n \geq 1$, $m \in \mathbb{Z}$, then the difference field (K, τ) is also generic.

One can show that no infinite definable subset of $k = \mathrm{Fix}(\sigma)$ is modular. The non-modularity of k is witnessed by the existence of a 2-dimensional family of curves on k^2, namely the curves $y = ax + b$ ($a, b \in k$). Modularity of a set D of rank 1 can be defined as the non-existence of a 2-dimensional family of curves (or rank 1 sets) on D^2. One can also show that definable sets of infinite rank are not modular.

In contrast with the fixed field case, assume that the characteristic of K is not 2, and consider the subgroup B of $\mathbb{G}_m(K)$ defined by the equation $\sigma(x) = x^2$. Note that if $a, b \in B$ and $f \in K[X, Y]$, then $f(a, b) \in B$ implies that $\sigma(f)(a^2, b^2) = f(a, b)^2$. Not very surprisingly, if such an f defines a map $B \times B \to B$, then $f(X, Y) = cX^i Y^j$ for some $c \in B$ and $i, j \in \mathbb{N}$. One can show that indeed B is modular.

Theorem 3 below, generalizing to our context an older result of Hrushovski-Pillay [11], will bring more information on the type of applications one can expect from the notion of modularity. But first we state the dichotomy theorem:

Theorem 2. *([4, 5]) Let K be a generic difference field, and let $D \subseteq K^n$ be definable over $E = \mathrm{acl}_\sigma(E)$, and of finite rank. Then either D is modular, or there is an infinite definable subset U of D, a subfield k of K defined by a formula $\sigma(x) = x$ or $\sigma^n(x) = x^{p^m}$ for some $n \geq 1$, $m \in \mathbb{Z}$ if the characteristic is $p > 0$, and a definable map $f\colon U \to k$, with $f(U)$ infinite.*

If the characteristic is 0 and D is modular, then D is fully stable, which implies that D with all the structure induced from K is superstable.

3.5. A criterion for modularity

In characteristic 0, there is a criterion for modularity of definable sets of rank 1: let $D \subseteq K^n$ be definable over $E = \mathrm{acl}_\sigma(E)$, and of rank 1. Then D is modular if and only for every $\bar{a} \in D$, $\bar{a} \notin E$, one of the following happens:

(i) $\mathrm{tr.\,deg}(E(\bar{a})_\sigma/E) > 1$, or
(ii) $\mathrm{tr.\,deg}(E(\bar{a})_\sigma/E) = 1$ and the set $\{[E(\bar{a},\sigma^k(\bar{a})) : E(\bar{a})] \mid k \in \mathbb{Z}\}$ is unbounded.

This criterion is slightly unsatisfactory for the following reason: given a definable set D of finite rank, it is sometimes difficult to determine whether D has rank 1 or higher rank. The criterion does not generalize properly to the positive characteristic case. Indeed, in characteristic $p > 0$, the set defined by the equation $\sigma^2(x) = x^p$ has rank 1 and satisfies (i) but is certainly non-modular. In case $\mathrm{tr.\,deg}(E(\bar{a})_\sigma/E) \leq 1$ for every $\bar{a} \in D$, then (ii) can be adapted to give a criterion of modularity of D.

Theorem 3. ([4, 5]) *Let K be a generic difference field, and G an algebraic group definable over K. Assume that $B \subseteq G(K)$ is a definable modular subgroup of $G(K)$, defined over $E = \mathrm{acl}_\sigma(E)$. Let X be a quantifier-free definable subset of $G(K)$. Then $X \cap B$ is a (finite) Boolean combination of cosets of σ-closed subgroups of B (i.e., subgroups of the form $B \cap H$, where H is σ-closed). If X is σ-closed, then $X \cap B$ is a finite union of cosets of σ-closed subgroups of B. The σ-closed subgroups of B are defined over E.*

Assume that the characteristic of K is 0. If X is a definable subset of $G(K)$, then $X \cap B$ is a (finite) Boolean combination of cosets of definable subgroups of B. The definable subgroups of B are defined over E.

Remark. *In positive characteristic, this last statement is not always true. We have examples of modular definable subgroups S and T of $\mathbb{G}_a(K)$, with $S \cap T$ finite, such that there is a definable non-degenerate bilinear map $q: S \times T \to \mathbb{F}_p$. If $b \in T$, then $\{a \in S \mid q(a,b) = 0\}$ is a definable subgroup of S, whose definition needs parameters from outside S. See [4].*

3.6.

Theorem 3 is not very useful unless one has a good criterion for modularity of definable subgroups of algebraic groups. Using model theory, one can reduce to the case where G is a simple abelian group. In the case of $\mathbb{G}_m$ and of a simple abelian variety, a good description is given by Hrushovski in [8]. In particular, he shows that if A is a simple abelian variety, and $A(K)$ has a definable non-modular subgroup, then A is isomorphic to an abelian variety A' defined over some fixed field. From this fact, the study of modularity of definable subgroups of semi-abelian varieties can be reduced to the case where $A = \mathbb{G}_m$ or A is a simple abelian variety defined over $\mathrm{Fix}(\sigma)$. Before stating his results in that simple case, let me say that he completely describes, up to commensurability, definable subgroups of abelian varieties and of tori. In characteristic 0, all definable subgroups of finite rank of $\mathbb{G}_a(K)$ are non-modular. In positive characteristic however, there are many examples of modular subgroups of $\mathbb{G}_a(K)$. As far as I know, however, we do not have as nice a description.

Theorem 4. ([8]) *Let K be a generic difference field of characteristic 0, A be a simple abelian variety defined over $\mathrm{Fix}(\sigma)$, and R its ring of algebraic endomorphisms.*

(1) *If B is a definable subgroup of $A(K)$, then there is $n \in \mathbb{N}$, and $a_0, \ldots, a_n \in R$ such that if $C = \{\bar{a} \in A(K) \mid \sum_{i=0}^{n} a_i(\sigma^i(\bar{a})) = 0\}$, then $C \cap B$ has finite index in B and in C. If B has infinite rank, then $B = A(K)$.*

(2) *Let B and C be defined as in (1). Then B is modular if and only if the polynomial $f(T) = \sum_{i=0}^{n} a_i T^i$ is relatively prime (in $\mathbb{Q} \otimes_{\mathbb{Z}} R[T]$) to all polynomials of the form $T^m - 1$, $m \geq 1$.*

This criterion can be suitably modified in characteristic $p > 0$. In characteristic 0, modularity is equivalent to having "no non-trivial connection" with $\mathrm{Fix}(\sigma)$, and the above criterion tells us when this happens. In positive characteristic, modularity means having no non-trivial connection with any of the fixed fields.

3.7.

Similar results hold for the multiplicative group $\mathbb{G}_m$, and because modularity is preserved under short exact sequences, one can obtain e.g.:

Theorem 5. ([8]) *Let K be a generic difference field of characteristic 0, A a semi-abelian variety defined over $\mathrm{Fix}(\sigma)$, and $f(T) \in \mathbb{Z}[T]$ a polynomial having no roots of unity among its roots. Then the subgroup of $A(K)$ defined by the equation $f(\sigma)(\bar{x}) = 0$ is modular.*

3.8.

Theorems 4 and 5 were used by Hrushovski in his proof of the Manin-Mumford conjecture over number fields [8]. Let me give the idea of his proof when A is a semi-abelian variety, and we are interested in the intersection of a subvariety X of A with the prime-to-p torsion of A, $\mathrm{Tor}_{p'}(A)$, for p a prime of good reduction. Using results of Weil on abelian varieties defined over finite fields and the functional equation of the Frobenius automorphism, one finds an automorphism σ of $\mathbb{Q}^{\mathrm{alg}}$, and a polynomial $f(T)$ with integral coefficients, of degree $\leq 2\dim(A)$, with no root of unity among its roots, and with absolute sum of the coefficients bounded by some function of p, and such that $f(\sigma)$ vanishes on $\mathrm{Tor}_{p'}(A)$. Let $B = \mathrm{Ker}\, f(\sigma) \subseteq A(K)$, in some generic difference field K extending $(\mathbb{Q}^{\mathrm{alg}}, \sigma)$. Then we look at the number of σ-irreducible components of $X \cap B$. By modularity of B, we know that $X \cap B$ is a union of translates of definable subgroups of B, which means that its Zariski closure is a union of translates of algebraic subgroups of A. Moreover, because we have a bound on the complexity of the σ-equations defining B, we get a bound on the number of σ-irreducible components of $X \cap B$, which is of the form $c\deg(X)^e$, where c, e depend only on A. Some simple arguments then give us that $X \cap \mathrm{Tor}_{p'}(A)$

is the union of m sets of the form $c_i + \mathrm{Tor}_{p'}(A_i)$, where each A_i is a subvariety of A and $m \leq c \deg(X)^e$.

3.9. The Jacobi conjecture for difference fields

We have n algebraic difference equations $u_1(\bar{x}) = 0, \ldots, u_n(\bar{x}) = 0$, in the variables $\bar{x} = (x_1, \ldots, x_n)$, with coefficients in some difference field E, which we will assume contained in a generic difference field K. For $k, i = 0, \ldots, n$, let h_k^i be the order of the equation $u_i(\bar{x}) = 0$ with respect to the variable x_k (if x_k does not occur at all, we let $h_k^i = -\infty$). We then define $H = \max_{\theta \in \mathrm{Sym}(n)} \sum_{k=1}^n h_k^{\theta(k)}$. The conjecture states that if the σ-closed set X defined by the equations $u_i(\bar{x}) = 0, i = 1, \ldots, n$, has finite rank, then the *order of X*, $\max_{\bar{a} \in X} \mathrm{tr.\,deg}(E(\bar{a})_\sigma / E)$, is bounded by H.

This conjecture is the transposition to difference fields of the original Jacobi conjecture for differential fields and differential equations. The Jacobi conjecture for differential fields remains open. Some cases of the Jacobi conjecture for difference fields have been established by Cohn and Lando. Hrushovski [10] establishes the conjecture for difference fields in all cases. His proof uses a strong form of Bezout's theorem:

For $i = 1, \ldots, n$, let S_i be a hypersurface in $(\mathbb{P}_1)^n$, of multi-degree $(d_1^i, \ldots, d_n^i)$ (see [8] or [7] for the definition of multi-degree). If Y denotes the intersection of the S_i, then the zero-dimensional components of Y have at most M points where $M = \sum_{\theta \in \mathrm{Sym}(n)} \prod_{k=1}^n d_k^{\theta(k)}$.

One then proves the conjecture by simply counting points in the structures F_q for q approaching ∞: if X_i is an irreducible component of X, then the order of X_i is given by $\log_q(|X_i(F_q)|)$ for $q \to \infty$. By Theorem 1, this gives the result in general. Note that his proof actually gives the bound on the order of any irreducible component of X of finite rank (i.e., X may also have components of infinite rank).

3.10.

Scanlon gives a positive solution of Denis's conjecture on Drinfeld modules in [18]. Let me explain the setting in a simple case:

Let K be an algebraically closed field of positive characteristic p and of positive transcendence degree. Consider the ring $\mathrm{End}_K(\mathbb{G}_a)$ of endomorphisms of $\mathbb{G}_a$ defined over K. Then $\mathrm{End}_K(\mathbb{G}_a)$ is isomorphic to the twisted polynomial ring $K[\sigma_p]$ (with $\sigma_p \colon x \mapsto x^p$). Let $A = \mathbb{F}_p[T]$ and view it as a subring of K, by identifying T with some transcendental $t \in K$. A *Drinfeld module (over A)* is given by a ring homomorphism $\varphi \colon A \to \mathrm{End}_K(A)$ so that if $\varphi(T) = \sum_{i=0}^n a_i \sigma_p^i$, then $a_0 = t$ and $a_n = 1$.

Theorem. ([18]) *Let φ be a Drinfeld module. Consider K^N as an A-module via φ. If X is a subvariety of K^N then the intersection of X with the A-torsion subgroup of K^N $(= \{\bar{x} \in K^N \mid \varphi(a)(\bar{x}) = 0$ for some non-zero $a \in A\})$ is a finite union of translates of A-torsion subgroups of algebraic subgroups of K^N.*

Again, this is proved by choosing an automorphism σ of some field L containing K, such that the functional equation satisfied by σ on the A-torsion subgroup of K defines a modular group in any generic difference field extending (L, σ).

3.11. Galois groups of difference equations

Let K be a generic difference field of characteristic 0, let A be an abelian variety defined over an algebraically closed subfield k of $\text{Fix}(\sigma)$, and let B be a modular subgroup of $A(K)$ defined by $f(\sigma)(\bar{x}) = 0$, for some $f(T) \in \text{End}(A)[T]$. Let $\bar{c}$ be transformally transcendental over k, let $F = k(\bar{c}, B)_\sigma$, and X the set of solutions of $f(\sigma)(\bar{x}) = \bar{c}$. Let $\bar{a} \in X$. By full stability of B, model theory tells us that the subgroup G of B consisting of those elements $\bar{g}$ such that there is an F-automorphism of the difference field $\text{acl}_\sigma(F, \bar{a})$ which sends $\bar{a}$ to $\bar{a} + \bar{g}$, is an intersection of definable subgroups of B of finite index in B. In particular, if B has no definable subgroup of finite index, then $G = B$.

References

[1] E. Bouscaren, *Théorie des modèles et conjecture de Manin-Mumford, [d'après Ehud Hrushovski]*, Séminaire N. Bourbaki exposé 870 (Mars 2000), to appear in Astérisque.

[2] Z. Chatzidakis, *Groups definable in ACFA*, in: Algebraic Model theory, B. Hart et al. ed., NATO ASI Series **C 496**, Kluwer Academic Publishers, 1997, 25–52.

[3] Z. Chatzidakis, *A survey on the model theory of difference fields*, to appear in: Proc. Workshop, MSRI Publications **39**, Cambridge University Press, 65–96.

[4] Z. Chatzidakis and E. Hrushovski, *Model theory of difference fields*, Trans. A.M.S. **351** (1999), 2997–3071.

[5] Z. Chatzidakis, E. Hrushovski and Y. Peterzil, *Model theory of difference fields, II: Periodic ideals and the trichotomy in all characteristics*, preprint, (1999).

[6] R. M. Cohn, *Difference algebra*, Tracts in Mathematics **17**, Interscience Pub., 1965.

[7] W. Fulton, *Intersection Theory*, Springer-Verlag New York, 1998.

[8] E. Hrushovski, *The Manin-Mumford conjecture and the model theory of difference fields*, to appear in Annals of Pure and Applied Logic.

[9] E. Hrushovski, *The first-order theory of the Frobenius*, preprint (1996).

[10] E. Hrushovski, Lecture at MSRI (June 98), private e-mail (Jan. 2000).

[11] E. Hrushovski and A. Pillay, *Weakly normal groups*, in: Logic Colloquium 85, North Holland, 1987, 233–244.

[12] E. Hrushovski and A. Pillay, *Groups definable in local fields and pseudo-finite fields*, Israel J. of Math. **85** (1994), 203–262.

[13] E. Hrushovski and A. Pillay, *Definable subgroups of algebraic groups over finite fields*, J. reine angew. Math. **462** (1995), 69–91.

[14] S. Lang and J.-P. Serre, *Sur les revêtements non-ramifiés des variétés algébriques*, Amer. J. of Maths. **79** (1957), 319–330.

[15] A. Macintyre, *Generic automorphisms of fields*, APAL **88** Nr 2–3 (1997), 165–180.

[16] A. Macintyre, *Nonstandard Frobenius*, in preparation.

[17] A. Pillay, *ACFA and the Manin-Mumford conjecture*, in: Algebraic Model theory, B. Hart et al. ed., NATO ASI Series **C 496**, Kluwer Academic Publishers, 1997, 195–205.

[18] T. Scanlon, *The diophantine geometry of the torsion of a Drinfeld module*, preprint 1999.

Université Paris 7
UFR de Mathématiques, Case 7012
2, place Jussieu
75251 Paris Cedex 05, France
E-mail address: zoe@logique.jussieu.fr

Geometric Aspects of Polynomial Interpolation in More Variables and of Waring's Problem

Ciro Ciliberto

Abstract. In this paper I treat the problem of determining the dimension of the vector space of homogeneous polynomials in a given number of variables vanishing with some of their derivatives at a finite set of general points in projective space. I will illustrate the geometric meaning of this problem and the main results and conjectures about it. I will finally point out its connection with the so-called Waring's problem for forms, of which I will also indicate the geometric meaning.

1. Introduction

The classical polynomial interpolation theory of functions in numerical analysis is based on the elementary fact, recalled in §2, that a polynomial of degree d in one variable is uniquely determined by its zeroes with their multiplicities. In the present expository paper I will deal with an extension of this property to polynomials in more variables. This extension has to do with linear systems of hypersurfaces in projective space with finitely many assigned base points of given multiplicities. The general setting is presented in §3, where I will also state the main problems. Among these, I will mainly discuss the *general dimensionality problem*, which, roughly speaking, can be stated as follows: given a linear system of hypersurfaces in projective space and finitely many general points with assigned multiplicities at each point, what is the dimension of the subsystem of the given one formed by all hypersurfaces having at the given points at least the assigned multiplicities? This problem, which is elementary to state and somehow basic in algebraic geometry, has been considered, in one form or another, since the beginnings of this discipline. It can be traced back in Bezout's work in the XVIII century, it is present in Plücker, Cremona, M. Noether, Bertini, C. Segre etc. in the XIX century and contributions have been given in the XX century by Castelnuovo, Enriques, Severi, Terracini, among others. Its relations with other topics have been considered by several authors, e.g. [63] and [57]. So far however the problem is still unsolved in its generality. I will discuss here what is known about it and what are the techniques involved, what are the conjectures, the open problems and the connections of this problem with others.

In §4 I will concentrate on the case of linear systems of plane curves. I will state and explain the main conjectures on the general dimensionality problem and in §5 I will illustrate the main results about it. In §6 I turn to the higher dimensional case and I will state and discuss the main result, which is a theorem of Alexander and Hirschowitz. In the last section §7 I will show the connection of Alexander-Hirschowitz's theorem with a famous algebraic problem, Waring's problem, of which I will illustrate the geometric meaning. This will lead us to the geometry of secant varieties and to the classification of defective varieties, the ones whose secant varieties have dimension smaller than expected, about which I will recall the main known results.

This is a survey paper, which contains and expands the material covered by my talk at the ECM of Barcelona 2000. For other interesting surveys on the subject I refer the reader to [38] and [61].

2. Polynomial Interpolation

A polynomial $f(x) = a_0 + a_1 x + \cdots + a_d x^d \in K[x]$ of degree at most d over a field K depends on $d + 1$ parameters, namely its coefficients $a_0, a_1, \ldots, a_d$. If we fix $d + 1$ distinct points $x_0, \ldots, x_d \in \mathbf{A}_K^1$ on the affine line over K and set the values:

$$f(x_i) = f_i \in K, \quad i = 0, \ldots, d \tag{1}$$

then, by linear algebra, there is some polynomial $f(x)$ satisfying the conditions (1). Moreover this polynomial is unique. The reason for this is that there is no non-zero polynomial of degree d with zeros at $x_0, \ldots, x_d$.

More generally, if we fix distinct points $x_1, \ldots, x_d \in \mathbf{A}_K^1$ and positive integers $m_1, \ldots, m_h$ such that $m_1 + \cdots + m_h = d + 1$ and set the values of the derivatives:

$$f^{(j-1)}(x_i) = f_{i,j}, \quad i = 1, \ldots, h, \quad j = 1, \ldots, m_i \tag{2}$$

again there is a unique polynomial $f(x)$ satisfying the conditions (2). This is because there is no non-zero polynomial of degree d with zeros of multiplicities $m_1, \ldots, m_h$ at $x_0, \ldots, x_h$.

In particular, the following happens. Let $F(x)$ be a differentiable function of a real variable. Fix $x_1, \ldots, x_h$ distinct points where $F(x)$ is defined, and positive integers $m_1, \ldots, m_h$ such that $m_1 + \cdots + m_h = d + 1$. Then there is a unique polynomial $f(x)$ of degree d satisfying (2) with $f_{i,j} = F^{(j-1)}(x_i)$, $i = 1, \ldots, h$, $j = 1, \ldots, m_i$. The polynomial $f(x)$ approximates $F(x)$ and of course the approximation is better and better as d increases. This approximating procedure of differentiable functions is called *polynomial interpolation*.

What is the situation in $n \geq 2$ variables? A polynomial $f(x_1, \ldots, x_n) \in K[x_1, \ldots, x_n]$ of degree at most d depends on $N_{n,d} + 1 := \binom{d+n}{n}$ parameters, namely its coefficients. Again we may fix points $p_i = (x_{1i}, \ldots, x_{ni}) \in \mathbf{A}_K^n$, $i =$

$1, \dots, h$, in the n-dimensional affine space over the field K and integers $m_1, \dots, m_h$ such that:

$$\sum_{i=1}^{h} \binom{m_i + n - 1}{n} = N_{n,d} + 1 \tag{3}$$

and we may impose $D^{(j-1)} f(x_i) = 0$, $i = 1, \dots, h$, $j = 1, \dots, m_i$, where $D^{(k)}$ is any derivative of order k. Notice that, according to (3), the number of conditions imposed equals the number of parameters on which the polynomials depend. In analogy with the one-variable case we may then ask: *is the resulting polynomial f identically zero?*

This is the question I will mainly deal with in this talk. Interestingly enough, there is yet no general answer to it and I will try to indicate the classical and recent developments on the subject and the techniques involved. It is useful to address the above question in a more general setting. This is what I will do next.

3. Linear Systems with Multiple Base Points

Let X be a smooth, irreducible, projective, complex variety of dimension n. Let $\mathcal{L}$ be a complete linear system of divisors on X. I will often abuse notation and denote with the same letter a linear system $\mathcal{L}$ and the corresponding line bundle on X. Fix $p_1, \dots, p_h$ distinct points on X and $m_1, \dots, m_h$ positive integers. I will denote by $\mathcal{L}(-\sum_{i=1}^{h} m_i p_i)$ the sublinear system of $\mathcal{L}$ formed by all divisors in $\mathcal{L}$ having multiplicity at least m_i at p_i, $i = 1, \dots, h$.

Having a point of multiplicity m at a fixed point p imposes $\binom{m+n-1}{n}$ linear conditions on the divisors of $\mathcal{L}$. Indeed this translates, for the equation $f(x_0, \dots, x_n) = 0$ of a divisor of $\mathcal{L}$ in local coordinates $(x_0, \dots, x_n)$ centered at p, in the vanishing of all monomials of degree at most $m - 1$ appearing in the Taylor expansion of $f(x_0, \dots, x_n)$. Thus, it makes sense to define the *expected dimension* of $\mathcal{L}(-\sum_{i=1}^{h} m_i p_i)$ as:

$$\operatorname{expdim}(\mathcal{L}(-\sum_{i=1}^{h} m_i p_i)) := \max\{\dim(\mathcal{L}) - \sum_{i=1}^{h} \binom{m_i + n - 1}{n}, -1\}$$

and one clearly has:

$$\dim(\mathcal{L}(-\sum_{i=1}^{h} m_i p_i)) \geq \operatorname{expdim}(\mathcal{L}(-\sum_{i=1}^{h} m_i p_i)). \tag{4}$$

The system $\mathcal{L}(-\sum_{i=1}^{h} m_i p_i)$ is said to be *non-special* if the equality holds in (4). Otherwise it is said to be *special*. Notice that, by definition, a system which is empty is non-special. For a non-empty system instead non-speciality means that the imposed conditions are independent. It is natural to expect that most systems are non-special. The *dimensionality problem* can be posed as follows: *classify all special systems.*

Put in this way, the problem is too complicated. Indeed one moment of reflection shows that the answer depends not only on the numerical data involved in it, but also on the position of the points $p_1, \ldots, p_h$ on X. However, $\dim(\mathcal{L}(-\sum_{i=1}^{h} m_i p_i))$ is an *upper-semicontinuous* function in the position of the points $p_1, \ldots, p_h$, hence it reaches its minimum for $p_1, \ldots, p_h$ in *general position* on X.

When the points $p_1, \ldots, p_h$ are sufficiently general on X, we set:

$$\mathcal{L}(-\sum_{i=1}^{h} m_i p_i) := \mathcal{L}(m_1, \ldots, m_h)$$

or equivalently:

$$\mathcal{L}(m_1, \ldots, m_h) := \mathcal{L}(m_1^{l_1}, \ldots, m_t^{l_t})$$

if $l_1 + \cdots + l_t = h$ and m_i is repeated l_i times. The case $t = 1$ is called *homogeneous*. The case $t = 2$, $l_1 = 1$ is called *quasi-homogeneous*.

Then we define the *general dimension* of the linear system $\mathcal{L}(-\sum_{i=1}^{h} m_i p_i)$ as:

$$\mathrm{gendim}(\mathcal{L}(-\sum_{i=1}^{h} m_i p_i)) := \dim(\mathcal{L}(m_1, \ldots, m_h))\,.$$

With this definition in mind, the dimensionality problem splits as:

(i) the *general dimensionality problem*: is $\mathrm{gendim}(\mathcal{L}(-\sum_{i=1}^{h} m_i p_i))$ equal to the expected dimension? Or rather, put in a different, but equivalent way: classify all systems $\mathcal{L}(m_1, \ldots, m_h)$ which are special;

(ii) the *hard dimensionality problem*: describe the stratification of X^h determined by the closures of the loci of h-tuples of points where

$$\dim(\mathcal{L}(-\sum_{i=1}^{h} m_i p_i)) > \mathrm{gendim}(\mathcal{L}(-\sum_{i=1}^{h} m_i p_i))\,.$$

The hard dimensionality problem has never been systematically explored. In the rest of this paper I will therefore stick to the general dimensionality problem.

Of course one can also ask more refined questions about systems of the type $\mathcal{L}(m_1, \ldots, m_h)$, like: describe its base locus, its general element, etc. Very little is known about this kind of questions and we will only touch upon some of them later on.

Notice that the general dimensionality problem is equivalent to the problem of determining the *Hilbert function* of the 0-dimensional subscheme of $\mathbf{P}^2$ given by the union of h general *fat points* of given multiplicities. One may of course ask for more refined questions like: what is the resolution of the ideal sheaf of this 0-dimensional scheme? This is also an active field of research, which I will not report on here, referring the reader, for example, to the recent paper [34] for information and references.

As in the interpolation problem, the general dimensionality problem is trivial in one variable, namely in the case of curves, i.e. if $n = \dim(X) = 1$. In this case all systems $\mathcal{L}(m_1, \ldots, m_h)$ are non-special. Another case which never causes speciality is when all the points have multiplicity one, i.e. $m_1 = \cdots = m_h = 1$: requiring the divisors of a non-empty linear system to contain a general point of the variety (e.g. a non-base point of the system) certainly imposes one condition. By contrast, the problem becomes quite difficult in more variables, namely as soon as $n = \dim(X) \geq 2$, and with higher multiplicities, namely $m_1, \ldots, m_h \geq 2$, the situation which we will consider from now on.

A first wise reduction of the problem is to consider, for the time being, particular varieties X and linear systems $\mathcal{L}$ on them. From this viewpoint, the first obvious choice is to take $X = \mathbf{P}^n$ and $\mathcal{L} = \mathcal{L}_{n,d} := |\mathcal{O}_{\mathbf{P}^n}(d)|$ the system of all hypersurfaces of degree d in $\mathbf{P}^n$. It should be clear to the reader that, in this setting, the problem essentially coincides with the original *interpolation problem* for polynomials in more variables considered in §2.

In this case

$$\operatorname{expdim}(\mathcal{L}_{n,d}(-\sum_{i=1}^{h} m_i p_i)) = \max\{\operatorname{virtdim}(\mathcal{L}_{n,d}(-\sum_{i=1}^{h} m_i p_i)), -1\}$$

where:

$$\operatorname{virtdim}(\mathcal{L}_{n,d}(-\sum_{i=1}^{h} m_i p_i)) := \binom{d+n}{n} - 1 - \sum_{i=1}^{h} \binom{m_i + n - 1}{n}$$

is the so-called *virtual dimension* of $\mathcal{L}_{n,d}(-\sum_{i=1}^{h} m_i p_i)$.

4. The Planar Case

I will consider now the case $X = \mathbf{P}^2$, in which there is a very precise conjecture about the general dimensionality problem, namely the Harbourne-Hirschowitz Conjecture 4.8. This section is devoted to stating this important conjecture.

The prototype of this conjecture goes back to B. Segre [78] who apparently was the first one to stress that speciality yields reducibility and even non-reducedness of the general curve of the involved linear systems with general multiple base points.

Conjecture 4.1. (B. Segre, 1961) *If a linear system of plane curves with general multiple base points $\mathcal{L}_{2,d}(m_1, \ldots, m_h)$ is special, then its general member is non-reduced, namely the linear system has, according to Bertini's theorem, some multiple fixed component.*

B. Segre's conjecture has been made more precise by A. Gimigliano [37] in 1987, on the basis of various examples, which we will partly mention in a moment:

Conjecture 4.2. (A. Gimigliano, 1987) *Consider a linear system of plane curves with general multiple base points $\mathcal{L}_{2,d}(m_1,\ldots,m_h)$. Then one has the following possibilities:*

(i) *the system is non-special and its general member is irreducible;*

(ii) *the system is non-special, its general member is reduced, reducible, its fixed components are all rational curves, but at most one (this may occur only if the system has dimension 0), the general member of its movable part is either irreducible or composed of rational curves in a pencil;*

(iii) *the system is non-special of dimension 0 and it consists of a unique multiple elliptic curve;*

(iv) *the system is special and it has some multiple rational curve as a fixed component.*

I want now to state the Harbourne-Hirschowitz conjecture. I will later explain, in §5, the relations among the various conjectures. In order to do so, let us consider the blow-up $\pi\colon \tilde{\mathbf{P}}^2 \to \mathbf{P}^2$ of the plane $\mathbf{P}^2$ at $p_1,\ldots,p_h$. Let $E_1,\ldots,E_h$ be the exceptional divisors corresponding to the blown-up points $p_1,\ldots,p_h$ and let H be the pull-back of a general line of $\mathbf{P}^2$ via π. The strict transform of the system $\mathcal{L}:=\mathcal{L}_{2,d}(-\sum_{i=1}^{h} m_i p_i)$ is the system $\tilde{\mathcal{L}} = |dH - \sum_{i=1}^{h} m_i E_i|$.

Consider two linear systems of this type $\mathcal{L}:=\mathcal{L}_{2,d}(-\sum_{i=1}^{h} m_i p_i)$ and $\mathcal{L}':=\mathcal{L}_{2,d}(-\sum_{i=1}^{h} m_i' p_i)$. We define their *intersection product* by using the intersection product of their strict transforms on $\tilde{\mathbf{P}}^2$, i.e. we set:

$$\mathcal{L} \cdot \mathcal{L}' := \tilde{\mathcal{L}} \cdot \tilde{\mathcal{L}}' = dd' - \sum_{i=1}^{h} m_i m_i'$$

as dictated by the classical Bezout's theorem. Also consider the *anticanonical class* $-K := -K_{\tilde{\mathbf{P}}}^2$ of $\tilde{\mathbf{P}}^2$ corresponding to the linear system $\mathcal{L}_{2,3}(-\sum_{i=1}^{h} p_i)$, which, by abusing notation, we also denote by $-K$.

The adjunction formula tells us that the arithmetic genus $p_a(\tilde{\mathcal{L}})$ of a curve in $\tilde{\mathcal{L}}$ is:

$$p_a(\tilde{\mathcal{L}}) = \frac{\mathcal{L} \cdot (\mathcal{L} + K)}{2} + 1 = \binom{d-1}{2} - \sum_{i=1}^{h} \binom{m_i}{2}$$

which one defines to be:

$$g_{\mathcal{L}} = \text{the geometric genus of } \mathcal{L}.$$

This is the classical *Clebsch's formula*. The theorem of Riemann-Roch then says that:

$$\begin{aligned}
\dim(\mathcal{L}) = \dim(\tilde{\mathcal{L}}) &= \frac{\mathcal{L} \cdot (\mathcal{L} - K)}{2} + h^1(\tilde{\mathbf{P}}^2, \tilde{\mathcal{L}}) - h^2(\tilde{\mathbf{P}}^2, \tilde{\mathcal{L}}) \\
&= \mathcal{L}^2 - g_{\mathcal{L}} + 1 + h^1(\tilde{\mathbf{P}}^2, \tilde{\mathcal{L}}) = \text{virtdim}(\mathcal{L}) + h^1(\tilde{\mathbf{P}}^2, \tilde{\mathcal{L}})
\end{aligned} \tag{5}$$

because clearly $h^2(\tilde{\mathbf{P}}^2, \tilde{\mathcal{L}}) = 0$. Hence:

$$\mathcal{L} \text{ is non-special} \Leftrightarrow h^0(\tilde{\mathbf{P}}^2, \tilde{\mathcal{L}}) \cdot h^1(\tilde{\mathbf{P}}^2, \tilde{\mathcal{L}}) = 0. \tag{6}$$

According to a recent terminology, this is expressed by saying that $\mathcal{L}$, or rather $\tilde{\mathcal{L}}$, has *natural cohomology*, meaning that at most one of the cohomology groups of the line bundle $\tilde{\mathcal{L}}$ is different from zero.

Now we can see how, in this setting, special systems naturally arise. Indeed let us look for an irreducible curve C on $\tilde{\mathbf{P}}^2$, corresponding to a linear system $\mathcal{L}$ on $\mathbf{P}^2$, which is *expected to exist* but *its double is not expected to exist*. However crazy this requirement may appear, it translates in the following set of inequalities:

$$\text{virtdim}(\mathcal{L}) \geq 0$$
$$g_\mathcal{L} \geq 0 \tag{7}$$
$$\text{virtdim}(2\mathcal{L}) \leq -1.$$

This system is equivalent to:

$$C^2 - C \cdot K \geq 0$$
$$C^2 + C \cdot K \geq -2 \tag{8}$$
$$2C^2 - C \cdot K \leq 0$$

and it has the only solution:

$$C^2 = C \cdot K = -1$$

which makes all the inequalities in (7) and (8) equalities. Accordingly C is a *rational curve*, i.e. a curve of genus 0, with *self-intersection* -1. Surface theorists call these curves (-1)-*curves*. A famous theorem of Castelnuovo's (see [10, pg. 27]) says that these are the only curves that can be contracted to smooth points via a birational morphism of the surface on which they lie to another surface. By abusing terminology the curve $\Gamma \subset \mathbf{P}^2$ corresponding to C is also called a (-1)-curve.

Example 4.3.

(i) *A line through two points $\mathcal{L}_{2,1}(-p-q)$ is a (-1)-curve. Hence $\mathcal{L}_{2,2}(-2p-2q)$, a conic with two double points, is special. Its virtual dimension is -1, namely it is expected not to exist, however it exists, and consists precisely of the line through the two points counted twice.*

(i′) *More generally $\mathcal{L} := \mathcal{L}_{n,2}(-\sum_{i=1}^h 2p_i)$ is special if $h \leq n$. Actually, quadrics in $\mathbf{P}^n$ singular at h independent points $p_1, \dots, p_h$ are cones with vertex the $\mathbf{P}^{h-1}$ spanned by $p_1, \dots, p_h$. Therefore the system is empty as soon as $h \geq n+1$, whereas, if $h \leq n$ one easily computes:*

$$\dim(\mathcal{L}) = \text{virtdim}(\mathcal{L}) + \binom{h}{2}.$$

(ii) *A conic through five general points $\mathcal{L}_{2,2}(-\sum_{i=1}^{5} p_i)$ is a (-1)-curve. Hence the system $\mathcal{L}_{2,4}(-\sum_{i=1}^{5} 2p_i)$ of quartics singular at five general points is special. Again its virtual dimension is -1, but it is non-empty, consisting of the conic through the five points counted twice.*

(ii′) *Similarly for $\mathcal{L} := \mathcal{L}_{n,4}(-\sum_{i=1}^{h} 2p_i)$. Speciality arises if:*

$$\mathrm{virtdim}(\mathcal{L}) < 0, \quad \mathrm{virtdim}(\mathcal{L}_{n,2}(-\sum_{i=1}^{h} p_i)) \geq 0$$

which occurs if and only if (n,h) is one of the pairs $(2,5)$, $(3,9)$, $(4,14)$.

More generally, one has special linear systems in the following situation. Let $\mathcal{L}$ be a linear system on $\mathbf{P}^2$ which is non-empty, let C be a (-1)-curve on $\tilde{\mathbf{P}}^2$ corresponding to a curve Γ on $\mathbf{P}^2$, such that $\tilde{\mathcal{L}} \cdot C = -N < 0$. Then C [resp. Γ] splits off with multiplicity N as a fixed component from all curves of $\tilde{\mathcal{L}}$ [resp. $\mathcal{L}$], and one has:

$$\tilde{\mathcal{L}} = NC + \tilde{\mathcal{M}}, \quad [\text{resp. } \mathcal{L} = N\Gamma + \mathcal{M}]$$

where $\tilde{\mathcal{M}}$ [resp. $\mathcal{M}$] is the residual linear system. Then one computes:

$$\dim(\mathcal{L}) = \dim(\mathcal{M}) \geq \mathrm{virtdim}(\mathcal{M}) = \mathrm{virtdim}(\mathcal{L}) + \binom{N}{2}$$

and therefore, if $N \geq 2$, then $\mathcal{L}$ is special.

Example 4.4. *One immediately finds examples of special systems of this type by starting from the (-1)-curves of Example 4.3.*
For instance consider $\mathcal{L} := \mathcal{L}_{2,2d}(-\sum_{i=1}^{5} dp_i)$ which is non-empty, consisting of the conic $\mathcal{L}_{2,2}(-\sum_{i=1}^{d} p_i)$ counted d times, though it has virtual dimension $-\binom{d}{2}$.

Even more generally, consider a linear system $\mathcal{L}$ on $\mathbf{P}^2$ which is non-empty, $C_1, \dots, C_k$ some (-1)-curves on $\tilde{\mathbf{P}}^2$ corresponding to curves $\Gamma_1, \dots, \Gamma_k$ on $\mathbf{P}^2$, such that $\tilde{\mathcal{L}} \cdot C_i = -N_i < 0$, $i = 1, \dots, k$. Then:

$$\mathcal{L} = \sum_{i=1}^{k} N_i \Gamma_i + \mathcal{M}, \quad \tilde{\mathcal{L}} = \sum_{i=1}^{k} N_i C_i + \tilde{\mathcal{M}}$$

and $\tilde{\mathcal{M}} \cdot C_i = 0$, for $i = 1, \dots, k$. As before, $\mathcal{L}$ is special as soon as there is an $i = 1, \dots, k$ such that $N_i \geq 2$. Furthermore $C_i \cdot C_j = \delta_{ij}$, because the union of two (-1)-curves meeting moves, according to the Riemann-Roch theorem, in a linear system of positive dimension on $\tilde{\mathbf{P}}^2$, and therefore it cannot be fixed for $\tilde{\mathcal{L}}$. In this situation, the reducible curve $C := \sum_{i=1}^{k} C_i$ [resp. $\Gamma := \sum_{i=1}^{k} N_i \Gamma_i$] is called a (-1)-*configuration* on $\tilde{\mathbf{P}}^2$ [resp. on $\mathbf{P}^2$].

Example 4.5. *Consider $\mathcal{L} := \mathcal{L}_{2,d}(-m_0 p_0 - \sum_{i=1}^{h} m_i p_i)$, with $m_0 + m_i = d + N_i$, $N_i \geq 1$. Let Γ_i be the line joining p_0, p_i. It splits off N_i times from $\mathcal{L}$. Hence $\mathcal{L} = \sum_{i=1}^{h} N_i \Gamma_i + \mathcal{L}_{2,d-\sum_{i=1}^{h} N_i}(-(m_0 - \sum_{i=1}^{h} N_i)p_0 - \sum_{i=1}^{h}(m_i - N_i)p_i)$. If we* *require the latter system to have non-negative virtual dimension, e.g. $d \geq \sum_{i=1}^{h} m_i$ if $m_0 = d$, and some $N_i > 1$ we have as many special systems as we want.*

With all this in mind we can now give a definition:

Definition 4.6. *A linear system $\mathcal{L}$ on $\mathbf{P}^2$ is (-1)-reducible if $\tilde{\mathcal{L}} = \sum_{i=1}^{k} N_i C_i + \tilde{\mathcal{M}}$, where $C = \sum_{i=1}^{k} C_i$ is a (-1)-configuration, $\tilde{\mathcal{M}} \cdot C_i = 0$, for all $i = 1, \ldots, k$, and $\mathrm{virtdim}(\mathcal{M}) \geq 0$.*

The system $\mathcal{L}$ is called (-1)-special if, in addition, there is an $i = 1, \ldots, k$ such that $N_i > 1$.

Remark 4.7. *This is an effective definition once one has an algorithm to produce (-1)-curves. We do not discuss this aspect here but refer the reader to [24] and [25], where the question is treated.*

We are finally ready to state the Harbourne-Hirschowitz conjecture (see also [40, 41, 45, 24] and [25]), which, with the terminology we just introduced, sounds very simple.

Conjecture 4.8. (Harbourne-Hirschowitz, 1989) *A linear system of plane curves $\mathcal{L} := \mathcal{L}_{2,d}(m_1, \ldots, m_h)$ with general multiple base points is special if and only if it is (-1)-special.*

This is a rather bold conjecture, whose basic motivation lies in the fact that, in more than a century of research on the subject, no special system has been discovered except (-1)-special systems. On the other hand, as we shall see in the next section, there are several recent results that make the conjecture rather plausible.

We can complement it and Gimigliano's one with another conjecture, mentioned in [58] and also attributed to Hirschowitz:

Conjecture 4.9. (Hirschowitz) *Consider a linear system of plane curves $\mathcal{L} := \mathcal{L}_{2,d}(m_1, \ldots, m_h)$ with general multiple base points $p_1, \ldots, p_h$ which is non-empty and non-special, with $g_{\mathcal{L}} \geq 0$. Then the general curve $C \in \mathcal{L}$ is irreducible, smooth off the multiple base points $p_1, \ldots, p_h$ where it has ordinary singularities of multiplicities exactly $m_1, \ldots, m_h$, unless $d = 3m$, $h = 9$, $m_1 = \cdots = m_9 = m \geq 2$, in which case $\mathcal{L}$ consists of the unique cubic through $p_1, \ldots, p_9$, counted with multiplicity m.*

I close this section with the following useful remark:

Remark 4.10. *Suppose that the Harbourne-Hirshowitz conjecture holds. Let $p_1, \ldots, p_h$ be general points of $\mathbf{P}^2$ and let C be an irreducible curve on the blow-up $\tilde{\mathbf{P}}^2$ at those points. Then one has $C^2 \geq p_a(C) - 1 \geq -1$ and $C^2 = -1$ if and only if C is a (-1)-curve. This is an immediate consequence of the Riemann-Roch theorem (5).*

5. Results on the Harbourne-Hirschowitz Conjecture

In this section I will present what is known about the conjectures introduced in §4, and I will try to briefly explain what are the techniques involved in the proofs. I will mainly follow the chronological development of the subject. As the reader will see, though the subject is so classical, almost all results do not go back more than twenty years and most of them are very recent.

As shown by (6), a non-empty linear system $\mathcal{L}$ in the plane $\mathbf{P}^2$ is non-special if and only if the corresponding linear system $\tilde{\mathcal{L}}$ on $\tilde{\mathbf{P}}^2$ has $h^1(\tilde{\mathbf{P}}^2, \tilde{\mathcal{L}}) = 0$. Any algebraic geometer knows that the literature is full of *vanishing theorems* for cohomology spaces. Think about Kodaira's or Kawamata-Vieweg's vanishing theorem (see [48]) and, in the surface case we are specifically dealing with, Mumford's and Franchetta-Ramanujam's (see [62, 70, 71]). If one tries to apply one of these theorems to the general dimensionality problem, one, unfortunately, does not go too far. It turns out that, in order to usefully apply them, one needs $-K$ to be effective on $\tilde{\mathbf{P}}^2$, hence all the points we blow-up have to lie on a cubic. If we blow-up h general points, this means that $h \leq 9$. In this way one proves a result already known to Castelnuovo [16], and later rediscovered by several authors like Nagata [64], Gimigliano [37] and Harbourne [41].

Theorem 5.1. (Castelnuovo, 1891; Nagata, 1960; Gimigliano, Harbourne, 1986) *The Harbourne-Hirschowitz conjecture holds for all linear systems with $h \leq 9$ general multiple base points.*

Recall that we are considering linear systems $\mathcal{L}_{2,d}(m_1, \dots, m_h)$ such that $m_1, \dots, m_h \geq 2$. The simplest case to look at is therefore the homogeneous case $m_1 = \cdots = m_h = 2$. This case was classically examined by Campbell [15], Palatini [66] and Terracini [84] in a wider context, as we will see later in §7. In recent times it has been reconsidered by Arbarello and Cornalba [7] in 1981. Their approach relies on the use of a classical *infinitesimal deformation technique* consisting in moving the base points of the system and computing the first order deformation of a curve which moves keeping its singularities. Arbarello-Cornalba's result, in the case they consider, is half-way between Harbourne-Hirschowitz Conjecture 4.8 and Hirschowitz Conjecture 4.9.

Theorem 5.2. (Arbarello-Cornalba, 1981) *Consider $\mathcal{L} := \mathcal{L}_{2,d}(2^h)$. Assume:*

(i) $\frac{d(d+3)}{2} \geq 3h$, *i.e.* $\mathrm{virtdim}(\mathcal{L}) \geq 0$;

(ii) $\binom{d-1}{2} \geq h$, *i.e.* $g_{\mathcal{L}} \geq 0$.

Then $\mathcal{L}$ is non-special, and $C \in \mathcal{L}$ general is irreducible, with nodes at the imposed general double points $p_1, \dots, p_h$, and elsewhere smooth, except for $\mathcal{L}_{2,6}(2^9)$ which is a double cubic.

The infinitesimal deformation computation performed by Arbarello and Cornalba is a particular case of a lemma which goes back to Terracini and we will come back to it later (see Lemma 6.3 and [82]). Unfortunately it works well only in the case of double points. In the higher multiplicity case infinitesimal deformation techniques have never been successfully used in this problem. However work in progress by C. Ciliberto, H. Clemens and R. Miranda, suggests that there are some chances in this direction.

A conceptually opposite approach, which is also natural to the problem, is to argue by *degeneration*, meaning with this that one specializes the base points of the linear system in order to be able to better compute the dimension of the linear system. Recall that the dimension of $\mathcal{L} := \mathcal{L}_{n,2}(-\sum_{i=1}^{h} m_i p_i)$ is upper-semicontinuous in the position of the points $p_1, \ldots, p_h$. Therefore if one finds a *particular* set of points $q_1, \ldots, q_h$ such that $\mathcal{L}_0 := \mathcal{L}_{n,2}(-\sum_{i=1}^{h} m_i q_i)$ is non-special, then also $\mathcal{L}$ is non-special. Unfortunately, this is often too naive: as soon as one puts the points $p_1, \ldots, p_h$ in a particular position, e.g. one puts them on some curve on which they should not lie, then the dimension of $\mathcal{L}$ tends to increase, and the method, in this crude form, does not work. However, there is still something which one can do: even if the dimension of $\mathcal{L}$ increases, one can actually compute the *limit* of $\mathcal{L}$ when $p_1, \ldots, p_h$ approach $q_1, \ldots, q_h$. There is no time here to enter in any detail about this idea, which is the one elaborated by A. Hirschowitz in his paper [44]. He called his degeneration technique *la méthode d'Horace*, i.e. *Horace's method*, consisting in successive specializations of the multiple base points on particular curves. Exploiting it, he has been able to prove

Theorem 5.3. (Hirschowitz, 1985) *The Harbourne-Hirschowitz conjecture holds in the homogeneous case $\mathcal{L}_{2,d}(m^h)$, $m \leq 3$.*

The application of Horace's method usually requires a deep geometric understanding of the problem and a special capability of guessing the right specializations to be performed. The unfortunate circumstance of this is that Horace's method seldom appears to be systematic, rather it seems ingenuous but too ad hoc to become a theory.

More recently a different specialization technique has been introduced, and successfully used, in this problem by Ciliberto and Miranda [24, 25]. The idea, which I will explain in some detail a few lines below, basically consists in using a degeneration technique worked out by Z. Ran [72] mainly for studying enumerative problems of families of plane nodal curves. It consists in *degenerating the plane* to a reducible surface and in following the linear system in the degeneration. The restriction of the limit linear system to the components of the reducible limit surface are *easier* than the system one starts with, so that one can hope to successfully use induction. The outcome of this method is the following substantial improvement of Hirschowitz's Theorem 5.3:

Theorem 5.4. (Ciliberto-Miranda, 1998) *The Harbourne-Hirschowitz conjecture holds in the quasi-homogeneous cases $\mathcal{L}_{2,d}(n, m^h)$, $m \leq 3$ and in the homogeneous cases $\mathcal{L}_{2,d}(m^h)$, $m \leq 12$.*

Remark 5.5. *It is worth mentioning, along the same lines, a recent result independently proved, with similar techniques, by J. Seibert [79] and A. Laface [50], to the*

effect that the Harbourne-Hirschowitz conjecture holds in the quasi-homogeneous case $\mathcal{L}_{2,d}(n, 4^h)$, and a result of Laface's [50], who proves that a suitable version of the Harbourne-Hirschowitz conjecture holds in the homogeneous case $\mathcal{L}(m^h)$, $m \leq 3$, for any linear system $\mathcal{L}$ on a Hirzebruch surface $\mathbf{F}_n$. Recall that $\mathbf{F}_n$ is the unique minimal rational ruled surface with an irreducible curve of self-intersection $-n$ (see [10, Chapter III]).

Note that a contribution to the case $\mathcal{L}_{2,d}(4^h)$, with different combinatorial methods coming from numerical analysis, is due to G. Lorentz and R. Lorentz [55] (see also [53, 54]). These methods however can be interpreted as an application of degeneration techniques rather similar to those used by L. Evain in [31], to which I will come back later on.

Let us now go back to the proof of Theorem 5.4. As promised, I want to give some details of the ideas involved. For more information, I refer to the original papers [24, 25].

First, let me describe Z. Ran's degeneration of the plane. Let Δ be a disc in $\mathbf{C}$ with centre the origin. Let $p \colon X \to \Delta$ be the flat family obtained by blowing-up $\Delta \times \mathbf{P}^2$ along a line L in the fibre of $0 \in \Delta$. The general fibre of the family is $X_t = p^{-1}(t) = \mathbf{P}^2$, for $t \neq 0$, whereas the *central fibre* is $X_0 = p^{-1}(0) = \mathbf{P} \cup \mathbf{F}$, where $\mathbf{P} = \mathbf{P}^2$, $\mathbf{F} = \mathbf{F}_1$ is the exceptional divisor of the blow-up, and $\mathbf{P} \cap \mathbf{F} = L$. Notice that X_0 thus appears as a flat limit of $\mathbf{P}^2$.

Next one takes to the limit the linear system. The natural map $\pi \colon X \to \mathbf{P}^2$ endows X with a line bundle $\mathcal{O}_X(d) := \pi^*(\mathcal{O}_{\mathbf{P}^2}(d))$ for any integer d. Of course $\mathcal{O}_X(d)_{|X_t} \simeq \mathcal{O}_{\mathbf{P}^2}(d)$ for all $t \neq 0$. But, for any integer k, one has also $\mathcal{O}_{\mathbf{P}^2}(d) = \mathcal{O}_X(d) \otimes \mathcal{O}_X(k\mathbf{P})_{|X_t}$. Hence each one of the line bundles $\mathcal{O}_{X_0}(d, k) := \mathcal{O}_X(d) \otimes \mathcal{O}_X(k\mathbf{P})_{|X_0}$ is a *limit* of $\mathcal{O}_{\mathbf{P}^2}(d)$ on the limit, reducible surface X_0. The failure of the uniqueness of the limit line bundle plays in our favour, inasmuch as the presence of the parameter k gives us more freedom in the numerical choices we will have to make next.

Send now $b < h$ of the h limiting points $q_1, \ldots, q_h$ on X_0 to $\mathbf{F}$ as general points, the remaining $h-b$ to $\mathbf{P}$ as general points, and consider the linear system $\mathcal{L}_0$ of all divisors in the linear system associated to $\mathcal{O}_{X_0}(d, k)$ having multiplicty at least m_i at q_i, $i = 1, \ldots, h$. This is a limit linear system of $\mathcal{L}_{2,d}(m_1, \ldots, m_h)$, which is called a (k, b)-*degeneration* $\mathcal{L}_0$ of $\mathcal{L}_{2,d}(m_1, \ldots, m_h)$. The usual upper-semicontinuity argument tells us that if the dimension of $\mathcal{L}_0$ equals the expected dimension of $\mathcal{L}$, then $\mathcal{L}$ is non-special.

Notice now that the two components $\mathbf{P}$ and $\mathbf{F}$ of X_0 are a plane and a plane blown-up at a point. Hence the restrictions of $\mathcal{L}_0$ to the two components $\mathbf{P}$ and $\mathbf{F}$ of X_0 are basically again linear systems of plane curves with general multiple base points. Thus one is in a position to use induction in order to estimate the dimension of $\mathcal{L}_0$. A basic ingredient in this computation is a *transversality lemma*, which, roughly speaking, tells us that the restrictions of $\mathcal{L}_0$ to $\mathbf{P}$ and $\mathbf{F}$ in turn restrict to $L = \mathbf{P} \cap \mathbf{F}$ in the most general possible way. A systematic use of (m, b)-degenerations leads then to the following result of independent interest:

Proposition 5.6. *There is a function $D(m) = \frac{m^2}{3} + o(m)$ such that if the Harbourne-Hirschowitz conjecture holds for every homogeneous system $\mathcal{L}_{2,d}(m^h)$ with $d \leq D(m)$, then the same conjecture holds for all homogeneous systems of the form $\mathcal{L}_{2,d}(m^h)$.*

The final part of the strategy is to try to prove the Harbourne-Hirschowitz conjecture for all homogeneous systems $\mathcal{L}_{2,d}(m^h)$, with $d \leq D(m)$, i.e. with d small with respect to m. In order to do so, one uses other (k,b)-degenerations, with other k's, but this does not work in all cases. For example Dixmier's example $\mathcal{L}_{2,19}(6^{10})$ worked out by Hirschowitz in [44] with Horace's method, cannot be attacked with (k,b)-degenerations. So one has to use also ad hoc geometric arguments or rely on the help of suitable computer programs. This is what is done in [25] in the cases $m \leq 12$.

Remark 5.7. *As a concluding remark on the proof of Theorem 5.4, I want to stress that a (k,b)-degeneration can be seen, ultimately, as a way of degenerating the set of points $p_1, \ldots, p_h$ by putting b of them on a line, and of letting the line split from the curves of the linear system k times. Thus, in principle, there is not so great a difference with Horace's method. However this approach seems quite systematic and has given so far very good results. Indeed, in principle, there is no reason why it should not work for higher values of m and in fact there is promising work in progress with F. Cioffi, R. Miranda and F. Orecchia on the algorithmic side mentioned a few lines above, in order to improve the bound $m \leq 12$ in Theorem 5.4. So far we have been able to work out a computer program which verifies the Harbourne-Hirschowitz conjecture for $\mathcal{L}_{2,d}(m^h)$. We tested the program and we have been able in this way to prove the conjecture for $m \leq 20$.*

I strongly believe that the method of (k,b)-degenerations can be still pushed further, to give better and better results along these lines.

Another aspect of the results in [24, 25] to be mentioned is the full classification of homogeneous (-1)-special systems, which is rather interesting and surprising in its own and plays an important role in the induction process described before.

First a little combinatorial analysis leads to the following:

Proposition 5.8. (Classification of homogeneous (-1)-configurations) *The only homogeneous linear systems $\mathcal{L}_{2,d}(m^h)$ which are (-1)-configurations are:*

$$\mathcal{L}_{2,1}(1^2) : \text{ a line through 2 points}$$

$$\mathcal{L}_{2,2}(1^5) : \text{ a conic through 5 points}$$

$$\mathcal{L}_{2,3}(2^3) : \text{ 3 lines each through 2 of 3 points}$$

$$\mathcal{L}_{2,12}(5^6) : \text{ 6 conics each through 5 of 6 points}$$

$$\mathcal{L}_{2,21}(8^7) : \text{ 7 cubics each through 6 points, double at another}$$

$$\mathcal{L}_{2,48}(17^8) : \text{ 8 sextics double at 7 points, triple at another}.$$

This leads to the following:

Theorem 5.9. (Classification of homogeneous (-1)-special systems) *The only homogeneous linear systems $\mathcal{L}_{2,d}(m^h)$ which are (-1)-special are:*

$$\mathcal{L}_{2,d}(m^2) \quad with \quad m \leq d \leq 2m - 2$$

$$\mathcal{L}_{2,d}(m^3) \quad with \quad \frac{3m}{2} \leq d \leq 2m - 2$$

$$\mathcal{L}_{2,d}(m^5) \quad with \quad 2m \leq d \leq \frac{5m - 2}{2}$$

$$\mathcal{L}_{2,d}(m^6) \quad with \quad \frac{12m}{5} \leq d \leq \frac{5m - 2}{2}$$

$$\mathcal{L}_{2,d}(m^7) \quad with \quad \frac{21m}{8} \leq d \leq \frac{8m - 2}{3}$$

$$\mathcal{L}_{2,d}(m^8) \quad with \quad \frac{48m}{17} \leq d \leq \frac{17m - 2}{6} \, .$$

As a remarkable consequence we have that the Harbourne-Hirschowitz conjecture for homogeneous system takes the form:

Conjecture 5.10. *Every homogeneous system of the form $\mathcal{L}_{2,d}(m^h)$ with $h \geq 10$ is non-special.*

This is probably the right moment for recalling another famous conjecture concerning singular plane curves. In [63] Nagata showed a counterexample to the fourteenth problem of Hilbert. In his construction, he proved that if the linear system $\mathcal{L}_{2,d}(m^{k^2})$ is non-empty for a integer $k \geq 4$, then one has $d > km$. He also conjectured that a similar result should hold for any, not necessarily a square, number of points in general position, namely he fomulated the following:

Conjecture 5.11. (Nagata, 1960) $\mathcal{L}_{2,d}(m^h)$ *is empty as soon as $h \geq 10$ and $d \leq \sqrt{h} \cdot m$.*

It is worth pointing out the following fact:

Remark 5.12. *Harbourne-Hirschowitz Conjecture 4.8 or 5.10, implies Nagata's Conjecture 5.11. Indeed, let $\mathcal{L} := \mathcal{L}_{2,d}(m^h)$, $h \geq 10$, be non-empty and let C be an irreducible component of the strict transform of the general element of $\mathcal{L}_{2,d}(m^h)$ on $\tilde{\mathbf{P}}^2$. By Remark 4.10 we have $C^2 \geq p_a(C) - 1$. On the other hand one cannot have $C^2 = -1, p_a(C) = 0$, since, by Proposition 5.8, there is no (-1)-configuration for $h \geq 10$. Thus $C^2 \geq 0$. Hence $\mathcal{L}^2 \geq 0$, which reads $d^2 \geq hm^2$.*

Before going back to our main topic, I cannot resist indicating the following connection of Nagata's conjecture, hence of the Harbourne-Hirschowitz conjecture, with another interesting subject.

Remark 5.13. *Let C be a curve of genus g. Curves on the product $C \times C$ are correspondences of the curve into itself. Similarly curves on the symmetric product $C(2)$*

are symmetric correspondences of C. A version of Petri's problem *(see* [8]*) for correspondences* is: describe the effective cone of the symmetric product $C(2)$ when C is a general curve of genus g. *It is known that, if C is general of genus g, then* $NS(C(2)) \simeq \mathbf{Z}\langle x, \frac{\delta}{2}\rangle$, *where x is the class of the curve and δ is the diagonal. The structure of the effective cone of $C(2)$ for C general of genus g is known when* $g \leq 3$. *Ciliberto and Kouvidakis' paper* [23] *(see also* [49]*), suggests the following conjecture:* if $g \geq 4$ there is no irreducible curve of negative self-intersection on $C(2)$ except the diagonal. *This conjecture would imply that the effective cone of $C(2)$ is bounded by the line spanned by the diagonal and the line of slope $-\frac{1}{\sqrt{g}-1}$ in the $(x, \frac{\delta}{2})$-plane, which is an open boundary line as soon as $g \geq 5$. One of the results in* [23] *is that:* Nagata's conjecture implies Ciliberto-Kouvidakis' conjecture. *The rather unexpected connection is provided by the fact that one may degenerate C to a rational g-nodal curve so that the curves on $C(2)$ degenerate to suitable plane curves.*

For more information on Nagata's conjecture and recent results on the subject, see [42, 74, 31].

Going back to the Harbourne-Hirschowitz conjecture, the following recent results are worth mentioning.

Theorem 5.14. (A. Bruno [14]**, 1998)** $\mathcal{L} = \mathcal{L}(m_1, \ldots, m_h)$ *is non-special if* virtdim$(\mathcal{L}) \geq 0$ *and* $g_{\mathcal{L}} \geq 0$ *and the general curve in $\mathcal{L}$ has ordinary m_i-tuple points at p_i, $i = 1, \ldots, h$.*

This result is quite interesting, though the hypothesis about the general curve in $\mathcal{L}$ is certainly too strong. The proof uses a bit of deformation theory, which reappears here after Arbarello-Cornalba's Theorem 5.2. However, the main tool in Bruno's proof is the use of the moduli space of curves, of stable reduction, and of the theory of limit linear series on reducible curves (for a general introduction to these ideas, see Harris-Morrison's book [43]). This is a really new idea in this setting and may possibly give further good results in the future.

The following theorem is due to T. Mignon in his thesis [58, 59] and it is based on the use of Horace's method:

Theorem 5.15. (T. Mignon, 1998) *Let $\mathcal{L} = \mathcal{L}(m_1, \ldots, m_h)$. Then:*

(i) *if $m_i \leq 4$ Harbourne-Hirschowitz Conjecture 4.8 holds;*
(ii) *if $g_{\mathcal{L}} \leq 4$ and* virtdim$(\mathcal{L}) \geq 0$ *then Harbourne-Hirschowitz Conjecture 4.8 and Hirschowitz Conjecture 4.9 both hold;*
(iii) *if $m_i \leq 3$, $d \geq 33$,* virtdim$(\mathcal{L}) \geq 0$ *and $g_{\mathcal{L}} \geq 0$ then Harbourne-Hirschowitz Conjecture 4.8 and Hirschowitz Conjecture 4.9 both hold.*

The interest of the next theorem, due to L. Evain [32], resides in the fact that it is the only evidence, so far, that the Harbourne-Hirschowitz conjecture holds for $\mathcal{L}_{2,d}(m^h)$ for infinitely many values of h.

Theorem 5.16. (L. Evain, 1998) $\mathcal{L}_{2,d}(m^h)$ *is never special if h is of the form $h = 4^k$.*

The proof uses a suitable version of Horace's method. Evain lets all the multiple points come together in a smart way in a unique singular point which gives independent conditions to curves of any degree. Some of these techniques go back to Hirschowitz [44] and to Caporaso-Harris (1996, unpublished) where they prove the Harbourne-Hirschowitz conjecture in the homogeneous case $m \leq 6$ under some restrictive hypotheses. Using Evain's ideas one can probably prove that $\mathcal{L}_{2,d}(m^h)$ is non-special in other situations, e.g. if $h = k^2 + l$, $l \leq 2k$ and either $d \leq km$ or $d \geq km + m + l - 3$, giving more information on Nagata's conjecture. Also other cases like $h = 4^k l^2$, $h = 9^k$ etc. can probably be analysed in the same way. There is also work in progress by Ciliberto and Miranda, who are able to give a rather easy proof of Evain's theorem using a Z. Ran's type of degeneration of the plane, as for the proof of Theorem 5.4. This also gives some hope of further extensions to other values of h.

A further important application of a refined version of Horace's method (what the authors call the *differential Horace method*, see [6]) is to get asymptotic results confirming the Harbourne-Hirschowitz conjecture. The prototype of results of this sort is the following theorem of Hirschowitz [45]:

Theorem 5.17. *The system $\mathcal{L}_{2,d}(m_1, \ldots, m_h)$ is non-special as soon as* $\left[\frac{(d+3)^2}{4}\right] > \sum_{i=1}^{h} \binom{m_i+2}{2}$.

A much deeper result is the following theorem of Alexander-Hirschowitz [6]:

Theorem 5.18. (Alexander-Hirschowitz, 1998) *Given any projective, reduced variety X and an ample line bundle $\mathcal{L}$ on it, there is a function $d(m)$ such that if $m_i < m$, $i = 1, \ldots, h$, and $d > d(m)$ then $\mathcal{L}^{\otimes d}(m_1, \ldots, m_h)$ is non-special.*

Note the independence of $d(m)$ by h the number of points: this makes Theorem 5.18 stronger than Theorem 5.17. More precise results about the function $d(m)$ in the planar case are due to other authors (see [9, 60, 39, 85]).

I will finish this section discussing the relations between the various Conjectures 4.1, 4.2, 4.8. While it is clear that both, the Harbourne-Hirschowitz conjecture and Gimigliano's conjecture, imply Segre's, it is rather surprising that the three conjectures are essentially equivalent. This is the content of the following theorem, whose proof will appear in [26].

Theorem 5.19. (Ciliberto-Miranda, 2000) *Segre's Conjecture 4.1 implies both Gimigliano's Conjecture 4.2 and the Harbourne-Hirschowitz Conjecture 4.8. In particular, given a linear system $\mathcal{L} := \mathcal{L}_{2,d}(m_1, \ldots, m_h)$ of plane curves with general multiple base points $p_1, \ldots, p_h$, if Segre's conjecture holds then:*

(i) *$\mathcal{L}$ is special if and only if it is (-1)-special;*

(ii) *if $\mathcal{L} \neq \emptyset$, then $C \in \mathcal{L}$ general has multiplicity m_i at p_i, $i = 1, \ldots, h$;*

(iii) *if $\mathcal{L}$ is non-special, then either $C \in \mathcal{L}$ general is irreducible, or $\mathcal{L}$ is (-1)-reducible, or $\mathcal{L}$ consists of a unique, maybe multiple, elliptic curve, or $\mathcal{L}$ is composed of a pencil of rational curves.*

The surprisingly easy proof is based on standard surface theory. I want also to mention, from the same paper [26], the following result, which goes in the direction of Hirschowitz Conjecture 4.9:

Corollary 5.20. (Ciliberto-Miranda, 2000) *Suppose Segre's conjecture is true. Consider a non-special, not (-1)-reducible, linear system $\mathcal{L}:=\mathcal{L}_{2,d}(m_1,\ldots,m_h)$ of plane curves with general multiple base points such that the general curve $C \in \mathcal{L}$ is reducible. Then there is a Cremona transformation sending $\mathcal{L}$ to one of the two systems:*

$$\mathcal{L}_{2,d}(d), \quad \textit{d-tuples of lines through a point } p$$

$$\mathcal{L}_{2,3d}(d^9), \quad \textit{the cubic through nine general points, counted d-times}.$$

The proof is a consequence of Theorem 5.19 and of the following classical fact:

Lemma 5.21. (Noether's lemma) *Let $\mathcal{L}$ be as in the statement of Corollary 5.20, let $C \in \mathcal{L}$ general be irreducible of genus $g \leq 1$. Then there is a Cremona transformation sending $\mathcal{L}$ to one of these systems:*

$$
\begin{array}{lll}
\textit{a linear system of lines,} & \dim(\mathcal{L}) \leq 2, & g = 0 \\
\mathcal{L}_{2,2}, & \dim(\mathcal{L}) = 5, & g = 0 \\
\mathcal{L}_{2,d}(d-1), & \dim(\mathcal{L}) = 2d, & g = 0 \\
\mathcal{L}_{2,d}(1, d-1), & \dim(\mathcal{L}) = 2d-1, & g = 0 \\
\mathcal{L}_{2,3}(1^h), h \leq 9, & \dim(\mathcal{L}) = 9 - h, & g = 1 \\
\mathcal{L}_{2,4}(2^2), & \dim(\mathcal{L}) = 8, & g = 1.
\end{array}
$$

6. Interpolation in More Variables

Little is known about the general dimensionality problem for linear systems in $\mathbf{P}^n$, $n \geq 3$. That little is mostly concentrated in the following beautiful result of Alexander-Hirschowitz which classifies the special linear systems $\mathcal{L}_{n,d}(2^h)$:

Theorem 6.1. (Alexander-Hirschowitz, 1996) $\mathcal{L}_{n,d}(2^h)$ *is non-special unless:*

$$
\begin{array}{lccccc}
n & any & 2 & 3 & 4 & 4 \\
d & 2 & 4 & 4 & 4 & 3 \\
h & 2,\ldots,n & 5 & 9 & 14 & 7
\end{array}
\tag{9}
$$

Remark 6.2. *The statement of the Alexander-Hirschowitz theorem was divined by Bronowski in [11], but he had only a plausibility argument rather than a proof of it. Terracini [83] instead has a proof for the case $n = 3$.*

Almost all the special systems shown in Table (9) have been met already in Examples 4.3, (i') and (i"). The only new one is $\mathcal{L}_{4,3}(2^7)$ whose virtual dimension is -1 whereas it is non-empty. In fact there is a unique rational normal quartic curve Γ through seven general points $p_1,\ldots,p_7$ in $\mathbf{P}^4$. The secant variety of Γ, *i.e. the variety described by all lines meeting Γ at two points, is a hypersurface of degree 3 and it is singular along Γ, hence it is singular at $p_1, \ldots, p_7$, thus it sits in $\mathcal{L}_{4,3}(2^7)$. This example was well known to Terracini [82] and it has been more recently rediscovered by Ciliberto-Hirschowitz [22].*

The original proof of the theorem of Alexander-Hirschowitz requires the full strength of Horace's method and it is long and difficult. Indeed it occupies a whole series of papers [1]–[5]. An easier proof has been recently provided by K. Chandler [18]. It still uses Horace's method but in a much simpler way, by subsequent specializations of part of the general double points of the linear system to a hyperplane. Work in progress by Ciliberto and Miranda indicates that an alternative and quite simple proof can be also obtained by using suitable Z. Ran's type of degenerations of $\mathbf{P}^n$. In essence, this approach is not so different from Chandler's, but, again, it looks more systematic and it gives some hope to extend the analysis to higher multiplicities.

The important feature of the special systems appearing in Table (9) is the following: *for each special $\mathcal{L}_{n,d}(2^h)$, the general member $D \in \mathcal{L}_{n,d}(2^h)$ is singular along a positive dimensional variety containing the general double base points of the system $\mathcal{L}_{n,d}(2^h)$.* Roughly speaking the phenomenon of speciality is not concentrated at the base points but somehow *propagates in space*! At least for double base points, this observation has a quite general meaning, and goes back to Terracini [81] (for a modern version see [22]):

Lemma 6.3. (Terracini, 1915; Ciliberto-Hirschowitz, 1991) *Let X be any projective variety, let $\mathcal{L}$ be a linear system on X, let $p_1, \ldots, p_h$ be general points of X. If $\mathcal{L}(-\sum_{i=1}^{h} 2p_i)$ is special then every $D \in \mathcal{L}(-\sum_{i=1}^{h} 2p_i)$ is singular along a positive dimensional variety containing $p_1, \ldots, p_h$.*

The proof of the lemma is based on an easy first-order infinitesimal computation, to the effect that any first-order deformation of a singular hypersurface D which preserves the singularities of D is a hypersurface D' containing the singular locus of D. As already mentioned before, this computation is basically the one needed for the proof of Arbarello-Cornalba's Theorem 5.2.

It is quite natural to conjecture that the phenomenon of *propagation in space* of speciality of linear systems with multiple general base points should take place for higher multiplicities too. This is the content of the following conjecture, which I share with R. Miranda:

Conjecture 6.4. (Ciliberto-Miranda) *Let $\mathcal{L} = \mathcal{L}_{n,d}(m_1, \ldots, m_h)$ be a linear system with multiple base points at $p_1, \ldots, p_h$. If the general member $D \in \mathcal{L}$ has isolated singularities at $p_1, \ldots, p_h$, then $\mathcal{L}$ is non-special.*

Remark 6.5. *Notice that the converse of the conjecture certainly does not hold. If $\mathcal{L}$ is non-special then the general member $D \in \mathcal{L}$ may very well have non-isolated singularities containing $p_1, \ldots, p_h$. An example is $\mathcal{L}_{3,4}(2^8)$, which, according to Alexander-Hirschowitz's theorem, is non-special, hence of dimension 2. Notice that* *there is a pencil of quadrics $\mathcal{P} = \mathcal{L}_{3,2}(1^8)$ through the eight base points, having a base locus Γ which is an elliptic quartic curve. Then $\mathcal{L}_{3,4}(2^8)$ is composed of all pairs of elements of $\mathcal{P}$ and therefore the general element of $\mathcal{L}_{3,4}(2^8)$ is singular along Γ.*

Conjecture 6.4 would be in perfect analogy with Segre's Conjecture 4.1. However one can be bolder, and try to make a conjecture which parallels Harbourne-Hirschowitz's. Let us try to do it now.

Let me start with a definition. Recall, in the course of it, that a theorem of Bellatalla-Grothendieck [10, pg. 43], asserts that any vector bundle on $\mathbf{P}^1$ splits in an essentially unique way as a direct sum of line bundles.

Definition 6.6. *Let X be a smooth, projective variety of dimension n, let C be a smooth, irreducible curve on X and let $\mathcal{N}_{C|X}$ be the normal bundle of C in X. We will say that C is a negative curve if there is a line bundle $\mathcal{N}$ of negative degree and a surjective map $\mathcal{N}_{C|X} \to \mathcal{N}$.*

The curve C is called a (-1)-curve of size a, with $1 \leq a \leq n - 1$, on X if $C \simeq \mathbf{P}^1$ and $\mathcal{N}_{C|X} \simeq \mathcal{O}_{\mathbf{P}^1}(-1)^{\oplus a} \oplus \mathcal{N}$, where $\mathcal{N}$ has no summands of negative degree.

We are now ready to make our general conjecture:

Conjecture 6.7. *Let X be the blow-up of $\mathbf{P}^n$ at general points $p_1, \ldots, p_h$ and let $\mathcal{L} = \mathcal{L}_{n,d}(m_1, \ldots, m_h)$ be a linear system with multiple base points at $p_1, \ldots, p_h$. Then:*

 (i) *the only negative curves on X are (-1)-curves;*
 (ii) *$\mathcal{L}$ is special if and only if there is a (-1)-curve C on X corresponding to a curve Γ on $\mathbf{P}^n$ containing $p_1, \ldots, p_h$ such that the general member $D \in \mathcal{L}$ is singular along Γ;*
 (iii) *if $\mathcal{L}$ is special, let B be the component of the base locus of $\mathcal{L}$ containing Γ according to Bertini's theorem. Then the codimension of B in $\mathbf{P}^n$ is equal to the size of C and B appears multiply in the base locus scheme of $\mathcal{L}$.*

Remark 6.8. *Of course, the above conjecture coincides with Conjecture 4.8 for dimension $n = 2$ (see also Remark 4.10 for part (i) of the conjecture).*

As a general warning, I should stress that there is not too much evidence for Conjecture 6.7, to the extent that I do not even know whether it is true for general multiple double points. In particular the case $\mathcal{L}_{4,4}(2^{14})$ remains rather difficult to attack, whereas in the other cases in Table (9) the conjecture holds.

For instance, in the case $\mathcal{L}_{3,4}(2^7)$, the rational quartic curve Γ through the seven double base points $p_1, \ldots, p_7$, corresponds to a curve $C \simeq \mathbf{P}^1$ on the blow-up of $\mathbf{P}^4$ at $p_1, \ldots, p_7$, whose normal bundle is $\mathcal{O}_{\mathbf{P}^1}(-1)^{\oplus 3}$, hence C is a (-1)-curve of size 3. This fits with Conjecture 6.7.

The other case in Table (9) is $\mathcal{L}_{3,4}(2^9)$ which consists of the unique quadric $B \in \mathcal{L}_{3,2}(1^9)$ counted twice. On B there is a unique rational quintic curve Γ through the nine base points, which corresponds to a curve C on the blow-up $X = \tilde{\mathbf{P}}^3$. Since *the normal bundle to $\Gamma \simeq \mathbf{P}^1$ is $\mathcal{O}_{\mathbf{P}^1}(8) \oplus \mathcal{O}_{\mathbf{P}^1}(10)$, then the normal bundle to C in X is $\mathcal{O}_{\mathbf{P}^1}(-1) \oplus \mathcal{O}_{\mathbf{P}^1}(1)$. Hence C is a (-1)-curve of size 1 on X, which again fits with the conjecture.*

It goes without saying that, at the present moment, I have no idea about the possible relations between the two Conjectures 6.4 and 6.7 above. Another related interesting question would be: describe the set of (-1)-curves on the blow-up X of $\mathbf{P}^n$ at general points $p_1, \dots, p_h$. If $n = 2$, for instance, it is known that this set is finite if and only if $h \leq 8$ and its behaviour under the action of the so-called Kantor group *is known also in the infinite case $h \geq 9$ (see [30]). Is there any extension of this to the case $n \geq 3$?*

7. Waring's Problem and Defective Varieties

The birthplace of Waring's problem is number theory. It can be stated as follows:

Problem 7.1. (Waring's problem) *Given positive integers d, h, may we write any positive integer as a sum of h non-negative d-th powers?*

One of the first instances of this problem is the famous *Gauss' four squares theorem* to the effect that: *every positive integer is a sum of four squares.* A theorem which is sharp, inasmuch as there are integers, like 7, which cannot be written as a sum of three squares.

I will not be interested here in the original, number theoretic Problem 7.1, but rather in the following transposition of it to the realm of polynomials:

Problem 7.2. (Waring's problems for forms) *Given positive integers d, h, n, may we write any homogeneous polynomial $f(x_0, \dots, x_n)$ of degree d as a sum of h d-th powers of linear forms $l_i(x_0, \dots, x_n)$, $i = 1, \dots, h$, i.e. as $f(x_0, \dots, x_n) = \sum_{i=1}^{h} l_i(x_0, \dots, x_n)^d$?*

The relations of this problem with the general dimensionality problem introduced in §3 will be clear in a while, once we give a geometric interpretation of it based on *secant varieties* and *Terracini's lemma*, ideas which have been developed by the classical Italian school of algebraic geometry. One word of warning before doing that: an equivalent interpretation can be given in a slightly different way using other concepts like differential operators and *inverse systems*. This approach, classically developed by Macaulay [56], has been in recent times taken up by various authors, starting with Iarrobino [46] and [47]. Via the classical and well-known notion of *apolarity*, the two approachs are basically equivalent, so that I feel free to present here only the first one, referring the reader to [46] or to the nice expository paper [61] for the other.

First I recall a basic definition:

Definition 7.3. *Let $X \subseteq \mathbf{P}^N$ be an irreducible, non-degenerate projective variety of dimension n and let k be a positive integer. Take $k+1$ independent points $p_0, \dots, p_k$ of X. The span $\langle p_0, \dots, p_k \rangle$ is a subspace of $\mathbf{P}^N$ of dimension k which is called* a $(k+1)$-secant $\mathbf{P}^k$ of X. By $\mathrm{Sec}_k(X)$ *we denote the closure of the union of all* $(k+1)$-*secant* $\mathbf{P}^k$*'s of* X. *This is an irreducible algebraic variety which is called the* k-th secant variety *of* X.

A basic parameter count shows that:

$$\dim(\mathrm{Sec}_k(X)) \leq \min\{(n+1)(k+1)-1, N\}; \tag{10}$$

the right-hand side of (10) is the expected dimension *of* $\mathrm{Sec}_k(X)$. *If the strict inequality holds in (10), then* X *is said to be* k-defective, *and its* k-defect *is* $\delta_k(X) := \min\{(n+1)(k+1)-1, N\} - \dim(\mathrm{Sec}_k(X))$.

Let now $\mathbf{P} := \mathbf{P}^{N_{n,d}}$ be the projective space $\mathbf{P}(\mathbf{C}[x_0,\dots,x_n]_d)$ associated to the vector space $\mathbf{C}[x_0,\dots,x_n]_d$ of all complex homogeneous polynomials of degree d in the variables $x_0,\dots,x_n$. Recall that $N_{n,d} = \binom{d+n}{n} - 1$. A remarkable subvariety of $\mathbf{P}$ is the so-called *dual d-th Veronese of* $\mathbf{P}^n$, namely the set $\mathcal{V}_{n,d}$ of all d-th powers of linear forms in $x_0,\dots,x_n$. Of course $f(x_0,\dots,x_n) \in \mathbf{C}[x_0,\dots,x_n]_d$ is a sum of h d-th powers of linear forms if and only if $[f] \in \mathbf{P}$ is contained in an h-secant $\mathbf{P}^{h-1}$ to $\mathcal{V}_{n,d}$. Thus, given the positive integers d, h, n, if Waring's problem has an affirmative answer for all forms $f(x_0,\dots,x_n) \in \mathbf{C}[x_0,\dots,x_n]_d$ then $\mathrm{Sec}_{h-1}(\mathcal{V}_{n,d}) = \mathbf{P}^{N_{n,d}}$, which implies that $h(n+1)-1 \geq N_{n,d}$, i.e. one has $h \geq \lceil \frac{1}{n+1} \cdot \binom{d+n}{n} \rceil$.

The main question is then: does the converse hold? It turns out that the answer is provided by Alexander-Hirschowitz's Theorem 6.1 In order to explain why this is the case, we need another bit of geometric information, namely the following result (see [81, 27]) which is the basic tool for understanding defective varieties:

Lemma 7.4. (Terracini's lemma) *Let* $X \subseteq \mathbf{P}^N$ *be an irreducible, non-degenerate projective variety of dimension* n *and let* $k \leq N$ *be a positive integer. Take* $k + 1$ *general points* $p_0,\dots,p_k$ *of* X *and let* $p \in \langle p_0,\dots,p_k \rangle$ *be a general point in* $\mathrm{Sec}_k(X)$. *Then the tangent space* $T_{\mathrm{Sec}_k(X),p}$ *to* $\mathrm{Sec}_k(X)$ *at* p *is given by* $T_{\mathrm{Sec}_k(X),p} = \langle \cup_{i=0}^k T_{X,p_i} \rangle$.

Hence, assume that $h \geq \lceil \frac{1}{n+1} \cdot \binom{d+n}{n} \rceil$ and that Waring's problem has no solution for $f(x_0,\dots,x_n) \in \mathbf{C}[x_0,\dots,x_n]_d$ general. Then $\mathrm{Sec}_{h-1}(\mathcal{V}_{n,d})$ is not equal to $\mathbf{P}^{N_{n,d}}$ and therefore, for $p_1,\dots,p_h \in \mathcal{V}_{n,d}$ general, one has that the subspace $\langle \cup_{i=1}^h T_{\mathcal{V}_{n,d},p_i} \rangle$ must be contained in a hyperplane of $\mathbf{P}^{N_{n,d}}$.

To connect all this with the general dimensionality problem and the Alexander-Hirschowitz theorem, dualize and look at $\mathcal{V}_{n,d}$ in $\mathbf{P}^{N_{n,d}}$ as $\mathbf{P}^n$ embedded via the complete linear system $\mathcal{L}_{n,d}$. Thus a hyperplane H in $\mathbf{P}^{N_{n,d}}$ cuts out on $\mathcal{V}_{n,d}$ a divisor whose pull-back to $\mathbf{P}^n$ is hypersurface D_H of degree d, i.e. a member of $\mathcal{L}_{n,d}$. If p is a point in $\mathcal{V}_{n,d}$, we abuse notation and denote by p also the corresponding point in $\mathbf{P}^n$. The hyperplane H is tangent to $\mathcal{V}_{n,d}$ at p, i.e. it contains $T_{\mathcal{V}_{n,d},p}$, if and only if the corresponding hypersurface D_H is singular at p.

In conclusion, if Waring's problem has no solution for a general $f(x_0,\dots,x_n) \in \mathbf{C}[x_0,\dots,x_n]_d$, then for $p_1,\dots,p_h \in \mathbf{P}^n$ general points, there is a hypersurface $D \in \mathcal{L}_{n,d}$ singular at $p_1,\dots,p_h$. Since $h(n+1)-1 \geq N_{n,d}$ is equivalent to virtdim$(\mathcal{L}_{n,d}(2^h)) \leq -1$, we see that the cases in which Waring's problem has no solution for $f(x_0, \dots, x_n) \in \mathbf{C}[x_0, \dots, x_n]_d$ general and $h(n+1) - 1 \geq N_{n,d}$ are exactly those listed in Table (9) from Alexander-Hirschowitz Theorem 6.1.

Remark 7.5. *The above cases of failure of a positive answer to Waring's problem were well known to the old geometers and invariant theorists. Besides the case of quadrics, which is trivial, the case of plane quartics was known to Clebsch (see the citation in [83]), who proved that in that case the sum of 6, instead of 5, as expected, fourtuple powers of linear forms is needed to obtain a general form of degree 4 in three variables. The remaining three cases were known to Palatini [69] and Terracini [83], who also proved that one more summand than expected is needed to represent a general form in each of the cases in question.*

In the old times an interesting extension of Waring's problem has also been considered. It goes back to Darboux [29], Reye [73], London [52], Palatini [65], Bronowski [12], Terracini [82] and is the following:

Problem 7.6. (Extended Waring's problem) *Given positive integers d, h, n, s, may we write any s homogeneous polynomial $f_j(x_0, \dots, x_n)$, $j = 1, \dots, s$, of degree d as linear combinations of the same h d-th powers of linear forms $l_i(x_0, \dots, x_n)$, $i = 1, \dots, h$?*

This leads right away to the following quite intriguing geometric problem:

Problem 7.7. *What is the dimension of the variety* $\mathrm{Sec}_{l,k}(X)$ *described by all* $\mathbf{P}^l$ *'s contained in* $(k+1)$*-secant* $\mathbf{P}^k$ *'s to a variety* X *of dimension* n *in* $\mathbf{P}^N$?

The expected dimension of $\mathrm{Sec}_{l,k}(X)$, which is contained in the grassmannian $\mathbf{G}(l, N)$, is $\min\{(l+1)(N-l), (k+1)n + (l+1)(k-l)\}$, but its actual dimension may be smaller. Thus a refined, grassmannian version of the concept of defect arises. Unfortunately there is no Terracini type lemma which helps in this situation. However there is recent interesting work of Chiantini-Coppens [19] on the case $n = 2$, $N = 5$, $l = 1$, $k = 2$ for Problem 7.7. It should also be noticed that Terracini claims in [82] that for $n = s = 2$ he has a complete solution to Problem 7.6: the only exception to the expected answer is for $d = 3$, i.e. the lines contained in a 5-secant $\mathbf{P}^4$ to $\mathcal{V}_{2,3} \subset \mathbf{P}^9$ are not all the lines of $\mathbf{P}^9$ as a parameter count suggests.

As the original Waring's problem is related to Alexander-Hirschowitz's theorem, the extended Waring's Problem 7.6 might lead to interesting extensions of Alexander-Hirschowitz's theorem. I believe this is an open, promising field of research.

In the same circle of ideas presented in this section, one is led in a natural way to the problem of the *classification of defective varieties*, which thus appears as another geometric counterpart of the general dimensionality problem introduced in §3. This is a classical, basic question concerning the extrinsic geometry of projective varieties, which has also applications to other branches of mathematics. For instance, computation of defects of Segre varieties, i.e. the products of projective spaces, is related to the linear algebra problem of determining the *rank* of a general tensor, which again is a relevant question in numerical analysis (see [17, 33, 51] and [80])

Among the main tools here are the two Terracini's Lemmas 6.3 and 7.4. It is not possible to enter now in too many details, but, before finishing, I will briefly recall, without any pretense of being exhaustive, some of the main results on the subject.

As in the general dimensionality problem, curves are never defective. Surfaces instead, can be defective. The subject of defective surfaces has been considered classically by Palatini [67] and [68], whose classification theorem contained a serious gap, and Terracini [84], who completed Palatini's classification (see also Scorza's [76] and Bronowski's [13] papers on the subject). Both Palatini and Terracini's papers are quite obscure and difficult to read. In more recent times Palatini-Terracini's classification of defective surfaces has been reconsidered, rediscovered and worked out again by M. Dale [28].

As for higher dimensional defective varieties, we are rather far away from a classification. In Zak's book [86] one finds several general properties. In particular a smooth, irreducible, non-degenerate, 1-defective variety $X \subset \mathbf{P}^N$ of dimension n is such that $N + 1 \leq \binom{n+1}{2}$ and if the equality holds then $X = \mathcal{V}_{n,2}$ is the Veronese variety of quadrics of $\mathbf{P}^n$ in which case the defect is 1 (see [86, Theorem 2.1, pg. 126]). Zak has similar, equally beautiful, theorems for 1-defective varieties with higher defect (see [86, Chapter VI]).

A weaker concept than the one of a k-defective variety, is the concept of a k-*weakly defective* variety. This is a variety $X \subset \mathbf{P}^N$ such that the general hyperplane which is tangent at $k + 1$ general points $p_0, \ldots, p_k$, is tangent along a positive dimensional variety containing $p_0, \ldots, p_k$. Recalling Lemma 6.3, it is clear that a k-defective variety is also k-weakly defective. The converse does not hold in general. It turns out that the classification of weakly defective varieties of dimension smaller than n matters in the classification of defective varieties of dimension n.

A full classification of weakly defective surfaces, which extends previous partial results by Terracini [84], has been recently obtained by Chiantini-Ciliberto [20]. This can be also seen as a wide extension of Arbarello-Cornalba's Theorem 5.2. In addition this gives some hope for the complete classification of defective 3-folds. This is a subject which has classically been studied by Scorza [75], who claims to have a classification of all 1-defective 3-folds. He also studied defective 4-folds in [77]. In more recent times the subject has been reconsidered by other authors, e.g. Zak [86], Fujita-Roberts [36] and Fujita [35], who essentially consider the case of smooth 3-folds. For any 3-fold, without any smoothness assumption, Ciliberto-Chiantini [21] have reworked Scorza's classification of 1-defective 3-folds. Their approach, easier and faster than Scorza's original one, is essentially based on a refined version of Lemma 6.3. The result is the following:

Theorem 7.8. *An irreducible, non-degenerate, projective 3-fold $X \in \mathbf{P}^N$ is 1-defective if and only if it is of one of the following types:*

(i) X *is a cone;*

(ii) X *sits in a 4-dimensional cone over a curve;*

(iii) $N = 7$ *and X is contained in a 4-dimensional cone over the Veronese surface $\mathcal{V}_{2,2}$ in $\mathbf{P}^5$;*

(iv) X *is the 2-Veronese embedding $\mathcal{V}_{3,3}$ of $\mathbf{P}^3$ in $\mathbf{P}^9$ or a projection of it in $\mathbf{P}^8$;*

(v) $N = 7$ *and X is a hyperplane section of the Segre embedding of $\mathbf{P}^2 \times \mathbf{P}^2$ in $\mathbf{P}^8$.*

Work in progress by Ciliberto-Chiantini indicates that along the same lines one may possibly obtain the full classification of defective 3-folds.

References

[1] J. Alexander, *Singularités imposables en positions générale à une hypersurface projective*, Compositio Math., **68** (1988), 305–354.

[2] J. Alexander and A. Hirschowitz, *La méthode d'Horace éclatée: application à l'interpolation en degré quatre*, Inventiones Math., **107** (1992), 585–602.

[3] J. Alexander and A. Hirschowitz, *Un lemme d'Horace différentiel: application aux singularités hyperquartiques de $\mathbf{P}^5$*, J. Algebraic Geom., **1** (1992), 411–426.

[4] J. Alexander and A. Hirschowitz, *Polynomial interpolation in several variables*, J. Algebraic Geom., **4** (1995), 201–222.

[5] J. Alexander and A. Hirschowitz, *Generic hypersurface singularities*, pre-print (1999).

[6] J. Alexander and A. Hirschowitz, *An asymptotic vanishing theorem for generic unions of multiple points*, pre-print (1999).

[7] E. Arbarello and M. Cornalba, *Footnotes to a paper of B. Segre*, Math. Ann., **256** (1981), 341–362.

[8] E. Arbarello, M. Cornalba, Ph. Griffiths and J. Harris, *Geometry of algebraic curves. Volume I*, Grundlehren der math. Wissenshaften, **267**, Springer-Verlag, New York, Berlin, Heidelberg, Tokyo (1984).

[9] E. Ballico, *Curves of minimal degree with prescribed planar singularities*, Illinois J. Math., **43** (1999), 672–676.

[10] A. Beauville, *Surfaces algébriques complexes*, Asterique, **54** (1978).

[11] J. Bronowski, *The sum of powers as canonical expression*, Proc. Camb. Phil. Soc., **29**, I (1933), 69–81.

[12] J. Bronowski, *The sum of powers as simultaneous canonical expressions*, Proc. Camb. Phil. Soc., **29**, II (1933), 245–256.

[13] J. Bronowski, *Surfaces whose prime sections are hyperelliptic*, J. London Math. Soc., **8**, (1933), 308–312.

[14] A. Bruno, *Degenerations of linear series and binary curves*, Ph. D. Thesis, Harvard Univ. (1997).

[15] J. E. Campbell, *Note on the maximum number of arbitrary points which can be double points on a curve, or a surface, of any degree*, The Messenger of Math., **21** (1891–92), 158–164.

[16] G. Castelnuovo, *Ricerche generali sopra i sistemi lineari di curve piane*, Mem. Accad. Sci. Torino, II **42** (1891).

[17] M. V. Catalisano, A. V. Geramita and A. Gimigliano, *On the secant varieties to embeddings of products of projective spaces*, to appear in Queen's papers in Pure and Applied Math. (2000).

[18] K. Chandler, *A brief proof of a maximal rank theorem for generic double points in projective space*, pre-print (1998).

[19] L. Chiantini and M. Coppens, *Grassmannians of secant varieties*, to appear on Forum Mathematicum.

[20] L. Chiantini and C. Ciliberto, *Weakly defective varieties*, to appear in Trans. Amer. Math. Soc.

[21] L. Chiantini and C. Ciliberto, *Threefolds with degenerate secant variety: on a theorem of G. Scorza*, in Geometric and Combinatorial Aspects of Commutative Algebra (Ed. J. Herzog and G. Restuccia), M. Dekker, Lect. Notes in Pure and Appl. Math., **217** (2001), 111–124.

[22] C. Ciliberto and A. Hirschowitz, *Hypercubique de* $\mathbf{P}^4$ *avec sept points singuliers génériques*, C. R. Acad. Sci. Paris, **313** I (1991), 135–137.

[23] C. Ciliberto and A. Kouvidakis, *On the symmetric product of a curve with general moduli*, Geometriae Dedicata, **78** (1999), 327–343.

[24] C. Ciliberto and R. Miranda, *Degenerations of planar linear systems*, J. reine und angew. Math., **501** (1998), 191–220.

[25] C. Ciliberto and R. Miranda, *Linear systems of plane curves with base points of equal multiplicity*, to appear in Transactions Amer. Math. Soc.

[26] C. Ciliberto and R. Miranda, *A deformation theory approach to linear systems with general triple point*, to appear in the special volume *Le Matematiche* in honour of Prof. S. Greco.

[27] M. Dale, *Terracini's lemma and the secant variety of a curve*, Proc. London Math. Soc., **49** (3) (1984), 329–339.

[28] M. Dale, *On the secant variety of an algebraic surface*, University of Bergen, Dept. of Math., Pre-print no. 33 (1984).

[29] G. Darboux, *Sur les systèmes linéaires de coniques et de surfaces du second ordre*, Bull. des Sciences Math., **1** (1870).

[30] P. Du Val, *On the Kantor group of a set of points in a plane*, Proceedings London Math. Soc., **42** (1936), 18–51.

[31] L. Evain, *Une minoration du degré des courbes planes à singularités imposées*, Bull. S.M.F., **126** (1998), 525–543.

[32] L. Evain, *La fonction de Hilbert de la réunion de* 4^h *points génériques de* $\mathbf{P}^2$ *de même multiplicité*, J. of Alg. Geom., **8** (1999).

[33] B. Fantechi, *Secanti di varietà proiettive e applicazioni*, Tesi di laurea, Università di Pisa (1988).

[34] S. Fitchett, B. Harbourne and S. Holay, *Resolutions of ideals of uniform fat points subschemes of* $\mathbf{P}^2$, pre-print (2000).

[35] T. Fujita, *Projective threefolds with small secant varieties*, Sci. Papers College Gen. Ed. Univ. Tokyo, **32** (1982), 33–46.

[36] T. Fujita and J. Roberts, *Varieties with small secant varieties: the extremal case*, Amer. J. of Math., **103** (1981) 1, 953–976.

[37] A. Gimigliano, *On linear systems of plane curves*, Thesis, Queen's University, Kingston (1987).

[38] A. Gimigliano, *Our thin knowledge of fat points*, Queen's Papers in Pure and Applied Math., **83**, The Curve Seminar at Queen's, Vol. IV, Queen's University, Kingston, Canada (1989).

[39] G.-M. Greuel, C. Lossen and E. Schustin, *Plane curves of minimal degree with prescribed singularities*, Inventiones Math., **133** (1998), 539–580.

[40] B. Harbourne, *The geometry of rational surfaces and Hilbert functions of points in the plane*, Proccedings of the 1984 Vancouver Conference in Algebraic Geometry, CMS Conf. Proc., **6** (1986) 95–111.

[41] B. Harbourne, *Points in good position in* $\mathbf{P}^2$, Zero-dimensional schemes (Ravello 1992), de Gruyter, Berlin, (1994), 213–229.

[42] B. Harbourne, *On Nagata's conjecture*, pre-print (1999).

[43] J. Harris and I. Morrison, *Moduli of curves*, Graduate Texts in Math., Springer-Verlag, **187** (1998).

[44] A. Hirschowitz, *La méthode d'Horace pour l'interpolation à plusieurs variables*, Manuscripta Math., **50** (1985), 337–388.

[45] A. Hirschowitz, *Une conjecture pour la cohomologie des diviseurs sur les surfaces rationelles génériques* , J. Reine Angew. Math., **397** (1989), 208–213.

[46] A. Iarrobino, *The inverse system of a symbolic power II: the Waring problem for forms*, J. Algebra, **174** (1995) 1091–1110.

[47] A. Iarrobino and V. Kanev, *Power sums, Gorestein algebras and determinantal loci*, Lecture Notes in Math., Springer-Verlag **1721** (1999).

[48] J. Kollár, *Vanishing theorems for cohomology groups*, Proceedings Symp. Pure Math., **46** (1) (1987), 233–243.

[49] A. Kouvidakis, *Divisors on symmetric products of curves*, Transactions Amer. Math. Soc., **337** (1) (1993), 117–128.

[50] A. Laface, *On the dimension of linear systems with fixed base points of given multiplicity*, Tesi di Dottorato, Univ. di Milano (1999).

[51] T. Lickteig, *Typical tensorial rank*, Linear algegra and its applications, **69** (1985), 95–120.

[52] F. London, *Ueber die Polarfiguren der ebenen Kurven dritter Ordnung*, Math. Ann., **36** (1890).

[53] G. Lorentz and R. Lorentz, *Solvability problems of bivariate interpolation. I.*, Constructive approximation, **2** (1986), 153–170.

[54] G. Lorentz and R. Lorentz, *Solvability problems of bivariate interpolation. II: applications*, Approximation Theory and applications, **3**, (1987), 79–97.

[55] G. Lorentz and R. Lorentz, *Bivariate Hermite interpolation and applications to algebraic geometry*, J. Num. Math., **57**, 6/7 (1990), 669–680.

[56] F. S. H. Macaulay, *The algebraic theory of modular systems*, Cambridge Univ. Press, Crambridge, UK (1916).

[57] D. McDuff and L. Polterovich, *Symplectic packings and algebraic geometry*, Inventiones Math., **115** (1994), 405–429.

[58] T. Mignon, *Systèmes linéaires de courbes planes*, Thèse, Univ. de Nice (1997).

[59] T. Mignon, *Systèmes linéaires de courbes planes à singularités ordinaires imposées*, C. R. Acad. Sci. Paris Sér. I Math., **327** (1998), 651–654.

[60] T. Mignon, *An asymptotic existence theorem for plane curves with prescribed singularities*, pre-print (2000).

[61] R. Miranda, *Linear systems of plane curves*, Notices of the Amer. Math. Soc., **46** (2) (1999), 192–202.

[62] D. Mumford, *Some footnotes to the work of C. P. Ramanujam*, in C. P. Ramanujam, A Tribute, Springer-Verlag, Berlin (1978), 247–262.

[63] M. Nagata, *On the 14-th problem of Hilbert*, Amer. J. of Math., **33** (1959), 766–772.

[64] M. Nagata, *On rational surfaces II*, Mem. Coll. Sci. Univ. Kyoto, Ser. A, Math., **33** (1960), 271–293.

[65] F. Palatini, *Sulla rappresentazione delle forme e in particolare della cubica quinaria con la somma di potenze di forme lineari*, Atti Accad. Sci. Torino, **38** (1903).

[66] F. Palatini, *Sulla rappresentazione delle forme ternarie mediante somme di potenze di forme lineari*, Rend. Accad. Lincei, (V) **12** (III) (1903), 378–384.

[67] F. Palatini, *Sulle superficie algebriche i cui S_h $(h+1)$-seganti non riempiono lo spazio ambiente*, Atti. Accad. Torino, (1906).

[68] F. Palatini, *Sulle varietà algebriche per le quali sono di dimensione minore dell'ordinario, senza riempire lo spazio ambiente, una o alcune delle varietà formate da spazi seganti*, Atti. Accad. Torino, **44** (1909) 362–374.

[69] F. Palatini, *Sulla rappresentazione delle coppie di forme ternarie mediante somme di potenze di forme lineari*, Ann. Mat. Pura e Appl., **24** (III) (1915), 1–10.

[70] C. P. Ramanujam, *Remarks on the Kodaira vanishing theorem*, J. Ind. Math. Soc., **36** (1972), 41–51.

[71] C. P. Ramanujam, *Supplement to the article: "Remarks on the Kodaira vanishing theorem"*, J. Ind. Math. Soc., **38** (1974), 121–124.

[72] Z. Ran, *Enumerative geometry of singular plane curves*, Inventiones Math., **97** (1989), 447–465.

[73] Th. Reye, *Ueber lineare Systeme und Gewebe von Flächen zweiten Grades*, Journ. für Math., **82** (1877).

[74] J. Roè, *On the existence of plane curves with imposed multiple points*, pre-print (1999).

[75] G. Scorza, *Determinazione delle varietà a tre dimensioni di S_r, $(r \geq 7)$, i cui S_3 tangenti si tagliano a due a due*, Rend. Circ. Mat. Palermo, **25** (1907), 193–204.

[76] G. Scorza, *Un problema sui sistemi lineari di curve appartenenti a una superficie algebrica*, Rend. R. Ist. Lombardo, (2) **41** (1908), 913–920.

[77] G. Scorza, *Sulle varietà a quattro dimensioni di S_r $(r \geq 9)$ i cui S_4 tangenti si tagliano a due a due*, Rend. Circ. Mat. Palermo, **27** (1909), 148–178.

[78] B. Segre, *Alcune questioni su insiemi finiti di punti in geometria algebrica*, Atti Convegno Intern. di Geom. Alg. di Torino, (1961), 15–33.

[79] J. Seibert, *The dimension of quasi-homogeneous linear systems with multiplicity four,* pre-print (1999).

[80] V. Strassen, *Rank and optimal computation of generic tensors,* Linear algebra and its applications, **52–53** (1983), 645–685.

[81] A. Terracini, *Sulle V_k per cui la varietà degli S_h $(h+1)$-seganti ha dimensione minore dell'ordinario,* Rend. Circ. Mat. Palermo, **31** (1911), 392–396.

[82] A. Terracini, *Sulla rappresentazione delle coppie di forme ternarie mediante somme di potenze di forme lineari,* Ann. Mat. Pura e Appl., **24** (1915), 91–100.

[83] A. Terracini, *Sulla rappresentazione delle forme quaternarie mediante somme di potenze di forme lineari,* Atti. Accad. Sci. Torino, **51** (1915–16).

[84] A. Terracini, *Su due problemi, concernenti la determinazione di alcune classi di superficie, considerati da G. Scorza e F. Palatini,* Atti Soc. Natur. e Matem. Modena, V, **6** (1921–22), 3–16.

[85] G. Xu, *Ample line bundles on smooth surfaces,* J. reine und angew. Math., **469** (1995), 199–209.

[86] F. Zak, *Tangents and secants of algebraic varieties,* Transl. Math. Monogr., **127** (1993).

Università di Roma Tor Vergata
Dipartimento di Matematica
Via della Ricerca Scientifica
00133 Roma, Italy
E-mail address: `cilibert@axp.mat.uniroma2.it`

The Calibration Method
for Free Discontinuity Problems

Gianni Dal Maso

Abstract. The calibration method is used to identify some minimizers of the Mumford-Shah functional. The method is then extended to more general free discontinuity problems.

1. Introduction

In [5] De Giorgi introduced the name *free discontinuity problems* to denote a wide class of minimum problems for functionals of the form

$$F(u) := \int_{\Omega \setminus S_u} f(x, u(x), \nabla u(x))dx + \int_{S_u} \psi(x, u^+(x), u^-(x), \nu_u(x))d\mathcal{H}^{n-1}, \quad (1)$$

where Ω is a given bounded domain in $\mathbf{R}^n$ with Lipschitz boundary, $f \colon \Omega \times \mathbf{R} \times \mathbf{R}^n \to [0, +\infty]$ and $\psi \colon \Omega \times \mathbf{R} \times \mathbf{R} \times \mathbf{S}^{n-1} \to [0, +\infty]$ are given Borel functions, $\mathbf{S}^{n-1} = \{v \in \mathbf{R}^n : |v| = 1\}$, $\mathcal{H}^{n-1}$ is the $(n-1)$-dimensional Hausdorff measure, and the unknown function $u \colon \Omega \to \mathbf{R}$ is assumed to be regular out of a (partially regular) singular set S_u of dimension $n-1$, with unit normal ν_u, on which u admits unilateral traces u^+ and u^-. The main feature of these problems is that the shape and location of the discontinuity set S_u are not prescribed. Thus minimizing F means optimizing both the function u and the singular set S_u, which is indeed often regarded as an independent unknown.

These problems have an increasing importance in many branches of applied analysis, such as image processing (Mumford-Shah functional for image segmentation) and fracture mechanics (Griffith's criterion and Barenblatt cohesive zone model).

The *Mumford-Shah functional* was introduced in [10] in the context of a variational approach to image segmentation problems (for which we refer to [9]). It can be written, in dimension n, as

$$F_g^{\alpha,\beta}(u) := \int_{\Omega \setminus S_u} |\nabla u(x)|^2 dx + \alpha \mathcal{H}^{n-1}(S_u) + \beta \int_{\Omega \setminus S_u} |u(x) - g(x)|^2 dx, \quad (2)$$

where g is a given function in $L^\infty(\Omega)$ (interpreted as the grey level of the image to be analysed), and $\alpha > 0$ and $\beta \geq 0$ are constants. When $n = 2$ (the only case considered in image processing), the singular set S_u of a minimizer u of $F_g^{\alpha,\beta}$ is interpreted as the set of the most relevant segmentation lines of the image.

Using different classes of infinitesimal variations, one can show that every minimizer must satisfy certain equilibrium conditions, which could be globally called *Euler-Lagrange equations* for $F_g^{\alpha,\beta}$. For instance, u must satisfy the equation $\Delta u = \beta(u - g)$ on $\Omega \setminus S_u$, with Neumann boundary conditions on $S_u \cup \partial\Omega$. Moreover, there is a link between the mean curvature of S_u (where defined) and the traces of u and ∇u on the two sides of S_u; for instance, when $\alpha = 1$ and $\beta = 0$, the mean curvature of S_u must be equal to the difference of the squares of the norms of the traces of ∇u. Additional conditions have been derived for the two-dimensional case. We refer the reader to [10] and [2] for a precise description of these equilibrium conditions.

However, since $F_g^{\alpha,\beta}$ is not convex, all conditions which can be derived by infinitesimal variations are necessary for minimality, but never sufficient. The purpose of this note is precisely to present a *sufficient condition* for minimality (Theorem 3.1 for $F_g^{\alpha,\beta}$ and Theorem 3.4 for F), and give a few applications (Examples 4.1–4.8). Detailed proofs and further results will be given in the forthcoming paper [1].

2. Notation and Preliminaries

For a complete mathematical treatment of the minimum problems for the functional F considered in (1), we use the space $SBV(\Omega)$ of *special functions of bounded variation*, introduced by De Giorgi and Ambrosio in [6]. A self-contained presentation of this space can be found in the recent book [2], which contains also the complete proof of the existence of a minimizer u of $F_g^{\alpha,\beta}$, and of the partial regularity of the corresponding singular set S_u (the regularity of u on $\Omega \setminus S_u$ follows from the standard theory of elliptic equations).

We recall that for every $u \in SBV(\Omega)$ the *approximate upper and lower limits* $u^+(x)$ and $u^-(x)$ at a point $x \in \Omega$ are defined by

$$u^{\pm}(x) := \pm \inf\{t \in \mathbf{R} : \lim_{\rho \to 0+} \rho^{-n} \mathcal{L}^n(\{\pm u > t\} \cap B_\rho(x)) = 0\},$$

where $B_\rho(x)$ is the open ball with centre x and radius ρ. The *singular set* (or *jump set*) of u is defined by $S_u := \{x \in \Omega : u^-(x) < u^+(x)\}$. It is known that S_u is countably $(\mathcal{H}^{n-1}, n-1)$-rectifiable and that there exists a Borel measurable function $\nu_u : S_u \to \mathbf{S}^{n-1}$ such that for $\mathcal{H}^{n-1}$-a.e. $x \in S_u$ we have

$$\lim_{\rho \to 0+} \frac{1}{\rho^n} \int_{B_\rho^{\pm}(x)} |u(y) - u^{\pm}(x)|\, dy = 0, \tag{3}$$

where $B_\rho^{\pm}(x) := \{y \in B_\rho(x) : \pm(y - x) \cdot \nu_u(x) > 0\}$ and $\cdot$ denotes the scalar product in $\mathbf{R}^n$ (see [7, Theorem 4.5.9]). Condition (3) says that $\nu_u(x)$ points from the side of S_u corresponding to $u^-(x)$ to the side corresponding to $u^+(x)$.

The gradient Du of u is a measure that can be decomposed as the sum of two measures $Du = D^a u + D^s u$, where $D^a u$ is absolutely continuous and $D^s u$ is singular with respect to the Lebesgue measure $\mathcal{L}^n$. The density of $D^a u$ with respect to $\mathcal{L}^n$ is denoted by ∇u. Since $u \in SBV(\Omega)$, for every Borel set B in Ω we have

$$(Du)(B) = \int_B \nabla u(x)\, dx + \int_{B \cap S_u} (u^+(x) - u^-(x))\nu_u(x)\, d\mathcal{H}^{n-1}\,.$$

The *graph* of u is defined as

$$\Gamma_u := \{(x,t) \in \Omega \times \mathbf{R} : u^-(x) \le t \le u^+(x)\}\,.$$

The characteristic function of the subgraph $\{(x,t) \in \Omega \times \mathbf{R} : t \le u(x)\}$ is denoted by 1_u. It is defined by $1_u(x,t) := 1$ if $t \le u(x)$, and $1_u(x,t) := 0$ if $t > u(x)$. It belongs to $SBV(\Omega \times \mathbf{R})$ and its gradient $D1_u$ is a measure concentrated on Γ_u.

3. The Main Results

We fix an open subset U of $\Omega \times \mathbf{R}$ of the form

$$U := \{(x,t) \in \Omega \times \mathbf{R} : \tau_1(x) < t < \tau_2(x)\}\,, \tag{4}$$

where τ_1 and τ_2 are two continuous functions on $\overline{\Omega}$ such that $-\infty \le \tau_1(x) \le \tau_2(x) \le +\infty$ for every $x \in \overline{\Omega}$.

Let F be the functional introduced in (1). We say that a function $u \in SBV(\Omega)$, with graph $\mathcal{H}^n$-contained in U (i.e., $\mathcal{H}^n(\Gamma_u \setminus U) = 0$), is a *Dirichlet U-minimizer* of F, if $F(u) \le F(v)$ for every $v \in SBV(\Omega)$ with the same trace as u on $\partial\Omega$ and with graph $\mathcal{H}^n$-contained in U. If the inequality $F(u) \le F(v)$ holds for every $v \in SBV(\Omega)$ with graph $\mathcal{H}^n$-contained in U, we say that u is a *U-minimizer* of F. We omit U when $U = \Omega \times \mathbf{R}$.

The symbol ϕ will always denote a bounded Borel measurable vectorfield defined on U with values in $\mathbf{R}^{n+1} = \mathbf{R}^n \times \mathbf{R}$, with components $\phi^x \in \mathbf{R}^n$ and $\phi^t \in \mathbf{R}$. The divergence of ϕ is then $\operatorname{div}\phi(x,t) = \operatorname{div}_x \phi^x(x,t) + \partial_t \phi^t(x,t)$.

We begin with a theorem concerning the functional $F_g^{\alpha,\beta}$ introduced in (2).

Theorem 3.1. *Let $u \in SBV(\Omega)$ with graph $\mathcal{H}^n$-contained in U. Assume that there exists a bounded vectorfield ϕ of class C^1 on U with the following properties:*

(a1) $\frac{1}{4}|\phi^x(x,t)|^2 \le \phi^t(x,t) + \beta|t - g(x)|^2$
       for $\mathcal{L}^n$-a.e. $x \in \Omega$ and for every $\tau_1(x) < t < \tau_2(x)$;

(a2) $\phi^x(x,u(x)) = 2\nabla u(x)$ *and* $\phi^t(x,u(x)) = |\nabla u(x)|^2 - \beta|u(x) - g(x)|^2$
       for $\mathcal{L}^n$-a.e. $x \in \Omega$;

(b1) $\left| \displaystyle\int_{t_1}^{t_2} \phi^x(x,t)\, dt \right| \le \alpha$
       for $\mathcal{H}^{n-1}$-a.e. $x \in \Omega$ and for every $\tau_1(x) < t_1 < t_2 < \tau_2(x)$;

320   G. Dal Maso

(b2) $\displaystyle\int_{u^-(x)}^{u^+(x)} \phi^x(x,t)\,dt = \alpha\,\nu_u(x)$

   $\qquad$ *for $\mathcal{H}^{n-1}$-a.e. $x \in S_u$;*

(c1) div $\phi(x,t) = 0$

   $\qquad$ *for every $(x,t) \in U$.*

Then u is a Dirichlet U-minimizer of $F_g^{\alpha,\beta}$. If, in addition, $\phi^x(x,t)$ satisfies the boundary condition

(c2) $\displaystyle\lim_{(y,s)\to(x,t)} \phi^x(y,s)\cdot\nu(x) = 0$

   $\qquad$ *for $\mathcal{H}^{n-1}$-a.e. $x \in \partial\Omega$ and for $\mathcal{L}^1$-a.e. $t \in [\tau_1(x),\tau_2(x)]$,*

where $\nu(x)$ is the outer unit normal to $\partial\Omega$, then u is a U-minimizer of $F_g^{\alpha,\beta}$.

A vectorfield ϕ which satisfies conditions (a1)–(c1) of Theorem 3.1 is called a *calibration* for the functional $F_g^{\alpha,\beta}$ on U. If ϕ satisfies also (c2), it is called a *Neumann calibration*. Theorem 3.1 is an immediate consequence of the following lemmas.

Lemma 3.2. *Let ϕ be a vectorfield which satisfies conditions (a1) and (b1) of Theorem 3.1. Then for every $u \in SBV(\Omega)$ with graph $\mathcal{H}^n$-contained in U we have*

$$F_g^{\alpha,\beta}(u) \geq \int_U \phi \cdot d(D1_u)\,. \tag{5}$$

Moreover, equality holds in (5) for a given u if and only if conditions (a2) and (b2) of Theorem 3.1 are satisfied.

The next lemma is a consequence of the divergence theorem.

Lemma 3.3. *Suppose that ϕ is of class C^1 and that $\mathrm{div}\,\phi = 0$ on U. Then*

$$\int_U \phi \cdot d(D1_u) = \int_U \phi \cdot d(D1_v) \tag{6}$$

for every pair of functions u, v in $BV(\Omega)$ with the same trace on $\partial\Omega$ and with graphs $\mathcal{H}^n$-contained in U. If, in addition, ϕ satisfies condition (c2) of Theorem 3.1, then (6) holds for every pair of functions u, v in $BV(\Omega)$ with graphs $\mathcal{H}^n$-contained in U.

As a matter of fact, the method of calibrations can be easily adapted to the functional F defined in (1).

Theorem 3.4. *Let $u \in SBV(\Omega)$ with graph $\mathcal{H}^n$-contained in U. Assume that there exists a bounded vectorfield ϕ of class C^1 on U with the following properties:*

(a1) $\phi^x(x,t)\cdot v \leq \phi^t(x,t) + f(x,t,v)$

   $\qquad$ *for $\mathcal{L}^n$-a.e. $x \in \Omega$, for every $\tau_1(x) < t < \tau_2(x)$, and for every $v \in \mathbf{R}^n$;*

(a2) $\phi^x(x,u(x))\cdot\nabla u(x) = \phi^t(x,u(x)) + f(x,u(x),\nabla u(x))$

   $\qquad$ *for $\mathcal{L}^n$-a.e. $x \in \Omega$;*

(b1) $\nu \cdot \displaystyle\int_{t_1}^{t_2} \phi^x(x,t)\,dt \leq \psi(x,t_1,t_2,\nu)$

for $\mathcal{H}^{n-1}$-a.e. $x \in \Omega$, for every $\tau_1(x) < t_1 < t_2 < \tau_2(x)$, and for every $\nu \in \mathbf{S}^{n-1}$;

(b2) $\nu_u(x) \cdot \displaystyle\int_{u^-(x)}^{u^+(x)} \phi^x(x,t)\,dt = \psi(x,u^-(x),u^+(x),\nu_u(x))$

for $\mathcal{H}^{n-1}$-a.e. $x \in S_u$;

(c1) $\operatorname{div}\phi(x,t) = 0$

for every $(x,t) \in U$.

Then u is a Dirichlet U-minimizer of F. If $\phi^x(x,t)$ satisfies also the boundary condition (c2) of Theorem 3.1, then u is a U-minimizer of F.

Remark 3.5. *We note that in Theorem 3.4 there is no regularity or convexity hypothesis on f or ψ. If $f^*(x,t,v^*)$ is the convex conjugate of $f(x,t,v)$ with respect to v, condition (a1) is equivalent to*

(a1') $f^*(x,t,\phi^x(x,t)) \leq \phi^t(x,t)$

for $\mathcal{L}^n$-a.e. $x \in \Omega$ and for every $\tau_1(x) < t < \tau_2(x)$.

If this condition is satisfied, and $f(x,t,v)$ is convex and differentiable with respect to v, then condition (a2) is equivalent to

(a2')
$\begin{cases} \phi^x(x,u(x)) = \partial_v f(x,u(x),\nabla u(x)) \\ \phi^t(x,u(x)) = f^*(x,u(x),\phi^x(x,u(x))) \end{cases}$
$\qquad$ *for* $\mathcal{L}^n$*-a.e.* $x \in \Omega$.

Remark 3.6. *In Theorems 3.1 and 3.4 the hypothesis that ϕ is of class C^1 is too strong for many applications. It is used only in Lemma 3.3 and it can be relaxed in several ways (see [1] for details). For instance, one may consider piecewise C^1 vectorfields, which may be discontinuous along sufficiently regular interfaces. In this case the divergence-free condition (c1) must be understood in the distributional sense, i.e., the pointwise divergence vanishes (where defined) and the normal component of ϕ is continuous across the discontinuity surfaces.*

4. Some Examples

The following examples show that the calibration method is very flexible, and can be used to prove the minimality of a given function u in many different situations. In the first examples we will consider only the "homogeneous" functional $F^\alpha := F_g^{\alpha,0}$, in which the lower order term $\beta \int_\Omega |u-g|^2 dx$ vanishes.

Example 4.1. (Affine function in one dimension) *Let $n := 1$, $\Omega := \,]0,a[$, and $u(x) := \lambda x$, with $\lambda > 0$. It is easy to see that u is a Dirichlet minimizer of F^α if and only if $a\lambda^2 \leq \alpha$. In this case a calibration is given by the piecewise constant* *function*

$$\phi(x,t) := \begin{cases} (2\lambda, \lambda^2), & \text{if } \frac{\lambda}{2}x \leq t \leq \frac{\lambda}{2}(x+a), \\ (0,0), & \text{otherwise}. \end{cases} \tag{7}$$

Another calibration is given by

$$\phi(x,t) := \begin{cases} \left(2\frac{t}{x}, \left(\frac{t}{x}\right)^2\right), & \text{if } 0 \leq t \leq \lambda x, \\ \left(2\frac{\lambda a - t}{a - x}, \left(\frac{\lambda a - t}{a - x}\right)^2\right), & \text{if } \lambda x \leq t \leq \lambda a, \\ (0,0), & \text{otherwise}. \end{cases} \tag{8}$$

If $a\lambda^2 > \alpha$, then the function $u(x) := \lambda x$ is not a Dirichlet minimizer of F^α, but it is still a Dirichlet U-minimizer with

$$U := \left\{ (x,t) \in \,]0,a[\, \times \mathbf{R} : \lambda x - \tfrac{\alpha}{4\lambda} < t < \lambda x + \tfrac{\alpha}{4\lambda} \right\}.$$

A calibration on U is given by $\phi(x,t) := (2\lambda, \lambda^2)$.

Example 4.2. (Jump in one dimension) *Let $n := 1$, $\Omega := \,]0,a[$, $u(x) := 0$ for $0 < x < c$, and $u(x) := h$ for $c < x < a$, with $0 < c < a$ and $h > 0$. It is easy to see that u is a Dirichlet minimizer of F^α if and only if $a\alpha \leq h^2$. In this case two different calibrations are given by (7) and (8) with $\lambda = \sqrt{\alpha}/\sqrt{a}$.*

Suppose now that $a\alpha > h^2$. Let $\varepsilon > 0$ be a constant such that $2\varepsilon + \sqrt{2\alpha\varepsilon} \leq h$, let

$$\tau_1(x) = \begin{cases} -\varepsilon, & \text{if } x \leq c, \\ -\varepsilon + \frac{h}{\varepsilon}(x - c), & \text{if } c \leq x \leq c + \varepsilon, \\ h - \varepsilon, & \text{if } c + \varepsilon \leq x, \end{cases}$$

let $\tau_2(x) = \tau_1(x + \varepsilon) + 2\varepsilon$, and let U be the open set defined by (4). Then u is a Dirichlet U-minimizer of F^α, and a calibration on U is given by the piecewise constant function

$$\varphi(x,t) := \begin{cases} (2\lambda, \lambda^2), & \text{if } c - \varepsilon < x < c + \varepsilon \text{ and} \\ & \varepsilon + \frac{\lambda}{2}(x - c + \varepsilon) < t < \varepsilon + \frac{\lambda}{2}(x - c + \varepsilon) + \frac{\alpha}{2\lambda}, \\ (0,0), & \text{otherwise}, \end{cases}$$

where $\lambda > 0$ is any constant such that $\varepsilon + \varepsilon\lambda + \frac{\alpha}{2\lambda} \leq h - \varepsilon$, for instance $\lambda = \sqrt{\alpha}/\sqrt{2\varepsilon}$.

Example 4.3. (Harmonic function) *Let Ω be a bounded domain in $\mathbf{R}^n$, n arbitrary, and let u be a harmonic function on Ω. As pointed out by Chambolle [3], u is a Dirichlet minimizer of F^α if*

$$\operatorname*{osc}_{\Omega} u \, \sup_{\Omega} |\nabla u| \leq \alpha, \tag{9}$$

where $\operatorname{osc}_\Omega u := \sup_\Omega u - \inf_\Omega u$. Note that for $n = 1$ this condition reduces to the constraint $a\lambda^2 \leq \alpha$ of Example 4.1. Inspired by the one dimensional case (see (7)), *we construct the calibration*

$$\phi(x,t) := \begin{cases} \big(2\nabla u(x), |\nabla u(x)|^2\big), & \text{if } \frac{1}{2}(u(x)+m) \le t \le \frac{1}{2}(u(x)+M)\,, \\ (0,0), & \text{otherwise}\,, \end{cases} \tag{10}$$

where $m := \inf_\Omega u$ *and* $M := \sup_\Omega u$. *Another calibration (see (8)) is given by*

$$\phi(x,t) := \begin{cases} \big(2\frac{t-m}{u(x)-m}\nabla u(x), (\frac{t-m}{u(x)-m})^2|\nabla u(x)|^2\big), & \text{if } m \le t \le u(x)\,, \\ \big(2\frac{M-t}{M-u(x)}\nabla u(x), (\frac{M-t}{M-u(x)})^2|\nabla u(x)|^2\big), & \text{if } u(x) \le t \le M\,, \\ (0,0), & \text{otherwise}\,. \end{cases}$$

If (9) is not satisfied, u is still a Dirichlet U-minimizer of F^α, for

$$U := \{(x,t) \in \Omega \times \mathbf{R} : u(x) - \tfrac{\alpha}{4}|\nabla u(x)|^{-1} < t < u(x) + \tfrac{\alpha}{4}|\nabla u(x)|^{-1}\}\,, \tag{11}$$

and a calibration in U is given by $\phi(x,t) := \big(2\nabla u(x), |\nabla u(x)|^2\big)$.

Example 4.4. (Pure jump) *Let $n \ge 2$ and let $\Omega :=]0,a[\times V$, where V is a bounded domain in $\mathbf{R}^{n-1}$ with Lipschitz boundary. Denoting the first coordinate of x by x_1, let $u(x) := 0$ for $0 < x_1 < c$, and $u(x) := h$ for $c < x_1 < a$, with $0 < c < a$ and $h > 0$. Using the results of Example 4.2 it is easy to see that u is a Dirichlet minimizer of F^α if $a\alpha \le h^2$. In this case two different calibrations can be constructed in the following way: the projection of these calibrations onto the (x_1,t)-plane are given by (7) and (8), with $\lambda = \sqrt{\alpha}/\sqrt{a}$ and x replaced by x_1, while all other components of these calibrations vanish.*

If $a\alpha > h^2$, it may happen that u is still a Dirichlet minimizer of F^α. For instance, if $n = 2$ and $V =]0,b[$, with $b\alpha\pi \le 2h^2$, a different calibration has been constructed in [1]. Therefore u is a Dirichlet minimizer of F^α even if $a\alpha$ is very large with respect to h^2, provided that $b\alpha$ is small enough.

Arguing as in the last part of Example 4.2 one can prove that for every a and V there exists an open set U of the form (4), containing Γ_u, such that u is a Dirichlet U-minimizer of F^α.

Example 4.5. (Triple junction) *Let $n := 2$, let $\Omega := B(0,r)$ be the open ball with radius $r > 0$ centered at the origin, and let u be given, in polar coordinates, by $u(\rho,\theta) := a$ for $0 \le \theta < \frac{2}{3}\pi$, $u(\rho,\theta) := b$ for $\frac{2}{3}\pi \le \theta < \frac{4}{3}\pi$, and $u(\rho,\theta) := c$ for $\frac{4}{3}\pi \le \theta < 2\pi$, where a, b, and c are distinct constants. Thus S_u is given by three line segments meeting at the origin with equal angles. If*

$$2\alpha r \le \min\{|a-b|^2, |b-c|^2, |c-a|^2\}\,, \tag{12}$$

then u is a Dirichlet minimizer of F^α. To construct a calibration, it is not restrictive to assume $a < b = 0 < c$. Inspired by the one dimensional case described in Example 4.2, we take $e_\pm := (\pm\sqrt{3}/2, -1/2)$, and $\lambda > 0$ such that $\frac{\lambda r}{2} + \frac{\alpha}{\lambda} \le \min\{-a, c\}$

(which is possible by (12)), and we define the calibration by

$$\phi(x,t) := \begin{cases} (\lambda e_+, \lambda^2/4), & \text{if } \frac{\lambda}{4}(r + x \cdot e_+) \leq t \leq \frac{\lambda}{4}(r + x \cdot e_+) + \frac{\alpha}{\lambda}, \\ (\lambda e_-, \lambda^2/4), & \text{if } \frac{\lambda}{4}(-r + x \cdot e_-) - \frac{\alpha}{\lambda} \leq t \leq \frac{\lambda}{4}(-r + x \cdot e_-), \\ (0,0), & \text{otherwise}. \end{cases} \tag{13}$$

If αr is much larger than $\min\{|a-b|^2, |b-c|^2, |c-a|^2\}$, it is easy to construct a comparison function v with the same boundary values as u and such that $F^\alpha(v) < F^\alpha(u)$. This shows that in this case u is not a Dirichlet minimizer.

However, for every value of the parameters α, r, a, b, c, one can construct a suitable neighbourhood U of the graph Γ_u, of the form (4), such that a variant of (13) is a calibration in U, and therefore u is a Dirichlet U-minimizer of F^α. We refer to [1] for the details.

We consider now the functional $F_g^{\alpha,\beta}$, with $\beta > 0$.

Example 4.6. (Solution of the Neumann problem) *Let Ω be a bounded open set in $\mathbf{R}^n$ with boundary of class $C^{1,\varepsilon}$ for some $\varepsilon > 0$, and let u be the solution of the Neumann problem*

$$\begin{cases} \Delta u = \beta(u - g) & \text{on } \Omega, \\ \frac{\partial u}{\partial \nu} = 0 & \text{on } \partial\Omega, \end{cases} \tag{14}$$

with $\beta > 0$ and $g \in L^\infty(\Omega)$. Assume that

$$\underset{\Omega}{\operatorname{osc}} g \, \underset{\Omega}{\sup} |\nabla u| \leq \alpha. \tag{15}$$

Then u is a minimizer of $F_g^{\alpha,\beta}$. If the strict inequality holds in (15), then u is the unique minimizer. A Neumann calibration $\phi(x,t)$ is given by

$$\begin{cases} \left(0, \beta|\frac{m}{2} - \frac{u(x)}{2}|^2 - \beta|\frac{m}{2} + \frac{u(x)}{2} - g(x)|^2\right), & \text{if } t - \frac{u(x)}{2} < \frac{m}{2}, \\ \left(2\nabla u(x), |\nabla u(x)|^2 - \beta|t - g(x)|^2 + \beta|t - u(x)|^2\right), & \text{if } \frac{m}{2} \leq t - \frac{u(x)}{2} \leq \frac{M}{2}, \\ \left(0, \beta|\frac{M}{2} - \frac{u(x)}{2}|^2 - \beta|\frac{M}{2} + \frac{u(x)}{2} - g(x)|^2\right), & \text{if } \frac{M}{2} < t - \frac{u(x)}{2}, \end{cases}$$

where $m := \inf_\Omega u$ and $M := \sup_\Omega u$.

If (15) is not satisfied, u is still a U-minimizer of $F_g^{\alpha,\beta}$, for a suitable neighbourhood U of Γ_u. A Neumann calibration on U is given by

$$\phi(x,t) := \left(2\nabla u(x), |\nabla u(x)|^2 - \beta|t - g(x)|^2 + \beta|t - u(x)|^2\right).$$

The hypothesis that $\partial\Omega$ is of class $C^{1,\varepsilon}$ is used only to obtain the boundary condition (c2) of Theorem 3.1, which, in this case, becomes

$$\lim_{y \to x} \nabla u(y) \cdot \nu(x) = 0 \quad \text{for } \mathcal{H}^{n-1}\text{-a.e. } x \in \partial\Omega. \tag{16}$$

It is clear that (16) is still true if for $\mathcal{H}^{n-1}$-a.e. $x \in \partial\Omega$ there exists an open neighbourhood V_x of x in $\mathbf{R}^n$ such that $V_x \cap \partial\Omega$ is a manifold of class $C^{1,\varepsilon}$ (see [2, Theorem 7.5.2]). Therefore the result of this example is true also when Ω is polyhedral.

In the next examples we construct a calibration for $F_g^{\alpha,\beta}$ when the parameter β is large enough.

Example 4.7. (Smooth g and large β) *Let Ω be a bounded open set in $\mathbf{R}^n$ with smooth boundary, and let $g \in C^2(\overline{\Omega})$. There exists a constant $\beta_0 \geq 0$, depending on g and α, such that for every $\beta > \beta_0$ the solution u of the Neumann problem (14) of Example 4.6 is the unique minimizer of $F_g^{\alpha,\beta}$. A Neumann calibration is constructed in* [1].

This shows that the minimizer of $F_g^{\alpha,\beta}$ is smooth, provided that g is smooth and β is large enough. Therefore the solution of the image segmentation problem $(n = 2)$ based on the minimization of $F_g^{\alpha,\beta}$ has an empty set of segmentation lines if the "grey level" function g is smooth and the parameter β in the fidelity term $\beta \int_\Omega |u - g|^2 dx$ is large.

Example 4.8. (Function g with only two values) *Let Ω be an open set in $\mathbf{R}^n$ and let E be a compact set contained in Ω with boundary of class C^2. Let $g(x) := a$ for $x \in E$ and $g(x) := b$ for $x \in \Omega \setminus E$, with $a \neq b$. There exists a constant $\beta_0 \geq 0$, depending on g and α, such that for every $\beta > \beta_0$ the function $u := g$ is the unique minimizer of $F_g^{\alpha,\beta}$. To construct a calibration, it is not restrictive to assume $a < b$. We take a C^1 vectorfield $v \colon \Omega \to \mathbf{R}^n$ with compact support in Ω such that $|v(x)| \leq 1$ for every $x \in \Omega$ and $v(x)$ is the outer unit normal to ∂E for every $x \in \partial E$. Then we set $\phi^x(x,t) = \sigma(t)v(x)$, where σ is a fixed positive smooth function with integral equal to α and support contained in $]a,b[$. We see that conditions (b1), (b2), and (c2) of Theorem 3.1 are satisfied by construction. It remains to choose ϕ^t so that (a1), (a2), and (c1) hold. Condition (a2) forces us to set $\phi^t(x,t) = 0$ for $t = g(x)$, while (c1) gives $\partial_t \phi^t(x,t) = -\sigma(t) \operatorname{div}_x v(x)$. These two conditions determine $\phi^t(x,t)$ at every point (x,t). It is then easy to see that (a1) holds if β is large enough. We refer to* [1] *for the details.*

This example shows that, if $g \in SBV(\Omega)$ has only two values, and S_g is smooth enough, then the minimizer of the Mumford-Shah functional $F_g^{\alpha,\beta}$ reconstructs g exactly, when β is large enough.

Recently the following question has been studied by using the calibration method: is it true that a function u is a (Dirichlet) minimizer of $F_g^{\alpha,\beta}$, if it satisfies the Euler-Lagrange equations and the domain Ω is *sufficiently small*? For the moment we have only a partial answer. In [4] we have considered the case where $n := 2$ and S_u is a line segment joining two points of the boundary of Ω. If u satisfies the Euler-Lagrange equations for the "homogeneous functional" $F^\alpha := F_g^{\alpha,0}$, then for every $x_0 \in S_u$ there exists an open neighbourhood Ω_0 of x_0, contained in Ω, such that u is a Dirichlet minimizer of F^α in Ω_0. The minimality is proved by constructing a complicated calibration on $\Omega_0 \times \mathbf{R}$.

This result has been extended in [8] to the case where S_u is an analytic curve joining two points of $\partial\Omega$. The (more difficult) construction of the calibration presented in this paper shows that one can take the same set Ω_0 for every $x_0 \in S_u$; in other words, one can take as Ω_0 a suitable tubular neighbourhood of S_u.

Moreover, it is proved in [8] that an additional condition on u and S_u implies that u is a Dirichlet U-minimizer for a suitable open neighbourhood U of the graph Γ_u. A counterexample (where S_u is a line segment joining two points of $\partial\Omega$) shows that this is not always true when u is just a solution of the Euler-Lagrange equations with $S_u \neq \emptyset$, in contrast to the case $S_u = \emptyset$ (see Example 4.6).

References

[1] G. Alberti, G. Bouchitté and G. Dal Maso: *The calibration method for the Mumford-Shah functional and free discontinuity problems*, Preprint SISSA, Trieste, 2001.

[2] L. Ambrosio, N. Fusco and D. Pallara: *Functions of Bounded Variation and Free Discontinuity Problems*, Oxford Mathematical Monographs, Oxford University Press, Oxford, 2000.

[3] A. Chambolle: *Personal communication*, Trieste, 1996.

[4] G. Dal Maso, M.G. Mora and M. Morini: *Local calibration for minimizers of the Mumford-Shah functional with rectilinear discontinuity sets*, J. Math. Pures Appl., **79** (2000), 141–162.

[5] E. De Giorgi: *Free discontinuity problems in calculus of variations*, in: Frontiers in Pure and Applied Mathemathics, a collection of papers dedicated to Jacques-Louis Lions on the occasion of his sixtieth birthday, R. Dautray ed., North Holland, Amsterdam (1991), 55–62.

[6] E. De Giorgi and L. Ambrosio: *Un nuovo funzionale del calcolo delle variazioni*, Atti Accad. Naz. Lincei Rend. Cl. Sci. Fis. Mat. Natur., **82** (1988), 199–210.

[7] H. Federer: *Geometric Measure Theory*, Springer-Verlag, Berlin, 1969.

[8] M. G. Mora and M. Morini: *Local calibration for minimizers of the Mumford-Shah functional with a regular discontinuity sets*, Ann. Inst. H. Poincaré, Anal. Non Linéaire, to appear.

[9] J.-M. Morel and S. Solimini: *Variational Methods in Image Segmentation*, Progr. Nonlinear Differential Equations Appl., **14** (1995), Birkhäuser, Boston.

[10] D. Mumford and J. Shah: *Optimal approximation by piecewise smooth functions and associated variational problems*, Comm. Pure Appl. Math., **42** (1989), 577–685.

SISSA
Via Beirut 4
34014 Trieste, Italy
E-mail address: `dalmaso@sissa.it`

Geometry on Arc Spaces of Algebraic Varieties

Jan Denef and François Loeser

Abstract. This paper is a survey on arc spaces, a recent topic in algebraic geometry and singularity theory. The geometry of the arc space of an algebraic variety yields several new geometric invariants and brings new light to some classical invariants.

1. Introduction

For an algebraic variety X over the field $\mathbb{C}$ of complex numbers, one considers the arc space $\mathcal{L}(X)$, whose points are the $\mathbb{C}[[t]]$-rational points on X, and the truncated arc spaces $\mathcal{L}_n(X)$, whose points are the $\mathbb{C}[[t]]/t^{n+1}$-rational points on X. The geometry of these spaces yields several new geometric invariants of X and brings new light to some classical invariants. For example, Denef and Loeser [13] showed that the Hodge spectrum of a critical point of a polynomial can be expressed in terms of geometry on arc spaces, yielding a new proof and a generalization [15] of the Thom-Sebastiani Theorem for the Hodge spectrum due to Varchenko [42] and Saito [33, 34]. In a different direction, Batyrev [6] used arc spaces to prove a conjecture of Reid [30] on quotient singularities (the McKay correspondence), and to construct his stringy Hodge numbers [5] appearing in mirror symmetry. All these developments are based on Kontsevich's construction [23] of a measure on the arc space $\mathcal{L}(X)$, the motivic measure, which is an analogue of the p-adic measure on a p-adic variety.

In Section 2 we define the arc spaces $\mathcal{L}(X)$ and $\mathcal{L}_n(X)$ of an algebraic variety over any field k of characteristic zero. The first question that appears is how $\mathcal{L}_n(X)$ and $\pi_n(\mathcal{L}(X))$ change with n, where π_n denotes the truncation map from $\mathcal{L}(X)$ to $\mathcal{L}_n(X)$. As a partial answer to this question we will see in Theorem 2.1 that the power series

$$J(T,\chi) := \sum_{n \geq 0} \chi(\mathcal{L}_n(X)) T^n, \quad P(T,\chi) = \sum_{n \geq 0} \chi(\pi_n(\mathcal{L}(X))) T^n$$

are rational (i.e. a quotient of two polynomials), for any reasonable generalized Euler characteristic χ. This is a direct consequence of results of Denef and Loeser [14]. Instead of working with particular generalized Euler characteristics, such as the topological Euler characteristic, the Hodge polynomial or the Hodge characteristic, it is more general to work with the universal Euler characteristic which associates to any algebraic variety X over k its class $[X]$ in the Grothendieck group $K_0(\mathrm{Var}_k)$ of algebraic varieties over k. This is the abelian group generated by symbols $[X]$, for X a variety over k, with the relations $[X] = [Y]$ if X and Y are isomorphic, and $[X] = [Y] + [X \setminus Y]$ if Y is Zariski closed in X. There is a natural ring structure on $K_0(\mathrm{Var}_k)$, the product of $[X]$ and $[Y]$ being equal to $[X \times Y]$. We denote by $\mathcal{M}_k$ the ring obtained from $K_0(\mathrm{Var}_k)$ by inverting the class of $\mathbb{A}^1_k$. The above rationality result applied to the universal Euler characteristic says that the power series

$$J(T) := \sum_{n \geq 0} [\mathcal{L}_n(X)]\, T^n, \quad P(T) := \sum_{n \geq 0} [\pi_n(\mathcal{L}(X))]\, T^n$$

in $\mathcal{M}_k[[T]]$ are rational.

Power series like $J(T)$ and $P(T)$, with coefficients in $\mathcal{M}_k$, are called "motivic", because they specialize to power series over the Grothendieck group $K_0(\mathrm{Mot}_k)$ of the category of Chow motives over k. Actually in several of our papers on arcs we work over $K_0(\mathrm{Mot}_k)$ instead of over $\mathcal{M}_k$.

In Section 3 we introduce the motivic zeta function $Z(T)$ associated to a morphism f from a nonsingular algebraic variety X of dimension d to the affine line, cf. [18]. A naive version of it is the power series over $\mathcal{M}_k$ defined by

$$Z^{\mathrm{naive}}(T) := \sum_{n \geq 1} [\mathfrak{X}_n]\, [\mathbb{A}^1_k]^{-nd}\, T^n\,.$$

Here $\mathfrak{X}_n$ denotes the set of arcs φ in $\mathcal{L}(X)$ with $f(\varphi)$ a power series of order n. The motivic zeta function of f contains a wealth of geometric information about f. For example the Hodge spectrum of any critical point of f can be expressed in terms of $\lim_{T \to \infty} Z(T)$. This limit is a well-defined element of $\mathcal{M}_k$, and can be considered as the "virtual motivic incarnation" of the Milnor fibers of f. All this is explained in Section 3.5. In Section 3.4 we also show that the topological zeta functions of Denef and Loeser [12] can be expressed in terms of the motivic zeta function.

We explain in Section 4 the notion of motivic integration on $\mathcal{L}(X)$, due to Kontsevich [23], and further developed by Batyrev [5, 6], Denef and Loeser [13]–[18], and Looijenga [27]. This notion plays a key role in the present paper. Kontsevich used it to prove that two birationally equivalent Calabi-Yau manifolds have the same Hodge numbers. This result, together with some other direct applications of motivic integration, is discussed in Section 4.4.

One of the most striking applications of arc spaces and motivic integration is Batyrev's proof [6] of the conjecture of Reid on the generalized McKay correspondence. We will not treat this material in the present paper, but refer to the Bourbaki report of Reid [30], see also [16] and [27].

In Section 5 we explain how the relation between the Hodge spectrum and the motivic zeta function yields a new proof of Varchenko's and Saito's Thom-Sebastiani Theorem which expresses the Hodge spectrum of $f(x) + g(y)$ in terms of the Hodge spectra of $f(x)$ and $g(y)$. Our method [15] actually yields a much stronger result which relates the "virtual motivic incarnations" of the Milnor fibers of these three functions. In our paper [15] we obtained this result only at the level of Chow motives, and it was Looijenga [27] who showed how to work at the level of the Grothendieck ring of algebraic varieties.

Finally in Section 6 we briefly discuss the connections with the p-adic case. Considering motivic integration as an analogue of p-adic integration, several arithmetical results in the p-adic case find their natural counterpart in complex geometry and in the theory of motives.

In the present paper we have avoided Chow motives (with one important exception in Section 6). Indeed, the more recent material in [18] and [27] shows that this is possible except for the functional equation in Section 3 of [13] and the results in [17].

2. The Arc Space of a Variety

We fix a base field k of characteristic zero. The reader may choose to consider only the case where k is the field $\mathbb{C}$ of complex numbers. Let X be an algebraic variety over k, not necessarily irreducible, i.e. X is a reduced separated scheme of finite type over k.

2.1. The arc space of X

For each natural number n we consider the *space $\mathcal{L}_n(X)$ of arcs modulo t^{n+1} on X*. This is an algebraic variety over k, whose K-rational points, for any field K containing k, are the $K[t]/t^{n+1}K[t]$-rational points of X. For example when X is an affine variety with equations $f_i(\vec{x}) = 0$, $i = 1, \ldots, m$, $\vec{x} = (x_1, \ldots, x_n)$, then $\mathcal{L}_n(X)$ is given by the equations, in the variables $\vec{a}_0, \ldots, \vec{a}_n$, expressing that $f_i(\vec{a}_0 + \vec{a}_1 t + \cdots + \vec{a}_n t^n) \equiv 0 \mod t^{n+1}, i = 1, \ldots, m$.

Taking the projective limit of these algebraic varieties $\mathcal{L}_n(X)$ we obtain the *arc space $\mathcal{L}(X)$ of X*, which is a reduced separated scheme over k. In general, $\mathcal{L}(X)$ is not of finite type over k (i.e. $\mathcal{L}(X)$ is an "algebraic variety of infinite dimension"). The K-rational points of $\mathcal{L}(X)$ are the $K[[t]]$-rational points of X. These are called *K-arcs on X*. For example when X is an affine complex variety with equations $f_i(\vec{x}) = 0, i = 1, \ldots m, \vec{x} = (x_1, \ldots, x_n)$, then the $\mathbb{C}$-rational points of $\mathcal{L}(X)$ are the sequences $(\vec{a}_0, \vec{a}_1, \vec{a}_2, \ldots) \in (\mathbb{C}^n)^{\mathbb{N}}$ satisfying $f_i(\vec{a}_0 + \vec{a}_1 t + \vec{a}_2 t^2 + \ldots) = 0$, for $i = 1, \ldots, m$. For any n, and for $m > n$, we have natural morphisms

$$\pi_n: \mathcal{L}(X) \to \mathcal{L}_n(X) \quad \text{and} \quad \pi_n^m: \mathcal{L}_m(X) \to \mathcal{L}_n(X),$$

obtained by truncation. Note that $\mathcal{L}_0(X) = X$ and that $\mathcal{L}_1(X)$ is the tangent bundle of X. For any arc γ on X (i.e. a K-arc for some field K containing k), we call $\pi_0(\gamma)$ *the origin of the arc γ*.

By a theorem of Greenberg [20], given an algebraic variety X over k, there exists a number $c > 0$, such that for any n and for any field K containing k we have

$$\pi_n(\mathcal{L}(X)(K)) = \pi_n^{cn}(\mathcal{L}_{cn}(X)(K)),$$

writing $Y(K)$ to denote the set of K-rational points on any variety Y over k. This implies that $\pi_n(\mathcal{L}(X))$ is a constructible subset of the algebraic variety $\mathcal{L}_n(X)$. If X is smooth, then we can take $c = 1$ and π_n is surjective. Moreover, in that case, π_n^m is a locally trivial fibration with fiber $\mathbb{A}_k^{(m-n)\dim X}$. Here $\mathbb{A}_k^d$ denotes the affine space of dimension d over k.

Probably Nash [28] was the first to study arc spaces in a systematic way (his paper was written in 1968, but published only recently). For a singular point P on X, he considered the space $\mathcal{L}_{\{P\}}(X) := \pi_0^{-1}(P)$ of arcs on X with origin P, and its subspace $\mathcal{N}_{\{P\}}(X)$ of arcs with origin P which are not contained in the singular locus of X. He proved that the number of irreducible components of the Zariski closure of $\pi_n(\mathcal{N}_{\{P\}}(X))$ stabilizes for n big enough, and associated to each irreducible component of this closure (when $n \gg 0$), in a canonical and injective way, an irreducible component of the preimage of P in any resolution of singularities of X. Moreover he conjectured that any irreducible component of the preimage of P in a given resolution of X, which "appears" in all resolutions of X, is obtained in that way. For other results related to this we refer to [24, 21, 25].

2.2. How do $\mathcal{L}_n(X)$ and $\pi_n(\mathcal{L}(X))$ change with n?

The work of Nash [28] is the first result towards the question of how the geometry of $\pi_n(\mathcal{L}_{\{P\}}(X))$ changes with n. Recently we investigated how the topological Euler characteristic χ_{top} (case $k = \mathbb{C}$), and generalized Euler characteristics of the spaces $\mathcal{L}_n(X), \pi_n(\mathcal{L}(X)), \pi_n(\mathcal{L}_{\{P\}}(X))$ change with n.

With a *generalized Euler characteristic* on the category Var_k of algebraic varieties over k, we mean a map χ from Var_k to some commutative ring R such that $\chi(X) = \chi(Y)$ when $X \cong Y$, $\chi(X) = \chi(Y) + \chi(X \setminus Y)$ when Y is a Zariski closed subvariety of X, and $\chi(X \times Y) = \chi(X) \cdot \chi(Y)$. Clearly $\chi = \chi_{\mathrm{top}}$ satisfies the above requirements with $R = \mathbb{Z}$, when $k = \mathbb{C}$. Another example of a generalized Euler characteristic, when $k = \mathbb{C}$, is given by $\chi_{hp} \colon \mathrm{Var}_{\mathbb{C}} \to \mathbb{Z}[u, v] \colon X \mapsto \sum_{i,p,q}(-1)^i h_{p,q}^i u^p v^q$, where $h_{p,q}^i$ is the dimension of the (p, q)-component of the mixed Hodge structure on $H_c^i(X, \mathbb{C})$. One calls $\chi_{hp}(X)$ the *Hodge polynomial of X*. For example, $\chi_{hp}(\mathbb{A}_{\mathbb{C}}^1) = uv$. We refer to 3.1.2 for the *Hodge characteristic* χ_h which takes values in the Grothendieck group $K_0(\mathrm{HS})$ of the abelian category of Hodge structures. There are many other examples of generalized Euler characteristics. For example, to mention an exotic one, when $k = \mathbb{Q}$, there is *the conductor $c(X)$ of X*, which yields a generalized Euler characteristic $c \colon Var_{\mathbb{Q}} \to \mathbb{Q}_{>0} \colon X \mapsto c(X) := \prod_i (c_i)^{(-1)^{i+1}}$, where c_i denotes the conductor of the ℓ-adic representation of $\mathrm{Gal}(\bar{\mathbb{Q}}, \mathbb{Q})$ on the étale cohomology $H_c^i(X_{\bar{\mathbb{Q}}}, \mathbb{Q}_\ell)$, where ℓ is a fixed prime and $\bar{\mathbb{Q}}$ an algebraic closure of $\mathbb{Q}$; see e.g. [36]. Here $\mathbb{Q}_{>0}$ is the multiplicative group of positive rational numbers with ring structure inherited by $Q_{>0} \cong \oplus_{p\,\mathrm{prime}} \mathbb{Z} \colon x \mapsto (\mathrm{ord}_p x)_p$. For example, when X is an elliptic curve, $c(X)$ is the usual conductor of X, related to the primes at which X has had reduction.

Theorem 2.1. ([14]) *Let $\chi\colon \mathrm{Var}_k \to R$ be a generalized Euler characteristic, and suppose that $\chi(\mathbb{A}_k^1)$ is not a zero divisor in R. Then the power series*

$$J(T,\chi) := \sum_{n\geq 0} \chi(\mathcal{L}_n(X))\, T^n, \quad P(T,\chi) = \sum_{n\geq 0} \chi(\pi_n(\mathcal{L}(X)))\, T^n$$

are rational (i.e. a quotient of two polynomials). Actually the denominators are products of polynomials of the form $1 - (\chi(\mathbb{A}_k^1))^a T^b$, with $b \in \mathbb{N} \setminus \{0\}, a \in \mathbb{Z}$.

In 2.3 below, we will construct the universal Euler characteristic $\mathrm{Var}_k \to K_0(\mathrm{Var}_k)\colon X \mapsto [X]$, where $K_0(\mathrm{Var}_k)$ denotes the Grothendieck group of varieties over k (see 2.3). Any generalized Euler characteristic on Var_k factorizes over this universal one. So it suffices to prove Theorem 2.1 with R the ring $\mathcal{M}_k$ obtained from $K_0(\mathrm{Var}_k)$ by inverting $[\mathbb{A}_k^1]$. The rationality of $J(T,\chi)$ follows from the material we will discuss in Section 3, when X is an affine hypersurface, see 3.3.1. The proof of the rationality of $P(T,\chi)$ is more complicated and uses a theorem of Pas [29] on quantifier elimination for power series rings. If $f\colon X \to Y$ is a morphism of algebraic varieties, then the image $f(A)$ in $\mathcal{L}(Y)$ of a constructible subset A (cf. 4.1 below) of $\mathcal{L}(X)$ is generally not constructible. The theorem of Pas implies that $f(A)$ still has a "simple description" and is in fact what is called a semi-algebraic subset of $\mathcal{L}(Y)$, cf. [14]. This plays a key role in the proof of the rationality of $P(T,\chi)$. For a survey on applications of quantifier elimination results for valued fields, see [11].

2.3. Grothendieck groups of varieties

Let S be an algebraic variety over k. By an S-variety we mean a variety X together with a morphism $X \to S$. The S-varieties from a category denoted by Var_S, the arrows are the morphisms that commute with the morphisms to S.

We denote by $K_0(\mathrm{Var}_S)$ *the Grothendieck group of S-varieties.* It is an abelian group generated by symbols $[X]$, for X an S-variety, with the relations $[X] = [Y]$ if X and Y are isomorphic in Var_S, and $[X] = [Y] + [X \setminus Y]$ if Y is Zariski closed in X. There is a natural ring structure on $K_0(\mathrm{Var}_S)$, the product of $[X]$ and $[Y]$ being equal to $[X \times_S Y]$. Sometimes we will also write $[X/S]$ instead of $[X]$, to emphasize the role of S. We write $\mathbb{L}$ to denote the class of $\mathbb{A}_k^1 \times S$ in $K_0(\mathrm{Var}_S)$, where the morphism from $\mathbb{A}_k^1 \times S$ to S is the natural projection. We denote by $\mathcal{M}_S$ the ring obtained from $K_0(\mathrm{Var}_S)$ by inverting $\mathbb{L}$. When A is a constructible subset of some S-variety, we define $[A/S]$ in the obvious way, writing A as a disjoint union of a finite number of locally closed subvarieties A_i. Indeed $[A/S] := \sum_i [A_i/S]$ does not depend on the choice of the subvarieties A_i.

When S consists of only one geometric point, i.e. $S = \mathrm{Spec}(k)$, then we will write $K_0(\mathrm{Var}_k)$ instead of $K_0(\mathrm{Var}_S)$ (to denote the Grothendieck group of algebraic varieties over k), and $\mathcal{M}_k$ instead of $\mathcal{M}_S$. Clearly the map $\mathrm{Var}_k \to K_0(\mathrm{Var}_k)$ is the universal generalized Euler characteristic, in the sense that any generalized Euler characteristic on Var_k factors through it.

In our papers [13]–[18], we always work with $K_0(\mathrm{Var}_k)$, but recently E. Looijenga, in his Bourbaki talk [27], introduced the relative Grothendieck ring

$K_0(\mathrm{Var}_S)$, stating some of our results in a stronger form. For example, considering $\mathcal{L}_n(X)$ as an X-variety through the morphism π_0^n, our proof of Theorem 2.1 actually yields the slightly stronger

Theorem 2.2. *Let X be a variety over k, then the power series*

$$J(T) := \sum_{n \geq 0} [\mathcal{L}_n(X)/X]\, T^n, \quad P(T) := \sum_{n \geq 0} [\pi_n(\mathcal{L}(X))/X]\, T^n$$

in $\mathcal{M}_X[[T]]$ are rational, with denominator a product of polynomials of the form $1 - \mathbb{L}^a T^b$, with $b \in \mathbb{N} \setminus \{0\}, a \in \mathbb{Z}$.

2.4. Equivariant Grothendieck groups

We need some technical preparation in order to take care of the monodromy actions in the next section.

For any positive integer n, let μ_n be the group of all n-th roots of unity (in some fixed algebraic closure of k). Note that μ_n is actually an algebraic variety over k, namely $\mathrm{Spec}(k[x]/(x^n - 1))$. The μ_n form a projective system, with respect to the maps $\mu_{nd} \to \mu_n \colon x \mapsto x^d$. We denote by $\hat{\mu}$ the projective limit of the μ_n. Note that the group $\hat{\mu}$ is not an algebraic variety. It is called a pro-variety.

Let X be an S-variety. A *good μ_n-action on X* is a group action $\mu_n \times X \to X$ which is a morphism of S-varieties, such that each orbit is contained in an affine subvariety of X. This last condition is automatically satisfied when X is a quasi-projective variety. A *good $\hat{\mu}$-action on X* is an action of $\hat{\mu}$ on X which factors through a good μ_n-action, for some n.

The *monodromic Grothendieck group* $K_0^{\hat{\mu}}(\mathrm{Var}_S)$ is defined as the abelian group generated by symbols $[X, \hat{\mu}]$ (also denoted by $[X/S, \hat{\mu}]$, or simply $[X]$), for X an S-variety with good $\hat{\mu}$-action, with the relations $[X, \hat{\mu}] = [Y, \hat{\mu}]$ if X and Y are isomorphic as S-varieties with $\hat{\mu}$-action, and $[X, \hat{\mu}] = [Y, \hat{\mu}] + [X \setminus Y, \hat{\mu}]$ if Y is Zariski closed in X with the $\hat{\mu}$-action on Y induced by the one on X, and moreover $[X \times V, \hat{\mu}] = [X \times \mathbb{A}_k^n, \hat{\mu}]$ where V is the n-dimensional affine space over k with any linear $\hat{\mu}$-action, and $\mathbb{A}_k^n$ is taken with the trivial $\hat{\mu}$-action. There is a natural ring structure on $K_0^{\hat{\mu}}(\mathrm{Var}_S)$, the product being induced by the fiber product over S. We write $\mathbb{L}$ to denote the class in $K_0^{\hat{\mu}}(\mathrm{Var}_S)$ of $\mathbb{A}_k^1 \times S$ with the trivial $\hat{\mu}$-action.

We denote by $\mathcal{M}_S^{\hat{\mu}}$ the ring obtained from $K_0^{\hat{\mu}}(\mathrm{Var}_S)$ by inverting $\mathbb{L}$. When A is a constructible subset of X which is stable under the $\hat{\mu}$-action, then we define $[A, \hat{\mu}]$ in the obvious way. When S consists of only one geometric point, i.e. $S = \mathrm{Spec}(k)$, then we will write $K_0^{\hat{\mu}}(\mathrm{Var}_k)$ instead of $K_0^{\hat{\mu}}(\mathrm{Var}_S)$. The group $K_0^{\hat{\mu}}(\mathrm{Var}_k)$ was first introduced in [18].

Note that for any $s \in S(k)$ we have natural maps $K_0^{\hat{\mu}}(\mathrm{Var}_S) \to K_0^{\hat{\mu}}(\mathrm{Var}_k)$ and $\mathcal{M}_S^{\hat{\mu}} \to \mathcal{M}_k^{\hat{\mu}}$ given by $[X, \hat{\mu}] \to [X_s, \hat{\mu}]$, where X_s denotes the fiber at s of $X \to S$.

Although $\mathcal{M}_k^{\hat{\mu}}$ is a very complicated ring, there are many interesting maps from it to simpler rings. For example when $k = \mathbb{C}$, for any character α of $\hat{\mu}$ (i.e.

a group homomorphism $\alpha\colon \hat{\mu} \to \mathbb{C}^\times$ with finite image), there is a natural group homomorphism

$$\chi_{\mathrm{top}}(-,\alpha)\colon \mathcal{M}_k^{\hat{\mu}} \longrightarrow \mathbb{Z}\colon \quad X \longmapsto \sum_{q\geq 0}(-1)^q \dim H^q(X,\mathbb{C})_\alpha\,,$$

where $H^*(X,\mathbb{C})_\alpha$ is the part of $H^*(X,\mathbb{C})$ on which $\hat{\mu}$ acts by multiplication by α.

3. The Motivic Zeta Function of a Regular Function

Let X be a nonsingular irreducible algebraic variety over k of dimension d and $f\colon X \to \mathbb{A}_k^1$ a non-constant morphism. In this section we introduce several new invariants of f. These are constructed using arc spaces. We first recall in 3.1 some classical invariants associated to f. In what follows we denote by X_0 the *locus of $f = 0$* in X.

3.1. The monodromy zeta function and the Hodge spectrum

In this Subsection 3.1 we suppose that $k = \mathbb{C}$. Let x be a point of $X_0 = f^{-1}(0)$. We fix a smooth metric on X.

3.1.1. MONODROMY We set $X_{\epsilon,\eta}^\times := B(x,\epsilon) \cap f^{-1}(D_\eta^\times)$, with $B(x,\epsilon)$ the open ball of radius ϵ centered at x and $D_\eta^\times := D_\eta \setminus \{0\}$, with D_η the open disk of radius η centered at 0. For $0 < \eta \ll \epsilon \ll 1$, the restriction of f to $X_{\epsilon,\eta}^\times$ is a locally trivial fibration, called the *Milnor fibration*, onto $D_\eta^\times$ with fiber F_x, *the Milnor fiber at x*. The action of a characteristic homeomorphism of this fibration on cohomology gives rise to the *monodromy operator*

$$M_x\colon H^{\cdot}(F_x,\mathbb{Q}) \to H^{\cdot}(F_x,\mathbb{Q})\,.$$

For any natural number n, we consider the *Lefschetz number*

$$\Lambda(M_x^n) := \sum_{q\geq 0}(-1)^q \operatorname{Trace}\left(M_x^n, H^q(F_x,\mathbb{Q})\right),$$

of the n-th iterate of M_x. These numbers are related to the *monodromy zeta function of f at x*

$$Z_x^{\mathrm{mon}}(T) := \prod_{q\geq 0} \operatorname{Det}\left(Id - TM_x, H^q(F_x,\mathbb{Q})\right)^{(-1)^q}$$

as follows: if one writes $\Lambda(M_x^n) = \sum_{i|n} s_i$ for $n \geq 1$, then $Z_x^{\mathrm{mon}}(T) = \prod_{i\geq 1}(1 - t^i)^{s_i/i}$. The monodromy zeta function of f at x (or equivalently the Lefschetz numbers) is an important topological invariant of f which has been studied intensively.

3.1.2. Hodge structures A *Hodge structure* is a finite dimensional $\mathbb{Q}$-vector-space H together with a bigrading $H \otimes \mathbb{C} = \oplus_{p,q \in \mathbb{Z}} H^{p,q}$, such that $H^{q,p}$ is the complex conjugate of $H^{p,q}$ and each *weight m summand*, $\oplus_{p+q=m} H^{p,q}$, is defined over $\mathbb{Q}$. The Hodge structures, with the evident notion of morphism, form an abelian category HS with tensor product. The elements of the Grothendieck group $K_0(\mathrm{HS})$ of this abelian category are representable as a formal difference of Hodge structures $[H] - [H']$, and $[H] = [H']$ iff $H \cong H'$. Note that $K_0(\mathrm{HS})$ becomes a ring with respect to the tensor product.

A *mixed Hodge structure* is a finite dimensional $\mathbb{Q}$-vector space V with a finite increasing filtration $W_\bullet V$, called the *weight filtration*, such that the associated graded vector space $\mathrm{Gr}_\bullet^W(V)$ underlies a Hodge structure having $\mathrm{Gr}_m^W(V)$ as weight m summand. Note that V determines in a natural way an element $[V]$ in $K_0(\mathrm{HS})$, namely $[V] := \sum_m [\mathrm{Gr}_m^W(V)]$.

When X is an algebraic variety over $k = \mathbb{C}$, the simplicial cohomology groups $H_c^i(X, \mathbb{Q})$ of X, with compact support, underlie a natural mixed Hodge structure, and the *Hodge characteristic* $\chi_h(X)$ of X (with compact support) is defined by

$$\chi_h(X) := \sum_i (-1)^i [H_c^i(X, \mathbb{Q})] \in K_0(\mathrm{HS}).$$

This yields a map $\chi_h \colon \mathrm{Var}_{\mathbb{C}} \to K_0(\mathrm{HS})$, which is a generalized Euler characteristic, and which factors through $\mathcal{M}_h$, because $\chi_k(\mathbb{A}_k^1)$ is actually invertible in the ring $K_0(\mathrm{HS})$. When X is proper and smooth, the mixed Hodge structure on $H_c^i(X, \mathbb{Q})$ is in fact a Hodge structure, the weight filtration being concentrated in weight i. We refer to [41] for an introduction to Hodge structures.

3.1.3. The Hodge spectrum The cohomology groups $H^i(F_x, \mathbb{Q})$ of the Milnor fiber F_x carry a natural mixed Hodge structure ([39, 31, 32]), which is compatible with the semi-simplification of the monodromy operator M_x. Hence we can define the Hodge characteristic $\chi_h(F_x)$ of F_x by

$$\chi_h(F_x) := \sum_i (-1)^i [H^i(F_x, \mathbb{Q})] \in K_0(\mathrm{HS}).$$

Actually by taking into account the monodromy action we can consider $\chi_h(F_x)$ as an element of the Grothendieck group $K_0(\mathrm{HS}^{\mathrm{mon}})$ of the abelian category $\mathrm{HS}^{\mathrm{mon}}$ of Hodge structures with an endomorphism of finite order. Again $K_0(\mathrm{HS}^{\mathrm{mon}})$ is a ring by the tensor product.

There is a natural linear map, called the Hodge spectrum

$$\mathrm{hsp} \colon K_0(\mathrm{HS}^{\mathrm{mon}}) \to \mathbb{Z}[t^{1/\mathbb{Z}}] := \cup_{n \geq 1} \mathbb{Z}[t^{1/n}, t^{-1/n}],$$

with $\mathrm{hsp}([H]) := \sum_{\alpha \in \mathbb{Q} \cap [0,1[} t^\alpha (\sum_{p,q \in \mathbb{Z}} \dim(H^{p,q})_\alpha) t^p$, for any Hodge structure H with an endomorphism of finite order, where $H_\alpha^{p,q}$ is the eigenspace of $H^{p,q}$ with respect to the eigenvalue $e^{2\pi \sqrt{-1} \alpha}$. Note that hsp is not a ring homomorphism, although it becomes one when we endow $K_0(\mathrm{HS}^{\mathrm{mon}})$ with a different ring multiplication, namely the one induced by the operation $*$ in Section 5.

We recall that $\mathrm{hsp}(f,x) := (-1)^{d-1}\mathrm{hsp}(\chi_h(F_x) - 1)$ is called the Hodge spectrum of f at x. It is a very important invariant, with remarkable properties, see [39, 40, 42].

3.2. The motivic zeta function

Let $n \geq 1$ be an integer. The morphism $f \colon X \to \mathbb{A}^1_k$ induces a morphism $f_n \colon \mathcal{L}_n(X) \to \mathcal{L}_n(\mathbb{A}^1_k)$.

Any point α of $\mathcal{L}(\mathbb{A}^1_k)$, resp. $\mathcal{L}_n(\mathbb{A}^1_k)$, yields a K-rational point, for some field K containing k, and hence a power series $\alpha(t) \in K[[t]]$, resp. $\alpha(t) \in K[[t]]/t^{n+1}$. This yields maps

$$\mathrm{ord}_t \colon \mathcal{L}(\mathbb{A}^1_k) \to \mathbb{N} \cup \{\infty\}, \quad \mathrm{ord}_t \colon \mathcal{L}_n(\mathbb{A}^1_k) \to \{0, 1, \ldots, n, \infty\},$$

with $\mathrm{ord}_t\alpha$ the largest e such that t^e divides $\alpha(t)$.

We set

$$\mathfrak{X}_n := \{\varphi \in \mathcal{L}_n(X) \mid \mathrm{ord}_t f_n(\varphi) = n\}.$$

This is a locally closed subvariety of $\mathcal{L}_n(X)$. Note that $\mathfrak{X}_n$ is actually an X_0-variety, through the morphism $\pi_0^n \colon \mathcal{L}_n(X) \to X$. Indeed $\pi_0^n(\mathfrak{X}_n) \subset X_0$, since $n \geq 1$. We consider the morphism

$$\bar{f}_n \colon \mathfrak{X}_n \to \mathbb{G}_{m,k} := \mathbb{A}^1_k \setminus \{0\},$$

sending a point φ in $\mathfrak{X}_n$ to the coefficient of t^n in $f_n(\varphi)$. There is a natural action of $\mathbb{G}_{m,k}$ on $\mathfrak{X}_n$ given by $a \cdot \varphi(t) = \varphi(at)$, where $\varphi(t)$ is the vector of power series corresponding to φ (in some local coordinate system). Since $\bar{f}_n(a \cdot \varphi) = a^n \bar{f}_n(\varphi)$ it follows that $\bar{f}_n$ is a locally trivial fibration.

We denote by $\mathfrak{X}_{n,1}$ the fiber $\bar{f}_n^{-1}(1)$. Note that the action of $\mathbb{G}_{m,k}$ on $\mathfrak{X}_n$ induces a good action of μ_n (and hence of $\hat{\mu}$) on $\mathfrak{X}_{n,1}$. Since $\bar{f}_n$ is a locally trivial fibration, the X_0-variety $\mathfrak{X}_{n,1}$ and the action of μ_n on it, completely determines both the variety $\mathfrak{X}_n$ and the morphism

$$(\bar{f}_n, \pi_0^n) \colon \mathfrak{X}_n \to \mathbb{G}_{m,k} \times X_0 .$$

Indeed it is easy to verify that $\mathfrak{X}_n$, as a $(\mathbb{G}_{m,k} \times X_0)$-variety, is isomorphic to the quotient of $\mathfrak{X}_{n,1} \times \mathbb{G}_{m,k}$ under the μ_n-action defined by $a(\varphi, b) = (a\varphi, a^{-1}b)$.

Definition 3.1. *The motivic zeta function of $f \colon X \to \mathbb{A}^1_k$, is the power series over* $\mathcal{M}^{\hat{\mu}}_{X_0}$ *defined by*

$$Z(T) := \sum_{n \geq 1} [\mathfrak{X}_{n,1}/X_0, \hat{\mu}] \, \mathbb{L}^{-nd} T^n .$$

Moreover we define the naive motivic zeta function of f as the power series over $\mathcal{M}_{X_0}$ *defined by*

$$Z^{\mathrm{naive}}(T) := \sum_{n \geq 1} [\mathfrak{X}_n/X_0] \, \mathbb{L}^{-nd} T^n .$$

Theorem 3.2 and Corollary 3.3 below show that $Z(T)$ and $Z^{\text{naive}}(T)$ are rational. In 3.4 and 3.5 we will see that $Z(T)$ and $Z^{\text{naive}}(T)$ give rise to interesting new invariants of f. The definition of $Z(T)$ goes back to [18] (in the non-relative version working in $\mathcal{M}_k^{\hat{\mu}}$). Many related motivic zeta functions, and $Z^{\text{naive}}(T)$, were first introduced in [13], inspired by work of Kontsevich [23]. The idea of rather working in the relative Grothendieck group was introduced by Looijenga [27].

3.3. A formula for the motivic zeta function

We recall that X_0 denotes the locus of $f = 0$ in X. Let (Y, h) be a resolution of f. By this, we mean that Y is a nonsingular and irreducible algebraic variety over k, $h \colon Y \to X$ is a proper morphism, that the restriction $h \colon Y \setminus h^{-1}(X_0) \to X \setminus X_0$ is an isomorphism, and that $h^{-1}(X_0)$ has only normal crossings as a subvariety of Y.

We denote by $E_i, i \in J$, the irreducible components (over k) of $h^{-1}(X_0)$. For each $i \in J$, denote by N_i the multiplicity of E_i in the divisor of $f \circ h$ on Y, and by $\nu_i - 1$ the multiplicity of E_i in the divisor of $h^* dx$, where dx is a local non-vanishing volume form at any point of $h(E_i)$, i.e. a local generator of the sheaf of differential forms of maximal degree. For $i \in J$ and $I \subset J$, we consider the nonsingular varieties $E_i^\circ := E_i \setminus \cup_{j \neq i} E_j$, $E_I = \cap_{i \in I} E_i$, and $E_I^\circ := E_I \setminus \cup_{j \in J \setminus I} E_j$.

Let $m_I = \gcd(N_i)_{i \in I}$. We introduce an unramified Galois cover $\tilde{E}_I^\circ$ of E_I°, with Galois group μ_{m_I}, as follows. Let U be an affine Zariski open subset of Y, such that, on U, $f \circ h = uv^{m_I}$, with u a unit on U and v a morphism from U to $\mathbb{A}_k^1$. Then the restriction of $\tilde{E}_I^\circ$ above $E_I^\circ \cap U$, denoted by $\tilde{E}_I^\circ \cap U$, is defined as

$$\{(z, y) \in \mathbb{A}_k^1 \times (E_I^\circ \cap U) | z^{m_I} = u^{-1}\}.$$

Note that E_I° can be covered by such affine open subsets U of Y. Gluing together the covers $\tilde{E}_I^\circ \cap U$, in the obvious way, we obtain the cover $\tilde{E}_I^\circ$ of E_I° which has a natural μ_{m_I}-action (obtained by multiplying the z-coordinate with the elements of μ_{m_I}). This μ_{m_I}-action on $\tilde{E}_I^\circ$ induces a $\hat{\mu}$-action on $\tilde{E}_I^\circ$ in the obvious way.

Theorem 3.2. ([18, 27]) *With the previous notations, the following relation holds in* $\mathcal{M}_{X_0}^{\hat{\mu}}[[T]]$

$$Z(T) = \sum_{\emptyset \neq I \subset J} (\mathbb{L} - 1)^{|I|-1} [\tilde{E}_I^\circ / X_0, \hat{\mu}] \prod_{i \in I} \frac{\mathbb{L}^{-\nu_i} T^{N_i}}{1 - \mathbb{L}^{-\nu_i} T^{N_i}}.$$

From the above theorem one easily deduces (using e.g. Lemma 5.1 in [27]) the following corollary, which is basically a special case of Theorem 2.2.1 in [13].

Corollary 3.3. *With the previous notations, the following relation holds in* $\mathcal{M}_{X_0}[[T]]$

$$Z^{\text{naive}}(T) = \sum_{\emptyset \neq I \subset J} (\mathbb{L} - 1)^{|I|} [E_I^\circ / X_0] \prod_{i \in I} \frac{\mathbb{L}^{-\nu_i} T^{N_i}}{1 - \mathbb{L}^{-\nu_i} T^{N_i}}.$$

3.3.1. PROOF OF THE RATIONALITY OF $J(T)$ We defined $J(T)$ in Theorem 2.2 above. We will now discuss the proof of the rationality of $J(T)$ in the special case when the variety X in Theorem 2.2 is the locus X_0 of a polynomial f in the affine space $\mathbb{A}_k^d$. Let $Z^{\text{naive}}(T)$ be the naive motivic zeta function of $f\colon \mathbb{A}_k^d \to \mathbb{A}_k^1$. It is straightforward to verify that $J(T\mathbb{L}^{-d}) = \frac{[X_0] - Z^{\text{naive}}(T)}{1-T}$. Hence the rationality of $J(T)$ is a direct consequence of Corollary 3.3.

3.4. The topological zeta functions

Let $\mathcal{M}_{S,\text{loc}}$ resp. $\mathcal{M}_{S,\text{loc}}^{\hat{\mu}}$, be the ring obtained from $\mathcal{M}_S$, resp. $\mathcal{M}_S^{\hat{\mu}}$, by inverting the elements $[\mathbb{P}_k^i] = 1 + \mathbb{L} + \mathbb{L}^2 + \cdots + \mathbb{L}^i$, for $i = 1, 2, 3, \ldots$, where $\mathbb{P}_k^i$ denotes the i-dimensional projective space over k.

We keep the notations of 3.3, but take $k = \mathbb{C}$. For any integer $s \geq 1$, evaluating $Z^{\text{naive}}(T)$ at $T = \mathbb{L}^{-s}$ yields a well-defined element of $\mathcal{M}_{X_0,\text{loc}}$, namely $\sum_{\emptyset \neq I \subset J}[E_I^\circ/X_0] \prod_{i \in I}[\mathbb{P}^{sN_i + \nu_i - 1}]^{-1}$. Applying the topological Euler characteristic χ_{top} we obtain

$$Z_{\text{top}}(s) := \chi_{\text{top}}(Z^{\text{naive}}(\mathbb{L}^{-s})) := \sum_{\emptyset \neq I \subset J} \chi_{\text{top}}(E_I^\circ) \prod_{i \in I} \frac{1}{sN_i + \nu_i}. \qquad (*)$$

We call $Z_{\text{top}}(s)$, considered as a rational function in the variable s, the *untwisted topological zeta function* of $f\colon X \to \mathbb{A}_k^1$.

Evaluating $(\mathbb{L} - 1)Z(T)$, instead of $Z^{\text{naive}}(T)$, at $T = \mathbb{L}^{-s}$, and applying the equivariant topological Euler characteristic $\chi_{\text{top}}(-, \alpha)$, with $\alpha\colon \hat{\mu} \to \mathbb{C}$ a character of order e, we obtain the *twisted topological zeta function* (for any integer $e \geq 1$)

$$Z_{\text{top}}^{(e)}(s) := \chi_{\text{top}}((\mathbb{L} - 1)Z(\mathbb{L}^{-s}), \alpha)$$

$$:= \sum_{\emptyset \neq I \subset J,\, e \mid m_I} \chi_{\text{top}}(E_I^\circ) \prod_{i \in I} \frac{1}{sN_i + \nu_i}. \qquad (**)$$

Note that, if we would define the topological zeta functions by the right-hand-side of $(*)$ and $(**)$, then it would be not at all clear that this is independent of the choosen resolution. It is the intrinsic definition using the motivic zeta function (which is based on the notion of arc spaces) that makes this independence obvious. The topological zeta functions were first introduced by Denef and Loeser in [12] using p-adic integration and the Grothendieck-Lefschetz trace formula to prove their independence of the choosen resolution. Our approach using arc spaces first appeared in [13].

The topological zeta functions are quite subtle invariants of f, and have been further investigated by Veys [45, 46]. There are some fascinating conjectures about them.

Conjecture 3.4. (Monodromy conjecture for $Z_{\text{top}}^{(e)}$) *If s is a pole of $Z_{\text{top}}^{(e)}(s)$ then $e^{2\pi\sqrt{-1}s}$ is an eigenvalue of the monodromy action on the cohomology of the Milnor fiber at some point of the locus of f.*

Conjecture 3.5. Holomorphy conjecture for $Z_{\mathrm{top}}^{(e)}$ *The function $Z_{\mathrm{top}}^{(e)}(s)$ is a polynomial in s, unless there is an eigenvalue with order divisible by e of the monodromy action on the cohomology of the Milnor fiber at some point of the locus of f.*

Loeser [26] and Veys [44] proved that these conjectures are true when $X = \mathbb{A}_{\mathbb{C}}^2$. A lot of experimental evidence has been obtained by Veys [43] when $X = \mathbb{A}_{\mathbb{C}}^3$. We refer to [46] and [47] for very interesting generalizations.

3.5. Relations with monodromy and the motivic Milnor fiber

The Lefschetz numbers $\Lambda(M_x^n)$ of f at x, which we recalled in 3.1.1 can be expressed in terms of a resolution of f, by the following formula of A'Campo.

Theorem 3.6. (A'Campo, [1]) *Let $k = \mathbb{C}$. Assume the notations of 3.1 and 3.3. Then for any integer $n \geq 0$ we have*

$$\Lambda(M_x^n) = \sum_{N_i \mid n} N_i \, \chi_{\mathrm{top}}(E_i^\circ \cap h^{-1}(x)) \,.$$

In particular we see that the right-hand-side of the above formula is independent of the chosen resolution h. Note that the material in 3.3 and 3.4 yields many other expressions which are independent of the chosen resolution, but A'Campo's result was probably the first in this direction.

Applying the natural map $\mathrm{Fiber}_x \colon \mathcal{M}_{X_0}^{\hat{\mu}} \to \mathcal{M}_k^{\hat{\mu}} \colon [A/X_0, \hat{\mu}] \mapsto [A \times_{X_0} \{x\}, \hat{\mu}]$, followed by the equivariant topological Euler characteristic $\chi_{\mathrm{top}}(-, 1)$, on the coefficients of $Z(T)$ and using Theorem 3.2, we obtain the following theorem.

Theorem 3.7. ([18]) *Let $k = \mathbb{C}$, then for any integer $n \geq 1$ we have $\Lambda(M_x^n) = \chi_{\mathrm{top}}(\mathfrak{X}_{n,1,x})$, where*

$$\mathfrak{X}_{n,1,x} := \mathfrak{X}_{n,1} \times_{X_0} \{x\} \,.$$

Thus we see that the monodromy zeta function of f at x is completely determined by the motivic zeta function $Z(T)$. Next, we will see that also the Hodge spectrum of f at x is determined by $Z(T)$.

Definition 3.8. ([13, 18]) *Expanding the rational function $Z(T)$ as a power series in T^{-1} and taking minus its constant term, yields a well-defined element of $\mathcal{M}_{X_0}^{\hat{\mu}}$, namely*

$$\mathcal{S} := -\lim_{T \to \infty} Z(T) := \sum_{\emptyset \neq I \subset J} (1 - \mathbb{L})^{|I|-1} [\tilde{E}_I^\circ] \,.$$

Moreover we set $\mathcal{S}_x := \mathrm{Fiber}_x(\mathcal{S}) \in \mathcal{M}_k^{\hat{\mu}}$. Instead of $\mathcal{S}$ and $\mathcal{S}_x$ we will also write $\mathcal{S}_f$ and $\mathcal{S}_{f,x}$. These definitions hold for any field k of characteristic zero.

Note again that the most right-hand-side of the above formula is independent of the choosen resolution (because of its relation to $Z(T)$), although a priori this is not at all evident.

We strongly believe that $\mathcal{S}_x$ is the correct virtual motivic incarnation of the Milnor fiber F_x of f at x (which is in itself not at all motivic). We will see below

(Theorem 3.10) that this is indeed true for the Hodge realization. A similar result holds for ℓ-adic cohomology, see [10]. Moreover we strongly believe that $\mathcal{S}$ is the virtual motivic incarnation of the so-called complex of nearby cycles ψ_f of f, which is a complex of sheaves on X_0. For the definition of ψ_f and the complex ϕ_f of vanishing cycles, we refer to [35], Exp. XIII; but we will not need these notions in the present survey. Inspired by the notation ϕ_f from the theory of vanishing cycles, we introduce the following

Notation 3.9. *We set* $\mathcal{S}_f^\phi := (-1)^{d-1}(\mathcal{S}_f - [X_0]) \in \mathcal{M}_{X_0}^{\hat\mu}$ *and* $\mathcal{S}_{f,x}^\phi := (-1)^{d-1}(\mathcal{S}_{f,x} - 1) \in \mathcal{M}_k^{\hat\mu}$.

We regard $\mathcal{S}_f^\phi$ as the virtual motivic incarnation of the complex $\phi_f[d-1]$.

Assume now again that $k = \mathbb{C}$. We denote by χ_h the canonical ring homomorphism (called the Hodge characteristic)

$$\chi_h \colon \mathcal{M}_k^{\hat\mu} \to K_0(\mathrm{HS}^{\mathrm{mon}}),$$

which associates to any complex algebraic variety Z, with a good μ_n-action, its Hodge characteristic as defined in 3.1.2, together with the endomorphism induced by $Z \to Z \colon z \mapsto e^{2\pi\sqrt{-1}/n} z$. (For the definition of $K_0(\mathrm{HS}^{\mathrm{mon}})$, see 3.1.3.)

Theorem 3.10. ([13]) *Assume the above notation with $k = \mathbb{C}$, and the notation of 3.1. Then we have the following equality in $K_0(\mathrm{HS}^{\mathrm{mon}})$*

$$\chi_h(F_x) = \chi_h(\mathcal{S}_x).$$

Moreover this theorem can be enhanced as an equality in the Grothendieck group of the abelian category of variations of Hodge structures with a finite-order endomorphism, when we replace $\mathcal{S}_x$ by $\mathcal{S}$, and F_x by ψ_f.

Theorem 3.10 yields that $\mathrm{hsp}(f, x) = \mathrm{hsp}(\chi_h(\mathcal{S}_{f,x}^\phi))$. Thus the motivic zeta function $Z(T)$ completely determines the Hodge spectrum of f at x.

4. Motivic Integration and the Proof of Theorem 3.2

The notion of motivic integration on $\mathcal{L}(X)$ is due to Kontsevich [23], who discovered its basic properties when X is nonsingular. This subject has been further developed by Batyrev [5, 6] and Denef-Loeser [13, 14, 15, 16, 17, 18]. See also the recent report by Looijenga [27] which contains some substantial improvements. Actually the best way to understand motivic integration is to consider it as being an analogue of p-adic integration, cf. Section 6.

Let X be an algebraic variety over k of pure dimension d, not necessarily nonsingular. Let X_{sing} denote the singular locus of X.

4.1. Naive motivic integration

A subset A of $\mathcal{L}(X)$ is called *constructible* if $A = \pi_n^{-1}(C)$ with C a constructible subset of $\mathcal{L}_n(X)$ for some integer $n \geq 0$. A subset A of $\mathcal{L}(X)$ is called *stable* if it is constructible and $A \cap \mathcal{L}(X_{\mathrm{sing}}) = \emptyset$. If $A \subset \mathcal{L}(X)$ is stable, then $[\pi_n(A)]\mathbb{L}^{-(n+1)d}$, considered as an element of $\mathcal{M}_k$, stabilizes for n big enough, and

$$\tilde{\mu}(A) := \lim_{n \to \infty} [\pi_n(A)]\mathbb{L}^{-(n+1)d} \in \mathcal{M}_k$$

is called the *naive motivic measure* of A. When X is nonsingular, this claim follows from the fact that the natural maps $\mathcal{L}_{n+1}(X) \to \mathcal{L}_n(X)$ are locally trivial fibrations with fiber $\mathbb{A}_k^d$. In the general case, the claim follows from [14], Lemma 4.1.

When $\theta \colon A \to \mathcal{M}_k$ is a map with finite image whose fibers are stable subsets of $\mathcal{L}(X)$, we define the integral $\int_A \theta d\tilde{\mu} := \sum_{c \in \mathrm{Image}\,\theta} c\tilde{\mu}(\theta^{-1}(c))$. The most fundamental result in the theory of arc spaces is the following change of variables formula, which was first obtained by Kontsevich [23] when X is nonsingular.

Theorem 4.1. ([23, 14, 16]) *Let $h \colon Y \to X$ be a morphism of algebraic varieties over k. Suppose that h is birational and proper. Let $A \subset \mathcal{L}(X)$ be stable and suppose that $\mathrm{ord}_t\mathrm{Jac}_h$ is bounded on $h^{-1}(A) \subset \mathcal{L}(Y)$. Then*

$$\tilde{\mu}(A) = \int_{h^{-1}(A)} \mathbb{L}^{-\mathrm{ord}_t\mathrm{Jac}_h} d\tilde{\mu}\,.$$

In the above theorem, $\mathrm{ord}_t\mathrm{Jac}_h(y)$, for $y \in \mathcal{L}(Y)$, denotes the t-order of the Jacobian of h at y. When X and Y are nonsingular this is the ord_t of the determinant of the Jacobian matrix of h at y with respect to any system of local coordinates on X and on Y. For the definition of $\mathrm{ord}_t\mathrm{Jac}_h$, in the general case, we refer to [14] and [16].

4.2. About the proof of Theorem 3.2

The proof of Theorem 3.2 consists of an explicit calculation of $[\mathfrak{X}_{n,1}/X_0, \hat{\mu}] \in \mathcal{M}_{X_0}^{\hat{\mu}}$ for each n. Note that in $\mathcal{M}_k$ we have the equality

$$[\mathfrak{X}_{n,1}] = \mathbb{L}^{(n+1)d}\tilde{\mu}(\pi_n^{-1}(\mathfrak{X}_{n,1}))\,.$$

Thus using the change of variables formula (Theorem 4.1), we see that $[\mathfrak{X}_{n,1}]$ is equal to an integral over a stable subset of $\mathcal{L}(Y)$, where $h \colon Y \to X$ is a resolution of f as in 3.3. Because $f \circ h$ is locally a monomial, that integral can be explicitly calculated and yields an explicit expression for $[\mathfrak{X}_{n,1}]$ as an element of $\mathcal{M}_k$. Taking into account the μ_n-action on $\mathfrak{X}_{n,1}$ and the natural map $\mathfrak{X}_{n,1} \to X_0$, one actually obtains a similar formula for $[\mathfrak{X}_{n,1}/X_0, \hat{\mu}]$, which yields Theorem 3.2.

4.3. Motivic integration

Let A be a constructible subset of $\mathcal{L}(X)$. When A is not stable, $[\pi_n(A)]\mathbb{L}^{-(n+1)d}$ will not always stabilize. However it is easy to prove (see [14]) that the limit

$$\mu(A) := \lim_{n \to \infty} [\pi_n(A)]\mathbb{L}^{-(n+1)d}$$

exists in the *completed Grothendieck group* $\hat{\mathcal{M}}_k$, which is the completion of $\mathcal{M}_k$ with respect to the filtration $F^m\mathcal{M}_k, m \in \mathbb{Z}$, where $F^m\mathcal{M}_k$ is the subgroup of $\mathcal{M}_k$ generated by the elements $[S]\mathbb{L}^{-i}$, with $S \in \mathrm{Var}_k$, $i - \dim S \geq m$. The completed Grothendieck ring $\mathcal{M}_k$ was first introduced by Kontsevich. (In a similar way one can define the completions $\hat{\mathcal{M}}_S$ and $\hat{\mathcal{M}}_S^{\hat{\mu}}$ of $\mathcal{M}_S$ and $\mathcal{M}_S^{\hat{\mu}}$.) The element $\mu(A)$ of $\hat{\mathcal{M}}_k$ is called the *motivic measure of A*. This yields a σ-additive measure μ on the Boolean algebra of constructible subsets of $\mathcal{L}(X)$. Actually all the above still works when A is a semi-algebraic subset of $\mathcal{L}(X)$, cf. [14]. It is even possible to define the notion of a measurable subset of $\mathcal{L}(X)$ and to integrate measurable functions on $\mathcal{L}(X)$, see [5, 16].

The change of variables formula (Theorem 4.1) remains true with $\tilde{\mu}$ replaced by μ, for any constructible (or measurable) subset of $\mathcal{L}(X)$, without assuming that $\mathrm{ord}_t\mathrm{Jac}_h$ is bounded on $h^{-1}(A)$.

It is not known whether the natural map $\mathcal{M}_k \to \hat{\mathcal{M}}_k$ is injective, but the topological Euler characteristic, the Hodge-Deligne polynomial, the Hodge characteristic, and many other important generalized Euler characteristics all factor through the image $\bar{\mathcal{M}}_k$ of $\mathcal{M}_k$ in $\hat{\mathcal{M}}_k$ (after inverting the image of $\mathbb{L}$ in the target ring).

We can consider the motivic volume of the whole arc space $\mathcal{L}(X)$, namely $\mu(\mathcal{L}(X))$. Clearly, when X is nonsingular, $\mu(\mathcal{L}(X)) = [X]\mathbb{L}^{-d}$ in $\hat{\mathcal{M}}_k$. Here and in what follows, we denote the image of $[X]$, resp. $\mathbb{L}$, in $\hat{\mathcal{M}}_k$ again by $[X]$, resp. $\mathbb{L}$. When X is not necessarily nonsingular, we can calculate $\mu(\mathcal{L}(X))$ using a suitable resolution of singularities $h\colon Y \to X$ of X. More precisely we have the following

Theorem 4.2. *Let $h\colon Y \to X$ be a proper birational morphism with Y nonsingular. Assume that the exceptional locus of h has normal crossings and that the image of $h^*(\Omega_X^d)$ in Ω_Y^d is an invertible sheaf, where Ω_X^d and Ω_Y^d denote the sheaf of differential forms of maximal degree. Let $E_j, j \in J$, be the k-irreducible components of the exceptional locus of h. For any subset I of J, set $E_I^\circ = (\cap_{i \in I} E_i) \setminus \cup_{j \in J\setminus I} E_j$. For $i \in I$, let $\nu_i - 1$ be the multiplicity along E_i of the divisor associated to $h^*(\Omega_X^d)$. Then, in $\hat{\mathcal{M}}_k$, we have*

$$\mu(\mathcal{L}(X)) = \mathbb{L}^{-d} \sum_{I \subset J} [E_I^\circ] \prod_{i \in I} [\mathbb{P}^{\nu_i - 1}]^{-1}.$$

In particular we see that $\mu(\mathcal{L}(X)) \in \bar{\mathcal{M}}_{k,\mathrm{loc}} \subset \hat{\mathcal{M}}_k$, where $\bar{\mathcal{M}}_{k,\mathrm{loc}}$ denotes the ring obtained from $\bar{\mathcal{M}}_k$ by inverting the elements $1 + \mathbb{L} + \cdots + \mathbb{L}^i$, for all $i = 1, 2, 3, \ldots$.

About the proof of this theorem, we remark that $\mu(\mathcal{L}(X)) = \int_{\mathcal{L}(Y)} \mathbb{L}^{-\mathrm{ord}_t\mathrm{Jac}_h} d\mu$, by the change of variables formula. Because Jac_h is locally a monomial, this integral can be easily calculated, which yields the theorem.

4.4. Applications

4.4.1. NEW INVARIANTS OF SINGULAR VARIETIES Suppose $k = \mathbb{C}$. Since χ_{top} and χ_{hp} factor through $\bar{\mathcal{M}}_k$, we have natural maps $\chi_{\mathrm{top}}\colon \bar{\mathcal{M}}_{k,\mathrm{loc}} \to \mathbb{Q}$ and

$\chi_{hp}\colon \bar{\mathcal{M}}_{k,\mathrm{loc}} \to \mathbb{Z}[[u,v]][u^{-1},v^{-1}]$. Hence we can consider $\chi_{\mathrm{top}}(\mathcal{L}(X)) \in \mathbb{Q}$ and $(uv)^d \chi_{\mathrm{hp}}(\mathcal{L}(X)) \in \mathbb{Z}[[u,v]]$, which are new invariants of X when X is singular. When X is nonsingular, these invariants equal $\chi_{\mathrm{top}}(X)$, resp. $\chi_{\mathrm{hp}}(X)$. We call the coefficients of $(uv)^d \chi_{\mathrm{hp}}(\mathcal{L}(X))$ (with an appropriate sign change) the arc-Hodge numbers of X. When X has only canonical Gorenstein singularities, Batyrev [5] introduced the so called *stringy Hodge numbers* of X, which are obtained in a similar way, replacing $\mu(\mathcal{L}(X))$ by $\int_{\mathcal{L}(X)} \mathbb{L}^{-\mathrm{ord}_t \omega_X} d\mu$, where ω_X denotes the canonical class of X. The stringy Hodge numbers play an important role in the work of Batyrev on mirror symmetry, see [5, 7, 3, 2]. Other fascinating related invariants were obtained by Veys [46, 47].

4.4.2. CALABI-YAU MANIFOLDS   Let X and Y be two Calabi-Yau manifolds, i.e. nonsingular proper complex algebraic varieties which admit a nonvanishing regular differential form of maximal degree, which we denote respectively by ω_X and ω_Y. Kontsevich [23] proved that X and Y have the same Hodge numbers and the same Hodge structure on their cohomology, when X and Y are birationally equivalent. The proof goes as follows There exists a nonsingular proper complex algebraic variety Z and birational morphisms $h_X\colon Z \to X$ and $h_Y\colon Z \to Y$. Note that $(h_Y \circ h_X^{-1})^*(\omega_Y)$ equals $c\,\omega_X$ for some $c \in \mathbb{C}^\times$ because ω_X has no zeroes. Hence $c\,h_X^*(\omega_X) = h_Y^*(\omega_Y)$. Thus $\mathrm{ord}_t\mathrm{Jac}_{h_X} = \mathrm{ord}_t\mathrm{Jac}_{h_Y}$ on $\mathcal{L}(Z)$, and by the change of variables formula both $\mu(\mathcal{L}(X))$ and $\mu(\mathcal{L}(Y))$ equal the same integral on $\mathcal{L}(Z)$. Because $\mu(\mathcal{L}(X)) = [X]\mathbb{L}^{-d}$ and $\mu(\mathcal{L}(Y)) = [Y]\mathbb{L}^{-d}$, this implies that $[X] = [Y]$ in $\bar{\mathcal{M}}_k$, which finishes the proof.

Actually Batyrev [4] first proved that X and Y have the same Betti numbers using p-adic integration and the Weil conjectures, and Kontsevich invented motivic integration to prove that X and Y have the same Hodge numbers.

4.4.3. EULER CHARACTERISTICS AND MODIFICATIONS   Let $h\colon Y \to X$ be a modification of nonsingular algebraic varieties over k, meaning that h is a proper birational morphism. Assume that the exceptional locus of h has normal crossings, and let J, E_i, E_I° and ν_i be as in Theorem 4.2. Because X is nonsingular, $\mu(\mathcal{L}(X)) = [X]\mathbb{L}^{-d}$ and Theorem 4.2 yields the following equality in $\bar{\mathcal{M}}_{k,\mathrm{loc}}$:

$$[X] = \sum_{I \subset J} [E_I^\circ] \prod_{i \in I} [\mathbb{P}^{\nu_i - 1}]^{-1}. \tag{$*$}$$

a) When $k = \mathbb{C}$, applying the topological Euler characteristic on $(*)$ yields $\chi_{\mathrm{top}}(X) = \sum_{I \subset J} \chi(E_I^\circ)/\prod_{i \in I} \nu_i$. This surprising formula about the Euler characteristic of modifications was first obtained in [12] using p-adic integration and the Grothendieck-Lefschetz trace formula.

b) When $k = \mathbb{Q}$, applying the conductor (with respect to ℓ-adic cohomology, see Section 2.2), yields the following remarkable formula for the conductor $c(X)$ of X

$$c(X) = \prod_{I \subset J} c(E_I^\circ)^{1/\prod_{i \in I} \nu_i}.$$

5. The Motivic Thom-Sebastiani Theorem

Let k be a field of characteristic zero, X and Y nonsingular irreducible algebraic varieties over k, $f : X \to \mathbb{A}_k^1$, $g : Y \to \mathbb{A}_k^1$ non-constant morphisms, and $x \in X(k), y \in Y(k)$. We denote by $f * g$ the morphism

$$f * g \colon X \times Y \to \mathbb{A}_k^1 \colon (x, y) \mapsto f(x) + g(y).$$

The following theorem was first proved by A. Varchenko [42], when x and y are isolated singular points of f and g, and by M. Saito [33, 34] in the general case. A similar but much weaker result for the eigenvalues of monodromy was first proved by Thom and Sebastiani [37].

Theorem 5.1. (Thom-Sebastiani Theorem for the Hodge spectrum) *Assume the notation of 3.1.3, with $k = \mathbb{C}$. We have the following equality in $\mathbb{Z}[t^{1/\mathbb{N}}]$:*

$$\mathrm{hsp}(f * g, (x, y)) = \mathrm{hsp}(f, x)\, \mathrm{hsp}(g, y) .$$

We recall that $\mathrm{hsp}(f, x) = \mathrm{hsp}(\chi_h(\mathcal{S}_{f,x}^{\phi}))$, with the notation of 3.9. We will see next that the above theorem is a direct consequence of a much stronger result which expresses $\mathcal{S}_{f*g,(x,y)}^{\phi}$ in terms of $\mathcal{S}_{f,x}^{\phi}$ and $\mathcal{S}_{g,y}^{\phi}$. Below, we define a binary operation $*$ on $\mathcal{M}_k^{\hat{\mu}}$ which yields an alternative ring structure on $\mathcal{M}_k^{\hat{\mu}}$, such that $\mathrm{hsp} \circ \chi_h$ becomes a homomorphism of rings (which is not true for the usual multiplication on $\mathcal{M}_k^{\hat{\mu}}$).

Using the theory of arc spaces and the definition of $\mathcal{S}_f$ in terms of the motivic zeta function $Z(T)$, work of Denef, Loeser and Looijenga yields the following theorem

Theorem 5.2. (Motivic Thom-Sebastiani Theorem) *Let k be a field of characteristic zero, containing all roots of unity. Then $\mathcal{S}_{f*g,(x,y)}^{\phi}$ and $\mathcal{S}_{f,x}^{\phi} * \mathcal{S}_{g,y}^{\phi}$ are equal in $\mathcal{M}_k^{\hat{\mu}}$, where the operation $*$ is defined below.*

Actually Denef and Loeser [15] first proved the above equality in the completed Grothendieck group of Chow motives. Later Looijenga [27] introduced the operation $*$ and proved, using basically the same method, an equality which is similar to Theorem 5.2. The proof of Theorem 5.2 uses arc spaces in a very essential way by deriving first a formula relating the motivic zeta functions of $f * g$, f and g, and taking afterwards the limit for $T \to \infty$. More precisely, set

$$Z_f^{\phi}(T) := (-1)^{d-1}\left[Z_f(T) + [X_0] + \frac{Z_f^{\mathrm{naive}}(T) - [X_0]}{1 - T}\right],$$

where $Z_f(T)$, resp. $Z_f^{\mathrm{naive}}(T)$, is the motivic, resp. naive motivic, zeta function of f, X_0 is the locus of $f = 0$ in X, and d is the dimension of X. Let $Z_{f,x}^{\phi}(T)$ be obtained from $Z_f^{\phi}(T)$ by applying the map Fiber_x to its coefficients. Clearly $-\lim_{T \to \infty} Z_f^{\phi}(T)$ is $\mathcal{S}_f^{\phi}$. We recently proved that

$$Z_{f*g,(x,y)}^{\phi}(T) = Z_{f,x}^{\phi}(T) * Z_{g,y}^{\phi}(T)$$

in $\mathcal{M}_k^{\hat{\mu}}$, where $*$ is defined coefficientswise. This implies Theorem 5.2 because $-\lim_{T\to\infty}$ commutes with $*$ on such power series without constant terms.

Finally, we explain the definition of the operation $*$ on $\mathcal{M}_k^{\hat{\mu}}$. We assume that k contains all roots of unity. Let X and Y be algebraic varieties over k with good μ_n-action, for some integer $n \geq 1$. Let J_n be the Fermat curve in $(\mathbb{A}_k^1 \setminus \{0\})^2$ defined by $u^n + v^n = 1$. There is an action of $\mu_n \times \mu_n$ on J_n given by $(\xi, \xi') \cdot (u, v) := (\xi u, \xi' v)$. We define $J_n(X, Y)$ in Var_k as the quotient of $J_n \times X \times Y$ under the equivalence relation given by $(\xi u, \xi' v, x, y) = (u, v, \xi x, \xi' y)$ for all $\xi, \xi' \in \mu_n$. We let μ_n act on $J_n(X, Y)$ by $\xi \cdot (u, v, x, y) := (\xi u, \xi v, x, y)$. This yields an element $[J_n(X, Y)]$ in $\mathcal{M}_k^{\hat{\mu}}$. If m is a divisor of n, and the action of μ_n on X and Y factors through μ_m, then $J_m(X, Y) = J_n(X, Y)$. Thus, in this way, we obtain a binary operation $J: \mathcal{M}_k^{\hat{\mu}} \times \mathcal{M}_k^{\hat{\mu}} \to \mathcal{M}_k^{\hat{\mu}}$, which was first introduced by Looijenga [27]. The operation J is commutative and bilinear over $\mathcal{M}_k$, considering $\mathcal{M}_k^{\hat{\mu}}$ as a module over $\mathcal{M}_k$ through the natural map $\mathcal{M}_k \to \mathcal{M}_k^{\hat{\mu}}$. One verifies that $J(a, 1) = (\mathbb{L} - 1)\bar{a} - a$, where $a \mapsto \bar{a}: \mathcal{M}_k^{\hat{\mu}} \to \mathcal{M}_k$ is the morphism induced by $[Z] \to$ [space of $\hat{\mu}$-orbits of Z], for any k-variety Z with good $\hat{\mu}$-action, cf. [27]. In particular we see that 1 is not a neutral element for the operation J. For this reason it is natural to introduce the operation $*$ on $\mathcal{M}_k^{\hat{\mu}}$ given by

$$a * b = -J(a, b) + (\mathbb{L} - 1)\overline{ab},$$

for a and b in $\mathcal{M}_k^{\hat{\mu}}$. Clearly the operation $*$ is commutative and bilinear over $\mathcal{M}_k$, and $a * 1 = a$ for all a in $\mathcal{M}_k^{\hat{\mu}}$. Moreover the operation $*$ is associative. One easily verifies that $\mathrm{hsp} \circ \chi_h$ is a ring homomorphism with respect to the alternative ring structure on $\mathcal{M}_k^{\hat{\mu}}$ given by $*$.

6. The Arithmetic Motivic Poincaré Series $P_{\mathrm{arith}}(T)$

6.1. The p-adic case

Assume that X is an algebraic variety over $\mathbb{Z}$, i.e. a reduced separated scheme of finite type over $\mathbb{Z}$. Let p be a prime number. We consider the Poincaré series

$$J_p(T) = \sum_{n \in \mathbb{N}} \#X(\mathbb{Z}/p^{n+1}\mathbb{Z}) \, T^n, \quad P_p(T) = \sum_{n \in \mathbb{N}} \#(\pi_n(X(\mathbb{Z}_p))) \, T^n,$$

where $\mathbb{Z}_p$ denotes the ring of p-adic integers and π_n is the natural projection $\pi_n: \mathbb{Z}_p \to \mathbb{Z}/p^{n+1}\mathbb{Z}$. Igusa [22], resp. Denef [8], proved that $J_p(T)$, resp. $P_p(T)$, is a rational function of T. The proofs are based on p-adic integration, resolution of singularities, and for $P_p(T)$ also the theory of p-adic semi-algebraic sets. Actually the proof of Theorem 2.2 about the rationality of $J(T)$ and $P(T)$ was very much inspired by the proofs of the rationality of $J_p(T)$ and $P_p(T)$, replacing p-adic integration by motivic integration. As a matter of fact, for all the material discussed in the previous sections, p-adic counterparts exist which were discovered first, see [9] for a survey.

6.2. Comparing $J(T)$ and $J_p(T)$

For any rational power series $G(T)$ over $K_0(\mathrm{Var}_{\mathbb{Q}})$ we choose representatives in $K_0(\mathrm{Var}_{\mathbb{Z}})$ for the coefficients in $K_0(\mathrm{Var}_{\mathbb{Q}})$ of numerator and denominator. In this way we find a power series over $K_0(\mathrm{Var}_{\mathbb{Z}})$, and, for any prime number p, we can apply to each coefficient the operation $N_p \colon K_0(\mathrm{Var}_{\mathbb{Z}}) \to \mathbb{Z} \colon [X] \mapsto \#X(\mathbb{Z}/p\mathbb{Z})$. This yields a power series over $\mathbb{Z}$ which we will denote by $N_p(G(T))$. If we choose other representatives in $K_0(\mathrm{Var}_{\mathbb{Z}})$, the resulting power series $N_p(G(T))$ will be the same for almost all p (i.e. for all but finitely many prime numbers p).

Comparing the proof of the rationality of $J(T)$ and $J_p(T)$ actually yields the following

Theorem 6.1. *Assume the notation of 6.1 and 6.2. For almost all p we have* $J_p(T) = N_p(J(T))$.

Also the motivic zeta functions $Z(T)$ and $Z^{\mathrm{naive}}(T)$, have similar arithmetic interpretations, related to Igusa's local zeta functions, see [13]. However it is not true in general that $N_p(P(T)) = P_p(T)$ for almost all p. Indeed $N_p(P(T))$ does not count the elements of $X(\mathbb{Z}/p^{n+1}\mathbb{Z})$ which can be lifted to $X(\mathbb{Z}_p)$, but counts (for almost all p) the elements of $X(\mathbb{Z}/p^{n+1}\mathbb{Z})$ which can be lifted to $X(\mathbb{Z}_p^{\mathrm{unram}})$, where $\mathbb{Z}_p^{\mathrm{unram}}$ is the maximal unramified extension of $\mathbb{Z}_p$. Note that the residue field of $\mathbb{Z}_p^{\mathrm{unram}}$ is the algebraic closure of $\mathbb{Z}/p\mathbb{Z}$.

6.3. The motivic Poincaré series $P_{\mathrm{arith}}(T)$

The above discussion leads to the question of defining in a canonical way a power series $P_{\mathrm{arith}}(T)$ over (some localization of) $K_0(\mathrm{Var}_Q)$ such that $N_p(P_{\mathrm{arith}}(T)) = P_p(T)$ for almost all p.

In our recent paper [17], we construct, for any algebraic variety over k, in a canonical way, a rational power series $P_{\mathrm{arith}}(T)$ over $K_0(\mathrm{Mot}_k) \otimes \mathbb{Q}$, such that if $k = \mathbb{Q}$ then $N_p(P_{\mathrm{arith}}(T)) = P_p(T)$ for almost all p. Here Mot_k denotes the category of Chow motives over k. We refer to [38] for the definition of this important category, and we only remark here that there is a natural ring morphism $K_0(\mathrm{Var}_k) \to K_0(\mathrm{Mot}_k)$, see [19]. Actually the coefficients of $P_{\mathrm{arith}}(T)$ are in the image of $K_0(\mathrm{Var}_k) \otimes \mathbb{Q}$. We need to work at the level of Chow motives, to make our construction canonical.

The proof of our result is rather complicated and uses several results from mathematical logic (quantifier elimination for valued fields and finite fields). For a survey on such relations between logic, geometry and arithmetic, we refer to [11]. Examples seem to suggest that $P_{\mathrm{arith}}(T)$ captures more geometric information than $P(T)$, but very little is presently known about it!

References

[1] N. A'Campo, *La fonction zêta d'une monodromie*, Comment. Math. Helv. **50** (1975), 233–248.

[2] V. Batyrev, L. Borisov, *Mirror duality and string-theoretic Hodge numbers*, Invent. Math. **126** (1996), 183–203.

[3] V. Batyrev, D. Dais, *Strong McKay correspondence, string-theoretic Hodge numbers and mirror symmetry*, Topology **35** (1996), 901–929.

[4] V. Batyrev, *Birational Calabi-Yau n-folds have equal Betti numbers*, in New trends in algebraic geometry, Klaus Hulek et al., eds., CUP (1999), 1–11.

[5] V. Batyrev, *Stringy Hodge numbers of varieties with Gorenstein canonical singularities*, in Integrable systems and algebraic geometry (Kobe/Kyoto, 1997), 1–32, World Sci. Publishing, River Edge, NJ, 1998.

[6] V. Batyrev, *Non-archimedian integrals and stringy Euler numbers of log terminal pairs*, Journal of European Math. Soc. **1** (1999), 5–33.

[7] V. Batyrev, *Mirror symmetry and toric geometry*, Doc. Math., J. DMV, Extra Vol. ICM Berlin 1998, **II** (1998), 239–248.

[8] J. Denef, *On the rationality of the Poincaré series associated to the p-adic points on a variety*, Invent. Math., **77** (1984), 1–23.

[9] J. Denef, *Report on Igusa's local zeta function*, in Séminaire Bourbaki, volume 1990/91, exposé 741. Astérisque **201-202-203** (1991), 359–386.

[10] J. Denef, *Degree of local zeta functions and monodromy*, Compositio Math., **89** (1993), 207-216.

[11] J. Denef, *Arithmetic and Geometric Applications of Quantifier Elimination for Valued Fields*, in Model theory, algebra, and geometry, D. Haskell (ed.) et al., Cambridge University Press. Math. Sci. Res. Inst. Publ. **39**, 173–198 (2000).

[12] J. Denef, F. Loeser, *Caractéristiques d'Euler-Poincaré, fonctions zêta locales et modifications analytiques*, J. Amer. Math. Soc., **5** (1992), 705-720.

[13] J. Denef, F. Loeser, *Motivic Igusa zeta functions*, J. Algebraic Geom., **7** (1998), 505–537.

[14] J. Denef, F. Loeser, *Germs of arcs on singular algebraic varieties and motivic integration*, Invent. Math. **135** (1999), 201–232.

[15] J. Denef, F. Loeser, *Motivic exponential integrals and a motivic Thom-Sebastiani Theorem*, Duke Math. J., **99** (1999), 285–309.

[16] J. Denef, F. Loeser, *Motivic integration, quotient singularities and the McKay correspondence*, Compositio Math. (to appear, 20 pages), available at math. AG/9903187.

[17] J. Denef, F. Loeser, *Definable sets, motives and P-adic integrals*, J. Amer. Math. Soc., **14** (2001), 429–469.

[18] J. Denef, F. Loeser, *Lefschetz numbers of iterates of the monodromy and truncated arcs*, Topology (to appear, 10 pages), available at math. AG/0001105.

[19] H. Gillet, C. Soulé, *Descent, motives and K-theory*, J. Reine. Angew. Math., **478** (1996), 127–176.

[20] M. Greenberg, *Rational points in discrete valuation rings*, Publ. Math. I.H.E.S., **31** (1996), 59–64.

[21] M. Hickel, *Fonction de Artin et germes de courbes tracées sur un germe d'espace analytique*, Amer. J. Math., **115** (1993), 1299-1334.

[22] J.-I. Igusa, *Lectures on forms of higher degree*, Tata Institute of Fundamental Research Lectures on Mathematics and Physics, **59** (1978).

[23] M. Kontsevich, Lecture at Orsay (December 7, 1995).

[24] M. Lejeune-Jalabert, *Courbes tracées sur un germe d'hypersurface*, Amer. J. Math., **112** (1990), 525–568.

[25] M. Lejeune-Jalabert, A. Reguera-López, *Arcs and wedges on sandwiched surface singularities*, Amer. J. Math., **121** (1999), 1191–1213.

[26] F. Loeser, *Fonctions d'Igusa p-adiques et polynômes de Bernstein*, Amer. J. Math., **110** (1988), 1–21.

[27] E. Looijenga, *Motivic Measures*, in Séminaire Bourbaki, exposé 874, Mars 2000.

[28] J. Nash Jr., *Arc structure of singularities*, Duke Math. J., **81** (1995), 31–38.

[29] J. Pas, *Uniform p-adic cell decomposition and local zeta functions*, J. Reine Angew. Math., **399** (1989), 137–172.

[30] M. Reid, *La correspondance de McKay*, in Séminaire Bourbaki, exposé 867, Novembre 1999, 20 p., available at math. AG/9911165 (en anglais).

[31] M. Saito, *Modules de Hodge polarisables*, Publ. Res. Inst. Math. Sci., **24** (1988), 849–995.

[32] M. Saito, *Mixed Hodge modules*, Publ. Res. Inst. Math. Sci., **26** (1990), 221–333.

[33] M. Saito, *Mixed Hodge modules and applications* in Proceedings of the International Congress of Mathematicians, Vol. I, II (Kyoto, 1990), Math. Soc. Japan, Tokyo (1991), 725–734.

[34] M. Saito, *Hodge filtration on vanishing cycles*, preprint, May 1998.

[35] Groupes de monodromie en géométrie algébrique, dirigé par A. GROTHENDIECK, avec la collaboration de M. RAYNAUD et D. S. RIM pour la partie I et par P. DELIGNE et N. KATZ pour la partie II, *Lecture Notes in Math.*, vol. 288 and 340, Springer-Verlag 1972 and 1973.

[36] J.-P. Serre, *Facteurs locaux des fonctions zêta des variétés algébriques (définitions et conjectures)*, Séminaire Delange-Pisot-Poitou, 1969/70, n° **19**.

[37] M. Sebastiani and R. Thom, *Un résultat sur la monodromie*, Invent. Math., **13** (1971), 90–96.

[38] A. Scholl, *Classical motives* in Motives (Seattle, Wash., 1991), Proc. Sympos. Pure Math., **55**, Part 1, Amer. Math. Soc., Providence, 1994, 163–187.

[39] J. Steenbrink, *Mixed Hodge structures on the vanishing cohomology*, in Real and Complex Singularities, Sijthoff and Noordhoff, Alphen aan den Rijn, 1977, 525–563.

[40] J. Steenbrink, *The spectrum of hypersurface singularities*, in Théorie de Hodge, Luminy 1987, Astérisque, **179-180** (1989), 163–184.

[41] J. Steenbrink, *A summary of mixed Hodge theory*, American Mathematical Society, Proc. Symp. Pure Math. **55**, Pt. 1, (1994), 31–41.

[42] A. Varchenko, *Asymptotic Hodge structure in the vanishing cohomology*, Math. USSR Izvestija, **18** (1982), 469–512.

[43] W. Veys, *Poles of Igusa's local zeta function and monodromy*, Bull. Soc. Math. France, **121** (1993), 545–598.

[44] W. Veys, *Holomorphy of local zeta functions for curves*, Math. Ann. **295** (1993), 631–641.

[45] W. Veys, *Embedded resolution of singularities and Igusa's local zeta function*, Academiae Analecta, to appear (survey paper, 46p).

[46] W. Veys, *The topological zeta function associated to a function on a normal surface germ*, Topology **38** (1999), 439–456.

[47] W. Veys, *Zeta functions and 'Kontsevich invariants' on singular varieties*, Canad. J. Math. (to appear, 29 pages).

J. Denef
Department of Mathematics
University of Leuven
Celestijnenlaan 200B
3001 Leuven, Belgium
E-mail address: Jan.Denef@wis.kuleuven.ac.be
URL: http://www.wis.kuleuven.ac.be/wis/algebra/denef.html

F. Loeser
École Normale Supérieure
Département de mathématiques et applications
45 rue d'Ulm
75230 Paris Cedex 05, France
(UMR 8553 du CNRS)
E-mail address: Francois.Loeser@ens.fr
URL: http://www.dma.ens.fr/~loeser/

Stacks for Everybody

Barbara Fantechi

Abstract. Let $\mathfrak{S}$ be a category with a Grothendieck topology. A stack over $\mathfrak{S}$ is a category fibered in groupoids over $\mathfrak{S}$, such that isomorphisms form a sheaf and every descent datum is effective. If $\mathfrak{S}$ is the category of schemes with the étale topology, a stack is algebraic in the sense of Deligne-Mumford (respectively Artin) if it has an étale (resp. smooth) presentation.

I will try to explain the previous definitions so as to make them accessible to the widest possible audience. In order to do this, we will keep in mind one fixed example, that of vector bundles; if you know what pullback of vector bundles is in some geometric context (schemes, complex analytic spaces, but also varieties or manifolds) you should be able to follow this exposition.

1. Introduction

Stacks (the french original name is *champs* [4]) have been part of algebraic geometry for several decades now; algebraic stacks were introduced by Deligne and Mumford in [3] in order to study the moduli space of curves, and their definition was later generalized by Artin [1]. Since then, algebraic stacks have become a very useful tool for algebraic geometers, but still not a very popular one: possible reasons are the lack of references (see however the recent book of Laumon and Moret-Bailly [5]) and the long and technical definitions, which can discourage the newcomer.

The idea of this exposition is to alternate rigorous, general definitions with the study of one concrete example: the classifying stack parametrizing rank r vector bundles. The "everybody" in the title means that you don't have to be an algebraic geometer: stacks, and even reasonable analogues of algebraic stacks, can be defined in the context of complex analytic spaces, manifolds (your favorite kind) and even topological spaces.

2. A Category with a Grothendieck Topology

2.1. The base category $\mathfrak{S}$

We want to talk of geometric objects, so as a first step we have to specify what kind of geometry we want to do. My favorite is a very small part of algebraic geometry, namely the study of quasiprojective schemes over the complex numbers.

You might prefer other kind of schemes, or maybe complex analytic spaces, or real (or complex) manifolds, or topological spaces. In any case, the objects of our study form a category, i.e. we are also interested in morphisms among them; regular morphisms for schemes, differentiable or analytic maps for manifolds, and so on.

In this paper we will consider the category we work on as fixed, and when needed will refer to it as $\mathfrak{S}$. Its objects will be called schemes, because "scheme" makes for easier reading then "object of $\mathfrak{S}$" or "quasiprojective scheme over the complex numbers". Feel free to replace scheme by manifold (or variety, or complex analytic space, etc) everywhere.

2.2. Cartesian diagrams and fiber products

A commutative diagram

$$
\begin{array}{ccc}
T' & \xrightarrow{\bar{f}} & T \\
{\scriptstyle p'}\downarrow & & \downarrow{\scriptstyle p} \\
S' & \xrightarrow{f} & S
\end{array}
\tag{1}
$$

is called *cartesian* if it induces all other commutative diagrams with the same lower-right corner; that is, for any other commutative diagram

$$
\begin{array}{ccc}
U & \xrightarrow{g} & T \\
{\scriptstyle q}\downarrow & & \downarrow{\scriptstyle p} \\
S' & \xrightarrow{f} & S
\end{array}
$$

there is a unique morphism $h\colon U \to T'$ such that $q = p' \circ h$ and $g = \bar{f} \circ h$.

Another way to express this is to say that T' is the *fiber product* of S' and T over S, and to write $T' = S' \times_S T$. In fact, given f and p, T', p' and $\bar{f}$ are unique up to canonical isomorphism. One can also say that p' is the *base change* of p induced by the morphism f.

A concrete way to construct a fiber product is to consider the morphism $(f, p)\colon S' \times T \to S \times S$ and take the fiber product to be the inverse image of the diagonal; you can check that this works for schemes over a fixed base (take the scheme-theoretic inverse image, given by the pullback ideal sheaf), topological spaces, or sets. It also works for manifolds in case, say, p is a submersion (in which case so is p').

2.3. The étale topology

We will assume that the category $\mathfrak{S}$ we work with has a *Grothendieck topology*, that is given a scheme S, it makes sense to say whether any given collection of morphisms $\{S_i \to S\}$ is an open covering. You can find a precise definition of Grothendieck topology in [2].

When objects of $\mathfrak{S}$ are topological spaces, we can take open coverings to be the usual ones. You can stick to this and proceed to the next section now if you want: we will use the convention that, for an open covering $\{S_i \to S\}$, we write S_{ij} for $S_i \cap S_j$ and similarly for S_{ijk}.

However, in order to define algebraic stacks when $\mathfrak{S}$ is a category of schemes, the Zariski topology is not appropriate, because it's too coarse; in particular, the analogue of the implicit function theorem does not hold. An étale morphism (for smooth schemes this means one whose differential is an isomorphism at every point) is not necessarily a local isomorphism.

Because of this, in the definition of algebraic stack one uses the *étale topology*; that is, define an open covering to be a collection of étale morphisms $\{S_i \to S\}$ such that $\cup S_i \to S$ is surjective. We will use the following notational convention: if $\{S_i \to S\}$ is an open covering, we write S_{ij} for the fiber product $S_i \times_S S_j$ and analogously for S_{ijk}. For fixed j, $\{S_{ij} \to S_j\}$ is an open covering of S_j because the property of being étale is invariant under base change. If each $S_i \to S$ is an open embedding, then S_{ij} is canonically isomorphic to $S_i \cap S_j$.

3. A Category Fibered in Groupoids

3.1. Our guiding example: the category V_r

I assume you know what a vector bundle over a scheme is (remember, if you want you can read manifold wherever I write scheme), and what the pullback of a vector bundle is. To fix notation, if E is a vector bundle over S, and $f \colon T \to S$ is a morphism of schemes, I will call a diagram

$$\begin{array}{ccc} F & \xrightarrow{\bar{f}} & E \\ \downarrow & & \downarrow \\ T & \xrightarrow{f} & S \end{array} \tag{2}$$

a *pullback diagram* if F is a vector bundle over T, and the diagram makes F into the pullback of E via f (hence, the diagram is cartesian and $\bar{f}$ induces a linear isomorphism on fibers). We will also say that $(F, \bar{f})$ is *a pullback of E via f*.

Pullback is essentially unique; that is, given another pullback $(F', \bar{f}')$, there exists a unique isomorphism $\alpha \colon F' \to F$ of vector bundles over T such that $\bar{f}' = \bar{f} \circ \alpha$. This uniqueness depends on having fixed not only the bundle F but also the morphism $\bar{f}$.

We define the category V_r as follows. Its objects are rank r vector bundles over schemes; its morphisms are pullback diagrams, i.e. diagram (2) defines a morphism from F to E. You can figure out for yourself how composition of morphisms is defined. There is a natural forgetful functor from V_r to $\mathfrak{S}$, which associates to every bundle its base scheme and to every pullback diagram (2) the morphism f of the bases.

3.2. Schemes as categories over $\mathfrak{S}$

A *category over* $\mathfrak{S}$ is a category X with a fixed covariant functor $\pi\colon X \to \mathfrak{S}$. We say that an object E of X is *over* a scheme S, or *lifts* S, or *is a lifting of* S, if $\pi(E) = S$, and similarly for morphisms. If S is a scheme, the *fiber* of X over S is the subcategory of objects over S, and morphisms over the identity of S.

For instance, V_r is a category over $\mathfrak{S}$; the fiber over S is the category whose objects are vector bundles over S, and whose morphisms are the isomorphisms among them.

To a scheme S we can associate a category $\mathfrak{S}/S$ (the category of S-schemes) over $\mathfrak{S}$ as follows: the objects are morphisms with target S in $\mathfrak{S}$; a morphism from $f\colon T \to S$ to $f'\colon T' \to S$ is a $g\colon T \to T'$ such that $f = f' \circ g$; the projection functor sends the object $T \to S$ to T and a morphism g to itself.

Pictorially,

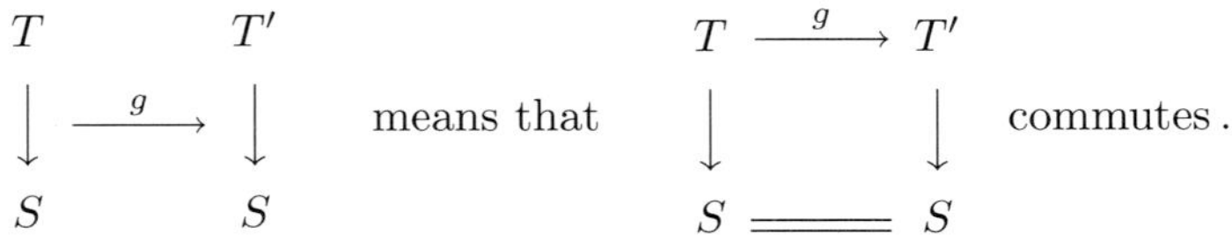

In the particular case where S is a point p (or a final object in the category $\mathfrak{S}$, if you find this clearer), the category $\mathfrak{S}/p$ is just the category $\mathfrak{S}$ itself, and the natural projection is the identity functor.

3.3. Morphisms of categories

A *morphism* of categories over $\mathfrak{S}$ is a covariant functor commuting with the projection to $\mathfrak{S}$.

Let S be a scheme, X a category over $\mathfrak{S}$, $f\colon \mathfrak{S}/S \to X$ a morphism of categories over $\mathfrak{S}$. To this morphism we can associate an object E of X over S, the image of $\mathrm{id}_S\colon S \to S$.

For instance, to every morphism $\mathfrak{S}/S \to V_r$ we can associate a vector bundle E over S. Conversely, given the associated vector bundle E, a morphism $\mathfrak{S}/S \to V_r$ is determined by the datum, for every $T \to S$, of a vector bundle F over T, together with a pullback diagram (2) (to prove this, use that every object $f\colon T \to S$ in $\mathfrak{S}/S$ has a unique morphism to id_S, namely f itself).

If S and T are schemes, and $f\colon \mathfrak{S}/T \to \mathfrak{S}/S$ is a morphism of categories over $\mathfrak{S}$, then the associated object is a morphism $g\colon T \to S$, and f is uniquely determined by g. Hence, morphisms of categories over $\mathfrak{S}$ from $\mathfrak{S}/T$ to $\mathfrak{S}/S$ are the same as morphisms of schemes $T \to S$: therefore, the category $\mathfrak{S}/S$ determines the scheme S up to isomorphism.

From now on we will use the same letter to indicate a scheme S and the stack $\mathfrak{S}/S$.

3.4. 2-morphisms and isomorphisms of categories

If X and Y are categories over $\mathfrak{S}$, and f, g are morphisms from X to Y, a *2-morphism* $f \to g$ is a natural transformation over the identity functor on $\mathfrak{S}$.

Because of the existence of 2-morphisms, categories over $\mathfrak{S}$ form a 2-category, i.e., morphisms can be isomorphic without being equal; the situation is analogous to that of homotopy theory, where two continuous maps can be homotopic without being equal.

As an example, define $f\colon V_r \to V_r$ by associating to each vector bundle its dual. Then f is a morphism of categories over $\mathfrak{S}$ and there is a 2-isomorphism between $f \circ f$ and the identity of V_r.

An *isomorphism* of categories over $\mathfrak{S}$ is a morphism which is an equivalence of categories, that is it induces bijections on morphisms and is surjective on objects up to isomorphism. An isomorphism has an inverse up to 2-isomorphisms, although to prove this one may need some form of the axiom of choice.

Let G be an algebraic group (or a Lie group, or a topological group, as you prefer); we can make a category BG over $\mathfrak{S}$ whose objects are principal G-bundles, and morphisms are pullback diagrams. We can define an isomorphism from V_r to $BGL(r)$ by associating to each vector bundle its frame bundle.

3.5. A category fibered in groupoids

The existence and uniqueness-up-to-isomorphism property for the pullback of vector bundles can be restated in categorical language by saying that V_r is a category fibered in groupoids over $\mathfrak{S}$.

Definition 3.1. *A category X over $\mathfrak{S}$ is called a* category fibered in groupoids, *or* groupoid fibration, *over $\mathfrak{S}$ if for any choice of a morphism of schemes $f\colon T \to S$ and of a lifting E of S to X, there exists a lifting $\bar{f}\colon F \to E$ of f to X, and the lifting is unique up to unique isomorphism: i.e., for any other lifting $\bar{f}'\colon F' \to E$ there is a unique isomorphism $\alpha\colon F' \to F$ over id_T such that $\bar{f}' = \bar{f} \circ \alpha$.*

As a partial motivation for the name, note that any morphism of X over an isomorphism of $\mathfrak{S}$ is also an isomorphism; in particular, for every scheme S, the *fiber* of X over S defined in 3.2, is a groupoid, a category where all morphisms are isomorphisms.

4. Stacks

As we saw, a category fibered in groupoids over $\mathfrak{S}$ is something that "pulls back like bundles". It is a stack if, moreover, it glues like bundles.

4.1. Notational conventions

Let X be a groupoid fibration over $\mathfrak{S}$. We assume that for every morphism $f\colon T \to S$, and every object E over S, we have chosen one lifting $f_E\colon f^*E \to E$ of f with target E. This can be achieved by direct construction, or by a suitable version of the axiom of choice. Note that it is not required, in this choice, that $g^*(f^*E) = (f \circ g)^*E$; the two are only canonically isomorphic. This choice of pullback is not logically necessary; it just makes it easier to write down the definition of stack (see also §6.3).

If E' is another object over S, and $\alpha\colon E' \to E$ is a morphism in the fiber (hence an isomorphism) there is a unique (iso)morphism $f^*\alpha\colon f^*E' \to f^*E$ such that the diagram

$$
\begin{array}{ccc}
f^*E' & \xrightarrow{\;f_{E'}\;} & E' \\
{\scriptstyle f^*\alpha}\downarrow & & \downarrow{\scriptstyle \alpha} \\
f^*E & \xrightarrow{\;f_E\;} & E
\end{array}
$$

commutes.

A further bit of convention: if the morphism $f\colon T \to S$ is clear from the context, we write $E \mid T$ or E over T instead of f^*E, and similarly for morphisms.

4.2. Descent data

Let $\{S_i \to S\}$ be an open covering of a scheme, and E a vector bundle on S; let E_i be a pullback of E to S_i. We cannot reconstruct E from knowing the E_i's only; for instance, we may have different bundles which become trivial on the same open covering.

However, the fact that E_i is the pullback of E means that we have induced isomorphisms $\alpha_{ij}\colon E_i \mid S_{ij} \to E_j \mid S_{ij}$, which satisfy the cocycle condition on S_{ijk}, and E can be recovered up to isomorphism by knowing E_i and α_{ij}. This motivates the following:

Definition 4.1. *Let X be a category fibered in groupoids over $\mathfrak{S}$. A descent datum for X over a scheme S is the following: an open covering $\{S_i \to S\}$; for every i, a lifting E_i of S_i to X; for every i, j an isomorphism $\alpha_{ij}\colon E_i \mid S_{ij} \to E_j \mid S_{ij}$ in the fiber which satisfies the cocycle condition $\alpha_{ik} = \alpha_{jk} \circ \alpha_{ij}$ over S_{ijk}.*

The descent datum is said to be effective *if there exists a lifting E of S to X together with isomorphisms $\alpha_i\colon E \mid S_i \to E_i$ in the fiber such that $\alpha_{ij} = \alpha_j \mid S_{ij} \circ (\alpha_i \mid S_{ij})^{-1}$.*

You can think of the covering $\{S_i\}$ as lying over S (after all, it covers it); we have a collection of bundles above, and we 'descend' them to a bundle over S.

4.3. Definition of stack

Before giving the definition of stack, we must introduce another 'technical' condition, the categorical expression of the fact that isomorphisms between bundles on the same scheme can be defined locally on an open covering and then glued in a unique way, once they agree on the overlaps.

Definition 4.2. *Let X be a category fibered in groupoids over $\mathfrak{S}$. We say that isomorphisms are a sheaf for X if, for any scheme S and any E, E' in the fiber over S, for every open covering $\{S_i \to S\}$ of S, and for every collection of isomorphisms $\alpha_i\colon E \mid S_i \to E' \mid S_i$ in the fiber over S_i such that $\alpha_i \mid S_{ij} = \alpha_j \mid S_{ij}$, there is a unique isomorphism $\alpha\colon E \to E'$ such that $\alpha \mid S_i = \alpha_i$.*

Definition 4.3. *A stack is a category fibered in groupoids over $\mathfrak{S}$ such that isomorphisms are a sheaf and every descent datum is effective.*

If X is a stack, then for every descent datum as in Definition 4.1, the (E, α_i) whose existence follows by effectivity are essentially unique: i.e., for any other possibility (E', α_i') there exists a unique isomorphism $\beta \colon E' \to E$ in the fiber such that $\alpha_i' = \alpha_i \circ \beta$.

Morphisms of stacks are defined to be morphisms of categories over $\mathfrak{S}$, and the same for 2-morphisms and isomorphisms.

4.4. Representable stacks and morphisms

The category $\mathfrak{S}/S$ is always fibered in groupoids over $\mathfrak{S}$, but not a priori a stack: this depends on the topology. It is true, and easy to prove, for each $\mathfrak{S}$ we mentioned (schemes, varieties, complex analytic spaces, manifolds, topological spaces, etc) with the usual topology. It is also true for schemes with the étale topology.

Definition 4.4. *A stack X over $\mathfrak{S}$ is* representable *if it is isomorphic to the stack $\mathfrak{S}/S$ induced by a scheme S (see also §6.3).*

Informally speaking, a representable morphism of stacks is one whose fibers are schemes.

Definition 4.5. *A morphism $X \to Y$ of stacks is* representable *if, for every morphism $S \to Y$ with S (the stack associated to) a scheme, the fiber product $S \times_Y X$ is representable.*

Oops! I didn't tell you what the fiber product of stacks is, did I? Well, it was done on purpose. The definition can be found in the last section; think of it for the moment as enjoying a universal property analogous to the one we saw for schemes. The proof of the next lemma is in §6.2.

Lemma 4.6. *Let W be a vector space of dimension r. Define a morphism $p \to V_r$ (recall that p is the category $\mathfrak{S}/p$) mapping every scheme T to the trivial bundle $T \times W$, and every morphism to the obvious pullback diagram. Then for every scheme S and every morphism $S \to V_r$ (with associated bundle E on S), the fiber product $S \times_{V_r} p$ is a representable stack, isomorphic to the frame bundle of E. Hence, the morphism $p \to V_r$ is representable.*

5. Algebraic Stacks and Groupoid Schemes

A property of morphisms of schemes is *invariant under base change* if, for any cartesian diagram (1), if p has the property then so has p'.

Definition 5.1. *Let P be a property of morphisms of schemes which is invariant under base change. A representable morphism $X \to Y$ of stacks has property P if, for every morphism $S \to Y$ with S a scheme, the induced morphism of schemes $S \times_Y X \to S$ has P.*

As an example, among such properties there are smooth, étale, proper, an open embedding, a closed embedding. If you're not an algebraic geometer, smooth is essentially the same as submersion, that is morphism of manifolds with surjective differential at every point.

From now on, $\mathfrak{S}$ is the category of quasiprojective schemes over the complex numbers, with the étale topology. Many other choices are possible in the context of algebraic geometry, but we are striving for clearness and not for generality. If you want to keep thinking that $\mathfrak{S}$ is complex analytic spaces, manifolds, etc, with the usual topology, you can do so, as most of what follows still makes sense: however, I don't know whether there is an official definition of analytic or differentiable stack.

Definition 5.2. *A stack X over $\mathfrak{S}$ is algebraic in the sense of Deligne and Mumford (resp. Artin) if there exists an étale (resp. smooth) and surjective representable morphism $S \to X$ where S is (the stack associated to) a scheme: we say $S \to X$ is a* presentation *of X.*

Lemma 4.6 implies that $p \to V_r$ is representable, smooth and surjective. Hence, the stack V_r is algebraic in the sense of Artin, and the morphism $p \to V_r$ is a presentation.

The stack V_r is a special case of quotient stack. Let G be an algebraic group acting on a scheme S on the left. Let $[S/G]$ be the following category over $\mathfrak{S}$: its objects are principal homogeneous G-bundles with a G-equivariant morphism to S. In analogy with V_r, its morphisms are those pullback diagrams which are compatible with the morphism to S. If S is a point p, with the trivial G action, then $[p/G]$ is BG defined before.

Theorem 5.3. *The category $[S/G]$ is an algebraic stack, there is a presentation $S \to [S/G]$ which is smooth and surjective of relative dimension $\dim G$.*

The proof of the theorem is analogous to that of Lemma 4.6: for any morphism $T \to [S/G]$ (associated to a G-equivariant morphism $P \to S$, where P is a principal G-bundle), the fiber product $T \times_{[S/G]} S$ is naturally isomorphic to P. Many moduli stacks arise as quotients; in fact, one of the key features of stacks is that for any group action you can take a quotient which behaves as if the action were fixed-point free.

6. Final Remarks

6.1. Fiber product of stacks

Definition 6.1. *If $f\colon X \to Y$ and $g\colon Z \to Y$ are morphisms of stacks over $\mathfrak{S}$, their fiber product can be defined as follows. Its objects are triples (x, z, α) where $\alpha\colon f(x) \to g(z)$ is a morphism in a fiber of Y; a morphism $(x, z, \alpha) \to (x', z', \alpha')$ is a pair of morphisms $(\beta_1\colon x \to x', \beta_2\colon z \to z')$ in fibers of X and Z respectively such that $g(\beta_2) \circ \alpha = \alpha' \circ f(\beta_1)\colon f(x) \to g(z').$*

This is indeed a stack and it enjoys 2-categorical properties similar to those of the usual fiber product; there are natural morphisms from $X \times_Y Z$ to X (mapping (x, z, α) to x and (β_1, β_2) to β_1) and to Z. The induced diagram

$$
\begin{array}{ccc}
X \times_Y Z & \longrightarrow & Z \\
\downarrow & & \downarrow{\scriptstyle g} \\
X & \xrightarrow{\ f\ } & Y
\end{array}
$$

does not commute, but only 2-*commute*, that is the two composition morphisms $X \times_Y Z \to Y$ are not the same but only canonically 2-isomorphic. In analogy with the case of schemes, the diagram 'induces' all other 2-commuting diagrams with the same lower right corner.

6.2. Outline of proof of Lemma 4.6

Recall that the frame bundle P of a vector bundle E over S has as fiber over $s \in S$ the set of possible bases for E_s. Fix a basis $B = (v_1, \ldots, v_r)$ of W. For a scheme T, there is a natural bijection between pullback diagrams

$$
\begin{array}{ccc}
T \times W & \xrightarrow{\ \bar{f}\ } & E \\
\downarrow & & \downarrow \\
T & \xrightarrow{\ f\ } & S
\end{array}
$$

and morphisms $g \colon T \to P$ (given the pullback diagram, for every $t \in T$ set $g(t) = \bar{f}(B)$, a basis of $E_{f(t)}$).

Let Z be the fiber product $S \times_{V_r} p$, and fix a scheme T. Spelling out Definition 6.1 in this case, an object of Z over T is a pullback diagram

$$
\begin{array}{ccc}
F & \xrightarrow{\ \bar{f}\ } & E \\
\downarrow & & \downarrow \\
T & \xrightarrow{\ f\ } & S
\end{array}
$$

(the image via $S \to V_r$ of f) together with an isomorphism $\alpha \colon F \to T \times W$ of vector bundles over T.

We can define a morphism $Z \to P$ sending such an object to the morphism $T \to P$ associated to the pullback diagram

$$
\begin{array}{ccc}
T \times W & \xrightarrow{\ \bar{f} \circ \alpha^{-1}\ } & E \\
\downarrow & & \downarrow \\
T & \xrightarrow{\ f\ } & S\,;
\end{array}
$$

we leave it to the reader to define the functor $Z \to P$ on morphisms and to check that it is an isomorphism (in fact, one can easily construct an explicit inverse).

6.3. Strictly speaking

Definitions 4.1 and 4.2 implicitly assume that $(E \mid S_i) \mid S_{ij}$ is the same bundle as $(E \mid S_j) \mid S_{ij}$. This is not true, but each of them is canonically isomorphic to $E_{ij} := E \mid S_{ij}$, and we have just omitted the canonical isomorphisms. Actually my favorite convention is that $f_E \colon f^*E \to E$ should mean *any* lifting of f with target E (after all, they're all canonically isomorphic).

In Definition 4.4 a representable stack should be one isomorphic to the stack associated to an algebraic space, not to a scheme, to be consistent with the existing literature. As every scheme is an algebraic space, this means that the definition of algebraic stack presented here is slightly narrower then the usual one.

6.4. Afterword

In analogy with the identification of a scheme S with the category $\mathfrak{S}/S$, a stack X "is" the collection of all morphisms from any scheme to X: the fiber of X over a scheme S is equivalent to the category of morphisms from S to X. The big difference with $\mathfrak{S}/S$ is that morphisms are allowed to be isomorphic without being equal. The stack condition ensures that, as in the case of schemes, morphisms can be defined locally and then glued.

The presentation can be used to define geometrical properties of an algebraic stack: e.g., V_r is smooth of dimension $-r^2$ (yes, it's negative) because p is smooth of dimension zero and $p \to V_r$ has relative dimension r^2. Of course one has to check that this does not depend on the choice of the presentation. In fact, many of the usual tools of algebraic geometry can be extended to algebraic stacks: coherent and locally free sheaves, cohomology, and even Riemann-Roch theorem and Chow groups. I would like to tell you more, but (to quote a favorite author) my paper reminds me to conclude.

Acknowledgements

I tend to learn more from people than from books. To all those who helped me I'm very thankful: a special mention goes to Kai Behrend, who patiently taught me most of what I know about stacks.

This paper is dedicated to Susanna, who gave me a very special start into the new millennium, and to Lothar who was with me in that occasion, as well as in many others.

References

[1] M. Artin, *Versal Deformations and Algebraic Stacks*, Invent. Math., **27**, (1974), 165–189.

[2] M. Artin, *Grothendieck Topologies*, Cambridge, Mass., 1962, mimeographed lecture notes.

[3] P. Deligne and D. Mumford, *The irreducibility of the space of curves of given genus*, Publ. Math. IHES, **36**, (1969), 75–109.

[4] J. Giraud, *Cohomologie non-abélienne*, Springer-Verlag, Berlin 1971, Grundlehren der mathematischen Wissenschaften, Band 179.

[5] G. Laumon and L. Moret-Bailly, *Champs algébriques*, Springer-Verlag, 1999.

Dipartimento di Matematica e Informatica
Università di Udine
Viale delle Scienze, 206
33100 Udine, Italy
E-mail address: `fantechi@dimi.uniud.it`

Multiple ζ-Values, Galois Groups, and Geometry of Modular Varieties

Alexander B. Goncharov

Abstract. We discuss two *arithmetical* problems, at first glance unrelated:
1) The properties of the multiple ζ-values

$$\zeta(n_1,\ldots,n_m) := \sum_{0<k_1<k_2<\cdots<k_m} \frac{1}{k_1^{n_1} k_2^{n_2} \cdots k_m^{n_m}} \qquad n_m > 1 \qquad (1)$$

and their generalizations, multiple polylogarithms at N-th roots of unity.
2) The action of the absolute Galois group on the pro-l completion

$$\pi_1^{(l)}(X_N) := \pi_1^{(l)}(\mathbb{P}^1 \backslash \{0, \mu_N, \infty\}, v)$$

of the fundamental group of $X_N := \mathbb{P}^1 \backslash \{0, \infty$ and all N-th roots of unity$\}$.
These problems are the Hodge and l-adic sites of the following one:
3) Study the Lie algebra of the image of motivic Galois group acting on the motivic fundamental group of $\mathbb{P}^1 \backslash \{0, \mu_N, \infty\}$.
We will discuss a surprising connection between these problems and *geometry* of the modular varieties

$$Y_1(m : N) := \Gamma_1(m; N) \backslash GL_m(\mathbb{R})/O_m \cdot \mathbb{R}_+^*$$

where $\Gamma_1(m; N)$ is the subgroup of $GL_m(\mathbb{Z})$ stabilizing $(0, \ldots, 0, 1) \bmod N$.
In particular using this relationship we get precise results about the Lie algebra of the image of the absolute Galois group in $\operatorname{Aut} \pi_1^{(l)}(X_N)$, and sharp estimates on the dimensions of the $\mathbb{Q}$-vector spaces generated by the multiple polylogarithms at N-th roots of unity, *depth m* and *weight $w := n_1 + \cdots + n_m$*.
The simplest case of the problem 3) is related to the classical theory of cyclotomic units. Thus the subject of this lecture is *higher cyclotomy*.

1. The Multiple ζ-Values

1.1. The algebra of multiple ζ-values and its conjectural description

Multiple ζ-values (1) were invented by L. Euler [8]. Euler discovered that the numbers $\zeta(m, n)$, when $w := m + n$ is odd, are $\mathbb{Q}$-linear combinations of $\zeta(w)$ and $\zeta(k)\zeta(w - k)$. Then these numbers were neglected. About 10 years ago they were resurrected as the coefficients of Drinfeld's associator [7], rediscovered by D. Zagier [31], appeared in works of M. Kontsevich on knot invariants and the author [10, 9] on mixed Tate motives over $\mathrm{Spec}(\mathbb{Z})$. More recently they showed up in quantum field theory [25, 3], deformation quantization and so on.

We say that $w := n_1 + \cdots + n_m$ is the *weight* and m is the *depth* of (1).

Let $\mathcal{Z}$ be the space of $\mathbb{Q}$-linear combinations of multiple ζ's. It is a commutative algebra over $\mathbb{Q}$. For instance

$$\zeta(m) \cdot \zeta(n) = \zeta(m,n) + \zeta(n,m) + \zeta(m+n) \tag{2}$$

because

$$\sum_{k_1,k_2>0} \frac{1}{k_1^m k_2^n} = \left(\sum_{0<k_1<k_2} + \sum_{0<k_1=k_2} + \sum_{k_1>k_2>0} \right) \frac{1}{k_1^m k_2^n}. \tag{3}$$

The only known results about the classical ζ-values are the following:

$$\zeta(2n) = (-1)^{n-1}(2\pi)^{2n} \cdot \frac{B_{2n}}{2 \cdot (2n)!} \quad \text{(Euler)}; \qquad \zeta(3) \notin \mathbb{Q} \quad \text{(Apery)}; \tag{4}$$

(were B_k are the Bernoulli numbers: $\frac{t}{e^t-1} = \sum B_k t^k/k!$) and recent results of T. Rivoal: dimension of the $\mathbb{Q}$-space spanned by $\zeta(3)$, $\zeta(5)$, $\ldots$, $\zeta(2n+1)$ grows at least as $C \log n$.

To describe the hypothetical structure of the algebra $\mathcal{Z}$ we introduce a free graded Lie algebra $\mathcal{F}(3,5,\ldots)_\bullet$, which is freely generated by elements e_{2n+1} of degree $-(2n+1)$ where $n \geq 1$. Let

$$U\mathcal{F}(3,5,\ldots)_\bullet^{\vee} := \oplus_{n\geq 1}\left(U\mathcal{F}(3,5,\ldots)_{-(2n+1)} \right)^{\vee}$$

be the graded dual to its universal enveloping algebra. It is $\mathbb{Z}_+$-graded.

Conjecture 1.1.
   a) *The weight provides a grading on the algebra $\mathcal{Z}$.*
   b) *One has an isomorphism of graded algebras over $\mathbb{Q}$*

$$\mathcal{Z}_\bullet = \mathbb{Q}[\pi^2] \otimes_{\mathbb{Q}} U\mathcal{F}(3,5,\ldots)_\bullet^{\vee} \qquad \deg \pi^2 := 2. \tag{5}$$

Part a) means that relations between ζ's of different weight, like $\zeta(5) = \lambda \cdot \zeta(7)$ where $\lambda \in \mathbb{Q}$, are impossible. For motivic interpretation/formulation of Conjecture 1.1 see Conjecture 17b in [10]. For its l-adic version see Conjecture 2.1 below.

Theorem 1.2. *One has* $\dim \mathcal{Z}_k \leq \dim(\mathbb{Q}[\pi^2] \otimes U\mathcal{F}(3,5,\ldots)^{\vee})_k$.

Both the origin of Conjecture 1.1 and proof of this theorem are based on theory of mixed Tate motives over $\mathrm{Spec}(\mathbb{Z})$: multiple ζ-values are periods of framed mixed Tate motives over $\mathbb{Z}$, and one can *prove* that the framed mixed Tate motives over $\mathbb{Z}$ form an algebra which is isomorphic to the one appearing on the right-hand side, (see [17]). This gives Theorem 1.2. Conjecture 1.1 just means that *every* such a period is given by multiple ζ-values.

For the definition of the *abelian* category of mixed Tate motives over a number field convenient for our approach see Chapter 5 in [15]. It has all the expected properties and is based on V. Voevodsky's construction of the triangulated category of motives [29]. Another approach to mixed motives has been developed by M. Levine [27, 28]. A construction of the abelian category of mixed Tate motives over the ring $\mathcal{O}_{F,S}$ of D-integers in a number field F see in ch. 3 of [17].

It is difficult to estimate $\dim \mathcal{Z}_k$ from below: we believe that $\zeta(5) \notin \mathbb{Q}$ but nobody can prove it.

One may reformulate Conjecture 1.1 as a hypothetical description of the $\mathbb{Q}$-vector space $\mathcal{PZ}$ of primitive multiple ζ's:

$$\mathcal{PZ}_{\bullet} := \frac{\mathcal{Z}_{>0}}{\mathcal{Z}_{>0} \cdot \mathcal{Z}_{>0}} \overset{?}{=} \langle \pi^2 \rangle \oplus \mathcal{F}(3, 5, \dots)_{\bullet}^{\vee}. \tag{6}$$

Here $\langle \pi^2 \rangle$ is a 1-dimensional $\mathbb{Q}$-vector space generated by π^2, and $\mathcal{Z}_{>0}$ is generated by π^2 and $\zeta(n_1, \dots, n_m)$.

Example 1.3. *There are 2^{10} convergent multiple ζ's of the weight 12. However according to Theorem 1.2 $\dim \mathcal{Z}_{12} \leq 12$. One should have $\dim \mathcal{PZ}_{12} = 2$ since $\mathcal{F}(3, 5, \dots)_{-12}$ is spanned over $\mathbb{Q}$ by $[e_5, e_7]$ and $[e_3, e_9]$. The $\mathbb{Q}$-vector space of decomposable multiple ζ's of the weight 12 is supposed to be generated by*

$$\pi^{12}, \quad \pi^6 \zeta(3)^2, \quad \pi^4 \zeta(3)\zeta(5), \quad \pi^4 \zeta(3,5), \quad \pi^2 \zeta(3)\zeta(7), \quad \pi^2 \zeta(5)^2, \quad \pi^2 \zeta(3,7),$$

$$\zeta(3)^4, \quad \zeta(5)\zeta(7), \quad \zeta(3)\zeta(9).$$

The algebra $U\mathcal{F}(3, 5, \dots)_{\bullet}^{\vee}$ is commutative. It is isomorphic to the space of noncommutative polynomials in variables f_{2n+1}, $n = 1, 2, 3, \dots$ with the algebra structure given by the shuffle product.

Let $\mathcal{F}(2, 3)_{\bullet}$ be the free graded Lie algebra generated by two elements of degree -2 and -3. Its graded dual $U\mathcal{F}(2, 3)_{\bullet}^{\vee}$ is isomorphic as a graded vector space to the space of noncommutative polynomials in two variables p and g_3 of degrees 2 and 3. There is canonical isomorphism of *graded vector spaces*

$$\mathbb{Q}[\pi^2] \otimes U\mathcal{F}(3, 5, \dots)_{\bullet}^{\vee} = U\mathcal{F}(2, 3)_{\bullet}^{\vee}.$$

The rule is clear from the pattern $(\pi^2)^3 f_3 (f_7)^3 (f_5)^2 \longrightarrow p^3 g_3 (g_3 p^2)^3 (g_3 p)^2$.

In particular if $d_k := \dim \mathcal{Z}_k$ then one should have $d_k = d_{k-2} + d_{k-3}$. This rule has been observed in computer calculations of D. Zagier for $k \leq 12$. Later on extensive computer calculations, confirming it, were made by D. Broadhurst [3].

1.2. The depth filtration

Conjecture 1.1, if true, would give a very simple and clear picture for the structure of the multiple ζ-values algebra. However this algebra has an additional structure: the depth filtration, and Conjecture 1.1 tells us nothing about it. The study of the depth filtration moved the subject in a completely unexpected direction: towards geometry of modular varieties for GL_m.

To formulate some results about the depth filtration consider the algebra $\overline{\mathcal{Z}}$ spanned over $\mathbb{Q}$ by the numbers

$$\overline{\zeta}(n_1, \dots, n_m) := (2\pi i)^{-w} \zeta(n_1, \dots, n_m).$$

364   A. B. Goncharov

It is filtered by the weight and depth. Since $\overline{\zeta}(2) = -1/24$, there is no weight grading anymore. Let $\mathrm{Gr}_{w,m}^{W,D} \mathcal{P}\overline{\mathcal{Z}}$ be the associated graded. We assume that 1 is of depth 0. Denote by $d_{w,m}$ its dimension over $\mathbb{Q}$.

Euler's classical computation of $\zeta(2n)$ (see (4)) tells us that $d_{2n,1} = 0$. Generalizing this it is not hard to prove that $d_{w,m} = 0$ if $w + m$ is odd.

Theorem 1.4.

a)
$$d_{w,2} \leq \left[\frac{w-2}{6}\right] \qquad \text{if } w \text{ is even}. \tag{7}$$

b)
$$d_{w,3} \leq \left[\frac{(w-3)^2 - 1}{48}\right] \qquad \text{if } w \text{ is odd}.$$

Part a) is due to Zagier; the dimension of the space of cusp forms for $SL_2(\mathbb{Z})$ showed up in his investigation of the double shuffle relations for the depth two multiple ζ's, ([31]). Part b) has been proved in [12]. Moreover we proved that, assuming some standard conjectures in arithmetic algebraic geometry, these estimates are exact, see also Corollary 2.5 and Theorem 7.5.

Problem 1.5. *Define explicitly a depth filtration on the Lie coalgebra $\mathcal{F}(3,5,\dots)^{\vee}$ which under the isomorphism (6) should correspond to the depth filtration on the space of primitive multiple ζ-values.*

The cogenerators of the Lie coalgebra $\mathcal{F}(3,5,\dots)^{\vee}$ correspond to $\zeta(2n+1)$. So a naive guess would be that the dual to the lower central series filtration on $\mathcal{F}(3,5,\dots)$ coincides with the depth filtration. However then one should have $d_{12,2} = 2$, while according to formula (7) $d_{12,2} = 1$. Nevertheless $\dim \mathcal{P}\mathcal{Z}_{12} = 2$, but the new transcendental number appears only in the depth 4.

1.3. A heuristic discussion

Conjecture 1.1 in the form (6) tells us that the space of primitive multiple ζ's should have a Lie coalgebra structure. How to determine its coproduct δ in terms of the multiple ζ's? Here is the answer for the depth 1 and 2 cases. (The general case later on.) Consider the generating series

$$\zeta(t) := \sum_{m>0} \zeta(m) t^{m-1}, \qquad \zeta(t_1, t_2) := \sum_{m,n>0} \zeta(m,n) t_1^{m-1} t_2^{n-1}.$$

Then $\delta\zeta(t) = 0$, i.e. $\delta\zeta(n) = 0$ for all n, and

$$\delta\zeta(t_1, t_2) = \zeta(t_2) \wedge \zeta(t_1) + \zeta(t_1) \wedge \zeta(t_2 - t_1) - \zeta(t_2) \wedge \zeta(t_1 - t_2). \tag{8}$$

To make sense out of this we have to go from the numbers $\zeta(n_1, \dots, n_m)$ to their more structured counterparts: framed mixed Tate motives $\zeta_{\mathcal{M}}(n_1, \dots, n_m)$, or their Hodge or l-adic realizations, (see [17]). The advantage is immediately seen: the coproduct $\delta_{\mathcal{M}}$ is well defined by the general formalism (see Section 10 in [10] or [17]), one easily proves not only that $\zeta_{\mathcal{M}}(2n) = 0$ (motivic version of Euler's theorem) as well as $\zeta_{\mathcal{M}}(1) = 0$, but also that $\overline{\zeta}_{\mathcal{M}}(2n+1) \neq 0$, and there are no linear relations between $\zeta_{\mathcal{M}}(2n+1)$'s! Hypothetically we lose no information:

linear relations between the multiple ζ's should reflect linear relations between their motivic avatars. Using $\zeta_{\mathcal{M}}(2n) = 0$ we rewrite formula (8) as

$$\delta_{\mathcal{M}} : \zeta_{\mathcal{M}}(t_1, t_2) \longmapsto \left(1 + U + U^2\right)\zeta_{\mathcal{M}}(t_1) \wedge \zeta_{\mathcal{M}}(t_2) \tag{9}$$

where U is the linear operator $(t_1, t_2) \longmapsto (t_1 - t_2, t_1)$. For example $\delta_{\mathcal{M}}$ sends the subspace of weight 12 double $\zeta_{\mathcal{M}}$'s to a one-dimensional $\mathbb{Q}$-vector space generated by $3\zeta_{\mathcal{M}}(3) \wedge \zeta_{\mathcal{M}}(9) + \zeta_{\mathcal{M}}(5) \wedge \zeta_{\mathcal{M}}(7)$. One can identify the cokernel of the map (9), restricted to the weight w subspace, with $H^1(GL_2(\mathbb{Z}), S^{w-2}V_2 \otimes \varepsilon_2)$ where V_2 is the standard GL_2-module, and $\otimes\varepsilon_2$ is the twist by the determinant, i.e. with the space of weight w cusp forms for $GL_2(\mathbb{Z})$. Moreover, one can prove that $\mathrm{Ker}\,\delta_{\mathcal{M}}$ is spanned by $\zeta_{\mathcal{M}}(2n + 1)$'s: this is a much more difficult result which uses all the machinery of mixed motives. Thus an element of the depth 2 associated graded of the space of primitive double ζ's is zero if and only if its coproduct is 0. So formula (9) provides a complete description of the space of double ζ's. In particular $d_{12,2} = 1$.

For the rest of this paper we suppress the motives working mostly with the l-adic side of the story and looking at the Hodge side for motivations.

2. Galois Symmetries of the pro-l Completion of the Fundamental Group of $\mathbb{P}^1\backslash\{0, \mu_N, \infty\}$

2.1. The Lie algebra of the image of the Galois group

Let X be a regular curve, $\overline{X}$ the corresponding projective curve, and v a tangent vector at a point $x \in \overline{X}$. According to Deligne [4] one can define the geometric profinite fundamental group $\widehat{\pi}_1(X, v)$ based at v. If X, x and v are defined over a number field F then the group $\mathrm{Gal}_F := \mathrm{Gal}(\overline{\mathbb{Q}}/F)$ acts by automorphisms of $\widehat{\pi}_1(X, v)$.

If $X = \mathbb{P}^1\backslash\{0, \mu_N, \infty\}$ there is a tangent vector v_∞ corresponding to the inverse t^{-1} of the canonical coordinate t on $\mathbb{P}^1\backslash\{0, \mu_N, \infty\}$. Denote by $\pi^{(l)}$ the pro-l-completion of the group π. We will investigate the map

$$\Phi_N^{(l)} : \mathrm{Gal}_{\mathbb{Q}} \longrightarrow \mathrm{Aut}\,\pi_1^{(l)}(\mathbb{P}^1\backslash\{0, \mu_N, \infty\}, v_\infty). \tag{10}$$

When $N = 1$ it was studied by Grothendieck [18], Deligne [4], Ihara (see [20, 22]), Drinfeld [7], and others (see [21]), but for $N > 1$ it was not investigated.

Denote by $H(m)$ the lower central series for the group H. Then the quotient $\pi_1^{(l)}(X_N)/\pi_1^{(l)}(X_N)(m)$ is an l-adic Lie group. Taking its Lie algebra and making the projective limit over m we get a pronilpotent Lie algebra over $\mathbb{Q}_l$:

$$\mathbb{L}_N^{(l)} := \varprojlim \mathrm{Lie}\left(\frac{\pi_1^{(l)}(X_N)}{\pi_1^{(l)}(X_N)(m)}\right).$$

Similarly one defines an l-adic pronilpotent Lie algebra $\mathbb{L}^{(l)}(X, v)$ corresponding to the geometric fundamental group $\widehat{\pi}_1(X, v)$ of a variety X with a base at v.

For topological reasons $\mathbb{L}_N^{(l)}$ is a free pronilpotent Lie algebra over $\mathbb{Q}_l$ with $n+1$ generators corresponding to the loops around 0 and N-th roots of unity.

Let $\mathbb{Q}(\zeta_n)$ be the field generated by n-th roots of unity. Set $\mathbb{Q}(\zeta_{l^\infty N}) := \cup\mathbb{Q}(\zeta_{l^a N})$. We restrict map (10) to the Galois group $G_{\mathbb{Q}(\zeta_{l^\infty N})}$. Passing to Lie algebras we get a homomorphism

$$\phi_N^{(l)}\colon\ \mathrm{Gal}_{\mathbb{Q}(\zeta_{l^\infty N})} \longrightarrow \mathrm{Aut}(\mathbb{L}_N^{(l)})\,.$$

Let us linearize the image of this map. Let $\mathbb{L}_N^{(l)}(m)$ be the lower central series for the Lie algebra $\mathbb{L}_N^{(l)}$. There are homomorphisms to l-adic Lie groups

$$\phi_{N;m}^{(l)}\colon\ \mathrm{Gal}_{\mathbb{Q}(\zeta_{l^\infty})} \longrightarrow \mathrm{Aut}\left(\mathrm{L}_N^{(l)}/\mathbb{L}_N^{(l)}(m)\right)\,.$$

The main hero of this story is the pronilpotent Lie algebra

$$\mathcal{G}_N^{(l)} := \varprojlim_m \mathrm{Lie}\left(\mathrm{Im}\,\phi_{N;m}^{(l)}\right) \hookrightarrow \mathrm{Der}\,\mathbb{L}_N^{(l)}\,.$$

When $N = 1$ we denote it by $\mathcal{G}^{(l)}$.

Conjecture 2.1. $\mathcal{G}^{(l)}$ *is a free Lie algebra with generators indexed by odd integers* ≥ 3.

It has been formulated, as a question, by Deligne [4] and Drinfeld [7].

2.2. The weight and depth filtration on $\mathbb{L}_N^{(l)}$

There are two increasing filtrations by ideals on the Lie algebra $\mathbb{L}_N^{(l)}$, indexed by negative integers.

The weight filtration $\mathcal{F}_\bullet^W$. It coincides with the lower central series for $\mathbb{L}_N^{(l)}$:

$$\mathbb{L}_N^{(l)} = \mathcal{F}_{-1}^W\mathbb{L}_N^{(l)}; \qquad \mathcal{F}_{-n-1}^W\mathbb{L}_N^{(l)} := [\mathcal{F}_{-n}^W\mathbb{L}_N^{(l)}, \mathbb{L}_N^{(l)}]\,.$$

The depth filtration $\mathcal{F}_\bullet^D$. The natural inclusion

$$\mathbb{P}^1\backslash\{0, \mu_N, \infty\} \hookrightarrow \mathbb{P}^1\backslash\{0, \infty\}$$

provides a morphism of the corresponding fundamental Lie algebras

$$p\colon \mathbb{L}_N^{(l)} \longrightarrow \mathbb{L}^{(l)}(\mathbb{P}^1\backslash\{0, \infty\}) = \mathbb{Q}_l(1)\,.$$

Let $\mathcal{I}_N$ be the kernel of this projection. Its powers give the depth filtration:

$$\mathcal{F}_0^D\mathbb{L}_N^{(l)} = \mathbb{L}_N^{(l)}, \quad \mathcal{F}_{-1}^D\mathbb{L}_N^{(l)} = \mathcal{I}_N, \quad \mathcal{F}_{-n-1}^D\mathbb{L}_N^{(l)} = [\mathcal{I}_N, \mathcal{F}_{-n}^D\mathbb{L}_N^{(l)}]\,.$$

2.3. The Galois Lie algebra and its shape

These filtrations induce two filtrations on the Lie algebra $\operatorname{Der}\mathbb{L}_N^{(l)}$ and hence on the Lie algebra $\mathcal{G}_N^{(l)}$. The associated graded Lie algebra $\operatorname{Gr}\mathcal{G}_{\bullet\bullet}^{(l)}(\mu_N)$, which we call the level N Galois Lie algebra, is bigraded by the weight $-w$ and depth $-m$. The weight filtration can be defined by a grading. Moreover one can define it in a way compatible with the depth filtration and the subspace $\mathcal{G}_N^{(l)} \subset \operatorname{Der}\mathbb{L}_N^{(l)}$. Therefore

$$\operatorname{Gr}\mathcal{G}_{\bullet\bullet}^{(l)}(\mu_N) \hookrightarrow \operatorname{Gr}_{\bullet\bullet}\operatorname{Der}\mathbb{L}_N^{(l)}.$$

The depth m quotients. For any $m \geq 1$ there is the depth $\geq -m$ (we will also say depth m) quotient Galois Lie algebra:

$$\operatorname{Gr}\mathcal{G}_{\bullet,\geq -m}^{(l)}(\mu_N) := \frac{\operatorname{Gr}\mathcal{G}_{\bullet,\bullet}^{(l)}(\mu_N)}{\operatorname{Gr}\mathcal{G}_{\bullet,<-m}^{(l)}(\mu_N)}. \tag{11}$$

It is a nilpotent graded Lie algebra of the nilpotence class $\leq m$.

The diagonal Lie algebra. Notice that $\operatorname{Gr}\mathcal{G}_{-w,-m}^{(l)}(\mu_N) = 0$ if $w < m$. We define the diagonal Galois Lie algebra $\operatorname{Gr}\mathcal{G}_{\bullet}^{(l)}(\mu_N)$ as the Lie subalgebra of $\operatorname{Gr}\mathcal{G}_{\bullet\bullet}^{(l)}(\mu_N)$ formed by the components with $w = m$. It is graded by the weight. It can be defined as the Lie subalgebra of $\operatorname{Gr}^W\mathcal{G}^{(l)}(\mu_N)$ by imposing the weight $=$ depth condition. Thus we do not need to take the associated graded for the depth filtration for its definition. Therefore the diagonal Galois Lie algebra is isomorphic, although non canonically, to a Lie subalgebra of $\mathcal{G}^{(l)}(\mu_N)$.

The picture below exhibits all possibly non zero components of the Galois Lie algebra and indicates its depth 2 quotient and the diagonal Lie subalgebra.

The shape of the Galois Lie algebra:

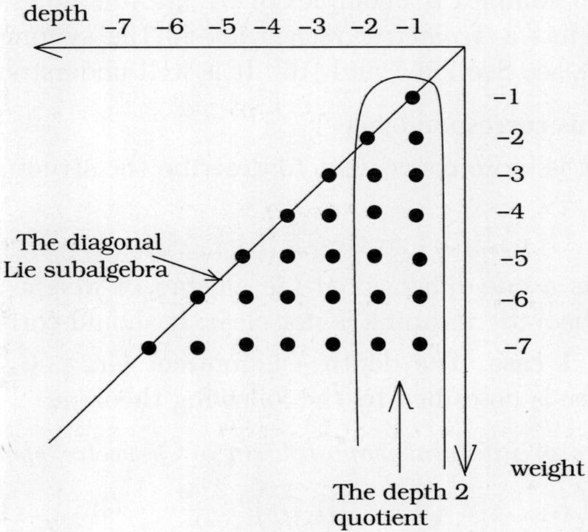

2.4. The mysterious correspondence

Let V_m be the standard m-dimensional representation of GL_m. Our key point ([9]–[13]) is that

The structure of Galois Lie algebra

$$\mathrm{Gr}\,\mathcal{G}^{(l)}_{\bullet,\geq -m}(\mu_N)$$

is related to

geometry of local systems with fibers $S^{\bullet -m}V_m$ over (the closure of) modular variety $Y_1(m; N)$, which is defined for $m > 1$ as

$$\Gamma_1(m; N)\backslash GL_m(\mathbb{R})/O_m \cdot \mathbb{R}^*.$$

The adelic approach to modular varieties shows that for $m = 1$ we have

$$Y_1(1; N) := S_N := \mathrm{Spec}\,\mathbb{Z}[\zeta_N][\frac{1}{N}]. \tag{12}$$

In particular the diagonal level N Galois Lie algebra is related to the geometry of the modular varieties $Y_1(m; N)$ themselves.

Both the dual to the Galois Lie algebras (11) and the modular varieties $\overline{Y}_1(m; N)$ form inductive systems with respect to m. The correspondence is compatible with these inductive structures.

Recall the standard cochain complex of a Lie algebra $\mathcal{G}$

$$\mathcal{G}^\vee \xrightarrow{\delta} \Lambda^2\mathcal{G}^\vee \xrightarrow{\delta} \Lambda^3\mathcal{G}^\vee \longrightarrow \cdots$$

where the first differential is dual to the commutator map $[,]: \Lambda^2\mathcal{G} \longrightarrow \mathcal{G}$, and the others are obtained using the Leibniz rule. The condition $\delta^2 = 0$ is equivalent to the Jacobi identity.

For a precise form of this correspondence see Section 6. It relates the depth m, weight w part of the standard cochain complex of the Lie algebra (11) with

$$(\text{rank } m \text{ modular complex}) \otimes_{\Gamma_1(m;N)} S^{w-m}V_m. \tag{13}$$

The rank m modular complex is a complex of $GL_m(\mathbb{Z})$-modules constructed purely combinatorially. It has a *geometric realization* in the symmetric space $\mathbb{H}_m := GL_m(\mathbb{R})/O(m) \cdot \mathbb{R}^*$, see Section 7 and [16]. It is well understood only for $m \leq 4$.

2.5. Examples of this correspondence

Strangely enough it is more convenient to describe the structure of the bigraded Lie algebra

$$\mathrm{Gr}\,\widehat{\mathcal{G}}^{(l)}_{\bullet\bullet}(\mu_N) := \mathrm{Gr}\,\mathcal{G}^{(l)}_{\bullet\bullet}(\mu_N) \oplus \mathbb{Q}_l(-1,-1)$$

where $\mathbb{Q}_l(-1,-1)$ is a one-dimensional Lie algebra of weight and depth -1. Its motivic or Galois-theoretic meaning is not clear: it should correspond to $\zeta(1)$.

a) The depth 1 case. The depth -1 quotient $\mathrm{Gr}_{\bullet,-1}\,\mathcal{G}^{(l)}_N$ is an abelian Lie algebra. Its structure is described by the following theorem.

Theorem 2.2. *There is a natural isomorphism of $\mathbb{Q}_l$-vector spaces*

$$\mathrm{Hom}\Big(K_{2n-1}(\mathbb{Z}[\zeta_N, N^{-1}]), \mathbb{Q}_l\Big) \xrightarrow{=} \mathrm{Gr}_{-n,-1}\,\mathcal{G}^{(l)}_N. \tag{14}$$

According to the Borel theorem one has

$$\dim K_{2n-1}(\mathbb{Z}) \otimes \mathbb{Q} = \begin{cases} 0 & n:\text{ even} \\ 1 & n > 1:\text{ odd} \end{cases} \tag{15}$$

$$\dim K_{2n-1}(\mathbb{Z}[\zeta_N, N^{-1}]) \otimes \mathbb{Q} = \begin{cases} \dfrac{\varphi(N)}{2} & N > 2,\, n > 1 \\[2mm] \dfrac{\varphi(N)}{2} + p(N) - 1 & N > 2,\, n > 1 \\[2mm] 1 & N = 2 \end{cases}, \tag{16}$$

where $p(N)$ is the number of prime factors of N.

Theorem 2.2 for $N = 1$ is known thanks to Soulé, Deligne [4], and Ihara [21]. The general case can be deduced from the motivic theory of classical polylogarithms developed by Deligne and Beilinson [4, 1]. In the case $n = 1$ there is a canonical isomorphism justifying the name "higher cyclotomy" for our story:

$$\operatorname{Gr} \mathcal{G}^{(l)}_{-1,-1}(\mu_N) = \operatorname{Hom}\Big(\text{group of the cyclotomic units in } \mathbb{Z}[\zeta_N][\tfrac{1}{N}],\, \mathbb{Q}_l\Big). \tag{17}$$

The level N modular variety for $GL_{1/\mathbb{Q}}$ is the scheme S_N, see (12). It has $\varphi(N)$ complex points parametrized by the primitive roots of unity ζ_N^α where $(\alpha, N) = 1$. Our correspondence for $m = 1$ is given by the natural isomorphism (where $+$ means invariants under the complex conjugation):

$$[\mathbb{Q}(n-1) - \text{valued functions on } S_N \otimes \mathbb{C}]^+ \xrightarrow{=} K_{2n-1}(S_N) \otimes \mathbb{Q}$$

provided by motivic classical polylogarithms: one associates to ζ_N^α the cyclotomic element $\{\zeta_N^\alpha\}_n \in K_{2n-1}(S_N)$, whose regulator is computed via $\operatorname{Li}_n(\zeta_N^\alpha)$.

b) The depth 2 case, $N = 1$. The structure of the depth ≥ -2 quotient of the Lie algebra $\operatorname{Gr} \widehat{\mathcal{G}}^{(l)}_{\bullet,\bullet}$ is completely described by the commutator map

$$[,]\colon \Lambda^2 \operatorname{Gr} \widehat{\mathcal{G}}^{(l)}_{-1,\bullet} \longrightarrow \operatorname{Gr} \mathcal{G}^{(l)}_{-2,\bullet}. \tag{18}$$

Construction of the dual to complex (18). Look at the classical modular triangulation of the hyperbolic plane $\mathbb{H}_2$ where the central ideal triangle has vertices at 0, 1, ∞:

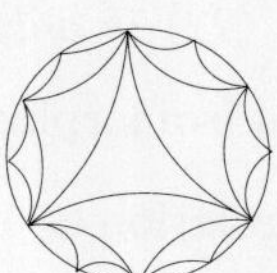

The group $GL_2(\mathbb{R})$, acting on $\mathbb{C}\backslash\mathbb{R}$ by $z \longmapsto \frac{az+b}{cz+d}$ commutes, with $z \longmapsto \bar{z}$. We let $GL_2(\mathbb{R})$ act on $\mathbb{H}_2$ by identifying $\mathbb{H}_2$ with the quotient of $\mathbb{C}\backslash\mathbb{R}$ by complex conjugation. The subgroup $GL_2(\mathbb{Z})$ preserves the modular picture. Consider the chain complex of the modular triangulation placed in degrees $[1, 2]$:

$$M^*_{(2)} := M^1_{(2)} \longrightarrow M^2_{(2)}. \tag{19}$$

It is a complex of $GL_2(\mathbb{Z})$-modules. The group $M^1_{(2)}$ is generated by the triangles, and $M^2_{(2)}$ by the geodesics. Let ε_2 be the one-dimensional GL_2-module given by the determinant.

Lemma 2.3. *Let Γ be a finite index subgroup of $GL_2(\mathbb{Z})$ and V a GL_2-module over $\mathbb{Q}$. Then the complex $M^*_{(2)} \otimes_\Gamma V[1]$ computes the cohomology $H^*(\Gamma, V \otimes \varepsilon_2)$.*

Theorem 2.4. *The weight w part of the dual to complex (18) is **canonically** isomorphic to the complex*

$$\left(M^*_{(2)} \otimes_{GL_2(\mathbb{Z})} S^{w-2}V_2 \right) \otimes \mathbb{Q}_l . \tag{20}$$

A motivic version of this theorem was obtained in Section 7 of [9]. Its Hodge side provides a refined version of the story told in Section 1.3.

According to the lemma, complex (20) computes $H^{*-1}(GL_2(\mathbb{Z}), S^{w-2}V_2 \otimes \varepsilon_2)$. Since these cohomology groups are known, we compute the Euler characteristic of the complex (18) and using Theorem 2.2 and formula (15) get the following result, the l-adic version of Theorem 1.4, (see related results of Ihara and Takao in [22]).

Corollary 2.5.

$$\dim \operatorname{Gr} \mathcal{G}^{(l)}_{-w,-2} = \begin{cases} 0 & w\colon odd \\ \left[\frac{w-2}{6}\right] & w\colon even \end{cases} . \tag{21}$$

c) $N = p$ is a prime, $w = m = -2$. The structure of the weight ≥ -2 quotient of the *diagonal* Galois Lie algebra $\operatorname{Gr} \mathcal{G}^{(l)}_\bullet(\mu_N)$ is described by the commutator map

$$[,]\colon \Lambda^2 \operatorname{Gr} \widehat{\mathcal{G}}^{(l)}_{-1,-1}(\mu_p) \longrightarrow \operatorname{Gr} \mathcal{G}^{(l)}_{-2,-2}(\mu_p) . \tag{22}$$

Projecting the modular triangulation of the hyperbolic plane onto the modular curve $Y_1(p) := \Gamma_1(p)\backslash \mathbb{H}_2$ we get the modular triangulation of $Y_1(p)$. The complex involution acts on the modular curve preserving the triangulation. Consider the following complex, where $+$ means invariants of the complex involution:

$$\left(\text{the chain complex of the modular triangulation of } Y_1(p)\right)^+ \otimes \mathbb{Q}_l . \tag{23}$$

Theorem 2.6. *The dual to complex (22) is naturally **isomorphic** to complex (23).*

In particular there is canonical isomorphism

$$\mathbb{Q}_l[\text{triangles of the modular triangulation of } Y_1(p)]^+ = \left(\operatorname{Gr} \mathcal{G}^{(l)}_{-2,-2}(\mu_p)\right)^\vee . \tag{24}$$

Computing the Euler characteristic of the complex (22) using (24) and (17) we get

$$\dim \operatorname{Gr} \mathcal{G}^{(l)}_{-2}(\mu_p) = \frac{(p-5)(p-1)}{12} .$$

Deligne proved [5] that the Hodge-theoretic version of $\mathcal{G}^{(l)}_N$ is free when $N = 2$, and recently extended the arguments to the case $N = 3, 4$ in [6]. The results above imply that it cannot be free for sufficiently big N. For instance it is not free for a prime $N = p$ if the genus of $Y_1(p)$ is positive, i.e. $p > 7$. Indeed,

The depth ≤ 2 part of $H^2(\mathcal{G}^{(l)}_p) = H^2_{(2)}(\mathcal{G}^{(l)}_\bullet(\mu_p)) = H^1(\Gamma_1(p), \varepsilon_2)$.

2.6. Our strategy

To describe the structure of the Galois Lie algebras (11) in general we need the dihedral Lie algebra of the group μ_N ([12, 13]) recalled in Section 4. To motivate to some extent its definition we turn in Section 3 to the Hodge side of higher cyclotomy. As explained in Section 5 both Galois and dihedral Lie algebra of μ_N act in a special way on the pronilpotent completion of $\pi_1(\mathbb{P}^1 \backslash \{0, \mu_N, \infty\}, v_\infty)$, and the Galois is contained in the dihedral Lie algebra ([13]). In Section 6 we relate the standard cochain complex of the dihedral Lie algebra of μ_N with the modular complex ([12]), whose canonical geometric realization in the symmetric space ([16]) is given in Section 7. Thus we related the structure of the Galois Lie algebras with geometry of modular varieties.

3. Multiple Polylogarithms and Higher Cyclotomy

3.1. Definition and iterated integral presentation

Multiple polylogarithms ([10, 9]) are defined as the power series

$$\mathrm{Li}_{n_1,\ldots,n_m}(x_1,\ldots,x_m) = \sum_{0<k_1<k_2<\cdots<k_m} \frac{x_1^{k_1} x_2^{k_2} \cdots x_m^{k_m}}{k_1^{n_1} k_2^{n_2} \cdots k_m^{n_m}} \tag{25}$$

generalizing both the classical polylogarithms $\mathrm{Li}_n(x)$ (if $m = 1$) and multiple ζ-values (if $x_1 = \cdots = x_m = 1$). These series are convergent for $|x_i| < 1$.

Recall a definition of iterated integrals. Let $\omega_1, \ldots, \omega_n$ be 1-forms on a manifold M and $\gamma \colon [0,1] \to M$ a path. The iterated integral $\int_\gamma \omega_1 \circ \cdots \circ \omega_n$ is defined inductively:

$$\int_\gamma \omega_1 \circ \cdots \circ \omega_n := \int_0^1 \left(\int_{\gamma_t} \omega_1 \circ \cdots \circ \omega_{n-1} \right) \gamma^* \omega_n . \tag{26}$$

Here γ_t is the restriction of γ to the interval $[0, t]$ and $\int_{\gamma_t} \omega_1 \circ \cdots \circ \omega_{n-1}$ is considered as a function on $[0, 1]$. We multiply it by the 1-form $\gamma_t^* \omega_n$ and integrate.

Denote by $I_{n_1,\ldots,n_m}(a_1 : \cdots : a_m : a_{m+1})$ the iterated integral

$$\int_0^{a_{m+1}} \underbrace{\frac{dt}{a_1 - t} \circ \frac{dt}{t} \circ \cdots \circ \frac{dt}{t}}_{n_1 \text{ times}} \circ \cdots \circ \underbrace{\frac{dt}{a_m - t} \circ \frac{dt}{t} \circ \cdots \circ \frac{dt}{t}}_{n_m \text{ times}} . \tag{27}$$

Its value depends only on the homotopy class of a path connecting 0 and a_{m+1} on $\mathbb{C}^* \backslash \{a_1, \ldots, a_m\}$. Thus it is a multivalued analytic function of $a_1, \ldots, a_{m+1}$. The following result provides an analytic continuation of multiple polylogarithms.

Theorem 3.1. $\mathrm{Li}_{n_1,\ldots,n_m}(x_1,\ldots,x_m) = I_{n_1,\ldots,n_m}(1 : x_1 : x_1 x_2 : \cdots : x_1 \cdots x_m)$.

The proof is easy: develop $dt/(a_i - t)$ into a geometric series and integrate. If $x_i = 1$ we get the Kontsevich formula. In particular in the depth 1 case we recover the classical Leibniz presentation for $\zeta(n)$:

$$\zeta(n) = \int_0^1 \underbrace{\frac{dt}{1-t} \circ \frac{dt}{t} \circ \cdots \circ \frac{dt}{t}}_{n \text{ times}} . \tag{28}$$

3.2. Multiple polylogarithms at roots of unity

Let $\overline{\mathcal{Z}}_{\leq w}(N)$ be the $\mathbb{Q}$-vector space spanned by the numbers

$$\overline{\mathrm{Li}}_{n_1,\dots,n_m}(\zeta_N^{\alpha_1},\dots,\zeta_N^{\alpha_m}) := (2\pi i)^{-w}\,\mathrm{Li}_{n_1,\dots,n_m}(\zeta_N^{\alpha_1},\dots,\zeta_N^{\alpha_m}); \qquad \zeta_N := e^{2\pi i/N} . \tag{29}$$

Here we may take any branch of $\mathrm{Li}_{n_1,\dots,n_m}(x_1,\dots,x_m)$. Similarly to (3) the space $\overline{\mathcal{Z}}(N) := \cup \overline{\mathcal{Z}}_{\leq w}(N)$ is an algebra *bifiltred* by the weight and by the depth. We want to describe this algebra and its associate graded for the weight and depth filtrations. Let us start from some relations between these numbers. Notice that

$$\mathrm{Li}_1(\zeta_N^\alpha) = -\log(1-\zeta_N^\alpha)$$

so the simplest case of this problem reduces to theory of cyclotomic units. By the Bass theorem all relations between the cyclotomic units $1 - \zeta_N^\alpha$ follow from the distribution relations and symmetry under $\alpha \to -\alpha$ valid modulo roots of unity.

3.3. Relations

The double shuffle relations. Consider the generating series

$$\mathrm{Li}(x_1,\dots,x_m \mid t_1,\dots,t_m) := \sum_{n_i \geq 1} \mathrm{Li}_{n_1,\dots,n_m}(x_1,\dots,x_m)t_1^{n_1-1}\cdots t_m^{n_m-1} .$$

Let $\Sigma_{p,q}$ be the subset of permutations of $p+q$ letters $\{1,\dots,p+q\}$ consisting of all shuffles of $\{1,\dots,p\}$ and $\{p+1,\dots,p+q\}$. Similarly to (2) multiplying power series (25) we immediately get

$$\mathrm{Li}(x_1,\dots,x_p \mid t_1,\dots,t_p) \cdot \mathrm{Li}(x_{p+1},\dots,x_{p+q} \mid t_{p+1},\dots,t_{p+q})$$

$$= \sum_{\sigma \in \Sigma_{p,q}} \mathrm{Li}(x_{\sigma(1)},\dots,x_{\sigma(p+q)} \mid t_{\sigma(1)},\dots,t_{\sigma(p+q)}) + \text{lower depth terms} . \tag{30}$$

To get the other set of the relations we multiply iterated integrals (27), and use Theorem 3.1 plus the following product formula for the iterated integrals:

$$\int_\gamma \omega_1 \circ \cdots \circ \omega_p \cdot \int_\gamma \omega_{p+1} \circ \cdots \circ \omega_{p+q} = \sum_{\sigma \in \Sigma_{p,q}} \int_\gamma \omega_{\sigma(1)} \circ \cdots \circ \omega_{\sigma(p+q)} . \tag{31}$$

For example the simplest case of formula (31) is derived as follows:

$$\int_0^1 f_1(t)dt \cdot \int_0^1 f_2(t)dt = \left(\int_{0\leq t_1 \leq t_2 \leq 1} + \int_{0\leq t_2 \leq t_1 \leq 1}\right) f_1(t_1)f_2(t_2)dt_1 dt_2 .$$

It is very similar in spirit to the derivation (3) of formula (2).

To get nice formulas consider the generating series

$$I^*(a_1 : \cdots : a_m : a_{m+1} \mid t_1, \ldots, t_m)$$

$$:= \sum_{n_i \geq 1} I_{n_1,\ldots,n_m}(a_1 : \cdots : a_m : a_{m+1})t_1^{n_1-1}(t_1 + t_2)^{n_2-1} \cdots (t_1 + \cdots + t_m)^{n_m-1}.$$

$$(32)$$

Theorem 3.2.

$$I^*(a_1 : \cdots : a_p : 1 \mid t_1, \ldots, t_p) \cdot I^*(a_{p+1} : \cdots : a_{p+q} : 1 \mid t_{k+1}, \ldots, t_{p+q})$$

$$= \sum_{\sigma \in \Sigma_{p,q}} I^*(a_{\sigma(1)}, \ldots, a_{\sigma(p+q)} : 1 \mid t_{\sigma(1)}, \ldots, t_{\sigma(p+q)}). \quad (33)$$

A sketch of the proof. It is not hard to prove the following formula

$$I^*(a_1 : \cdots : a_m : 1 \mid t_1, \ldots, t_m) = \int_0^1 \frac{s^{-t_1}}{a_1 - s} ds \circ \cdots \circ \frac{s^{-t_m}}{a_m - s} ds. \quad (34)$$

The theorem follows from this and product formula (31) for the iterated integrals.

For multiple ζ's these are precisely the relations of Zagier, who conjectured that, properly regularized, they provide all the relations between the multiple ζ's.

Distribution relations. From the power series expansion we immediately get

Proposition 3.3. *If $|x_i| < 1$ and l is a positive integer then*

$$\mathrm{Li}(x_1, \ldots, x_m \mid t_1, \ldots, t_m) = \sum_{y_i^l = x_i} \mathrm{Li}(y_1, \ldots, y_m \mid lt_1, \ldots, lt_m). \quad (35)$$

If $N > 1$ the double shuffle plus distribution relations do not provide all relations between multiple polylogarithms at N-th roots of unity. However I conjecture they do give all the relations if N is a prime and we restrict to the weight = depth case.

3.4. Multiple polylogarithms at roots of unity and the cyclotomic Lie algebras

Denote by $UC_\bullet$ the universal enveloping algebra of a graded Lie algebra $C_\bullet$. Let $UC_\bullet^\vee$ be its graded dual. It is a commutative Hopf algebra.

Conjecture 3.4.

a) *There exists a graded Lie algebra $C_\bullet(N)$ over $\mathbb{Q}$ such that one has an iso-morphism $\overline{\mathcal{Z}}(N) = UC_\bullet(N)^\vee$ of filtered by the weight on the left and by the degree on the right algebras.*

b) $H^1_{(n)}(C_\bullet(N)) = K_{2n-1}(\mathbb{Z}[\zeta_N][\frac{1}{N}]) \otimes \mathbb{Q}.$

c) $C_\bullet(N) \otimes \mathbb{Q}_l = \mathcal{G}_N^{(l)}$ *as filtered by the weight Lie algebras.*

Here $H_{(n)}$ is the degree n part of H. Notice that $H^1_{(n)}(C_\bullet(N))$ is dual to the space of degree n generators of the Lie algebra $C_\bullet(N)$.

Examples 3.5.

  i) *If $N = 1$ the generators should correspond to $\overline{\zeta}(2n+1)$.*

  ii) *If $N > 1$ the generators should correspond to $\overline{\mathrm{Li}}_n(\zeta_N^\alpha)$ where $(\alpha, N) = 1$.*

A construction of the Lie algebra $C_\bullet(N)$ using the Hodge theory see in [17]. Similarly to Theorem 1.2 one proves ([17]) that the algebra $\overline{\mathcal{Z}}(N)$ is a subalgebra of the universal enveloping algebra of the motivic Tate Lie algebra of $\mathrm{Spec}(\mathbb{Z}[\zeta_N][\frac{1}{N}])$, i.e. free graded Lie algebra generated by $K_{2n-1}(\mathbb{Z}[\zeta_N][\frac{1}{N}]) \otimes \mathbb{Q}$ in degrees $n \geq 1$. However unlike our expectation for $N = 1$ this estimate is not exact for sufficiently big N.

3.5. The coproduct

The iterated integral

$$I(a_0; a_1, \ldots, a_m; a_{m+1}) := \int_{a_0}^{a_{m+1}} \frac{dt}{t - a_1} \circ \cdots \circ \frac{dt}{t - a_m} \tag{36}$$

provides a framed mixed Hodge-Tate structure, denote by $I_{\mathcal{H}}(a_0; a_1, \ldots, a_m; a_{m+1})$, see [10, 11, 17]. The set of equivalence classes of framed mixed Hodge-Tate structures has a structure of the graded Hopf algebra over $\mathbb{Q}$ with the coproduct Δ.

Theorem 3.6. *For the framed Hodge-Tate structure corresponding to (36) we have:*

$$\Delta I_{\mathcal{H}}(a_0; a_1, a_2, \ldots, a_m; a_{m+1})$$

$$= \sum_{0=i_0<i_1<\cdots<i_k<i_{k+1}=m} I_{\mathcal{H}}(a_0; a_{i_1}, \ldots, a_{i_k}; a_{m+1}) \otimes$$

$$\otimes \prod_{p=0}^{k} I_{\mathcal{H}}(a_{i_p}; a_{i_p+1}, \ldots, a_{i_{p+1}-1}; a_{i_{p+1}}). \tag{37}$$

This formula also provides an explicit description of the variation of mixed Hodge-Tate structures whose period function is given by (36), see [11, 17]. Specializing it we get explicit formulas for the coproduct of all multiple polylogarithms (27). When a_i are N-th roots of unity and the lower depth terms are suppressed the result has a particularly nice form. It is described, in an axiomatized form of the coproduct for dihedral Lie algebras, in the next section.

4. The Dihedral Lie Coalgebra of a Commutative Group G

Let G and H be two commutative groups or, better, commutative group schemes. Then, generalizing a construction given in [12, 13] one can define a graded Lie coalgebra $\mathcal{D}_\bullet(G \mid H)$, called the dihedral Lie coalgebra of G and H ([17]). In the special case when $H = \mathrm{Spec}\,\mathbb{Q}[[t]]$ is the additive group of the formal line it is a bigraded Lie coalgebra $\mathcal{D}_{\bullet\bullet}(G)$ called the dihedral Lie coalgebra of G. (The second grading comes from the natural filtration on $\mathbb{Q}[[t]]$.) We recall its definition below. The construction of $\mathcal{D}_\bullet(G \mid H)$ is left as an easy exercise.

4.1. Formal definitions ([12, 13])

Let G be a commutative group. We will define a bigraded Lie coalgebra $\mathcal{D}_{\bullet\bullet}(G) = \oplus_{w \geq m \geq 1} \mathcal{D}_{w,m}(G)$. The $\mathbb{Q}$-vector space $\mathcal{D}_{w,m}(G)$ is generated by the symbols

$$I_{n_1,\ldots,n_m}(g_1 : \cdots : g_{m+1}), \qquad w = n_1 + \cdots + n_m, \quad n_i \geq 1. \tag{38}$$

To define the relations we introduce the generating series

$$\{g_1 : \cdots : g_{m+1} \mid t_1 : \cdots : t_{m+1}\} :=$$

$$:= \sum_{n_i > 0} I_{n_1,\ldots,n_m}(g_1 : \cdots : g_{m+1})(t_1 - t_{m+1})^{n_1 - 1} \cdots (t_m - t_{m+1})^{n_m - 1}. \tag{39}$$

We will also need two other generating series:

$$\{g_1 : \cdots : g_{m+1} \mid t_1, \ldots, t_{m+1}\} :=$$

$$:= \{g_1 : \cdots : g_{m+1} \mid t_1 : t_1 + t_2 : \cdots : t_1 + \cdots + t_m : 0\} \tag{40}$$

where $t_1 + \cdots + t_{m+1} = 0$, and

$$\{g_1, \ldots, g_{m+1} \mid t_1 : \cdots : t_{m+1}\} := \{1 : g_1 : g_1 g_2 : \cdots : g_1 \cdots g_m \mid t_1 : \cdots : t_{m+1}\} \tag{41}$$

where $g_1 \cdot \ldots \cdot g_{m+1} = 1$.

4.2. Relations

i) *Homogeneity.* For any $g \in G$ one has

$$\{g \cdot g_1 : \cdots : g \cdot g_{m+1} \mid t_1 : \cdots : t_{m+1}\} = \{g_1 : \cdots : g_{m+1} \mid t_1 : \cdots : t_{m+1}\}. \tag{42}$$

(Notice that the homogeneity in t is true by the very definition (39).)

ii) *The double shuffle relations.* $(p + q = m, \ p \geq 1, \ q \geq 1)$.

$$\sum_{\sigma \in \Sigma_{p,q}} \{g_{\sigma(1)} : \cdots : g_{\sigma(m)} : g_{m+1} \mid t_{\sigma(1)}, \ldots, t_{\sigma(m)}, t_{m+1}\} = 0, \tag{43}$$

$$\sum_{\sigma \in \Sigma_{p,q}} \{g_{\sigma(1)}, \ldots, g_{\sigma(m)}, g_{m+1} \mid t_{\sigma(1)} : \cdots : t_{\sigma(m)} : t_{m+1}\} = 0. \tag{44}$$

iii) *The distribution relations.* Let $l \in \mathbb{Z}$. Suppose that the l-torsion subgroup G_l of G is finite and its order is divisible by l. Then if $x_1, \ldots, x_m$ are l-powers

$$\{x_1 : \cdots : x_{m+1} \mid t_1 : \cdots : t_{m+1}\}$$

$$- \frac{1}{|G_l|} \sum_{y_i^l = x_i} \{y_1 : \cdots : y_{m+1} \mid l \cdot t_1 : \cdots : l \cdot t_{m+1}\} = 0$$

except the relation $I_1(e : e) = \sum_{y^l = e} I_1(y : e)$ which is not supposed to hold.

iv) $I_1(e : e) = 0$.

Denoted by $\widehat{\mathcal{D}}_{\bullet\bullet}(G)$ the bigraded space defined just as above except condition iv) is dropped, so $\widehat{\mathcal{D}}_{\bullet\bullet}(G) = \mathcal{D}_{\bullet\bullet}(G) \oplus \mathbb{Q}_{(1,1)}$ where $\mathbb{Q}_{(1,1)}$ is of bidegree $(1,1)$.

Theorem 4.1. (See Theorem 4.1 in [13].) *If $m \geq 2$ the double shuffle relations imply the dihedral symmetry relations, which include the cyclic symmetry*

$$\{g_1 : g_2 : \cdots : g_{m+1} \mid t_1 : t_2 : \cdots : t_{m+1}\} = \{g_2 : \cdots : g_{m+1} : g_1 \mid t_2 : \cdots : t_{m+1} : t_1\},$$

the reflection relation

$$\{g_1 : \cdots : g_{m+1} \mid t_1 : \cdots : t_{m+1}\} =$$
$$= (-1)^{m+1}\{g_{m+1} : \cdots : g_1 \mid -t_m : \cdots : -t_1 : -t_{m+1}\},$$

and the inversion relations

$$\{g_1 : \cdots : g_{m+1} \mid t_1 : \cdots : t_{m+1}\} = \{g_1^{-1} : \cdots : g_{m+1}^{-1} \mid -t_1 : \cdots : -t_{m+1}\}.$$

4.3. Pictures for the definitions

We think about generating series (39) as a function of $m + 1$ pairs $(g_1, t_1), \ldots,$ (g_{m+1}, t_{m+1}) located cyclically on an oriented circle as follows. The oriented circle has slots, where the g's sit, and in between the consecutive slots, dual slots, where t's sit:

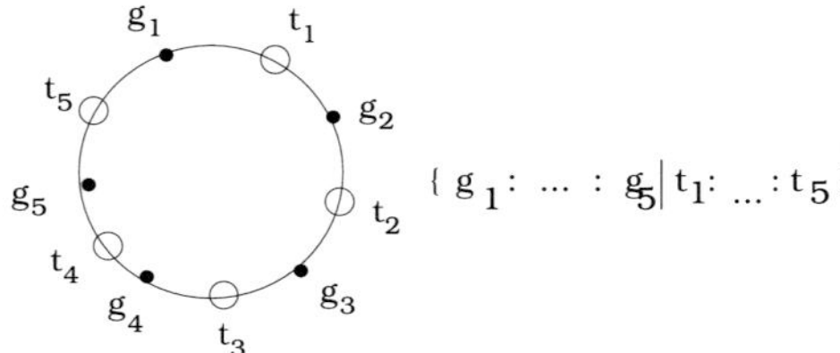

To make definitions (40) and (41) more transparent set $g_i' := g_i^{-1} g_{i+1}$, $t_i' := -t_{i-1} + t_i$ and put them on the circle together with g's and t's as follows:

Then it is easy to check that

$$\{g_1 : \cdots : g_{m+1} \mid t_1 : \cdots : t_{m+1}\} = \{g_1 : \cdots : g_{m+1} \mid t_1', \ldots, t_{m+1}'\}$$
$$= \{g_1', \ldots, g_{m+1}' \mid t_1 : \cdots : t_{m+1}\}. \qquad (45)$$

To picture any of three generating series (45) we leave on the circle only the two sets of variables among g's, g''s, t's, t''s which appear in this generating series:

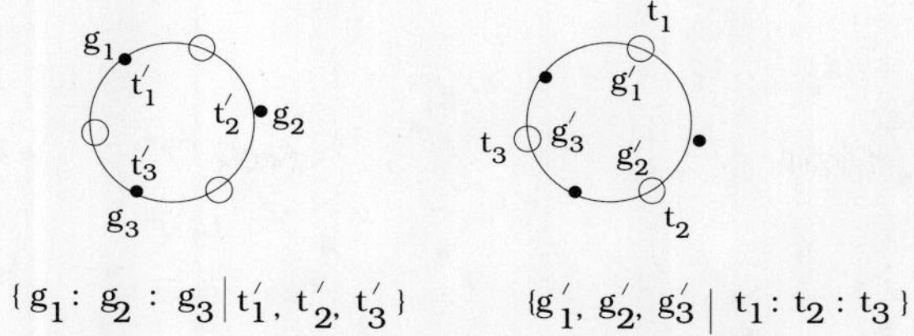

$$\{ g_1 : g_2 : g_3 \,|\, t_1', t_2', t_3' \} \qquad \{ g_1', g_2', g_3' \,|\, t_1 : t_2 : t_3 \}$$

The "$\{:\}$"-variables are outside, and the "$\{,\}$"-variables are inside of the circle.

4.4. Relation with multiple polylogarithms when $G = \mu_N$

Theorem 4.2. *(See* [12], [17]*) There is a well-defined homomorphism of the $\mathbb{Q}$-vector spaces*

$$\mathcal{D}_{w,m}(\mu_N) \longrightarrow \mathrm{Gr}_{w,m}^{W,D} \, \mathcal{P}\overline{\mathcal{Z}}(N)$$

defined on the generators by

$$I_{n_1,\dots,n_m}(a_1 : a_2 : \cdots : a_{m+1}) \longrightarrow (2\pi i)^{-w} \ integral \ (27). \qquad (46)$$

Here we consider iterated integrals (27) modulo the lower depth integrals and products of similar integrals. Formula (41) reflects Theorem 3.1. Formula (40) reflects definition (32) of the generating series I^*. The shuffle relations (43) (resp. (44)) correspond to relations (30) (resp. (33)).

Remark 4.3. *Notice an amazing symmetry between g's and t's in the generating series* (45), *completely unexpected from the point of view of iterated integrals* (27).

4.5. The cobracket $\delta\colon \mathcal{D}_{\bullet\bullet}(G) \longrightarrow \Lambda^2\mathcal{D}_{\bullet\bullet}(G)$

It will be defined by

$$\delta\{g_1 : \cdots : g_{m+1} \mid t_1 : \cdots : t_{m+1}\}$$

$$= -\sum_{k=2}^{m} \mathrm{Cycle}_{m+1}\left(\{g_1 : \cdots : g_{k-1} : g_k \mid t_1 : \cdots : t_{k-1} : t_{m+1}\}\wedge \right. \qquad (47)$$

$$\left. \wedge\{g_k : \cdots : g_{m+1} \mid t_k : \cdots : t_{m+1}\}\right)$$

where indices are modulo $m+1$ and $\mathrm{Cycle}_{m+1}f(v_1,\dots,v_m):=\sum_{i=1}^{m+1}f(v_i,\dots,v_{i+m})$.

Each term of the formula corresponds to the following procedure: choose a slot and a dual slot on the circle. Cut the circle at the chosen slot and dual slot and make two oriented circles with the data on each of them obtained from the initial data. It is useful to think about the slots and dual slots as of little arcs, not points, so cutting one of them we get the arcs on each of the two new circles marked by the corresponding letters. The formula reads as follows:

$$\delta(47) = -\sum_{\mathrm{cuts}} \ (\text{start at the dual slot}) \ \wedge \ (\text{start at the slot})$$

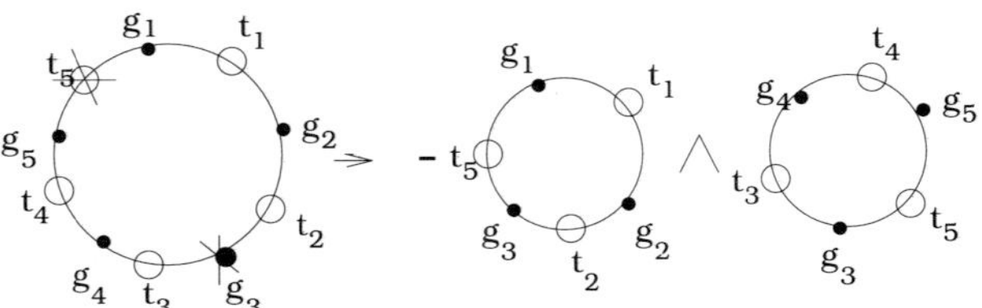

Theorem 4.4. *There exists a unique map* $\delta \colon \mathcal{D}_{\bullet\bullet}(G) \longrightarrow \Lambda^2 \mathcal{D}_{\bullet\bullet}(G)$ *for which* (47) *holds, providing a bigraded Lie coalgebra structure on* $\mathcal{D}_{\bullet\bullet}(G)$.

A similar result is true for $\widehat{\mathcal{D}}_{\bullet\bullet}(G)$. *Moreover there is an isomorphism of bigraded Lie algebras* $\widehat{\mathcal{D}}_{\bullet\bullet}(G) = \mathcal{D}_{\bullet\bullet}(G) \oplus \mathbb{Q}_{(1,1)}$.

5. The Dihedral Lie Algebra of μ_N and Galois Action on $\pi_1^{(l)}(X_N)$

5.1. Constraints on the image of the Galois group

Recall the homomorphism

$$\phi_N^{(l)} \colon \ \mathrm{Gal}(\overline{\mathbb{Q}}/\mathbb{Q}) \longrightarrow \mathrm{Aut}\,\mathbb{L}_N^{(l)}\,. \tag{48}$$

It has the following properties.

i) *The action of* μ_N. The group μ_N acts on X_N by $z \longmapsto \zeta_N z$. This action does not preserve the base vector v_∞. However one can define an action of μ_N on $\pi_1^{(l)}(X_N, v_\infty) \otimes \mathbb{Q}_l$, and hence on $\mathbb{L}^{(l)}(X_N)$, commuting with the action of the Galois group $\mathrm{Gal}_{\mathbb{Q}(\zeta_{l^\infty N})}$ [13].

ii) *"Canonical generator" at* ∞. Recall the projection

$$\mathbb{L}^{(l)}(X_N) \longrightarrow \mathbb{L}^{(l)}(\mathbb{G}_m) = \mathbb{Z}_l(1)\,. \tag{49}$$

Let X be a regular curve over $\overline{\mathbb{Q}}$, $\overline{X}$ the corresponding projective curve and v a tangent vector at $x \in \overline{X}$. Then there is a natural map of Galois modules ([4]):

$$\mathbb{Z}_l(1) = \pi_1^{(l)}(T_x\overline{X}\backslash 0, v) \longrightarrow \pi_1^{(l)}(X, v)\,.$$

For $X = \mathbb{G}_m$, $x = \infty$, $v = v_\infty$ it is an isomorphism. So for $X = X_N$ it provides a splitting of (49):

$$X_\infty \colon \mathbb{Z}_l(1) = \mathbb{L}^{(l)}(\mathbb{G}_m) \hookrightarrow \mathbb{L}^{(l)}(X_N)\,. \tag{50}$$

iii) *Special equivariant generators for* $\mathbb{L}_N^{(l)}$. For topological reasons there are well-defined conjugacy classes of "loops around 0 or $\zeta \in \mu_N$" in $\pi_1(X_N, v_\infty)$. It turns out that there are μ_N-equivariant representatives of these classes providing a set of the generators for the Lie algebra $\mathbb{L}_N^{(l)}$:

Lemma 5.1. *There exist maps* $X_0, X_\zeta \colon \mathbb{Q}_l(1) \longrightarrow \mathbb{L}_N^{(l)}$ *which belong to the conjugacy classes of the "loops around* $0, \zeta$*" such that* $X_0 + \sum_{\zeta \in \mu_N} X_\zeta + X_\infty = 0$ *and the action of* μ_N *permutes* X_ζ's *(i.e.* $\xi_* X_\zeta = X_{\xi\zeta}$) *and fixes* X_0, X_∞.

To incorporate these constraints we employ a more general setup.

5.2. The Galois action and special equivariant derivations

Let G be a commutative group written multiplicatively. Let $L(G)$ be the free Lie algebra with the generators X_i where $i \in \{0\} \cup G$ (we assume $0 \notin G$). Set $X_\infty := -X_0 - \sum_{g \in G} X_g$.

A derivation D of the Lie algebra $L(G)$ is called special if there are elements $S_i \in L(G)$ such that

$$D(X_i) = [S_i, X_i] \quad \text{for any } i \in \{0\} \cup G, \quad \text{and} \quad D(X_\infty) = 0. \tag{51}$$

The special derivations of $L(G)$ form a Lie algebra, denoted $\operatorname{Der}^S L(G)$. Indeed, if $D(X_i) = [S_i, X_i]$, $D'(X_i) = [S'_i, X_i]$, then

$$[D, D'](X_i) = [S''_i, X_i], \quad \text{where} \quad S''_i := D(S'_i) - D'(S_i) + [S'_i, S_i]. \tag{52}$$

The group G acts on the generators by $h \colon X_0 \longmapsto X_0$, $X_g \longmapsto X_{hg}$. So it acts by automorphisms of the Lie algebra $L(G)$. A derivation D of $L(G)$ is called equivariant if it commutes with the action of G. Let $\operatorname{Der}^{SE} L(G)$ be the Lie algebra of all special equivariant derivations of the Lie algebra $L(G)$.

Denote by $\mathbb{L}(\mu_N)$ the pronilpotent completion of the Lie algebra $L(\mu_N)$. Lemma 5.1 provides a (non-canonical) isomorphism $\mathbb{L}(\mu_N) \otimes \mathbb{Q}_l \longrightarrow \mathbb{L}_N^{(l)}$. Then it follows that $\mathcal{G}_N^{(l)}$ acts by special equivariant derivations of the Lie algebra $\mathbb{L}_N^{(l)}$, i.e.

$$\mathcal{G}_N^{(l)} \hookrightarrow \operatorname{Der}^{SE} \mathbb{L}_N^{(l)}. \tag{53}$$

5.3. Incorporating the two filtrations

The Lie algebra $L(G)$ is bigraded by the weight and depth. Namely, the free generators X_0, X_g are bihomogeneous: they are of weight -1, X_0 is of depth 0 and the X_g's are of depth -1. Each of the gradings induces a filtration of $L(G)$.

The Lie algebra $\operatorname{Der} L(G)$ is bigraded by the weight and depth. Its Lie subalgebras $\operatorname{Der}^S L(G)$ and $\operatorname{Der}^{SE} L(G)$ are compatible with the weight grading. However they are *not* compatible with the depth grading. Therefore they are graded by the weight, and *filtered* by the depth. A derivation (51) is of depth $-m$ if each $S_j \bmod X_j$ is of depth $-m$, i.e. there are at least m X_i's different from X_0 in $S_j \bmod X_j$. The depth filtration is compatible with the weight grading. Let $\operatorname{Gr} \operatorname{Der}^{SE}_{\bullet\bullet} L(G)$ be the associated graded for the depth filtration.

Consider the following linear algebra situation. Let $W_\bullet L$ be a filtration on a vector space L. A splitting $\varphi \colon \operatorname{Gr}^W L \longrightarrow L$ of the filtration leads to an isomorphism $\varphi^* \colon \operatorname{End}(L) \longrightarrow \operatorname{End}(\operatorname{Gr}^W L)$. The space $\operatorname{End}(L)$ inherits a natural filtration, while $\operatorname{End}(\operatorname{Gr}^W L)$ is graded. The map φ^* respects the corresponding filtrations. The map $\operatorname{Gr} \varphi^* \colon \operatorname{Gr}^W(\operatorname{End} L) \longrightarrow \operatorname{End}(\operatorname{Gr}^W L)$ does not depend on the choice of the splitting. Therefore if $L = \mathbb{L}_N$ we get a *canonical* isomorphism

$$\operatorname{Gr}^W(\operatorname{Der}^{SE} \mathbb{L}_N) \cong \operatorname{Der}^{SE} L(\mu_N) \tag{54}$$

respecting the weight grading. Thus there is a *canonical* injective morphism

$$\operatorname{Gr} \mathcal{G}^{(l)}_{\bullet\bullet}(\mu_N) \hookrightarrow \operatorname{Gr} \operatorname{Der}^{SE}_{\bullet\bullet} \mathbb{L}^{(l)}_N \cong \operatorname{Gr} \operatorname{Der}^{SE}_{\bullet\bullet} L(\mu_N) \otimes \mathbb{Q}_l. \tag{55}$$

The Lie algebra $\mathcal{G}^{(l)}_N$ is isomorphic to $\operatorname{Gr}^W_\bullet \mathcal{G}^{(l)}_N$, but this isomorphism is not canonical. The advantage of working with $\operatorname{Gr}^W_\bullet \mathcal{G}^{(l)}_N$ is that, via isomorphism (54), it became a Lie subalgebra of $\operatorname{Der}^{SE} L(\mu_N) \otimes \mathbb{Q}_l$, which has natural generators provided by the canonical generators of $L(\mu_N)$. This gives canonical "coordinates" for description of $\operatorname{Gr}^W_\bullet \mathcal{G}^{(l)}_N$. The benefit of taking its associated graded for the *depth* filtration is an unexpected relation with the geometry of modular varieties for GL_m, where m is the depth.

5.4. Cyclic words and special differentiations ([7, 23])

Denote by $A(G)$ the free associative algebra generated by elements X_i where $i \in \{0\} \cup G$. Let $\mathcal{C}(A(G)) := A(G)/[A(G), A(G)]$ be the space of cyclic words in X_i. Consider a map of linear spaces $\partial_{X_i} \colon \mathcal{C}(A(G)) \longrightarrow A(G)$ given on the generators by the following formula (the indices are modulo m):

$$\partial_{X_j} \mathcal{C}(X_{i_1} \cdots X_{i_m}) := \sum_{X_{i_k} = X_j} X_{i_{k+1}} \cdots X_{i_{k+m-1}}.$$

For example $\partial_{X_1} \mathcal{C}(X_1 X_2 X_1 X_2^2) = X_2 X_1 X_2^2 + X_2^2 X_1 X_2$.

Define special derivations just as in (51), but with $S_i \in A(G)$. There is a map

$$\kappa \colon \mathcal{C}(A(G)) \longrightarrow \operatorname{Der}^S A(G), \quad \kappa \mathcal{C}(X_{i_1} \cdots X_{i_m})(X_j) := [\partial_{X_j} \mathcal{C}(X_{i_1}, \ldots, X_{i_m}), X_j].$$

It is easy to check that it is indeed a special derivation. Denote by $\widetilde{\mathcal{C}}(A(G))$ the quotient of $\mathcal{C}(A(G))$ by the subspace generated by the monomials X_i^n. Then one can show that the map κ provides an isomorphism of vector spaces

$$\kappa \colon \widetilde{\mathcal{C}}(A(G)) \longrightarrow \operatorname{Der}^S A(G).$$

5.5. The dihedral Lie algebra as a Lie subalgebra of special equivariant derivations

Let G be a finite commutative group. We will use a notation Y for the generator X_0 of $A(G)$. So $\{Y, X_g\}$ are the generators of the algebra $A(G)$. Set

$$\mathcal{C}(X_{g_0} Y^{n_0-1} \cdot \ldots \cdot X_{g_m} Y^{n_m-1})^G := \sum_{h \in G} \mathcal{C}(X_{hg_0} Y^{n_0-1} \cdot \ldots \cdot X_{hg_m} Y^{n_m-1}).$$

Consider the following formal expression:

$$\xi_G := \sum \frac{1}{|\operatorname{Aut} \mathcal{C}|} I_{n_0,\ldots,n_m}(g_0 : \cdots : g_m) \otimes \mathcal{C}(X_{g_0} Y^{n_0-1} \cdot \ldots \cdot X_{g_m} Y^{n_m-1})^G \tag{56}$$

where the sum is over all G-orbits on the set of cyclic words $\mathcal{C}$ in X_g, Y. The weight $1/|\operatorname{Aut} \mathcal{C}|$ is the order of automorphism group of the cyclic word $\mathcal{C}$.

Applying the map $\operatorname{Id} \otimes \operatorname{Gr}(\kappa)$ we get a bidegree $(0,0)$ element

$$\xi_G \in \mathcal{D}_{\bullet\bullet}(G) \widehat{\otimes}_{\mathbb{Q}} \operatorname{Gr} \operatorname{Der}^{SE}_{\bullet\bullet} A(G).$$

Let $D_{-w,-m}(G) = \mathcal{D}_{w,m}(G)^{\vee}$. Then $D_{\bullet\bullet}(G) := \oplus_{w,m \geq 1} D_{-w,-m}(G)$ is a bigraded Lie algebra. Consider ξ_G as a map of bigraded spaces:

$$\xi_G \in \mathrm{Hom}_{\mathbb{Q}\text{-Vect}}(D_{\bullet\bullet}(G), \mathrm{Gr\,Der}_{\bullet\bullet}^{SE} A(G)). \tag{57}$$

Notice that $\mathrm{Gr\,Der}_{\bullet\bullet}^{SE} L(G)$ is a Lie subalgebra of $\mathrm{Gr\,Der}_{\bullet\bullet}^{SE} A(G)$.

Theorem 5.2. *The map ξ_G provides an injective Lie algebra morphism*

$$\xi_G \colon D_{\bullet\bullet}(G) \hookrightarrow \mathrm{Gr\,Der}_{\bullet\bullet}^{SE} L(G). \tag{58}$$

Theorem 5.3. $\mathrm{Gr}\,\mathcal{G}_{\bullet\bullet}^{(l)}(\mu_N) \hookrightarrow \xi_{\mu_N}(D_{\bullet\bullet}(\mu_N)) \otimes_{\mathbb{Q}} \mathbb{Q}_l$.

If G is a trivial group we set $D_{\bullet\bullet} := D_{\bullet\bullet}(\{e\})$, and $\xi := \xi_{\{e\}}$. Denote by $D_{\bullet}(G)$ the quotient of $D_{\bullet\bullet}(G)$ by the components with $w \neq m$.

Conjecture 5.4.
   a) *One has $\xi(D_{\bullet\bullet}) \otimes \mathbb{Q}_l = \mathrm{Gr}\,\mathcal{G}_{\bullet\bullet}^{(l)}$.*
   b) *Let p be a prime number. Then $\xi_{\mu_p}(D_{\bullet}(\mu_p)) = \mathrm{Gr}\,\mathcal{G}_{\bullet}^{(l)}(\mu_p)$.*

Summarizing we see the following picture: both Lie algebras $D_{\bullet\bullet}(\mu_N)$ and $\mathrm{Gr}\,\mathcal{G}_{\bullet\bullet}^{(l)}(\mu_N)$ are realized as Lie subalgebras of the Lie algebra of special equivariant derivations $\mathrm{Gr\,Der}_{\bullet\bullet}^{SE} L(G)$. The (image of) dihedral Lie algebra contains the (image of) Galois. Hypothetically they coincide when $N = 1$, or when weight = depth and N is prime. In general the gap between them exists, but should not be big.

Theorem 5.5. *Conjecture 5.4 is true for $m = 1, 2, 3$.*

The proof of this theorem is based on the following two ideas: the standard cochain complex of $D_{\bullet\bullet}(\mu_N)$ is related to the modular complex, and the modular complex has a geometric realization. We address them in the last two sections.

6. Modular Complexes and Galois Symmetries of $\pi_1^{(l)}(X_N)$

6.1. The modular complexes

Let L_m be a rank m lattice. The rank m modular complex $M^{\bullet}(L_m) = M_{(m)}^{\bullet}$ is a complex of $GL_m(\mathbb{Z})$-modules

$$M_{(m)}^1 \xrightarrow{\partial} M_{(m)}^2 \xrightarrow{\partial} \cdots \xrightarrow{\partial} M_{(m)}^m.$$

If $m = 2$ it is isomorphic to complex (19). In general it is defined as follows.

i) *The group $M_{(m)}^1$.* An extended basis of a lattice L_m is an $(m + 1)$-tuple of vectors $v_1, \ldots, v_{m+1}$ of the lattice such that $v_1 + \cdots + v_{m+1} = 0$ and $v_1, \ldots, v_m$ is a basis. The extended bases form a principal homogeneous space over $GL_m(\mathbb{Z})$.

The abelian group $M^1(L_m) = M_{(m)}^1$ is generated by extended bases. Denote by $\langle e_1, \ldots, e_{m+1} \rangle$ the generator corresponding to the extended basis $e_1, \ldots, e_{m+1}$.

Let $u_1, \ldots, u_{m+1}$ be elements of the lattice L_m such that the set of elements $\{(u_i, 1)\}$ form a basis of $L_m \oplus \mathbb{Z}$. The lattice L_m acts on such sets by

$l\colon \{(u_i, 1)\} \longmapsto \{(u_i + l, 1)\}$. We call the coinvariants of this action *homogeneous affine bases* of L_m and denote them by $\{u_1 : \cdots : u_{m+1}\}$.

To list the relations we need another set of the generators corresponding to the homogeneous affine bases of L_m (compare with (40)):

$$\langle u_1 : \cdots : u_{m+1}\rangle := \langle u_1', u_2', \ldots, u_{m+1}'\rangle; \qquad u_i' := u_{i+1} - u_i.$$

We will also employ the notation

$$[v_1, \ldots, v_k] := \langle v_1, \ldots, v_k, v_{k+1}\rangle, \quad v_1 + \cdots + v_k + v_{k+1} = 0.$$

Relations. One has $\langle v, -v\rangle = \langle -v, v\rangle$. For any $1 \le k \le m$, $m \ge 2$ one has (compare with (43)–(44)):

$$\sum_{\sigma \in \Sigma_{k, m-k}} \langle v_{\sigma(1)}, \ldots, v_{\sigma(m)}, v_{m+1}\rangle = 0, \tag{59}$$

$$\sum_{\sigma \in \Sigma_{k, m-k}} \langle u_{\sigma(1)} : \cdots : u_{\sigma(m)} : u_{m+1}\rangle = 0. \tag{60}$$

ii) *The group $M_{(m)}^k$.* It is the sum of the groups $M^1(L^1) \wedge \cdots \wedge M^1(L^k)$ over all unordered lattice decompositions $L_m = L^1 \oplus \cdots \oplus L^k$. Thus it is generated by the elements $[A_1] \wedge \cdots \wedge [A_k]$ where A_i is a basis of the sublattice L_i and $[A_i]$'s anticommute. Define a map $\partial\colon M_{(m)}^1 \longrightarrow M_{(m)}^2$ by setting (compare with (47))

$$\partial\colon \langle v_1, \ldots, v_{m+1}\rangle \longmapsto -\operatorname{Cycle}_{m+1}\left(\sum_{k=1}^{m-1} [v_1, \ldots, v_k] \wedge [v_{k+1}, \ldots, v_m]\right)$$

where indices are modulo $m + 1$. We get the differential in $M_{(m)}^\bullet$ by Leibniz' rule:

$$\partial([A_1] \wedge [A_2] \wedge \cdots) := \partial([A_1]) \wedge [A_2] \wedge \cdots - [A_1] \wedge \partial([A_2]) \wedge \cdots + \cdots$$

6.2. The modular complexes and the cochain complex of $D_{\bullet\bullet}(\mu_N)$

Denote by $\Lambda_{(m,w)}^* D_{\bullet\bullet}(\mu_N)$ the depth m, weight w part of the standard cochain complex of the Lie algebra $D_{\bullet\bullet}(\mu_N)$.

Theorem 6.1.

a) *For $m > 1$ there exists a canonical surjective map of complexes*

$$\mu_{m;w}^*\colon M_{(m)}^* \otimes_{\Gamma_1(m;N)} S^{w-m} V_m \longrightarrow \Lambda_{(m,w)}^* D_{\bullet\bullet}(\mu_N). \tag{61}$$

b) *Let $N = 1$, or $N = p$ be a prime and $w = m$. Then this map is an isomorphism.*

The map (61) was defined in [12]. Here is the definition when $w = m$. Notice that

$$M_{(m)}^* \otimes_{\Gamma_1(m;N)} \mathbb{Q} = M_{(m)}^* \otimes_{GL_m(\mathbb{Z})} \mathbb{Z}[\Gamma_1(m; N)\backslash GL_m(\mathbb{Z})].$$

The set $\Gamma_1(m; N)\backslash GL_m(\mathbb{Z})$ is identified with the set $\{(\alpha_1, \ldots, \alpha_m)\}$ of all nonzero vectors in the vector space over $\mathbb{Z}/p\mathbb{Z}$. Then

$$\mu_{m;m}^1\colon [v_1, \ldots, v_m] \otimes (\alpha_1, \ldots, \alpha_m) \longrightarrow I_{1,\ldots,1}(\zeta_N^{\alpha_1}, \ldots, \zeta_N^{\alpha_m}).$$

The other components $\mu_{m;m}^*$ are the wedge products of the maps $\mu_{k;k}^1$.

6.3. Modular complexes and cochain complexes of Galois Lie algebras

Combining Theorems 6.1a) and 5.3 we get a surjective map of complexes

$$M^*_{(m)} \otimes_{\Gamma_1(m;N)} S^{w-m} V_m \longrightarrow \Lambda^*_{(m,w)} \operatorname{Gr} \mathcal{G}^{(l)}_{\bullet\bullet}(\mu_N)^\vee . \tag{62}$$

So using Theorem 6.1b) we reformulate Conjecture 5.4 as follows:

Conjecture 6.2. *Let $N = 1$, or $N = p$ is a prime and $w = m$. Then the map (62) is an isomorphism.*

According to Theorem 5.3 the cochain complex of $D_{\bullet\bullet}(\mu_N)$ projects onto the cochain complex of the level N Galois Lie algebra. Combining this with Theorem 5.5 (and Conjecture 6.2) we describe the structure of the Galois Lie algebras via the modular complexes. Since the modular complexes are defined very explicitly this leads to a precise description of Galois Lie algebras. However to get the most interesting results about them we need the geometric realization of modular complexes, which allows us to express the structure of the Galois Lie algebra in terms of the *geometry and topology* of modular varieties.

7. Geometric Realization of Modular Complexes in Symmetric Spaces

As was emphasized before the rank m modular complex is a purely combinatorial object. Surprisingly it has a *canonical* realization in the symmetric space $\mathbb{H}_n$. In the simplest case $m = 2$ it identifies the rank two modular complex with the chain complex of the modular triangulation of the hyperbolic plane.

7.1. Voronoi's cell decomposition of the symmetric space for $GL_m(\mathbb{R})$

Let

$$\mathbb{H}_m := GL_m(\mathbb{R})/O(m) \cdot \mathbb{R}^* = \frac{> 0 \text{ definite quadratic forms on } V_m^*}{\mathbb{R}_+^*} .$$

Let $L_m \subset V_m$ be a lattice. Any vector $v \in V_m$ defines a nonnegative definite quadratic form $\varphi(v) := \langle v, \cdot \rangle^2$ on V_m^*. The convex hull of the forms $\varphi(l)$, when l runs through all non-zero primitive vectors of the lattice L_m, is an infinite polyhedra. Its projection into the closure of $\mathbb{H}_m$ defines a polyhedral decomposition of $\mathbb{H}_m$ invariant under the symmetry group of the lattice L_m.

Example 7.1. *If $m = 2$ we get the modular triangulation of the hyperbolic plane.*

Denote by $(V^{(m)}_\bullet, d)$ the chain complex of the Voronoi decomposition.

7.2. The relaxed modular complex

Consider a version $\widehat{M}^{\bullet}_{(m)}$ of the modular complex, called the relaxed modular complex, where the group $\widehat{M}^{1}_{(m)}$ is defined using the same generators $[v_1, \dots, v_m]$ satisfying only the first shuffle relations (59) and the dihedral symmetry relations

$$\langle v_2, \dots, v_{m+1}, v_1 \rangle = \langle v_1, \dots, v_{m+1} \rangle = (-1)^{m+1} \langle v_{m+1}, \dots, v_1 \rangle .$$

The other groups are defined in a similar way. The differential is as before.

7.3. The geometric realization map

Denote by $\varphi(v_1, \dots, v_k)$ the convex hull of the forms $\varphi(v_1), \dots, \varphi(v_k)$ in the space of quadratic forms. Let $v_1, \dots, v_{n_1}$ and $v_{n_1+1}, \dots, v_{n_1+n_2}$ be two sets of lattice vectors such that the lattices they generate have zero intersection. Define the join $*$ by

$$\varphi(v_1, \dots, v_{n_1}) * \varphi(v_{n_1+1}, \dots, v_{n_1+n_2}) := \varphi(v_1, \dots, v_{n_1+n_2})$$

and extend it by linearity. Make a homological complex $\widehat{M}^{(m)}_{\bullet}$ out of $\widehat{M}^{\bullet}_{(m)}$ by

$$\widehat{M}^{(m)}_{i} := \widehat{M}^{2m-1-i}_{(m)} .$$

Theorem 7.2. *There exists a canonical morphism of complexes*

$$\widehat{\psi}^{(m)}_{\bullet} : \widehat{M}^{(m)}_{\bullet} \longrightarrow V^{(m)}_{\bullet} \qquad such\ that$$

$$\widehat{\psi}^{(m)}_{\bullet} \left([A_1] \wedge \cdots \wedge [A_k] \right) := \widehat{\psi}^{(m)}_{\bullet}([A_1]) * \cdots * \widehat{\psi}^{(m)}_{\bullet}([A_k]) . \tag{63}$$

In particular $\widehat{\psi}^{(m)}_{m-1} \left([v_1] \wedge \cdots \wedge [v_m] \right) = \varphi(v_1, \dots, v_m)$.

To define such a morphism $\widehat{\psi}^{(m)}_{\bullet}$ one needs only to define $\widehat{\psi}^{(m)}_{2m-2}([v_1, \dots, v_m])$ for vectors $v_1, \dots, v_m$ forming a basis of the lattice L_m in such a way that the dihedral and the first shuffle relations go to zero and

$$d\widehat{\psi}^{(m)}_{2m-2}([v_1, \dots, v_m]) = \widehat{\psi}^{(m)}_{2m-3}(\partial[v_1, \dots, v_m])$$

where the right-hand side is computed by (63) and the formula for ∂.

7.4. Construction of the map $\widehat{\psi}^{(m)}_{2m-2}$

A *plane tree* is a tree without self intersections located on the plane. The edges of a tree consist of legs (external edges) and internal edges. Choose a lattice L_m. A *colored tree* is a plane tree whose legs are in a bijective correspondence with the elements of an affine basis of the lattice L_m. In particular a colored 3-valent tree has $2m - 1$ edges. We visualize it as follows:

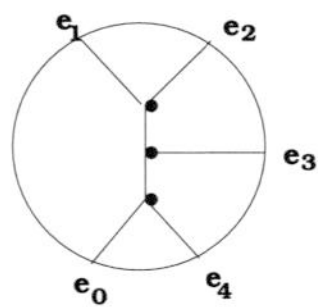

The vectors $e_0, \ldots, e_m$ of an affine basis are located cyclically on an oriented circle and the legs of the tree end on the circle and are labelled by $e_0, \ldots, e_m$. (The circle itself is not a part of the graph.)

Construction. Each edge E of the tree T provides a vector $f_E \in L_m$ defined up to a sign. Namely, the edge E determines two trees rooted at E, as in the picture.

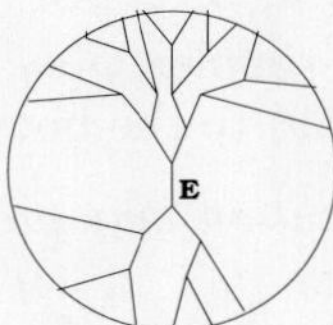

The union of the incoming (i.e. different from E) legs of these rooted trees coincides with the set of all legs of the initial tree. Take the sum of all the vectors e_i corresponding to the incoming legs of one of these trees. Denote it by f_E. If we choose the second rooted tree the sum will change the sign. So the degenerate quadratic form $\varphi(f_E)$ is well defined. Set

$$\widehat{\psi}^{(m)}_{2m-2}(\langle e_0, e_1, \ldots, e_m \rangle) :=$$

$$:= \sum_{\text{plane 3-valent trees}} \text{sgn}(E_1 \wedge \cdots \wedge E_{2m-1}) \cdot \varphi(f_{E_1}, \ldots, f_{E_{2m-1}}). \quad (64)$$

Here the sum is over all plane 3-valent trees colored by $e_0, \ldots, e_m$. The sign is defined as follows. Let $V(E)$ be the $\mathbb{R}$-vector space generated by the edges of a tree. An orientation of a tree is a choice of the connected component of $\det(V(E))\backslash 0$. A plane 3-valent tree has a canonical orientation. Indeed, the orientation of the plane provides orientations of links of each of the vertices. The sign in (64) is taken with respect to the canonical orientation of the plane 3-valent tree. Then one proves ([16]) that this map has all the required properties, so we get Theorem 7.2.

Examples 7.3.

a) *For $m = 2$ there is one plane 3-valent tree colored by e_0, e_1, e_2, so we get a modular triangle $\varphi(e_0, e_1, e_2)$ on the hyperbolic plane. The geometric realization in this case leads to an isomorphism of complexes $M_\bullet^{(2)} \longrightarrow V_\bullet^{(2)}$:*

$$[e_1, e_2] \longmapsto \varphi(e_0, e_1, e_2); \quad [e_1] \wedge [e_2] \longmapsto \varphi(e_1) * \varphi(e_2) = \varphi(e_1, e_2).$$

b) *Let $f_{ij} := e_i + e_j$. For $m = 3$ there are two plane 3-valent trees colored by e_0, e_1, e_2, e_3, see the picture, so the chain is*

$$\widehat{\psi}^{(3)}_4([e_1, e_2, e_3]) := \varphi(e_0, e_1, e_2, e_3, f_{01}) - \varphi(e_0, e_1, e_2, e_3, f_{12})$$

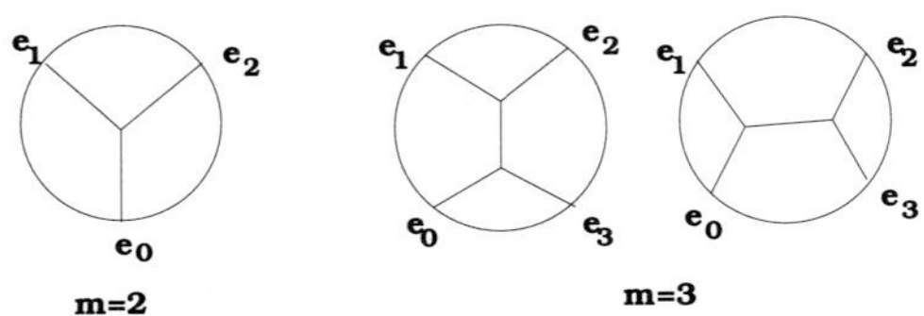

The symmetric space $\mathbb{H}_3$ has dimension 5. The Voronoi decomposition consists of the cells of dimensions 5, 4, 3, 2. All Voronoi cells of dimension 5 are $GL_3(\mathbb{Z})$-equivalent to the Voronoi simplex $\varphi(e_0, e_1, e_2, e_3, f_{01}, f_{12})$. The map $\widehat{\psi}_4^{(3)}$ sends the second shuffle relation (59) to the boundary of a Voronoi 5-simplex.

Theorem 7.4. *The geometric realization map provide quasiisomorphisms*

$$M_\bullet^{(3)} \longrightarrow \tau_{[4,2]}(V_\bullet^{(3)}); \qquad M_\bullet^{(4)} \longrightarrow \tau_{[6,3]}(V_\bullet^{(4)}).$$

7.5. Some corollaries

a) Let $N = p$ be a prime. Take the geometric realization of the rank 3 relaxed modular complex. Project it onto the modular variety $Y_1(3; p)$. Take the quotient of the group of the 4-chains generated by $\widehat{\psi}_4^{(3)}(v_1, v_2, v_3)$ on $Y_1(3; p)$ by the subgroup generated by the boundaries of the Voronoi 5-cells. Then the complex we get is *canonically isomorphic* to the weight = depth 3 part of the standard cochain complex of the level p Galois Lie algebra. Therefore

$$H_{(3)}^i(\mathrm{Gr}\,\widehat{\mathcal{G}}_\bullet^{(l)}(\mu_p)) = H^i(\Gamma_1(3; p)) \quad i = 1, 2, 3.$$

In particular we associate to each of the numbers $\mathrm{Li}_{1,1,1}(\zeta_p^{\alpha_1}, \zeta_p^{\alpha_2}, \zeta_p^{\alpha_3})$, or to the corresponding Hodge, l-adic or motivic avatars of these numbers, a certain 4-cell on the 5-dimensional orbifold $Y_1(3; p)$. The properties of the framed motive encoded by this number, like the coproduct, can be read from the geometry of this 4-cell.

Similarly the map $\widehat{\psi}_{2m-2}^{(m)}$ provides a canonical $(2m-2)$-cell on $Y_1(m; p)$ which "knows everything" about the framed motive with the period $\mathrm{Li}_{1,\dots,1}(\zeta_p^{\alpha_1}, \dots, \zeta_p^{\alpha_m})$.

b) $N = 1$. Theorems 5.5, 6.1a) and 7.4 lead to the following

Theorem 7.5.

$$\dim \mathrm{Gr}\,\mathcal{G}_{-w,-3}^{(l)} = \begin{cases} 0 & w\text{: even} \\ \left[\frac{(w-3)^2-1}{48}\right] & w\text{: odd} \end{cases}. \tag{65}$$

Since, according to standard conjectures, this number should coincide with $d_{w,3}$ the estimate given in Theorem 1.4 should be exact.

8. Multiple Elliptic Polylogarithms

The story above is related to the field $\mathbb{Q}$. I hope that for an arbitrary number field F there might be a similar story. The Galois group $\mathrm{Gal}(\overline{F}/F)$ should have a remarkable quotient Gal_F given by an extension of the maximal abelian quotient of $\mathrm{Gal}(\overline{F}/F)$ by a prounipotent group U_F:

$$0 \longrightarrow U_F \longrightarrow \mathrm{Gal}_F \longrightarrow \mathrm{Gal}(\overline{F}/F)^{\mathrm{ab}} \longrightarrow 0\,.$$

Its structure should be related to modular varieties for $GL_m/_F$, for all m.

The group $\mathrm{Gal}_{\mathbb{Q}}$ is obtained from the motivic fundamental group of $\mathbf{G}_m -$ {"all" roots of unity}. It turns out that for an imaginary quadratic filed K one can get a similar picture by taking the motivic fundamental group of the CM elliptic curve $E_K := \mathbb{C}/\mathcal{O}_K$ punctured at the torsion points. Below we construct the periods of the corresponding mixed motives, *multiple Hecke L-values*, as the values at torsion points of multiple elliptic polylogarithms. We define the multiple polylogarithms for arbitrary curves as correlators for certain Feynman integrals. We make sense out of these Feynman integrals by using the perturbation expansion via Feynman diagrams, which in this case are plane 3-valent trees. Unlike the Feynman integrals the coefficients of the perturbative expansion are given by convergent finite dimensional integrals, and so well defined. I leave to the reader the pleasure to penetrate the analogy between this construction and the geometric realization of modular complexes described in section 7.4.

8.1. The classical Eisenstein-Kronecker series

Let E be an elliptic curve over $\mathbb{C}$ with the period lattice Γ, so that $E(\mathbb{C}) = \mathbb{C}/\Gamma$. The intersection form $\Lambda^2\Gamma \longrightarrow 2\pi i\mathbb{Z}$ leads to the pairing $\chi\colon E(\mathbb{C}) \times \Gamma \longrightarrow S^1$. So for $a \in E(\mathbb{C})$ we get a character $\chi_a\colon \Gamma \longrightarrow S^1$.

Consider the generating function for the classical Eisenstein-Kronecker series

$$G(a \mid t) := \frac{\mathrm{vol}(\Gamma)}{\pi} \sideset{}{'}\sum_{\gamma \in \Gamma} \frac{\chi_a(\gamma)}{|\gamma - t|^2}$$

where $\sum'$ means the summation over all non zero vectors γ of the lattice. It depends on a point a of the elliptic curve and an element t in a formal neighborhood of zero in $H_1(E, \mathbb{R})$. It is invariant under the involution $a \longmapsto -a$, $t \longmapsto -t$. Expanding it into power series in t and $\overline{t}$ we get, as the coefficients, the classical Eisenstein-Kronecker series:

$$G(a \mid t) = \sum_{p,q \geq 1} \left(\frac{\mathrm{vol}(\Gamma)}{\pi} \sideset{}{'}\sum_{\gamma \in \Gamma} \frac{\chi_a(\gamma)}{\gamma^p \overline{\gamma}^q} \right) t^{p-1}\overline{t}^{q-1}\,.$$

When E is a CM curve their special values at the torsion points of E provide the special values of the Hecke L-series with Groessencharacters.

8.2. Multiple Eisenstein-Kronecker series: a description

We define them as the coefficients of certain generating functions. The generating function for the depth m multiple Eisenstein-Kronecker series is a function

$$G(a_1 : \cdots : a_{m+1} \mid t_1, \ldots, t_{m+1}), \quad t_1 + \cdots + t_{m+1} = 0$$

where a_i are points on the elliptic curve E and t_i are elements in a formal neighborhood of zero in $H_1(E, \mathbb{R})$. It is invariant under the shift $a_i \longrightarrow a_i + a$. Decomposing this function into the series in t_i, $\bar{t}_i$ we get the depth m multiple Eisenstein-Kronecker series.

Construction. Consider a plane trivalent tree T colored by $m + 1$ pairs consisting of points a_i on the elliptic curve E and formal elements $t_i \in H_1(E, \mathbb{R})$:

$$(a_1, t_1), \ldots, (a_{m+1}, t_{m+1}); \qquad t_1 + \cdots + t_{m+1} = 0 \tag{66}$$

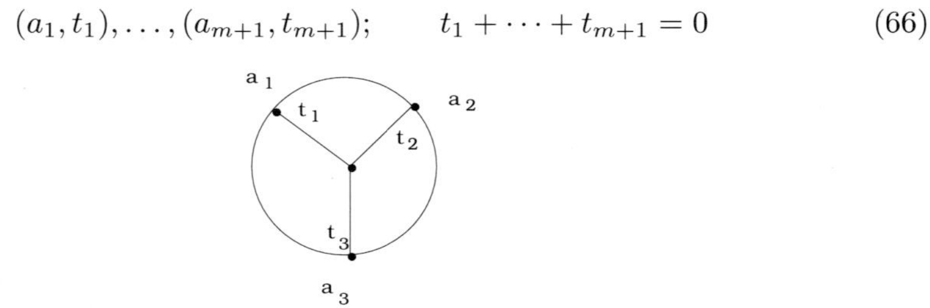

Each oriented edge $\vec{E}$ of the tree T provides an element $t_{\vec{E}} \in H_1(E, \mathbb{R})$. Namely, as explained in section 7.4 the edge E determines two trees rooted at E. An orientation of the edge E corresponds to the choice of one of them: take the tree obtained by going in the direction of the orientation of the edge. Then $t_{\vec{E}}$ is the sum of all t_i's corresponding to the legs of this tree different from E. Since $\sum t_i = 0$ changing the orientation of the edge E we get $-t_{\vec{E}}$.

Let X be a manifold and $\mathcal{A}^i(X)$ the space of i-forms on X. We define a map

$$\omega_m \colon \Lambda^{m+1} \mathcal{A}^0(X) \longrightarrow \mathcal{A}^m(X),$$

$$\omega_m \colon \varphi_1 \wedge \cdots \wedge \varphi_{m+1} \longmapsto$$

$$\longmapsto \frac{1}{(m+1)!} \operatorname{Alt}_{m+1} \left(\sum_{j=0}^{m+1} (-1)^j \varphi_1 \partial \varphi_2 \wedge \cdots \wedge \partial \varphi_k \wedge \bar{\partial} \varphi_{k+1} \wedge \cdots \wedge \bar{\partial} \varphi_{m+1} \right).$$

If $\varphi_i = \log |f_i|$ it is the form used to define the Chow polylogarithms in [14].

Every edge E of the tree T defines a function G_E on $E(\mathbb{C})^{\{\text{vertices of } T\}}$ depending on t_i. Namely, let v_1^E, v_2^E be the vertices of the edge E. Their order orients the edge E. Consider the natural projection

$$p_E \colon E(\mathbb{C})^{\{\text{vertices of the tree } T\}} \longrightarrow E(\mathbb{C})^{\{v_1^E, v_2^E\}} = E(\mathbb{C}) \times E(\mathbb{C}). \tag{67}$$

Then $G_E := p_E^* G \left(x_1^E - x_2^E \mid t_{\vec{E}} \right)$ where (x_1^E, x_2^E) is a point at the right of (67). This function does not depend on the orientation of the edge E.

Definition 8.1.

$$G(a_1 : \cdots : a_{m+1} \mid t_1, \ldots, t_{m+1}) :=$$

$$:= \sum_{\text{plane 3-valent trees } T} \operatorname{sgn}(E_1 \wedge \cdots \wedge E_{2m-1}) \cdot$$

$$\cdot \int_{S^{m-1}E(\mathbb{C})} \operatorname{sym}^* \omega_{2m-2} \left(G_{E_1} \wedge \cdots \wedge G_{E_{2m-1}} \right). \quad (68)$$

The sum is over all plane 3-valent trees whose legs are cyclically labelled by (66). The correspondence sym: $S^{m-1}E(\mathbb{C}) \longrightarrow E(\mathbb{C})^{\{\text{internal vertices of } T\}}$ is given by the sum of all $(m-1)!$ natural maps $E(\mathbb{C})^{m-1} \longrightarrow E(\mathbb{C})^{\{\text{internal vertices of } T\}}$.

Recall the CM elliptic curve E_K. Let $\mathcal{N}$ be an ideal of $\operatorname{End}(E_K)$. Denote by $K_{\mathcal{N}}$ the field generated by the $\mathcal{N}$-torsion points of E_K. If a_i are $\mathcal{N}$-torsion points of E_K we view the numbers obtained in the $t, \bar{t}$-expansion (68) as multiple Hecke L-values related to K. They are periods of mixed motives over the ring of integers in $K_{\mathcal{N}}$, with $\operatorname{Norm}(\mathcal{N})$ inverted. These are the motives which appear in the motivic fundamental group of $E - \{\mathcal{N} - \text{torsion points}\}$.

9. Multiple Polylogarithms on Curves, Feynman Integrals and Special Values of L-Functions

9.1. Polylogarithms on curves and special values of L-functions

Let X be a regular complex algebraic curve. We assume for simplicity only, that it is a projective curve of genus $g \leq 1$. Choose a volume form on $X(\mathbb{C})$, and let $G(x, y)$ be the corresponding Green function. Set

$$\mathcal{H} := H_1(X, \mathbb{R}); \qquad \mathcal{H}_{\mathbb{C}} := \mathcal{H} \otimes \mathbb{C} = \mathcal{H}^{-1,0} \oplus \mathcal{H}^{0,-1}.$$

For each integer $n \geq 1$ we define a 0-current $G_n(x, y)$ on $X \times X$ with values in

$$\operatorname{Sym}^{n-1} \mathcal{H}_{\mathbb{C}}(1) = \oplus_{a+b=n-2} S^{a-1}\mathcal{H}^{-1,0} \otimes S^{b-1}\mathcal{H}^{0,-1}. \quad (69)$$

Then $G_1(x, y) := G(x, y)$. For $n > 1$ it is a function on $X(\mathbb{C}) \times X(\mathbb{C})$.

To define the function $G_n(x, y)$ we proceed as follows. Let

$$\overline{\Omega}_a = \omega_{\alpha_1} \cdot \ldots \cdot \omega_{\alpha_{a-1}} \in S^{a-1}\overline{\Omega}^1, \quad \Omega_b = \omega_{\beta_1} \cdot \ldots \cdot \omega_{\beta_{b-1}} \in S^{b-1}\Omega^1; \qquad \omega_* \in \Omega^1.$$

Then $\overline{\Omega}_a \otimes \Omega_b$ is an element of the dual to (69). We are going to define the pairing $\langle G_n(x, y), \overline{\Omega}_a \otimes \Omega_b \rangle$. Denote by $p_i \colon X^{n-1} \longrightarrow X$ the projection on i-th factor.

Definition 9.1. *The n-th polylogarithm function on the curve X is defined by*

$$\langle G_n(x,y), \overline{\Omega}_a \otimes \Omega_b \rangle := \operatorname{Alt}_{\{z_1,\ldots,z_{n-1}\}}$$

$$\times \left(\int_{X^{n-1}(\mathbb{C})} \omega_{n-1} \Big(G(x,z_1) \wedge G(z_1,z_2) \wedge \cdots \wedge G(z_{n-2}-z_{n-1}) \wedge G(z_{n-1},y) \Big) \wedge \right.$$

$$\left. \wedge \Lambda_{i=1}^{a-1} p_i^* \overline{\omega}_{\alpha_i} \wedge \Lambda_{j=1}^{b-1} p_{a-1+j}^* \omega_{\beta_j} \right).$$

We skewsymmetrized the integrand with respect to $z_1, \ldots, z_{n-1}$.

These functions provide a variation of $\mathbb{R}$-mixed Hodge structures on $X \times X - \Delta$ of motivic origin. If X is an elliptic curve it is given by Beilinson-Levin theory of elliptic polylogarithms [1].

In particular for a pair of distinct points x, y on X we get an $S^{n-1}\mathcal{H}(1)$-framed mixed motive (see [14] for the background) denoted $\{x,y\}_n$, whose period is given by $G_n(x,y)$. Its coproduct δ is given by $\{x,y\}_n \longmapsto \{x,y\}_{n-1} \wedge (x-y)$ where $(x-y)$ is the point of the Jacobian of X corresponding to the divisor $\{x\} - \{y\}$. If X is defined over a number field this leads to a very precise conjecture expressing the special value $L(S^{n-1}H^1(X), n)$ via the polylogarithms $G_n(x,y)$ —an analog of Zagier's conjecture. If X is an elliptic curve we are in the situation considered in [14, 30]. An especially interesting example appears when x, y are cusps on a modular curve. Then $(x-y)$ is a torsion point in the Jacobian, so $\delta\{x,y\}_n = 0$ and thus $G_n(x,y)$ is the regulator of an element of motivic $\operatorname{Ext}^1(\mathbb{Q}(0), S^{n-1}\mathcal{H}(1))$!

In particular we can apply this construction in the case when $X = \mathbb{G}_m$, $G(x,y) = \log|x-y|$ and $\omega = d\log(z)$. Then it boils down to the Chow polylogarithm [19] corresponding to the element

$$(x-z_1) \wedge (z_1-z_2) \wedge \cdots \wedge (z_{n-2}-z_{n-1}) \wedge (z_{n-1}-y) \wedge z_1 \wedge \cdots \wedge z_{n-1} \in \Lambda^{2n-1}\mathcal{O}^*(\mathbb{G}_m^{n-1})$$

9.2. Multiple polylogarithms on curves

We package the polylogarithms $G_n(x,y)$ into the generating series

$$G(x,y \mid t_1, t_2), \quad t_i \in \mathcal{H}, \; t_1 + t_2 = 0$$

so that $G_n(x,y)$ emerges as the weight $-n-1$ component of the power series decomposition into t_1, $\overline{t}_1$. Then $G(x,y \mid t_1,t_2) = G(y,x \mid t_2,t_1)$. The construction of the previous section provides multiple polylogarithms $G(a_1, \ldots, a_{m+1} \mid t_1, \ldots, t_{m+1})$ on X, where $t_1 + \cdots + t_{m+1} = 0$. Indeed, for an edge E of a plane 3-valent tree T set $G_E := p_E^* G(x_1^E, x_2^E \mid t_{\overrightarrow{E}}, -t_{\overrightarrow{E}})$ and repeat the construction. We call the constant term in t's the multiple Green function on X.

Applying this construction in the case when $X = \mathbb{G}_m$ we get a single-valued version of mutiple polylogarithms written as Chow polylogarithms. One can also interpret them as Grassmannian polylogarithms ([19]) on certain strata.

9.3. Feynman integral for multiple Green functions

Let φ be a function and ψ a $(1,0)$-form on $X(\mathbb{C})$ with values in $N \times N$ complex skew symmetric matrices. We denote by $\overline{\varphi}$ and $\overline{\psi}$ the result of the action of complex conjugation. Then the multiple Green function $G(a_1, \ldots, a_{m+1})$ emerges as the leading term of the asymptotic when $N \to \infty$ of the following correlator:

$$\int \mathrm{Tr}\Big((\varphi + \overline{\varphi})(a_1) \cdot \ldots \cdot (\varphi + \overline{\varphi})(a_{m+1}) \Big) e^{iS(\varphi,\psi)} \mathcal{D}\varphi \mathcal{D}\psi$$

where

$$S(\varphi, \psi) := \int_{X(\mathbb{C})} \mathrm{Tr}\Big(\varphi \overline{\partial} \psi + \overline{\varphi} \partial \overline{\psi} + \psi \overline{\psi} + \varphi[\psi, \overline{\psi}] + \overline{\varphi}[\overline{\psi}, \psi] \Big).$$

I conjecture that the special values $L(S^n H^1(X), n + m)$ can be expressed via the depth m multiple polylogarithms on X. So Feynman integrals provide construction of (periods of) mixed motives, which are in particular responsible for special values of L-functions. I hope this reflects a very general phenomena.

Acknowledgment

This work was partially supported by the NSF grant DMS-9800998.

References

[1] A. A. Beilinson and P. Deligne, *Motivic polylogarithms and Zagier's conjecture*, Unfinished manuscript.

[2] A. A. Beilinson and A. M. Levin, *The elliptic polylogarithms*, Proc. Symp. in Pure Math., vol. 55, (1994), part 2, 126–129.

[3] D. J. Broadhurst, *On the enumeration of irreducible k-fold sums and their role in knot theory and field theory*, Preprint hep-th/9604128.

[4] P. Deligne, *Le group fondamental de la droite projective moins trois points*, In: Galois groups over $\mathbb{Q}$. Publ. MSRI, no. 16 (1989) 79–298.

[5] P. Deligne, *A letter to D. Broadhurst*, June 1997.

[6] P. Deligne, *Letter to the author*, July 2000.

[7] V. G. Drinfeld, *On quasi-triangular quasi-Hopf algebras and some group related to closely associated with* $\mathrm{Gal}(\overline{\mathbb{Q}}/\mathbb{Q})$, Leningrad Math. Journal, 1991. (In Russian).

[8] L. Euler, "Opera Omnia," Ser. 1, Vol XV, Teubner, Berlin 1917, 217–267.

[9] A. B. Goncharov, *Multiple ζ-numbers, hyperlogarithms and mixed Tate motives*, Preprint MSRI 058–93, June 1993.

[10] A. B. Goncharov, *Polylogarithms in arithmetic and geometry*, Proc. ICM-94, Zurich, 374–387.

[11] A. B. Goncharov, *The double logarithm and Manin's complex for modular curves*, Math. Res. Letters, vol. 4. N 5 (1997), pp. 617–636.

[12] A. B. Goncharov, *Multiple polylogarithms, cyclotomy and modular complexes*, Math. Res. Letters, vol. 5. (1998), pp. 497–516.

[13] A. B. Goncharov, *The dihedral Lie algebras and Galois symmetries of* $\pi_1^{(l)}(\mathbb{P}^1 \setminus 0, \mu_N, \infty)$, Preprint MPI–1998–131 (1998); To appear in Duke Math. J. Math. AG 0009121 (2001).

[14] A. B. Goncharov, *Mixed elliptic motives*, in London Math. Soc. Lect. Note Series, 254, Cambridge Univ. Press, Cambridge, 1998, 147–221.

[15] A. B. Goncharov, *Volumes of hyperbolic manifolds and mixed Tate motives*, J. Amer. Math. Soc. 12 (1999) N 2, 569–618.

[16] A. B. Goncharov, *Galois groups, geometry of modular varieties and graphs*, Arbeitstagung, June 1999, Preprint MPI 1999–50-f (http//www.mpim-bonn.mpg.de/).

[17] A. B. Goncharov, *Multiple polylogarithms and mixed Tate motives*. Math. AG 0103059, and *Multiple polylogarithms and mixed Tate motives II* (to appear).

[18] A. Grothendieck, *Esquisse d'un programme*, Mimeographed note (1984).

[19] R. Hain and M. Matsumoto, *Weighted completion of Galois groups and some conjectures of Deligne*, Preprint June 2000.

[20] Y. Ihara, *Profinite braid groups, Galois representations and complex multiplications*, Ann. Math. 123 (1986) 43–106.

[21] Y. Ihara, *Braids, Galois groups, and some arithmetic functions*, Proc. ICM-90, Kyoto, (1990).

[22] Y. Ihara, *Some arithmetical aspects of Galois action on the pro-p fundamental group of* $\widehat{\pi}_1(\mathbb{P}^1 \backslash \{0, 1, \infty\})$, Preprint RIMS-1229, 1999.

[23] M. Kontsevich, *Formal (non)commutative symplectic geometry*, The Gelfand mathematical seminars, Birkhäuser, 1993, pp. 173–187.

[24] M. Kontsevich, *Operads and motives in deformation quantization*, Lett. Math. Phys., 48 (1999) N 1, 35–72.

[25] D. Kreimer, *Renormalization and knot theory*, J. Knot Theory Ramifications, 6, (1997) N 4, 479–581.

[26] A. Levin, *Kronecker double series and the dilogarithm*, Preprint MPI 2000–35 (2000).

[27] M. Levine, *Tate motives and the vanishing conjectures for algebraic K-theory*, In Algebraic K-theory and Algebraic topology, 167–188, NATO Adv. Sci. Inst. Ser. C Math. Phys. Sci., 407, Kluver, (1993).

[28] M. Levine, *Mixed motives*, Mathematical Surveys and Monographs, 57, AMS, Providence, RI, 1998.

[29] V. Voevodsky, *Triangulated category of motives over a field*, Cycles, transfers, and motivic homology theories, 188–238. Ann. of Math. Stud. 143, Princeton Univ. Press, Princeton, NJ, 2000.

[30] J. Wildeshaus, *On an elliptic analogue of Zagier's conjecture* 87 (1997), 355–407.

[31] D. Zagier, *Values of zeta functions and their applications*, Proc. ECM-92, vol. 2, 497–512, In Progr. Math., 120, Birkhäuser, Basel, 1994.

[32] D. Zagier, *Periods of modular forms, traces of Hecke operators, and multiple ζ-values*, Kokyuroku No. 843 (1993), 162–170. (In Japanese).

Department of Mathematics
Brown University
Providence, RI 02912, USA
E-mail address: sasha@math.brown.edu

Heat Kernels on Manifolds, Graphs and Fractals

Alexander Grigor'yan

Abstract. We consider heat kernels on different spaces such as Riemannian manifolds, graphs, and abstract metric measure spaces including fractals. The talk is an overview of the relationships between the heat kernel upper and lower bounds and the geometric properties of the underlying space. As an application, some estimate of higher eigenvalues of the Dirichlet problem is considered.

1. Introduction

This is a brief survey of heat kernel long-time estimates on various underlying spaces such as Riemannian manifolds, graphs and fractals. We have selected generic results which, when properly modified, remain true in all cases, although each particular result is presented here in one of the settings which is most convenient. In Section 2 we give necessary definitions. In Sections 3 and 4, we consider heat kernel on-diagonal estimates in relation with the first eigenvalue estimates. In Sections 5 and 6 we consider off-diagonal upper and lower estimates, of Gaussian and sub-Gaussian types, respectively. In Section 7, we discuss relations to isoperimetric properties of higher eigenvalues.

For simplicity, we restrict our consideration to *uniform* estimates of the heat kernel. For further results and for a detailed account of various related aspects of heat kernels and eigenvalues, we refer the reader to books and surveys [2, 6, 7, 9, 10, 13, 21, 29, 30, 31, 33, 34, 36, 40, 41].

Throughout the paper, C and c normally denote large and small positive constants, respectively, which may be different in different occurrences. A relation $f \simeq g$ means that the ratio of the functions f and g remains bounded between two positive constants for a specified range of their arguments.

2. The Notion of Heat Kernel

2.1. Manifolds

Let M be a smooth connected Riemannian manifold, and let $d(x, y)$ be a geodesic distance on M. Assume that a Borel measure μ is defined on M, which has a smooth density m with respect to a Riemannian measure (in particular, μ may be the Riemannian measure if $m = 1$). The couple (M, μ) is called *a weighted manifold*.

A natural Laplace operator Δ_μ is associated with (M, μ), namely,

$$\Delta_\mu = m^{-1}\text{div}\,(m\nabla)$$

where ∇ and div are the Riemannian gradient and divergence, respectively. An energy form of Δ_μ is given by

$$\mathcal{E}_\mu(f) = \int_M |\nabla f|^2\, d\mu\,.$$

It is well known that the heat equation

$$\frac{\partial u}{\partial t} = \Delta_\mu u\,,$$

(where $u(t, x)$ is defined for $t > 0$ and $x \in M$) has *a heat kernel*, which is denoted by $p_t(x, y)$ and can be defined in various alternative ways. Here are some equivalent definitions:

1. For any $y \in M$, the function $(t, x) \mapsto p_t(x, y)$ is the smallest positive fundamental solution to the heat equation with a source at y.
2. The function $p_t(x, y)$ is an integral kernel of *the heat semigroup* $P_t := e^{t\Delta_\mu}$ that is defined by using the spectral theorem (indeed, the operator Δ_μ with domain $C_0^\infty(M)$ is essentially self-adjoint in $L^2(M, \mu)$ and negative definite).
3. The function $p_t(x, y)$ is the transition density of *the Brownian motion* X_t on (M, μ), which is by definition a diffusion process generated by Δ_μ.

Here are some examples of exact heat kernels.

If $(M, \mu) = \mathbb{R}^N$ then the heat kernel is given by the Gauss-Weierstrass formula

$$p_t(x, y) = \frac{1}{(4\pi t)^{N/2}}\exp\left(-\frac{d^2(x, y)}{4t}\right)\,.$$

If $(M, \mu) = \mathbb{H}^3$ — the 3-dimensional hyperbolic space of constant negative curvature -1, then

$$p_t(x, y) = \frac{1}{(4\pi t)^{3/2}}\exp\left(-t - \frac{d(x, y)^2}{4t}\right)\frac{d(x, y)}{\sinh d(x, y)}\,.$$

If M is a compact manifold then the operator Δ_μ^{-1} is compact. Let $\{\varphi_k\}_{k=0}^\infty$ be an orthonormal basis in $L^2(M, \mu)$ of eigenfunctions of $-\Delta_\mu$ with eigenvalues

$$0 = \lambda_0 < \lambda_1 \le \lambda_2 \le \ldots$$

(here $\varphi_0 = \text{const}$). Then the heat kernel on (M, μ) is determined by

$$p_t(x, y) = \sum_{k=0}^\infty e^{-\lambda_k t}\varphi_k(x)\varphi_k(y)\,.$$

2.2. Graphs

Consider now a discrete version of the heat kernel. Let Γ be a connected graph and let $d(x, y)$ be a combinatorial distance on Γ, that is, $d(x, y)$ is the smallest number of edges in a path connecting $x, y \in \Gamma$.

Let μ_{xy} be a weight on edges of Γ. More precisely, if vertices $x, y \in \Gamma$ are connected by an edge then we write $x \sim y$, denote the edge by $\overline{xy}$ and assign to it a positive number μ_{xy}. In particular, it may happen that $\mu_{xy} = 1$ for all edges, in which case we say that μ_{xy} is *a standard weight*. It is convenient to extend μ_{xy} by zero to those x, y which are not neighbors.

Any weight μ_{xy} gives rise to a function on vertices by

$$\mu(x) = \sum_{y \sim x} \mu_{xy}$$

and then to a measure on all finite sets $\Omega \subset \Gamma$ by

$$\mu(\Omega) = \sum_{x \in \Omega} \mu(x).$$

The couple (Γ, μ) is called a weighted graph (here μ refers both to the weight μ_{xy} and to the measure). There is a natural Laplace operator Δ_μ on (Γ, μ) which acts on functions on Γ by

$$\Delta_\mu f(x) = \frac{1}{\mu(x)} \sum_{y \in \Gamma} (\nabla_{xy} f) \, \mu_{xy}$$

where

$$\nabla_{xy} f = f(y) - f(x).$$

It is easy to verify that Δ_μ is a bounded self-adjoint operator in $L^2(\Gamma, \mu)$. Its energy form is given by

$$\mathcal{E}_\mu(f) = \sum_{x, y \in \Gamma} |\nabla_{xy} f|^2 \, \mu_{xy}.$$

The heat kernel $p_n(x, y)$ is defined for non-negative integers n and for all $x, y \in \Gamma$ as a kernel with respect to μ of the operator $(I + \Delta)^n$. An operator $P = I + \Delta$ acts by

$$Pf(x) = \frac{1}{\mu(x)} \sum_{y \in \Gamma} f(y) \mu_{xy} = \sum_{y \in \Gamma} P(x, y) f(y)$$

where

$$P(x, y) = \frac{\mu_{xy}}{\mu(x)}.$$

The operator P is Markov and defines a nearest neighborhood random walk X_n on Γ by the rule

$$\mathbb{P}\left(X_{n+1} = y \mid X_n = x\right) = P(x, y).$$

Denote by $P_n(x, y)$ the transition function of X_n, that is

$$P_n(x, y) = \mathbb{P}\left(X_n = y \mid X_0 = x\right).$$

Then the heat kernel p_n is a density of P_n with respect to measure μ:

$$p_n(x, y) = \frac{P_n(x, y)}{\mu(y)}.$$

There are practically no explicit formulas for heat kernels on graphs, even in simple situations. If $\Gamma = \mathbb{Z}^N$ and μ is a standard weight then $p_n(x, y)$ admits the following Gaussian estimates[1]

$$p_n(x, y) \simeq \frac{1}{n^{N/2}} \exp\left(-\frac{d^2(x, y)}{cn}\right) \tag{1}$$

provided

$$n \geq d(x, y) \quad \text{and} \quad n \equiv d(x, y) \bmod 2.$$

If $n < d(x, y)$ then always $p_n(x, y) = 0$. If n and $d(x, y)$ have different parities then $p_n(x, y) = 0$ for all bipartite graphs, in particular for $\mathbb{Z}^N$.

2.3. Metric-measure-energy spaces

Let (M, d) be a locally compact separable metric space. Suppose that a Radon measure μ with full support is defined on M. Assume also that (M, μ) admits an energy form $\mathcal{E}$ which is a regular Dirichlet form on $L^2(M, \mu)$.

In general, any energy form $\mathcal{E}$ has a generator $-\Delta$ which is a self-adjoint operator on $L^2(M, \mu)$ with a dense domain, so that

$$\mathcal{E}(f) = -\int_M f \Delta f \, d\mu,$$

for all $f \in dom(\Delta)$ (see for example [14, Theorem 4.4.2], [16]). It is natural to say that Δ is the Laplace operator of the space $(M, \mu, \mathcal{E})$.

The heat semigroup P_t is defined as a one-parameter family of operators $\left\{e^{t\Delta}\right\}_{t \geq 0}$ in $L^2(M, \mu)$. If P_t has an integral kernel $p_t(x, y)$ with respect to measure μ then it is called the heat kernel of Δ (or $\mathcal{E}$). There are various conditions for existence of a heat kernel and its continuity. With $\mathcal{E}$ one associates a Hunt process X_t on M with generator Δ and transition density $p_t(x, y)$.

Weighted manifolds are simple examples of metric-measure-energy spaces. Graphs fit also apart from the fact that on graphs it is more natural to consider the heat kernel with discrete time, which arises from the semi-group $(I + \Delta)^n$ rather than $e^{t\Delta}$, although the latter can also be considered.

An example of different nature arises from fractal sets. Let M be a fractal set in $\mathbb{R}^N$ such as a Sierpinski gasket or a Sierpinski carpet (see [2, 35]). A distance d on M is inherited from the ambient space. A measure μ is a Hausdorff measure of a proper dimension α, namely, α is just the Hausdorff dimension of M. Definition

---

[1]The constant c in the exponential term of (1) may be different for the upper and lower bounds. This remark applies to all Gaussian estimates to be considered below.

of an energy form is highly non-trivial. This is done by using the self-similarity structure of M and a limiting procedure, and we refer the reader to [2]. In any case, $\mathcal{E}$ can be defined and, moreover, the corresponding heat kernel $p_t(x,y)$ is jointly continuous in $x, y \in M$ and $t > 0$.

The heat kernel estimate which is described below is due to Barlow-Perkins [4] for a Sierpinski gasket and Barlow-Bass [3] for a Sierpinski carpet. This estimate requires two parameters which describe the geometry of the underlying space. The first one is α which is the Hausdorff dimension of M, and the second one is "a walk dimension" which we denote by β. It can be defined, for example, as follows. Denote by $B(x,r)$ a metric ball

$$B(x,r) = \{y \in M : d(x,y) < r\} \tag{2}$$

and by $T_{x,r}$ the first exit time from the ball $B(x,r)$, that is

$$T_{x,r} = \inf\{t \geq 0 : X_t \notin B(x,r)\} . \tag{3}$$

Then β is defined by the relation

$$\mathbb{E}_x T_{x,r} \simeq r^\beta ,$$

for all $x \in M$ and $r < r_0$ (here r_0 is either finite or infinite depending on whether M is bounded or not). It has been proved that $\beta > 2$ on the fractals, and the heat kernel admits the following *sub-Gaussian estimate*

$$p_t(x,y) \simeq \frac{1}{t^{\alpha/\beta}} \exp\left(-\left(\frac{d^\beta(x,y)}{ct}\right)^{\frac{1}{\beta-1}}\right), \tag{4}$$

for all $x, y \in M$ and $t < t_0 = t_0(r_0)$.

Observe that the Gaussian heat kernel in $\mathbb{R}^N$ satisfies (4) with $\alpha = N$ and $\beta = 2$. Hence, we can say that fractals extend the family of Euclidean spaces in two ways: first, allowing fractional values of α, and second, introducing a second parameter β so that the potential theory on such spaces is determined by *two* parameters.

3. On-Diagonal Upper Bounds and Faber-Krahn Inequalities

We will distinguish various types of heat kernel estimates, and start with an *on-diagonal upper bound*

$$p_t(x,x) \leq \frac{C}{f(ct)}, \tag{5}$$

where $f(t)$ is an increasing function of t. For simplicity, let us restrict to the case when the underlying space is a complete non-compact weighted manifold. Then (5) is supposed to hold for all $t > 0$ and $x \in M$.

The necessary and sufficient condition for (5) can be stated in terms of *a Faber-Krahn inequality*. For any precompact region $\Omega \subset M$, let $\lambda_1(\Omega)$ be the first Dirichlet eigenvalue of $-\Delta_\mu$, that is

$$\lambda_1(\Omega) = \inf_{\varphi \in C_0^\infty(\Omega), \varphi \not\equiv 0} \frac{\int_\Omega |\nabla \varphi|^2 \, d\mu}{\int_\Omega \varphi^2 d\mu} \, .$$

Given a decreasing non-negative function Λ on $(0, \infty)$, we say that (M, μ) admits a Faber-Krahn inequality with function Λ if, for any precompact region Ω,

$$\lambda_1(\Omega) \geq \Lambda(\mu(\Omega)) \, . \tag{6}$$

For example, $\mathbb{R}^N$ admits a Faber-Krahn inequality with function $\Lambda(v) = cv^{-2/N}$.

Theorem 3.1. ([18]) *If (M, μ) admits a Faber-Krahn inequality with function Λ then the heat kernel upper bound (5) holds with function f defined by*

$$t = \int_0^{f(t)} \frac{dv}{v\Lambda(v)} \, , \tag{7}$$

assuming that the integral converges at 0.

    Conversely, if (5) holds with a function f satisfying certain regularity condition then (M, μ) admits a Faber-Krahn inequality with function $c\Lambda(C\cdot)$ where Λ is defined by (7).

Consider some examples. Let (M, μ) be an N-dimensional Riemannian manifold of bounded geometry. Then it admits a Faber-Krahn inequality with function

$$\Lambda(v) = c \left\{ \begin{array}{ll} v^{-2/N}, & v < 1, \\ v^{-2}, & v \geq 1, \end{array} \right.$$

which implies the estimate

$$p_t(x, x) \leq \frac{C}{\min\left(t^{N/2}, t^{1/2}\right)}$$

(see [8, 19]). This estimate is sharp if $M = \mathbb{R} \times K$ where K is an $(N-1)$-dimensional compact manifold.

Let (M, μ) admit a discrete group G of isometries with a compact fundamental domain. Then, by [12] (see also [21, Section 7.6]) a Faber-Krahn function is determined by the volume function $V(x, r) = \mu(B(x, r))$ as follows: fix a point x_0, denote $V(r) = V(x_0, r)$ and define

$$\Lambda(v) = \left(\frac{c}{V^{-1}(Cv)} \right)^2 . \tag{8}$$

For example, if $V(r) \simeq e^{cr}$, for large r, then we obtain, for large v,

$$\Lambda(v) \simeq \frac{1}{\log^2 v}$$

whence, for large t,

$$p_t(x, x) \leq C \exp\left(-ct^{1/3}\right). \tag{9}$$

Similar relation between a volume growth and a Faber-Krahn function Λ takes place on manifolds of non-negative Ricci curvature. Indeed, if M is such a manifold and in addition $V(x, r) \geq cr^\alpha$ for all $r \geq 1$ and $x \in M$, then M admits a Faber-Krahn inequality with function Λ defined by (8) for $V(r) = cr^\alpha$, which yields for large v

$$\Lambda(v) \simeq v^{-2/\alpha}$$

(see [22, p.198]). Therefore, we obtain by Theorem 3.1, for large t,

$$p_t(x, x) \leq Ct^{-\alpha/2}.$$

This estimate follows also from [28].

4. On-Diagonal Lower Bounds and Anti-Faber-Krahn Inequalities

Here we consider the case of graphs and lower bounds of the following type:

$$\sup_{x \in \Gamma} p_{2n}(x, x) \geq \frac{c}{f(Cn)}. \tag{10}$$

We say that a graph (Γ, μ) admits *anti-Faber-Krahn inequality* with a function Λ, if there exists a sequence $\{\Omega_k\}$ of non-empty finite sets in Γ and a numerical sequence $\{v_k\}$ such that

$$v_{k+1} \leq Cv_k, \quad \lim_{k \to \infty} v_k = +\infty,$$

and

$$\mu(\Omega_k) \leq v_k \quad \text{and} \quad \lambda_1(\Omega_k) \leq \Lambda(v_k).$$

Theorem 4.1. ([11]) *If M admits anti-Faber-Krahn inequality with function Λ then the heat kernel lower estimate (10) holds with function f defined by (7) provided f possesses certain regularity.*

A particularly nice application occurs if Γ is a Cayley graph of a finitely generated group G. In this case we take μ to be a standard measure, which implies that $p_n(x, x)$ does not depend on x. On certain classes of groups, one can directly construct the sets $\{\Omega_k\}$ with controlled volumes and eigenvalues. To obtain an upper bound for $\lambda_1(\Omega_k)$, it suffices to have a smaller set $\Omega'_k \subset \Omega_k$ such that

$$\mu(\Omega'_k) \geq c\mu(\Omega_k) \quad \text{and} \quad d(\Omega'_k, \complement\Omega_k) \geq c\rho_k$$

with some numerical sequence $\{\rho_k\}$. Then it is easy to show that

$$\lambda_1(\Omega_k) \leq C\rho_k^{-2}.$$

For example, if G is a polycyclic group then, using structure results, it is possible to construct sequences of sets Ω_k and Ω_k' so that $v_k = C^k$ and $\rho_k = k$ whence

$$\lambda_1(\Omega_k) \leq \frac{C}{k^2} = \frac{C}{\log^2 v_k} \, .$$

Therefore, we can take

$$\Lambda(v) = \frac{C}{\log^2 v} \, ,$$

which implies by (7) and (10)

$$p_{2n}(x,x) \geq c \exp\left(-Cn^{1/3}\right) \, .$$

This estimate was first proved by Alexopoulos [1] by a different method. Together with a discrete version of the upper bound (9), it provides a sharp on-diagonal heat kernel decay on polycyclic groups of exponential volume growth.

5. Gaussian Estimates

Let us return to the setting of manifolds.

Theorem 5.1. ([39, 20]) *Assume that, for two points x, y on an arbitrary weighted manifold (M, μ),*

$$p_t(x,x) \leq \frac{1}{f(t)} \quad \text{and} \quad p_t(y,y) \leq \frac{1}{g(t)}, \quad \text{for all } 0 < t < t_0 \,,$$

where f and g are monotone increasing functions possessing certain regularity. Then, for all $0 < t < t_0$,

$$p_t(x,y) \leq \frac{C}{\sqrt{f(ct)g(ct)}} \exp\left(-\frac{d^2(x,y)}{Ct}\right) \, . \tag{11}$$

Here t_0 may be either finite or infinite. In particular, the Faber-Krahn inequality (6) implies (11) for all $x, y \in M$ and $t > 0$ with functions $f = g$ defined by (7). As we see, the Gaussian upper bound does *not* require any additional geometric assumptions on top of those which already provide the on-diagonal estimates.

The question of Gaussian lower bound is less understood. However, for two-sided Gaussian estimates there is the following result. Denote

$$V(x,r) = \mu(B(x,r)) \, .$$

Also, let $\lambda_1^{(N)}(\Omega)$ be the first non-zero Neumann eigenvalue of $-\Delta_\mu$ in a region Ω. The following result can be extracted from [32, 17].

Theorem 5.2. *Let (M, μ) be a complete non-compact weighted manifold, and let $f(t)$ be a positive increasing function on $(0, \infty)$ satisfying the doubling property*

$$f(2t) \leq Cf(t), \quad \forall t > 0 \, .$$

Then a two-sided heat kernel bound

$$p_t(x,y) \simeq \frac{1}{f(t)} \exp\left(-\frac{d^2(x,y)}{ct}\right),\tag{12}$$

for all $x,y \in M$ and $0 < t < t_0$, is equivalent to the following two conditions (valid for all $x \in M$ and $0 < r < c\sqrt{t_0}$):

1. *the volume growth condition*

$$V(x,r) \simeq f(r^2)$$

2. *and the Poincaré inequality*

$$\lambda_1^{(N)}(B(x,r)) \geq cr^{-2}.$$

Here t_0 may be either finite or infinite. A similar result holds on graphs [15] and on local Dirichlet spaces [37]. For example, a uniform volume growth $V(x,r) \simeq r^\alpha$ together with the Poincaré inequality is equivalent on any of those spaces to the estimate

$$p_t(x,y) \simeq \frac{1}{t^{\alpha/2}} \exp\left(-\frac{d^2(x,y)}{ct}\right).$$

A discrete version of Theorem 5.2 applies on a Cayley graph (Γ, μ) of any finitely generated group G with polynomial volume growth (see [26]), in which case α is exactly the exponent of the volume growth of G.

Note also that the Poincaré inequality holds on any complete Riemannian manifold of non-negative Ricci curvature (see [5]). For such manifolds, the estimate (12) was first proved by Li and Yau [28].

6. Sub-Gaussian Estimates

We state the result of this section for graphs although with certain modifications it holds also on manifolds and fractals. We would like to find conditions for a weighted graph (Γ, μ) which would ensure the heat kernel sub-Gaussian estimate similar to (4). By sub-Gaussian heat kernel estimates on graphs we mean the following inequalities:

$$p_n(x,y) \leq Cn^{-\alpha/\beta} \exp\left(-\left(\frac{d(x,y)^\beta}{Cn}\right)^{\frac{1}{\beta-1}}\right)\tag{13}$$

and

$$p_n(x,y) + p_{n+1}(x,y) \geq cn^{-\alpha/\beta} \exp\left(-\left(\frac{d(x,y)^\beta}{cn}\right)^{\frac{1}{\beta-1}}\right), \quad n \geq d(x,y),\tag{14}$$

where x,y are arbitrary points on Γ and n is a positive integer. The necessity of considering $p_n + p_{n+1}$ for the lower bound arises from a possible parity problem (in general, either p_n or p_{n+1} may vanish or be very small but a priori we cannot say which one is small).

To state the result, we need the volume function

$$V(x,r) = \mu(B(x,r)) = \mu\left\{y \in \Gamma : d(x,y) < r\right\},$$

as well as the Green function

$$G(x,y) = \sum_{n=0}^{\infty} P_n(x,y),$$

which, alternatively, is the infimum of all positive fundamental solutions to Δ_μ. It may happen that $G \equiv \infty$ in which case the random walk X_n is recurrent.

Theorem 6.1. ([24]) *Let (Γ, μ) be a weighted graph, and assume that, for some positive constant p_0,*

$$P(x,y) \geq p_0 \quad \forall x \sim y. \tag{15}$$

Given two numbers $\alpha > \beta > 1$, the sub-Gaussian estimates (13) and (14) are equivalent to the following two conditions:

1. *the polynomial volume growth, for all $x \in M$ and $r \geq 1$,*

$$V(x,r) \simeq r^\alpha \tag{16}$$

2. *and the polynomial Green function decay, for all $x \neq y$,*

$$G(x,y) \simeq d(x,y)^{-(\alpha-\beta)}. \tag{17}$$

The Green kernel uniform decay (17) implies a *uniform Harnack inequality*

$$\begin{aligned}\sup_{B(x,r)} u &\leq C \inf_{B(x,r)} u, \\ \text{provided } \Delta_\mu u &= 0 \text{ and } u \geq 0 \text{ in } B(x,2r),\end{aligned} \tag{18}$$

for all $x \in \Gamma$ and $r > 0$. Conversely, assuming the Harnack inequality (18) and the volume growth (16), one can deduce the Green function estimate (17) from one of the following conditions:

1. The estimate of the first Dirichlet eigenvalue of $-\Delta_\mu$ in a ball

$$\lambda_1(B(x,r)) \simeq r^{-\beta} \tag{19}$$

2. The capacity estimate

$$\text{cap}(B(x,r), B(x,2r)) \simeq r^{\alpha-\beta} \tag{20}$$

3. The mean exit time estimate

$$\mathbb{E}_x T_{x,r} \simeq r^\beta. \tag{21}$$

All conditions are assumed to be true for all $x \in \Gamma$ and $r \geq 1$. The capacity is defined to be the infimum of $\mathcal{E}(\varphi)$ over test functions φ which are equal to 1 on $B(x,r)$ and vanish outside $B(x,2r)$. The first exit time $T_{x,r}$ is defined by (3).

The following theorem covers also the recurrent case $G \equiv \infty$ when Theorem 6.1 is not applicable.

Theorem 6.2. ([23]) *Assume that (15) holds on (Γ, μ). Given two numbers $\alpha > 0$ and $\beta > 1$, the sub-Gaussian estimates (13) and (14) are equivalent to the following three conditions:*

1. *the Harnack inequality (18)* (which provides the homogeneity of the graph in question)
2. *the volume growth (16)* (which determines the parameter α)
3. *and any one of the conditions (20), (21)* (which determines the second parameter β).

Although a priori we assume only $\alpha > 0$ and $\beta > 1$, the hypothesis (16) and any of the conditions (20), (21) imply $2 \leq \beta \leq \alpha + 1$ (see [38]).

A geometric background for the Harnack inequality (18) is yet to be understood.

7. Higher Eigenvalues

Let (M, μ) be a weighted manifold. For any precompact domain $\Omega \subset M$, denote by $\lambda_k(\Omega)$ the k-th Dirichlet eigenvalue of $-\Delta_\mu$ in Ω.

Theorem 7.1. ([18]) *Assume that (M, μ) admits a Faber-Krahn inequality with function Λ, that is, for any precompact Ω,*

$$\lambda_1(\Omega) \geq \Lambda(\mu(\Omega)).$$

If Λ possesses a certain regularity property then, for all integers $k > 1$ and precompact Ω,

$$\lambda_k(\Omega) \geq c\Lambda\left(C\frac{\mu(\Omega)}{k}\right).$$

The proof goes through the heat kernel upper bound given by Theorem 3.1. In particular, we have

$$\lambda_1(\Omega) \geq c\mu(\Omega)^{-\delta} \implies \lambda_k(\Omega) \geq c\left(\frac{k}{\mu(\Omega)}\right)^\delta.$$

This results admits a generalization as follows. Assume that, apart from the measure μ, there is another Radon measure σ on M, and consider the following quadratic form associated with σ:

$$\mathcal{E}_\sigma(f) = \int_M |\nabla f|^2 \, d\sigma.$$

Assuming that in any precompact region $\Omega \subset M$ the form $\mathcal{E}_\sigma$ with domain $C_0^\infty(\Omega) \subset L^2(\Omega, \mu)$ is closable and has a discrete spectrum, we denote its k-th eigenvalue by $\lambda_k(\Omega, \mathcal{E}_\sigma)$. Note that the associated Rayleigh quotient is

$$\frac{\int_\Omega |\nabla f|^2 \, d\sigma}{\int_\Omega f^2 \, d\mu}.$$

The proof of the following theorem is based on the heat kernel techniques as well as on ideas from [27].

Theorem 7.2. ([25]) *Assume that for a Radon measure ν on M, for any precompact domain Ω and some $\delta > 0$,*

$$\lambda_1(\Omega, \mathcal{E}_\sigma) \geq \nu(\Omega)^{-\delta}.$$

Then, for all integers $k > 1$ and precompact Ω,

$$\lambda_k(\Omega, \mathcal{E}_\sigma) \geq c \left(\frac{k}{\nu(\Omega)} \right)^\delta.$$

References

[1] G. K. Alexopoulos, *A lower estimate for central probabilities on polycyclic groups,* Can. J. Math., **44** (1992), no. 5, 897–910.

[2] M. T. Barlow, *Diffusions on fractals,* in: Lectures on Probability Theory and Statistics, Ecole d'été de Probabilités de Saint-Flour XXV – 1995, Lecture Notes Math. 1690, Springer-Verlag, **1998**. 1–121.

[3] M. T. Barlow, R. F. Bass, *Brownian motion and harmonic analysis on Sierpinski carpets,* Canad. J. Math., **54** (1999), 673–744.

[4] M. T. Barlow, A. Perkins, *Brownian motion on the Sierpinski gasket,* Probab. Th. Rel. Fields, **79** (1988), 543–623.

[5] P. Buser, *A note on the isoperimetric constant,* Ann. Scient. Ec. Norm. Sup., **15** (1982), 213–230.

[6] I. Chavel, *Eigenvalues in Riemannian geometry,* Academic Press, New York, **1984.**

[7] I. Chavel, *Isoperimetric inequalities and heat diffusion on Riemannian manifolds,* Lecture notes **1999.**

[8] I. Chavel, E. A. Feldman, *Modified isoperimetric constants, and large time heat diffusion in Riemannian manifolds,* Duke Math. J., **64** (1991), no. 3, 473–499.

[9] F. R. K. Chung , *Spectral Graph Theory,* CBMS Regional Conference Series in Mathematics 92, AMS publications, **1996.**

[10] T. Coulhon, *Heat kernels on non-compact Riemannian manifolds: a partial survey,* Séminaire de théorie spectrale et géométrie, 1996–1997, Institut Fourier, Grenoble, **15** (1998), 167–187.

[11] T. Coulhon, A. Grigor'yan, Ch. Pittet, *A geometric approach to on-diagonal heat kernel lower bounds on groups,* to appear in Ann. Inst. Fourier

[12] T. Coulhon, L. Saloff-Coste, *Isopérimétrie pour les groupes et les variétés,* Revista Matemática Iberoamericana, **9** (1993), no. 2, 293–314.

[13] E. B. Davies, *Heat kernels and spectral theory,* Cambridge University Press, **1989.**

[14] E. B. Davies, *Spectral theory and differential operators,* Cambridge University Press, **1995.**

[15] T. Delmotte, *Parabolic Harnack inequality and estimates of Markov chains on graphs,* Revista Matemática Iberoamericana, **15** no. 1, (1999), 181–232.

[16] M. Fukushima, Y. Oshima, M. Takeda, *Dirichlet forms and symmetric Markov processes,* Studies in Mathematics 19, De Gruyter, **1994.**

[17] A. Grigor'yan, *The heat equation on non-compact Riemannian manifolds,* (in Russian) Matem. Sbornik, **182** (1991), no. 1, 55–87. Engl. transl. Math. USSR Sb., **72** (1992), no. 1, 47–77.

[18] A. Grigor'yan, *Heat kernel upper bounds on a complete non-compact manifold,* Revista Matemática Iberoamericana, **10** (1994), no. 2, 395–452.

[19] A. Grigor'yan, *Heat kernel on a manifold with a local Harnack inequality,* Comm. Anal. Geom., **2** (1994), no. 1, 111–138.

[20] A. Grigor'yan, *Gaussian upper bounds for the heat kernel on arbitrary manifolds,* J. Diff. Geom., **45** (1997), 33–52.

[21] A. Grigor'yan, *Estimates of heat kernels on Riemannian manifolds,* in: Spectral Theory and Geometry. ICMS Instructional Conference, Edinburgh 1998, ed. B.Davies and Yu.Safarov, London Math. Soc. Lecture Note Series 273, Cambridge Univ. Press, **1999.** 140–225.

[22] A. Grigor'yan, *Analytic and geometric background of recurrence and non-explosion of the Brownian motion on Riemannian manifolds,* Bull. Amer. Math. Soc., **36** (1999), 135–249.

[23] A. Grigor'yan, A. Telcs, *Harnack inequalities and sub-Gaussian estimates for random walks,* preprint.

[24] A. Grigor'yan, A. Telcs, *Sub-Gaussian estimates of heat kernels on infinite graphs,* to appear in Duke Math. J.

[25] A. Grigor'yan, S.-T. Yau, *On isoperimetric properties of higher eigenvalues of elliptic operators,* in preparation.

[26] W. Hebisch, L. Saloff-Coste,, *Gaussian estimates for Markov chains and random walks on groups,* Ann. Prob., **21** (1993), 673–709.

[27] P. Li, S.-T. Yau, *On the Schrödinger equation and the eigenvalue problem,* Comm. Math. Phys., **88** (1983), 309–318.

[28] P. Li, S.-T. Yau, *On the parabolic kernel of the Schrödinger operator,* Acta Math., **156** (1986), no. 3–4, 153–201.

[29] V. G. Maz'ya, *Sobolev spaces,*(in Russian) Izdat. Leningrad Gos. Univ. Leningrad, **1985.** Engl. transl. Springer-Verlag, **1985.**

[30] Ch. Pittet, L. Saloff-Coste, *A survey on the relationship between volume growth, isoperimetry, and the behavior of simple random walk on Cayley graphs, with examples,* preprint.

[31] D. W. Robinson, *Elliptic operators and Lie groups,* Oxford Math. Mono., Clarenton Press, Oxford New York Tokyo, **1991.**

[32] L. Saloff-Coste, *A note on Poincaré, Sobolev, and Harnack inequalities,* Duke Math J., Internat. Math. Res. Notices, **2** (1992), 27–38.

[33] L. Saloff-Coste, *Parabolic Harnack inequality for divergence form second order differential operators,* Potential Analysis, **4** (1995), 429–467.

[34] R. Schoen, S.-T. Yau, *Lectures on Differential Geometry,* Conference Proceedings and Lecture Notes in Geometry and Topology 1, International Press, **1994.**

[35] R. S. Strichartz, *Analysis on fractals,* Notices of AMS, **46** (1999), no. 10, 1199–1208.

[36] D. W. Stroock, *Estimates on the heat kernel for the second order divergence form operators,* in: Probability theory. Proceedings of the 1989 Singapore Probability Conference held at the National University of Singapore, June 8–16 1989, ed. L.H.Y. Chen, K.P. Choi, K. Hu and J.H. Lou, Walter De Gruyter, **1992.** 29–44.

[37] K.-Th. Sturm, *Analysis on local Dirichlet spaces III. The parabolic Harnack inequality,* Journal de Mathématiques Pures et Appliquées, **75** (1996), no. 3,   273–297.

[38] A. Telcs, *Random walks on graphs, electric networks and fractals,* J. Prob. Theo. and Rel. Fields, **82** (1989),   435–449.

[39] V. I. Ushakov, *Stabilization of solutions of the third mixed problem for a second order parabolic equation in a non-cylindric domain,* (in Russian) Matem. Sbornik, **111** (1980),   95–115. Engl. transl. Math. USSR Sb., **39** (1981),   87–105.

[40] N. Th. Varopoulos, L. Saloff-Coste, T. Coulhon, *Analysis and geometry on groups,* Cambridge University Press, Cambridge, **1992.**

[41] W. Woess, *Random walks on infinite graphs and groups,* Cambridge Tracts in Mathematics 138, Cambridge Univ. Press., **2000.**

Department of Mathematics
Imperial College
London SW7 2BZ, England
E-mail address: a.grigoryan@ic.ac.uk

Local Langlands Correspondences and Vanishing Cycles on Shimura Varieties

Michael Harris

Abstract. We report on the results and techniques of the author's recent joint work with Richard Taylor, which analyzes in detail the bad reduction of certain Shimura varieties in order to prove the compatibility of local and global Langlands correspondences, obtaining as a consequence the local Langlands conjecture for $GL(n)$ of a p-adic field. These Shimura varieties have natural models over p-adic integer rings, as moduli spaces for abelian varieties with additional structure. The starting point of the work with Taylor is the stratification of the special fiber of an integral model in minimal level, according to the isogeny type of the universal family of p-divisible groups attached to these abelian varieties. Similar stratifications can conjecturally be constructed for any Shimura variety, and indeed are known to exist for most PEL types. We discuss a series of conjectures regarding the behavior of vanishing cycles along these strata, with the aim of extending Kottwitz' conjectures on the cohomology of Shimura varieties to the case of bad reduction.

1. Introduction

Let F be a local field and n a positive integer. Let $\mathcal{A}(n, F)$ denote the set of equivalence classes of irreducible admissible representations of $GL(n, F)$, $\mathcal{A}_0(n, F)$ the subset of supercuspidal representations. Let $\mathcal{G}(n, F)$ denote the set of equivalence classes of n-dimensional complex representations of the Weil-Deligne group $WD(F)$ on which Frobenius acts semisimply, $\mathcal{G}_0(n, F)$ the subset of irreducible representations. A *local Langlands correspondence* (for general linear groups), a non-abelian generalization of local class field theory, is a family of bijections $\pi \mapsto \sigma(\pi)$ from $\mathcal{A}(n, F)$ to $\mathcal{G}(n, F)$, for all n, identifying $\mathcal{A}_0(n, F)$ with $\mathcal{G}_0(n, F)$, and satisfying a list of properties recalled below.

The existence of local Langlands correspondences, previously known in various special cases, has now been established in full generality. For F of positive characteristic, this was proved by Laumon, Rapoport and Stuhler [35], using a variant of Drinfeld's modular varieties; in particular, the techniques of [35] are

1991 *Mathematics Subject Classification.* Primary classification: 11S37, 22E50, 14G35 Secondary classification: 14L05.

global. For p-adic fields, the first proof was given in joint work with Richard Taylor [20], again using global methods, this time involving the geometry of certain Shimura varieties, together with cases of non-Galois automorphic induction proved in [19] (also using Shimura varieties). Shortly after distribution of the first version of [20], Henniart found a much simpler proof [25], obtaining the local Langlands correspondence directly from the results of [19]. All of these proofs rely crucially on a weaker version of the correspondence, the *numerical local Langlands correspondence*, proved by Henniart in [23].

The present article is a report on the results and techniques of [20]. Many of these techniques appear to apply to a more general class of Shimura varieties than those considered in [20]. Shimura varieties[1] are conjectured to be moduli spaces for certain kinds of motives. This was proved by Shimura for many Shimura varieties attached to classical groups, the motives in this case arising from abelian varieties with additional structure (PEL types). In this way many Shimura varieties, together with their Hecke correspondences, acquire natural models over p-adic integer rings. The varieties considered in [20] are of PEL type. The starting point of [20] is the stratification of the special fiber of the integral model, according to the isogeny type of the universal family of *p-divisible groups* (with additional structure) attached to the moduli problem. Such stratifications can be constructed for any Shimura variety realized as a moduli space for motives.

Let $Sh(G, X)$ be a Shimura variety, with G a connected reductive group over $\mathbb{Q}$, and X a $G(\mathbb{R})$-conjugacy class of homomorphisms $h \colon R_{\mathbb{C}/\mathbb{R}}(\mathbb{G}_{m,\mathbb{C}}) \to G_{\mathbb{R}}$, satisfying a familiar list of axioms. Then $Sh(G, X)$ has a canonical model over the reflex field $E = E(G, X)$. Fix a rational prime p and a prime v of E above p with residue field $k(v)$, and assume $Sh(G, X)$ has a model over the v-adic integer ring $\mathcal{O}_v$. If π is a cohomological automorphic representation of (the adele group of) G, with finite part π_f, let $r_\ell(\pi)$ denote the virtual ℓ-adic representation of $\mathrm{Gal}(\bar{E}/E)$ on the π_f-isotypic component of $\sum_i (-1)^i H_c^i(Sh(G, X)_{\bar{E}}, \mathbb{Q}_\ell)$; more generally, $\mathbb{Q}_\ell$ can be replaced by an ℓ-adic local system. Following a technique introduced by Ihara and Langlands, and developed systematically by Kottwitz [29, 30], we study the local behavior at v of $r_\ell(\pi)$ by comparing the Grothendieck-Lefschetz trace formula on the cohomology of the special fiber at v of $Sh(G, X)^2$ with the Arthur-Selberg trace formula for G. Unlike Kottwitz, however, we do not assume $Sh(G, X)$ to have good reduction at v. Thus $r_\ell(\pi)$ is ramified in general, and the Lefschetz formula is applied to the cohomology with coefficients in the nearby cycle complex. In [20] it is proved that the sheaves of nearby cycles are locally constant on each stratum in the étale topology. This is unlikely to be true in general, but the stalks can be predicted in terms of the local uniformization theory of Rapoport and Zink [40]. In this way, we arrive at a conjectural expression for the contribution of each stratum to $r_\ell(\pi)$. Up to semisimplification, this contribution depends only on the local component of π at p.

---

[1] For experts: we always assume the weight morphism is rational over $\mathbb{Q}$.
[2] More precisely, the Shimura variety $_K Sh(G, X)$ at an appropriate finite level K.

2. The Local Langlands Correspondence

In what follows, p is a prime number. The local Langlands conjecture for $GL(n)$ is best stated for all positive integers n and all p-adic fields F simultaneously. Notation is as in the introduction.

Local Langlands Conjecture. *Let F be a finite extension of $\mathbb{Q}_p$.*

 (i) *For every integer $n \geq 1$, there exists a bijection $\pi \to \sigma(\pi)$ between $\mathcal{A}(n, F)$ and $\mathcal{G}(n, F)$ that identifies $\mathcal{A}_0(n, F)$ with $\mathcal{G}_0(n, F)$.*

 (ii) *Let χ be a character of $F^\times$, which we identify with a character of $WD(F)$ via the reciprocity isomorphism of local class field theory. Then $\sigma(\pi \otimes \chi \circ \det) = \sigma(\pi) \otimes \chi$. In particular, when $n = 1$, the bijection is given by local class field theory.*

 (iii) *If $\pi \in \mathcal{A}(n, F)$ with central character $\xi_\pi \in \mathcal{A}(1, F)$, then $\xi_\pi = \det(\sigma(\pi))$.*

 (iv) *$\sigma(\check{\pi}) = \check{\sigma}(\pi)$, where $\check{\ }$ denotes contragredient.*

 (v) *Let $\alpha \colon F \to F_1$ be an isomorphism of local fields. Then α induces bijections $\mathcal{A}(n, F) \to \mathcal{A}(n, F_1)$ and $\mathcal{G}(n, F) \to \mathcal{G}(n, F_1)$ for all n, and we have $\sigma(\alpha(\pi)) = \alpha(\sigma(\pi))$. In particular, if F is a Galois extension of a subfield F_0, then the bijection σ respects the $\mathrm{Gal}(F/F_0)$-actions on both sides.*

 (vi) *Let F'/F denote a cyclic extension of prime degree d. Let $BC \colon \mathcal{A}(n, F) \to \mathcal{A}(n, F')$ and $AI \colon \mathcal{A}(n, F') \to \mathcal{A}(nd, F')$ denote the local base change and automorphic induction maps [1, 26]. Let $\pi \in \mathcal{A}(n, F)$, $\pi' \in \mathcal{A}(n, F')$. Then $\sigma(BC(\pi)) = \sigma(\pi)|_{WD(F')}$, $\sigma(AI(\pi')) = \mathrm{Ind}_{F'/F}\, \sigma(\pi')$, where $\mathrm{Ind}_{F'/F}$ denotes induction from $WD(F')$ to $WD(F)$.*

 Let n and m be positive integers, $\pi \in \mathcal{A}(n, F)$, $\pi' \in \mathcal{A}(m, F)$. Then

 (vii) *$L(s, \pi \otimes \pi') = L(s, \sigma(\pi) \otimes \sigma(\pi'))$.*

(viii) *For any additive character ψ of F, $\epsilon(s, \pi \otimes \pi', \psi) = \epsilon(s, \sigma(\pi) \otimes \sigma(\pi'), \psi)$.*

Here the terms on the left of (vii) and (viii) are as in [27, 43] and are compatible with the global functional equation for Rankin-Selberg L-functions. The right-hand terms are given by Artin and Weil (for (vii)) and Langlands and Deligne (for (viii)) and are compatible with the functional equation of L-functions of representations of the global Weil group. In particular both sides have Artin conductors and (viii) implies that $a(\sigma(\pi)) = a(\pi)$.

I refer to Carayol's Bourbaki exposés [11, 12], and the introduction to [20], for more details on the history of this conjecture. A theorem of Henniart [24] implies that σ is uniquely determined by these properties. A version of the local Langlands conjecture for general connected p-adic reductive groups is recalled in §5 below, in connection with Conjecture 5.3.

The proof of this conjecture in [20] is based on the following considerations. The logical first step is the

Theorem 2.1. *There is a family of maps $\sigma_0^{\mathrm{van}} \colon \mathcal{A}_0(n, F) \to \mathcal{G}(n, F)$, for all positive integers n and all p-adic fields F, satisfying (iii–vi).*

Theorem 2.1 summarizes the contents of Theorem VIII.1.3 and Lemma VIII.2.6 of [20]. The map, whose existence was conjectured by Carayol [10], is realized on a geometric model arising from the deformation theory of p-divisible groups. Its construction is global, however (see (17), below). A theorem of Henniart [23, 7] implies that any such family of maps is in fact a family of *bijections* $\mathcal{A}_0(n, F) \to \mathcal{G}_0(n, F)$ that preserves conductors and satisfies (i–vii).

It therefore remains to prove (viii). The proof is global in nature. We work over a CM field E, a quadratic extension of a totally real subfield E^+ of degree d. Let Π be a cuspidal automorphic representation of $GL(n)_E$, unramified outside a finite set S. For any finite set S' containing S, let $L^{S'}(s, \Pi)$ denote the partial standard L-function of π with the Euler factors at S' removed. Let L be a number field, $\{\lambda\}$ the set of finite places of L. Let $\sigma = \{\sigma_\lambda\}$ be a compatible family of n-dimensional λ-adic representations of $\mathrm{Gal}(\overline{E}/E)$. We say σ is *weakly associated* to Π if

$$L^{S'}(s, \sigma) = L^{S'}(s, \Pi) \tag{1}$$

as Euler products for some finite set S' containing S. Here $L^{S'}(s, \sigma)$ denotes the partial L-function of $\{\sigma\}$. Both sides of (1) are normalized so that the functional equation is symmetric around the line $\mathrm{Re}(s) = \frac{1}{2}$. Let $Reg(n, E)$ denote the set of Π for which Π_∞ is of cohomological type. We assume the coefficient system to be trivial, for simplicity, but this is not necessary. The following theorem is mainly due to Clozel, with improvements due to Taylor and Blasius, and depends crucially on Kottwitz' study of points on Shimura varieties over finite fields [13, 30, 31, 5]; cf. [20, Theorem VIII.1.9]:

Theorem 2.2. ([13, Théorème 5.7], [14]) *Let c denote complex conjugation on E. Let $\Pi \in Reg(n, E)$. Suppose (i) The local component Π_v at at least one finite place v is square-integrable (supercuspidal or generalized Steinberg); (ii) Π is dual to Π^c.*

Then there exists a compatible family $\sigma(\Pi) = \{\sigma_\lambda(\Pi)\}$ of semi-simple n-dimensional λ-adic representations of $\mathrm{Gal}(\overline{E}/E)$, weakly associated to Π.

To any $\pi \in \mathcal{A}(n, F)$ we associate its *supercuspidal support* $\mathrm{Supp}(\pi)$, consisting of a partition $n = n_1 + \cdots + n_r$ and, for each $i \in \{1, \ldots, r\}$, a $\pi_i \in \mathcal{A}_0(n_i, F)$, such that π is an irreducible constituent of the (normalized) induced representation $I_P^{GL(n,F)}(\pi_1 \otimes \cdots \otimes \pi_r)$, where $P \subset GL(n, F)$ is any parabolic subgroup with Levi factor $\prod_i(GL(n_i, F))$. Suppose we have bijections $\sigma_0 \colon \mathcal{A}_0(m, F) \leftrightarrow \mathcal{G}_0(m, F)$ for all m and all F. We extend these bijections to all π in two steps. Let $\mathcal{G}_{ss}(n, F) \subset \mathcal{G}(n, F)$ denote the subset of representations of $WD(F)$ factoring through the Weil group $W(F)$. If $\pi \in \mathcal{A}(n, F)$, $\mathrm{Supp}(\pi) = \{(n_i, \pi_i)\}$, let

$$\sigma_{ss}(\pi) = \oplus_{i=1}^r \sigma(\pi_i) \in \mathcal{G}_{ss}(n, F). \tag{2}$$

The set of $\pi \in \mathcal{A}(n, F)$ with support $\{(n_i, \pi_i)\}$ was classified by Zelevinski, who proved the existence of a natural extension of σ_{ss} to a set of bijections $\sigma \colon \mathcal{A}(m, F) \leftrightarrow \mathcal{G}(m, F)$ [45]. If σ_0 satisfies (iii), (iv), (vi), (vii), and (viii) for all m, then so does σ [24, 3.4]. The situation for (vi) is a bit more complicated, since BC and AI do not preserve supercuspidality in general, but allowing for this complication, it also suffices to verify (vi) for σ_0.

The main purpose of [20] is to remove the modifier "weakly" from Theorem 2.2:

Theorem 2.3. ([20, Theorem VIII.1.9]) *Let* $\Pi \in Reg(n, E)$. *Assume* Π *satisfies (i) and (ii) of Theorem 2.2. Then for all primes v of E not dividing the characteristic of λ, the following relation holds:*

$$\sigma_\lambda(\Pi)_{v,ss} \xrightarrow{\sim} \sigma_{ss,\lambda}(\Pi_v). \tag{3}$$

Here $\Pi_v \in \mathcal{A}(n, F)$ *is the local component at v of Π, and $\sigma_\lambda(\Pi)_{v,ss}$ is the semi-simplification of the restriction of $\sigma_\lambda(\Pi)$ to $W(F)$.*

Remarks 2.4.

 (i) *When Π_v is unramified, this comes down to the equality of local Euler factors asserted in Theorem 2.2.*

 (ii) *The article [18] uses rigid-analytic uniformization of slightly different Shimura varieties to obtain maps $\sigma_0^{\mathrm{rig}} : \mathcal{A}_0(n, F) \to \mathcal{G}(n, F)$, with the properties indicated in Theorem 2.1. As above, Henniart's theorem implies that these maps define a family of bijections $\mathcal{A}_0(n, F) \to \mathcal{G}_0(n, F)$. Both σ_0^{rig} and σ_0^{van} satisfy Theorem 2.3 when π_v is supercuspidal. A posteriori, it follows that the two correspondences coincide.*

 (iii) *For $n = 2$, Carayol proved the stronger result [9] that $\sigma_\lambda(\Pi)_v \xrightarrow{\sim} \sigma_\lambda(\Pi_v)$ as representations of the full Weil-Deligne group; T. Saito has proved the analogous result without restriction on the residue characteristic of λ [42][3]. Removal of the subscript ss in Theorem 2.3 seems to require proof of Deligne's conjecture on the purity of the monodromy weight filtration.*

The reduction of the local Langlands conjecture to Theorem 2.3 is the subject of [19]. The point is to show that, as E varies over CM fields, the set of representations $Reg(n, E) \times Reg(m, E)$, for varying n and m, contains sufficiently many pairs (Π, Π') whose global L-functions are known *a priori* to satisfy two functional equations, one involving the automorphic local constants $\epsilon(s, \Pi \otimes \Pi', \psi)$, the other involving the Galois-theoretic local constants of Langlands and Deligne. A technique originating with Deligne then permits identification of the corresponding local constants at the prime v of interest.

An argument involving Brauer's theorem shows that it suffices to construct $\Pi = \Pi(\chi) \in Reg(n, E)$ such that the associated $\sigma(\Pi)$ are induced from appropriate Hecke characters χ, of CM extensions E'/E, with fixed local behavior at primes dividing v, such that the Galois closure of E' over E is solvable. It follows that $\sigma(\Pi)$

---

[3] *Note added in proof.* Saito's work concerns the representation $D(\sigma_\lambda(\Pi)_v)$ of $WD(F)$ defined by Fontaine when v divides the characteristic of λ. Saito has recently announced joint work with K. Kato implying that, for general n, $D(\sigma_\lambda(\Pi)_v)$ and $\sigma_\lambda(\Pi_v)$ coincide after restriction to the wild ramification subgroup.

is the compatible system associated to a complex representation of the global Weil group of E, hence that its L-function satisfies a functional equation of Artin-Weil type.

This version of non-Galois automorphic induction, carried out in §4 of [19], relies on Clozel's theorem 2.2, and thereby on Kottwitz' analysis of the good reduction of Shimura varieties. Henniart's simple proof of the local Langlands conjecture in [25] proceeds in the opposite direction. Starting with the $\pi(\chi)$ constructed in [19], he defines maps $\pi_n \colon \mathcal{G}_0(n, F) \to \oplus_{n \geq 1} \mathbb{Z} \cdot \mathcal{A}(n, F)$, where the target of π_n is a formal direct sum. The maps π_n are defined globally, but using properties of L-functions Henniart shows that they are well defined, and that the image of π_n is contained in $\mathcal{A}_0(n, F)$. The results of [23, 24] then suffice to prove that π_n are independent of all choices and have properties (i)–(viii).

3. Shimura Varieties Attached to Twisted Unitary Groups

If G is a connected reductive group over a global or local field F, we denote by $\hat{G}$ its Langlands dual group, viewed as the points of a reductive group over an algebraically closed field $\mathbf{F}$. Denote by $^L G$ the L-group of G, a semi-direct product of $\hat{G}$ with either the Weil group $W(F)$ or the Galois group $\mathrm{Gal}(\bar{F}/F)$, depending on context.

Let (G, X) be a Shimura datum, as in the introduction. Let $\mu \colon \mathbb{G}_{m,\mathbb{C}} \to G_{\mathbb{C}}$ be the cocharacter attached to a point $h \in X$. The G-conjugacy class of μ is independent of h and its field of definition is the reflex field $E(G, X)$; we will write μ_X for any point in this conjugacy class. This is a cocharacter of some maximal torus of G, hence a character of a maximal torus $\hat{T} \subset \hat{G}$. We let r_μ be the irreducible representation of $\hat{G}$ with extreme weight μ. The weight μ is necessarily minuscule. Langlands has defined an extension of r_μ, also denoted r_μ, to a representation of the L-group $^L G$ of G over the base field $E(G, X)$. The Shimura variety $Sh(G, X)$ has a canonical model over the field $E(G, X)$.

The article [20] is concerned with a specific family of Shimura varieties. Let p be a rational prime, E an imaginary quadratic field in which p splits, F^+ a totally real field of degree d, and $F = F^+ \cdot E$; let $c \in \mathrm{Gal}(F/F^+)$ denote complex conjugation. Choose a prime u of E above p, and let $w = w_1, w_2, \ldots, w_r$ be the primes of F above u. Choose a distinguished embedding $\tau_0 \colon F \to \mathbb{C}$. Let σ_0 (resp. τ_E) denote the restriction of τ_0 to F^+ (resp. to E), and let Σ denote the set of complex embeddings of F restricting to τ_E on E.

Let B be a central division algebra of dimension n^2 above F, admitting an anti-automorphism $\#$ restricting to c on the center F. Assume B is split at w and at every prime that does not split over F^+, and that at every place B is either split or a division algebra.

Define a connected reductive $\mathbb{Q}$-algebraic group G by

$$G(R) = \{ g \in (B^{\mathrm{op}} \otimes_{\mathbb{Q}} R)^\times \mid g \cdot g^\# = \nu(g) \in R^\times \} \tag{4}$$

for any $\mathbb{Q}$-algebra R. The kernel G_1 of the map $\nu\colon G \to \mathbb{G}_m$ is the restriction of scalars to $\mathbb{Q}$ of a group G^+ over F^+. Under a certain parity condition [13, §2]; [20, Lemma II.7.1], which we assume, we can choose $\#$ so that G is quasi-split at all rational primes that do not split in $E/\mathbb{Q}$ and so that $G^+_{\sigma_0}$ is isomorphic to $U(1, n-1)$ but G^+_σ is a compact unitary group for all real places $\sigma \neq \sigma_0$.

Since p splits in E, we can identify

$$G(\mathbb{Q}_p) \xrightarrow{\sim} GL(n, F_w) \times \prod_{i>1} B^{\mathrm{op}, \times}_{w_i} \times \mathbb{Q}^\times_p \,, \tag{5}$$

where the map $G(\mathbb{Q}_p) \to \mathbb{Q}^\times_p$ is given by ν.

Choose an $\mathbb{R}$-algebra homomorphism $h_0\colon \mathbb{C} \to B^{\mathrm{op}} \otimes_\mathbb{Q} \mathbb{R}$ such that $h_0(z)^\# = h_0(\bar{z})$ for all $z \in \mathbb{C}$. The image is contained in G and we may assume it is centralized by a maximal compact subgroup of $G(\mathbb{R})$. Let (G, X) be the Shimura datum for which X is the $G(\mathbb{R})$-conjugacy class containing h_0. Then the reflex field $E(G, X)$ is isomorphic to F, identified with its image in $\mathbb{C}$ under τ_0.

3.1. The moduli problem

If A is an abelian scheme over a base scheme S, let $T_f(A)$ denote the direct product of the Tate modules $T_\ell(A)$ over all primes ℓ, $V_f(A) = \mathbb{Q} \otimes T_f(A)$. Let $K \subset G(\mathbf{A}_f)$ be a compact open subgroup. Consider the functor $\mathcal{A}_K(B, *)$ on schemes over F, which to S associates the set of equivalence classes, for the usual equivalence relation, of quadruples (A, λ, i, η), where A is an abelian scheme over S of dimension dn^2, $\lambda\colon A \to \hat{A}$ is a polarization, $i\colon B \hookrightarrow \mathrm{End}(A) \otimes \mathbb{Q}$ is an embedding, and $\eta\colon V \otimes_\mathbb{Q} \mathbf{A}_f \xrightarrow{\sim} V_f(A)$ an isomorphism of $B \otimes_\mathbb{Q} \mathbf{A}_f$-modules, modulo K [30, p. 390]. These data are assumed to satisfy the standard compatibilites. More importantly, i induces an action i_F of the center F of B on the $\mathcal{O}_S$-module $\mathrm{Lie}(A)$. For each embedding $\tau\colon F \to \mathbb{C}$, we let $\mathcal{O}_{S,\tau} = \mathcal{O}_S \otimes_{F,\tau} \mathbb{C}$, and let $\mathrm{Lie}(A)_\tau = \mathrm{Lie}(A) \otimes_{F,\tau} \mathbb{C}$. We then assume that

(i) $\mathrm{Lie}(A)_\tau = 0$, $\tau \in \Sigma$, $\tau \neq \tau_0$; module of rank n^2, $\tau \neq \tau_0$;

(ii) $\mathrm{Lie}(A)_{\tau_0}$ is a projective $\mathcal{O}_{S,\tau}$ module of rank n.

For K sufficiently small, $\mathcal{A}_K(B, *)$ is represented by a smooth projective scheme over F, also denoted $\mathcal{A}_K(B, *)$, isomorphic to $|\ker^1(\mathbb{Q}, G)|$ copies of the canonical model of ${}_K Sh(G, X)$, where $\ker^1(\mathbb{Q}, G)$ measures the defect of the Hasse principle for $H^1(\mathbb{Q}, G)$.

Assume K factors as $K_p \times K^p$, with K^p sufficiently small, $K_p = \prod_i K_{w_i} \times \mathbb{Z}^\times_p$, with respect to (5), and $K_w = K_{w_1} = GL(n, \mathcal{O}_w)$. Then $\mathcal{A}_K(B, *)$ has a smooth model over $\mathrm{Spec}(\mathcal{O}_w)$, also denoted $\mathcal{A}_K(B, *)$, that represents a slightly modified version of the functor considered above: in conditions (i)–(ii), the $\mathcal{O}_{S,\tau}$'s are replaced by $\mathcal{O}_S \otimes_{\mathcal{O}_w} \mathcal{O}_{\tilde{w}}$, where $\tilde{w}$ run through the primes of F above p. As above, $\mathcal{A}_K(B, *)$ is the union of $|\ker^1(\mathbb{Q}, G)|$ copies of a smooth model $\mathcal{O}_w$-model $S_K(G, X)$ of ${}_K Sh(G, X)$. We let $\bar{S} = \bar{S}_K(G, X)$ denote the special fiber of this model.

The moduli space $\mathcal{A}_K(B, *)$ admits a universal abelian scheme (with PEL structures) denoted $A = A_K(B, *)$. Let $A[w^\infty]$ denote the corresponding p-divisible $\mathcal{O}_w$-module. The action of a maximal order in $B_w \xrightarrow{\sim} M(n, F_w)$ breaks up $A[w^\infty]$ as a sum of n-copies of a p-divisible $\mathcal{O}_w$ module $\mathcal{G}$. Conditions (i)–(ii) imply that $\mathcal{G}$ is a one-dimensional height n divisible $\mathcal{O}_w$ module. The Serre-Tate theorem implies that the infinitesimal local structure of $\mathcal{A}_K(B, *)$ near a point s of the special fiber is controlled by the deformation theory of the fiber $\mathcal{G}_s$ at s. This is the basis of the stratification of the special fiber $\bar{S}$, discussed in the following section.

4. Stratifications of Shimura Varieties

We will work with a general Shimura variety $Sh(G, X)$, as in the introduction; to avoid complications, we assume the derived subgroup of G to be simply connected. Fix a prime p and a level subgroup $K = K_p \times K^p \subset G(\mathbf{A}_f)$. Assume G is quasi-split at p and K^p is sufficiently small, so that $_K Sh(G, X)$ is smooth. Let v be a prime of the reflex field E dividing p, $F = E_v$. Let W denote the ring of Witt vectors of $k(v)$, $\mathcal{K} = \mathrm{Frac}(W)$, $\mathcal{L}$ the compositum of F and $\mathcal{K}$, σ the (arithmetic) Frobenius automorphism of $\mathcal{L}$ over F. We assume that $_K Sh(G, X)$ has a smooth model S over $\mathrm{Spec}(\mathcal{O}_v)$, to which the Hecke correspondences extend; we let $\bar{S}$ denote the special fiber. The usual hypothesis is that p be unramified in E and that K_p be a hyperspecial maximal compact subgroup. This hypothesis is sufficient when $Sh(G, X)$ is of PEL type ([34, 30]), but is certainly stronger than necessary. Results of Labesse [33, Prop. 3.6.4] suggest it may suffice to take K_p to be a "very special" maximal compact subgroup, provided G splits over F; this is true in the cases considered in [20].

For any algebraic torus T, let $X^*(T)$ and $X_*(T)$ denote the group of its characters and cocharacters, respectively. Let $P_0 \subset G$ be a $\mathbb{Q}_p$-rational minimal parabolic subgroup, with Levi factor T_0 and unipotent radical N_0. This determines an order on the root lattice of G and, dually, on that of $\hat{G}$. The prime p being fixed, we let $B(G)$ denote the set of σ-conjugacy classes in $G(\mathcal{L})$, and let $\kappa \colon B(G) \to X^*(Z((\hat{G}))^{\Gamma_p})$ be the invariant defined in [28], with $\Gamma_p = \mathrm{Gal}(\overline{\mathbb{Q}}_p/\mathbb{Q}_p)$. Let $A \subset T_0$ be the maximal split torus, $\mathfrak{a} = X_*(A) \otimes_{\mathbb{Z}} \mathbb{R}$, $\mathfrak{a}_{\mathbb{Q}} = X_*(A) \otimes_{\mathbb{Z}} \mathbb{Q}$. Let $\bar{C} \subset \mathfrak{a}$ be the closed positive chamber corresponding to N_0, $\bar{C}_{\mathbb{Q}} = \bar{C} \cap \mathfrak{a}_{\mathbb{Q}}$. The *Newton map* $\bar{\nu} \colon B(G) \to \bar{C}_{\mathbb{Q}}$ is defined in [39, 32]; it is known that

$$\bar{\nu} \times \kappa \colon B(G) \to \bar{C}_{\mathbb{Q}} \times X^*(Z((\hat{G}))^{\Gamma_p})$$

is injective [32, 4.13]. The class b is *basic* if and only if $\bar{\nu}(b)$ is in the intersection of all root hyperplanes. In that case, b defines an inner twist of G [28, 4.4].

On the other hand, the cocharacter μ can be interpreted as a character of the dual torus $\hat{T}_0$; let $\mu^\# \in X^*(Z(\hat{G})^{\Gamma_p})$ denote the restriction of this character. Following Kottwitz [32, §6], we let $B(G, \mu) = B(G_{\mathbb{Q}_p}, \mu)$ be the set of $b \in B(G)$ satisfying $\kappa(b) = \mu^\#$ and such that $\bar{\nu}(b) \leq \mu_{\mathfrak{a}}$ with $\leq$ the usual lexicographic order.

Consider the Langlands representation r_μ of $^L G$. Let $P = LU \subset G$ be a standard parabolic. The representation r_μ decomposes, upon restriction to $^L L$, as a sum of irreducible components $\mathcal{C}_0(L, \mu)$, each intervening with multiplicity one. Indeed, μ is a minuscule weight, with stabilizer $W_\mu = W_{Q_\mu}$ for a certain parabolic subgroup $Q = Q_\mu \subset G$ defined over $\bar{\mathbb{Q}}$. Here W_{Q_μ} is the Weyl group of any Levi factor of Q_μ. The irreducible components of r_μ are indexed by $(W_P \backslash W_{\hat{G}} / W_{Q_\mu})$ where W_P is the Weyl group of the Langlands dual $\hat{L}$ of L and $W_{\hat{G}}$ is the absolute Weyl group of G. The highest weight of the component corresponding to w, relative to the standard ordering induced by P_0, is the one in the orbit containing $w\mu$. We identify two elements $\lambda, \lambda' \in \mathcal{C}_0(L, \mu)$ if they are associate; i.e., if there is an element of the (relative) Weyl group W_G that normalizes L that takes λ to λ'. Let $\mathcal{C}(L, \mu)$ be the set of equivalence classes for this relation, and let $\mathcal{C}(\mu) = \coprod_L \mathcal{C}(L, \mu)$, where L runs through the classes of standard Levi subgroups of G. The set $\mathcal{C}(\mu)$ of pairs $(L, w\mu)$ is partially ordered by inclusion on the first factor. If $L \subset G$ is an F-rational Levi factor, let $i_{LG} : B(L) \to B(G)$ denote the natural map.

Proposition 4.1.

(i) *There is a canonical surjective map* $\mathrm{Strat} : \mathcal{C}(\mu) \to B(G, \mu)$ *such that, for any* $(L, w\mu) \in \mathcal{C}(\mu)$, $\mathrm{Strat}(L, w\mu) = i_{LG}(b_L(w\mu))$, *where* $b_L(w\mu) \in B(L)$ *is the unique basic class such that* $\kappa(b_L(w\mu)) = (w\mu)^\#$ *for* L.

(ii) *Let* $b \in B(G, \mu)$, *and let* $\mathrm{Rep}(b) = \mathrm{Strat}^{-1}(b) \subset \mathcal{C}(\mu)$. *The set* $\mathrm{Rep}(b)$ *contains a unique maximal element* $(M = M(b), w_b \mu)$. *Here* $M(b)$ *is the centralizer of the slope morphism attached to* b [39, 32] *and* b *is the image of a basic* σ-*conjugacy class* $b_M \in B(M)$ *under the natural map* $B(M) \to B(G)$.

(iii) *There is a bijection between* $\mathrm{Rep}(b)$ *and the set of* $\mathcal{P}(b)$ *of standard* $\mathbb{Q}_p$-*rational parabolics* $P \subset M = M(b)$ *that transfer to the inner form* $J(b)$ *of* M *defined by the basic* σ-*conjugacy class* b. *(We call such parabolics* b-*relevant.)*

The proof of this proposition makes use of simple properties of minuscule weights. It is an amusing exercise to work it out explicitly when $G = GL(n)$ and μ is any minuscule weight.

We assume $\bar{S}$ admits a stratification by locally closed reduced $k(v)$-rational subschemes

$$\bar{S} = \coprod_{b \in B(G, \mu)} S(b), \tag{6}$$

with each $S(b)$ stable under Hecke correspondences. This is conjectured to be true in general. In the PEL case, the existence of such a stratification follows from Theorem 3.6 of [39]. This defines a map from $\bar{S}(\bar{k}(v))$ to $B(G)$, with image necessarily in $B(G, \mu)$ (by Mazur's theorem). We define $S(b)_{\mathrm{geom}}$ to be the inverse image of b with respect to this map. Theorem 3.6 of [39] asserts that $S(b)_{\mathrm{geom}}$ is the set of geometric points of a locally closed reduced subscheme $S(b)$. Stability under Hecke correspondences follows from invariance of isocrystals with respect to isogeny. The same theorem of [39] asserts moreover that $S(b') \subset \bar{S}(b)$ if and only if $\bar{\nu}(b') \le \bar{\nu}(b)$. We assume our stratification to have this property as well.

416   M. Harris

4.1. Example (notation as in §3)

With respect to the factorization (5), we can write $\mu = (\mu_1, \ldots, \mu_r; \mu_0)$, where μ_0 is the $\mathbb{Q}_p^\times$-factor; likewise, we write $b = (b_1, \ldots, b_r; b_0)$. With our conventions, $\mu_i = 0$ for $i > 1$, and μ_1 is the minuscule coweight $(1, 0, \ldots, 0)$ in the usual coordinates; i.e., r_{μ_1} is the standard n-dimensional representation of $GL(n)$. Then $b \in B(G, \mu)$ if and only if $b_1 \in B(GL(n)_{F_w}, \mu_1)$, $b_i = 0$ for $i > 1$, and $\kappa(b_0) = \mu_0^\# \in X^*(\hat{\mathbb{G}}_m)$. Let $M_{\frac{r}{s}}$ denote a simple isocrystal with slope $\frac{r}{s}$. We can write $B(GL(n)_{F_w}, \mu_1) = \{b_1(h), 0 \leq h \leq n - 1\}$, where $b_1(h)$ corresponds to the height n F_{w_1}-isocrystal M with slope decomposition $M_{\frac{1}{n-h}} \oplus M_0^h$. Let $b(h)$ be the element of $B(G, \mu)$ corresponding to $b_1(h)$. Then the group $J(b(h))$ is isomorphic to

$$D_{\frac{1}{n-h}}^\times \times GL(h, F_w) \times \prod_{i>1} B_{w_i}^{\mathrm{op}, \times} \times \mathbb{Q}_p^\times, \tag{7}$$

where $D_{\frac{1}{n-h}}$ is the division algebra over F_w with invariant $\frac{1}{n-h}$.

Let $\bar{S}$ denote the special fiber of the smooth $\mathcal{O}_w$-model $S_K(G, X)$ of $_K Sh(G, X)$. Then the stratification (6) holds with $S(b(h))$ defined as follows. As mentioned at the end of §3, $S_K(G, X)$ carries a natural family $\mathcal{G}$ of p-divisible $\mathcal{O}_w$-modules of height n and dimension 1. For each dimension $h = 0, 1, \ldots, n - 1$, there is a unique stratum $S(h) = S(b(h)) \subset \bar{S}$ of dimension h, defined by the property that, over every geometric point $x \in S(h)$, the maximal étale quotient $\mathcal{H}_x^{et}$ of $\mathcal{H}_x$ is of height h (i.e., of height $h[F : \mathbb{Q}_p]$ as p-divisible group). It is proved in [20] that each stratum $S(h)$ is smooth. The proof makes use of Drinfeld's explicit deformation theory for one-dimensional formal $\mathcal{O}_w$-modules.

For general Shimura varieties, $S(b)$ is almost never expected to be smooth (see, e.g., [37]). Langlands and Rapoport have formulated a conjecture describing $\bar{S}(\bar{k}(v))$ as a disjoint union of subsets $S(\phi)$, each stable under Hecke correspondences. Here ϕ runs through the set $\Phi(G, X)$ of *Langlands-Rapoport parameters*: admissible homomorphisms $\phi \colon \mathfrak{P} \to \mathfrak{G}_G$, where $\mathfrak{P}$ is the pseudomotivic groupoid and $\mathfrak{G}_G$ is the neutral groupoid attached to G [34, 36].

There is a map $b \colon \Phi(G, X) \to B(G, \mu)$, defined by restricting ϕ to the Dieudonné groupoid $\mathcal{D}_p$ (cf. [36, p. 181]), so that $S(\phi) \subset S(b(\phi))(\bar{k}(v))$. Moreover, every point in $S(b)(\bar{k}(v))$ belongs to exactly one $S(\phi)$ with $b(\phi) = b$. We add the assumption that, to each admissible ϕ, we can associate a locally finite disjoint union $\tilde{S}(\phi)$ of closed reduced subschemes such that $S(\phi)$ is the set of $\bar{k}(v)$ points of $\tilde{S}(\phi)$. Under these conditions, the formal completion $S_{/\tilde{S}(\phi)}$ of the scheme S along the subset $\tilde{S}(\phi)$ of its special fiber can be defined as in [40, 6.22].

Let I_ϕ denote the automorphism group of the admissible homomorphism ϕ, in the sense of [34]. I_ϕ is a reductive algebraic group over $\mathbb{Q}$, an inner form of the centralizer in G of the torus $T(\phi) := \phi(\mathfrak{P}^+)$, where $\mathfrak{P}^+$ is the identity component of $\mathfrak{P}$. There is a natural homomorphism $I_\phi(\mathbb{Q}) \to J(b(\phi))$, well defined up to conjugacy. At all finite primes $\ell \neq p$, the inner twist is trivial, hence there is a natural map $I_\phi(\mathbb{Q}) \to G(\mathbf{A}_f^p)$, well defined up to conjugacy. For any $b \in B(G)$, we let $G^b(\mathbf{A}_f) = G(\mathbf{A}_f^p) \times J(b)$. The superscript ^{rig} denotes the rigid-analytic generic fiber of a formal scheme.

Conjecture 4.2.

(i) *For each $b \in B(G, \mu)$, there is a formal scheme $\breve{\mathcal{M}}(b, \mu)$ over $Spf(\mathcal{O}_{\mathcal{L}})$, with a Weil descent datum [40, 3.45] over $Spf(\mathcal{O}_v)$ and a compatible action of $J = J(b)$.*

(ii) *There is an étale $J(b)$-equivariant surjective rigid-analytic morphism*

$$\pi\colon \breve{\mathcal{M}}^{\mathrm{rig}}(b, \mu) \to \breve{\mathcal{F}}^{\mathrm{wa}}(b, \mu) \times \Delta',$$

where $\breve{\mathcal{F}}^{\mathrm{wa}}(b, \mu)$ is the rigid-analytic open subset of weakly admissible flags (relative to b) in the flag variety G/Q_μ [40, §1] and Δ' is a discrete homogenous space for $G(\mathbb{Q}_p)$. The morphism π is compatible with Weil descent data on both sides.

(iii) *For any open subgroup $K' \subset K_p$, there is a rigid-analytic covering $\pi_{K'}\colon \mathcal{M}_{K'}(b, \mu) \to \breve{\mathcal{M}}^{\mathrm{rig}}(b, \mu)$, and for any pair $K'' \subset K'$ of open subgroups of K_p, a morphism $\pi_{K'',K'}\colon \mathcal{M}_{K''}(b, \mu) \to \mathcal{M}_{K'}(b, \mu)$, such that $\pi_{K''} = \pi_{K'} \circ \pi_{K'',K'}$. The projective system $\mathcal{M}_{K'}(b, \mu)$ thus inherits a continuous action of $G(\mathbb{Q}_p)$, covering the natural action on the second factor of $\breve{\mathcal{F}}^{\mathrm{wa}}(b, \mu) \times \Delta'$.*

(iv) *Let ϕ be a Langlands-Rapoport parameter, and let $b = b(\phi)$. Let $x \in \tilde{S}(\phi)$ be a basepoint. There is an isomorphism of formal schemes (*local uniformization*):*

$$u_{\phi,x}\colon [\breve{\mathcal{M}}(b, \mu) \times (I_\phi(\mathbb{Q}) \backslash G^{b(\phi)}(\mathbf{A}_f)/K^p)]/J(b(\phi)) \xrightarrow{\sim} S_{/\tilde{S}(\phi)}.$$

Here $J(b(\phi))$ acts diagonally, on the first factor as in (i) and on the second via the inclusion of $J(b(\phi))$ in $G^{b(\phi)}$. The Weil descent datum on $\breve{\mathcal{M}}(b, \mu)$ induces an effective descent datum on the left-hand side, and $u_{\phi,x}$ is an isomorphism of formal schemes over $Spf(\mathcal{O}_v)$. Moreover $u_{\phi,x}$ is equivariant with respect to the actions of the Hecke correspondences (relative to K) on both sides.

(v) *More generally, for any open subgroup $K' \subset K_p$, there is an isomorphism of rigid-analytic spaces*

$$u_{\phi,x,K'}\colon [\mathcal{M}_{K'}(b, \mu) \times (I_\phi(\mathbb{Q}) \backslash G^{b(\phi)}(\mathbf{A}_f)/K^p)]/J(b(\phi))$$
$$\xrightarrow{\sim} (\Pi_{K'})^{-1}([S_{/\tilde{S}(\phi)}]^{\mathrm{rig}}.$$

Here $\Pi_{K'}\colon {}_{K' \times K^p}Sh(G, X) \to {}_K Sh(G, X)$ is the standard étale covering in characteristic zero. The isomorphism $u_{\phi,x,K'}$ is rational over E and equivariant with respect to Hecke correspondences (relative to $K' \times K^p$).

Sections 5 and 6 of [40] are largely devoted to proving a version of conjecture 4.2 for the PEL-type Shimura varieties considered there, in which Langlands-Rapoport parameters are replaced by isogeny classes of PEL abelian varieties[4]. In particular, the conjecture is true for the Shimura varieties considered in [20]. However, much more is true.

---

[4]Assertion (ii) is proved in [40] assuming a conjecture of Fontaine, recently proved by Breuil.

Theorem 4.3. *Let $Sh(G, X)$ be the Shimura variety considered in §3. Then*

(i) *Conjecture 4.2 is valid for $Sh(G, X)$, with Langlands-Rapoport parameters replaced by isogeny classes for the moduli problem 3.1.*

(ii) *For any isogeny class ϕ for $Sh(G, X)$, the connected components of $\tilde{S}(\phi)$ are closed points.*

(iii) *For any open subgroup $K' \subset K_p$, and any $b \in B(G, \mu)$, there is a formal scheme $\check{\mathcal{M}}_{K'}(b, \mu)$ over $Spf(\mathcal{O}_{\mathcal{L}})$, with $J(b)$-action, a Weil descent datum over $Spf(\mathcal{O}_v)$, and a $J(b)$-equivariant isomorphism $\check{\mathcal{M}}_{K'}^{\mathrm{rig}}(b, \mu) \xrightarrow{\sim} \mathcal{M}_{K'}(b, \mu)$. These isomorphisms are compatible with inclusions $K'' \subset K'$ of open subgroups, and the $G(\mathbb{Q}_p)$-action on the projective system $\{\check{\mathcal{M}}_{K'}^{\mathrm{rig}}(b, \mu)\}$ induced by the isomorphism with $\{\mathcal{M}_{K'}(b, \mu)\}$ extends to a continuous $G(\mathbb{Q}_p)$-action on the projective system $\{\check{\mathcal{M}}_{K'}(b, \mu)\}$ of formal schemes. Each $\check{\mathcal{M}}_{K'}(b, \mu)$ is regular.*

(iv) *Similarly, there is a $G(\mathbf{A}_f)$-equivariant system of regular $\mathcal{O}_v$-schemes $S_{K' \times K^p}$ with $G(\mathbf{A}_f)$-equivariant isomorphisms*

$$S_{K' \times K^p} \otimes_{\mathrm{Spec}(\mathcal{O}_v)} \mathrm{Spec}(F) \xrightarrow{\sim} {}_{K' \times K^p} Sh(G, X).$$

(v) *Let ϕ denote an isogeny class for the moduli problem (?), and let $b = b(\phi)$. Let $\tilde{S}_{K' \times K^p}(\phi)$ denote the scheme-theoretic inverse image of $\tilde{S}(\phi)$ in $S_{K' \times K^p}$. The rigid uniformization of 4.2(v) extends to a Hecke equivariant isomorphism of formal schemes over $Spf(\mathcal{O}_v)$:*

$$u : [\check{\mathcal{M}}_{K'}(b, \mu) \times (I_\phi(\mathbb{Q}) \backslash G^{b(\phi)}(\mathbf{A}_f) / K^p)] / J(b(\phi))$$

$$\xrightarrow{\sim} [S_{K' \times K^p}]_{/ \tilde{S}_{K' \times K^p}(\phi)}.$$

As noted above, (i) is a special case of the results of [40]. Assertion (iii) is due to Drinfeld [16], and uses his theory of level structures for one-dimensional divisible $\mathcal{O}_v$-modules (Drinfeld bases); (ii) is a consequence of the explicit deformation theory of [16]. Assertion (iv) is proved using Drinfeld bases, and (v) is a formal consequence of the corresponding assertion when $K = K_p$.

5. Vanishing Cycles

Let E, v, and F be as above. Henceforward, we assume G to be anisotropic modulo its center; then $Sh(G, X)$ is a projective limit of projective varieties. For any open $K' \subset K_p$, the direct image $\pi_{K', *} \mathbb{Q}_\ell$ of the constant sheaf on ${}_{K' \times K^p} Sh(G, X)$ is a locally constant étale sheaf on ${}_K Sh(G, X)$. We use the same notation for the corresponding étale sheaf on ${}_K Sh(G, X)^{\mathrm{rig}}$. We let $R\Psi_{K'}$ denote the nearby cycles complex $R\Psi(\pi_{K', *} \mathbb{Q}_\ell)$ on the special fiber $\bar{S}$. We view $R\Psi_{K'}$ as an object in the bounded derived category of constructible ℓ-adic complexes on $\bar{S}$. For $K'' \subset K'$, pullback via $\pi_{K'',K'}$ induces a canonical morphism $R\Psi_{K'} \to R\Psi_{K''}$. Define

$$R\Gamma(\bar{S}, R\Psi) := \varinjlim_{K',K^p} R\Gamma(\bar{S}, R\Psi_{K'});$$

$$R\Gamma(Sh(G,X), \mathbb{Q}_\ell) = \varinjlim_{K',K^p} R\Gamma(_{K' \times K^p} Sh(G,X), \mathbb{Q}_\ell).$$

Here the subscript K^p is omitted from $\bar{S}$ for convenience. There is a canonical spectral sequence

$$E_2^{p,q} = H_c^p(\bar{S}, \Psi^q) \Rightarrow H_c^{p+q}(Sh(G,X), \mathbb{Q}_\ell) \qquad (8)$$

of $G(\mathbf{A}_f) \times \mathrm{Gal}(\bar{F}/F)$-modules, with $\Psi^q = R^q \Psi$.

We work in a modified Grothendieck group $\mathrm{Groth}(G)$ of equivalence classes of admissible $G(\mathbf{A}_f) \times W(F)$-modules [6, 20].

If M is an admissible $G(\mathbf{A}_f) \times W(F)$-module, $[M]$ denotes the corresponding object of $\mathrm{Groth}(G)$. Let $[H(Sh(G,X), \mathbb{Q}_\ell)] = \sum_i (-1)^i H^i(Sh(G,X), \mathbb{Q}_\ell) \in \mathrm{Groth}(G)$. Then (8) corresponds to an equality in $\mathrm{Groth}(G)$:

$$[H(Sh(G,X), \mathbb{Q}_\ell)] = \sum_{p,q} (-1)^{p+q} [H_c^p(\bar{S}, \Psi^q)]. \qquad (9)$$

Here and below, the constant sheaf $\mathbb{Q}_\ell$ can be replaced by the ℓ-adic local system $\mathcal{L}_\xi$ attached to a finite-dimensional absolutely irreducible representation ξ of G with coefficients in $\bar{\mathbb{Q}}_\ell$.

On the other hand, the stratification $\bar{S} = \coprod_{b \in B(G,\mu)} S(b)$ gives rise to a spectral sequence of dévissage, for each term on the right-hand side of (9). Let Ψ_b^q denote the pullback to $S(b)$ of Ψ^q. When Π_f is an irreducible admissible representation of $G(\mathbf{A}_f)$, let $[\Pi_f]$ denote the Π_f isotypic component, which makes sense in the Grothendieck group. Then

$$[H(Sh(G,X), \mathbb{Q}_\ell)][\Pi_f] = \sum_{p,q,b} (-1)^{p+q} [H_c^p(S(b), \Psi_b^q)][\Pi_f], \qquad (10)$$

for any Π_f.

Now let (b, μ) be a general weakly admissible pair for $G(\mathbb{Q}_p)$, in the sense of [40]. Consider the $G(\mathbb{Q}_p) \times J(b) \times W(F)$-equivariant system $\mathcal{M}_{K'}(b, \mu)$ of rigid-analytic converings of $\breve{\mathcal{M}}^{\mathrm{rig}}(b, \mu)$. Let $\mathrm{Groth}(G, b)$ be the Grothendieck group, in the above sense, now of smooth representations of $G(\mathbb{Q}_p) \times J(b) \times W(F)$. Define

$$[H_{(b,\mu)}] = \sum_i (-1)^i [\varinjlim_{K'} H_c^i(\mathcal{M}_{K'}(b, \mu), \mathbb{Q}_\ell)].$$

The cohomology is ℓ-adic étale cohomology, in the sense of Berkovich [2]. It is smooth as a representation of $G(\mathbb{Q}_p) \times J(b)$ [3]. It is reasonable to assume (cf. [18, 20]) it is a direct limit of finite-dimensional representations of $W(F)$. Let $R\Theta_{(b,\mu)} = \varinjlim_{K'} R\Theta(\pi_{K',*} \mathbb{Q}_\ell)$, where $\pi_{K',*} \mathbb{Q}_\ell$ is viewed as an étale sheaf on $\breve{\mathcal{M}}^{\mathrm{rig}}(b, \mu)$ and $R\Theta$ is Berkovich's nearby cycle functor for formal schemes [3, I., §4]: $R\Theta_{(b,\mu)}$ belongs to the derived category of constructible ℓ-adic sheaves on the special fiber $\bar{\mathcal{M}}(b,\mu)$ of the formal $Spf(\mathcal{O}_v)$-scheme $\mathcal{M}(b,\mu)$. Writing $\Theta^j = R^j\Theta$, and letting $\check{\Theta}^j$ denote the $G(\mathbb{Q}_p)$-smooth contragredient, we then have

$$[H_{(b,\mu)}] = \sum_{i,j}(-1)^{i+j}[H_c^i(\bar{\mathcal{M}}(b,\mu),\check{\Theta}^j_{(b,\mu)})]\,. \tag{11}$$

Conjecture 5.1. *Let ϕ be a Langlands-Rapoport parameter for the special fiber $\bar{S}$ of $Sh(G,X)$ at v. Let $b = b(\phi)$, and let $\mu = \mu_X$. Let $\Psi^q(\phi)$ be the pullback of Ψ^q (or equivalently of Ψ_b^q) to the locally finite scheme $\tilde{S}(\phi)$. For every j, there is a $G(\mathbf{A}_f) \times W(F)$-equivariant isomorphism of ind-constructible sheaves:*

$$u_{\phi,x}^*\colon [\Theta^q_{(b,\mu)} \times (I_\phi(\mathbb{Q})\backslash G^{b(\phi)}(\mathbf{A}_f)/K^p)]/J(b(\phi)) \xrightarrow{\sim} \Psi^q(\phi)\,,$$

covering the uniformization map $u_{\phi,x}$.

When $Sh(G,X)$ is one of the PEL type Shimura varieties treated in [40], the analogue of this conjecture for torsion coefficients has been proved by L. Fargues.

5.1. Reduction to basic classes

Suppose $b \in B(G,\mu)$ is the image under the map Strat of a pair $(L,w\mu)$, as in Proposition 4.1. Let $b_L = b_L(w\mu) \in B(L)$ be the corresponding basic class. Then there is (conjecturally!) a Rapoport-Zink space $\check{\mathcal{M}}(b_L,w\mu)$, and thus a cohomology representation $[H_{(b_L,w\mu)}] \in \mathrm{Groth}(L,b_L)$. Let $P \subset G(\mathbb{Q}_p)$ be a parabolic subgroup with Levi factor L. Note that $J(b_L) = J(b)$. There is a thus a natural map (non-normalized parabolic induction)

$$\mathrm{Ind}_P^G\colon \mathrm{Groth}(L,b_L) \to \mathrm{Groth}(G,b)\,.$$

Suppose $(M(b),w\mu)$ is the maximal element in $\mathrm{Strat}^{-1}(b)$; we write b_M instead of $b_{M(b)}$ for the corresponding basic element of $B(M(b),w\mu)$. Then one parabolic P is better than the others. On any representation (ρ,V) of G, the class b defines the structure of an isocrystal. We may thus define the *slope filtration* on V to be the decreasing filtration F_a, indexed by $a \in \mathbb{Q}$, such that $F_a(V)$ is the sum of the isoclinic subspaces of slope $\geq a$. Let $P(b) \subset G$ be the parabolic stabilizing the slope filtration attached to b in any finite-dimensional representation of G. Since $M(b)$ is maximal, it is the centralizer of the slope morphism attached to b, hence $M(b)$ is a Levi factor of $P(b)$. On global grounds, it seems reasonable to propose the following

Conjecture 5.2. *There is an isomorphism*

$$[H_{(b,\mu)}] = \mathrm{Ind}_{P(b)}^G(\mathbb{Q}_p)[H_{(b_M,w\mu)}]\,.$$

There is apparently no geometric relation between the two sides, but Strauch's methods [44] may suffice to establish the corresponding identity of distribution characters.

Now assume b basic, so that $M(b) = G$. Write $J = J(b)$. Kottwitz has proposed a conjecture [38, §5] for the contribution to $[H_{(b,\mu)}]$ of discrete series representations of $G \times J(b)$, generalizing Carayol's conjectures in [10]. Kottwitz' conjecture is expressed in terms of the conjectural classification of L-packets by Langlands parameters, here denoted $\psi\colon W_{\mathbb{Q}_p} \to {}^L G$, and the characters of their centralizers S_ψ. Assume ψ is discrete; i.e., that the identity component S_ψ^0 of S_ψ is contained in $Z(\hat{G})^{\Gamma_p}$. Let $\hat{S}_\psi$ denote the set of equivalence classes of irreducible representations of S_ψ. We let

$$\Pi_\psi(G) = \{\tau \in \hat{S}_\psi | \tau|_{Z(\hat{G})^{\Gamma_p}} = 1\}; \ \Pi_\psi(J) = \{\tau' \in \hat{S}_\psi | \tau'|_{Z(\hat{G})^{\Gamma_p}} = \kappa(b)\}\,.$$

The *local Langlands conjecture* for G identifies elements of $\Pi_\psi(G)$ (resp. $\Pi_\psi(J)$) with irreducible discrete series representations of $G(\mathbb{Q}_p)$ (resp. J), and conjectures that, as ψ varies over discrete Langlands parameters and τ varies over characters of $\hat{S}_\psi$ as above, this gives a complete parametrization of the discrete series.

For $\pi \in \Pi_\psi(J)$, let

$$[H_{(b,\mu)}][\pi'] = \sum_{i,j,k} (-1)^{i+j+k}[\mathrm{Ext}^k_J(H^i_c(\bar{\mathcal{M}}(b,\mu), \check{\Theta}^j_{(b,\mu)}), \pi')]\,. \tag{12}$$

The Ext groups are taken in the category of smooth J-modules. The result $[H_{(b,\mu)}][\pi]$ is a virtual representation of $G(\mathbb{Q}_p) \times W(F)$, and should belong to the Grothendieck group of admissible $G(\mathbb{Q}_p) \times W(F)$-modules in the naive sense (i.e., modules of finite length).

Recall that r_μ is taken to be a representation of $\hat{G} \rtimes W(F)$. In what follows, $\check{\pi}$ and $\check{\tau}$ denote the contragredients of π and τ, respectively.

Conjecture 5.3. (Kottwitz) *Let ψ be a discrete Langlands parameter for $G(\mathbb{Q}_p)$, and let $\pi' \in \Pi_\psi(J)$. Then (up to sign)*

$$[H_{(b,\mu)}][\pi'] = \sum_{\pi \in \Pi_\psi(G)} [\check{\pi} \otimes \mathrm{Hom}_{S_\psi}(\tau_\pi \otimes \check{\tau}_{\pi'}, r_\mu \circ \psi)]$$

as virtual representation of $G(\mathbb{Q}_p) \times W(F)$.

The global theory of Shimura varieties, in conjunction with proposition 4.1(iii), suggests a generalization of Conjecture 5.3 to accomodate Langlands parameters induced from b-relevant parabolic subgroups of G. Let $P \subset G$ be a b-relevant parabolic, with Levi subgroup L, and let $P_J \subset J$ be a transfer of P, with Levi subgroup L_J. There is a basic $b_L \subset B(L)$ such that $i_{LG}(b_L) = b$. The bijection of Proposition 4.1(iii) realizes b_L as $\mathrm{Strat}(L, \mu(b_L))$ for some minuscule character $\mu(b_L)$ of (a maximal torus of) $\hat{L}$, corresponding to a minuscule representation r_{μ, b_L}. We consider discrete Langlands parameters ψ for $L(\mathbb{Q}_p)$, and define the L-packets $\Pi_\psi(L)$ and $\Pi_\psi(J_L)$ as above. Let I^G_P, $I^J_{P_J}$ denote normalized induction. For $\pi \in \Pi_\psi(L)$, let

$$[H_{(b,\mu)}][\pi] = \sum_{i,j,k} (-1)^{i+j+k}[\mathrm{Ext}^k_{G(\mathbb{Q}_p)}(H^i_c(\bar{\mathcal{M}}(b,\mu), \check{\Theta}^j_{(b,\mu)}), I^G_P\pi)]\,. \tag{13}$$

We make the analogous definition for $\pi' \in \Pi_\psi(L_J)$.

Conjecture 5.4. *Let ψ be a discrete Langlands parameter for $L(\mathbb{Q}_p)$, and let $\pi' \in \Pi_\psi(L_J)$. Suppose $I^J_{P_J}\pi'$ is irreducible. Then (up to sign)*

$$[H_{(b,\mu)}][\pi'] = \sum_{\pi \in \Pi_\psi(L)} [I^G_P(\check{\pi}) \otimes \operatorname{Hom}_{S_\psi}(\tau_\pi \otimes \check{\tau}_{\pi'}, r_{\mu,b_L} \circ \psi)]$$

as virtual representation of $L(\mathbb{Q}_p) \times W(F)$.

Remark 5.5. *There are analogous conjectures with the roles of G and J exchanged.*

It is not clear whether the formula in (5.4) should still hold when $I^G_P\pi$ is reducible. For example, is it compatible with Conjecture 5.3 when $P = B$ is a Borel subgroup and $I^G_P\pi$ contains the Steinberg representation as subquotient?

5.2. The twisted unitary case

In the remainder of this section, we restrict attention to the Shimura varieties $Sh(G, X)$ of §3. In that case, Conjecture 5.2 holds for straightforward geometric reasons, and nearly all the conjectures considered above are established in [20]. The starting point is Theorem 4.3. Note that the conjectures simplify considerably. As in (4.1), we can factor

$$\bar{M}(b,\mu) \xrightarrow{\sim} \bar{M}(b_1,\mu) \times \prod_{i \neq 1} \bar{M}(b_i, \mu_i). \tag{14}$$

For $i > 1$ the factors $\bar{M}(b_i, \mu_i)$ are discrete sets with trivial Galois action. Similarly, $\bar{M}(b_0, \mu_0) \xrightarrow{\sim} \mathbb{Q}_p^\times/\mathbb{Z}_p^\times$; the Weil group action on this factor is non-trivial but elementary. Finally, let $L(h) \subset GL(n, F_w)$ be the Levi factor corresponding to $b_1(h) \in B(GL(n, F_w), \mu_1)$. Then

$$\bar{M}(b_1(h), \mu_1) \xrightarrow{\sim} GL(n, F_w) \times_{L(h)} [\bar{M}(h) \times GL(h, F_w)/GL(h, \mathcal{O}_w)]. \tag{15}$$

In the first place, the schemes $\bar{M}(b,\mu)$ are *zero-dimensional* (more precisely, are inductive limits of zero-dimensional reduced schemes) for all $b \in B(G,\mu)$. Thus $i = 0$ in (12) and $\check{\Theta}^j_{(b,\mu)}$ is just a vector space. Next, one sees easily from (14) and (15), and from compactness of $D^\times_{\frac{1}{n-h}}$ modulo its center, that $\check{\Theta}^j_{(b,\mu)}$ is a projective object in the category of smooth $J(b)$-modules. Thus (when G is replaced by J) the index k in (12) can also be taken to be zero. Finally, the L-packets Π_ψ are all singletons. Let $J = J(b) = J(b(h))$. There is a map $JL \colon \Pi_\psi(J) \to \Pi_\psi(G)$ defined as follows. We write $\Pi_\psi(J) = \pi' = [\pi'_1(n-h) \otimes \pi'_1(h)] \otimes \prod_{i>1} \pi'_i \otimes \pi'_0$, with respect to the factorization (7). Here $\pi'_1(n-h)$ (resp. $\pi'_1(h)$) is an irreducible representation of $D^\times_{\frac{1}{n-h}}$ (resp. $GL(h, F_w)$), and the other factors are clear. Let $P(h) = P(b(h))$. Define

$$JL(\pi') = \operatorname{Ind}^{GL(n,F_w)}_{P(h)}[JL(\pi'_1(n - h)) \otimes \pi'_1(h)] \otimes \prod_{i \neq 1} \pi'_i \otimes \pi'_0 \tag{16}$$

where the JL on the right-hand side is the Jacquet-Langlands correspondence
[41, 15] between representations of $D^\times_{\frac{1}{n-h}}$ and discrete series representations of
$GL(n-h, F_w)$. Thus Conjecture 5.4 asserts

$$\pm \sum_j (-1)^j [\mathrm{Hom}_{J(b)}(\check{\Theta}^j_{(b,\mu)}, \pi')] = JL(\check{\pi}') \otimes \sigma(JL(\check{\pi}')). \tag{17}$$

This formula, a version of Carayol's conjecture [10], is the substance of Theorem
VIII.1.3 of [20], at least in the supercuspidal case.

6. Local Terms in the Lefschetz Formula

We return to the general setting of §§4 and 5, and assume the truth of Conjectures 4.2 and 5.1. The identity (10) reduces study of the representation of
$G(\mathbf{A}_f) \times W(F)$ on $[H(Sh(G, X), \mathbb{Q}_\ell)]$ to the determination of the representations
on the individual strata. For fixed $b \in B(G, \mu)$, we write

$$[H(S(b), R\Psi)] = \sum_{p,q} (-1)^{p+q} [H^p_c(S(b), \Psi^q(b))].$$

Fix an admissible irreducible representation Π_f of $G(\mathbf{A}_f)$, and suppose the
p-adic component $\Pi_{f,p}$ is induced from a discrete representation π of the Levi component L of a parabolic subgroup $P \subset G(\mathbb{Q}_p)$. Let ψ be the corresponding discrete
Langlands parameter for L. Roughly speaking, one expects that the map Strat of
Proposition 4.1 identifies the set of b for which $[H(S(b), R\Psi)][\Pi_f] \neq 0$ with the
set $\mathcal{C}(L, \mu)$. Conjectures 5.3 and 5.4 suggest that the semisimplified local Galois
representation on $\sum_b [H(S(b), R\Psi)][\Pi_f]$ then should be something like

$$\sum_{w \in W_P \backslash W_G / W_{Q_\mu}} r_{w\mu} \circ \psi. \tag{18}$$

This natural generalization of (3) needs to be modified [29] when Π_f is attached
to an endoscopic automorphic representation, but it provides a heuristic interpretation of Proposition 4.1 as well as Conjectures 5.3 and 5.4.

It is plausible that Conjectures 5.3 and 5.4 can be established by global methods, using a trick due to Boyer [6]. Assuming Conjecture 5.2 is true, as in [20] (and
[6]), identity (10) shows that, if $\Pi_{f,p}$ is supercuspidal, then $[H(S(b), R\Psi)][\Pi_f] \neq 0$
only for b basic. If $b = b(\phi)$ is basic, I_ϕ is a twisted inner form of G. Conjecture 5.3
should then follow by comparing stable trace formulas for G and I_ϕ, at least when
all $\pi \in \Pi_\psi$ are supercuspidal.

Remark 6.1. *Boyer's trick should also provide a means to apply the argument of
Taylor-Wiles to study deformations of* (mod ℓ) *Galois representations in the part
of the cohomology of $Sh(G, X)$ supercuspidal at p. See [21] for an application of
this type in a case where the whole special fiber at p is basic.*

Assuming Conjectures 5.2, 5.3, and 5.4, one obtains a formula similar to (18) for all strata, by purely local means, except that the terms $r_{w\mu}$ occur with undetermined multiplicities. It appears that the multiplicities can only be calculated by global means, generalizing Kottwitz' techniques for "counting points" in [29, 30].

Let $\mathcal{H}(G(\mathbf{A}_f))$ denote the big Hecke algebra of locally constant compactly supported functions on $G(\mathbf{A}_f)$, and define $\mathcal{H}(G(\mathbb{Q}_p))$, $\mathcal{H}(J(b))$, and $\mathcal{H}(G(\mathbf{A}_f^p))$ analogously. For $K' \subset G(\mathbf{A}_f)$ an open compact subgroup, we let $\mathcal{H}_{K'} \subset \mathcal{H}(G(\mathbf{A}_f))$ be the subalgebra of K'-biinvariant functions. These define correspondences on $_{K'}Sh(G, X)$, and we suppose these extend compatibly to cohomological correspondences on $[H(S(b), R\Psi)]^{K'} \subset [H(S(b), R\Psi)]$, covering an action by correspondences on $S(b)$. We will consider functions $f = f^p \otimes f_p \in \mathcal{H}_{K'}$, with $f^p \in \mathcal{H}(G(\mathbf{A}_f^p))$ and $f_p \in \mathcal{H}(G(\mathbb{Q}_p))$ such that the fixed point set $\mathrm{Fix}(f) \in S(b)(\bar{k}(v))$ is finite. As K' varies, such functions generate $\mathcal{H}(G(\mathbf{A}_f))$, hence their traces determine the admissible virtual $G(\mathbf{A}_f)$-module $[H(S(b), R\Psi)]$ up to isomorphism.

Since the actions of $G(\mathbf{A}_f)$ and $W(F)$ commute, the trace $\mathrm{Tr}(f|[H(S(b), R\Psi)])$ takes values in $\overline{\mathbb{Q}}_\ell \otimes \mathrm{Groth}(W(F))$, where $\mathrm{Groth}(W(F))$ is the Grothendieck group of virtual $W(F)$-modules. Indeed, if $f \in \mathcal{H}_{K'}$, for each p, q, and $a \in \overline{\mathbb{Q}}_\ell$, the generalized eigenspace $H_c^p(S(b), \Psi_b^q)^{K'}((f - a))$ of f with eigenvalue a is a finite-dimensional $W(F)$-module. We let

$$W\,\mathrm{Tr}(f|[H(S(b), R\Psi)]) = \sum_{p,q,a}(-1)^{p+q}a \otimes [H_c^p(S(b), \Psi_b^q)^{K'}((f - a))], \qquad (19)$$

where $[\,]$ denotes passage to $\mathrm{Groth}(W(F))$.

Similarly, if $f_p \in \mathcal{H}(G(\mathbb{Q}_p))$, and π' is any admissible representation of $J(b)$, we can define a virtual trace

$$\mathrm{Loc}(f_p, b, \pi') = W\,\mathrm{Tr}(f_p|[H_{(b,\mu)}][\pi']) = \sum_k (-1)^k W\,\mathrm{Tr}(f_p|\,\mathrm{Ext}^k_{J(b)}([H_{(b,\mu)}]), \pi')\,.$$

$$(20)$$

Here $[H_{(b,\mu)}][\pi']$, defined as in (13), is an admissible virtual representation of $G(\mathbb{Q}_p) \times W(F)$, and $\mathrm{Loc}(f_p, b, \pi') \in \overline{\mathbb{Q}}_\ell \otimes \mathrm{Groth}(W(F))$. The following conjecture is presumably a straightforward application of the Paley-Wiener theorem [4].

Conjecture 6.2. *For any $f_p \in \mathcal{H}(G(\mathbb{Q}_p))$ there is a function $F_b(f_p) \in \mathcal{H}(J(b)) \otimes \mathrm{Groth}(W(F))$ such that*

$$\mathrm{Loc}(f_p, b, \pi') = \mathrm{Tr}(\pi')(F_b(f_p))\,.$$

For $\gamma \in I_\phi(\mathbb{Q})$, let $I_\gamma \subset I_\phi$ denote its centralizer, and let $v(\gamma) = \mathrm{vol}(I_\gamma(\mathbb{Q})\backslash I_\gamma(\mathbf{A}_f))$. The choice of measure on $I_\gamma(\mathbf{A}_f)$ is irrelevant at this stage. The point-counting argument in [20] begins with a formula of the following type:

$$W\,\mathrm{Tr}(f|[H(S(b), R\Psi)]) = \sum_{\{\phi|b(\phi)=b\}} \sum_{\gamma \in I_\phi(\mathbb{Q})/\cong} c(\gamma)v(\gamma)O_\gamma(f^p \cdot F_b(f_p))\,. \qquad (21)$$

Here $I_\phi(\mathbb{Q})/\cong$ is the set of conjugacy classes in $I_\phi(\mathbb{Q})$, $O_\gamma(f^p \cdot f_{\pi'})$ is the orbital integral of f^p over the conjugacy class of γ in $G^b(\mathbf{A}_f)$, and $c(\gamma)$ is a "dimensionless constant" involving signs and the like. Moreover, it is assumed f belongs to a certain class of "acceptable" functions, sufficiently large to determine $[H(S(b), R\Psi)]$ up to isomorphism, for which Fujiwara's simple version of the Lefschetz formula [17] applies. Theorem 2.3 is obtained by relating this formula to the global trace formula on cohomology and combining the result with (10) and (17).

It is tempting to hope that a formula like (21) holds for general groups. However, it is not clear whether the coefficients $c(\gamma)$ can be defined to make this formula comparable to the global trace formula without assuming the fundamental lemma for endoscopy.

Acknowledgements

Work on this report began in the summer of 1999, during a visit to the Sonderforschungsbereich in Münster. I thank Christopher Deninger for making that visit so enjoyable, and Matthias Strauch for encouraging me to lecture on [20].

References

[1] J. Arthur and L. Clozel, *Simple algebras, base change, and the advanced theory of the trace formula*, Annals of Math. Studies, **120**, Princeton: Princeton University Press (1989).

[2] V. G. Berkovich, *Étale cohomology for non-archimedean analytic spaces*, Publ. Math. I.H.E.S., **78**, 5–161 (1993).

[3] V. G. Berkovich, *Vanishing cycles for formal schemes*, Invent. Math., **115**, 549–571 (1994); *II*, Invent. Math., **125**, 367–390 (1996).

[4] J. Bernstein, P. Deligne and D. Kazhdan, *Trace Paley-Wiener theorem for reductive p-adic groups*, J. Analyse Math., **47**, 180–192 (1986).

[5] D. Blasius, *Automorphic forms and Galois representations: Some examples*, in L. Clozel and J. S. Milne, eds., Automorphic Forms, Shimura varieties, and L-functions, New York: Academic Press, vol II, 1–14 (1990).

[6] P. Boyer, *Mauvaise réduction de variétés de Drinfeld et correspondance de Langlands locale*, Invent. Math., **138**, 573–629 (1999).

[7] C. Bushnell, G. Henniart and P. Kutzko, *Correspondance de Langlands locale pour GL_n et conducteurs de paires*, Ann. Sci. E. N.S., **31**, 537–560 (1998).

[8] H. Carayol, *Sur la mauvaise réduction des courbes de Shimura*, Compositio Math., **59**, 151–230 (1986).

[9] H. Carayol, *Sur les représentations ℓ-adiques associées aux formes modulaires de Hilbert*, Ann. scient. Ec. Norm. Sup, **19**, 409–468 (1986).

[10] H. Carayol, *Non-abelian Lubin-Tate theory*, in L. Clozel and J. S. Milne, eds., Automorphic Forms, Shimura varieties, and L-functions, New York: Academic Press, vol II, 15–39 (1990).

[11] H. Carayol, *Variétés de Drinfeld compactes, d'après Laumon, Rapoport, et Stuhler,* Séminaire Bourbaki exp. 756 (1991–1992), Astérisque, **206** (1992), 369–409.

[12] H. Carayol, *Preuve de la conjecture de Langlands locale pour GL_n: Travaux de Harris-Taylor et Henniart,* Séminaire Bourbaki exp. 857 (1998–1999).

[13] L. Clozel, *Représentations Galoisiennes associées aux représentations automorphes autoduales de GL(n),* Publ. Math. I.H.E.S., **73**, 97–145 (1991).

[14] L. Clozel and J.-P. Labesse, *Changement de base pour les représentations cohomologiques de certains groupes unitaires,* appendix to [33].

[15] P. Deligne, D. Kazhdan and M.-F. Vigneras, *Représentations des algèbres centrales simples p-adiques,* in J.-N. Bernstein, P. Deligne, D. Kazhdan, M.-F. Vigneras, Représentations des groupes réductifs sur un corps local, Paris: Hermann (1984).

[16] V. G. Drinfeld, *Elliptic modules,* Math. USSR Sbornik, **23**, 561–592 (1974).

[17] K. Fujiwara, *Rigid geometry, Lefschetz-Verdier trace formula and Deligne's conjecture,* Invent. Math., **127**, 489–533 (1997).

[18] M. Harris, *Supercuspidal representations in the cohomology of Drinfel'd upper half spaces; elaboration of Carayol's program,* Invent. Math., **129**, 75–119 (1997).

[19] M. Harris, *The local Langlands conjecture for $GL(n)$ over a p-adic field, $n < p$,* Invent. Math., **134**, 177–210 (1998).

[20] M. Harris and R. Taylor, *On the geometry and cohomology of some simple Shimura varieties,* Princeton: Annals of Math. Studies (in press).

[21] M. Harris and R. Taylor, *Deformations of automorphic Galois representations,* manuscript (1998).

[22] G. Henniart, *On the local Langlands conjecture for $GL(n)$: the cyclic case,* Ann. of Math., **123**, 145–203 (1986).

[23] G. Henniart, *La conjecture de Langlands locale numérique pour $GL(n)$,* Ann. scient. Ec. Norm. Sup, **21**, 497–544 (1988).

[24] G. Henniart, *Caractérisation de la correspondence de Langlands locale par les facteurs ϵ de paires,* Invent. Math., **113**, 339–350 (1993).

[25] G. Henniart, *Une preuve simple des conjectures de Langlands pour $GL(n)$ sur un corps p-adique,* Invent. Math., **139**, 439–455 (2000).

[26] G. Henniart and R. Herb, *Automorphic induction for $GL(n)$ (over local non-archimedean fields),* Duke Math. J., **78**, 131–192 (1995).

[27] H. Jacquet, I. I. Piatetski-Shapiro and J. Shalika, *Rankin-Selberg convolutions,* Am. J. Math., **105**, 367–483 (1983).

[28] R. Kottwitz, *Isocrystals with additional structure,* Compositio Math., **56** (1985), 201–220.

[29] R. Kottwitz, *Shimura varietes and λ-adic representations,* in L. Clozel and J. S. Milne, eds., Automorphic Forms, Shimura varieties, and L-functions, New York: Academic Press, vol I, 161–209 (1990).

[30] R. Kottwitz, *Points on some Shimura varieties over finite fields,* Jour. of the AMS, **5** (1992), 373–444.

[31] R. Kottwitz, *On the λ-adic representations associated to some simple Shimura varieties,* Invent. Math., **108**, 653–665 (1992).

[32] R. Kottwitz, *Isocrystals with additional structure II,* Compositio Math., **109** (1997), 255–339.

[33] J.-P. Labesse, *Cohomologie, stabilisation, et changement de base,* Astérisque, **257** (1999).

[34] R. P. Langlands, Jr. and M. Rapoport, *Shimuravarietäten und Gerben,* J. reine angew. Math., **378** (1987), 113–220.

[35] G. Laumon, M. Rapoport and U. Stuhler, $\mathcal{D}$-*elliptic sheaves and the Langlands correspondence,* Invent. Math., **113**, 217–338 (1993).

[36] J. S. Milne, *The points on a Shimura variety modulo a prime of good reduction,* in R. P. Langlands and D. Ramakrishnan, eds., The Zeta Functions of Picard Modular Surfaces, Montréal: Les publications CRM (1992), 151–253.

[37] F. Oort, *Which abelian surfaces are products of elliptic curves?,* Math. Ann., **214**, 35–47 (1975).

[38] M. Rapoport, *Non-archimedean period domains,* Proceedings of the International Congress of Mathematicians, Zürich, 1994, 423–434 (1995).

[39] M. Rapoport et M. Richartz, *On the classification and specialization of F-isocrystals with additional structure,* Compositio Math., **103** 153–181 (1996).

[40] M. Rapoport et T. Zink, *Period Spaces for p-divisible Groups,* Princeton: Annals of Mathematics Studies **141** (1996).

[41] J. Rogawski, *Representations of GL(n) and division algebras over a p-adic field,* Duke Math. J., **50**, 161–196 (1983).

[42] T. Saito, *Hilbert modular forms and p-adic Hodge theory,* preprint (1999).

[43] F. Shahidi, *Local coefficients and normalization of intertwining operators for GL(n),* Comp. Math., **48**, 271–295 (1983).

[44] M. Strauch, *On the Jacquet-Langlands correspondence in the cohomology of the Lubin-Tate deformation tower,* Preprintreihe SFB 478 (Münster), **72**, (1999).

[45] A. V. Zelevinsky, *Induced representations of reductive p-adic groups II: on irreducible representations of GL(n),* Ann. Sci. E.N.S., **13**, 165–210 (1980).

UFR de Mathématiques
Université Paris 7
2, Pl. Jussieu
75251 Paris cedex 05, France
E-mail address: harris@math.jussieu.fr

Burgers Turbulence and Dynamical Systems

Renato Iturriaga and Konstantin Khanin

Abstract. We discuss a dynamical system approach to a problem of Burgers turbulence. It is shown that there exists a unique stationary distribution for solutions to spatially periodic inviscid random forced Burgers equations in arbitrary dimension. The construction is based on analysis of minimizing orbits for time-dependent random Lagrangians on a d-dimensional torus. We also discuss how dynamical properties of minimizing trajectories lead to quantitative predictions for physically important universal critical exponents.

1. Introduction

In this paper we discuss recent progress in the understanding of Burgers turbulence. A significant part of this development is connected with applications of the theory of dynamical systems. Certainly dynamical systems were used in hydrodynamics for a very long time. However, the main features of the approach which we describe below are connected with the fact that dynamical systems methods are used not only to construct stationary distributions for solutions of hydrodynamic equations, but also to obtain mathematical results on structure of singularities for typical stationary solutions. The resulting geometrical picture leads to important quantitative predictions for universal scaling exponents related to the pdf (probability distribution function) for the velocity gradients. It also allows us to analyse analytic properties of the structure functions. In this short article it is not possible to comment on all the important aspects of mathematical hydrodynamics. We discuss here only problems which are related to turbulence driven by a random force. The proofs of the results which we discuss here will be published in [22]. Results in the one-dimensional case are based on joint papers of one of us with Weinan E, Alex Mazel and Yakov Sinai (see [9, 10]).

Most of the results which we discuss below are obtained for Burgers equation:

$$\partial_t u + (u \cdot \nabla)u = \nu \Delta u + f(y, t) \tag{1}$$

where $u(y, t) = (u_i(y, t), 1 \leq i \leq d)$ is a velocity field, $y = (y_i, 1 \leq i \leq d) \in \mathbb{R}^d$, and $f(y, t) = (f_i(t, y), 1 \leq i \leq d)$ is an external force. However, for general discussion we shall also refer to the Navier–Stokes equations:

$$\partial_t u + (u \cdot \nabla)u = \nu \Delta u - \nabla p + f(y, t), \quad \text{div } u = 0. \tag{2}$$

A general setting involves one of the hydrodynamics equations listed above being driven by random force. The forcing term is responsible for the constant pumping of energy into the system. The energy also dissipates through different dissipation mechanisms. The balance between pumping and dissipation should in principle lead to the establishment of a stationary probability distribution for solutions. This is a general philosophy and the first task of a mathematical theory of turbulence consists in establishing of the existence and, preferably, uniqueness of this stationary distribution, or, in dynamical systems terminology, the invariant measure. The difficulty of the problem depends on the mechanism of dissipation involved. This mechanism obviously depends on the type of equation and the dimension d. In general the difficulty increases when one passes from a Burgers equation to a Navier–Stokes equation, from viscous case $\nu > 0$ to inviscid limit $\nu \to 0$, and from dimension $d = 1$ (for Burgers) or $d = 2$ (for Navier–Stokes) equation to higher dimensions.

It is important to mention that the construction of the stationary distribution or the invariant measure corresponds to the limit $t \to \infty$. At the same time, the inviscid case, which is mostly interesting for turbulence theory, requires another limiting process $\nu \to 0$. It is tempting, first, to construct an invariant measure μ_ν for $\nu > 0$, that is first to take the limit as $t \to \infty$ and then study the limit of μ_ν as $\nu \to 0$. However, it is extremely difficult to control the last limiting process. Experience of the random Burgers equation, for which an invariant measure can be constructed for both viscous and inviscid cases, suggests that it is more productive to consider from the very beginning $\nu = 0$, construct unique invariant measure μ, and only later prove that $\mu_\nu \to \mu$ as $\nu \to 0$.

As we have mentioned above, a mathematical theory starts with establishing the existence and uniqueness of an invariant measure. However, from the physical point of view the most interesting problem is the analysis of statistical properties of stationary solutions. In other words, how does a typical solution looks like? What are the leading singularities? How does the power spectrum decay? What are the asymptotic properties of pdfs and the structure functions? Those questions are much harder to answer. They form the core of the problem of turbulence. It is probably fair to say that the problem of turbulence splits into a "soft part" of existence-uniqueness statements and a "hard part" of analysing properties of typical solutions. In the rest of the paper we discuss only the Burgers equation, where one can complete the "soft part" in a very general situation, and also answer many of the "hard part" questions (especially in the one-dimensional case). On the contrary, despite recent progress (see [23, 14, 5]), even the situation with a two-dimensional Navier–Stokes equation remains largely open. In the most important case of a finite number of modes being excited by the random force, one can prove uniqueness of the stationary distribution only for large enough viscosity ν. Correspondingly, in the case of the Navier–Stokes equation there are no mathematical results on the asymptotic behaviour in the limit $\nu \to 0$.

2. Random Setting

To make the general discussion more precise we have to specify assumptions on the random force $f(y,t)$. The most natural, from the turbulence theory point of view, is a setting where $f(y,t)$ is a smooth function in space variable y and quite irregular in time t. This brings us to a popular model, where

$$f(y,t) = -\nabla F(y,t), \;\; F(y,t) = \sum_{k=1}^{N} F_k(y)B_k(t). \tag{3}$$

Here $F_k(y)$ are smooth potentials and $B_k(t) = \dot{W}_k(t)$ are independent white noises, corresponding to independent Wiener processes $W_k(t)$. Summation in (3) can be finite or infinite. In the latter case one has to make sure that potentials $F_k(y)$ decay fast enough so that the series converges and defines a smooth forcing term $f(y,t)$. Under assumption (3) the hydrodynamics equations (1), (2) become stochastic PDEs. However, one can essentially eliminate the stochastic part by considering another popular model for a random force, namely the random kicking force. Let

$$f(y,t) = -\nabla F(y,t), \;\; F(y,t) = \sum_{j\in\mathbb{Z}} F^j(y)\delta(t - t_j). \tag{4}$$

The force in (4) corresponds to the situation when the system evolves without any force between moments of kicks t_{j-1} and t_j, so that $u(y,t_j-)$ is obtained from $u(y,t_{j-1}+)$ as a result of free evolution of a system. Then, suddenly, we apply a kicking force which changes the solution discontinuously, i.e.,

$$u(y,t_j+) = u(y,t_j-) - \nabla F^j(y). \tag{5}$$

Smooth potentials $\{F^j(y), j \in \mathbb{Z}\}$ form a realization of some stationary potential-valued random process. The simplest situation corresponds to a Bernoulli process, when different F^j are chosen independently according to some probability distribution $\chi \in \mathcal{P}(C^\infty(T^d))$ in a space of smooth potentials. Here and below we denote by $\mathcal{P}(M)$ the set of all probability distributions on the measure space M. One can also make different assumptions on kicking times t_j. We shall just indicate two natural cases. The first one corresponds to a periodic sequence $t_j = \bar{t}j, \; j \in \mathbb{Z}$. As we shall see later, this assumption in the case of a Burgers equation leads to a problem, which is closely related to random Aubry–Mather theory. In the second situation kicking times are determined by the realization of a stationary random process which is independent of the random process $F^j(y)$. For example, $\{t_j, j \in \mathbb{Z}\}$ can be a realization of a Poisson process. In this paper we consider only the simplest case when kicking potentials form a Bernoulli sequence, and kicking times are periodic with $\bar{t} = 1$.

In both "white" and "kicked" settings the force is given by a stationary random process. Denote by $(\Omega, \mathcal{B}, P)$ the probability space corresponding to this random process. Sometimes we shall use the notation $f^\omega(y,t), F^\omega(y,t), F^{j,\omega}, \; \omega \in \Omega$, for the forcing term and the potentials, indicating their randomness. The time shift

$f(y,t) \to f(y, t + \tau)$ corresponds to the flow of automorphisms of the probability space Ω :

$$\theta^\tau : \Omega \to \Omega, \ \theta^\tau P = P, \ F^\omega(y, t + \tau) = F^{\theta^\tau \omega}(y, t), \ \tau \in \mathbb{R}. \tag{6}$$

Equality $\theta^\tau P = P$ in (6) expresses the stationarity of the random force. Obviously, in the case of periodic kicking times one should consider only the values $\tau = j, j \in \mathbb{Z}$ in (6). We can define now a standard skew-product structure which leads to the definition of invariant measure. Consider the Cauchy problem for one of the equations (1), (2). Denote by $\mathcal{U}$ some natural functional space, which is invariant under the evolution given by (1), (2), or an invariant set in such space. This means that $u(y,t) \in \mathcal{U}$ for all $t > 0$, provided $u(y,0) \in \mathcal{U}$. Denote by $S^\omega(\tau)$ a nonlinear random transformation on $\mathcal{U}$ corresponding to the solution of the Cauchy problem on the time interval $[0, \tau]$ with the random force $f^\omega(y, t)$. In other words, $u(y, \tau) = S^\omega(\tau)u(y,0)$. Clearly, solution of the equation for longer times implies iteration of $S^\omega(\tau)$ for shifted ω. It is easy to see that this process corresponds to non-random transformation $\hat{S}$ acting on the product space $\mathcal{U} \times \Omega$:

$$\hat{S}(\tau)(u, \omega) = (S^\omega(\tau)u, \theta^\tau \omega). \tag{7}$$

Definition 2.1. *A probability measure $\nu \in \mathcal{P}(\mathcal{U} \times \Omega)$ is called an invariant measure for the random equation if*

1. *For any τ, $\hat{S}(\tau)$ preserves ν : $\hat{S}(\tau)\nu = \nu$.*
2. *Marginal distribution of ν on Ω is given by P, i.e.,*

$$\nu(du, d\omega) = \nu^\omega(du)P(d\omega),$$

   where $\nu^\omega(du)$ are the conditional distributions on $\mathcal{U}$ under the condition of fixed ω.
3. *The conditional distributions ν^ω are measurable with respect to the σ - algebra $\mathcal{B}^0_{-\infty}$ generated by the random process on the time interval $(-\infty, 0]$.*

The last condition means that ν^ω does not depend on the future distribution of a random force $f^\omega(y, t), t > 0$. It may look a bit artificial but one may say that all physically relevant invariant measures always satisfy this condition.

We shall describe now a slightly different point of view which also leads to the notion of a stationary distribution for solutions of the random equation. It is easy to see that $S^\omega(\tau)$ defines a stationary Markov process on $\mathcal{U}$ with transition probabilities given by:

$$p_\tau(A \mid u_0) = P(\omega \colon u(y, \tau) = S^\omega(\tau)u_0(y) \in A). \tag{8}$$

Definition 2.2. *A probability measure $\mu \in \mathcal{P}(\mathcal{U})$ is called a stationary distribution for the random equation if it is a stationary distribution for the Markov process (8), i.e.,*

$$\mu(A) = \int_{\mathcal{U}} p_\tau(A \mid u_o)\mu(du_0), \ \tau > 0.$$

The notions of invariant measure and stationary distribution for a random equation are closely connected. Namely, if ν is an invariant measure, then its marginal distribution $\mu(du) = \int_\Omega \nu^\omega(du)P(d\omega)$ is a stationary distribution for the corresponding Markov process. Notice that this property holds only if an invariant measure ν satisfies Condition 3 in Definition 2.1.

3. Random Hamilton–Jacobi Equation and Minimizers

Our main results which we shall formulate in the next section are proven for the Burgers equation on a compact connected Riemannian manifold. For simplicity we consider here only the case of the torus $T^d = \mathbb{R}^d/\mathbb{Z}^d$ which corresponds to periodic potentials F_k in (3), or F^j in (4). Everywhere below we assume that potentials F_k, F^j in (3), (4) belong to $C^\infty(\mathbb{R}^d)$ and are $\mathbb{Z}^d$-periodic functions:

$$F_k(y+m) = F_k(y), \ \ F^j(y+m) = F^j(y), m \in \mathbb{Z}^d, y \in \mathbb{R}^d. \tag{9}$$

We shall denote points of T^d by x and points of the universal cover $\mathbb{R}^d$ by y.

It is well known that due to the creation of shocks the inviscid Burgers equation has no strong, or, in other words, smooth solutions. However, there exists a unique "physical" weak solution for the Cauchy problem, which is called the "viscosity" solution. Given initial data $u(x,0) = u_0(x)$ one can construct the "viscosity" solutions by considering the viscous Burgers equation and taking the limit as $\nu \to 0$. We shall always assume that u_0 is a gradient-like function, i.e., $u_0 = \nabla\phi_0$, where ϕ_0 is a Lipschitz function. It is easy to see that the Burgers equation with a potential force is an evolution equation in a space of gradient-like functions. That means that the set of gradient-like functions is an invariant set. Taking $u(x,t) = \nabla\phi(x,t)$ we obtain the Hamilton–Jacobi equation for ϕ:

$$\partial_t \phi(y,t) + \frac{1}{2}(\nabla\phi(y,t))^2 + F^\omega(y,t) = 0. \tag{10}$$

As in the case of the Burgers equation there are many weak solutions to the Hamilton–Jacobi equation. However, there exists a unique "viscosity" solution corresponding to the Burgers viscosity solution. The exact expression for the viscosity solution is given by the Hopf–Lax–Oleinik variational principle. That is, the "viscosity" solution of the Cauchy problem for (10) in $y \in \mathbb{R}^d, t \in [0,T]$ with the initial condition $\phi(\cdot,0) = \phi_0$ is given by:

$$\phi(y,t) = \inf\left[\phi_0(\gamma(0)) + \int_0^t \left(\frac{1}{2}\dot\gamma^2 - F^\omega(\gamma(\tau),\tau)\right)d\tau\right], \ t \in [0,T], \tag{11}$$

where the infimum is taken over all absolutely continuous curves $\gamma\colon [0,t] \to \mathbb{R}^d$ such that $\gamma(t) = y$ ([21, 24, 31]). Notice, that if potentials F^ω are $\mathbb{Z}^d$-periodic functions, and $\phi_0(y) = b \cdot y + \psi_0(y)$, where $b \in \mathbb{R}^d$ and $\psi_0(y)$ is $\mathbb{Z}^d$-periodic, then the solution (11) has the same form. Namely,

$$\phi(y,t) = b \cdot y + \psi(y,t), t > 0 \tag{12}$$

where $\psi(y,t)$ is a $\mathbb{Z}^d$-periodic function of y. The linear form $b \cdot y$ can be considered as a first integral of the Hamilton–Jacobi equation. For the Burgers equation b corresponds to the average velocity

$$b = \int_{T^d} u(y,t)dx \, ,$$

which is well known to be a first integral. From now on we shall assume that the value of the first integral b is fixed.

Using formula (11) we get the variational principle for $\psi(x,t)$, which we consider now as a function on the torus T^d. For fixed b denote the *action* of a curve $\gamma \colon [s,t] \to T^d$ by

$$A_{s,t}^{\omega,b}(\gamma) = \int_s^t \left(\frac{1}{2}(\dot\gamma(\tau) - b)^2 - F^\omega(\gamma(\tau),\tau) - \frac{b^2}{2} \right) d\tau. \tag{13}$$

Then,

$$\psi(x,t) = \inf_{\gamma \in AC(0;x,t)} (\psi_0(\gamma(0)) + A_{0,t}^{\omega,b}(\gamma)), \tag{14}$$

where $AC(s;x,t)$ is the set of all absolutely continuous curves $\gamma \colon [s,t] \to T^d$ on the torus T^d such that $\gamma(t) = x$. We shall also use the following notations. Denote $AC(x_1,s;x_2,t)$ the set of absolutely continuous curves $\gamma \colon [s,t] \to T^d$ such that $\gamma(s) = x_1$ and $\gamma(t) = x_2$; $AC(x,t)$ the set of absolutely continuous curves $\gamma \colon (-\infty,t] \to T^d$ such that $\gamma(t) = x$; AC the set of all absolutely continuous curves $\gamma \colon (-\infty,+\infty) \to T^d$. Finally, we define the Lax operator $\mathcal{L}_{s,t}^{\omega,b}$:

$$\mathcal{L}_{s,t}^{\omega,b}\psi(x) = \inf_{\gamma \in AC(s;x,t)} \psi(\gamma(s)) + A_{s,t}^{\omega,b}(\gamma). \tag{15}$$

Obviously, a function $\phi(y,t)$, $y \in T^d$, $t \in [t_1,t_2]$ is a "viscosity" solution of the Hamilton–Jacobi equation if $\phi(y,t) = b \cdot y + \psi(y,t)$, where ψ is $\mathbb{Z}^d$-periodic and for all $t_1 \leq s < t \leq t_2$:

$$\mathcal{L}_{s,t}^{\omega,b}\psi(\cdot,s) = \psi(\cdot,t).$$

The variational principle (11, 14) can be easily extended to the "kicked" case. Let $u(y,n\pm) = \nabla\phi(y,n\pm), \phi(y,n\pm) = b \cdot y + \psi(y,n\pm), \psi(y,n+) = \psi(y,n-) - F^{n,\omega}(y)$. Denote by $\mathcal{A}_{m,n}^{\omega,b}(\mathbf{X})$ the "kicked" action of the sequence $\mathbf{X} = \{x_i \in T^d, m \leq i \leq n\}$:

$$\mathcal{A}_{m,n}^{\omega,b}(\mathbf{X}) = \sum_{i=m}^{n-1} \left(\frac{1}{2}\rho_b^2(x_{i+1},x_i) - F^{i,\omega}(x_i) - \frac{b^2}{2} \right), \tag{16}$$

where $\rho_b(x,x') = \min_{k \in \mathbb{Z}^d}(\|x - x' - b - k\|)$. Then,

$$\psi(x,n) = \inf_{\mathbf{X} \in S(0;x,n)} \left[\psi(x_0,0-) + \mathcal{A}_{m,n}^{\omega,b}(\mathbf{X}) \right], \tag{17}$$

where $S(m;x,n)$ is the set of sequences $\mathbf{X} = \{x_i \in T^d, m \leq i \leq n\}$ such that $x_n = x$.

In analogy with the continuous case we shall also use the following notations: $S(x', m; x, n)$ is the set of all sequences $\{x_i \in T^d, m \leq i \leq n\}$ such that $x_m = x', x_n = x$; $S(x, n)$ is the set of infinite sequences $\{x_i \in T^d, -\infty < i \leq n\}$ such that $x_n = x$; S is the set of all double-infinite sequences $\{x_i \in T^d, -\infty < i - \infty\}$.

We next define an important notion of a minimizer.

Definition 3.1.

1. *A curve* $\gamma \colon [s, t] \to T^d$ *is called a minimizer if*

$$A_{s,t}^{\omega,b}(\gamma) = \min_{\sigma \in AC(\gamma(s), s; \gamma(t).t)} A_{s,t}^{\omega,b}(\sigma).$$

2. *A curve* $\gamma \colon [s, t] \to T^d$ *is called a ψ-minimizer if*

$$\psi(\gamma(s)) + A_{s,t}^{\omega,b}(\gamma) = \min_{\sigma \in AC(s; \gamma(t).t)} \Big(\psi(\sigma(s)) + A_{s,t}^{\omega,b}(\sigma)\Big).$$

3. *A curve* $\gamma \colon (-\infty, t_0] \to T^d$ *is called a one-sided minimizer if it is a minimizer for all* $(s, t), -\infty < s < t \leq t_0$.

4. *A curve* $\gamma \colon (-\infty, \infty) \to T^d$ *is called a global minimizer if it is a minimizer for all* $(s, t), -\infty < s < t < \infty$.

Similar definitions apply to the "kicked" case. One has just to replace curves γ by sequences $\mathbf{X}$, the action $A_{s,t}^{\omega,b}$ by the "kicked" action $\mathcal{A}_{m,n}^{\omega,b}$ and pairs (s, t) by the pairs of integer moments of time (m, n).

Clearly, minimizers are special trajectories of the Lagrangian flow $L_s^\omega, s \in \mathbb{R}$ which is generated by solutions of the Euler–Lagrange system of equations:

$$\begin{aligned} \dot{x} &= v \\ \dot{v} &= -\nabla F^\omega(x, t). \end{aligned} \tag{18}$$

The Lagrangian flow L_s^ω is a stochastic flow of diffeomorphisms of $T^d \times \mathbb{R}^d$:

$$L_s^\omega \colon (x_0, v_0) \mapsto (x_s, v_s), \ s \in \mathbb{R}. \tag{19}$$

We shall also define the skew-product extension $\hat{L}_s$ of the Lagrangian flow. Namely, we consider the flow of non-random transformations of $T^d \times \mathbb{R}^d \times \Omega$ given by:

$$\hat{L}_s \colon (x_0, v_0, \omega) \mapsto (L_s^\omega(x_0, v_0), \theta^s \omega) = (x_s, v_s, \theta^s \omega), \ s \in \mathbb{R}. \tag{20}$$

In the "kicked" case the Lagrangian dynamics is given by a random Standard-like map (instead of an Euler–Lagrange equation):

$$\begin{aligned} x_{n+1} &= x_n + v_n - \nabla F^{n,\omega}(x_n) \\ v_{n+1} &= v_n - \nabla F^{n,\omega}(x_n). \end{aligned} \tag{21}$$

Then the Lagrangian flow L_n^ω and its skew-product extension is defined exactly in the same way as above

$$L_n^\omega \colon (x_0, v_0) \mapsto (x_n, v_n), \ \hat{L}_n \colon (x_0, v_0, \omega) \mapsto (x_n, v_n, \theta^n \omega), \ n \in \mathbb{Z}. \tag{22}$$

We shall see in the next section that global minimizers correspond to non-trivial invariant measures of $\hat{L}$.

4. Formulation of The Main Results

Consider the random Hamilton–Jacobi equation (10, 3) in a semi-infinite interval of time $(-\infty, t_0]$. It turns out that for any $b \in \mathbb{R}^d$ there exists a unique (up to a constant) "viscosity" solution. More precisely, the following theorem holds.

Theorem 4.1.

1. *For almost all ω and all $b \in \mathbb{R}^d$, $t_0 \in \mathbb{R}$ there exists a unique (up to an aditive constant) function $\phi_b^\omega(y, t)$, $y \in \mathbb{R}^d, t \in (-\infty, t_0]$, such that*

$$\phi_b^\omega(y, t) = b \cdot y + \psi_b^\omega(y, t)$$

where ψ_b^ω is a $\mathbb{Z}^d$-periodic function and for all $-\infty < s < t \leq t_0$

$$\psi_b^\omega(y, t) = \mathcal{L}_{s,t}^\omega \psi_b^\omega(y, s).$$

2. *The function ϕ_b^ω is Lipschitz in y. If $x \in T^d$ is a point of differentiability of $\phi_b^\omega(x, t)$, then there exists a unique one-sided minimizer $\gamma_{x,t}^\omega$ at (x, t) and its velocity is given by the gradient of ϕ: $\dot{\gamma}_{x,t}^\omega(t) = \nabla \phi_b^\omega(x, t)$.*

3. *For almost all ω and an arbitrary family of continuous functions $\zeta_s(y) = b \cdot y + \eta_s(y)$, where η_s are $\mathbb{Z}^d$-periodic functions, the following convergence holds:*

$$\lim_{s \to -\infty} \mathcal{L}_{s,t}^\omega \eta_s = \psi_b^\omega(x, t) \quad (mod \quad constant).$$

Moreover, if x is a point of differentiability of $\phi_b^\omega(x, t)$ then

$$\lim_{s \to -\infty} \dot{\gamma}_{\eta_s, s; x, t}^\omega(t) = \dot{\gamma}_{x,t}^\omega(t) = \nabla \phi_b^\omega(x, t),$$

where $\gamma_{\eta_s, s; x, t}^\omega$ is an η_s-minimizer.

A similar theorem holds in the "kicked" case. Suppose the probability distribution χ for the kicking potentials $F^{j,\omega}(x)$ satisfy the following assumption.

Assumption 4.2. *For any $x \in T^d$ there exists a potential F belonging to the support of χ and such that F has a unique global non-degenerate maximum at x.*

Theorem 4.3. *Suppose probability distribution $\chi \in \mathcal{P}(C^\infty(T^d))$ satisfies Assumption 4.2. Then all the statements of Theorem 4.1 remain true with replacement of t by $n+$ and $n-$, $n \in \mathbb{Z}$.*

The following statement is a simple consequence of Theorems 4.1, 4.3.

Corollary 4.4.

1. *Consider the random Burgers equation (1, 3) or (1, 4) in a semi-infinite interval of time $(-\infty, t_0]$. Assume that probability distribution $\chi \in \mathcal{P}(C^\infty(T^d))$ satisfies Assumption 4.2. Then in both "white" and "kicked" cases, almost*

surely for all $b \in \mathbb{R}^d$ there exists a unique "viscosity" solution $u_b^\omega(x,t)$ with the average velocity b.

2. *For Lebesgue-almost all x,*

$$u_b^\omega(x,t) = \nabla \phi_b^\omega(x,t) = \dot{\gamma}_{x,t}^\omega(t),$$

where $\gamma_{x,t}^\omega$ is the unique one-sided minimizer at (x,t).

3. *Almost surely for all all $b \in \mathbb{R}^d$, $t \in (\infty, t_0]$ and an arbitrary family of gradient-like functions $w_s(x)$, such that $\int_{T^d} w_s(x)dx = b$, the following convergence holds for Lebesgue-almost all $x \in T^d$:*

$$\lim_{s \to -\infty} w_s(x,t) = u_b^\omega(x,t),$$

where $w_s(x,t)$ is the solution of the Cauchy problem for the forced Burgers equation with the initial data at time s given by $w_s(x)$.

Corollary 4.4 implies uniqueness of the stationary distribution for the random Burgers equation. Denote by $\mathcal{U}_b = \{u(x) : \ u(x) = b + \nabla\psi(x), \psi \in Lip(T^d)\}$. Obviously, $\mathcal{U}_b$ is an invariant set for the Burgers dynamics. We shall regard $\mathcal{U}_b$ as a subset of $L^p(T^d, dx), p > 0$. This embedding defines a measurable structure on $\mathcal{U}_b$. It is easy to see that a mapping $\omega \mapsto u_b^\omega(x,0)$ (or, $u_b^\omega(x,0+)$ in the "kicked" case) gives a measurable transformation $U: \Omega \to \mathcal{U}_b$. Denote by $\delta_b^\omega(du)$ the atomic measure on $\mathcal{U}_b$ with the atom at $u_b^\omega(x,0)$ (or at the $u_b^\omega(x,0+)$ in the "kicked case"), and define a probability measure ν_b on $\mathcal{P}(\mathcal{U}_b \times \Omega)$ by $\nu_b(du, d\omega) = \delta_b^\omega(du)P(d\omega)$. We shall also consider the marginal distribution of ν_b on $\mathcal{U}_b$. Denote this marginal distribution by $\mu_b : \mu_b(du) = \int_\Omega \delta_b^\omega(du)P(d\omega)$. As in the previous theorem we assume that probability distribution $\chi \in \mathcal{P}(C^\infty(T^d))$ satisfies Assumption 4.2.

Theorem 4.5. *Consider the evolution given by the random Burgers equation (1, 3), or (1, 4). Then for all $b \in \mathbb{R}^d$ the following statements hold.*

1. *The probability measure $\nu_b \in \mathcal{P}(\mathcal{U}_b \times \Omega)$ is the unique invariant measure for the skew-product dynamics (7) on $\mathcal{P}(\mathcal{U}_b \times \Omega)$.*

2. *The probability measure $\mu_b \in \mathcal{P}(\mathcal{U}_b)$ is the unique stationary distribution for the Markov process (8) on $\mathcal{U}_b$.*

We next discuss the properties of minimizers. It follows from Theorem 4.1 that for a fixed time t and for Lebesgue-almost all points x there exists a unique one-sided minimizer. In fact, an easy compactness argument gives the existence of at least one one-sided minimizer for all $x \in T^d$. Points of non-uniqueness are called shock waves, or simply shocks. Every one-sided minimizer γ_{x,t_0} is a trajectory of the Lagrangian flow corresponding to the Euler–Lagrange system (18), or the random Standard map (21). It can be continued as a trajectory of Lagrangian flow for all t. However, its continuation is likely not to be a one-sided minimizer for large enough $t > t_0$. Global minimizers correspond to exactly those one-sided minimizers that can be continued as minimizers for all t. It turns out that in a very general situation there exists a unique global minimizer. We shall first formulate a theorem in the "white" force case.

Theorem 4.6. *Suppose that the mapping*

$$(F_1(x), \ldots, F_N(x)) \colon T^d \to \mathbb{R}^N$$

is an embedding. Then with probability 1 there exists only one global minimizer.

In the "kicked" case we shall assume that the kicking potentials $F^{j,\omega}(x)$ are given by linear combination of the smooth non-random potentials $F_i, 1 \le i \le M$ with random coefficients.

Assumption 4.7.

$$F^{j,\omega}(x) = \sum_{i=1}^{M} \xi_i^j(\omega) F_i(x), \ \ j \in \mathbb{Z},$$

where the random vectors $\xi^j(\omega) = \{\xi_i^j(\omega), 1 \le i \le M\}$ are independent identically distributed vectors in $\mathbb{R}^M$ with an absolutely continuous distribution.

Theorem 4.8. *Assume that Assumptions 4.2, 4.7 hold, and in addition the potentials $F_i(x), 1 \le i \le M$, are such that the mapping*

$$(F_1(x), \ldots, F_M(x)) \colon T^d \to \mathbb{R}^M$$

is one-to-one. Then with probability 1 there exists only one global minimizer.

Denote by x_g^ω the position of the global minimizer at time $t = 0$. For fixed $t > 0$ consider all one-sided minimizers $\gamma_{x,t}^\omega$, $x \in T^d$. Denote by $A^\omega(t) = \{x_0 = \gamma_{x,t}^\omega(0)\}$ the set of points at which one-sided minimizers cross T^d at $t = 0$. Notice that $x_g^\omega \in A^\omega(t)$ for all $t \ge 0$. The following statement follows immediately from the uniqueness of the global minimizer.

Corollary 4.9. *If there exists a unique global minimizer, then $\operatorname{diam} A^\omega(t) \to 0$ as $t \to \infty$.*

Corollary 4.9 implies that $\gamma_{x,t}^\omega(0)$ approaches x_g^ω as $t \to \infty$. However, in the multi-dimensional case we cannot control the rate of convergence. At the same time, in the one-dimensional case convergence to the global minimizer is exponentially fast. It follows from the hyperbolicity of the global minimizer. Denote by v_g^ω the velocity of the global minimizer at time $t = 0$. Define a probability measure $\kappa \in \mathcal{P}(T^d \times \mathbb{R}^d \times \Omega)$ by

$$\kappa(dx, dv, d\omega) = \delta_{x_g^\omega, v_g^\omega}(dx, dv) P(d\omega),$$

where $\delta_{x_g^\omega, v_g^\omega}(dx, dv)$ is an atomic measure in $\mathcal{P}(T^d \times \mathbb{R}^d)$ with the atom at (x_g^ω, v_g^ω). The following theorem holds if $d = 1$.

Theorem 4.10. (see [9])

1. *The probability measure κ is a hyperbolic ergodic invariant measure for the skew-product extension of the Lagrangian flow. In particular, it implies that with probability 1 the global minimizer is a hyperbolic trajectory of the Lagrangian flow with non-random positive Lyapunov exponent $\lambda > 0$.*

2. *With probability 1 there exists a one-dimensional C^∞-smooth unstable manifold passing through (x_g^ω, v_g^ω), which consists of all (x, v) whose Lagrangian trajectories are asymptotic to the global minimizer as $t \to -\infty$.*

3. *With probability 1 for all b the function $u_b^\omega(x, 0)$ is piecewise C^∞-smooth with a finite number of shocks (jump discontinuities). The graph of each smooth piece of $u_b^\omega(x, 0)$ belongs to the unstable manifold of the global minimizer.*

4. *All the shocks except one correspond to double-folds of the unstable manifold. The only special shock, which is called the main shock, has a topological nature. It is the only shock which can be traced ever in the past. All other shocks have finite history and have appeared from preshocks at some negative time.*

Remark 4.11. *Part of the statement 1 of Theorem 4.10 remains valid in the multi-dimensional case. Namely, it is still true that the probability measure κ is an ergodic invariant measure for the skew-product extension of the Lagrangian flow. However, hyperbolicity of the measure κ is still an open problem.*

The proof of all the theorems (except the last one) will be published in [22]. The results of the last theorem have been proved in [9] for the "white" force case. The proof requires an additional assumption (see nondegeneracy condition (A1) in [9, p. 900]), which is of the same nature as the condition of Theorem 4.6 above.

5. Concluding Remarks

1. Most of the results which we have formulated in the previous section are related to the multi-dimensional Burgers equation driven by a random force. Similar results in the one-dimensional case have been proven in [9]. However, the methods in [9] were purely one-dimensional. The new approach which we apply here is more general and straightforward. The advantage of the method is connected with a systematic use of Lagrangian formalism and the Hamilton–Jacobi equation. This allows us to prove results on the uniqueness of stationary distribution in any dimension. Our method also gives a proof of the uniqueness of global the minimizer, which is a fundamental property of random Lagrangian systems. In fact, one can consider the work presented here as a study of minimizing orbits of random Lagrangian systems. This problem is quite interesting on its own without any connection with the hydrodynamics and the Burgers equation. The theory of minimizing orbits for Lagrangian systems with autonomous or time-periodic potentials has a long history. It goes back to the works of M. Morse and G. Hedlund (see [30, 20]). In the last 15 years this subject became a very active research area in connection with Aubry–Mather theory (see [1, 28, 29, 26, 27, 15, 8]). R. Mañé has conjectured that for a generic time-periodic potential there exists a unique global minimizer which is a hyperbolic periodic trajectory of the Lagrangian system. This conjecture remains open. However, a related result on the uniqueness of a minimizing measure has been proven in [26]. It was also shown in [7] that the invariant measure is hyperbolic provided it is supported by a single periodic orbit. In the random case we can prove not only the uniqueness of a minimizing measure, but also show that this measure is supported by the unique global minimizer. We hope that the uniqueness of the global minimizer will make it possible to prove that it is a hyperbolic trajectory with probability 1. There is an interesting connection between our hyperbolicity conjecture in the random case and Mane's conjecture. Indeed, random Lagrangian systems are generic with probability 1 in the measure-theoretic sense, while Mane's conjecture is related to topologically generic Lagrangian systems.

2. The case of "kicking" potentials corresponds to a random generalisation of Aubry–Mather theory (see [1, 28]). The Lagrangian in this case can be considered as a random Frenkel–Kontorova type model, which describes an infinite chain of particles placed in periodic potentials and connected by elastic springs. In the random situation the potentials are chosen independently for each particle, contrary to the classical case when one periodic potential serves all of them. In the non-random case the structure of global minimizers is very complicated. It crucially depends on the arithmetic properties of the rotation vector, which in turn depends non-trivially on the vector b in the linear form $b \cdot y$. In the case of fixed Diophantine rotation number ($d = 1$) it is believed that one has transition from elliptic minimizers for small potentials to hyperbolic global minimizers when potentials are large enough. This transition is related to destruction of invariant KAM curves which correspond to elliptic minimizers. All this complicated picture should be compared with the uniqueness of the global minimizer for all b in the random case (see Theorems 4.6, 4.8). In the random one-dimensional case one can also prove (see Theorem 4.10) the hyperbolicity of the global minimizer, while in the classical Aubry–Mather theory the hyperbolicity of minimizers after destruction of KAM curves remains an outstanding open problem.

3. The uniqueness of the "viscosity" solutions of the random Hamilton–Jacobi and Burgers equations for $-\infty < t \leq t_0$ (Theorems 4.1, 4.3, Corollary 4.4) demonstrates the extreme stability of those equations. The uniqueness result can be formulated as the "one force – one solution" principle. We just mention that in the case of the Navier–Stokes equations the situation is much more complicated. The "one force - one solution" principle is valid only for large enough values of the viscosity.

4. Finally, we shall discuss the quantitative predictions which we have mentioned at the beginning of the article. One-dimensional random Burgers equations with large scale forcing have been discussed very intensively in the physical literature in the last five years (see [11, 32, 6, 19, 2, 18, 3, 4, 17]). The methods used vary from numerical simulations to quantum field theoretic closure techniques. It was shown that the pdf g(z) for the velocity gradient $z = \frac{du^\omega}{dx}$ decays super exponentially for large positive values of z, and algebraically for negative values, i.e., $g(z) \sim \frac{C}{|z|^\alpha}$ as $z \to -\infty$. The critical exponent α is universal. It means that α is independent of the precise formula for a driven force. In ([32, 2]) it was predicted that α is equal to $\frac{5}{2}$ or 3. However, as it was shown in ([10]) the geometrical picture formulated above (see Theorem 4.10) suggests that the main contribution to $g(z)$ for large negative values of z comes from small neighbourhoods of the preshock events. Then an easy calculation implies that $\alpha = \frac{7}{2}$ (see also [12, 13]. This prediction, although not rigorous, seems to be more and more accepted by the physical community. The prediction for α is based on the hyperbolic geometrical picture which implies that the number of shocks is finite (rather than infinite with shocks being everywhere dense on $S^1 = \mathbb{R}/\mathbb{Z}$). This suggests that the preshocks events form a set of isolated points on the (x, t) space, and, hence, leads to the predicted value $\alpha = \frac{7}{2}$. We hope that in the multidimensional case one will be able to get nontrivial quantitative results, provided that the hyperbolicity is established.

Acknowledgement

We are grateful to J. Bec, A. Fahti, U. Frisch, G. Contreras, V. Sedykh and Ya. Sinai for helpful discussions. The work was supported by EPSRC research grant GR/M45610, with further support for R. Iturriaga from Conacyt grant 28489-E.

References

[1] S. Aubry, *The twist map, the extended Frenkel–Kontorova model and the devil's staircase,* Physica D, **7** , 1983, 240–258.

[2] S. Boldyrev, *Velocity-difference probability density for Burgers turbulence,* Phys. Rev. E, **55** (1997), 6907–6910.

[3] E. Balkovsky, G. Falkovich, I. Kolokolov, and V. Lebedev, *Intermittency of Burgers turbulence,* Phys. Rev. Lett., **78** (1997), 1452–1455.

[4] J. Bec, U. Frisch and K. Khanin, *Kicked Burgers Turbulence,* J. Fluid Mech., **416** (2000), 239–267.

[5] J. Bricmont, A. Kupiainen and R. Lefevere *Ergodicity of the 2D Navier–Stokes equations with random forcing,* Preprint (2000).

[6] J. P. Bouchaud, M. Mezard and G. Parisi, *Scaling and intermittency in Burgers turbulence,* Phys. Rev. E, **52** (1995), 3656–3674.

[7] G. Contreras and R. Iturriaga, *Convex Hamiltonians without conjugate points,* Ergod. Th and Dynam. Sys., **19** (1999), 901–952.

[8] G. Contreras and R. Iturriaga, *Global Minimizers of Autonomous Lagrangians,* 22 Colloqio Brasileiro de Matematica, 1999.

[9] W. E, K. Khanin, A. Mazel and Ya. Sinai, *Invariant measures for Burgers equation with stochastic forcing,* Annals of Math., **151** (2000), 877–960.

[10] W. E, K. Khanin, A. Mazel and Ya. Sinai, *Probability distribution functions for the random forced Burgers equations,* Phys. Rev. Lett. **78** (1997), 1904–1907.

[11] A. Chekhlov and V. Yakhot, *Kolmogorov Turbulence in a random force-driven Burgers equation* Phys. Rev E, **51** (1995), R2739–R2749.

[12] W. E and E. Vanden Eijnden, *Asymptotic theory for the probability density functions in Burgers turbulence* Phys. Rev. Lett., **83** (1999), 2572–2575.

[13] W. E and E. Vanden Eijnden *Statistical theory of the stochastic Burgers equation in the inviscid limit,* Comm. Pure Appl. Math., **53** (2000), 852–901.

[14] W. E, J.C. Mattingly and Ya. Sinai, *Gibbsian dynamics and ergodicity for the stochastically forced Navier–Stokes Equation,* Preprint (2000).

[15] A. Fathi, *Théorème KAM faible et Théorie de Mather sur les systems Lagrangiens,* C.R. Acad. Sci. Paris, t. 324, Série I (1997), 1043–1046.

[16] F. Flandoli, B. Maslowski, *Ergodicity of the 2-D Navier–Stokes equation under random perturbation* Comm. in Math Physics, **171** (1995), 119–141.

[17] U. Frisch, J. Bec and B. Villone, *Singularities and the distribution of density in the Burgers/adhension model,* Physica D, **152–153** (1–4) 2001, 620–635.

[18] T. Gotoh and R. H. Kraichnan, *Burgers turbulence with large scale forcing,* Phys. Fluids A, **10** (1998), 2859–2866.

[19] V. Gurarie and A. Migdal, *Instantons in Burgers equations,* Phys. Rev. E, **54** (1996), 4908–4914.

[20] G. Hedlund *Geodesics on a two dimensional Riemannian manifold with periodic coefficients,* Annals of Math., **33** (1932), 719–739.

[21] E. Hopf *The partial differrential equation* $u_t + uu_x = \mu u_{xx}$, Comm. Pure Appl. Math., **3** (1950), 201–230.

[22] R. Iturriaga and K. Khanin, *Burgers Turbulence and Random Lagrangian Systems,* Preprint 2001.

[23] S. Kuksin and A. Shirikyan, *Stochastic dissipative PDEs and Gibbs measures,* Preprint (2000).

[24] P. D. Lax, *Hyperbolic systems of conservation laws,* Comm. Pure Appl. Math., **10** (1957), 537–566.

[25] P. L. Lions, *Generalized Solutions of Hamilton-Jacobi Equations,* Research Notes in Math., **69**, Pitman Advanced Publishing Program, Boston, 1982.

[26] R. Mañé, *Generic properties and problems of minimizing measures of Lagrangian systems,* Nonlinearity, **9** (1996), 273–310.

[27] R. Mañé, *Lagrangian flows: the dynamics of globally minimizing orbits,* International Congress on Dynamical Systems in Montevideo (a tribute to Ricardo Mañé), F. Ledrappier, J. Lewowicz, S. Newhouse eds, Pitman Research Notes in Math., **362**, Pitman Advanced Publishing Program, Boston, 1996, 120–131. Reprinted in Bol. Soc. Bras. Mat., Vol **28** (1997), 141–153.

[28] J. Mather, *Existence of quasi-periodic orbits for twist homeomorphisms of the annulus,* Topology, **21** (1982), 457–467.

[29] J. Mather, *Action minimizing measures for positive definite Lagrangian systems,* Math. Zeitschrift, **207** (1991), 169–207.

[30] M. Morse, *Calculus of variations in the large,* Amer. Math. Soc. Colloquium Publications, **XVIII**, 1934.

[31] O. A. Oleinik, *Discontinuous solutions of nonlinear differential equations,* Uspekhi Mat. Nauk, **12** (1957), 3–73.

[32] A. Polyakov, *Turbulence without pressure,* Phys. Rev E, **52** (1995), 6183–6188.

[33] Ya. Sinai, *Two results concerning asymptotic behavior of solutions of the Burgers equation with force,* J. Stat. Phys., **64** (1992), 1–12.

R. Iturriaga:
Centro de Investigación en Matemática
Guanajuato, Mexico
E-mail address: renato@cimat.mx

Isaac Newton Institute for Mathematical Sciences
University of Cambridge
20 Clarkson Road
Cambridge CB3 0EH, United Kingdom

K. Khanin:
Isaac Newton Institute for Mathematical Sciences
University of Cambridge
20 Clarkson Road
Cambridge CB3 0EH, United Kingdom
E-mail address: K.Khanin@newton.cam.ac.uk

BRIMS
Hewlett-Packard Laboratories, Bristol

Department of Mathematics
Heriot-Watt University, Edinburgh

Landau Institute for Theoretical Physics
Moscow

Random Growth and Random Matrices

Kurt Johansson

Abstract. We give a survey of some of the recent results on certain two-dimensional random growth models and their relation to random matrix theory, in particular to the Tracy-Widom distribution for the largest eigenvalue. The problems are related to that of finding the length of the longest increasing subsequence in a random permutation. We also give a new approach to certain results for the Schur measure introduced by Okounkov.

1. Random Growth Models in the Plane

1.1. Eden-Richardson growth

During the last twenty years there has been a lot of interest in models where an object grows by some rule involving randomness. Basically, there are two types of models, non-local models like diffusion-limited aggregation (DLA) and local models, which is our concern here. There are many types of local random growth models in the plane, see [24] for a review and more background. As an example consider the *Eden-Richardson growth model*, [13, 30], which is defined as follows. The shape Ω_t at time t of the growing object is a connected set, which is the union of unit squares centered at points in $\mathbb{Z}^2$. Let $\partial\Omega_t$ denote the set of all unit squares, centered at integer points, which are adjacent to Ω_t. In the continuous time version each square in $\partial\Omega_t$ is added to Ω_t independently of each other and with exponential waiting times; i.e. as soon as a square joins $\partial\Omega_t$ it's clock starts to tick and the square is added to Ω_t after a random time T with the exponential distribution, $P[T > s] = e^{-s}$. In the discrete time version, at each time $t \in \mathbb{Z}^+$, each square in $\partial\Omega_{t-1}$, is added to Ω_{t-1} with probability $p = 1 - q$ independently of each other, and the resulting set is Ω_t. At time $t = 0$ we take $\Omega_0 = [-1/2, 1/2]^2$. In both cases the object grows linearly in time, [21], $\Omega_t/t \to A$, the asymptotic shape, as $t \to \infty$. We are interested in the roughness of Ω_t, the fluctuations of Ω_t around tA. This growth model is equivalent with a certain *first-passage site percolation* model. With each site $(i, j) \in \mathbb{Z}^2$ we associate a random variable $\tau(i, j)$, which we think of as a random time. Variables associated with different sites are independent. A *path* π from $(0,0)$ to (M, N) is a sequence $\{p_r\}_{r=0}^R \subseteq \mathbb{Z}^2$ with $|p_r - p_{r-1}| = 1$, $p_0 = (0,0)$ and $p_R = (M, N)$. The *first-passage time* from $(0,0)$ to (M, N) is

$$T(M, N) = \min_\pi \sum_{p_r \in \pi} \tau(p_r)\,. \tag{1}$$

If we take $\tau(0,0) = 0$ and $P[\tau(i,j) > s] = e^{-s}$ (continuous time) or $\tau(i,j) = w(i,j) + 1$ with $P[w(i,j) = s] = (1-q)q^s$, $s \geq 0$, $0 < q < 1$ (discrete time), then

$$\Omega_t = \{(M,N) \in \mathbb{Z}^2; T(M,N) \leq t\} + [-\tfrac{1}{2}, \tfrac{1}{2}]^2. \tag{2}$$

Thus, fluctuations in Ω_t can be translated into fluctuations of $T(M,N)$. It is conjectured that the standard deviation $SD(T(N,N)) \sim N^\chi$, as $N \to \infty$ with $\chi = 1/3$, [24, 27]. Since $T(N,N) \sim cN$ as $N \to \infty$, we see that the standard deviation grows as $(\text{mean})^{1/3}$. This means that the *longitudinal fluctuations* of Ω_t are of order $t^{1/3}$, which is conjectured to be true generally for local two-dimensional random growth models. We can also consider the *transversal fluctuations*. Let d_N be the maximal deviation of all paths π, which are minimizers in (1) with $M = N \geq 0$, from the straight line $x = y$. It is conjectured that d_N is of order N^ξ with $\xi = 2/3$, [25]. To prove that $\chi = 1/3$ and $\xi = 2/3$ in the Eden-Richardson growth model is an open problem. Below we will consider related models, where this, and more, can be rigorously proved.

1.2. The corner growth model

We can modify the model above by allowing growth only upwards or to the right. In (1) this corresponds to allowing only *up/right paths* π in (1), i.e. $p_r - p_{r-1} = (1,0)$ or $(0,1)$, which gives a so-called *directed first-passage percolation* model. We can add one more restriction by allowing growth only in corners. In this model $\Omega_0 = \mathbb{R}^2 \setminus \mathbb{R}_+^2$ and we can add a square $Q = [m-1,m] \times [n-1,n]$ to Ω_t only if both $[m-2,m-1] \times [n-1,n] \subseteq \Omega_t$ and $[m-1,m] \times [n-2,n-1] \subseteq \Omega_t$, i.e. Q lies in a "corner" of Ω_t. For this model we have to replace (1) by

$$G(M,N) = \max_\pi \sum_{(i,j)\in\pi} w(i,j), \tag{3}$$

where the maximum is over all up/right paths from $(1,1)$ to (M,N). Thus, we get instead what can be called a *last-passage directed percolation model*, and the random shape is given by

$$\Omega_t = \{(M,N) \in \mathbb{Z}^2; G(M,N) + M + N - 1 \leq t\} + [-1,0]^2, \tag{4}$$

since all up/right paths from $(1,1)$ to (M,N) contain the same number of points $M + N - 1$. The continuous time case can be obtained by taking the limit $q \to 1$, see [17]. Note that if $q = 1 - 1/L$, then $L^{-1}w(i,j)$ converges, as $L \to \infty$ to an exponentially distributed random variable. As explained in [31], the corner growth model is equivalent with the discrete or continuous time *totally asymmetric simple exclusion process* (TASEP) and results for the growth model can be translated into results for the TASEP, [17].

In order to state the results we have to define the Tracy-Widom distribution. Let $\text{Ai}(x)$ denote the *Airy function* and define the *Airy kernel*,

$$A(x,y) = \int_0^\infty \text{Ai}(x+t)\,\text{Ai}(y+t)dt = \frac{\text{Ai}(x)\,\text{Ai}'(y) - \text{Ai}'(x)\,\text{Ai}(y)}{x-y}. \tag{5}$$

The *Tracy-Widom distribution* is defined by the Fredholm determinant

$$F(t) = \det(I - A)_{L^2(t,\infty)}, \quad t \in \mathbb{R}. \tag{6}$$

Let M be an $N \times N$ matrix from the *Gaussian Unitary Ensemble* (GUE), [26], where we put the measure $Z_N^{-1} \exp(-\operatorname{Tr} M^2)dM$ on the space of Hermitian matrices. The space of $N \times N$ Hermitian matrices is isomorphic to R^{N^2} and dM is Lebesgue measure on this space. If $\lambda_{\max}$ is the largest eigenvalue of M, then

$$\lim_{N \to \infty} P[\sqrt{2}N^{1/6}(\lambda_{\max} - \sqrt{2N}) \le t] = F(t), \tag{7}$$

see [36].

We can now state the main theorem for the corner growth model.

Theorem 1.1. [17]. *For each $q \in (0,1)$, $\gamma \ge 1$ and $s \in \mathbb{R}$,*

$$\lim_{N \to \infty} P[\frac{G([\gamma N], N) - N\omega(\gamma, q)}{\sigma(\gamma, q)N^{1/3}} \le s] = F(s), \tag{8}$$

where

$$\omega(\gamma, q) = \frac{(1 + \sqrt{q\gamma})^2}{1 - q} - 1 \tag{9}$$

and

$$\sigma(\gamma, q) = \frac{q^{1/6}\gamma^{-1/6}}{1 - q}(\sqrt{\gamma} + \sqrt{q})^{2/3}(1 + \sqrt{q\gamma})^{2/3}. \tag{10}$$

From $\omega(\gamma, q)$ we can compute the asymptotic shape A for the corner growth model, which was also done in [33], and we obtain $A \cap \mathbb{R}_+^2 = \{(x, y) \in \mathbb{R}_+^2; y + 2\sqrt{qxy} + x \le 1 - q\}$. Note that in this model the standard deviation grows as $\sim (\text{mean})^{1/3}$, so we have a proof of $\chi = 1/3$. Also we see from (8) that $G(M, N)$, for M and N large, behaves like the largest eigenvalue of a big random hermitian matrix. There are analogous results for the continuous time case and the TASEP, see [17].

Theorem 1.1 is proved using the following representation of the distribution function for $G(M, N)$, $M \ge N$,

$$P[G(M, N) \le n] = \frac{1}{Z_{M,N}} \sum_{h \in \{0, \ldots, n+N-1\}^N} \Delta_N(h)^2 \prod_{j=1}^{N} \binom{h_j + M - N}{h_j} q^{h_j}, \tag{11}$$

where $\Delta_N(h) = \prod_{1 \le i < j \le N}(h_i - h_j)$ is the Vandermonde determinant and $Z_{M,N}$ is a normalization constant. How this formula is obtained is described in the next section. The formula (11) should be compared with the formula for the distribution function for the largest eigenvalue of a GUE-matrix,

$$P[\lambda_{\max} \le t] = \frac{1}{Z_N} \int_{(-\infty,t]^N} \Delta_N(x)^2 \prod_{j=1}^{N} e^{-x_j^2} d^N x. \tag{12}$$

Note that in the continuous time case we obtain, by taking a limit $q \to 1$ in (11), a similar integral giving the largest eigenvalue in the Laguerre ensemble, [17].

The right-hand side of (12) can also be written as

$$\frac{1}{Z_N} \int_{(-\infty,t]^N} \exp\left[-\left(\sum_{j \neq k} \log|x_j - x_k|^{-1} + \sum_j x_j^2\right)\right] d^N x, \tag{13}$$

which leads to Dyson's Coulomb gas interpretation. We see that $\lambda_{\max}$ is the position of the rightmost charge in a logarithmic Coulomb gas on the line confined by a quadratic external potential. In (11) we get instead a discrete Coulomb gas on $\mathbb{N}$, also confined by an external potential. At the edge of the support of the charges the density is low, so we expect the discrete Coulomb gas to be well approximated by a continuous Coulomb gas in this region, and thus, heuristically, $G(M, N)$ and $\lambda_{\max}$ should behave similarly.

To establish this rigorously one can proceed as follows. A standard computation in random matrix theory, [26, ch. 5], shows that the right-hand side of (12) equals a certain Fredholm determinant and this is used to prove (7). Carrying out the same computation for (11), [17], gives

$$P[G(M, N) \leq n] = \det(I - K_{M,N})_{\ell(\{n+N, n+N+1, \dots\})}, \tag{14}$$

where $K_{M,N}$ is the operator on $\ell^2(\mathbb{N})$ with kernel

$$K_{M,N}(x,y) = \frac{\kappa_{N-1}}{\kappa_N} \frac{M_N(x)M_{N-1}(y) - M_{N-1}(x)M_N(y)}{x - y}$$

$$\times \left[\binom{x + M - N}{x}\binom{y + M - N}{y}q^{x+y}\right]^{1/2}, \tag{15}$$

for $x, y \in \mathbb{N}$. Here $M_n(x) = \kappa_n x^n + \dots$ are the normalized, discrete orthogonal polynomials with respect to the weight $\binom{x+M-N}{x}q^x$ on $\mathbb{N}$. They are multiples of the standard Meixner polynomials, [11]. Theorem 1.1 is proved by analysing the asymptotics of the Fredholm determinant (14), using the fact that the asymptotics of the *Meixner kernel*, (15) can be analysed using the integral formula for the Meixner polynomials, [17].

1.3. Hammersley's model

Consider a Poisson process with intensity α in the unit square $[0, 1]^2$. An *up/right path* from $(0, 0)$ to $(1, 1)$ through the points is a sequence $\{(x_k, y_k)\}_{k=1}^L$ of Poisson points such that $x_k \leq x_{k+1}$ and $y_k \leq y_{k+1}$ for each k. Let $L(\alpha)$ denote the maximum number of points in such a path. This random variable was introduced in [16] to study random permutations. In fact, if we condition the number of points in the square to be N, then $L(\alpha)$ has the same distribution as the length of the longest increasing subsequence in a random permutation from S_N with uniform distribution, see also [1]. Hammersley's model can also be obtained as a limit of the corner growth model by taking $q = \alpha/N^2$ and letting N go to infinity. With this choice of q the $N \times N$ matrix $(w(i, j))_{i,j=1}^N$ has, with probability going to 1 as

$N \to \infty$, at most one 1 in each row and column, and all other elements are 0. From this it is not hard to see that $G(N, N)$ converges to $L(\alpha)$. Thus one approach to the results for $L(\alpha)$ is to take $M = N$ and $q = \alpha/N^2$ in (14) and then compute the limit as $N \to \infty$, see [19] and [20] for a review.

The distribution function for $L(\alpha)$ also has another representation, [15], as a Toeplitz determinant,

$$P[L(\alpha) \leq n] = D_n(e^{2\sqrt{\alpha}\cos\theta}) \doteq \det\left(\frac{1}{2\pi}\int_{-\pi}^{\pi} e^{2\sqrt{\alpha}\cos\theta - i(j-k)\theta} d\theta\right)_{j,k=1}^{N}. \quad (16)$$

The asymptotics of the Toeplitz determinant in (16) can be analysed using the steepest descent method for Riemann-Hilbert problems, [12], and this leads to a proof of

Theorem 1.2. [5]. *For all $t \in \mathbb{R}$,*

$$\lim_{\alpha \to \infty} P[L(\alpha) \leq 2\sqrt{\alpha} + t\alpha^{1/6}] = F(t), \quad (17)$$

where $F(t)$ is the Tracy-Widom distribution (6), and also all moments of $L(\alpha)$ converge.

Note that the standard deviation of $L(\alpha)$ is $\sim c\alpha^{1/6} = c(\sqrt{\alpha})^{1/3}$ and the mean is $\sim 2\sqrt{\alpha}$, so we have the same exponent $\chi = 1/3$ as above. This theorem was the first case where this exponent was rigorously verified. In the proof of (17) in [5] another expression for the Tracy-Widom distribution is obtained, namely

$$F(t) = \exp\left(-\int_t^{\infty}(x-t)u(x)^2 dx\right), \quad (18)$$

where $u(x)$ is the solution of the Painlevé II equation $u'' = xu + 2u^3$, which satisfies $u(x) \sim \mathrm{Ai}(x)$ as $x \to \infty$. The fact that (6) and (18) are equal is proved in [36]. It is also possible to consider random permutations (or the Hammersley model) with symmetry restrictions and restrictions on the number of fixed points, [3, 4]. For these models one can also obtain the GOE and GSE largest eigenvalue distributions, [36], as limits besides the GUE distribution discussed above, see [4].

Hammersley's model has an interesting interpretation as a two-dimensional growth model called *polynuclear growth* (PNG), see [29]. Some of the symmetrized models mentioned above also have an interpretation in this context, [29]. In Hammersley's model we can also obtain the exponent ξ discussed above, by looking at the largest deviation of any maximal path from the diagonal. Using estimates from [5] and ideas from [25, 38] it is possible to prove that $\xi = 2/3$ in this model, see [18].

The rigorous results discussed above are of course proved for very special models. On the other hand one would expect the results to hold more generally, [23], [17, Conjecture 1.9], [29], since limiting laws should have some degree of universality. To understand this universality rigorously is an interesting open problem.

450   K. Johansson

2. Analysis of the Corner Growth Model

Let $\lambda = (\lambda_1, \ldots, \lambda_n)$ be a partition of k, i.e. $\lambda_1 \geq \lambda_2 \geq \cdots \geq \lambda_n \geq 0$, $\lambda_j \in \mathbb{N}$ and $\sum_j \lambda_j = k$. The *Young* or *Ferrers diagram* corresponding to λ is the set $\cup_{j=1}^n \{(i,j); 1 \leq i \leq \lambda_{n+1-j}\}$, which we also denote by λ. A *semistandard Young tableaux* T of shape λ, $\mathrm{sh}(T) = \lambda$, with elements in $\{1, \ldots, N\}$ is a map $T: \lambda \to \{1, \ldots, N\}$ such that $T(i,j) \leq T(i+1,j)$ and $T(i,j) < T(i,j+1)$. Let $m_r(T)$ denote the number of points $(i,j) \in \lambda$, such that $T(i,j) = r$, see [14, 32] or [35] for more details. The *Robinson-Schensted-Knuth (RSK) correspondence*, [22], sets up a bijection between integer $M \times N$ matrices $W = (w(i,j))$ and pairs (T,S) of semistandard Young tableaux of the same shape λ, where T has elements in $\{1, \ldots, N\}$ and S has elements in $\{1, \ldots, M\}$. This bijection has the property that $G(M,N)$ defined by (3) is equal to λ_1, see [17], and furthermore $\sum_j w(i,j) = m_i(S)$, $\sum_i w(i,j) = m_j(T)$.

Consider a generalization of the corner growth model defined above, where $w(i,j)$ is geometrically distributed with parameter $x_i y_j$, i.e. $P[w(i,j) = s] = (1 - x_i y_j)(x_i y_j)^s$, $s \in \mathbb{N}$. Here x_i, y_i, $i \geq 1$, are given numbers in $[0,1)$. From the facts above we obtain

$$P[G(N,N) \leq n] = \sum_{W;G(N,N)\leq n} P[W]$$

$$= \prod_{i,j=1}^N (1 - x_i y_j) \sum_{W;G(N,N)\leq n} \prod_{i=1}^N x_i^{\sum_j w(i,j)} \prod_{j=1}^N y_j^{\sum_i w(i,j)}$$

$$= \prod_{i,j=1}^N (1 - x_i y_j) \sum_{\lambda;\lambda_1 \leq n} \left(\sum_{S;\mathrm{sh}(S)=\lambda} \prod_{i=1}^N x_i^{m_i(S)} \right) \left(\sum_{T;\mathrm{sh}(T)=\lambda} \prod_{j=1}^N y_j^{m_j(T)} \right). \quad (19)$$

In (19) we can recognize one possible definition of the Schur polynomial. Given $\lambda = (\lambda_1, \ldots, \lambda_N)$, a partition of k, the *Schur polynomial* $s_\lambda(x_1, \ldots, x_N)$ is a homogeneous symmetric polynomial of degree k, defined by

$$s_\lambda(x_1, \ldots, x_N) = \sum_{S;\mathrm{sh}(S)=\lambda} \prod_{i=1}^N x_i^{m_i(S)} = \frac{1}{\Delta_N(x)} \det(x_i^{\lambda_j+N-j})_{i,j=1}^N. \quad (20)$$

The second equality is the Jacobi-Trudi identity, [32, 35]. Combining (19) and (20), we obtain

$$P[G(N,N) \leq n] = \prod_{i,j=1}^N (1 - x_i y_j) \sum_{\lambda;\lambda_1 \leq n} s_\lambda(x) s_\lambda(y), \quad (21)$$

[17, 3]. In [3] many variations of this formula are given. Note that we can think of

$$P_S[\lambda] = \prod_{i,j=1}^N (1 - x_i y_j) s_\lambda(x) s_\lambda(y) \quad (22)$$

as a probability measure on all partitions with at most N non-zero parts. This is the *Schur measure* introduced in [28], where it is defined on the set of all partitions and x_i, y_i are allowed to be complex. As discussed in [8], the Schur measure generalizes many other measures on partitions motivated by representation theory, see [7], including the classical *Plancherel measure*, which corresponds to random permutations, [9]. The Poissonized Plancherel measure can be obtained as a limit of (22) with $x_i = y_i = \sqrt{\alpha}/N$, see [19, 20].

Note that if we put $y_i = 0$, $M < i \leq N$, $y_i = \sqrt{q}$, $1 \leq i \leq M$ and $x_i = \sqrt{q}$, $1 \leq j \leq N$ in (22) we get the distribution function for $G(N, M)$ of Section 1. Using the second equality in (20) we can rewrite the right-hand side of (22) and obtain (11), [17]. The crucial fact in going from (11) to (14) is the fact that the measure in (11) has determinantal correlation functions, [26], see [34] for review on random point fields with determinantal correlation functions. In fact, this is true also for the Schur measure (22) as proved in [28]. In the next section we will give a new proof of this, and actually we will see that it is possible to go directly from (21) to a Fredholm determinant.

3. The Schur Measure

Our analysis of the Schur measure is based on [37], which is an alternative to the approach in [26], see also [6]. We start by outlining some results in [37]. Let μ be a measure on the space Ω and let f, ϕ_j, ψ_j, $j \geq 1$ be integrable functions from Ω to $\mathbb{C}$. The following identity, [2], is central.

$$Z_N[f] \doteq \frac{1}{N!} \int_{\Omega^N} \det(\phi_j(t_k))_{j,k=1}^N \det(\psi_j(t_k))_{j,k=1}^N \prod_{j=1}^N f(t_j) d\mu(t_j)$$

$$= \det(\int_\Omega \phi_j(t)\psi_k(t)f(t)d\mu(t))_{j,k=1}^N. \tag{23}$$

Let

$$A = (\int_\Omega \phi_j(t)\psi_k(t)d\mu(t))_{j,k=1}^N \tag{24}$$

and assume that A is invertible with inverse $A^{-1} = (\mu_{jk})_{j,k=1}^N$. Set

$$K_N(u, v) = \sum_{j,k=1}^N \psi_k(u)\mu_{kj}\phi_j(v). \tag{25}$$

The arguments in [37] show that, writing $f = 1 + g$,

$$\frac{Z_N[1 + g]}{Z_N[1]} = \det(I + K_N g)_{L^2(\Omega,\mu)}, \tag{26}$$

where the right-hand side is a Fredholm determinant on $L^2(\Omega, \mu)$ and $K_N g$ is the operator which is first multiplication by g and the application of the operator with kernel $K_N(u, v)$. Think of

$$\rho_N(t_1, \ldots, t_N)d\mu^N(t) = \frac{1}{N!Z_N[1]} \det(\phi_j(t_k))_{j,k=1}^N \det(\psi_j(t_k))_{j,k=1}^N d\mu^N(t) \quad (27)$$

as a complex measure on Ω^N. It follows from (26), [37], that $\rho_N d\mu^N$ has marginal distributions given by

$$\rho_k(t_1, \ldots, t_k) \doteq \int_{\Omega^{N-k}} \rho_N(t)d\mu(t_{k+1}) \ldots d\mu(t_N)$$

$$= \frac{(N-k)!}{N!} \det(K_N(t_i, t_j))_{i,j=1}^k. \quad (28)$$

Let $A^{(jk)}$ be the matrix obtained by deleting row j and column k in the matrix A. Then, by (25) and the standard formula for the inverse of a matrix,

$$K_N(u, v) = \sum_{j,k=1}^N (-1)^{j+k}\psi_k(u)\frac{\det A^{(jk)}}{\det A}\phi_j(v). \quad (29)$$

If we take $\phi_j(v) = y_j^v$, $\psi_k(u) = x_k^u$ and $d\mu$ to be the counting measure on $\mathbb{N}$, then (27) becomes the Schur measure. To see this use the second equality in (20), the Jacobi-Trudi identity, set $t_j = \lambda_j + N - j$, and note that the Schur measure is a symmetric function of the t_i :s. Hence we can regard the Schur measure, (22) as a measure on $\mathbb{N}^N$, which we denote by $\rho_N^{Sc}(t_1, \ldots, t_N)$, $t \in \mathbb{N}^N$. Note that with this choice,

$$Z_N[1] = \det A = \det(\sum_{t=0}^\infty y_j^t x_k^t) = \det(\frac{1}{1 - y_j x_k})$$

$$= \frac{\Delta_N(x)\Delta_N(y)}{\prod_{j,k=1}^N (1 - x_j y_k)}, \quad (30)$$

by the formula for a Cauchy determinant, so we really get (22) with the right normalization. We see that $\det A^{(jk)}$ is also a Cauchy determinant and (29) becomes, after cancellation of common factors,

$$K_N(u, v) = \sum_{r,s=1}^N y_s^v x_r^u \frac{\prod_j (1 - x_j y_s)(1 - x_r y_j)}{(1 - x_r y_s) \prod_{j \neq r}(x_r - x_j) \prod_{j \neq s}(y_s - y_j)}. \quad (31)$$

We can rewrite this using the residue theorem. Let γ_2 be the circle $|w| = 1$ and γ_1' the circle $|\zeta| = (1 + \epsilon)^{-1}$, where $\epsilon > 0$ is chosen so that $|y_j| < (1 + \epsilon)^{-1}$ for $1 \leq j \leq N$. Both curves have positive orientation. An application of the residue theorem shows that the right-hand side of (31) is

$$K_N(u, v) = \frac{1}{(2\pi i)^2} \int_{\gamma_1'} d\zeta \int_{\gamma_2} dw \frac{w^u \zeta^v}{1 - \zeta w} \prod_{j=1}^N \left(\frac{1 - wy_j}{w - x_j}\right)\left(\frac{1 - x_j\zeta}{\zeta - y_j}\right). \quad (32)$$

If γ_1 is the circle $|z| = 1 + \epsilon$ and we put $z = 1/\zeta$, we get from (32)

$$K_N(u,v) = \frac{1}{(2\pi i)^2} \int_{\gamma_1} d\zeta \int_{\gamma_2} dw \frac{w^u}{z^{v+1}} \frac{1}{z-w} \prod_{j=1}^{N} \left(\frac{1 - wy_j}{w - x_j} \right) \left(\frac{z - x_j}{1 - y_j z} \right) , \quad (33)$$

and we have rederived the result in [28].

By (21), the definition of the Schur measure and the remarks above, we obtain

$$P[G(N,N) \leq n] = \sum_{t \in \mathbb{N}^N} P_N^{\mathrm{Sc}}(t_1, \ldots, t_N) \prod_{j=1}^{N} (1 - \chi_{[n+N,\infty)}(t_j)) . \quad (34)$$

Hence, by (26),

$$P[G(N,N) \leq n] = \det(I - K_N)_{\ell(\{n+N, n+N+1, \ldots\})} \quad (35)$$

with K_N given by (33). If we put $y_i = 0$, $M < i \leq N$, $y_i = \sqrt{q}$, $1 \leq i \leq M$ and $x_i = \sqrt{q}$, $1 \leq j \leq N$, this formula can be used to prove Theorem 1.1. The asymptotics can be analysed by a saddle-point argument of the integral in the right-hand side of (33). Note that in this approach we do not need to know anything about Meixner polynomials. After some manipulation, where we use the integral formula for Meixner polynomials, it is possible to obtain the representation (15) for the kernel, see [8].

Let λ' denote the *conjugate partition* to λ, i.e. $\lambda'_j = $ the length of the j:th column in λ, and let $\ell(\lambda)$ denote the number of non-zero-parts in λ. We have

$$\sum_{\lambda; \lambda_1 \leq n} s_\lambda(x) s_\lambda(y) = \sum_{\mu; \ell(\mu) \leq n} s_{\mu'}(x) s_{\mu'}(y) . \quad (36)$$

Let $e_k(x)$ denote the k:th elementary symmetric function, i.e.

$$\prod_{j=1}^{N} (1 + zx_j) = \sum_{k=0}^{N} e_k(x) z^k , \quad (37)$$

and $e_k(x) = 0$ if $k < 0$, $k > N$. Then, there is a third formula for the Schur polynomial, [32, 35],

$$s_{\mu'}(x) = \det(e_{t_i - j}(x))_{i,j=1}^{N} , \quad (38)$$

where $\mu = (\mu_1, \ldots, \mu_N)$ and $t_i = \mu_i + N - i$. As described in [3], we can now use formula (23) to derive Gessel's formula. Inserting (38) into (36) and using (23) gives

$$\sum_{\lambda; \lambda_1 \leq n} s_\lambda(x) s_\lambda(y) = \det(\sum_{m=0}^{\infty} e_{m-j}(x) e_{m-k}(y))_{j,k=1}^{n} . \quad (39)$$

Using (21) we find,

$$P[G(N,N) \leq n] = \prod_{j,l=1}^{N} (1 - x_j y_k) \det(\sum_{m=0}^{\infty} e_m(x) e_{m+j-k}(y))_{j,k=1}^{n} . \quad (40)$$

454 K. Johansson

Note that the determinant in the right-hand side of (40) is a Toeplitz determinant. If we set

$$f(z) = \prod_{j=1}^{N} (1 + x_j/z)(1 + y_j z),$$ (41)

then, by (37) and (40),

$$P[G(N, N) \le n] = \prod_{j,l=1}^{N} (1 - x_j y_k) D_n(f(e^{i\theta})).$$ (42)

If we take $x_j = y_j = \sqrt{\alpha}/N$ and let $N \to \infty$ we obtain (16). Combining (35) and (42) we find,

$$D_n(f(e^{i\theta})) = \prod_{j,l=1}^{N} \frac{1}{(1 - x_j y_k)} \det(I - K_N)_{\ell^2(\{n+N, n+N+1, \dots \})}.$$ (43)

This is the Borodin-Okounkov identity, which can be extended to more general f by taking appropriate limits, [8], see also [10] for a completely different derivation.

References

[1] D. Aldous, P. Diaconis, *Longest increasing subsequences: From patience sorting to the Baik-Deift-Johansson theorem,* Bull. AMS, **36** (1999), 199–213.

[2] C. Andréief, *Note sur une relation les intégrales définies des produits des fonctions,* Mém. de la Soc. Sci. Bordeaux, **2** (1883), 1–14.

[3] J. Baik, E. Rains, *Algebraic aspects of increasing subsequences,* math.CO/9905083.

[4] J. Baik, E. Rains, *The asymptotics of monotone subsequences of involutions,* math.CO/9905084.

[5] J. Baik, P. A. Deift, K. Johansson, *On the distribution of the length of the longest increasing subsequence in a random permutation,* J. Amer. Math. Soc., **12**, (1999), 1119–1178.

[6] A. Borodin, *Biorthogonal ensembles,* Nucl. Phys. B, **536** (1999), 704–732.

[7] A. Borodin, G. Olshanski, *Z-measures on partitions, Robinson-Schensted-Knuth correspondence, and $\beta = 2$ random matrix ensembles,* math.CO/9905189.

[8] A. Borodin, A. Okounkov, *A Fredholm determinant formula for Toeplitz determinants,* Integral Eq. Oper. Th., **37** (2000), 386–396.

[9] A. Borodin, A. Okounkov, G. Olshanski, *Asymptotics of Plancherel measures for symmetric groups,* J. Amer. Math. Soc., **13** (2000), 481–515.

[10] E. L. Basor, H. Widom, *On a Toeplitz determinant identity of Borodin and Okounkov,* preprint, Integral Eq. Oper. Th., **37** (2000), 397–401.

[11] T. S. Chihara, *An Introduction to Orthogonal Polynomials,* New York, Gordon and Breach, 1978.

[12] P. A. Deift, X. Zhou, *A Steepest descent method for oscillatory Riemann-Hilbert problems,* Ann. Math., **137**, (1993), 295–368.

[13] M. Eden, in *Proceedings of the Fourth Berkeley Symposium on Mathematical Statistics and Probability,* Ed. F. Neyman, vol. IV, University of California Press, Berkeley, 1961.

[14] W. Fulton, *Young Tableaux,* London Mathematical Society, Student Texts 35, Cambridge Univ. Press, 1997.

[15] I. M. Gessel, *Symmetric functions and P-recursiveness,* J. Combin. Theory Ser. A, **53**, (1990), 257–285.

[16] J. M. Hammersley, A few seedlings of research, In *Proc. Sixth Berkeley Symp. Math. Statist. and Probability,* Volume 1, pp. 345–394, University of California Press, 1972.

[17] K. Johansson, *Shape fluctuations and random matrices,* Commun. Math. Phys., **209**, (2000), 437–476.

[18] K. Johansson, *Transversal fluctuations for increasing subsequences on the plane,* Probab. Th. and Rel. Fields, **116** (2000), 445–456.

[19] K. Johansson, *Discrete orthogonal polynomial ensembles and the Plancherel measure,* Annals of Math., **153** (2001), 259–296.

[20] K. Johansson, *Random permutations and the discrete Bessel kernel,* to appear in the MSRI volume on random matrices, Cambridge University Press.

[21] H. Kesten, Aspects of first passage percolation, in *Lecture Notes in Mathematics vol 1180,* Springer-Verlag, Berlin, 1986.

[22] D. E. Knuth, *Permutations, Matrices and Generalized Young Tableaux,* Pacific J. Math., **34**, (1970), 709–727.

[23] J. Krug, P. Meakin, T. Halpin-Healy, *Amplitude universality for driven interfaces and directed polymers in random media,* Phys. Rev. A, **45** (1992), 638–653.

[24] J. Krug, H. Spohn, Kinetic Roughening of Growing Interfaces. In: *Solids far from Equilibrium: Growth, Morphology and Defects,* Ed. C Godrèche, Cambridge, Cambridge University Press, 1992, 479–582.

[25] C. Licea, C. M. Newman, M. S. T. Piza, *Superdiffusivity in first-passage percolation,* Probab. Th. and Rel. Fields, **106** (1996), 559–591.

[26] M. L. Mehta, *Random Matrices,* 2nd ed., Academic Press, San Diego, 1991.

[27] C. M. Newman, M. S. T. Piza, *Divergence of shape fluctuations in two dimensions,* Ann. Prob., **23** (1995), 977–1005.

[28] A. Okounkov, *Infinite wedge and measures on partitions,* math.RT/9907127.

[29] M. Prähofer, H. Spohn, *Universal distributions for growth processes in $1 + 1$ dimensions and random matrices,* Phys. Rev. Lett., **84** (2000), 4882–4885.

[30] D. Richardson, *Random growth in a tesselation,* Proc. Camb. Phil. Soc., **74** (1973), 515–528.

[31] H. Rost, *Non-Equilibrium Behaviour of a Many Particle Process: Density Profile and Local Equilibria,* Zeitschrift f. Wahrsch. Verw. Geb., **58** (1981), 41–53.

[32] B. Sagan, *The Symmetric Group,* Brooks/Cole Publ. Comp., 1991.

[33] T. Seppäläinen, *Coupling the totally asymmetric simple exclusion process with a moving interface,* Markov Process. Rel. Fields, **4** (1998), 592–628.

[34] A. Soshnikov, *Determinantal random point fields,* math.PR/0002099.

[35] R. P. Stanley, *Enumerative Combinatorics,* Vol. 2, Cambridge University Press, 1999.

[36] C. A. Tracy, H. Widom, *Level Spacing Distributions and the Airy Kernel*, Commun. Math. Phys., **159**, (1994), 151–174.

[37] C. A. Tracy, H. Widom, *Correlation Functions, Cluster Functions, and Spacing Distributions for Random Matrices*, J. Statist. Phys., **92**, (1998), 809–835.

[38] M. V. Wüthrich, *Scaling identity for crossing Brownian motion in a Poissonian potential*, Probab. Th. and Rel. Fields, **112** (1998), 299–319.

Department of Mathematics
Royal Institute of Technology
S-100 44 Stockholm, Sweden
E-mail address: kurtj@math.kth.se

Sobolev Spaces and Quasiconformal Mappings on Metric Spaces

Pekka Koskela

Abstract. Heinonen and I have recently established a theory of quasiconformal mappings on Ahlfors regular Loewner spaces. These spaces are metric spaces that have sufficiently many rectifiable curves in a sense of good estimates on moduli of curve families. The Loewner condition can be conveniently described in terms of Poincaré inequalities for pairs of functions and upper gradients. Here an upper gradient plays the role that the length of the gradient of a smooth function has in the Euclidean setting. For example, the Euclidean spaces and Heisenberg groups and the more general Carnot groups admit the type of a Poincaré inequality we need. We describe the basics and discuss the associated Sobolev spaces that, for example, allow for a very abstract setting for variational integrals. We also discuss the concept of a Sobolev mapping between two metric spaces.

1. Introduction

Let X and Y be metric spaces and $f : X \to Y$ a homeomorphism. Then the distortion of f at a point $x \in X$ is

$$H(x) := \lim\,\sup_{r \to 0} \frac{L(x,r)}{l(x,r)}, \tag{1}$$

where

$$L(x,r) := \sup\{d_Y(f(x), f(y)) : d_X(x,y) \leq r\},$$
$$l(x,r) := \inf\{d_Y(f(x), f(y)) : d_X(x,y) \geq r\}.$$

We say that f is quasiconformal if there is a constant H so that $H(x) \leq H$ for every $x \in X$. This infinitesimal condition is easy to state but not easy to use. For instance, it is not clear from the definition if the inverse mapping is quasiconformal as well. It is thus desirable to find conditions on X and Y that would guarantee the stronger, global requirement:

$$H(x,r) := \frac{L(x,r)}{l(x,r)} \leq H' < \infty$$

for all $x \in X$ and all $r > 0$ whenever f is a quasiconformal mapping of X onto Y. We call this global condition quasisymmetry and a homeomorphism satisfying it a quasisymmetric mapping. Our notion of quasisymmetry is equivalent to the concept of a quasisymmetric function introduced by Beurling and Ahlfors in [2] when $X = Y$ is the real line and to the concept of quasisymmetry defined by Tukia and Väisälä in [28] for many metric spaces X and Y, especially for all metric spaces discussed below in Theorem 1.1. It is a fundamental fact that quasiconformal mappings between Euclidean spaces of dimension at least 2 are quasisymmetric. This fails in dimension 1; consider, for example, $f(x) = x + e^x$. It is immediate that the inverse mapping of a quasisymmetric mapping is also quasisymmetric and many other important properties of quasiconformal mappings also follow easily from quasisymmetry.

This infinitesimal-to-global principle was shown to hold on the Heisenberg group by Koranyi and Reimann, [17], and it holds for mappings between spaces that occur as conformal boundaries of rank-1 symmetric spaces by results of Mostow and Pansu, [23, 24]. This extends to the case of Carnot groups by results of Heinonen and Koskela in [11] and to the more general case of Carnot-Carathéodory spaces by the work of Margulis and Mostow, [22].

A natural metric setting that covers all the above cases turns out to be that of an Ahlfors regular metric space that supports a suitable Poincaré inequality. Here X is Ahlfors Q-regular if X is equipped with a Borel measure μ and there is a constant $C_\mu \geq 1$ such that

$$C_\mu^{-1} r^Q \leq \mu(B(x,r)) \leq C_\mu r^Q \,,$$

for all balls $B(x,r) \subset X$ of radius $r < \mathrm{diam} X$. We also assume that X is proper: each closed ball in X is compact.

Theorem 1.1. *Let $Q > 1$. If X is a proper, Q-regular metric space that supports a Q-Poincaré inequality, then each quasiconformal self-homeomorphism of X is quasisymmetric.*

For the assumption concerning Poincaré inequalities see Section 2. This result of Heinonen and Koskela is from [13] where versions with more general targets can be found as well. We arrived at Theorem 1.1 from results more intrinsic to quasiconformal mappings. This includes moduli of curve families and so-called Loewner spaces. The quasisymmetry gives Hölder continuity and other pleasant properties. If one assumes a p-Poincaré inequality for some $p < Q$, then the volume derivative of a quasiconformal mapping is a Muckenhoupt weight. For all this see [13]. We will now leave quasiconformal mappings with the following philosophical point of view: for quasiconformal mappings to be quasisymmetric, the infinitesimal distortion of shapes needs to integrate to global control and a Poincaré inequality tells us that the infinitesimal control of a function by its "gradient" results in global estimates.

In this paper we concentrate on metric spaces that support a Poincaré inequality. Section 2 deals with the basics of such spaces. One of the crucial concepts here is the notion of an upper gradient which will be our substitute for the length of the gradient of a smooth function. In Section 3 we discuss Sobolev spaces and notice that one obtains a rich theory as soon as a Poincaré inequality is available. The issue of a Sobolev mapping between metric spaces gets also briefly addressed. In the final short section, Section 4, we comment on the the consequences related to calculus of variations in the metric setting and point out some other applications.

2. Spaces that Support Poincaré Inequalities

To simplify the things we will assume that the spaces we consider are relatively nice. Our standing assumption will be that X is a proper (recall the definition from Introduction), *doubling* metric space: X is equipped with a Borel measure μ so that

$$\mu(2B) \leq C_d \mu(B) \tag{2}$$

whenever B is a ball and $2B$ is the ball with the same center as B and with radius twice that of B. Notice that each Ahlfors regular metric space is automatically doubling. Moreover, a simple iteration argument shows that the doubling condition (2) is equivalent to the existence of a constant $C > 0$ and an exponent $s > 0$ such that

$$\frac{\mu(B)}{\mu(B_0)} \geq C \left(\frac{r}{r_0}\right)^s \tag{3}$$

whenever B_0 is an arbitrary ball of radius r_0 and $B = B(x,r)$, $x \in B_0$, $r \leq r_0$.

2.1. Upper gradients

In the abstract setting of a metric space we cannot talk about partial derivatives. However, the length of a gradient of a smooth function has a rather natural generalization.

Definition 2.1. *Let* $u\colon A \to \overline{\mathbf{R}}$, $A \subset X$. *An* upper gradient *of* u *on* A *is a Borel function* $g\colon A \to [0,\infty]$ *such that for each rectifiable path* $\gamma\colon [0,l] \to A$

$$|u(\gamma(l)) - u(\gamma(0))| \leq \int_\gamma g\,ds. \tag{4}$$

This definition is from [12, 13] except that there functions g as in (4) were not called upper gradients. The current terminology comes from [20].

If u is Lipschitz:

$$|u(x) - u(y)| \leq Ld(x,y)$$

for all x,y, then trivially $g \equiv L$ is an upper gradient of u, but the local Lipschitz constant provides us with a smaller upper gradient than the global Lipschitz constant: Given u, define

$$\operatorname{Lip} u(x) = \liminf_{r \to 0^+} \sup_{\{y:d(x,y)\leq r\}} \frac{|u(x) - u(y)|}{r}.$$

See e.g. [6].

2.2. Further Examples

1. $g \equiv \infty$ is always an upper gradient.
2. If there are no rectifiable, non-constant curves in A, then $g \equiv 0$ is an upper gradient.
3. Each function u in the usual Sobolev class $W^{1,p}(\mathbf{R^n})$ has a representative that has a p-integrable upper gradient. See e.g. [19, 26].

Notice that the existence of an upper gradient that is integrable on many curves gives good control on u: If $\int_\gamma g\, ds < \infty$, then u is continuous and real-valued on γ.

To save space we mention here only that upper gradients behave quite the way one expects, see [6, 19, 25, 26].

2.3. Poincaré inequalities

Definition 2.2. *Let (X, d, μ) be as before. X is said to support a p-Poincaré inequality, $p \geq 1$, if there exist $C > 0$ and $\lambda \geq 1$ such that*

$$\fint_B |u - u_B|\, d\mu \leq C\, diam\,(B) \left(\fint_{\lambda B} g^p\, d\mu \right)^{1/p}, \tag{5}$$

for all balls B, for all continuous functions u on λB and for every upper gradient g of u on λB. Here u_B is the average of u over the ball B and the barred integrals mean averaged integrals.

When compared with the usual Poincaré inequalities in the Euclidean setting, two differences appear: instead of the (p, p)-inequality we ask for a $(1, p)$-inequality and the ball that g gets integrated over is larger (when $\lambda > 1$) than the corresponding ball for u. It follows from Theorem 2.4 that we could as well assume a (p, p)-inequality. It is however immediate from the $(1, p)$-formulation (using Hölder's inequality) that (5) becomes weaker as p increases whereas this is substantially harder to see from the (p, p)-formulation. In fact, (5) becomes strictly weaker when p decreases [10, 13]. Regarding the size of balls, we can omit λ if the geometry of balls is sufficiently nice, e.g. if the metric is a length metric, but not in general, see [10].

In the Euclidean setting the Poincaré inequality for Sobolev functions follows from that inequality for smooth functions. In the abstract setting we cannot speak about smooth functions but Lipschitz functions make sense. It is then natural to ask if a Poincaré inequality for Lipschitz functions guarantees that the space supports a Poincaré inequality. This turns out to be true in great generality.

Theorem 2.3. *Assume that X is path-wise connected. Then X supports a p-Poincaré inequality if and only if (5) holds for each ball B and for all Lipschitz functions u and their upper gradients. In fact, (5) will then hold for all measurable functions u.*

Here our standing assumption that X be proper is essential (see [10]). For a proof of Theorem 2.3 see [14, 19].

Recall that we required inequality (5) for every continuous function u and each upper gradient of u. It is then natural to inquire how good the functions u are for which (5) holds for some L^p-function g. According to the following result, such functions share many of the good properties of Euclidean Sobolev functions. The various constants C below are not necessarily the same as the constant in (5) but they depend only on the given data, of which u, g are not part.

Theorem 2.4. *Let X be a doubling space and s be an exponent as in (3). Assume that $g \in L^p(X)$, $p \geq 1$, and a locally integrable u satisfy inequality (5) for all balls B.*

1. *If $p < s$, and $q < \frac{ps}{s-p}$, then*

$$\left(\fint_B |u - u_B|^q \, d\mu \right)^{1/q} \leq C \, diam \, (B) \left(\fint_{5\lambda B} g^p \, d\mu \right)^{1/p}$$

    for all balls B.

2. *If X is connected and $p = s > 1$, then*

$$\fint_B \exp \left((t|u - u_B|)^{\frac{s}{s-1}} \right) \, d\mu \leq C_2 \tag{6}$$

    for each ball B, where

$$t = \frac{C_1 \mu(B)^{1/s}}{diam \, (B) \|g\|_{L^s(5\lambda B)}}.$$

3. *If $p > s$, then u (after redefinition in a set of measure zero) is locally Hölder continuous:*

$$|u(x) - u(y)| \leq C r_0^{s/p} d(x,y)^{1-s/p} \left(\fint_{5\lambda B_0} g^p \, d\mu \right)^{1/p} \tag{7}$$

    for all $x, y \in B_0$, where B_0 is an arbitrary ball of radius r_0.

4. *Let $x \in X$. If $p > s - 1$, then u (after redefinition in a set of measure zero) is uniformly Hölder continuous with exponent $1 - (s-1)/p$ for almost every $r > 0$ on the set $\{y : d(y, x) = r\}$.*

5. *Suppose that a sequence of pairs $(u_i, g_i)_{i \in \mathbb{N}}$ satisfies inequality (5) with uniform constants. If $\|u_i\|_{L^1(B)} + \|g_i\|_{L^p(5\lambda B)} \leq C$ for each i for a ball B, then there exists a subsequence $(u_{i_k})_{k \in \mathbb{N}}$ that converges in $L^q(B)$ to a function u. Here one can take any $1 \leq q < ps/(s-p)$ when $p < s$ and any finite $q \geq 1$ when $p \geq s$. Moreover, when $p > 1$, there is a function g in $L^p(5\lambda B)$ so that (5) holds for the pair (u, g) (with a possibly different constant C).*

For a proof of these results see [10] (Part 1 can be found already in [9]). The last conclusion in Part 5 is however not covered by [10] but it easily follows by selecting a weakly convergent subsequence of the subsequence g_{i_k} and using Mazur's lemma to get a strongly convergent sequence of convex combinations of these functions.

By Theorem 2.4, the Poincaré inequality (5) implies versions of Sobolev-Poincaré and Trudinger inequalities, a version of the Sobolev embedding theorem on spheres and of the Rellich-Kondrachov theorem. Besides the Sobolev-Poincaré inequality and the Sobolev embedding on spheres, we obtain as good results as in the Euclidean setting. If we assume that X supports a p-Poincaré inequality, then we have the full analog of the Sobolev-Poincaré inequality in (1) with $q = \frac{ps}{s-p}$ for function-upper gradient pairs (see [10]).

When does X then support a Poincaré inequality? A look at (4) should soon convince the reader that this is the case whenever the concept of a smooth function makes sense and one has a "usual" Poincaré inequality. This covers the case of Carnot groups and the vector field setting (c.f. [10]). Regarding necessary conditions, a Poincaré inequality implies the existence of "many" short curves, in particular, that any pair of points in X can be joined with a curve whose length is no more than a fixed constant times the distance between the points. Spaces with this property are called *quasiconvex*. For sufficient conditions and "exotic" examples see [4, 10, 13, 21, 25, 27]. We wish to stress that the spaces we consider need not be integer dimensional. Indeed, Laakso [21] constructs for each $Q > 1$ Ahlfors Q-regular spaces that support even a 1-Poincaré inequality. The first non-integer-dimensional examples were constructed by Bourdon and Pajot ([4]) as the Gromov boundaries of certain hyperbolic buildings.

2.4. Stability

One good property of the concept of a metric space supporting a p-Poincaré inequality is that it is stable both under bi-Lipschitz changes of the metric and under pointed, measured Gromov-Hausdorff convergence. The first stability is immediate; recall that $f \colon X \to Y$ is bi-Lipschitz if there is a constant L so that

$$d_X(x,y)/L \le d_Y(f(x), f(y)) \le L d_X(x,y)$$

for all $x, y \in X$. The second result is more complicated and essentially due to Cheeger [6]. Recall that all the spaces we consider are assumed to be proper and doubling.

Theorem 2.5. *Suppose that (X_i, d_i, μ_i) is a sequence of spaces that all support a p-Poincaré inequality with fixed C, λ and are all doubling with a fixed doubling constant C_d. If this sequence converges (as a subsequence always does) in the pointed, measured Gromov-Hausdorff sense to a space (X, d, μ), then X also supports a p-Poincaré inequality.*

Cheeger obtains only a q-Poincaré inequality for X for all $q > p$ but relying on his work one can verify even a p-Poincaré inequality, as was done in [19].

Above the Poincaré inequality, doubling and the properness of the spaces guarantee that each of the spaces X_i is quasiconvex with a fixed constant. Thus, using approriate bi-Lipschitz changes of metric, each X_i can be assumed to be a length space, and it follows that also X carries a length space metric. The requirement that the convergence be measured then means that, whenever $x_i \in X_i$

converge to $x \in X$, the measures of the balls $B(x_i, r)$, when normalized by the measure of $B(x_i, 1)$, converge to the measure of $B(x, r)$ for $r > 0$; here the balls are given in the length space metrics.

3. Sobolev Spaces

3.1. The real-valued case

Following Shanmugalingam [26] we define the Newtonian space $N^{1,p}(X)$ by

$$N^{1,p}(X) = \{u \in L^p(X) : \exists \text{ an upper gradient } g_u \in L^p(X)\}$$

and equip it with the norm

$$\|u\|_{1,p} = \|u\|_p + \inf_{g_u} \|g_u\|_p .$$

Naturally, we need to identify u and v if $\|u - v\|_{1,p} = 0$. Motivated by the Euclidean setting we further define

$$H^{1,p}(X) = \overline{\mathrm{Lip}\,(X)}^{\|\cdot\|_{1,p}} .$$

Finally we let

$$P^{1,p}(X) = \{u \in L^p(X) : \exists g \in L^p(X) \text{ such that (5) holds for the pair } u,\, g\} .$$

If X supports a p-Poincaré inequality, then all these spaces coincide, and the Newtonian space has many of the properties of the classical Sobolev spaces.

Theorem 3.1. *Assume that X supports a p-Poincaré inequality, $p > 1$. Then*

$$N^{1,p}(X) = H^{1,p}(X) = P^{1,p}(X)$$

coincide as sets. Moreover, $N^{1,p}(X)$ is a reflexive Banach space, and the usual Sobolev inequalities hold for functions in $N^{1,q}(X)$, $q \geq p$.

By Sobolev inequalities we mean the results from Theorem 2.4 with the exponent p there being any exponent larger than or equal to the exponent p in Theorem 3.1, except that for $p < s$ we allow $q = ps/(s - p)$ in the first of the inequalities. Other than for this improvement, the Sobolev inequalities directly follow from Theorem 2.4. For the above endpoint result, one uses truncation arguments, see [10]. The reflexivity is not easy to establish. This is due to Cheeger [6]; his definition for the Sobolev space is slightly different from the above definitions but, as shown by Shanmugalingam [26], the resulting space is isometrically equivalent to $N^{1,p}(X)$. For the coincidence of the three spaces above see [7, 10, 26]. There are also other possible definitions for a Sobolev class. The class introduced by Korevaar and Schoen [18] coincides with the above Sobolev spaces under the Poincaré inequality assumption but the space introduced by Hajłasz only for exponents $q > p$, see [20].

We began by considering upper gradients that form a substitute for the length of a gradient. Very surprisingly, the Poincaré inequality guarantees that even the differential of a function makes sense. This is a remarkable result of Cheeger [6] and it is a crucial part of his proof for the reflexivity. Another amazing conclusion of Cheeger is that the "minimal upper gradient" of a Sobolev function can be realized as the point-wise Lipschitz constant. Unfortunately, we have no space to discuss this in detail here, and we have to confine ourselves to referring to his paper.

3.2. Sobolev mappings

We originally got interested in Sobolev functions on metric spaces because of the need for tools suitable to handle quasiconformal mappings. For the basic questions, like Theorem 1.1, the real-valued theory was sufficient: the trick to be used was to replace "locally" a mapping f with $u(x) = d_Y(f(x), f(x_0))$, where x_0 is a "locally" fixed base point. The limits in this approach showed in that we were not able to obtain an analytic definition for quasiconformality under the Q-Poincaré inequality assumption. The above switch from a mapping f to a real-valued function u can be used to define the class of (local) Sobolev mappings from X to Y, but a better definition is obtained using post-composition with Lipschitz functions. This replaces the mapping with a family of real-valued functions. Our obstacle was the lack of Lipschitz approximations to our quasiconformal mapping.

The key for us was to embed Y isometrically into $\ell^\infty(Y)$ to gain linear structure; this can be done for any metric space Y. This allows us to conveniently invoke the vector-valued integration theory of Bochner and Pettis. The nice thing here is that the validity of a real-valued p-Poincaré inequality on X is equivalent with that for the $\ell^\infty(X)$-valued case, see [15]. Even though usual convolutions cannot be used, one can still use certain "discrete" convolutions to approximate our quasiconformal mapping to show that the mapping belongs to the desired Sobolev class. As an application to the theory of quasiconformal mappings we have the following consequence.

Theorem 3.2. *Let X be a proper, Q-regular metric space that supports a Q-Poincaré inequality and f a self-homeomorphism of X. Then f is quasiconformal if and only if $f \in N_{loc}^{1,Q}(X, X)$ and*

$$Lip\, f(x)^Q \leq K\mu_f(x)$$

for a.e. $x \in X$.

Here $Lip\, f(x)$ is a local Lipschitz constant of f, this time defined as

$$\mathrm{Lip}\, f(x) = \limsup_{r \to 0^+} \sup_{\{y : d(x,y) \leq r\}} \frac{d(f(x), f(y))}{r}$$

and

$$\mu_f(x) = \limsup_{r \to 0^+} \frac{\mu(fB(x,r))}{\mu(B(x,r))}$$

is the volume derivative of f. For this result see [15].

4. Minimization Problems and other Related Topics

By the reflexivity of the Sobolev space $N^{1,p}(X)$ under the p-Poincaré inequality assumption, direct methods from the calculus of variations are available. Thus one has the existence and uniqueness for solutions to the Dirichlet problem (c.f. [6]).

What about the regularity of the solutions? It seems plausible that one could use the Moser iteration scheme to obtain Harnack's inequalities for positive solutions and as a consequence Hölder continuity of the solutions. There is, however, a hidden danger which surfaces when trying to establish Caccioppoli-type inequalities: the norm on the differentials need not necessarily be smooth. This problem can be overcome by using De Giorgi's method, and the solutions are indeed Hölder continuous; see [16]. There are also recent results on the boundary continuity [3].

Another area where Poincaré inequalities in our sense have appeared is geometric measure theory. This relates to the borderline case $p = 1$ of Sobolev functions where functions of bounded variation occupy the theory. Ambrosio [1] studies properties of sets of finite perimeter in Ahlfors regular metric spaces that support a 1-Poincaré inequality. For applications to the setting of the Heisenberg groups see [8]. It is not a big surprise that the 1-Poincaré inequalities surface here: there is an intimate connection between isoperimetric and 1-Poincaré inequalities as observed already years ago by Maz'ya, Federer and Fleming.

Let us close this note by going back to the origin for our motivation to look at quasiconformal mappings and Sobolev functions in the metric setting. The quasiconformal mappings appeared for the first time in non-Riemannian setting in the work of Mostow on rigidity of symmetric spaces. It would thus be desireable that the metric theory of quasiconformal mappings also would result in rigidity results. This is indeed the case: Bourdon and Pajot [5] have very recently obtained ridigity results for hyperbolic buildings using the metric theory of quasiconformal mappings.

References

[1] L. Ambrosio, *Some fine properites of sets of finite perimeter in Ahlfors regular metric measure spaces*, Adv. Math. **159** (2001), 51–67.

[2] A. Beurling and L. V. Ahlfors, *The boundary correspondence under quasiconformal mappings*, Acta Math. **96** (1956), 125–142

[3] J. Björn, *Boundary continuity for quasi-minimizers on metric spaces*, Linköping University, Department of Mathematics, Preprint #1 (2000).

[4] M. Bourdon and H. Pajot, *Poincaré inequalities and quasiconformal structure on the boundaries of some hyperbolic buildings*, Proc. Amer. Math. Soc. **127** (1999), 2315–2324.

[5] M. Bourdon and H. Pajot, *Rigidity of quasi-isometries for some hyperbolic buildings*, Comment. Math. Helv. **75** (2000), 701–736.

[6] J. Cheeger, *Differentiability of Lipschitz functions on metric spaces*, Geom. Funct. Anal. **9** (1999), 428–517.

[7] B. Franchi, P. Hajłasz and P. Koskela, *Definitions of Sobolev classes on metric spaces*, Ann. Inst. Fourier (Grenoble) **49** (1999), 1903–1924.

[8] B. Franchi, R. Serapioni and F. Serra Cassano, *Rectifiability and perimeter in the Heisenberg group*, preprint (1999).

[9] P. Hajłasz and P. Koskela, *Sobolev meets Poincaré*, C. R. Acad. Sci. Paris **320** (1995), 1211–1215.

[10] P. Hajłasz and P. Koskela, *Sobolev met Poincaré*, Mem. Amer. Math. Soc. **145** (2000), no. 688, x+101 pp.

[11] J. Heinonen and P. Koskela, *Definitions of quasiconformality*, Invent. math. **120** (1995), 61–79

[12] J. Heinonen and P. Koskela, *From local to global in quasiconformal structures*, Proc. Nat. Acad. Sci. U.S.A. **93** (1996), 554–556.

[13] J. Heinonen and P. Koskela, *Quasiconformal maps on metric spaces with controlled geometry*, Acta Math. **181** (1998), 1–61.

[14] J. Heinonen and P. Koskela, *A note on Lipschitz functions, upper gradients and the Poincaré inequality*, New Zealand J. Math. **28** (1999), 37–42.

[15] J. Heinonen, P. Koskela, N. Shanmugalingam and J. T. Tyson, *Sobolev classes of Banach space-valued functions and quasiconformal mappings*, to appear in J. Analyse Math.

[16] J. Kinnunen and N. Shanmugalingam, *Quasi-minimizers on metric spaces*, to appear in Manuscripta Math.

[17] A. Korányi and H. M. Reimann, *Foundations for the theory of quasiconformal mappings on the Heisenberg group*, Adv. Math **111**, (1995), 1–87.

[18] N. J. Korevaar and R. M. Schoen, *Sobolev spaces and harmonic maps for metric space targets*, Comm. Anal. Geom. **1** (1993), 561–659.

[19] P. Koskela, *Upper gradients and Poincaré inequalities*, lectures in Trento in June, 1999.

[20] P. Koskela and P. MacManus, *Quasiconformal mappings and Sobolev spaces*, Studia Math. **131** (1998), 1–17.

[21] T. Laakso, *Ahlfors Q-regular spaces with arbitrary Q admitting weak Poincaré inequality*, Geom. Funct. Anal. **10** (2000), 111–123.

[22] G. A. Margulis and G. D. Mostow, *The differential of a quasiconformal mapping of a Carnot-Carathéodory space*, Geom. Funct. Anal. **5** (1995), 402–433

[23] G. D. Mostow, *A remark on quasiconformal mappings on Carnot groups*, Michigan Math. J. **41** (1994), 31–37.

[24] P. Pansu, *Métriques de Carnot-Carathéodory et quasiisométries des espaces symétriques de rang un*, Ann. Math. **129** (1989), 1–60.

[25] S. Semmes, *Finding curves on general spaces through quantitative topology, with applications to Sobolev and Poincaré inequalities*, Selecta Math. **2**, (1996), 155–295.

[26] N. Shanmugalingam, *Newtonian spaces: an extension of Sobolev spaces*, Rev. Mat. Iberoamericana **16** (2000), 243–279.

[27] K. T. Sturm, *Diffusion processes and heat kernels on metric spaces*, Ann. Probab. **26** (1998), 1–55.

[28] P. Tukia and J. Väisälä, *Quasisymmetric embeddings of metric spaces*, Ann. Acad. Sci. Fenn. Math. **5** (1980), 97–114.

Department of Mathematics
University of Jyväskylä
P.O. Box 35
Fin-40351 Jyväskylä, Finland
E-mail address: pkoskela@math.jyu.fi

Understanding Skyrmions Using Rational Maps

Nicholas S. Manton and Bernard M. A. G. Piette

Abstract. We discuss an ansatz for Skyrme fields in three dimensions which uses rational maps between Riemann spheres, and produces shell-like structures of polyhedral form. Houghton, Manton and Sutcliffe showed that a single rational map gives good approximations to the minimal energy Skyrmions up to baryon number of order 10. We show how the method can be generalized by using two or more rational maps to give a double-shell or multi-shell structure. Particularly interesting examples occur at baryon numbers 12 and 14.

1. Introduction

The Skyrme model is a nonlinear theory of pions in $\mathbb{R}^3$, with an $SU(2)$ valued scalar field $U(\mathbf{x}, t)$, the Skyrme field, satisfying the boundary condition $U \to 1$ as $|\mathbf{x}| \to \infty$. Static fields obey the equation

$$\partial_i(R_i - \frac{1}{4}[R_j, [R_j, R_i]]) = 0 \tag{1}$$

where R_i is the $su(2)$ valued current $R_i = (\partial_i U)U^{-1}$. Such fields are stationary points (either minima or saddle points) of the energy function

$$E = \int \{-\frac{1}{2}\,\text{Tr}(R_i R_i) - \frac{1}{16}\,\text{Tr}([R_i, R_j][R_i, R_j])\}\ d^3\mathbf{x}. \tag{2}$$

Associated with a Skyrme field is a topological integer, the baryon number B, defined as the degree of the map $U\colon \mathbb{R}^3 \mapsto SU(2)$. It is well defined because of the boundary condition at infinity.

Solutions of equation (1) are known for several values of B, but they can only be obtained numerically. Many of these solutions are stable, and probably represent the global minimum of the energy for given B. We shall refer to the solutions believed to be of lowest energy for each B as Skyrmions.

There is a nine-dimensional symmetry group of the equation and boundary condition. It consists of translations and rotations in $\mathbb{R}^3$ and the $SO(3)$ isospin transformations $U \mapsto \mathcal{O}U\mathcal{O}^{-1}$ where $\mathcal{O}$ is a constant element of $SU(2)$.

Skyrme found the spherically symmetric $B = 1$ Skyrmion. The $B = 2$ Skyrmion is toroidal. A substantial numerical search for Skyrmion solutions was undertaken by Braaten, Townsend and Carson [4], and minimal energy solutions up to $B = 5$ were found. (Their solution for $B = 6$ was rather inaccurate.) Surprisingly, the $B = 3$ solution has tetrahedral symmetry T_d, and the $B = 4$ solution has cubic symmetry O_h. Battye and Sutcliffe [2] subsequently found all Skyrmions up to $B = 8$. (Their solution for $B = 9$ is probably a saddle point.) The $B = 7$ Skyrmion has icosahedral symmetry Y_h. Recently, with new methods, they have found candidate solutions up to $B = 22$ [3].

2. Skyrme Fields from Rational Maps

Let us denote a point in $\mathbb{R}^3$ by its coordinates (r, z) where r is the radial distance from the origin and $z = \tan(\theta/2) \exp i\varphi$ specifies the direction from the origin. Houghton, Manton and Sutcliffe [6] showed that one can understand the structure of the known Skyrmions in terms of an ansatz using rational maps

$$R(z) = \frac{p(z)}{q(z)} \tag{3}$$

where p and q are polynomials, and a radial profile function $f(r)$. One identifies the target S^2 of the rational maps with spheres of latitude on $SU(2)$, that is, spheres at a fixed distance from the identity element.

Recall that the direction z corresponds to the Cartesian unit vector

$$\widehat{\mathbf{n}}_z = \frac{1}{1 + |z|^2} \left(2 \operatorname{Re}(z), 2 \operatorname{Im}(z), 1 - |z|^2 \right). \tag{4}$$

Similarly the value of the rational map R is associated with the unit vector

$$\widehat{\mathbf{n}}_R = \frac{1}{1 + |R|^2} \left(2 \operatorname{Re}(R), 2 \operatorname{Im}(R), 1 - |R|^2 \right). \tag{5}$$

The ansatz is

$$U(r, z) = \exp(i f(r)\, \widehat{\mathbf{n}}_{R(z)} \cdot \boldsymbol{\sigma}) \tag{6}$$

where $\boldsymbol{\sigma} = (\sigma_1, \sigma_2, \sigma_3)$ are Pauli matrices. For this to be well defined at the origin, $f(0) = k\pi$, for some integer k. The boundary value $U = 1$ at $r = \infty$ requires that $f(\infty) = 0$. The baryon number of this field is $B = Nk$, where $N = \max(\deg p, \deg q)$ is the degree of R. We consider only the case $k = 1$ here, so $B = N$. Note that an $SU(2)$ Möbius transformation of the rational map

$$R(z) \mapsto \frac{\alpha R(z) + \beta}{-\bar{\beta} R(z) + \bar{\alpha}} \tag{7}$$

with $|\alpha|^2 + |\beta|^2 = 1$ acts as an isospin transformation.

An attractive feature of the ansatz (6) is that it leads to a simple energy expression which can be minimized with respect to the rational map R and the profile function f to obtain close approximations to the true Skyrmions. A generalized rational map ansatz has also proved useful in the construction of solutions to Skyrme models with fields having values in $SU(N)$ [7].

The energy for a Skyrme field of the form (6) is

$$E = \int \left[f'^2 + 2(f'^2 + 1) \frac{\sin^2 f}{r^2} \left(\frac{1 + |z|^2}{1 + |R|^2} \left| \frac{dR}{dz} \right| \right)^2 \right. \\ \left. + \frac{\sin^4 f}{r^4} \left(\frac{1 + |z|^2}{1 + |R|^2} \left| \frac{dR}{dz} \right| \right)^4 \right] \frac{2i \, dz d\bar{z} r^2 dr}{(1 + |z|^2)^2} \, . \tag{8}$$

Now

$$\left(\frac{1 + |z|^2}{1 + |R|^2} \left| \frac{dR}{dz} \right| \right)^2 \frac{2i \, dz d\bar{z}}{(1 + |z|^2)^2} \tag{9}$$

is the pull-back of the area form $2i \, dR d\bar{R}/(1 + |R|^2)^2$ on the target sphere of the rational map; therefore its integral is 4π times the degree N. So the energy simplifies to

$$E = 4\pi \int \left(r^2 f'^2 + 2N(f'^2 + 1) \sin^2 f + \mathcal{I} \frac{\sin^4 f}{r^2} \right) dr \tag{10}$$

where $\mathcal{I}$ denotes the integral

$$\mathcal{I} = \frac{1}{4\pi} \int \left(\frac{1 + |z|^2}{1 + |R|^2} \left| \frac{dR}{dz} \right| \right)^4 \frac{2i \, dz d\bar{z}}{(1 + |z|^2)^2} \, . \tag{11}$$

Applying a Bogomolny-type argument to the expression (10), one can show that

$$E \geq 4\pi^2 (2N + \sqrt{\mathcal{I}}) \, . \tag{12}$$

This lower bound on the energy, which applies to the rational map ansatz, is higher than the Fadeev-Bogomolny bound satisfied by any Skyrme field, $E \geq 12\pi^2 B$. This is because the Schwarz inequality implies $\mathcal{I} \geq N^2$.

To minimize the energy E, one should first minimize $\mathcal{I}$ over all maps of degree N. Then the profile function f minimizing (10) is found by solving a second order differential equation with N and $\mathcal{I}$ as parameters. In [6] only rational maps of a given symmetric form were considered, with symmetries corresponding to a known Skyrmion. If these symmetric maps still contained a few free parameters, $\mathcal{I}$ was minimized with respect to these, and then f was calculated. A rational map, $R \colon S^2 \mapsto S^2$, is symmetric under a subgroup $G \subset SO(3)$ if there is a set of Möbius transformation pairs $\{g, D_g\}$ with $g \in G$ acting on the domain S^2 and D_g acting on the target S^2, such that

$$R(g(z)) = D_g R(z) \, . \tag{13}$$

Some rational maps also possess additional reflection or inversion symmetry. In their recent work, Battye and Sutcliffe have systematically sought the rational map that minimizes $\mathcal{I}$, up to $N = 22$ [3]. This work has confirmed that the choice of maps in [6], up to $N = 8$, was optimal.

The zeros of the Wronskian

$$W(z) = p'(z)q(z) - q'(z)p(z) \tag{14}$$

of a rational map $R(z)$ give interesting information about the shape of the corresponding Skyrme field. Where W is zero, the derivative dR/dz is zero, so the baryon density vanishes. The energy density is also low. The Skyrme field baryon density contours therefore look like a polyhedron with holes in the directions given by the $2N - 2$ zeros of W.

3. Symmetric Rational Maps

In this section, we present the symmetric rational maps of degrees 1 to 8, determined in reference [6]. Table 1 gives the energy of the resulting approximate Skyrmions, and also the energy of the true Skyrmions. All numerical values for the energies are the real energies divided by $12\pi^2 B$, and hence close to unity. Figure 1 shows a surface of constant baryon density for most of the approximate Skyrmions. The true solutions have very similar shapes [2].

For $B = 1$ the basic map is $R(z) = z$, for which the integral $\mathcal{I} = 1$, and (6) reduces to Skyrme's hedgehog field

$$U(r, \theta, \varphi) = \cos f + i \sin f (\sin \theta \cos \varphi \; \sigma_1 + \sin \theta \sin \varphi \; \sigma_2 + \cos \theta \; \sigma_3). \qquad (15)$$

This is $SO(3)$ invariant, since $R(g(z)) = g(z)$ for any $g \in SU(2)$. It gives the standard exact spherically symmetric Skyrmion with its usual profile $f(r)$, and with energy $E = 1.232$.

The rational map which gives the toroidal $B = 2$ Skyrmion is

$$R(z) = z^2. \qquad (16)$$

Using this, one finds $\mathcal{I} = \pi + 8/3$ and after determining the profile $f(r)$ one obtains $E = 1.208$, an energy 3% higher than that of the true solution.

The $B = 3$ Skyrmion has tetrahedral symmetry. A rational map with this symmetry is obtained by imposing

$$R(-z) = -R(z), \quad R(1/z) = 1/R(z) \qquad (17)$$

$$R\left(\frac{iz + 1}{-iz + 1}\right) = \frac{iR(z) + 1}{-iR(z) + 1}. \qquad (18)$$

This gives the degree 3 maps

$$R(z) = \frac{\sqrt{3}az^2 - 1}{z(z^2 - \sqrt{3}a)} \qquad (19)$$

with $a = \pm i$. Note that $z \mapsto (iz + 1)/(-iz + 1)$ sends $0 \mapsto 1 \mapsto i \mapsto 0$ and hence generates the 120° rotation cyclically permuting the Cartesian axes. The sign of a can be changed by the 90° rotation $z \mapsto iz$. For these maps $\mathcal{I} = 13.58$. Solving for the profile $f(r)$, one finds an energy $E = 1.184$. The Wronskian of maps of the form (19) is proportional to $z^4 \pm 2\sqrt{3}iz^2 + 1$, a tetrahedral Klein polynomial [8].

The $B = 4$ Skyrmion has cubic symmetry. The cubically symmetric rational map of degree 4 is the ratio of tetrahedral Klein polynomials

$$R(z) = \frac{z^4 + 2\sqrt{3}iz^2 + 1}{z^4 - 2\sqrt{3}iz^2 + 1}. \tag{20}$$

The $90°$ rotation is a symmetry, because $R(iz) = 1/R(z)$. Using (20) in the ansatz (6) gives an energy $E = 1.137$.

The $B = 5$ Skyrmion of minimal energy has symmetry D_{2d}, which is somewhat surprising. A nearby cubically symmetric solution exists but is a saddle point. The D_{2d}-symmetric family of rational maps is

$$R(z) = \frac{z(z^4 + bz^2 + a)}{az^4 - bz^2 + 1} \tag{21}$$

with a and b real. If $b = 0$ then $R(z)$ has D_4 symmetry, the symmetry of a square. There is cubic symmetry if, in addition, $a = -5$. This value ensures the $120°$ rotational symmetry (18) and the Wronskian is then proportional to $z^8 + 14z^4 + 1$, the face polynomial of an octahedron. When $b = 0$, $a = -5$, the integral $\mathcal{I} = 52.05$. However, $\mathcal{I}$ is minimized when $a = 3.07$, $b = 3.94$, taking the value $\mathcal{I} = 35.75$. This is consistent with the structure of the $B = 5$ Skyrmion, a polyhedron made from four pentagons and four quadrilaterals. With the optimal profile function $f(r)$, the energy is $E = 1.147$. The octahedral saddle point has $E = 1.232$. There is a further, much higher saddle point at $a = b = 0$, where the map (21) simplifies to $R(z) = z^5$, and gives a toroidal field.

The $B = 6$ Skyrmion has symmetry D_{4d}. The rational maps

$$R(z) = \frac{z^4 + a}{z^2(az^4 + 1)} \tag{22}$$

have this symmetry, and the minimal energy occurs at $a = 0.16$, giving $E = 1.137$. The Skyrme field has a polyhedral shape consisting of a ring of eight pentagons capped by squares above and below.

In a sense, the $B = 7$ case is similar to the case $B = 6$, but the Skyrmion has a dodecahedral shape. A dodecahedron is a ring of ten pentagons capped by pentagons above and below. Among the degree 7 rational maps with D_{5d} symmetry

$$R(z) = \frac{z^5 - a}{z^2(az^5 + 1)}, \tag{23}$$

the one with icosahedral symmetry has $a = -1/7$ (not $a = -3$ as stated in [6]). The Wronskian is then proportional to $z(z^{10} + 11z^5 - 1)$, the face polynomial of a dodecahedron. In another orientation, tetrahedral symmetry T is manifest. There is a one-parameter family of maps with the symmetries (17) and (18),

$$R(z) = \frac{bz^6 - 7z^4 - bz^2 - 1}{z(z^6 + bz^4 + 7z^2 - b)} \tag{24}$$

where b is complex. For real b, the symmetry extends to T_h and for b imaginary it extends to T_d. When $b = 0$ there is cubic symmetry O_h, and when $b = \pm 7/\sqrt{5}$ there is icosahedral symmetry Y_h. Using (24) in our ansatz, one finds the minimal energy at $b = \pm 7/\sqrt{5}$, which gives a dodecahedral Skyrme field, with energy $E = 1.107$. This is particularly close to the energy of the true solution. There is a saddle point at $b = 0$ with a cubic shape.

The $B = 8$ Skyrmion has symmetry D_{6d}, as do the rational maps

$$R(z) = \frac{z^6 - a}{z^2(az^6 + 1)} \, . \tag{25}$$

This time the minimal energy is $E = 1.118$ when $a = 0.14$. The shape is now a ring of twelve pentagons capped by hexagons above and below.

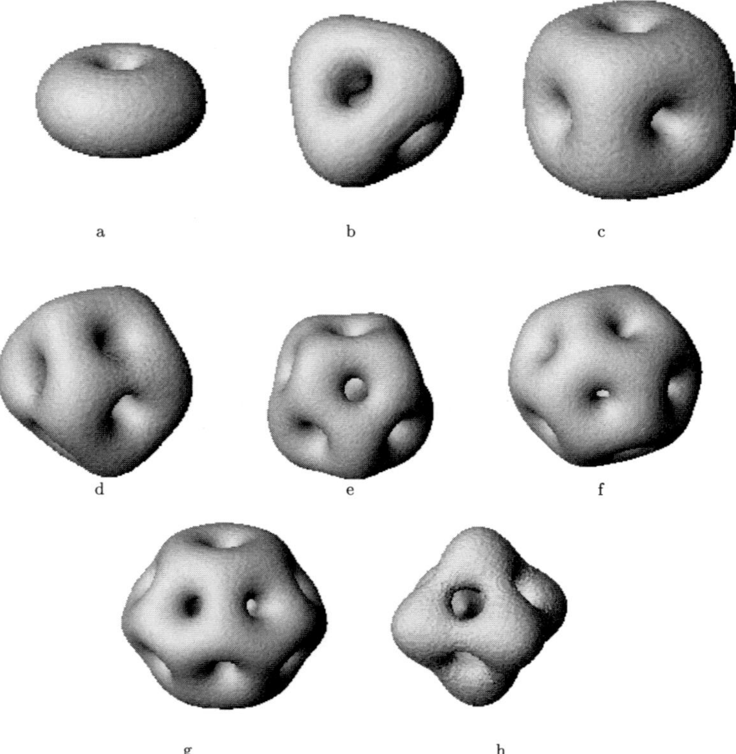

FIGURE 1. Surfaces of constant baryon density for the following approximate Skyrmions, constructed using rational maps: a) $B = 2$ torus; b) $B = 3$ tetrahedron; c) $B = 4$ cube; d) $B = 5$ with D_{2d} symmetry; e) $B = 6$ with D_{4d} symmetry; f) $B = 7$ dodecahedron; g) $B = 8$ with D_{6d} symmetry; h) $B = 5$ octahedron (saddle point).

B	$\mathcal{I}$	APPROX	TRUE	SYM
1	1.00	1.232	1.232	$O(3)$
2	5.81	1.208	1.171	$O(2) \times \mathbb{Z}_2$
3	13.58	1.184	1.143	T_d
4	20.65	1.137	1.116	O_h
5	35.75	1.147	1.116	D_{2d}
6	50.76	1.137	1.109	D_{4d}
7	60.87	1.107	1.099	Y_h
8	85.63	1.118	1.100	D_{6d}
5*	52.05	1.232	1.138	O_h

Table 1 : The energies of approximate Skyrmions generated from rational maps, and of true Skyrmions. The table gives the value of the angular integral $\mathcal{I}$, and the associated Skyrme field energy (APPROX), together with the energy of the true solution (TRUE), as determined in references [6, 2], and the symmetry (SYM) of the solution. A $*$ denotes a saddle point configuration.

4. Multi-Shell Rational Maps

The minimal energy solution of the Skyrme equation (1) with infinite baryon number is a three-dimensional cubic crystal. It is obtained by relaxing a face-centred cubic array of Skyrmions [5]. For finite, increasing B, the single-shell polyhedral structures we have discussed so far are therefore unlikely to remain the minimal energy solutions. Skyrmions will probably look more like part of the crystal. An approximate construction of Skyrmions as part of the crystal was carried out by Baskerville, for some special values of B, but the resulting energies were rather high [1]. Here we try a rational map ansatz with a two-shell structure. This is easily generalized to a multi-shell structure. The connection with the crystal will emerge below.

The simplest version of this generalized ansatz is (6) itself, with the profile function $f(r)$ having boundary values $f(0) = 2\pi$, $f(\infty) = 0$. However, this does not give a low energy. More promising is to use two different rational maps, $R_1(z)$ of degree N_1 for the inner shell, and $R_2(z)$ of degree N_2 for the outer shell. Let $r_0 > 0$ denote the radius where the inner and outer shells join. The ansatz is now

$$U(r,z) = \begin{cases} \exp(if_1(r)\,\widehat{\mathbf{n}}_{R_1(z)} \cdot \boldsymbol{\sigma}) & 0 \le r \le r_0, \\ \exp(if_2(r)\,\widehat{\mathbf{n}}_{R_2(z)} \cdot \boldsymbol{\sigma}) & r_0 \le r, \end{cases} \tag{26}$$

where the profiles $f_1(r)$ and $f_2(r)$ have boundary values $f_1(0) = 2\pi$, $f_1(r_0) = f_2(r_0) = \pi$, $f_2(\infty) = 0$. The field is continuous at $r = r_0$, but derivatives jump there. Note that $U = 1$ at the centre.

The baryon number of the Skyrme field (26) is easily seen to be $B = N_1 + N_2$. Its energy is the obvious generalization of (10),

$$
\begin{aligned}
E \;=\; & 4\pi \int_0^{r_0} \left(r^2 f_1'^2 + 2N_1(f_1'^2 + 1)\sin^2 f_1 + \mathcal{I}_1 \frac{\sin^4 f_1}{r^2} \right) dr \\
& + 4\pi \int_{r_0}^{\infty} \left(r^2 f_2'^2 + 2N_2(f_2'^2 + 1)\sin^2 f_2 + \mathcal{I}_2 \frac{\sin^4 f_2}{r^2} \right) dr \,.
\end{aligned}
\tag{27}
$$

There is considerably more choice than before in how to minimize this. One should consider all pairs N_1, N_2 whose sum is B, then find rational maps that minimize $\mathcal{I}_1$ and $\mathcal{I}_2$, and finally find the profiles $f_1(r)$ and $f_2(r)$, allowing r_0 to be variable. We have not carried out such a systematic analysis. Instead, we have considered those pairs of maps $R_1(z)$ and $R_2(z)$ which have a high degree of symmetry, and which appear to fit well together. Our aim is to obtain a field with a low, but possibly not optimal, value of the energy (27). We have then relaxed this field numerically to find a true solution of the Skyrme equation, usually with the same symmetry. We have used this approach for baryon numbers $B = 12$, 13 and 14. In each of these cases there is the possibility of a cubically symmetric solution, rather similar to part of the Skyrme crystal. We describe these in turn.

$B = 12$

There are various attractive choices for N_1 and N_2. Choosing $N_1 = N_2 = 6$, with the rational map (22), gives a rather low symmetry. More interesting is $N_1 = 3$ and $N_2 = 9$, where there are tetrahedrally symmetric maps. However, the most successful choice is $N_1 = 5$ and $N_2 = 7$. One could use the optimal single-shell maps given earlier (for $B = 5$, 7), but they have low combined symmetry. Better is to combine the maps with cubic symmetry mentioned earlier

$$
R_5(z) = \frac{z(z^4 - 5)}{-5z^4 + 1}, \quad R_7(z) = \frac{-7z^4 - 1}{z^3(z^4 + 7)} \,.
\tag{28}
$$

These maps both have the tetrahedral symmetries (17) and (18), as well as the $90°$ rotation symmetry $R(iz) = iR(z)$.

We have calculated the optimal profile functions and optimal r_0 for this pair of maps, obtaining an energy $E = 1.30$. Then, with this as a starting point, we have numerically relaxed the field to obtain a cubically symmetric, smooth solution of the Skyrme equation with energy $E = 1.15$. Its shape is shown in Figure 2a). Note that the figure does not exhibit a two-shell structure. After relaxation, the inner and outer shells coalesce. We shall return to this below.

Battye and Sutcliffe have also studied the $B = 12$ Skyrmion [3]. They have found the optimal single-shell rational map to use in (6). This map has only tetrahedral symmetry T_d, and gives an energy $E = 1.102$. They also relax their field to seek a true solution. This has the same tetrahedral symmetry, and energy $E = 1.086$. The cubically symmetric solution is not the true Skyrmion, but probably a saddle point.

$B = 13$

To construct a cubically symmetric $B = 13$ Skyrme field one might try a three-shell structure, with baryon numbers $1 + 7 + 5$ from the centre outward, combining the map $R_1(z) = z$ with the maps in (28). However, this cannot easily be implemented, because of the large size and energy of the initial configuration. Instead, we have constructed the solution, which resembles part of the Skyrme crystal, by relaxing a configuration made from a single Skyrmion and twelve nearest neighbours. It is shown in Figure 2b), and has energy $E = 1.09$.

Battye and Sutcliffe have also investigated the $B = 13$ case, using the single-shell ansatz. The rational map minimizing $\mathcal{I}$ has O symmetry. However, there is an O_h symmetric map with a slightly larger value of $\mathcal{I}$, and using this gives a field which looks almost identical to Figure 2b). We conclude that there is an O_h symmetric $B = 13$ solution, with $U = -1$ at the centre, which can be found starting in several ways. However, it appears that the solution with O symmetry is the true Skyrmion.

$B = 14$

Again we seek a cubically symmetric solution. We do this by taking a two-shell ansatz, using the dodecahedral rational maps of degree 7. The inner shell map is (24) with $b = 7/\sqrt{5}$, the outer shell uses the same map with $b = -7/\sqrt{5}$. Together, these maps (at different radii) possess only T_h symmetry, but if they can be made to coalesce, then there is cubic symmetry, because a $90°$ rotation transforms one into the other. Optimizing the profile functions gives an energy $E = 1.39$. Further relaxation produces a solution with T_h symmetry, and nearly cubically symmetric, with $E = 1.14$. Its form is shown in Figure 2c). This again looks like part of the Skyrme crystal; this time what one would obtain by taking the six nearest neighbours and eight next-nearest neighbours surrounding a hole in the crystal, and relaxing the field.

The optimal single-shell structure with $B = 14$ is quite different [3]. The rational map has only D_2 symmetry, and gives an energy $E = 1.103$. The field also differs because $U = -1$ at the centre. Relaxation of the solution will give a lower energy, but this has not yet been done.

5. Interpretation of the Two-Shell Ansatz

For certain profile functions f_1 and f_2, the two-shell rational map ansatz describes a Skyrmion of baryon number N_1 inside a Skyrmion of baryon number N_2, approximately. However, as the field relaxes, it changes its character. Consider a radial line (z fixed), and the field values U at the points along it where $f_1 = 3\pi/2$ and where $f_2 = \pi/2$. If these values are close, then the field between can be relaxed to be approximately constant, which makes the energy low in this direction. Conversely, if the field values are antipodal (on $SU(2)$), then the field gradient between them, and hence the energy, is large in this direction. In fact the winding of the field along this radial line indicates that there is a $B = 1$ Skyrmion in this direction.

Now, antipodal field values occur on this line if $R_1(z) = R_2(z)$. (The rational map values are the same, but $\sin f_1 = -1$, $\sin f_2 = 1$.) Thus the two-shell rational map ansatz produces a configuration which can be interpreted as a superposition of $B = 1$ Skyrmions located at $r = r_0$ and in those directions z which solve the equation

$$R_1(z) = R_2(z) \,. \tag{29}$$

Writing $R_1(z) = \frac{p_1(z)}{q_1(z)}$ and $R_2(z) = \frac{p_2(z)}{q_2(z)}$, this becomes

$$p_1(z)q_2(z) - p_2(z)q_1(z) = 0 \tag{30}$$

which is a polynomial equation of degree $N_1 + N_2$, with $N_1 + N_2$ solutions. So the number of Skyrmions one finds by solving (30) is precisely the total baryon number. The relative orientation of these Skyrmions has not yet been determined.

Equation (30) has a particularly symmetric form for maps we have been considering. For $B = 12$ it reduces to

$$z^{12} - 33z^8 - 33z^4 + 1 = 0 \,, \tag{31}$$

the Klein polynomial for the edges of a cube. For $B = 14$ it reduces to

$$z(z^4 - 1)(z^8 + 14z^4 + 1) = 0 \,, \tag{32}$$

the product of the Klein polynomials for the faces and vertices of a cube (one root is $z = \infty$). This is what one anticipates based on the analogy with the Skyrme crystal.

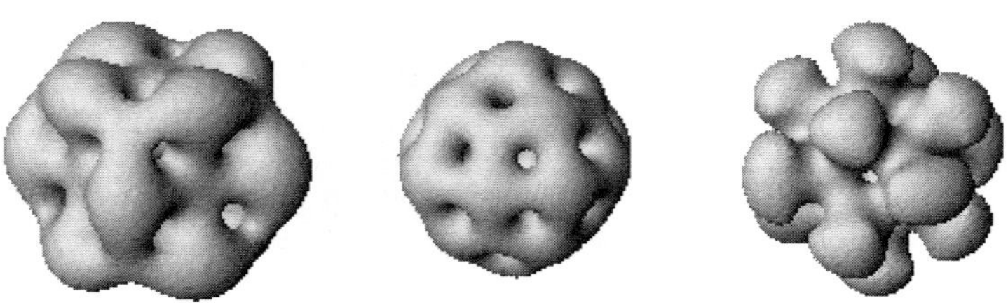

FIGURE 2. Surfaces of constant baryon density for the following solutions: a) $B = 12$ with cubic symmetry; b) $B = 13$ with cubic symmetry; c) $B = 14$ two-shell with near cubic symmetry.

6. Conclusions

We have discussed an ansatz for Skyrme fields, based on rational maps, which allows the construction of good approximations to several Skyrmions. We have also discussed a two-shell rational map ansatz as an approach to construct multi-shell

Skyrmion solutions. These ansätze are good starting points to construct solutions with certain symmetries. We have studied other examples of the two-shell rational map ansatz than the ones described here and in most cases the configuration relaxes to a single shell solution. Two-shell solutions were found only for $B = 14$, as we have shown, and also when using a rational map of degree 17, together with another of lower degree. To construct two-shell solutions with a relatively low energy, the outside shell must be large enough to contain a smaller shell inside it. Although the solutions we have found using the two-shell ansatz are not minimal energy Skyrmions, their energies are not much greater, and we believe that for higher baryon numbers the minimal energy solutions will exhibit a multi-shell structure.

Acknowledgements

We thank Richard Battye, Paul Sutcliffe and Wojtek Zakrzewski for useful discussions.

References

[1] W. K. Baskerville, Nucl. Phys. A596, 611 (1996).

[2] R. A. Battye and P. M. Sutcliffe, Phys. Rev. Lett. 79, 363 (1997).

[3] R. A. Battye and P. M. Sutcliffe, 'Skyrmions, Fullerenes and Rational Maps', to appear in Reviews in Math. Phys,; Phys. Rev. Lett. 86, 3989 (2001).

[4] E. Braaten, S. Townsend and L. Carson, Phys. Lett. B 235, 147 (1990).

[5] L. Castillejo, P. S. J. Jones, A. D. Jackson, J. J. M. Verbaarschot and A. Jackson, Nucl. Phys. A 501, 801 (1989).

[6] C. J. Houghton, N. S. Manton and P. M. Sutcliffe, Nucl. Phys. B 510, 507 (1998).

[7] T. Ioannidou, B. M. A. G. Piette and W. J. Zakrzewski, J. Math. Phys. 40, 6223 (1999); J. Math. Phys. 40, 6353 (1999).

[8] F. Klein, 'Lectures on the icosahedron', London, Kegan Paul, 1913.

Nicholas S. Manton
Department of Applied Mathematics and Theoretical Physics
University of Cambridge
Wilberforce Road, Cambridge CB3 0WA, England
E-mail address: N.S.Manton@damtp.cam.ac.uk

Bernard M. A. G. Piette
Department of Mathematical Sciences
University of Durham
South Road, Durham DH1 3LE, England
E-mail address: B.M.A.G.Piette@durham.ac.uk

Models for the Leaf Space of a Foliation

Ieke Moerdijk

The aim of this talk is to explain and compare some approaches to the leaf space (or "transverse structure") of a foliation. A foliation is a certain partition $\mathcal{F}$ of a manifold M into immersed submanifolds, the leaves of the foliation. Identifying each of the leaves to a single point yields a very uninformative, "coarse" quotient space, and the problem is to define a more refined quotient $M/\mathcal{F}$, which captures aspects of that part of the geometric structure of the foliation which is constant and/or trivial along the leaves.

It is possible to distinguish (at least) three approaches to this problem. One is in the spirit of non-commutative geometry [4], and uses the duality between the manifold M and the ring $C_c^\infty(M)$ of compactly supported smooth functions on M. The *quotient* $M/\mathcal{F}$ is then modelled, dually, by an extension of this ring $C_c^\infty(M)$, the so-called convolution algebra of the foliation. Completion of such convolution algebras leads one into C^*-algebras. Important invariants are the cyclic type (i.e. Hochschild, cyclic, periodic cyclic) homologies and the K-theory of these convolution and C^*-algebras.

A second approach, which predates non-commutative geometry, is to construct a quotient "up to homotopy". Like all such homotopy colimits in algebraic topology, this construction takes the form of a classifying space. This approach goes back to Haefliger, who constructed a classifying space $B\Gamma_q$ for foliations of codimension q, as the leaf space of the "universal" foliation [9, 2]. Important invariants are the cohomology groups of these classifying spaces, in particular the universal or characteristic classes coming from the cohomology of the universal leaf space $B\Gamma_q$.

A third approach, even older, is due to Grothendieck. Not surprisingly, Grothendieck uses the 'duality' between the space M and the collection of all its sheaves, which form a topos $\mathrm{Sh}(M)$. The quotient $M/\mathcal{F}$ can then be constructed as a suitable topos "$\mathrm{Sh}(M/\mathcal{F})$", consisting of sheaves on M which are invariant along the leaves in a suitable sense. One can then apply the whole machinery of [17], and study the Grothendieck fundamental group of $\mathrm{Sh}(M/\mathcal{F})$, its sheaf cohomology groups, etc. etc..

Central to all these approaches is the construction [18] of a smooth groupoid out of the foliated manifold $(M, \mathcal{F})$, called the holonomy groupoid and denoted $\mathrm{Hol}(M, \mathcal{F})$. The three approaches above then become special instances of the general procedure of associating to a smooth (or "Lie") groupoid G a convolution algebra $C_c^\infty(G)$, a classifying space BG, or a classifying topos $\mathrm{Sh}(G)$. Of these, the last one is intuitively closest to a manifold. For example, there are immediate natural constructions of the differential forms on such a topos, of its tangent bundle (another topos mapping to $\mathrm{Sh}(G)$), and so on for almost any construction of differential topology and geometry I can think of. This is partly caused by the fact that the Lie groupoids arising in this context are all (equivalent to) étale groupoids. The diagram below provides a schematic summary of the situation. In this lecture, I will first give more precise definitions and references for the notions occurring in this diagram and then explain some relations between the three legs.

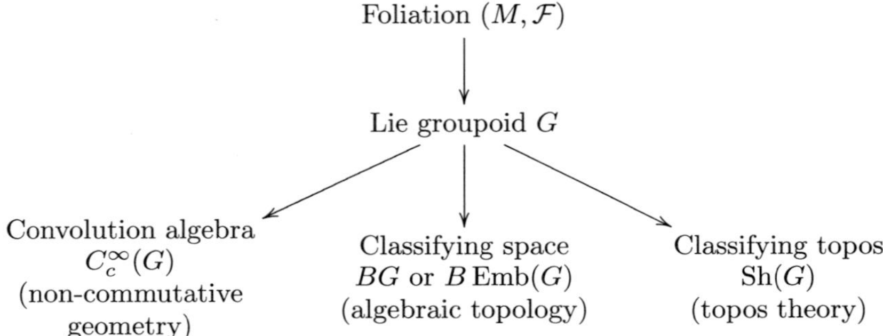

1. Foliations

Let M be a manifold of dimension n. A foliation $\mathcal{F}$ of M is an integrable subbundle $\mathcal{F} \subseteq TM$ of the tangent bundle. Integrability means that if two vector fields on M belong to $\mathcal{F}$ then so does their Lie bracket. If $\mathcal{F}$ is of rank p, the foliation is said to be of dimension p and of codimension $q = n - p$. Integrability implies that through each point $x \in M$ there is a unique connected p-dimensional immersed submanifold L_x which is everywhere tangent to $\mathcal{F}$, called the leaf of $\mathcal{F}$ through x. These leaves form a partition of M. This partition is locally trivial in the sense that at each point x there is a chart $\varphi \colon \mathbb{R}^n \to U$ (where U is a neighborhood of x) such that for $\mathbb{R}^n = \mathbb{R}^p \times \mathbb{R}^q$ the plaques $\varphi(\mathbb{R}^p \times \{t\})$ are exactly the connected components of the intersections of U with the leaves. (A specific leaf may pass through U in different plaques.) Here are two easy and well-known examples of foliations.

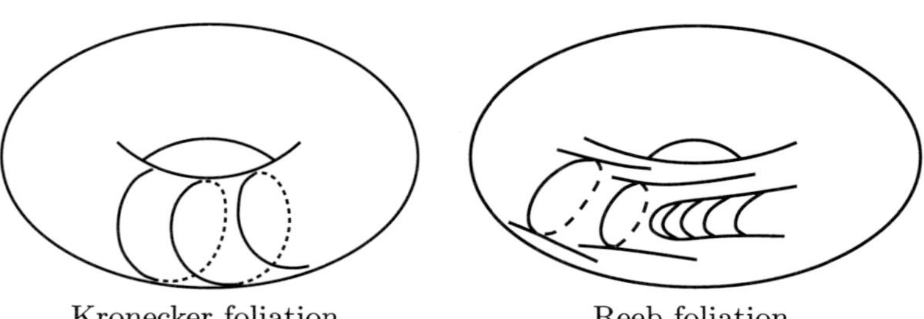

Kronecker foliation.   Reeb foliation.

The Kronecker foliation $\mathcal{K}$ of the torus $T = S^1 \times S^1$ is the foliation given by the 1-dimensional subbundle of vectors in $\mathbb{R}^2$ with a fixed irrational slope. The leaves are immersed copies of the real line which wrap around the torus infinitely often, each leaf being dense. The Reeb foliation $\mathcal{R}$ of the solid torus has one compact boundary leaf, and its other leaves are planes. If you imagine the solid torus as obtained from the solid cylinder $\mathbb{R} \times D$ by identifying (t, x) and $(t + 1, x)$ for every point x on the disk D, then the interior of $\mathbb{R} \times D$ is foliated by planes which are stacked upon each other as infinitely deep salad bowls, and the Reeb foliation is the quotient. To get rid of the boundary, one can construct S^3 as the union of two solid tori. Then the union of Reeb foliations is the Reeb foliation of S^3, with one compact leaf.

The theory of foliations is a vast subject, for which there are many good introductions, e.g. the books by Camacho and Neto, Godbillon, Hector and Hirsch, Tondeur, and others.

2. Lie Groupoids

A groupoid G is a small category all of whose arrows are isomorphisms. It thus has a set G_0 of objects $x, y, \ldots$ and a set G_1 of arrows $g, h, \ldots$. Each arrow has a source $x = s(g)$ and a target $y = t(g)$, written $g \colon x \to y$. Two arrows g and h with $s(g) = t(h)$ can be composed as $gh \colon s(h) \to t(g)$. This composition is associative, has a unit $1_x \colon x \to x$ at each object x, and has a two-sided inverse $g^{-1} \colon t(g) \to s(g)$ for each arrow g. All the structure is contained in a diagram

$$G_2 \xrightarrow{\ m\ } G_1 \xrightarrow{\ i\ } G_1 \overset{s}{\underset{t}{\rightleftarrows}} \xleftarrow{u} G_0 \tag{1}$$

(s = source, t = target, i = inverse, u = units, m = composition, defined on $G_2 = \{(g, h) \in G_1 \times G_1 : s(g) = t(h)\}$. The groupoid G is said to be *smooth* or *Lie* if G_0 and G_1 are smooth manifolds, each of the structure maps in (1) is smooth, and s, t are submersions so that G_2 is a smooth manifold as well. The classical reference for Lie groupoids is [11].

For a Lie groupoid G and a point $x \in G_0$, the arrows $g \colon x \to x$ form a Lie group G_x, called the *isotropy group* at x. A Lie groupoid G is called a *foliation groupoid* if each of its isotropy groups is discrete. All the groupoids arising from foliations have this property.

There is an obvious notion of *smooth functor* or *homomorphism* $\varphi \colon H \to G$ between Lie groupoids. It consists of two smooth maps (both) denoted $\varphi \colon H_0 \to G_0$ and $\varphi \colon H_1 \to G_1$, together commuting with all the structure maps in (1). Such a homomorphism is said to be an *essential equivalence* if (i), φ induces a surjective submersion $(y, g) \mapsto t(g)$ from the space $H_0 \times_{G_0} G_1 = \{(y, g) | \varphi(y) = s(g)\}$ onto H_0; and (ii), φ induces a diffeomorphism $h \mapsto (s(h), \varphi(h), t(h))$ from H_1 to the pullback $H_0 \times_{G_0} G_1 \times_{G_0} H_0$. Two Lie groupoids G and G' are said to be (*Morita*) *equivalent* if there are essential equivalences $G \leftarrow H \rightarrow G'$ from a third Lie groupoid H. (This notion is also often formulated in terms of principal bundles.)

A Lie groupoid G is said to be *étale* (or *r*-discrete) if all the structure maps in (1) are local diffeomorphisms (it is enough to require this for s). The relevance of étale groupoids for foliations is based on the following proposition [6].

Proposition 2.1. *A Lie groupoid is a foliation groupoid iff it is equivalent to an étale groupoid.*

If G is an étale groupoid, each arrow $g\colon x \rightarrow y$ in G uniquely determines the germ of a diffeomorphism $\tilde{g}\colon (G_0, x) \rightarrow (G_0, y)$, namely $\tilde{g} = \operatorname{germ}_x(t \circ \sigma)$ where σ is a section of $s\colon G_1 \rightarrow G_0$ on a neighbourhood U of x, with $\sigma(x) = g$ and U so small that $t \circ \sigma$ is a diffeomorphism from U onto its image. This construction gives in particular a group homomorphism $G_x \rightarrow \operatorname{Diff}_x(G_0)$. If this homomorphism is injective for each $x \in G_0$ the groupoid G is said to be *effective*. Effectivity is preserved under equivalence of groupoids, so we can define a foliation groupoid to be effective if it is Morita equivalent to an effective étale groupoid.

A groupoid G is called *proper* if $(s,t)\colon G_1 \rightarrow G_0 \times G_0$ is a proper map. An *orbifold groupoid* is a proper effective foliation groupoid. This notion is again invariant under equivalence. It can be shown that a Lie groupoid is an orbifold groupoid iff it is equivalent to the action groupoid associated to an infinitesimally free action of a compact Lie group on a manifold (see e.g. [15]).

3. The Holonomy Groupoid of a Foliation

Let $(M, \mathcal{F})$ be a foliated manifold. The holonomy groupoid $H = \operatorname{Hol}(M, \mathcal{F})$ is a smooth groupoid with $H_0 = M$ as space of objects. If $x, y \in M$ are two points on different leaves there are no arrows from x to y in H. If x and y lie on the same leaf L, an arrow $h\colon x \rightarrow y$ in H (i.e. a point $h \in H_1$ with $s(h) = x$ and $t(h) = y$) is an equivalence class $h = [\alpha]$ of smooth paths $\alpha\colon [0,1] \rightarrow L$ with $\alpha(0) = x$ and $\alpha(1) = y$. To explain the equivalence relation, let T_x and T_y be small q-disks through x and y, transverse to the leaves of the foliation. If $x' \in T_x$ is a point sufficiently close to x on a leaf L' , then α can be "copied" inside L' to give a path α' near α with endpoint $y' \in T_y$, say. In this way one obtains the germ of a diffeomorphism from T_x to T_y, sending x to y and x' to y'. This germ is called the holonomy of α and denoted $\operatorname{hol}(\alpha)$. By definition, two paths α and β from x to y in L are equivalent, i.e. define the same arrow $x \rightarrow y$ in H, iff $\operatorname{hol}(\alpha) = \operatorname{hol}(\beta)$. For example, if α and β are homotopic (inside L and relative endpoints) then $\operatorname{hol}(\alpha) = \operatorname{hol}(\beta)$. Composition and inversion of paths respects the equivalence relation, so that one obtains a well defined groupoid $H = \operatorname{Hol}(M, \mathcal{F})$, which can be shown to be smooth [18]. This groupoid is a foliation groupoid, and the (discrete) isotropy group H_x at x is called the holonomy group of the leaf through x. If $T \subseteq M$ is an embedded q-manifold transverse to the leaves and hitting each leaf at least once, then the restriction of H to T defines an étale groupoid

$H_T = \mathrm{Hol}_T(M, \mathcal{F})$, equivalent to H. We refer to H_T as "the" étale (model for the) holonomy groupoid of $(M, \mathcal{F})$.

Morally, every étale groupoid G is the holonomy groupoid of a foliation (for a precise formulation, see [16, p. 21]). Orbifold groupoids are exactly the groupoids which arise as holonomy groupoids of foliations with compact leaves and finite holonomy groups [15].

The reader is urged to work out the various étale models for the holonomy groupoids of the Kronecker and Reeb foliations. He will notice that in the second case the space H_1 is a non-Hausdorff manifold.

The Haefliger groupoid Γ_q has $\mathbb{R}^q$ for its space of objects, while the arrows $x \to y$ are the germs of diffeomorphisms $(\mathbb{R}^q, x) \to (\mathbb{R}^q, y)$. When this space of arrows is equipped with the sheaf topology, Γ_q becomes an étale smooth groupoid. If $(M, \mathcal{F})$ is a foliation of codimension q, there is an essentially unique map of étale groupoids $\mathrm{Hol}_T(M, \mathcal{F}) \to \Gamma_q$ (for a suitable choice of T).

4. The Classifying Space

For a smooth groupoid G, the nerve of G is the simplicial set whose n-simplices are strings of composable arrows in G,

$$x_0 \xleftarrow{\ g_1\ } x_1 \longleftarrow \cdots \xleftarrow{\ g_n\ } x_n .$$

This set is denoted G_n, consistent with the earlier notation for $n = 0, 1, 2$. The space G_n is a fibered product $G_1 \times_{G_0} \cdots \times_{G_0} G_1$, hence has the natural structure of a smooth manifold. Thus $G_{\textbf{.}}$ is a simplicial manifold. Its geometric realization is denoted BG and is called the classifying space of G. This construction respects Morita equivalence. In fact, an essential equivalence $H \to G$ induces a weak homotopy equivalence $BH \to BG$.

For a foliated manifold $(M, \mathcal{F})$ with holonomy groupoid H, there is a canonical 'quotient' map $M \to BH$, and BH models the space of leaves $M/\mathcal{F}$. One can also construct maps, canonical up to homotopy, $M \to BM_T \to B\Gamma_q$.

One of the problems with the space BH is that it is usually non-Hausdorff. There is another model for BH which doesn't have this defect. This model is based on a small (discrete) category $\mathrm{Emb}(G)$ constructed for any étale groupoid G. (In this context, $G = H_T$ is the étale model for the holonomy groupoid.) The objects of this new category $\mathrm{Emb}(G)$ are the members of a fixed basis of contractible open sets for the topology on G_0. For two such basic open U and V, each section $\sigma \colon U \to G_1$ of the source map, with the property that $t \circ \sigma \colon U \to G_0$ defines an embedding into V, defines an arrow $\hat{\sigma} \colon U \to V$ in the category $\mathrm{Emb}(G)$. Composition is defined by $\hat{\tau} \circ \hat{\sigma} = \hat{\rho}$ where $\rho(x) = \tau(t\sigma(x))\sigma(x)$ (multiplication in G). The nerve of this category $\mathrm{Emb}(G)$ is a simplicial set, whose geometric realization is denoted $B\,\mathrm{Emb}(G)$.

Theorem 4.1. [12] *For any étale groupoid G the spaces BG and $B\,\mathrm{Emb}(G)$ are weakly homotopy equivalent.*

For the special case where $G = \Gamma_q$, the category $\mathrm{Emb}(G)$ is categorically equivalent to the (discrete) monoid M_q of smooth embedding of $\mathbb{R}^q$ into itself, and one recovers Segal's theorem, $B\Gamma_q \simeq BM_q$.

We remark that, unlike BG, the classifying space $B\mathrm{Emb}(G)$ is a CW-complex, hence within the scope of the usual methods of algebraic topology. It is also very well suited for the explicit geometric construction of characteristic classes of foliations [6].

5. The Classifying Topos

Let G be an étale (or foliation) groupoid. A G-sheaf of sets is a sheaf on G_0 equipped with a continuous right action by G. When $\pi\colon S \to G_0$ is the étale space of the sheaf, the action can be described as a continuous map $S \times_{G_0} G_1 \to S$. For $\xi \in S_y = \pi^{-1}(y)$ and $g\colon x \to y$, the result of the action is denoted $\xi \cdot g \in S_x$. The usual identities $\xi \cdot 1_y = \xi$ and $(\xi \cdot g) \cdot h = \xi \cdot (gh)$ are required to hold. With the obvious notion of action preserving map, these sheaves form a category $\mathrm{Sh}(G)$. This category is a topos [17], called the classifying topos of G, and discussed in [13]. A homomorphism of groupoids $G \to H$ induces a topos map $\mathrm{Sh}(G) \to \mathrm{Sh}(H)$. The construction preserves Morita equivalence. (In fact, $\mathrm{Sh}(G)$ and $\mathrm{Sh}(H)$ are equivalent toposes, i.e. are equivalent as categories, if *and only if* G and H are Morita equivalent as topological —rather than smooth— groupoids.)

If follows that the category $\mathrm{Ab}\,\mathrm{Sh}(G)$ of sheaves of abelian groups has enough injectives, and one obtains for each sheaf A the sheaf cohomology groups $H^n(G, A)$ as those of the topos, i.e. $H^n(G, A) = H^n(\mathrm{Sh}(G), A)$ for $n \geq 0$ by definition. These cohomology groups are then automatically contravariant in G and invariant under Morita equivalence, and satisfy all the usual general properties of [17] (Leray spectral sequence, Čech spectral sequence, hypercover description, relation of $H^1(G, A)$ to the fundamental group of $\mathrm{Sh}(G)$, etc. etc.).

This approach is compatible with the classifying space, as follows.

Theorem 5.1. *Any abelian G-sheaf A induces in a natural way a sheaf $\tilde{A}$ on the classifying space BG, and one has a canonical isomorphism*

$$H^n(G, A) \xrightarrow{\sim} H^n(BG, \tilde{A}), \quad n \geq 0.$$

This isomorphism was conjectured by Haefliger and proved in [14].

There is also a dual *homology theory* for étale groupoids, introduced and studied in [7]. For an abelian G-sheaf A, we construct homology groups

$$H_n(G, A), \quad n \geq -\dim(G),$$

again invariant under Morita equivalence of groupoids and having good general properties. In particular, there is a Verdier type duality between the cohomology just described and this homology.

It seems difficult to give a description of this homology theory in terms of the classifying space BG, although in some special cases this is possible.

6. The Convolution Algebra and Cyclic Homology

For an algebra A, one can define the Hochschild, cyclic and periodic cyclic homology groups, denoted $HH_n(A), HC_n(A)$ $(n \in \mathbb{N})$ and $HP_\nu(A)$ $(\nu = 0, 1)$. The definition is based on iterated tensor products $A \otimes \cdots \otimes A$. In the special case where $A = C_c^\infty(M)$ is the ring of smooth compactly supported functions on a manifold, a well-known result of Connes' [5], which played a central role in the development of cyclic homology, provides the following relation to the De Rham cohomology,

$$HH_n(C_c^\infty(M)) = \Omega_c^n(M), \quad HP_\nu(M) = H_c^\nu(M). \tag{2}$$

Here $\Omega_c^n(M)$ is the vector space of compactly supported n-forms on M, and $H_c^\nu(M)$ is the product of the even $(\nu = 0)$ or odd $(\nu = 1)$ compactly supported De Rham cohomology groups.

It is important to note that for this result, the algebraic tensor product $A \otimes B$ is replaced by a completed topological tensor product (the inductive one) $A \hat\otimes B$ having the property that $C_c^\infty(M) \hat\otimes C_c^\infty(N) = C_c^\infty(M \times N)$ for two manifolds M and N.

Connes' result extends to generalized manifolds such as leaf spaces of foliations. More specifically, let G be an étale groupoid. The convolution algebra $C_c^\infty(G)$ is the algebra of compactly supported smooth functions a, b, ... on G_1, with "convolution" product

$$(a * b)(g) = \sum_{g=hk} a(h)b(k),$$

exactly as for the group ring. (The sum here makes sense, because it ranges over a space which is discrete because G is étale and finite because of compact supports.) Using the inductive topological tensor product $\hat\otimes$, one can then define the "cyclic type" homology groups $HH_*(C_c^\infty(G))$, $HC_*(C_c^\infty(G))$ and $HP_*(C_c^\infty(G))$. The construction of the convolution algebra $C_c^\infty(G)$ is not functorial in G, and the invariance under Morita equivalence of these cyclic type homology groups is established in a rather indirect way, by relating them to the homology groups mentioned above, as follows.

For an étale groupoid G, the 'loop groupoid' $\Lambda(G)$ has as its objects the arrows $g: x \to x'$ in G with $x = x'$. Arrows $g \to h$ in $\Lambda(G)$, from $(g: x \to x)$ to $(h: y \to y)$, are arrows $\alpha: x \to y$ in G with $h\alpha = \alpha g$. This groupoid $\Lambda(G)$ is a topological étale groupoid. This construction is functorial in G, and preserves Morita equivalence. There is also an evident retraction $\pi: \Lambda(G) \to G$.

Let $\mathcal{A}_n$ be the pullback along the diagonal $G_0 \to G_0^{n+1}$ of the sheaf of smooth functions on G_0^{n+1}. The stalk of $\mathcal{A}_n$ at x is the ring $C_x^\infty(G_0) \hat\otimes \cdots \hat\otimes C_x^\infty(G_0)$ and consists of germs of functions $f(x_0, \ldots, x_n)$. The usual Hochschild boundary of the complex $\mathcal{A}_0 \xleftarrow{b} \mathcal{A}_1 \leftarrow \cdots$ can be twisted by loops to give a complex $\pi^*(\mathcal{A}_0) \xleftarrow{b} \pi^*(\mathcal{A}_1) \leftarrow \cdots$ on $\Lambda(G)$, and one has the following comparison,

$$HH_*(C_c^\infty(G)) = H_*(\Lambda(G), \pi^*(\mathcal{A})), \tag{3}$$

expressing the Hochschild homology of the convolution algebra in terms of the (hyper-)homology of the étale groupoid $\Lambda(G)$. An immediate consequence is that $HH(C_c^\infty(G))$ is invariant under Morita equivalence of étale groupoids. (The invariance of HC and HP follows by the usual SBI-argument.) If G is a manifold, i.e. $G_0 = G_1 = M$, then $\Lambda(G) = G = M$ also, and one recovers Connes' isomorphism (2) from (3). If $G = \Gamma$ is a discrete group, $\Lambda(\Gamma)$ is Morita equivalent to the sum over all conjugacy classes of centralizer subgroups,

$$\Lambda(\Gamma) \cong \sum_{(\gamma)} Z_\gamma \,,$$

and one recovers the well-known description of the Hochschild homology of the group ring ([3, 10]). If $G = M \rtimes \Gamma$ is the action groupoid associated to the action of a discrete group Γ on a manifold M, then $\Lambda(G)$ is Morita equivalent to $\bigoplus_{(\gamma)} M^\gamma \rtimes Z_\gamma$, giving a familiar decomposition of $HH_*(C_c^\infty(M \rtimes \Gamma))$.

There are isomorphisms similar to (3) for the cyclic and periodic cyclic homology groups of étale groupoids, all relating these groups to the homology of étale groupoids (or categories) like $\Lambda(G)$. For precise formulations and computations I refer to [1, 8, 7]. Here as an illustration, I just single out the special case where G is an étale groupoid with finite isotropy groups (e.g. an orbifold groupoid). In this case, one has a natural isomorphism

$$HP_\nu(C_c^\infty(G)) = \prod_k H_{2k+\nu}(\Lambda(G), \mathbb{R}), \quad (\nu = 0, 1)\,,$$

extending Connes' isomorphism (2) to such groupoids.

References

[1] J.-L. Brylinski, V. Nistor, *Cyclic cohomology of étale groupoids*, K-theory **8** (1994), 341–365.

[2] R. Bott, *Lectures on characteristic classes and foliations*, Springer LNM **279**, 1–94.

[3] D. Burghelea, *The cyclic cohomology of group rings*, Comm. Math. Helv. **60** (1985), 354–365.

[4] A. Connes, *Noncommutative Geometry*, Academic Press, 1994.

[5] A. Connes, *Noncommutative differential geometry*, Publ. IHES **62** (1985), 41–144.

[6] M. Crainic, I. Moerdijk, *Cech-De Rham theory for leaf spaces of foliations*, in prep.

[7] M. Crainic, I. Moerdijk, *A homology theory for étale groupoids*, to appear in Crelle's J. f. reine und angewandte Mathematik.

[8] M. Crainic, *Cyclic cohomology of étale groupoids: The general case*, K-Theory **17** (1999), 319–362.

[9] A. Haefliger, *Feuilletages sur les variétés ouvertes*, Topology **9** (1970), 183–194.

[10] M. Karoubi, *Homologie cyclique des groupes et des algèbres*, CRAS **297** (1983), 381–384.

[11] K. Mackenzie, *Lie Groupoids and Lie Algebroids in Differential Geometry*, Cambridge U.P., 1987.

[12] I. Moerdijk, *On the weak homotopy type of étale groupoids*, in "Integrable Systems and Foliations", Birkhäuser (1997), 147–156.

[13] I. Moerdijk, *Classifying toposes and foliations*, Ann. Inst. Fourier **41** (1991),189–209.

[14] I. Moerdijk, *Proof of a Conjecture of A. Haefliger*, Topology **37** (1998), 735–741.

[15] I. Moerdijk, D. Pronk, *Orbifolds, Sheaves and Groupoids*, K-Theory **12** (1997), 3–21.

[16] J. Mrcun, *Stability and invariance of Hilsum-Skandaliz maps*, PhD Thesis, Utrecht, 1996.

[17] M. Artin, A. Grothendieck, J.-L. Verdier, *Théorie de topos et cohomologie étale des schémas ("SGA IV")*, Springer Lecture Notes 269, 270 (1972).

[18] H. Winkelnkemper, *The graph of a foliation*, Ann. Global Anal. Geom. **1** (1983), 613–665.

Mathematical Institute
University of Utrecht
PO Box 80.010
3508 TA Utrecht, The Netherlands
E-mail address: moerdijk@math.uu.nl

Multivariable Hypergeometric Functions

Eric M. Opdam

Abstract. The goal of this lecture is to present an overview of the modern developments around the theme of multivariable hypergeometric functions. The classical Gauss hypergeometric function shows up in the context of differential geometry, algebraic geometry, representation theory and mathematical physics. In all cases it is clear that the restriction to the one variable case is unnatural. Thus from each of these contexts it is desirable to generalize the classical Gauss function to a class of multivariable hypergeometric functions. The theories that have emerged in the past decades are based on such considerations.

1. The Classical Gauss Hypergeometric Function

The various interpretations of Gauss' hypergeometric function have challenged mathematicians to generalize this function. Multivariable versions of this function were proposed already in the 19th century by Appell, Lauricella, and Horn. Reflecting developments in geometry, representation theory and mathematical physics, a renewed interest in multivariable hypergeometric functions took place from the 1980s. Such generalizations have been initiated by Aomoto [1], Gelfand and Gelfand [14], and Heckman and Opdam [20], and these theories have been further developed by numerous authors in recent years.

The best introduction to this story is a recollection of the role of the Gauss function itself. So let us start by reviewing some of the basic properties of this classical function. General references for this introductory section are [25, 12], and [40].

The *Gauss hypergeometric series* with parameters $a, b, c \in \mathbf{C}$ and $c \notin \mathbf{Z}_{\leq 0}$ is the following power series in z:

$$F(a, b, c; z) := \sum_{n=0}^{\infty} \frac{(a)_n (b)_n}{(c)_n n!} z^n . \tag{1}$$

The *Pochhammer symbol* $(a)_n$ is defined by $(a)_n = a(a+1)\ldots(a+n-1)$ for $n \geq 1$, and $(a)_0 = 1$. This series is easily seen to be convergent when $|z| < 1$.

Gauss proved a number of remarkable facts about this function. He showed that

Proposition 1.1. *The hypergeometric series $F(a, b, c; z)$ and any two additional hypergeometric series whose 3-tuples of parameters are equal to (a, b, c) modulo $\mathbf{Z}^3$, satisfy a nontrivial linear relation with coefficients in the ring of polynomials in a, b, c, and z.*

A hypergeometric series whose parameters are $(a \pm 1, b, c)$, $(a, b \pm 1, c)$ or $(a, b, c \pm 1)$ is called *contiguous* to $F(a, b, c; z)$. Gauss worked out the basic cases of the relations between $F(a, b, c; z)$ and two of its contiguous functions, known as the contiguity relations of Gauss. Using such relations, he proved the famous "Gauss summation formula":

Lemma 1.2. *When $c \notin \{0, -1, -2, \dots\}$, and $\operatorname{Re}(c - a - b) > 0$, then*

$$F(a, b, c; 1) = \frac{\Gamma(c)\Gamma(c - a - b)}{\Gamma(c - a)\Gamma(c - b)}\,.$$

When we differentiate the series (1) we obtain

$$\frac{d}{dz}F(a, b, c; z) = \frac{ab}{c}F(a + 1, b + 1, c + 1; z)\,. \tag{2}$$

As a special case of Proposition 1.1 there exists a linear second order differential equation with polynomial coefficients for the series (1). By an easy direct computation one finds:

Proposition 1.3. *The Gauss series $F(a, b, c; z)$ satisfies the equation*

$$z(1 - z)f'' + (c - (1 + a + b)z)f' - abf = 0\,. \tag{3}$$

This equation is of Fuchsian type on the projective line $\mathbf{P}^1(\mathbf{C})$, and it has its *singular* points at $z = 0, 1$ and ∞. Locally in a neighborhood of any regular point $z_0 \in \mathbf{C}\backslash\{0, 1\}$ the space of holomorphic solutions to (3) will be two dimensional. This shows that we can continue any locally defined holomorphic solution of (3) holomorphically to any simply connected region in $\mathbf{C}\backslash\{0, 1\}$. In particular, the series (1) has such holomorphic continuations. This leads us in a natural way to consider the *monodromy representation* of the Gauss hypergeometric function. Choose a regular base point z_0, and consider the associated two dimensional complex vector space V_{z_0} of solutions to (3). For each element $\gamma \in \Pi_1(\mathbf{C}\backslash\{0, 1\}, z_0)$ consider the operator $\mu(\gamma) \in \operatorname{End}(V_{z_0})$ representing the effect in V_{z_0} of analytic continuation of a local solution along a closed loop representing γ. This is easily seen to be a *representation*

$$\Pi_1(\mathbf{C}\backslash\{0, 1\}, z_0) \to \operatorname{GL}(V_{z_0})\,. \tag{4}$$

This representation is very fundamental to the subject. The monodromy representation has important interpretations in algebraic geometry (Picard-Schwarz map) and representation theory (quantum Schur-Weyl duality), as we will see later.

1.1. Behavior at the singular points and monodromy

We can compute the monodromy representation of the hypergeometric function explicitly. This is based on the summation Lemma 1.2. We need to study the behavior of the solutions of (3) near the singular points. By substitution into the hypergeometric equation (3) we find that apart from

$$w_{0,1}(z) = F(a, b, c; z),\tag{5}$$

also the expression

$$w_{0,2}(z) = z^{1-c}F(1 - c + b, 1 - c + a, 2 - c; z)\tag{6}$$

gives us a solution of (3), locally defined in sectors of a punctured disk centered at $z = 0$. Treating the other singular points similarly, we obtain

$$w_{1,1}(z) = z^{-a}F(a, a - c + 1, a + b - c + 1; 1 - z^{-1}),$$
$$w_{1,2}(z) = z^{-b}(1 - z^{-1})^{c-a-b}F(c - a, 1 - a, c - a - b + 1; 1 - z^{-1})\tag{7}$$

at $z = 1$, and

$$w_{\infty,1}(z) = z^{-a}(1 - z^{-1})^{-a}F(a, c - b, a - b + 1; (1 - z)^{-1}),$$
$$w_{\infty,2}(z) = z^{-b}(1 - z^{-1})^{-b}F(b, c - a, b - a + 1; (1 - z)^{-1})\tag{8}$$

at $z = \infty$. This gives us a basis of local solutions in the vicinity of each of the singular points, at least when we assume that the parameters a, b and c do not differ by integers. Each of these six solutions can be expressed in four ways in terms of hypergeometric series (1), and together these constitute Kummer's 24 solutions of the hypergeometric differential equation. When the numbers a, b and c have integer differences, logarithmic terms are usually necessary to describe the local solutions at some of the singular points. This is an important phenomenon called *resonance*. We shall ignore this phenomenon for sake of simplicity.

When we want to understand the monodromy in terms of the local basis $w_{0,1}$, $w_{0,2}$, it is sufficient to find the relations with the other local bases (7) and (8) on a common domain. So let us write

$$w_{0,1} = c_1 w_{1,1} + c_2 w_{1,2}.\tag{9}$$

Since, when $\mathrm{Re}(c - a - b) > 0$, we have $w_{1,2}(1) = 0$, we obtain from the summation formula (2) that

$$c_1 = \frac{\Gamma(c)\Gamma(c - a - b)}{\Gamma(c - a)\Gamma(c - b)}.\tag{10}$$

By application of the Kummer transformation rules one can similarly deduce that

$$c_2 = \frac{\Gamma(c)\Gamma(a + b - c)}{\Gamma(a)\Gamma(b)}.\tag{11}$$

We can similarly deal with the transition to the basis (8).

All this shows us how the Kummer transformations together with the Gauss summation formula make it possible to obtain explicitly the matrices of the monodromy representation. It is a very special feature of the hypergeometric equation.

1.2. The Euler integral

There is another, more geometric way of thinking about the monodromy representation. It is based on the representation of local solutions by means of integrals over *twisted cycles*. The basic form of such a representation is the Euler integral:

Theorem 1.4. *When* $\mathrm{Re}(c) > \mathrm{Re}(a) > 0$, *and* $|z| < 1$, *then*

$$F(a,b,c;z) = \frac{\Gamma(c)}{\Gamma(a)\Gamma(c-a)} \int_0^1 t^{a-1}(1-t)^{c-a-1}(1-tz)^{-b}dt.$$

A proof of this theorem can be given by using the binomial expansion of $(1-tz)^{-b}$, and applying the Euler beta-integral formula.

The Euler integral gives rise to a new understanding of what we saw in the previous subsection. Let us first of all remark that we can replace the integration domain $[0,1]$ by any closed cycle C in $\mathbf{C}\backslash\{0,1,\frac{1}{z}\}$, provided that the integrand

$$t^{a-1}(1-t)^{c-a-1}(1-tz)^{-b} \tag{12}$$

is univalued on C. Such a cycle is called a twisted cycle for the coefficient system defined by (12). A famous example of a twisted cycle is the Pochhammer contour (see Figure 1) around the points 0 and 1.

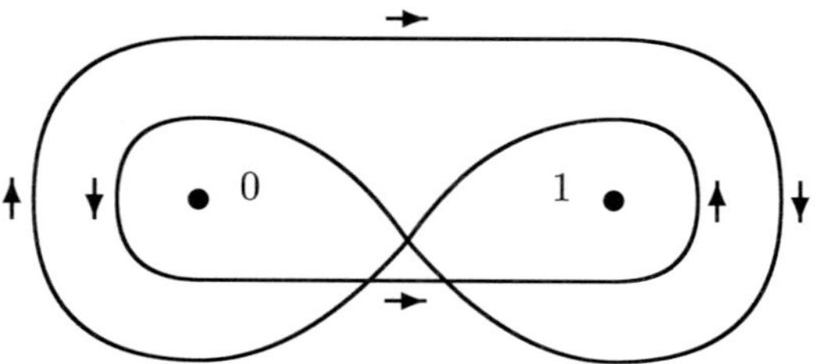

FIGURE 1. The Pochhammer contour

This has the advantage that we can remove the condition $\mathrm{Re}(c) > \mathrm{Re}(a) > 0$. Moreover, we obtain a linear map from the space of homology classes of twisted cycles in $Y_z = \mathbf{C}\backslash\{0,1,\frac{1}{z}\}$, to the space V_z of germs of local solutions at z:

$$H_1^{\mathrm{twist}}(Y_z) \to V_z$$

$$C \to \int_C t^{a-1}(1-t)^{c-a-1}(1-tz)^{-b}dt. \tag{13}$$

For generic parameters, $H_1^{\mathrm{twist}}(Y_z)$ is two dimensional and this map is an isomorphism.

Put $X = \mathbf{C}\backslash\{0,1\}$ and $Y = \mathbf{C}^2\backslash\{t = 0, t = 1, zt = 1\}$, and consider the projection $\pi\colon Y \to X$, $\pi(z,t) = z$ on the first coordinate.

$$
\begin{array}{ccc}
Y & \ni & (z,t) \\
\pi\Big\downarrow & & \Big\downarrow \\
X & \ni & z
\end{array}
\tag{14}
$$

This projection is a fibration with fiber $\pi^{-1}(z) = Y_z$. We define a vector bundle $H_1^{\text{twist}}(Y/X)$ over X whose fiber at z is the twisted homology group $H_1^{\text{twist}}(Y_z)$. An element $C_{z_0} \in H_1^{\text{twist}}(Y_{z_0})$ naturally defines a twisted cycle in every fiber $H_1^{\text{twist}}(Y_z)$ if z is sufficiently close to z_0. Such local sections of $H_1^{\text{twist}}(Y/X)$ are called *flat*, and this natural notion of flat local sections defines an integrable connection on the bundle $H_1^{\text{twist}}(Y/X)$. This is the Gauss Manin connection of the fibration π (with respect to the twisting by the local coefficient system). The "flat continuation" of elements of $H_1^{\text{twist}}(Y_{z_0})$ defines a monodromy representation of $\Pi_1(X, z_0)$ in $\mathrm{GL}(H_1^{\text{twist}}(Y_{z_0}))$.

In short, for generic parameters the isomorphism (13) interprets the monodromy action on the local solution space of the hypergeometric differential equation as the monodromy of the (twisted) Gauss-Manin connection of the fibration (14).

When the parameters a, b and c are rational, we have a projection of the space of 1-cycles of the Riemann surface Z_z of (12) to the space of twisted 1-cycles in the fiber Y_z. Variation of z in the base space X should be thought of as a variation of moduli of the surface Z. The hypergeometric functions are now interpreted as period integrals, considered as functions of the moduli of Z. This point of view gives rise to modular interpretations of X (or certain local compactifications of it) via the *Schwarz map S*. This is the *multivalued* map on X defined by taking the projective ratio

$$
S(z) := (\phi_1(z) : \phi_2(z))
\tag{15}
$$

of two linearly independent solutions of the hypergeometric differential equation. Its branches are related to each other by the action of the projective monodromy group Γ^+. The S-image of the upper half plane $X^+ \subset X$ is a circular triangle T called the Schwarz triangle. The vertices of T are the S-images of 0, 1, and ∞, and by (5), (7) and (8) its angles are

$$
(1 - c)\pi, \quad (c - a - b)\pi, \quad \text{and} \quad (a - b)\pi
\tag{16}
$$

respectively. In order to avoid degeneracies we now assume that (a, b, c) is such that contiguous parameters give equivalent monodromy representations (this is true when $a, b \not\equiv 0, c$ modulo $\mathbf{Z}$). Applying contiguity relations repeatedly we can reduce T so that its angles are nonnegative, and that the sum of two angles is at most π. This ensures that the Schwarz map is a bijection from X^+ to T.

By Schwarz' reflection principle, Γ^+ is realized explicitly as the normal subgroup of index two of holomorphic maps in the group Γ generated by the inversions in the edges of the Schwarz triangle T. By proper choice of the basis ϕ_1, ϕ_2 in (15),

T can be realized as a geodesic triangle in one of the three standard geometries. If the angle sum σ of T exceeds π, we can realize T as a geodesic triangle in $\mathbf{D}^+ := \mathbf{P}^1(\mathbf{C})$ (spherical case). When $\sigma < \pi$, we can realize T as a geodesic triangle in the upper half plane $\mathbf{D}^- := \mathbf{H}$ (hyperbolic case). Finally, when $\sigma = \pi$, we can realize T as a Euclidean triangle in $\mathbf{D}^0 := \mathbf{C}$.

We call T elementary when its angles are of the form $\frac{\pi}{n}$ with $n \in \{2, 3, \dots\}$. By elementary geometry in the natural geometric domain $\mathbf{D}^\epsilon$ ($\epsilon = \pm, 0$) of T, the group Γ^+ is a *discrete* subgroup of the group of isometries $\mathrm{Aut}(\mathbf{D}^\epsilon)$ if and only if T is finitely tesselated by copies of an elementary Schwarz triangle. When T is elementary, then its closure in its geometric domain $\mathbf{D}^\epsilon$ will be a fundamental domain for the action of Γ on $\mathbf{D}^\epsilon$.

When T is elementary, we can therefore find a holomorphic inverse J of S that extends to $\mathbf{D}^\epsilon$ by adding the points of finite branching order. The map J is automorphic for Γ^+ and realizes an isomorphism

$$J : \Gamma^+ \backslash \mathbf{D}^\epsilon \xrightarrow{\sim} \tilde{X} \tag{17}$$

where $\tilde{X}$ is obtained from X by adding the points corresponding to the points of finite branching order.

In the simplest case we consider $a = b = \frac{1}{2}$ and $c = 1$. All angles of T are 0 now. In this case the Euler integral solutions of the hypergeometric differential equation are in fact the classical *elliptic integrals*. We find that Z_z is a double cover of $\mathbf{P}^1(\mathbf{C})$ branched in 0, 1, ∞ and $\frac{1}{z}$, and this is an elliptic curve (with marked point of order two). The projective monodromy group is

$$\Gamma(2) = \left\{ g \in \mathrm{PSL}(2, \mathbf{Z}) \,\middle|\, g \equiv \begin{pmatrix} 1 & 0 \\ 0 & 1 \end{pmatrix} \text{ modulo } 2 \right\}. \tag{18}$$

In this case, the inverse J is the *lambda invariant* that maps the quotient $\Gamma(2) \backslash \mathbf{H}^+$ isomorphically to $X = \mathbf{C} \backslash \{0, 1\}$. It is an isomorphism between two natural models of the moduli space of elliptic curves with marked points of order two.

We should think of the base space $X = \mathbf{C} \backslash \{0, 1\}$ as the space of configurations of four points in $\mathbf{P}^1(\mathbf{C})$, i.e. the space of positions of four distinct points on the projective line, up to the simultaneous action of projective transformations on these points. In this way it is natural to generalize the above interpretation of the Euler integral to more general configuration spaces of geometric objects. This is a fruitful point of view for generalizing hypergeometric functions, which has led to applications in algebraic geometry.

Configuration spaces are also natural to consider in mathematical physics, of course, and the hypergeometric functions in mathematical physics arise in this way also as functions of configurations. I shall return to these matters later.

One-variable generalizations like the functions $_pF_q$ and q-deformations like the basic hypergeometric series are certainly important, but we shall restrict ourselves to discussing multivariable generalizations in this overview.

2. Generalizations of Euler's Hypergeometric Integral

The first generalization of hypergeometric functions that comes to mind when we consider the Euler integral is the Lauricella F_D function. Let $X(n)$ denote the space of $n \geq 4$ distinct, marked points $x_1, \ldots, x_n$ in $\mathbf{P}^1(\mathbf{C})$, modulo the action of $\mathrm{PGL}(2, \mathbf{C})$. The space $X(4)$ is nothing but the base space $X = \mathbf{C} \backslash \{0, 1\}$ we considered in the previous section, since we can send the first three points to 0, 1 and ∞ by a uniquely determined fractional linear map, leaving the fourth point as a free variable in X.

Let μ be an n tuple of complex numbers with $\sum \mu_i = 2$. Let w_μ denote the multivalued $(1, 0)$-form

$$w_\mu = \prod_{i=1}^{n} (t - x_i)^{-\mu_i} dt \tag{19}$$

on $Y_x := \mathbf{P}^1(\mathbf{C}) - \{x_1, \ldots, x_n\}$. For any twisted cycle C in Y_x with respect to this form, we define the following hypergeometric integral

$$I_C(x) := \int_C w_\mu. \tag{20}$$

These integrals are solutions of the Lauricella hypergeometric equations of "type D" when we fix $x_{n-2} = 0$, $x_{n-1} = 1$, and $x_n = \infty$ and think of the $I_C(x)$ as functions of the remaining $n - 3$ variables. It is known that this is an $n - 2$ dimensional space $V(\mu)$ of multivalued functions on $X(n)$. Choose a base point $b \in X(n)$. The map $C \to I_C(x)$ defines an isomorphism $H_1^{\mathrm{twist}}(Y_b)$ and the space $V(\mu)(U)$ where $x \in U$ and U is a suitable neighborhood of b.

The analog of the Schwarz map in this context was studied by Picard ($n = 5$) [36], Terada [38], Deligne and Mostow [8, 9] and others. It was shown that when $\mu_i \in (0, 1)$, the space of twisted cycles $H_1^{\mathrm{twist}}(Y_x)$ carries an hermitian intersection form M of signature $(1, n - 3)$, invariant for monodromy. Hence, for a suitable choice of basis C_i of $H_1^{\mathrm{twist}}(Y_b)$, the image of the Picard-Schwarz map

$$PS \colon X(n) \quad \to \quad \mathbf{P}^{n-3}(\mathbf{C}) \tag{21}$$

$$x \quad \to \quad (I_{C_1}(x) : \cdots : I_{C_{n-2}}(x)) \tag{22}$$

is inside the set $B = \{z = (z_1 : \cdots : z_{n-2}) | M(z, z) > 0\}$. The space B is isomorphic to the unit ball in $\mathbf{C}^{n-3}$.

The main theorem of [8] asserts that the projective monodromy group $\Gamma(\mu) \subset PU(1, n - 3)$ is discrete if there exist $m_{i,j} \in \mathbf{N} \cup \infty$ such that $1 - \mu_i - \mu_j = \frac{1}{m_{i,j}}$ or $\frac{2}{m_{i,j}}$ when $\mu_i = \mu_j$. Moreover, the image of PS is dense in B, and PS^{-1} extends holomorphically to B and gives an isomorphism

$$PS^{-1} \colon \Gamma(\mu) \backslash B \to \Sigma \backslash \tilde{X}(n) \tag{23}$$

where Σ is the group of permutations of points x_i with equal weights μ_i, and $\tilde{X}(n)$ is some quasi-projective local compactification of $X(n)$.

This is a delightful generalization of the theory of the Schwarz map. At the same time it is clear that it is not the end of the story! Other generalizations of the hypergeometric function can be obtained easily by considering hypergeometric integrals associated with configuration spaces of hyperplanes in $\mathbf{P}^n(\mathbf{C})$. For example, Yoshida obtained the modular interpretation of the configuration space $X(3,6)$ of six lines in $\mathbf{P}^2(\mathbf{C})$ in his book [40]. Other work in this direction was done by Couwenberg [7], working with the root system type hypergeometric functions that will be discussed in Section 3. Recently, a lot of progress was made on this subject by Heckman and Looijenga, see [19].

2.1. The Gelfand-Kapranov-Zelevinskii-hypergeometric function

This hypergeometric function (sometimes called A-hypergeometric function) was introduced in [15]. It is in fact a very general class of hypergeometric functions that resembles the case of Lauricella functions. The classical generalizations of the Gauss hypergeometric function like $_pF_q$, the Lauricella type functions, and Horn's hypergeometric functions all occur as special cases of the GKZ-systems.

The GKZ-hypergeometric functions are defined by means of a deceptively simple system of differential equations. Let $A \subset \mathbf{Z}^n$ be a finite generating subset of $\mathbf{Z}^n$. Assume that A lies inside a rational hyperplane. In other words, there exists a linear function $h\colon \mathbf{Z}^n \to \mathbf{Z}$ such that $h(A) = 1$. Let $L \subset \mathbf{Z}^A$ denote the lattice of relations in A, thus

$$L := \{(a_\omega) \in \mathbf{Z}^A | \sum_{\omega \in A} a_\omega \omega = 0\}\,. \tag{24}$$

For $a \in L$, define a constant coefficient partial differential operator $\square_a$ on $\mathbf{C}^A$ by

$$\square_a := \prod_{a_\omega > 0} \left(\frac{\partial}{\partial x_\omega}\right)^{a_\omega} - \prod_{a_\omega < 0} \left(\frac{\partial}{\partial x_\omega}\right)^{a_\omega}\,. \tag{25}$$

Note that $\square_a$ is homogeneous, since $\sum a_\omega = 0$ for every $a \in L$ (apply h to the relation defined by a).

Also define, for every $i = 1, \ldots, n$,

$$Z_i = \sum_{\omega \in A} \omega_i x_\omega \left(\frac{\partial}{\partial x_\omega}\right)\,. \tag{26}$$

When $(\gamma_1, \ldots, \gamma_n) \in \mathbf{C}^n$ is given, we define the following system of differential equations for functions on $\mathbf{C}^A$:

Definition 2.1. (GKZ-system of equations)

$$(1)\ \square_a f = 0\ \forall a \in L, \quad (2)\ Z_i f = \gamma_i f\ \forall i = 1, \ldots, n\,. \tag{27}$$

It is known that the system is *holonomic*, i.e. the system has finite dimensional local solutions spaces. It was shown in [15] that the dimension of the local solution space at a regular point is at least equal to the volume of the convex hull of A inside the rational hyperplane containing A, with equality in the non-resonant case. However, an exact general formula for the dimension doesn't seem to be known [17]. The monodromy representation is also not known in general. Gelfand, Kapranov and Zelevinskii [16, 17] have shown that in the non-resonant case the solutions of the system can be represented by generalized Euler integrals. The GKZ-hypergeometric function contains the hypergeometric functions on Grassmannians which were defined previously by Gelfand and Gelfand [14]. Also the hypergeometric integrals studied by Aomoto [1] are of GKZ-type. It was shown by Batyrev [2] that the period integrals of Calabi-Yau hypersurfaces satisfy a system of GKZ-hypergeometric equations.

3. Analogs of Spherical Functions on Symmetric Spaces

In this section we shall discuss a different kind of multivariable hypergeometric function, the hypergeometric function associated to root systems. This theory is based on other aspects of the Gauss hypergeometric function, namely its role in the representation theory of groups like $SL(2, \mathbf{R})$.

It is well known that Bessel functions of the half integer order $n/2$ show up as the radial eigenfunctions of the Laplace operator Δ of the Euclidean space $\mathbf{R}^n$. This has a generalization to hypergeometric functions, and this provides the basis of a theory of multivariable hypergeometric functions that is natural in relation to representation theory of reductive algebraic groups. The hypergeometric functions of this kind are called "hypergeometric functions associated to root systems" [20, 18, 21, 34], and are closely related to what is called "Macdonald-Cherednik theory" nowadays.

Let us review the basic construction of these functions. A *Riemannian symmetric space* X is a Riemannian manifold such that at every point p of X, the local geodesic inversion $i_p \colon \exp(tv) \to \exp(-tv)$ extends to a global isometry of X. With this assumption it follows simply that X is complete, and that the group G of isometries of X acts transitively on X. We choose a base point x_0 in X, and denote by K the stabilizer group of x_0 in G. The Lie group G acts transitively on X, and K is a compact subgroup of G which is pointwise fixed for the involution $g \to i_{x_0} g i_{x_0}$ of G.

The Euclidean spaces $\mathbf{R}^n$ are the simplest examples of such spaces. These are examples of *flat* symmetric spaces, by which we mean that the sectional curvature of these spaces is 0. Any simply connected Riemannian symmetric space is a product of factors with constant sectional curvature.

Let us assume from now on that X has sectional curvature -1. The local structure of X can be described by introducing "polar coordinates". Let A be a maximal flat totally geodesic submanifold through x_0. Then $A \simeq \mathbf{R}^r$ for some positive integer r. This dimension r is called the *rank* of X. It turns out that the group

$$W := \frac{\{k \in K \mid k \text{ stabilizes } A\}}{\{k \in K \mid k \text{ fixes } A \text{ pointwise}\}} \tag{28}$$

is a finite crystallographic reflection group acting on A. All K orbits intersect A, and this intersection is an orbit of W. The local structure of X is determined completely by the behavior of the function $\delta(x) := \mathrm{Vol}(Kx)$ on A. This function takes the form (see [22])

$$\delta(x) = \prod_{\alpha \in R} |\sinh \alpha(x)|^{\frac{1}{2} m_\alpha} \tag{29}$$

for a certain finite set of linear functions R on A, and certain non-negative integer labels m_α for the elements of R. Clearly δ has to be W invariant, implying that set R is W-stable and the labels are W-invariant. In fact, the orthogonal reflections in the hyperplanes $H_\alpha := \{x \mid \alpha(x) = 0\}$ ($\alpha \in R$) generate the reflection group W. Hence the local structure of X is determined by R and the labels m_α. We call R the *root system* of X, and the m_α are called the root multiplicities.

The analogs of the Laplace operator on $\mathbf{R}^n$ are the G invariant differential operators on X. The algebra of such operators is denoted by $\mathbf{D}(X)$. In the case $\mathbf{R}^n$ this is the operator algebra generated by the Laplace operator Δ. Although $\mathbf{D}(X)$ is in general no longer generated by a single operator, its structure is amazingly simple (cf. [22]):

Theorem 3.1. $\mathbf{D}(X)$ *is a polynomial algebra of rank r over* $\mathbf{C}$.

The analogs on X of the Bessel function of half integer order are the *elementary spherical functions* on X. An elementary spherical function ϕ on X (with origin x_0) is an eigenfunction of the algebra $\mathbf{D}(X)$ which is moreover K-invariant. In other words, there is an algebra homomorphism $\lambda \colon \mathbf{D}(X) \to \mathbf{C}$ such that

$$\Delta \phi = \lambda(\Delta) \phi \ \forall \Delta \in \mathbf{D}(X). \tag{30}$$

Such a function depends on the "radial" variables $x \in A \simeq \mathbf{R}^n$ only. As in the case of spherical waves on $\mathbf{R}^n$, we derive the differential equations for $\phi(x)$ by separation of the radial and rotational variables. This reduces the equations (30) to a system of W-invariant equations on $A \simeq \mathbf{R}^n$. The simplest equation of this type is the second order equation that is derived from the Laplace-Beltrami operator $\Delta_{\mathrm{LB}} \in \mathbf{D}(X)$ of X. Its radial part $L = L(R, m_\alpha)$ has the following form on A:

$$L(R, m_\alpha)\phi = \Delta_A \phi + \sum_{\alpha \in R} m_\alpha \frac{\cosh(\alpha)}{\sinh(\alpha)} \alpha(\nabla_A \phi), \tag{31}$$

where Δ_A is the Laplace operator of the Euclidean space A, and $\nabla_A \phi$ denotes the gradient vector of ϕ in A. When the rank r of X equals 1, the eigenfunction equations (30) reduce to the eigenfunction equation for L. In this case, the root system R has the form $R = \{\pm\beta, \pm 2\beta\}$. When we use $z = -\sinh^2(\beta(x))$ as a new coordinate, we obtain the hypergeometric differential equation (3) whose parameters a, b and c can be expressed in terms of m_β, $m_{2\beta}$, and the eigenvalue λ. In order to obtain the full three parameter family of hypergeometric functions we have to abandon the rank 1 symmetric spaces altogether, and allow arbitrary complex values for the labels m_β and $m_{2\beta}$. The situation is similar to the case of spherical waves on $\mathbf{R}^n$: only Bessel functions of half-integer order allow such a geometric interpretation.

How can we imitate this step in higher rank situations? The operator L is a certain W-invariant deformation of Δ_A, which has the remarkable property that it defines a *completely integrable system*. That is to say, the algebra of W-invariant differential operators commuting with L contains a polynomial algebra of rank r (the radial parts of the operators $\Delta \in \mathbf{D}(X)$).

The crucial step towards the theory of hypergeometric functions for root systems is the insight that this property of complete integrability is not lost when we choose arbitrary complex coefficients m_α in (31) instead of the positive integer labels dictated by the local structure of X (see [21, 35] and the references therein):

Theorem 3.2. *In the algebra of linear partial differential operators with polynomial coefficients in the unknowns m_α, the commutant algebra of the operator L given by (31) is isomorphic to a polynomial algebra $\mathbf{D}(R, m_\alpha)$ of rank r.*

This theorem is not merely an interpolation from the classical cases based on the theory of Riemannian symmetric spaces X. In general, for a given root system R, "nature" has only given us finitely many symmetric spaces X with a root system of type R. Theorem 3.2 is therefore rather surprising, and points at something new. We will explore this in the next subsection.

Anyway, it is now clear how we should define the hypergeometric functions associated to a root system:

Theorem 3.3. *Let $m = (m_\alpha)$ be a set of complex root labels. Given a character λ of the algebra $\mathbf{D}(R, m_\alpha)$, the system of hypergeometric differential equations is defined on the complexification $A_{\mathbf{C}}$ by*

$$\Delta\phi = \lambda(\Delta)\phi \ \forall \Delta \in \mathbf{D}(R, m_\alpha). \tag{32}$$

The system is invariant for W and for the lattice of translations T on which all the roots take values in $2\pi i$.

The system is regular at the regular points of the action of the affine reflection group $W \ltimes T$, and locally its solution space has dimension $|W|$ at regular points. There is a unique holomorphic solution $F_R(\lambda, m_\alpha; x)$ defined in a neighborhood of the origin in $A_{\mathbf{C}}$, normalized in the origin by $F_R(\lambda, m_\alpha; 0) = 1$. It is called *the hypergeometric function associated with R.*

This function has very elegant properties. Its monodromy representation has been determined, at least for generic parameters. The fundamental group of the regular orbit space of $W \ltimes T$ acting on $A_{\mathbf{C}}$ is the affine braid group associated with W. The monodromy representations factors through an affine Hecke algebra quotient of the group algebra of the braid group [20].

Considering its origin, it is not surprising that the root system hypergeometric function can be used as the kernel of a deformation of the Fourier transform, generalizing the harmonic analysis of zonal spherical functions [34, 5]. This harmonic analysis contains a lot of combinatorial information about root systems, already in the polynomial case (corresponding to zonal polynomials on the compact form of X). We return to this issue in the next subsection. The spectral analysis using these functions is also related to the dynamics of the integrable models of "Calogero-Moser" type in mathematical physics.

There is no Euler type integral representation known in general for the hypergeometric function associated with a root system, except for $W = S_n$. This has to be considered as a missing link.

However, recent work of Heckman and Looijenga shows that, for special parameter values, the monodromy of these functions has a deep geometric meaning [19].

3.1. The Cherednik-Macdonald theory

Complete integrability of a system of differential operators is a rare and delicate property. The integrability of the Laplace-Beltrami operator L of X as in the previous subsection is indeed very special, as it reflects the geometry of X. The fact that the deformations of L in (30) do not destroy the integrability is therefore remarkable, and it indicates that there should exist a more fundamental structure than the symmetric space X itself. On the algebraic level this structure is well understood. It is Ivan Cherednik's *double affine Hecke algebra* [4]. It simultaneously captures the so-called *spherical convolution algebras* of the p-adic symmetric spaces $X(\mathbf{Q}_p)$ and the algebras of G invariant differential operators $\mathbf{D}(X)$ (with X a real form of the symmetric space) as in the previous subsection.

Let us briefly look at this interesting object when the rank of X equals 1. We follow the nice presentation from [32]. As before, we put $R = \{\pm\beta, \pm 2\beta\}$. The double affine Hecke algebra $\mathcal{H}$ has generators $T_0, T_1, T_1^\vee$ and $T_0^\vee$ over the field K of rational functions in five indeterminates $t_i, t_i^\vee$ $(i = 0, 1)$ and q, with relations:

$$
\begin{aligned}
(T_i - t_i)(T_i + t_i^{-1}) &= 0, \\
(T_i^\vee - t_i^\vee)(T_i^\vee + t_i^{\vee\,-1}) &= 0, \\
T_0 T_1 T_1^\vee T_0^\vee &= q.
\end{aligned}
$$

The subalgebra $K\langle T_0, T_1\rangle$ of $\mathcal{H}$ generated by T_0 and T_1 is an ordinary affine Hecke algebra, which has a one dimensional representation ρ defined by

$$
\rho(T_0) = q_0, \quad \rho(T_1) = q_1. \tag{33}
$$

The induced module $\mathrm{Ind}^{\mathcal{H}}_{K\langle T_0, T_1\rangle}(\rho)$ is naturally isomorphic to the Laurent polynomial ring $K[X, X^{-1}]$, where $X = T_1 T_1^\vee$. This defines a faithful representation π of $\mathcal{H}$, as operator algebra on $K[X, X^{-1}]$.

For example, the operator $\pi(T_1)$ is a "Lusztig operator"

$$
\pi(T_1) = t_1 s_1 + (a_1 + a_1^\vee X^{-1})\frac{1}{1 - X^{-2}}(1 - s_1) \tag{34}
$$

where s_1 is the involutive automorphism of $K[X, X^{-1}]$ defined by $s_1(X^n) = X^{-n}$, $a_i = t_i - t_i^{-1}$ and $a_i^\vee = t_i^\vee - t_i^{\vee -1}$. Similarly we have

$$\pi(T_0) = t_0 s_0 + (a_0 + q^{-1} a_0^\vee X) \frac{1}{1 - q^{-2} X^2}(1 - s_0),\qquad(35)$$

where $s_0(X^n) = q^{2n} X^{-n}$. These formulas can be checked by some disciplined direct computations. The analog of the radial part of the Laplace-Beltrami operator $L(R, m_\beta)$ of (31) is now given by the operator

$$\Lambda(R, t_i, t_i^\vee, q) := (\pi(Y) - 1)(1 - \pi(Y^{-1}))\qquad(36)$$

on $K[X, X^{-1}]$, where $Y = T_1 T_0$. The relation with the previous subsection is achieved by a limiting procedure $q \to 1$, after the specializations $t_0 = t_0^\vee = q^{-m_\beta/2}$, $t_1 = q^{-m_{2\beta}}$, and finally $t_1^\vee = 1$. When we formally write $X = e^\beta$, we find by direct computation that the relation with the operator (31) is given by

$$L(R, m_\alpha) = \lim_{q \to 1} \frac{\Lambda(R, t_i, t_i^\vee, q)}{q - q^{-1}}.\qquad(37)$$

Let us return to the general rank case now. Intelligible sources for this material are [30] and [24]. The double affine Hecke algebra $\mathcal{H}$ can be defined without too much difficulty. It consists of two dual affine Hecke algebras, whose finite dimensional Hecke-subalgebras are identified. As in the rank 1 case, this algebra has a faithful representation in a Laurent polynomial algebra $K[X_1^{\pm 1} \ldots, X_r^{\pm r}]$, and it contains a rank r polynomial subalgebra

$$\mathcal{L} = K[\Lambda_1(R, t_i, t_i^\vee, q), \ldots, \Lambda_r(R, t_i, t_i^\vee, q)],\qquad(38)$$

the "q-analog" of the algebra $\mathbf{D}(R, m_\alpha)$ of Theorem 3.2.

The elementary spherical functions on the compact real form X^{comp} of a Riemannian symmetric space X, are the so-called zonal polynomials. In the rank 1 case these are the well-known Jacobi-polynomials. A far reaching generalization of the zonal spherical polynomial is the Macdonald-Koornwinder polynomial [30, 28]. These are by definition the W-invariant polynomial eigenfunctions of the algebra $\mathcal{L}$, suitably normalized. In the one-variable case, these are the polynomials of the q-Askey-Wilson scheme. The original zonal polynomials are obtained by the limit transition as described above. Hall-Littlewood polynomials, in their role of the elementary spherical functions on a p-adic symmetric space $X(\mathbf{Q}_p)$, arise as the limit for $q \to 0$ when we put $t_\alpha = \frac{1}{p}$.

The Macdonald polynomials have many interpretations in algebraic combinatorics, in mathematical physics and in representation theory. The polynomials have been instrumental in the solution of various conjectures on the combinatorial properties of reflection groups [33, 4]. Macdonald's "constant term conjectures" [29] are the most prominent among these. Figure 2 (also due to Macdonald) gives an overview of their representation theoretic significance. In addition, one knows in the case of the reflection group $W = S_n$, the symmetric group on n letters, that the polynomials are also the elementary spherical functions of the compact quantum group $U_q(n)$.

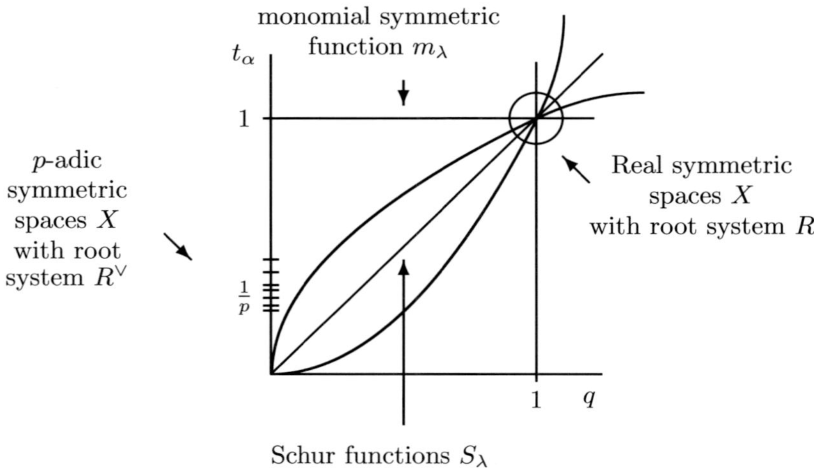

FIGURE 2. Macdonald polynomials and symmetric spaces

Much of the theory of Macdonald polynomials is related to the spectral analysis of the double affine Hecke algebra that generalizes the spherical harmonic analysis related to the compact real form of the symmetric space X.

It is a major open problem to find the spectral theory of the operator algebra $\mathcal{L}$ introduced in (38) that generalizes the spherical Harish-Chandra transform on real non-compact symmetric spaces and Macdonald's p-adic spherical transform. It has been achieved in the differential limit for q tends to 1 [34, 5], and for the rank 1 double affine Hecke algebra in [26].

4. Integrable Models and Hypergeometric Functions

We saw that the differential equations for the hypergeometric function associated to a root system is a completely integrable system. This system is also known in mathematical physics as the trigonometric *Calogero-Moser* system. When $W = S_n$, the symmetric group on n letters, this system describes the dynamics of a quantum mechanical system of n particles moving on the real line under the influence of a pair potential that is proportional to the inverse square of the hyperbolic sine of the distance of the particles. The generalization to the algebra of difference equations $\mathcal{L}$ of (38) has the interpretation of making the quantum system relativistic. For $W = S_n$ such a relativistic model was found explicitly by Ruijsenaars [37], and this was extended to general classical root systems by Van Diejen [10]. The spectral problem for Cherednik's operator realization of affine Hecke algebras, discussed in the previous subsection, is of course very important for the dynamics of these related integrable models from mathematical physics.

4.1. The Knizhnik-Zamolodchikov equations

There is another interpretation of hypergeometric functions, in mathematical physics, via the so-called Knizhnik-Zamolodchikov equations of conformal field theory [39, 13]. These equations are the differential equations for the n-point correlation functions $\psi(z_1, \ldots, z_n)$ of conformal field theory for a Kac-Moody algebra $\hat{\mathfrak{g}}$. The points $z_1, \ldots, z_n$ are distinct points on the complex line $\mathbf{C}$, and the correlation function takes values in an n-fold tensor product $V_1 \otimes \cdots \otimes V_n$ of representations of the finite dimensional simple Lie algebra $\mathfrak{g}$. The equations have the form

$$(k + h^\vee) \frac{d\psi}{dz_i} = \left(\sum_{\substack{j=1 \\ j \neq i}}^{n} \frac{\Omega_{i,j}}{z_i - z_j} \right) \psi \ (\forall i = 1, \ldots, n), \tag{39}$$

where Ω is the symmetric "Casimir tensor" $\Omega = \sum x_i \otimes x^i \in \mathfrak{g} \otimes \mathfrak{g}$ corresponding to the invariant scalar product on $\mathfrak{g}$, and $\Omega_{i,j}$ denotes the action of Ω on the i-th and j-th slot of the n-fold tensor ψ. The number $h^\vee$ is the dual Coxeter number of $\hat{\mathfrak{g}}$, and the complex number k is called the central charge.

These KZ-equations have been studied intensively by mathematicians because of their interesting relation with quantum groups. The system of differential equations is integrable and is invariant for the $\mathfrak{g}$ action on ψ. Hence it defines a monodromy representation of the fundamental group B_n of the regular orbit space of S_n acting on $\mathbf{C}^n$. This monodromy representation has values in the space of $\mathfrak{g}$-intertwiners between tensor products of $\mathfrak{g}$-modules. This defines the structure of a braided tensor category on the tensor category of $\mathfrak{g}$-modules. For generic values of k, this is the representation category of the quantum universal enveloping algebra $U_q(\mathfrak{g})$, where $q = \exp(\frac{\pi i}{k+h^\vee})$ (Drinfeld-Kohno theorem) [11, 27, 23]. When $\mathfrak{g} = \mathfrak{gl}_n$ this is the quantum group version of the classical Schur-Weyl duality.

It is fully justified to view the KZ-equations as generalized hypergeometric equations. In simple examples the equations reduce to (3) and the solutions of (39) allow integral representations that are generalizations of Euler's integral representation 1.4. This gives a beautiful geometric interpretation of the monodromy representation of the KZ-equation, analogous to the isomorphism (13). I refer the reader to the books [39] and [13] for a detailed exposition of this point of view.

The Knizhnik-Zamolodchikov equations have a number of variations and generalizations (allowing more complicated coefficient functions and discretization for example). For these complicated matters the reader is referred to [13].

In the case of trigonometric coefficient functions, we have the following relation to the Macdonald-Cherednik theory. Take $\mathfrak{g} = \mathfrak{gl}_n$ and let ψ take it values in the zero-weight space of $V^{\otimes n}$, where V is the defining representation of $\mathfrak{g}$. The resulting system of equations can be shown to be equivalent to 3.3 when $W = S_n$. Cherednik has introduced the root system analogs of these (very special) Knizhnik-Zamolodchikov equations [3]. He found an explicit system of first order differential equations, and these equations were shown to be equivalent to (32) in [31].

The representation theoretic meaning via the quantum Schur-Weyl duality and the geometric interpretation of the monodromy representation give the KZ-equation a very rich structure in the case of $W = S_n$.

References

[1] K. Aomoto, *On the structure of integrals of power products of linear functions*, Sci. Papers, Coll. Gen. Education, Univ. Tokyo **27** (1977), 49–61.

[2] V. V. Batyrev, *Variations of the mixed Hodge structure of affine hypersurfaces in algebraic tori*, Duke Math. J. **69(2)** (1993), 349–409.

[3] I. Cherednik, *A unification of Knizhnik-Zamolodchikov equations and Dunkl operators via affine Hecke algebras*, Inv. Math. **106** (1991), 411–432.

[4] ———, *Double affine Hecke algebras and Macdonald conjectures*, Ann. Math. **141** (1995), 191–216.

[5] ———, *Inverse Harish-Chandra transform and difference operators*, Internat. Math. Res. Notices **15** (1997), 733–750.

[6] ———, *Lectures on affine Knizhnik-Zamolodchikov equations, quantum many-body problems, Hecke algebras, and Macdonald theory*, RIMS-1144 (1997).

[7] W. J. Couwenberg, *Complex reflection groups and hypergeometric functions*, Thesis University of Nijmegen (1994).

[8] P. Deligne and G. D. Mostow, *Monodromy of hypergeometric functions and non-lattice integral monodromy*, I.H.E.S. Publ. Math. **63** (1986), 5–89.

[9] ———, *Commensurabilities among lattices in $PU(1, n)$*, Annals of Math studies **132**, Princeton UP (1993).

[10] J. F. van Diejen, *Self-dual Koornwinder-Macdonald polynomials*, Inv. Math. **126** (1996), 319–341.

[11] V. G. Drinfeld, *Quasi-Hopf algebras*, Algebra i Analiz **1(6)** (1989), 114–148, Engl. transl. in Leningrad Math. J. **1** (1990), 1419–1457.

[12] A. Erdéli (eds.), *Higher transcendental functions*, vol. **1**, McGraw-Hill (1953).

[13] P. I. Etingof, I. B. Frenkel and A. A. Kirillov, Jr., *Lectures on Representation Theory and Knizhnik-Zamolodchikov Equations*, Mathematical Surveys and Monographs **58**, A.M.S. (1998).

[14] I. M. Gelfand and S. I. Gelfand, *Generalized hypergeometric functions*, Dokl. Akad. Nauk. SSSR **228 (2)** (1986), 279–283.

[15] I. M. Gelfand, M. M. Kapranov and A. V. Zelevinskii, *Hypergeometric functions and toral manifolds*, Funct. Anal. and its Appl. **23** (1989), 94–106.

[16] ———, *Generalized Euler integrals and A-hypergeometric functions*, Adv. in Math. **84** (1990), 255–271.

[17] ———, *A correction to the paper "Hypergeometric functions and toral manifolds"*, Funct. Anal. and its Appl. **27** (1993), 295–295.

[18] G. J. Heckman, *Dunkl operators*, Séminaire BOURBAKI 49ème année, 1996–97, n° 828 (1997).

[19] G. J. Heckman and E. Looijenga, *The moduli space of rational elliptic surfaces*, arXiv math.AG/001002D (2001).

[20] G. J. Heckman and E. M. Opdam, *Root systems and hypergeometric functions I*, Comp. Math. **64** (1987), 329–352.

[21] G. J. Heckman, and H. Schlichtkrull, *Harmonic Analysis and Special Functions on Symmetric Spaces*, Academic Press (1994).

[22] S. Helgason, *Groups and Geometric Analysis*, Perspectives in Mathematics **16**, Academic Press, New York (1984).

[23] D. Kazhdan and G. Lusztig, *Tensor structures arising from affine Lie algebras I-IV*, J. A.M.S. **6** (1993), 905–947, 949–1011; **7** (1994), 335–381, 383–453.

[24] A. A. Kirillov, Jr., *Lectures on affine Hecke algebras and Macdonald's conjectures*, Bull. AMS **34**(3) (1997), 251–292.

[25] F. Klein, *Vorlesungen ueber die hypergeometrische Funktion*, Grundlehren der mathematischen Wissenschaften, Springer-Verlag (1933).

[26] E. Koelink and J. V. Stokman, *The Askey-Wilson transform*, preprint xxx.lanl.gov/ abs/ math.CA/0004053 (2000).

[27] T. Kohno, *Monodromy representations of braid groups and Yang-Baxter equations*, Ann. Inst. Fourier **37(4)** (1987), 139–160.

[28] T. H. Koornwinder, *Askey-Wilson polynomials for root systems of type BC*, Contemp. Math. **138** (1992), 189–204.

[29] I. G. Macdonald, *Some conjectures for root systems*, SIAM J. of Math. An. **13** (1982) 988–1007.

[30] ______, *Affine Hecke algebras and orthogonal polynomials*, Séminaire Bourbaki, Vol. 1994–1995 n° 797.

[31] A. Matsuo, *Integrable connections related to zonal spherical functions*, Invent. Math. **110** (1992), 95–121.

[32] M. Noumi and J. V. Stokman, *Askey-Wilson polynomials: an affine Hecke-algebraic approach*, preprint (2000).

[33] E. M. Opdam, *Some applications of hypergeometric shift operators*, Inv. Math. **98** (1989), 1–18.

[34] ______, *Harmonic analysis for certain representations of graded Hecke algebras*, Acta. Math. **175** (1995), 75–121.

[35] ______, *Lecture notes on Dunkl operators for real and complex reflection groups*, MSJ Memoirs **8** (2000), Math. Soc. of Japan.

[36] E. Picard, *Sur une extension aux fonctions de deux variables du problème relatif aux fonctions hypergéométriques*, Ann. E.N.S. **10** (1881), 305–321.

[37] S. N. M. Ruijsenaars, *Complete integrability of relativistic Calogero-Moser systems and elliptic function identities*, Comm. Math. Phys. **110** (1987), 191–213.

[38] T. Terada, *Problème de Riemann et fonctions automorphes provenant des fonctions hypergéométriques de plusieurs variables*, J. of Math. Kyoto Univ. **13-3** (1973), 557–578.

[39] A. Varchenko, *Multidimensional hypergeometric functions and representation theory of Lie algebras and quantum groups*, Advanced Series in Mathematical Physics **21**, World Scientific (1995).

[40] M. Yoshida, *Hypergeometric function, my love*, Aspects of Mathematics **E32**, Vieweg, (1997).

KdV Institute for Mathematics
University of Amsterdam
Plantage Muidergracht 24
1018 TV Amsterdam, The Netherlands
E-mail address: opdam@wins.uva.nl

Contact Structures, Rational Curves
and Mori Theory

Thomas Peternell

Abstract. This paper intends to report on the recent progress in the classification of contact structures on complex projective and compact Kähler manifolds. In particular we explain how to apply the geometry of rational curves and Mori theory to investigate projective contact manifolds whose second Betti number is at least 2.

1. Introduction

Given a projective or compact Kähler manifold X, a *special* subbundle or coherent subsheaf $E \subset T_X$ in the holomorphic tangent bundle T_X often carries significant geometric information about the underlying manifold. One important special property is certainly integrability. Integrable subbundles E, i.e. subbundles closed under the Lie bracket, define a foliation and one might hope – under additional assumptions – to find compact leaves and therefore to obtain strong geometric information on X. At the moment there are however only a very few results when the leaves are compact. In sharp contrast are contact structures, i.e. subbundles of corank 1 which are maximally non-integrable. Contact structures play an important role in real differential geometry and attracted rather recently interest in complex geometry via the theory of quaternionic-Kähler manifolds: given a quaternionic Kähler manifold (M, g) with positive scalar curvature, the twistor space Z is a Fano manifold (i.e. has a metric with positive Ricci curvature) with a contact structure and classifying those Fano manifolds means to classify the quaternionic-Kähler manifolds with positive scalar curvature (Salamon, LeBrun).

Therefore one is very much interested in classifying projective contact manifolds, and apart from the connection with quaternionic Kähler manifolds, the interest in complex geometry really lies in being "opposite" to foliations.

This article mainly intends to report the recent progress in this topic and to discuss the main open problems. It is written in such a way that it should be accessible also to differential geometers.

2. Complex Contact Structures: Basic Facts, State of the Art and Conjectures

Given a complex manifold X with sheaf $\mathcal{O}_X$ of holomorphic functions we shall denote T_X its tangent bundle, and denote by Ω^1_X the sheaf of holomorphic 1-forms, i.e. the sheaf of holomorphic sections of T^*_X. Moreover $K_X = \det T^*_X$ is the canonical bundle on X and $c_1(X) = c_1(T_X)$ denotes the first Chern class of X. A line bundle L on a (usually compact) manifold is *positive* or *ample* if it carries a metric of positive curvature. A line bundle L on a *projective* manifold X is called *nef*, if $c_1(L) \cdot C = c_1(L \mid C) \geq 0$ for any irreducible compact curve $C \subset X$. If L carries a metric of semipositive curvature, then L is nef but the converse is false. Finally we denote by $\kappa(L)$ the Kodaira dimension of L and let $\kappa(L) = \kappa(K_X)$ be the Kodaira dimension of X. So $\kappa(X) = -\infty$ if no multiple $mL = L^{\otimes m}$ has a section; $\kappa(L) = 0$, if some mL has a section but $\dim H^0(X, mL) \leq 1$ for all m and $\kappa(L) = k > 0$, if $\dim H^0(X, mL)$ grows as m^k.

Definition 2.1. *A compact complex manifold X of dimension $2n + 1$ together with a subbundle $F \subset T_X$ of rank $2n$ is a* contact manifold *if the pairing $\omega \colon F \times F \to T_X/F =: L$ induced by the Lie bracket is everywhere non-degenerate.*

An equivalent definition is as follows: we require that K_X is divisible by $n + 1$ ($\dim X = 2n + 1$); if we write $-K_X = (n + 1)L$, then we must have a section $\theta \in H^0(X, \Omega^1_X \otimes L)$ and the non-degeneracy condition can be reformulated as follows: an elementary calculation shows that $\theta \wedge (d\theta)^{\wedge k}$, although computed in a local trivialisation of L, gives a global section of $\Omega^{2k+1}_X \otimes L^k$. Then $\theta \wedge (d\theta)^n$ has no zeroes. The subbundle $F \subset T_X$ is now given by $F = \ker(\theta)$.

Remark 2.2. *As already mentioned, the anticanonical bundle is a multiple of the contact line bundle L: we have $-K_X = (n+1)L$. This already restricts the class of possible contact manifolds significantly. E.g. a quadric hypersurface in projective space can never carry a contact structure.*

For a detailed discussion of contact manifolds we refer to [12] and [4].

Examples 2.3.
(1) *Let $X = \mathbb{P}_{2n+1}$ and take a non-zero form $\omega \in H^0(\Omega^1_X \otimes \mathcal{O}(2))$. Then ω has no zeroes and therefore defines a bundle epimorphism $T_X \to \mathcal{O}(2)$. The kernel now defines a contact structure on $\mathbb{P}_{2n+1}$. In the theory of vector bundles on projective space, this kernel is known (up to a twist with a line bundle) as a null-correlation bundle, see [17].*
(2) *Let Y be any complex manifold of dimension $n+1$ and let $X = \mathbb{P}(T_Y)$. Then X has a contact form as follows.*
$H^0(X, \pi^*(\Omega^1_Y \otimes \mathcal{O}_X(1))) = H^0(Y, \Omega^1_Y \otimes T_Y)$ has a canonical element, namely the identity endomorphism on T_Y. So letting $L = \mathcal{O}_X(1)$ we obtain an element $\omega \in H^0(X, \Omega^1_X \otimes \mathcal{O}(1))$ which easily is seen to give a contact structure.
(3) *Let G be a complex simple Lie group with Lie algebra $\mathcal{G}$. Then there is a unique closed orbit $X_{\mathcal{G}}$ for the adjoint action of G on $\mathbb{P}(\mathcal{G})$. Now $X_{\mathcal{G}}$ can be*

seen to be a Fano contact manifold; for details see [3] *and* [4]. *We say that the contact structure is induced by a simple Lie group. Of course projective space can be rediscovered in that way* $(G = \mathrm{Sp}(2n+2))$.

(4) *Let* (M, g) *be a quaternion-Kähler manifold of positive scalar curvature. Then its twistor space* $p\colon Z \to M$, *a* S^2-*submersion over* M, *is a Fano contact manifold (and even Kähler-Einstein). We refer to* [19] *and* [12].

Now a standard conjecture says that these should be the only examples.

Conjecture 2.4. *Let* X *be a compact complex contact Kähler manifold. Then* X *is induced by a simple Lie group or* $X = \mathbb{P}(T_Y)$ *for some compact manifold* Y.

Notice that $\mathbb{P}(T_{\mathbb{P}_{n+1}})$ occurs in both lists $(G = \mathrm{Sl}(n+1))$, but all other manifolds coming from simple Lie groups have second Betti number $b_2(X) = 1$. There is a lot of evidence that the conjecture should be true in the projective category (and then also in the Kähler category), but in the Kähler some necessary tools are not yet available.

In particular this conjecture would imply a classification of quaternion-Kähler manifolds via the twistor construction.

In dimension 3 the conjecture has been proved by Ye [21], whereas in dimension 5 it is proved by Druel [7] up to possible contact 5-folds with nef canonical bundles. We next state the general results known so far to support the conjecture.

Proposition 2.5. (Druel [7]**)** *A compact Kähler contact manifold* X *has Kodaira dimension* $\kappa(X) = -\infty$, *i.e.* $H^0(X, \mathcal{O}(mK_X)) = 0$ *for all positive* m.

The Kähler assumption is needed to conclude that a holomorphic 1-form on X is closed.

Proposition 2.6. (Ye [21]**)** *A compact Kähler contact manifold* X *cannot have a numerically trivial canonical bundle.*

In fact, such a manifold has $\kappa(X) = 0$ by virtue of the decomposition theorem [2].

Very recently, Demailly [6] has generalized these two results and made the following important contribution.

Proposition 2.7. (Demailly) *Let* X *be a compact Kähler contact manifold. Then* K_X *is never pseudo-effective, i.e.* K_X *does not carry a (possibly singular) hermitian metric whose curvature current is non-negative.*

If X is projective, then K_X is pseudo-effective if and only if K_X is contained in the closure of the cone generated by the (classes of the) effective divisors, i.e. of the irreducible hypersurfaces in X. See Theorem 4.1 for a more general version of Proposition 2.7.

We shall distinguish the cases $b_2(X) = 1$ and $b_2(X) \geq 2$. If $b_2(X) = 1$, then X, being Kähler, is automatically projective and Propositions 2.5/2.6 imply that X is a Fano manifold, i.e. $-K_X$ is ample. In that case X should come from a simple Lie group according to the conjecture. In fact, one has

Theorem 2.8. (Beauville [3]**)** *Let X be a Fano contact manifold with $b_2(X) = 1$ and – as usual – with contact line bundle L. Assume that*
  (a) *the rational map defined by the linear system $|L|$ (i.e. by the global sections of L) is generically finite;*
  (b) *the group G of contact automorphisms is reductive.*
  Then the Lie algebra $\mathcal{G}$ of G is simple and X is of the form $X_\mathcal{G}$ (Example 2.3(3)).

The contact automorphisms are the automorphisms of X preserving F; notice also that the Lie algebra of G can be identified naturally with $H^0(X, L)$. In order to prove the conjecture, one now has to construct sufficiently many sections in L, which might be quite hard. In [4] Beauville removed the assumption that G is reductive but instead had to assume that L is very ample, i.e. the map given by the global sections of L is an embedding.

We next turn to the case $b_2(X) \geq 2$. From now on we will assume X projective and say only a few words on the Kähler case later. The aim is to rediscover the projective bundle structure X is expected to have according to the conjecture. In that case the canonical bundle K_X is negative on the projective spaces, in particular K_X is not nef, i.e. $K_X \cdot C < 0$ for some irreducible curve C. Therefore it is natural to use Mori theory for the investigation of X. We refer to [9, 15] for basic information on Mori theory. In general, Mori theory is applicable if K_X is not nef, this being guaranteed by Proposition 2.7. Therefore the main result of [10], discussed at length in Section 3, yields.

Theorem 2.9. *Let X be a projective contact manifold with $b_2(X) \geq 2$. Then X is of the form $X = \mathbb{P}(T_Y)$ with a projective manifold Y.*

The case "K_X nef" is potentially also ruled out by standard conjectures in minimal model theory.

Abundance Conjecture. *Let X be a projective manifold with K_X nef. Then mK_X is spanned by global sections for suitable large m. In particular $\kappa(X) \geq 0$.*

This conjecture has been proved by Miyaoka [13, 15] and Kawamata [8] in dimension 3, but it is completely open in higher dimensions. Of course we need only a "small" part, namely that $\kappa(X) \geq 0$, i.e. some multiple of K_X has a section.

Remark 2.10. *In the Kähler case we would basically need the corresponding results on Mori theory, which do not exist yet, to prove the conjecture. In dimension 3 how there are some results which suffice to settle Conjecture 2.4 except for the mysterious case of "simple threefolds with $\kappa(X) = -\infty$". We refer to [18].*

It would be nice to have a direct proof of the following conjecture which of course would be a direct consequence of Conjecture 2.4 and which also would rule out the nef case immediately.

Conjecture 2.11. *Let X be a compact complex contact Kähler manifold of dimension $2n + 1$. Then $c_1(X)^n \neq 0$.*

Notice that one cannot do better: let Y be a 2-dimensional complex torus and let X be the projectivised tangent bundle: $X = \mathbb{P}(T_Y)$ (we take hyperplanes in the fibers!). Then $X \simeq \mathbb{P}_1 \times Y$ and clearly $c_1(X)^2 = 0$.

Given a contact manifold X, it is natural to ask how many contact structures can exist on X. Of course two contact forms define the same contact structure if and only if they differ only by a scalar. In [10] it is proved that,

Theorem 2.12. *The contact structures on a complex manifold $X = \mathbb{P}(T_Y)$ are in natural $1 : 1$-correspondence with the space $H^0(Y, \mathrm{End}(\Omega_Y^1))$ of endomorphisms of Ω_Y^1.*

If Ω_Y^1 is simple, i.e. there are no endomorphisms except for multiples of the identity, then it follows that the contact structure is unique. This is in particular the case when X is projective and stable with respect to some ample polarisation. For example the Fano manifold $\mathbb{P}(T_{\mathbb{P}_{n+1}})$ has only one contact structure.

Concerning Fano manifolds with $b_2(X) = 1$, LeBrun proved in [12] that a Fano manifold which admits a Kähler-Einstein metric carries more than one contact structure if and only if X is projective space. It is natural to conjecture that this should be true without the Kähler-Einstein assumption; in fact Conjecture 2.4 implies that a Fano contact manifold is Kähler-Einstein, because it is rational-homogeneous.

It is interesting to notice that the contact sequence never splits in the Kähler case:

Proposition 2.13. *Let X be a compact Kähler contact manifold. Then the contact sequence $0 \to F \to T_X \to L \to 0$ never splits.*

Proof. Suppose the contact sequence splits. Consider the (proportional) Chern classes

$$c_1(L), c_1(F) \in H^1(X, \Omega_X^1) = H^1(X, L^*) \oplus H^1(X, F^*).$$

Then, not surprisingly, actually $c_1(L) \in H^1(X, L^*)$ and $c_1(F) \in H^1(X, F^*)$ (see [5]). Hence the proportionality $c_1(F) = nc_1(L)$ yields $c_1(X) = c_1(L) = c_1(F) = 0$ which is impossible as we already saw.

LeBrun ([12]) already noticed that the contact sequence never splits in case X is Fano. He also proved that the image of $c_1(L)$ under the canonical morphism

$$H^1(X, \Omega_X^1) \to H^1(X, F^*) \to H^1(X, F \otimes L^*)$$

is (up to $2\pi i$) the extension class of the contact sequence (the second map comes from the isomorphism $F^* \simeq F \otimes L^*$ provided by the non-degenerate map $F \times F \to L$).

3. The Use of Mori Theory

In this section we explain the methods of the proof of Theorem 2.9. So we fix for this section a projective contact manifold X of dimension $2n+1$ with contact line bundle L; we also suppose $b_2(X) \geq 2$. The final goal is to show that X is of the form $X = \mathbb{P}(T_Y)$. The main problem is to rediscover a $\mathbb{P}_n$-bundle structure on X: once we have this, it is not so difficult to show that X is a projectivised tangent bundle.

Let us suppose first that K_X is not nef. By Mori theory we obtain a surjective holomorphic map

$$\phi: X \to Y$$

to a normal projective variety Y such that $b_2(X) = b_2(Y) + 1$ and such that

$$-K_X \cdot C > 0$$

for one and hence for all irreducible algebraic curves C contracted by ϕ. Now we shall study this map ϕ and we would like to see that ϕ defines a $\mathbb{P}_n$-bundle structure. There are three main steps.

Lemma 3.1. $\dim Y < \dim X$, *i.e.* ϕ *is not birational.*

Lemma 3.2. *The general fiber of ϕ is $\mathbb{P}_n$.*

Lemma 3.3. *All fibers of ϕ are smooth of the same dimension.*

The **proof of Lemma 3.1** is very technical and relies on a detailed study of rational curves of small degree with respect to L, actually of rational curves C with $L \cdot C = 1$. Such curves exist: by Mori's breaking technique a projective manifold X of dimension m with K_X not nef always carries a rational curve C with $0 > K_X \cdot C \geq -m - 1$. Via $-K_X = (n+1)L$ in our situation, we obtain a rational curve C with $L \cdot C = 1$ or $L \cdot C = 2$. If we cannot find C with $L \cdot C = 1$, then one can see that X must be Fano with $b_2 = 1$ (we have large families of these lines covering X), so we always have "L-lines". Now one main point is to study the deformations of such a curve. In this context a main point is to study the restriction of T_X to such lines.

We postpone (some of) the technical details to the sketch of proof of Lemma 3.3; instead we present here a simplified version (not using any projectivity assumption):

Simplified Lemma. *Let X be a compact complex manifold, $\phi: X \to Y$ the blow-up of a submanifold B in the compact manifold Y. Then X cannot carry a contact structure.*

Proof. Take any non-trivial fiber X_y of ϕ. Let $k = \dim X_y$. Then $T_X \mid l$ is of the following form

$$T_X \mid l = \mathcal{O}(2) \oplus \mathcal{O}(1)^{\oplus k-1} \oplus \mathcal{O}^{\oplus n-k-1} \oplus \mathcal{O}(-1).$$

Here the factor $\mathcal{O}(2)$ comes from the tangent bundle T_l of l. In particular $c_1(X) \cdot l = k$. Now $c_1(X)$ is divisible by $n+1$, hence $k = n+1$. Thus $c_1(L) \cdot l = 1$. The tangent map $T_l \to T_X \mid l$ composed with the contact map $T_X|l \to L|l$ yields a map

$$\mu: T_l \to L \mid l.$$

But T_l has degree 2, whereas $L \mid l$ has degree 1, so $\mu = 0$. We conclude that therefore $T_l \subset F_l$ for all l. Since the tangents of the lines in X_y generate the tangent space of X_y at every point, it follows that $T_{X_y} \subset F_{X_y}$. Since $\dim X_y = n + 1$, this contradicts the non-integrability of F.

In some sense, birational Mori contractions are the proper generalizations of blow-ups in classification theory, so that the simplified lemma at least gives strong evidence for Lemma 3.1.

Next we discuss the **proof of Lemma 3.2**. We claim that the general fiber X_y must be $\mathbb{P}_n$. First notice that the contact structure defines an isomorphism $F^* \otimes L \to F$ which can be extended to a map (with 1-dimensional kernel)

$$\alpha \colon T_X^* \otimes L \to T_X \,.$$

We then consider the composition

$$\beta \colon \phi^* T_Y^* \otimes L \mid X_y \to T_X^* \otimes L \to T_X \mid X_y \,,$$

where the first arrow is given by the differential of ϕ. Observe that clearly $\beta \neq 0$ and that $\phi^* T_Y^* \otimes L \mid X_y = L^{\oplus m} \mid Y_y$, which is ample (having in mind that ϕ is a Mori contraction so that $-K_X$ is positive on the fibers of ϕ, hence L is positive on the fibers of L). Composing β with $T_{X_y} \to N_{X_y/X}$ (where N denotes the normal bundle) we obtain a map $\gamma \colon L^{\oplus m} \mid X_y \to N_{X_y}$. Since N_{X_y} is trivial, $\gamma = 0$. Hence we obtain an injective map $L_{X_y} \to T_{X_y}$. But a remarkable theorem of Wahl [20] says that then X_y must be projective space. By divisibility reasons, $X_y \simeq \mathbb{P}_n$. The theorem of Wahl, in a slightly weaker version due to Mori and Sumihiro [16], says that a projective manifold admitting a vector field which vanishes on an ample divisor, must be projective space.

The **proof of Lemma 3.3** is again rather technical. The main point is to prove that all fibers of ϕ have the same dimension n, then one can apply a result of Fujita to see that ϕ is a $\mathbb{P}_n$-bundle, then again it is not so difficult to conclude $X = \mathbb{P}(T_Y)$. We describe one important proposition (= (2.9) in [10]). We consider the space $\mathrm{Hom}(\mathbb{P}_1, X))$ of holomorphic maps $f \colon \mathbb{P}_1 \to X$. Geometrically any (non-constant) map f determines an irreducible rational curve in X. Now consider a component V of $\mathrm{Hom}(\mathbb{P}_1, X)$. We say that V is unsplit if the curves from V cannot be deformed into a sum of rational curves (possibly with multiplicities). Now if V is unsplit and if $\deg f^*(L) = 1$, for one (hence for all) $f \in V$, then the curves from V fill up X (up to closure) and the curves from V passing through a fixed point form a subvariety of dimension n in X_{2n+1}. This is based on a careful analysis of the differential of the map

$$F \colon \mathbb{P}_1 \times \mathrm{Hom}(\mathbb{P}_1, X) \to X, \; F(x, f) = f(x) \,.$$

By Mori theory V always exist and so one can apply these considerations: they yield immediately that φ cannot be birational (Lemma 3.1); equidimensionality requires further considerations.

4. The "Nef" Case

Demailly's contribution to the classification of contact Kähler manifolds (Proposition 2.7) is a special case of his more general

Theorem 4.1. *Let X be a compact Kähler manifold carrying a pseudo-effective line bundle L. Let $\eta \in H^0(X, \Omega_X^p \otimes L^*)$ be a non-zero L^*-valued holomorphic p-form for some $1 \leq p \leq \dim X$. Let $S \subset T_X$ be the coherent subsheaf of vector fields v such that the contraction $i_v(\eta)$ vanishes. Then S is integrable.*

The proof uses of course the theory of currents together with some ("singular") integration by parts. Specializing to $p = 1$, S defines a meromorphic foliation of codimension 1, i.e. $\eta \wedge d\eta = 0$. In the contact situation this is applied to $S = F$ to show that L^* and hence K_X cannot be pseudo-effective.

The partial results in [10] to exclude projective contact manifolds with K_X nef lead to some interesting questions on nef subsheaves in Ω_X^1. In principle one is interested in the problem of how positive a subsheaf in Ω_X^1 can be. We shall restrict ourselves to rank 1 subsheaves $\mathcal{E}$. Then Bogomolov has shown that $\kappa(\mathcal{E}) \leq \dim X - 1$. We will now make the following assumptions.

(*) X_n is a projective manifold, $\mathcal{E} \subset \Omega_X^1$ locally free of rank 1, $\mathcal{E}$ is nef and there exists a positive rational number α such that $\alpha \mathcal{E} = K_X$ (as $\mathbb{Q}$-divisors).

Proposition 4.2. *Assuming (*), we have $c_1(X)^2 = K_X^2 = 0$.*

Proof. We choose general hyperplane sections $H_1, \ldots, H_{n-1}$ to obtain a smooth surface $S = H_1 \cap \ldots \cap H_{n-1}$. Then consider the restricted sequence

$$0 \to \mathcal{E}_S \to \Omega_X^1 \mid S \to Q_S \to 0 \, .$$

Via the map $\Omega_X^1 \mid S \to \Omega_S^1$, we obtain a map $\varphi \colon \mathcal{E} \mid S \to \Omega_S^1$. This map is non-zero, hence injective; in fact, otherwise we would have a map $\mathcal{E} \mid S \to N_S^*$ which has to vanish since $\mathcal{E} \mid S$ is nef and the normal bundle N_S is a direct sum of ample line bundles (actually one can see that the general choice of S already enforces $\varphi \neq 0$). Now the already mentioned theorem of Bogomolov yields $\kappa(\mathcal{E} \mid S) \leq 1$. On the other hand, the nefness of $\mathcal{E} \mid S$ implies that $c_1(\mathcal{E} \mid S) \geq 0$. If however $c_1(\mathcal{E} \mid S)^2 > 0$, then the Riemann-Roch theorem gives $\kappa(\mathcal{E} \mid S) = 2$; so we must have $c_1(\mathcal{E} \mid S)^2 = 0$. Since $\mathcal{E} \mid S$ is proportional to $K_X \mid S$, we obtain

$$K_X^2 \cdot H_1 \cdot \ldots \cdot H_{n-2} = 0$$

for any choice of ample line bundles H_i. Now some standard considerations (see [10]) show that actually $K_X^2 = 0$.

As a consequence, it is easily shown that if $\kappa(X) \geq 0$, then either $K_X \equiv 0$ or $\alpha \geq 1$. The abundance conjecture predicts that $\kappa(X) \geq 0$ should always hold since K_X is nef. Assuming $\kappa(X) \geq 0$, we are reduced to study two cases: $K_X \equiv 0$ and $\alpha \geq 1$. If $K_X \not\equiv 0$, then $K_X^2 = 0$ predicts that we should have $\kappa(X) = 1$. If this is really true, then we have

Theorem 4.3. *Suppose (*) and $\kappa(X) = 1$. Then K_X is semi-ample, i.e. some mK_X is spanned. Let $f\colon X \to C$ be the Iitaka fibration and let B denote the divisor part of the zeroes of $f^*(\Omega^1_C) \to \Omega^1_X$. Then there exists an effective divisor D such that*

$$\mathcal{E} = f^*(\Omega^1_C) \otimes \mathcal{O}_X(B - D).$$

Turning to the case that $\kappa(X) = 0$, one of course expects that $K_X \equiv 0$, and therefore the decomposition theorem [2] says that there exists a finite étale cover $f\colon \tilde{X} \to X$ such that $\tilde{X} = A \times Y$ with A abelian and Y simply connected and that $f^*(\mathcal{E}) = \mathcal{O}_{\tilde{X}}$.

There is the following well-known *"conjecture K"* of Ueno

Conjecture 4.4. *Let X be a projective manifold with $\kappa(X) = 0$. Then the Albanese map is birational to an étale fiber bundle over its Albanese torus which is trivialised by an étale base change.*

Then we have [11]

Theorem 4.5. *In the situation of (*) suppose that $\kappa(X) = 0$. If Conjecture 4.4 holds, then $K_X \equiv 0$, and therefore there exists a finite étale cover $f\colon \tilde{X} \to X$ such that $\tilde{X} = A \times Y$ with A abelian and Y simply connected and that $f^*(\mathcal{E}) = \mathcal{O}_{\tilde{X}}$.*

Since Conjecture K is known to be true for $q(X) \geq \dim X - 2$, we conclude that if in (*) we have $\kappa(X) = 0$ and if $\dim X \leq 4$, then $K_X \equiv 0$.

As a conclusion, nef rank 1 subsheaves $\mathcal{E} \subset \Omega^1_X$ which are proportional to K_X will give some very precise geometric information on X. If we introduce

$$\mathcal{F} = (\Omega^1_X/\mathcal{E})^* \subset T_X,$$

then in case $\kappa(X) = 1$, the leaves of $\mathcal{F}$ are just the fibers of the Iitaka fibration $f\colon X \to C$, and in case $\kappa(X) = 0$, possibly after finite étale cover, $\mathcal{F} = pr^*(T_Y,)$ in the notation of Theorem 4.5, i.e. the leaves of $\mathcal{F}$ are the fibers of the projection $X \to A$.

References

[1] W. Barth, C. Peters and A. Van de Ven, *Compact complex surfaces*, Springer-Verlag 1984.

[2] A. Beauville, *Variétés dont la première classe de Chern est nulle*, J. Diff. Geom., **18** (1983), 755–782.

[3] A. Beauville, *Fano contact manifolds and nilpotent orbits*, Comm. Math. Helv., **73** (1998), 566–583.

[4] A. Beauville, *Riemannian holonomy and Algebraic Geometry*, Duke/alg-geom Preprint 9902110 (1999).

[5] A. Beauville, *Complex manifolds with split tangent bundle*, in: *Complex analysis and algebraic geometry*, volume in honour of M. Schneider, ed. T. Peternell, F. O. Schreyer, 61–70. Birkhäuser, 2000.

[6] J. P. Demailly, *Frobenius integrability of certain holomorphic p-forms*, Preprint (2000).

[7] S. Druel, *Contact structures on 5-dimensional manifolds*, C. R. Acad. Sci. Paris, **327** (1998), 365–368.

[8] Y. Kawamata, *Abundance theorem for minimal threefolds*, Inv. Math., **108** (1992), 229–246.

[9] Y. Kawamata, K. Matsuda and K. Matsuki, *Introduction to the minimal model problem*, Adv. Stud. Pure Mat., **10** (1987), 283–360.

[10] S. Kebekus, T. Peternell, A. J. Sommese and J. Wisniewski, *Projective contact manifolds*, Inv. Math. **142** (2000), 1–15.

[11] S. Kebekus, T. Peternell, A. J. Sommese and J. Wisniewski, *Nef subsheaves in the cotangent bundle*, Preprint (2000).

[12] J. LeBrun, *Fano manifolds, contact structures and quaternionic geometry*, Int. Journ. Math., **6** (1995), 419–437.

[13] Y. Miyaoka, *On the Kodaira dimension of minimal threefolds*, Math. Ann., **281** (1988), 325–332.

[14] Y. Miyaoka, *Abundance conjecture for 3-folds – case $\nu = 1$*, Comp. Math., **68** (1988), 203–220.

[15] Y. Miyaoka and T. Peternell, *Geometry of higher dimensional algebraic varieties*, DMV Seminar, vol. 26. Birkhäuser, 1997.

[16] S. Mori and Y. Sumihiro, *On Hartshorne's conjecture*, J. Math. Kyoto Univ., **19** (1978), 523–533.

[17] C. Okonek, M. Schneider and H. Spindler, *Vector bundles on complex projective spaces*; Birkhäuser, 1980.

[18] T. Peternell, *Towards a Mori theory on compact Kähler threefolds, III*, to appear in Bull. Soc. Math. France.

[19] S. M. Salamon, *Quaternionic-Kähler manifolds*, Inv. Math., **67** (1982), 143–171.

[20] J. Wahl, *A cohomological characterisation of $\mathbb{P}_n$*, Inv. Math., **72** (1983), 315–322.

[21] Y.-G. Ye., *A note on complex projective threefolds admitting holomorphic contact structures*, Inv. Math., **115** (1994), 311–314.

Added in proof: Recently S. Kebekus has proved (Uniqueness of complex contact structures, to appear in J. f. d. reine u. angew. Mathematik; math. AG 0004103) that a Fano manifold X with $b_2(X) = 1$ has at most one contact structure unless X is projective space.

Mathematisches Institut
Universität Bayreuth
D-95440 Bayreuth, Germany
E-mail address: thomas.peternell@uni-bayreuth.de

Analytic Topology

Alexander Reznikov

Abstract. This is an extended written version of an address to the European Congress of Mathematics in Barcelona. It reviews my work of the last 10 years in Analytic Topology and contains several new theorems.

1. Area Estimates and Geometry of Three-Manifolds

Volume estimates for maps and sections of flat bundles bear topological significance because they provide bounds for topological invariants such as secondary characteristic classes. In [33] it is proved that any map of a closed oriented surface Σ^g of genus $g \geq 2$ to a complete hyperbolic three-manifold can be homotoped to a map of area $\leq 2\pi(2g - 2)$. Thurston used his technique of pleated surfaces. A full generalization to negative curvature has been obtained in [21, Theorem C1]:

Theorem 1.1. *Let N be a complete simply-connected Riemannian manifold with the curvature satisfying $-K \leq K(N) \leq -k < 0$. Let $ISO(N)$ be the isometry group of N. Consider a flat N-bundle $N \to E \to \Sigma^g$ with holonomy in $ISO(N)$. Assume that the action of $\pi_1(\Sigma^g)$ on the sphere at infinity of N is fixed-point free. Then there exists a section s such that*

$$\text{Area} \ (s) \leq \frac{4\pi(g - 1)}{k} \ .$$

It is shown in [21] how to deduce from this theorem a well-known result of Goldman, identifying a Teichmüller space T_{6g-6} with a component of the character variety

$$\text{Hom} \ (\pi_1(\Sigma^g), PSL_2(\mathbb{R}))/PSL_2(\mathbb{R}) \ ,$$

corresponding to the top Euler number [12]. A different proof of Goldman's theorem had been earlier established in [15] and [32].

The theorem above has been established by "hard" techniques (harmonic sections). There are at least two "soft" techniques designed for similar purposes. The first is a twisted version of Thurston's straightening procedure, see, for example, [21, Lemma E12]. The second, which works only for sections of flat bundles over a base of dimension ≥ 3, is an explicit formula of Patterson type, given in [2], see a survey of [9]. Both techniques provide bounds for secondary characteristic classes.

We now turn to the *lower* bounds for the area in terms of the genus. The following result has been proven in [22, Theorem 1].

Theorem 1.2. *Let N be a compact oriented three-dimensional homologically ato-roidal Riemannian manifold, let $R(x)$ be the scalar curvature function and let*

$$0 < R(N) = \sup_N(-R(x)).$$

Then for any $z \in H_2(N, \mathbb{Z})/Jm\pi_2(N)$ one has

$$||z||_a \geq \frac{2\pi}{R(N)}||z||_g^i,$$

where $||z||_a$ is the least area of a cycle representing z and $||z||_g^i$ is the least absolute value of an Euler characteristic of a singular surface, representing z.

Corollary 1.3. *([22, Corollary 2]) Let N be a compact three-manifold satisfying*

$$-K \leq K(N) \leq -k < 0.$$

Then for any $z \in H_2(N, \mathbb{Z})$,

$$\frac{1}{k}||z||_g^i \geq \frac{1}{2\pi}||z||_a \geq \frac{1}{3K}||z||_g^i. \qquad (*)$$

Remark 1.4. *Let $||z||_g^e$ be the minimal absolute value of an Euler characteristic of an* embedded surface *representing z. By GMT there is a stable minimal surface $\Sigma^{g'}$, $g' \geq g$, representing z. The proof of [22, Theorem 1] applies to give*

$$||z||_a \geq \frac{2\pi}{R(N)}(2g' - 2) \geq \frac{2\pi}{R(N)}||z||_g^e$$

so that we can improve (*) as follows:

$$\frac{1}{k}||z||_g^i \geq \frac{1}{2\pi}||z||_a \geq \frac{1}{3K}||z||_g^e. \qquad (*')$$

Choosing a hyperbolic metric on N, which exists since N is irreducible and atoroidal and $H_2(N, \mathbb{Z}) \neq 0$ [34] we find

$$||z||_g^i \geq \frac{1}{2\pi}||z||_a \geq \frac{1}{3}||z||_g^e,$$

which of course implies

$$||z||_g^e \leq 3||z||_g^i.$$

This is, with slightly weaker constant, a well-known result of Gabai, see [7] or a survey in [8]. Gabai's approach is purely topological.

We will now describe an application to conformal geometry of three-mani-folds. Let N be a compact three-manifold and C a conformal class of metrics on N. Assume that N is irreducible, atoroidal and $b_1(N, \mathbb{Q}) > 0$ so that N admits a hyperbolic metric. Define $\text{Vol}(C)$ to be a volume of the Yamabe metric in C (a metric in C with scalar curvature -1). Then for any $w \in H^1(N, \mathbb{Z})$, the following inequality holds [22, Proposition 5]:

$$\text{Vol}^{2/3}(C) \cdot ||w||_{L^3} \geq 4\pi,$$

where $||w||_{L^3}$ is computed with respect to any metric in C.

2. Quadratic Equations in Groups

Let G be a finitely generated group. A solution of a homogeneous quadratic equation in G is simply a homotopy class of maps of a closed surface Σ^g to $K(G,1)$. If Σ^g is oriented, we say that a solution is free if the map factors through a handlebody.

Thurston [33] proved his famous result that in the fundamental group of a hyperbolic three-manifold satisfying $\operatorname{inj}(x) \to \infty$ as x escapes all compact sets, there are but finitely many copies of a given surface group. This was extended to any f.g. group, which is not a free product, in a given word-hyperbolic group [13]. The case of a surface group is more interesting however, because the statement can be much strengthened [21, 23]:

Theorem 2.1. *Let M be a manifold of pinched negative curvature such that $\operatorname{inj}(x) \to \infty$ as x escapes all compact sets (e.g. compact). Let Σ be a closed oriented surface. There are, up to conjugacy in $\pi_1(M)$ and an automorphism of $\pi_1(\Sigma)$, only finitely many homomorphisms $f \colon \pi_1(\Sigma) \to \pi_1(M)$, such that $f([\gamma]) \neq 1$ for any simple closed loop γ in S.*

This result immediately yields a classification of solutions of any homogeneous quadratic equation in $\pi_1(M)$ by genus reduction. A similar theorem holds for minimal genus solution of inhomogeneous equations [23]. This classification had been earlier obtained by Lysionok [17, 18] using the methods of combinatorial group theory.

In case of hyperbolic three-manifolds M^3 the theorem above is uniform on the class of manifolds with $\operatorname{inj}(M) > \varepsilon > 0$ (ε fixed).

Conjecture 2.2. *Consider homomorphisms $f \colon \pi_1(\Sigma) \to \pi_1(M)$ where M runs through hyperbolic three-manifolds with $\operatorname{inj}(M) > \varepsilon > 0$ and $\operatorname{inj}(x) \to \infty$ as x escapes all compact sets, and $f([\gamma]) \neq 1$ for any simple closed loop γ. Then a composite representation*

$$\pi_1(\Sigma) \to \pi_1(M) \to PSL_2(\mathbb{C}) \tag{*}$$

lies in a compact set $V(\varepsilon)$ of the representation variety

$$\operatorname{Hom}(\pi_1(S), PSL_2(\mathbb{C}))/PSL_2(\mathbb{C}),$$

after a conjugation by an element of $\operatorname{Aut}(\pi_1(\Sigma))$.

Sketch of the proof. We can represent f by an immersed minimal surface $F \colon \Sigma \to M$. Then the second quadratic form of Σ is a holomorphic quadratic differential $\varphi \in H^0(K^2)$ on Σ. The Gauss-Bonnet formula gives

$$\operatorname{area}(\Sigma) + \int_{\Sigma} |\varphi|^2 = -2\pi\chi(\Sigma),$$

so that

$$\text{area}\,(\Sigma) \leq -2\pi\chi(\Sigma)\,,$$

$$\int_{\Sigma} |\varphi|^2 \leq -2\pi\chi(\Sigma)\,.$$

The first inequality implies [21, 23] that the conformal class of the induced metric on Σ lies in a compact set of the moduli space of Riemann surfaces. Since F is a minimal immersion, the Higgs bundle corresponding to (*) [15] is an extension

$$0 \to K^{-1/2} \to E \to K^{1/2} \to 0\,,$$

with an extension class $\psi \in H^0(K^2) = H^1(\mathrm{Hom}\,(K^{1/2}, K^{-1/2}))$ and a Higgs field $\theta = \left(\begin{smallmatrix} 0 & 1 \\ 0 & 0 \end{smallmatrix}\right)$. It seems possible to prove that, since φ is L^2-bounded, then so is ψ, because φ controls the geometry of the minimal immersion. Now (E, θ) lies in a compact set of a moduli space of Higgs bundles, and the result follows from [15].

For every oriented closed hyperbolic three-manifold M there is a map $F\colon \Sigma^g \to M$ of (possibly non-oriented) surface Σ^g such that $F_*([\gamma]) \neq 1$ for any essential simple two-sided loop γ. One can classify hyperbolic three-manifold manifolds by the condition $g \leq G$, $\mathrm{inj}\,(M) \geq \varepsilon > 0$. The conjecture above seems to imply that the number of closed hyperbolic non-virtually Haken three-manifolds supporting an essentially injective map of a surface, such that these inequalities hold, is finite. $\square$

We now turn to an amazing application of analytic technique to quadratic equations in *finite* groups.

Theorem 2.3. ([23, Theorem 4.3]) *Let G be a free product $*G_i$ where G_i is either $\mathbb{Z}$, or a finite group which acts freely and isometrically on (S^3, can) or $(\mathbb{R}P^3, can)$. Then any solution of an oriented quadratic equation in G is free.*

The reader is invited to provide an algebraic proof that all solutions of an equation

$$[x_1, y_1] \ldots [x_{2000}, y_{2000}] = 1$$

in A_5 are free. The proof of [23] essentially uses minimal surfaces.

Let now X be a compact Kähler manifold, which is Kobayashi hyperbolic.

Lemma 2.4. *The classes of closed Riemann surfaces which admit a holomorphic immersion $F\colon S \to X$ such that $F_*([\gamma]) \neq 1$ in $\pi_1(X)$ for any* simple *closed loop, lie in a compact set of the moduli space.*

Proof. The proof is identical to the argument of [21, 23] once one has noticed that area (S) in the induced metric is bounded.

It follows that if S moves, then there is a projective subvariety Y of the moduli space such that a pull-back of the universal curve, say Z, maps to X (because a deformation of S cannot degenerate). We ignore here the difficulties in defining the universal curve. If S does not move, then we deduce immediately as in [20] that the number of such curves is finite. $\square$

Theorem 2.5. *Let X be a compact Kähler manifold, which is Kobayashi hyperbolic. Consider holomorphic immersions of Riemann surfaces (of a given genus g) $F\colon S \to X$ such that $F_*([\gamma]) \neq 1$ in $\pi_1(X)$ for any simple closed loop. If all such immersions are rigid, then they are finite in number. If they are not all rigid, then there is a fibration by smooth curves over a smooth projective base, which maps to X.*

Indeed, we can resolve the singularities of Y and pull the fibration back.

This theorem generalizes the solution of the Mordell conjecture for function fields to the case of *essentially* π_1-injective holomorphic immersions of *variable* Riemann surfaces.

3. Vanishing and Rationality of Characteristic Classes for CM Fields and Kähler Manifolds

Let M be a compact manifold and let

$$\rho\colon \pi_1(M) \to SL_n(\mathbb{C})$$

be a representation. Then one has a series of classes in $H^{2i-1}(M, \mathbb{C}/\mathbb{Z})$ [6, 5]. We will denote the $\mathbb{R}/\mathbb{Z}$-part $ChS_i(\rho)$ and $i\mathbb{R}$-part $b_i(\rho), i \geq 2$.

Lemma 3.1. (Rationality lemma, special case of [24, Fundamental Lemma]**)** *If ρ is defined over a number field and for all Galois conjugates $\sigma_i \circ \rho$,*

$$b_k(\sigma_i \circ \rho) = 0, \qquad k \geq 2,$$

then

$$ChS_k(\rho) \in H^{2k-1}(M, \mathbb{Q}/\mathbb{Z}).$$

As a corollary, one gets [24, Proof of Theorem D, page 688]:

Theorem 3.2. *Let K be a CM-field and let*

$$\rho\colon \pi_1(M) \to SL_n(K).$$

Then $ChS_k(\rho) \in H^{2k-1}(M, \mathbb{Q}/\mathbb{Z}), k \geq 2$.

This theorem is stated and proved in [24] only for $k = 2$, however, the proof applies informally to all dimensions. The lemma cited above becomes

$$ChS_k(\rho) = \sum_{\substack{i_1+\cdots+i_s=k \\ i_j\, odd}} \sum_i a_{i_1\ldots i_s,i} b_{i_1} \ldots b_{i_s}(\sigma_i \circ \rho) \qquad (\mathrm{mod}\ \mathbb{Z}\frac{1}{M}))$$

with some $a_{i_1\ldots i_s} \in \mathbb{R}$. The proof is the same as [25, page 376]. Then the proof goes as in [24, page 688].

For a Kähler base we have a fundamental vanishing theorem:

Theorem 3.3. (Vanishing Theorem, [24, 4.5] and [25, Theorem 1.4]) *Let M be compact Kähler. Then for any ρ,*

$$b_k(\rho) = 0, \quad k \geq 2.$$

As a corollary we get

Theorem 3.4. (Rationality Theorem, [24, Main Theorem] and [25, Theorem 1.1]) *Let M be compact Kähler. Then for any ρ,*

$$ChS_k(\rho) \in H^{2k-1}(M, \mathbb{Q}/\mathbb{Z}), \quad k \geq 2.$$

As explained in the above-cited papers, this implies the Bloch conjecture [3].

The proof uses a combination of rational algebraic K-theory and a theory of harmonic maps from Kähler manifolds to manifolds of nonpositive curvature.

4. Secondary Classes and Volume Estimates in Symplectic Topology

Given a closed submanifold L^q of a compact Riemannian manifold M^m, can we make its volume small by a smooth deformation? If $[L] \in H_q(M, \mathbb{Q})$ is nonzero, the answer is no, because if ω is a closed q-form such that $[w]([L]) \neq 0$, then for any deformation L',

$$0 < |[w]([L])| \leq ||w||_\infty \cdot \mathrm{Vol}\,(L').$$

If the class $[L] \in H_q(M, \mathbb{Z})$ is torsion but nonzero, one can appeal to integral geometry, as in the case $M = \mathbb{C}P^{2n}$, $L = \mathbb{R}P^{2n}$. The class

$$[\mathbb{R}P^{2n}] \in H_*(\mathbb{C}P^{2n}, \mathbb{Z}_2)$$

is nonzero, therefore any deformation L' intersects any totally geodesic $\mathbb{R}P^{2n}$ in $\mathbb{C}P^{2n}$, which gives a (sharp) lower bound on the volume of L'. In the context of symplectic topology, one assumes M to be symplectic, L Lagrangian and the deformation to be Lagrangian or even Hamiltonian. In case $M = \mathbb{C}P^{2n+1}$, $L = \mathbb{R}P^{2n+1}$ it was shown in [11] that L' still intersects all totally geodesic $\mathbb{R}P^{2n+1} \subset \mathbb{C}P^{2n+1}$ (though the group $H_{2n+1}(\mathbb{C}P^{2n+1}, \mathbb{Z})$ is 0). So the volume of L' with respect to any fixed metric is bounded away from zero. If the cohomology class of the symplectic form is integral, one can interpret symplectic actions

$$\mu_k \colon H_{2k-1}(L, \mathbb{Z}) \to \mathbb{R}/\mathbb{Z}$$

as Chern-Simons classes ([26, Section 3]). Using the Cheeger-Simons rigidity, one shows the following result.

Theorem 4.1. ([26, Theorem 3.5]) *Let $X \subset \mathbb{P}^n$ be a smooth projective variety defined over $\mathbb{R}$, $\dim X = 2k - 1$, and let $M = X(\mathbb{C})$ and $L = X(\mathbb{R})$. Assume the homomorphism $H_{2k-1}(L) \to H_{2k-1}(\mathbb{R}P^n) = \mathbb{Z}_2$ is nontrivial and $H_{2k-1}(M) = 0$. Then for any metric on M and any Lagrangian homotopy L' of L in M, $\mathrm{Vol}\,(L')$ stays bounded away from zero.*

The main technical point in the proof is the existence of isoperimetrical films [14].

5. Cohomology of $\mathrm{Sympl}\,(M)$ and $\mathrm{Diff}^{\,1,\alpha}(S^1)$

The group of symplectomorphisms of a compact symplectic manifold resembles on one hand, the compact Lie groups and on the other hand, very surprisingly, the group of diffeomorphisms of a "rough" circle $\mathrm{Diff}^{\,1,\alpha}(S^1)$. The Lie algebra $\mathrm{Lie}(\mathrm{Sympl}(M))$ (or $\mathrm{Sympl}_{\mathrm{Hamiltonian}}(M)$ if $H_1(M,\mathbb{R})\neq 0$) is simply $C^\infty(M)/\mathrm{const}$ and admits invariant polynomials and bi-invariant closed forms, introduced in [26]:

$$f \to \int_M f^k \qquad \text{where} \qquad \int f = 0\,,$$

and

$$f_1,\ldots,f_{2k-1} \to \mathrm{Alt}\sum \int \{f_1,f_2\}f_3\ldots f_{2k-1}\,. \tag{*}$$

One therefore defines classes in $H^{2k-1}(\mathrm{Sympl}^{\,\mathrm{top}}(M),\mathbb{R})$ and $H^{2k}(B\,\mathrm{Sympl}\,(M),\mathbb{R})$, modelled by Chern-Weil theory. The situation with secondary classes is more interesting. Let A be a group of periods of the forms $(*)$ on the Hurewitz image of $\pi_{2k-1}(\mathrm{Sympl}^{\,\mathrm{top}}(M))$ in $H_{2k-1}(\mathrm{Sympl}^{\,\mathrm{top}}(M),\mathbb{Z})$. Then one defines a class ([26, Section 2])

$$K^{\mathrm{alg}}_{2k-1}(\mathrm{Sympl}\,(M)) = \pi_{2k-1}(B\,\mathrm{Sympl}^{\,\delta}(M))^+ \to \mathbb{R}/A\,,$$

however to lift it to $H^{2k-1}(\mathrm{Sympl}^{\,\delta}(M))$, one needs information on the topology of $\mathrm{Sympl}\,(M)$, available only in low dimensions (like $M = (\mathbb{C}P^2,can)$). The classes just described are symplectic analogs of Chern-Simons classes. The situation with $\mathbb{R}$-valued classes is much more satisfactory: there is a series of classes in $H^{4k+2}(\mathrm{Sympl}^{\,\delta}(M),\mathbb{R})$ constructed in [27] using the action on twistor varieties. Some groups of central importance, like the mapping class groups Map_g admit a symplectic action

$$\mathrm{Map}_g \to \mathrm{Sympl}\,(\mathcal{M}_g)\,,$$

where $\mathcal{M}_g = \mathrm{Hom}^-(\pi_1(\Sigma^g),SO_3)/SO_3$ (the representations with Stiefel-Whitney class 1 to make $\mathcal{M}_g$ smooth). The class in $H^2(\mathrm{Sympl}\,(\mathcal{M}_g),\mathbb{R})$ detects the generator of $H^2(\mathrm{Map}_g,\mathbb{R})$ ([27, Theorem 3.6]).

A series of characteristic classes in $H^{4k+2}(\mathrm{Diff}^{\,1,\alpha}(S^1))$ is defined in [29, Chapter II]. The definition uses the action on an infinite-dimensional Siegel half-plane.

6. Secondary Classes, Domination and Representation Varieties in 3-Dimensional Topology

If M^3 is a closed oriented three-manifold, then it can dominate at most finitely many closed hyperbolic oriented three-manifolds N. The domination means that there is a map of nonzero degree. This is a recent result of [31]. During a conversation with the author (Toulouse, July 1997) Joan Porti noticed that this theorem follows immediately from the argument of [24, 28]. In fact, for a given M, the volume of a representation

$$\pi\colon \pi_1(M) \to PSL_2(\mathbb{C})$$

can assume only finitely many values by the rigidity of volume and a remark of Lusztig on representation varieties [24, Sections 5.16.1 and 5.17]. Now, if $f\colon M \to N$ has nonzero degree and

$$\rho\colon \pi_1(N) \to PSL_2(\mathbb{C})$$

is the uniformization representation, then by naturality of volume [24, Proof of Proposition 5.9] we have

$$\mathrm{Vol}\,(\rho \circ f_*) = \deg f \cdot \mathrm{Vol}\,(N)\,,$$

where $f_*\colon \pi_1(M) \to \pi_1(N)$ is induced by f. So we have finitely many possibilities for $\deg f \cdot \mathrm{Vol}\,(N)$, therefore for $\mathrm{Vol}\,(N)$ (since $\deg f$ is bounded by Gromov-Thurston, or, alternatively, since $\mathrm{Vol}\,(N)$ is bounded below by Margulis), and finally for N (by Jørgensen).

This argument was originally used in [24, Proposition 5.17] to prove that there are hyperbolic manifolds M such that the representation variety

$$V_{PSL_2(\mathbb{C})}^{\pi_1(M)} = \mathrm{Hom}\,(\pi_1(M), PSL_2(\mathbb{C}))/PSL_2(\mathbb{C})$$

may have arbitrarily many components.

7. How to Geometrize Group Theory?

A finitely-generated group G can be made geometric in two ways:
  A. **Time-geometry.** This means that G acts on a compact space with some additional structure:
   1) circle
   2) tree/building
   3) compact symplectic manifold
   4) compact manifold with volume form
      (a tree is, formally speaking, non-compact).
  B. **Space-geometry.** This means that G is a fundamental group of a compact manifold with additional structure:
   1) negative curvature
   2) Kähler metric
   3) quaternionic Kähler metric.

A principle of fundamental importance is that groups with time-geometry tend to be non-Kazhdan, whereas groups with space geometry of types 2), 3) tend to be Kazhdan (Space geometry of type 1) is indifferent to the Kazhdan property):

Theorem 7.1. ([30, Main Theorem]) *Let X be a compact Kähler manifold, let $G = \pi_1(X)$. Assume that G is not Kazhdan. Then $H^2(G, \mathbb{Q}) \neq 0$. If $G \xrightarrow{f} K \to 1$ is a central extension of a non-Kazhdan group, then $f^* \colon H^2(K, \mathbb{Q}) \to H^2(G, \mathbb{Q})$ is nonzero.*

In addition to this one shows [29, Chapter V] that there is a nonzero polynomially bounded class in $H^2(G, \mathbb{Q})$. This theorem essentially solves a conjecture of Carlson-Toledo [16].

Theorem 7.2. ([29, Theorem VI. 2.2]) *Let X be a compact quaternionic Kähler manifold of negative scalar curvature. Let $G = \pi_1(X)$. Then G is Kazhdan. Of course, if the scalar curvature is nonnegative, G is virtually abelian.*

On the contrary, groups acting nontrivially on a tree are not Kazhdan by a result of [1] and [35]. For other time geometries, one has the following:

Theorem 7.3. ([29, Theorem II. 1.7]) *Let G be a subgroup of $\mathrm{Diff}^{\,1,\alpha}(S^1)$, $\alpha > 1/2$. Suppose that either*
A. *The representation of G in half-densities on S^1 is irreducible or a finite direct sum of irreducibles, or*
B, C. *G contains a conjugate in $\mathrm{Diff}^{\,1,\alpha}(S^1)$ of a nonelliptic fractional linear transformation, or*
D.

$$\sup_{g \in G} \iint_{S^1} \left[\frac{\sqrt{g'(\varphi)g'(\eta)}(\varphi - \eta) - (g(\varphi) - g(\eta))}{(g(\varphi) - g(\eta))(\varphi - \eta)} \right]^2 d\varphi \, d\eta = \infty \,.$$

Then G is not Kazhdan.

During my lecture course on Analytic Topology in Göttingen in June, 2000, I noticed that a small variation of the method used in [29] gives a refined version of this theorem, namely, that any action of a Kazhdan group on S^1 in the same smoothness class factors through a finite group. In fact, such action commutes with a Hilbert-Schmidt deformation of the unbounded version of the Hilbert transform, introduced in [29] and called there R. The spectrum of R is discrete and the eigenspaces are finite-dimensional, and the same will be true for the deformation. We have therefore an invariant finite-dimensional space of half-densities on S^1. Choosing an orthonormal basis and taking the sum of squares of its elements, we arrive to an invariant density. Reparametrization of S^1 by this density gives a factorization through a Kazhdan abelian group, hence finite.

This general theorem was preceded by special results concerning the nonexistence of actions on S^1 of lattices in semi-simple Lie groups of rank ≥ 2 (see references in the cited paper). It also solves a conjecture of Ghys-Sergiesku [10].

The method of the proof also yields an important corollary. Let G be a surface group and let

$$\rho \colon G \to \mathrm{Diff}\,(S^1)$$

be a representation. We will say that ρ is generic, if there are four elements $g_i \in G$, such that there is a finite number of attracting fixed points $x_{i,j}^+$ and repelling fixed points $x_{i,j}^-$ of $\rho(g_i)$ with the property that for arbitrary neighbourhoods U_i^+ and U_i^- of $\{x_{i,j}^+\}$ and $\{x_{i,j}^-\}$, $\rho^N(g_i)(S^1 \setminus U_i^-) \subseteq U_i^+$ for $N \gg 1$. Recall that there exists a measurable equivariant map

$$S^1 \to \text{probability measures on } S^1 \,, \tag{$*$}$$

since S^1 is the stochastic boundary of G. The following result is a quantization of it:

Theorem 7.4. ([29, Theorem II 2.1]) *Let ρ be generic. Let $\mathcal{H}$ be the Hilbert space of half-densities on S^1. Then there exists a measurable equivariant map*

$$S^1 \xrightarrow{\psi} \mathcal{B}(\mathcal{H}) \,, \tag{$**$}$$

such that for $\theta \in S^1$, $(\psi(\theta) - H)$ is a Hilbert-Schmidt operator, where H is the Hilbert transform.

Problem Relate to $(*)$ to $(**)$.

Theorem 7.5. ([29, Theorem IV. 3.2]) *Let M be a compact symplectic manifold. Then there is a cocycle $l(\phi) \in Z^1(\text{Sympl}\,(M), L^2(M))$ with the following properties:*
  A. *$l(\phi)$ is canonically defined by a choice of deformation quantization*

$$f * g = f \cdot g + \{f, g\}h + c_2(f, g)h^2 + \dots \,;$$

  B. *$l(\phi)$ depends only on a second jet of ϕ;*
  C. *For a flat torus T^2 and $\phi = (\phi_1, \phi_2)$;*

$$l(\phi) = \frac{\partial^2 \phi_1}{\partial x_2^2} \frac{\partial^2 \phi_2}{\partial x_1^2} - \frac{\partial^2 \phi_1}{\partial x_1^2} \frac{\partial^2 \phi_2}{\partial x_2^2} \,;$$

  D. *If G is a subgroup of $\text{Sympl}\,(M)$ and $\sup_{g \in G} \|l(g)\| = \infty$, then G is not Kazhdan.*

This is a definition of l: notice that $c_2(f \circ \phi, g \circ \phi) \circ \phi^{-1} - c_2(f, g)$ is a Hochschild 2-cocycle therefore a section of $\Lambda^2 TM$. Identify it with a 2-form and multiply by a power of symplectic structure to get a function. In a forthcoming paper, we use Theorem 7.5 to construct a unitary representation of $\text{Sympl}\,(M)$ in the Fok space, modelled on $L^2(M)$, which we call mathematical quantization.

The results for volume-preserving actions are less definite (see Chapter V of the cited paper).

8. Cocycle Growth

A fine classification of non-Kazhdan groups may be based on the growth rate for unitary cocycles. A foundational result based on a deep work of Makarov on Bloch functions [19] is given below.

Theorem 8.1. [29, Theorem III. 3.1] *Let G be a surface group. Let $\rho\colon G \to U(\mathcal{H})$ be a unitary representation without almost fixed vectors and let $l \in Z^1(G, \mathcal{H})$. Then*

$$\|l(g)\| \leq c(\theta)\sqrt{\text{length }(g)\log\log\text{length }(g)}$$

for almost all $\theta \in S^1 = \partial G$ and $g \to \theta$ nontangentially.

This result extends immediately to cocompact complex hyperbolic lattices. There is a better estimate for primitives in l^p-cohomology:

Theorem 8.2. [29, Theorem I.5.1] *Let G be a surface group. Let $\mathcal{F}\colon G \to \mathbb{R}$ be such that $L_g\mathcal{F} - \mathcal{F} \in l^p(G)$ for all $g \in G$. Then $|\mathcal{F}(g)| < c[\text{length }(g)]^{1/p'}$, where $1/p + 1/p' = 1$.*

9. Analytic Topology of Negative Curvature

A problem of fundamental importance in topology is the following. Let $M^m \overset{\varphi}{\hookrightarrow} N^n$ be a π_1-injective immersion of compact manifolds of negative curvature. Lift it to universal covers: $\widetilde{\varphi}\colon \widetilde{M} \to \widetilde{N}$. Does there exist a boundary map $\partial\widetilde{\varphi}\colon S^{m-1} = \partial\widetilde{M} \to \partial\widetilde{N} = S^{n-1}$ and if there does, how "good" is it? For example, if N^3 is a fibration over a circle with pseudo-Anosov monodromy and M^2 is a fiber, then $\partial\widetilde{\varphi}\colon S^1 \to S^2$ is a continuous *surjective* map with amazing properties [4]. It is easy to show [29, Remark I.8.3] that if $\varphi_*\colon \pi_1(M) \to \pi_1(N)$ satisfies

$$\text{length }\varphi_*(g) \geq \text{const length }(g),$$

then $\partial\widetilde{\varphi}$ exists as a Hölder continuous map, that is, for some $\alpha > 0$,

$$\rho(\partial\widetilde{\varphi}x, \partial\widetilde{\varphi}y) \leq \text{const } \rho(x, y)^\alpha.$$

Without this assumption, we have the following general theorem.

Theorem 9.1. ([29, Theorem I.6.1]) *Let $p_0 \in \widetilde{M}$, $q_0 \in \widetilde{N}$ and $\pi\colon \widetilde{N}\setminus\{q_0\} \to S^{n-1}(T_{q_0}\widetilde{N})$ be the radial projection. Then:*

1. *For almost every $v \in S^{m-1}(T_{p_0}\widetilde{M})$, the restriction of $\pi \circ \widetilde{\varphi}$ on the geodesic $\gamma(p_0, v)$ starting at p_0 with velocity v, has an L^1-derivative as a map from $\mathbb{R}_+$ to $\mathbb{R}^n$; therefore $\lim_{t\to\infty}\pi \circ \widetilde{\varphi}(\gamma(p_0, v, t))$ exists a.e. on S^{m-1}. The resulting measurable map $\partial\widetilde{\varphi}\colon S^{m-1} \to S^{n-1}$ does not depend on the choice of p_0, q_0.*
2. *For $p \gg 1$, $\partial\widetilde{\varphi}$ induces a linear operator*

$$C^\infty(S^{n-1}) \to W_p^{(m-1)/p}(S^{m-1}),$$

   if $K(M) = -1$.
3. *If $K(M) = K(N) = -1$, then $\partial\widetilde{\varphi}$ induces a bounded operator*

$$W_p^{(n-1)/p}(S^{n-1})/\text{const} \to W_p^{(m-1)/p}(S^{m-1})/\text{const}.$$

   for all $p > (n-1)$.
   Here $W_p^s(S^m)$ is the Sobolev-Slobodecki space.

In particular, the coordinate functions of the Cannon-Thurston curve are in the class $W_p^{1/p}(S^1)$ for all $p > 2$. Morover:

Theorem 9.2. ([29, Theorem I.9.3]) *Let* $\varphi \in \mathrm{Map}_{g,1}(\Sigma^g)$ *be a pseudo-Anosov automorphism of the surface group* $G = \pi_1(\Sigma^g)$. *Let*

$$A_\varphi \in \mathrm{Aut}\,(W_p^{1/p}(S^1)/\,\mathrm{const}\,) \qquad (p > 2)$$

be the corresponding automorphism of the Banach space

$$W_p^{1/p}(S^1)/\,\mathrm{const} \simeq H^1(G, l^p(G))\,.$$

Then for any $v \in W_p^{1/p}(S^1)/\,\mathrm{const}$ *in the image of the natural map*

$$W_p^{2/p}(S^2)/\,\mathrm{const} \to W_p^{1/p}(S^1)/\,\mathrm{const}\,,$$

induced by the Cannon-Thurston curve, one has

$$\sum_{n \in \mathbb{Z}} \|A_\varphi^n v\|^p < \infty\,.$$

The last inequality is wildly exotic: I don't know of any "explicit" invertible operator in any Banach space which possesses such vectors $v \neq 0$.

10. Quantization of the Mapping Class Group

For a discrete group G one defines a bicohomology space ([29, I.10]) as

$$\mathcal{H}_p(G) = H^1(G_l, H^1(G_r, l^p(G)))\,,$$

where l and r stand for the left and right action, respectively. The outer automorphism group $\mathrm{Out}\,(G)$ acts on $\mathcal{H}_p(G)$. For an "average" group G, $\mathcal{H}_p(G) = 0$. However, for a surface group one has a very rich theory:

Theorem 10.1. ([29, Theorem I.11.2]) *Let* G *be a surface group of genus* g. *Then for* $\mathcal{H}_{p,g} = \mathcal{H}_p(G)$ *one has:*

1. *For* $p > 1$, $\mathcal{H}_{p,g}$ *is a nontrivial Banach space.*
2. *For* $p = 2$, $\mathcal{H}_{2,g}$ *is an infinite dimensional Hilbert space.*
3. *There exists a nondegenerate pairing*

$$\mathcal{H}_{p,g} \otimes \mathcal{H}_{p',g} \to \mathbb{R}\,,$$

   which for $p = 2$ *defines a nondegenerate symmetric scalar product in* $\mathcal{H}_{2,g}$ *of signature* (m, n), $m + n = \infty$, $n > 0$.
4. Map_g *acts in* $\mathcal{H}_{p,g}$ *by invertible operators and the pairing above is* Map_g-*equivariant.*

Theorem 10.2. ([29, Theorem I.11.12]) *Realize* G *as a lattice in* $PSL_2(\mathbb{R})$. *Let* $X = \mathcal{H}^2 \times \mathcal{H}^2/G$, $Y = \mathcal{H}^2 \times \overline{\mathcal{H}^2}/G$, *where* $\overline{\mathcal{H}^2}$ *means the hyperbolic plane with the opposite orientation. Let* $\mathcal{H}_+$ *be a space of* L^2 *holomorphic 2-forms on* X *(respectively,* Y*). Then*

$$\mathcal{H}_{2,g} = \mathcal{H}_+ \oplus \mathcal{H}_-\,,$$

and the restriction of the scalar product of $\mathcal{H}_{2,g}$ on $\mathcal{H}_+$ (respectively $\mathcal{H}_-$) is positive (respectively, negative).

Identifying $H^1(G_r, l^p(G))$ with $W_p^{1/p}(S^1)/\text{const}$, one defines a vacuum vector by an explicit cocycle

$$\phi \to \text{Arg}\ (\phi^{-1}\beta) - \text{Arg}\ \beta \quad (\text{mod const})\,,$$

where $\phi \in G$, $\beta \in S^1$. The vacuum vector is a nontrivial vector in $\mathcal{H}_{p,g}$ which is fixed under Map_g-action. For a pseudo-Anosov automorphism $\varphi \in \text{Map}_{g,1}$, or rather its reduction $\overline{\varphi} \in \text{Map}_g$, one has another fixed vector, which comes from the hyperbolic three-manifold, fibered over the circle with monodromy φ ([29, Theorem I.13.2]). To an extent, the bicohomology theory "linearizes" three-dimensional topology. All details are to be found in [29].

References

[1] R. Alperin, *Locally compact groups acting on trees and property T*, Mh. Math. **93** (1982), 262–265.

[2] G. Besson, G. Courtois and S. Gallot, *Entropies et rigidités des espaces localement symétriques de courboure strictement négative*, GAFA, **5** (1995), 731–799.

[3] S. Bloch, *Applications of the dilogarithm functions in algebraic K-theory and algebraic geometry*, Proc. Int. Symp. Alg. Geom. Kyoto, Kinokuniya, 1977, 103–114.

[4] J. Cannon and W. Thurston, *Equivariant Peano curves*, Preprint, (1986).

[5] J. Cheeger and J. Simons, *Differential characters and geometric invariants*, Geometry and Topology, Lectures Notes in Math., **1167** (1980), Springer, 50–80.

[6] S. S. Chern, and J. Simons, *Characteristic forms and geometric invariants*, Ann. of Math., **99** (1974), 48–69.

[7] D. Gabai, *Foliations and the topology of 3-manifolds*, J. Diff. Geom. **18** (1983), 445–503.

[8] D. Gabai, *Foliations and 3-manifolds*, Proc. Int. Congr. Math. Kyoto, 609–619 (1990).

[9] S. Gallot, *Curvature-decreasing maps are volume-decreasing*, Proc. Int. Congr. Math., Berlin, 1998, vol. II, 339–348.

[10] E. Ghys, V. Sergiesku, *Sur un groupe remarquable de difféomorphismes du cercle*, Comment. Math. Helv. **62** (1987), 185–239.

[11] A. Givental, *Periodic maps in symplectic topology*, Funct. Anal. Appl., **23** (1981), 37–52.

[12] W. Goldman, Berkeley Thesis.

[13] M. Gromov, *Filling Riemannian manifolds*, Journ. Diff. Geom. **18** (1983), 1–147.

[14] M. Gromov and Ya. Eliashberg, *Construction of nonsingular isoperimetric films*, Trudy Steklov Inst., **116**, 18–33.

[15] N. Hitchin, *The self-duality equation on a Riemann surface*, Proc. London Math. Soc., **55** (1987), 59–126.

[16] J. Kollár, *Shafarewich maps and automorphic forms*, UP, 1995.

[17] I. G. Lysionok, *On some algorithmic properties of hyperbolic groups*, Math. USSR Izv., **35** (1989), 145–163.

[18] I. G. Lysionok, *Algorithmic problems and quadratic equations in hyperbolic groups*, Thesis, Moscow, 1989, (Russian).

[19] N. G. Makarov, *On the radial behaviour of Bloch functions*, Soviet Math. Dokl. **40** (1990), 505–508.

[20] A. N. Parshin, *Finteness theorems and hyperbolic manifolds*, The Grothendieck Festschrift, Vol. III, Progr. Math. Birkhäuser Boston, 1990, 163–178.

[21] A. Reznikov, *Harmonic maps, hyperbolic cohomology and higher Milnor inequalities*, Topology, **32** (1993), 899–907.

[22] A. Reznikov, *Yamabe spectra*, Duke Math. Journ, **89** (1997), 87–94.

[23] A. Reznikov, *Quadratic equations in groups from the global geometry viewpoint*, Topology, **36** (1997), 849–865.

[24] A. Reznikov, *Rationality of secondary classes*, J. Diff. Geom., **43** (1996), 674–692.

[25] A. Reznikov, *All regulators of flat bundles are torsion*, Annals of Math., **141** (1995), 373–386.

[26] A. Reznikov, *Characteristic classes in symplectic topology*, Sel. Math., **3** (1997), 601–642.

[27] A. Reznikov, *Continuous cohomology of the groups of volume-preserving and symplectic diffeomorphism, measurable transfer and higher asymptotic cycles*, Sel. Math., **5** (1999), 181–198.

[28] A. Reznikov, *Volume of discrete groups and topological complexity of homology spheres*, Math. Ann., **306** (1996), 547–554.

[29] A. Reznikov, *Analytic topology of groups, actions, strings and varieties*, Preprint (August, 1999), Math. DG. 0001135.

[30] A. Reznikov, *Structure of Kähler groups: second cohomology*, Preprint (May, 1998), Math. DG. 9903023, to appear.

[31] T. Soma, *Non-zero degree maps to hyperbolic 3-manifolds*, J. Diff. Geom. **49** (1998), 517–546.

[32] D. Toledo, *Representations of surface groups in complex hyperbolic space*, J. Diff. Geom., **29** (1989), 125–133.

[33] W. Thurston, *Geometry and Topology of Three-Manifolds*, Princeton lecture notes.

[34] W. Thurston, *Hyperbolic structures on 3-manifolds*, I: Deformations of acylindrical manifolds, Ann. of Math. **124** (1986), 203–246, II: Deformations of 3-manifolds with incompressible boundary, Preprint, 1980/1986.

[35] Y. Watatani, *Property (T) of Kazhdan implies property (FA) of Serre*, Math. Japon. **27** (1981), 97–103.

Department of Mathematical Sciences
University of Durham
South Road
Durham DH1 3LE, United Kingdom
E-mail address: Alexander.Reznikov@durham.ac.uk

Towards Ryser's Conjecture

Bernhard Schmidt

Abstract. Ryser's conjecture asserts that there is no (v, k, λ)-difference set with $\gcd(v, k - \lambda) > 1$ in any cyclic group. We survey what is known on this conjecture and obtain progress towards it by improving the exponent bound for difference sets in [12]. As a consequence, with three possible exceptions, Ryser's conjecture is true for all parameters of known (v, k, λ)-difference sets with $k \leq 5 \cdot 10^{10}$. In particular, the circulant Hadamard matrix conjecture holds for orders $\leq 10^{11}$, also with only three possible exceptions. Finally, we obtain the first necessary and sufficient condition known in the literature for the existence of an infinite class of difference sets not relying on the self-conjugacy assumption.

1. Introduction

In 1938, Singer discovered that the desarguesian projective geometry $\mathrm{PG}(n, q)$ admits a cyclic regular automorphism group, nowadays called the **Singer cycle** of $\mathrm{PG}(n, q)$. Such an automorphism group is equivalent to a certain *difference set* in a cyclic group. Here a (v, k, λ)**-difference set** means a k-subset D of a group G of order v such that every nonidentity element g of G has exactly λ representations $g = d_1 d_2^{-1}$ with $d_1, d_2 \in D$. We call D **abelian**, **cyclic** etc. if G has this property. The parameter $n := k - \lambda$ is called the **order** of D. For convenience, n sometimes is added to the parameters, and we speak of a (v, k, λ, n)-difference set. Difference sets with $n = 0, 1$ are called **trivial** and will be excluded from our considerations.

Singer's discovery of an infinite family of difference sets inspired the development of an existence theory for these objects. First only cyclic groups were considered, later the theory was extended to noncyclic finite groups. In this paper, we mainly will be interested in the cyclic case. Until recently, only two main methods for the study of difference sets were known: Hall's multiplier theorem [3, 1947] and Turyn's self-conjugacy approach [16, 1965]. Both methods work well for *small* parameters (v, k, λ, n). Thus it was possible to settle the existence problem for cyclic difference sets with $k \leq 100$ already in the 60s [2, 1969]. These results and the fact that no cyclic (v, k, λ, n)-difference set with $\gcd(v, n) > 1$ has ever been found to motivate the following conjecture of Ryser [11] from 1963.

Conjecture 1.1. (Ryser's conjecture) *There is no cyclic (v, k, λ, n)-difference set with $\gcd(v, n) > 1$.*

Ryser's conjecture is still open, only some partial results are known. Despite many efforts, there had not been any new results since Turyn's work [16, 1965] until substantial progress was obtained in [12]. In the present paper, we will improve upon [12].

Ryser's conjecture implies two further longstanding conjectures, namely, the Barker and the circulant Hadamard matrix conjecture. A **circulant Hadamard matrix of order** v is a matrix of the form

$$
H = \begin{pmatrix}
a_1 & a_2 & \cdots & a_v \\
a_v & a_1 & \cdots & a_{v-1} \\
\cdots & \cdots & \cdots & \cdots \\
a_2 & a_3 & \cdots & a_1
\end{pmatrix}
$$

with $a_i = \pm 1$ and $HH^t = vI$ where I is the identity matrix. It is conjectured that no circulant Hadamard matrix of order $v > 4$ exists. A sequence $(a_i)_{i=1}^v$, $a_i = \pm 1$, is called a **Barker sequence of length** v if $|\sum_{i=1}^{v-j} a_i a_{i+j}| \leq 1$ for $j = 1, \ldots, v-1$. The **Barker conjecture** asserts that there are no Barker sequences of length $v > 13$. Storer and Turyn [15] proved the Barker conjecture for all odd v. The following is well known, see [1, Remark VI.14.13].

Result 1.2. *Ryser's conjecture implies the circulant Hadamard matrix conjecture which in turn implies the Barker conjecture.*

In [12], the Barker conjecture was verified for $v \leq 4 \cdot 10^{12}$. In the present paper, we will show that the circulant Hadamard matrix conjecture holds for $v \leq 10^{11}$ with only three possible exceptions. We will also show that, again with only three possible exceptions, Ryser's conjecture is true for all parameters of known difference sets with $k \leq 5 \cdot 10^{10}$.

Let us briefly discuss the previous work related to Ryser's conjecture. Hall's table of difference sets [3, 1956] shows that Ryser's conjecture is true for $k \leq 50$ with eleven possible exceptions. These eleven cases were eliminated by the independent and overlapping work of Mann [6], Rankin [10], Turyn [16], and Yamamoto [18]. Thus, in 1965, it was known that Ryser's conjecture is true for $k \leq 50$. Baumert [2, 1969] extended the table for difference sets in cyclic groups to $k \leq 100$ and, in particular, showed that Ryser's conjecture holds in this range. The two most difficult cases $(v, k, \lambda) = (441, 56, 7)$ and $(891, 90, 9)$ were excluded by H. Rumsey (unpublished) by extensive computations, see [2]. Baumert's table was extended by Vera Lopez and Garcia Sanchez [17, 1997] to $k \leq 150$. Together with the previously known results, their table shows the following.

Result 1.3. *Ryser's conjecture is true for $k \leq 107$.*

To my knowledge, the following are the only open cases for Ryser's conjecture with $k \leq 150$: $(v, k, \lambda) = (429, 108, 27)$, $(715, 120, 20)$, $(351, 126, 45)$, $(465, 150, 50)$, see [17]. We will be able to exclude the third of these cases later.

The most important result on Ryser's conjecture aside from [12] is the following due to Turyn [16]. We recall that a prime p is called **self-conjugate** modulo an integer w if -1 is a power of p modulo the p-free part of w. A composite integer m is called self-conjugate modulo w if every prime divisor of m has this property.

Result 1.4. *Assume the existence of a cyclic (v, k, λ, n)-difference set. Let m and w be positive integers with $(m, w) > 1$ such that m^2 divides n, w divides v, and m is self-conjugate modulo w. Then*

$$m \le \frac{2^{r-1}v}{w}$$

where r is the number of prime divisors of (m, w).

Turyn's result shows that Ryser's conjecture is true in the case of self-conjugacy:

Corollary 1.5. *If there is a prime p dividing (v, n) which is self-cojugate modulo v, then there is no cyclic (v, k, λ, n)-difference set.*

We note that the self-conjugacy assumption is very rarely satisfied if v has many prime divisors. In the present paper, we will obtain a result which does not need severe assumptions like self-conjugacy and thus is of broader applicability.

2. Characters

The standard method for the study of difference sets in abelian groups is the use of complex characters. We summarize the necessary facts here, see [7] for proofs. Let G be a finite abelian group. A complex character of G is a homomorphism $\chi \colon G \to \mathbb{C}^*$. The character χ_0 defined by $\chi_0(g) = 1$ for all $g \in G$ is called the **trivial** character. The set of characters of G forms a group G^* isomorphic to G where the group operation is defined by $\chi_1\chi_2(g) = \chi_1(g)\chi_2(g)$. If χ is a character of G of order e, then $\chi(g)$ is a complex eth root of unity for all $g \in G$. Any character of G can be extended to the group ring $\mathbb{Z}[G]$ by linearity. A subset D of G will be identified with $\sum_{d \in D} d \in \mathbb{Z}[G]$. For $X = \sum_{g \in G} a_g g \in \mathbb{Z}[G]$ we write $X^{(-1)} := \sum_{g \in G} a_g g^{-1}$. We use the notation $\xi_t = e^{2\pi i/t}$.

Lemma 2.1. ([16]) *Let D be a (v, k, λ, n)-difference set in an abelian group G. Let χ be a character of G of order e. Then $\chi(D) \in \mathbb{Z}[\xi_e]$ and*

$$|\chi(D)|^2 = n.$$

3. The Field Descent

Lemma 2.1 has been used in dozens of papers for the study of difference sets. Almost all of these results rely on the self-conjugacy assumption. The main merit of [12] is to provide a method free from this restrictive condition. The key to this method is the so-called "field descent", see [12, Theorem 3.5]. For the formulation of the field descent, we need a definition.

Definition 3.1. *Let m, n be positive integers, and let $m = \prod_{i=1}^{t} p_i^{c_i}$ be the prime power decomposition of m. For each prime divisor q of n let*

$$m_q := \begin{cases} \prod_{p_i \neq q} p_i & \text{if } m \text{ is odd or } q = 2\,, \\ 4 \prod_{p_i \neq 2, q} p_i & \text{otherwise}. \end{cases}$$

Let $\mathcal{D}(n)$ be the set of prime divisors of n. We define $F(m, n) = \prod_{i=1}^{t} p_i^{b_i}$ to be the minimum multiple of $\prod_{i=1}^{t} p_i$ such that for every pair (i, q), $i \in \{1, \ldots, t\}$, $q \in \mathcal{D}(n)$, at least one of the following conditions is satisfied.

   (a) $q = p_i$ *and* $(p_i, b_i) \neq (2, 1)$,

   (b) $b_i = c_i$,

   (c) $q \neq p_i$ *and* $q^{\text{ord}_{m_q}(q)} \not\equiv 1 \pmod{p_i^{b_i+1}}$.

Result 3.2. (Field descent) *Assume $|X|^2 = n$ for $X \in \mathbb{Z}[\xi_m]$ where n and m are positive integers. Then*

$$X \xi_m^j \in \mathbb{Z}[\xi_{F(m,n)}]$$

for some j.

In [12], Result 3.2 was used to obtain a general exponent bound for difference sets. In Section 5, we will improve this bound considerably.

4. Bounding the Absolute Value

In this section, we will use result 3.2 to obtain an upper bound on the absolute value of cyclotomic integers. This bound is an improvement [12, Theorem 4.2]. In the present paper, we will only give the applications of our new bound to abelian difference sets. Similar applications to relative difference sets, planar functions, and group invariant weighting matrices can be given [13, Chapter 3].

As a preparation for the proof of our bound, we need a simple lemma on conjugate characters. Two characters χ and τ of order e of an abelian group G are called **conjugate** if there is $\sigma \in \text{Gal}(\mathbb{Q}(\xi_e)/\mathbb{Q})$ with $\chi(g) = \tau(g)^\sigma$ for all $g \in G$. Let φ denote the Euler totient function. The following is well known and easy to prove, see [9, p. 6], for instance.

Lemma 4.1. *Let χ be a character of order e of an abelian group G. Then χ has exactly $\varphi(e)$ distinct conjugates. Furthermore, if $\chi(A) \in \mathbb{Q}$ for some $A \in \mathbb{Z}[G]$, then $\tau(A) = \chi(A)$ for all conjugates τ of χ.*

The following theorem on cyclotomic integers of prescribed absolute value is the main result of this paper.

Theorem 4.2. *Let $X \in \mathbb{Z}[\xi_m]$ be of the form*

$$X = \sum_{i=0}^{m-1} a_i \xi_m^i$$

with $0 \leq a_i \leq C$ for some constant C and assume that $n := X\overline{X}$ is an integer. Then

$$n \leq \frac{C^2 F(m,n)^2}{4\varphi(F(m,n))}.$$

Proof. By Theorem 3.2, we can assume $X \in \mathbb{Z}[\xi_f]$ where $f := F(m,n)$. Since $1, \xi_m, \ldots, \xi_m^{m/f-1}$ are independent over $\mathbb{Q}(\xi_f)$, we have $X = \sum_{i=0}^{f-1} b_i \xi_f^i$ where $b_i := a_{im/f}$. Now we view X also as an element of the group ring $\mathbb{Z}[G]$ where $G = \langle \xi_f \rangle$. Since $X\overline{X} = n \in \mathbb{Q}$, we have

$$\chi(X)\overline{\chi(X)} = n \tag{1}$$

for all $\varphi(f)$ characters χ of G of order f by Lemma 4.1. Write $l := \sum_{i=0}^{f-1} b_i$. The coefficient of 1 in $XX^{(-1)}$ is $\sum_{i=0}^{f-1} b_i^2$. From the Fourier inversion formula, we get $f \sum_{i=0}^{f-1} b_i^2 = \sum_{\tau \in G^*} |\tau(X)|^2$. Using (1) and $\chi_0(X) = l$ for the trivial character χ_0 of G, we get

$$f \sum b_i^2 \geq l^2 + \varphi(f)n. \tag{2}$$

Since $0 \leq b_i \leq C$, we have $\sum b_i^2 \leq Cl$. Thus $f \sum b_i^2 - l^2 \leq fCl - l^2 \leq f^2 C^2/4$. Combining this with (2) gives the assertion. $\qquad\square$

5. A Field Descent Exponent Bound

We now apply Theorem 4.2 to obtain a general exponent bound on abelian groups containing difference sets. By φ we denote the Euler totient function.

Theorem 5.1. *Assume the existence of a (v, k, λ, n)-difference D set in an abelian group G. Then*

$$\exp G \leq \frac{vF(v,n)}{2\sqrt{n\varphi(F(v,n))}}.$$

In particular, if G is cyclic, then

$$n \leq \frac{F(v,n)^2}{4\varphi(F(v,n))}.$$

Proof. Let χ be a character of G of order $e := \exp G$. By Lemma 2.1, we have $|\chi(D)|^2 = n$. Also, since the kernel of χ on G has order v/e, we have

$$\chi(D) = \sum_{i=0}^{e-1} a_i \xi_e^i$$

with $0 \leq a_i \leq v/e$. Thus, from Theorem 4.2, we get the assertion. $\qquad\square$

6. Application to Ryser's Conjecture

The most interesting test cases for our exponent bound are the parameter series corresponding to known families of difference sets. In this section, we apply Theorem 5.1 to all parameter series corresponding to known difference sets with $\gcd(v, n) > 1$. The following is a complete list of these series, see [4, 5] or [1].

(i) Hadamard parameters:

$(v, k, \lambda, n) = (4u^2, 2u^2 - u, u^2 - u, u^2)$
where u is any positive integer.

(ii) McFarland parameters:

$(v, k, \lambda, n) = (q^{d+1}[\frac{q^{d+1}-1}{q-1} + 1], q^d \frac{q^{d+1}-1}{q-1}, q^d \frac{q^d-1}{q-1}, q^{2d})$
where $q = p^f \neq 2$ and p is a prime.

(iii) Spence parameters:

$(v, k, \lambda, n) = (3^{d+1} \frac{3^{d+1}-1}{2}, 3^d \frac{3^{d+1}+1}{2}, 3^d \frac{3^d+1}{2}, 3^{2d})$
where d is any positive integer.

(iv) Chen/Davis/Jedwab parameters:

$(v, k, \lambda, n) = (4q^{2t} \frac{q^{2t}-1}{q^2-1}, q^{2t-1}[\frac{2(q^{2t}-1)}{q+1} + 1], q^{2t-1}(q - 1)\frac{q^{2t-1}+1}{q+1}), q^{4t-2})$
where $q = p^f$, p is a prime, and t is any positive integer.

We do not allow $q = 2$ for the McFarland parameters since then $(v, k, \lambda, n) = (2^{2d+2}, 2^{2d+1} - 2^d, 2^{2d} - 2^d, 2^{2d})$, and these are Hadamard parameters with $u = 2^d$. Hadamard difference sets are known to exist for every u of the form $u = 2^a 3^b r^2$ where $a, b \geq 0$ and r is any positive integer, see [1, 5]. Here we will consider arbitrary positive integers u. McFarland and Spence difference sets are known for any prime power q and any positive integer d, see [1, 5]. Difference sets of type (iv) are known to exist only if f is even or $p \leq 3$, see [1, 5]. However, here we will consider arbitrary f and p.

Now we come to the application of Theorem 5.1. The next theorem shows that Ryser's conjecture is true for most parameters of known difference sets.

Theorem 6.1.
a) *If there is a Hadamard difference set in a cyclic group of order $v = 4u^2$ then $F(v, u)^2 / \varphi(F(v, u)) \geq v$.*
b) *If there is a difference set with McFarland parameters in a cyclic group of order $q^{d+1}[\frac{q^{d+1}-1}{q-1} + 1]$, $q = p^f$, then $p > 2$, $d = f = 1$ and*

$$\frac{p+2}{\varphi(p+2)} \geq 4 - \frac{12}{p+2}. \tag{3}$$

In particular, $p + 2$ has at least 20 distinct prime divisors and $p > 2 \cdot 10^{28}$.

c) *There are no difference sets with Spence or Chen/Davis/Jedwab parameters in any cyclic groups.*

Proof. a) This is immediate from Theorem 5.1.

c) See [12, Theorem 6.3].

b) Assume the existence of a difference set with McFarland parameters in a cyclic group G of order $v = q^{d+1}[\frac{q^{d+1}-1}{q-1}+1]$ where $q = p^f$, and p is a prime. We first show $f = d = 1$.

If we take $p_1 = p$ in Definition 3.1 then $b_1 = 1$ if p is odd and $b_1 = 2$ if $p = 2$. In both cases $f := F(v, n)$ divides $p(\frac{q^{d+1}-1}{q-1}+1)$ since $\frac{q^{d+1}-1}{q-1}+1$ is even for $p = 2$. Thus

$$f \leq 2p^{fd+1} \, . \tag{4}$$

Since $2 \cdot 3 \cdot 5/(1 \cdot 2 \cdot 4) < 4$, and since $r^{2/25} > r/(r-1)$ for all $r \geq 7$, we have

$$\frac{x}{4\varphi(x)} < x^{2/25}$$

for all integers $x > 1$. From Theorem 5.1 and (4) we thus get

$$p^{2fd} < (2p^{fd+1})^{27/25} \, .$$

This implies $fd = 1$ or $fd = 2$ and $p = 2$. In the latter case we have $f = 2$ and $d = 1$ since we assumed $q = p^f \neq 2$ for McFarland parameters. A direct application of Theorem 5.1 shows that this case cannot occur. Thus we have shown $fd = 1$.

Now let $fd = 1$. Then $p \neq 2$ since $q \neq 2$, and we have $v = p^2(p+2)$. Thus $f := F(v, n)$ divides $p(p+2)$. Theorem 5.1 gives $p^2 \leq p^2(p+2)^2/[4\varphi(p(p+2))]$ proving (3). Let $Y = 3 \cdot 5 \cdots 73$ be the product of the 20 smallest odd primes. Then $Y/\varphi(Y) < 3.97$ and $Y > 2 \cdot 10^{28}$. This implies the remaining assertions of part b.

$\square$

Remark 6.2.
  a) *Theorem 6.1 eliminates the open case $(v, k, \lambda) = (351, 126, 45)$ mentioned in the introduction since these are Spence parameters with $d = 2$. The nonexistence of a cyclic difference set with these parameters also follows directly from Theorem 5.1 since $F(351, 81) = 39$.*
  b) *A heuristic argument [12, Remark 3.6] shows that the order of magnitude of $F(v, n)$ "usually" is the product of the primes dividing v. This indicates that Theorem 6.1a should rule out "almost all" cyclic Hadamard difference sets. The next result, in particular, supports this claim.*

Corollary 6.3. *For $k \leq 5 \cdot 10^{10}$, Ryser's conjecture is true for all parameters (v, k, λ) of known difference sets (see the list (i)–(iv) above) with the possible exception of $(v, k, \lambda) = (4u^2, 2u^2 - u, u^2 - u)$ with $u \in \{165, 11715, 82005\}$.*

Proof. For McFarland, Spence and Chen/Davis/Jedwab parameters, this immediately follows from Theorem 6.1. For Hadamard parameters, the result follows from a computer search using result 1.4 and Theorem 6.1a.

$\square$

540　　　　　　　　　　　　　　　B. Schmidt

Corollary 6.4. *There is no circulant Hadamard matrix of order v, $4 < v \leq 10^{11}$, with the possible exceptions $v = 4u^2$, $u \in \{165, 11715, 82005\}$.*

Proof. The existence of a circulant Hadamard matrix of order v implies the existence of a Hadamard difference set in the cyclic group of order v, see [1, Remark VI.14.13]. Thus the assertion follows from Corollary 6.3. $\square$

We conclude this paper with a necessary and sufficient condition for the existence of McFarland difference sets with $f = d = 1$. It is the first necessary and sufficient condition for a (presumably) infinite family of difference sets known in the literature which does not rely on the self-conjugacy argument.

Corollary 6.5. *Let p be an odd prime such that $p + 2$ is squarefree and*

$$\frac{p+2}{\varphi(p+2)} < 4 - \frac{12}{p+2}. \tag{5}$$

Then a $(p^2(p+2), p(p+1), p+1)$-difference set in an abelian group G exists if and only if

$$G \cong (\mathbb{Z}/p\mathbb{Z})^2 \times (\mathbb{Z}/(p+2)\mathbb{Z}).$$

Proof. The existence of a $(p^2(p+2), p(p+1), p+1, p^2)$-difference set in $(\mathbb{Z}/p\mathbb{Z})^2 \times (\mathbb{Z}/(p+2)\mathbb{Z})$ is due to McFarland [8]. The necessary part follows directly from Theorem 6.1b. $\square$

References

[1] T. Beth, D. Jungnickel and H. Lenz: *Design Theory* (2nd edition), Cambridge University Press, Cambridge 1999.

[2] L. D. Baumert: Difference sets. *SIAM J. Appl. Math.* **17** (1969), 826–833.

[3] M. Hall: Cyclic projective planes. *Duke Math. J.* **14** (1947), 1079–1090.

[4] D. Jungnickel: Difference Sets. *In: Contemporary Design Theory: A Collection of Surveys. Eds. J. H. Dinitz and D. R. Stinson,* Wiley, New York 1992, 241–324.

[5] D. Jungnickel and B. Schmidt: Difference Sets: An Update. *In: Geometry, Combinatorial Designs and Related Structures. Proceedings of the First Pythagorean Conference. Eds. J. W. P. Hirschfeld et al.,* Cambridge University Press 1997, 89–112.

[6] H. B. Mann: Balanced incomplete block designs and abelian difference sets. *Illinois J. Math.* **8** (1964), 252–261.

[7] H. B. Mann: *Addition Theorems.* Wiley, New York 1965.

[8] R. L. McFarland: A family of difference sets in non-cyclic groups. *J. Comb. Theory Ser. A* **15** (1973), 1–10.

[9] R. L. McFarland: Difference sets in abelian groups of order $4p^2$. *Mitt. Math. Sem. Giessen* **192** (1989), 1–70.

[10] R. A. Rankin: Difference sets. *Acta Arith.* **9** (1964), 161–168.

[11] H. J. Ryser: *Combinatorial Mathematics.* Wiley, New York 1963.

[12] B. Schmidt: Cyclotomic Integers and Finite Geometry. *J. Am. Math. Soc.* **12** (1999), 929–952.

[13] B. Schmidt: Characters and cyclotomic fields in finite geometry. Habilitation thesis, Universität Augsburg. In preparation.

[14] J. Singer: A theorem in finite projective geometry and some applications to number theory. *Trans. Amer. Math. Soc.* **43** (1938), 377–385.

[15] J. Storer and R. Turyn: On binary sequences. *Proc. Amer. Math. Soc.* **12** (1961), 394–399.

[16] R. J. Turyn: Character sums and difference sets. *Pacific J. Math.* **15** (1965) 319–346.

[17] A. Vera Lopez and M. A. Garcia Sanchez: On the existence of abelian difference sets with $100 < k \leq 150$. *J. Comb. Math. Comb. Comp.* **23** (1997), 97–112.

[18] K. Yamamoto: Decomposition fields of difference sets. *Pacific J. Math.* **13** (1963), 337–352.

Department of Mathematics
University of Augsburg
Universitätsstraße 14
86135 Augsburg, Germany
E-mail address: schmidt@math.uni-augsburg.de

The Dynamics of Algebraic $\mathbb{Z}^d$-Actions

Klaus Schmidt

Abstract. An *algebraic $\mathbb{Z}^d$-action* is a $\mathbb{Z}^d$-action by automorphisms of a compact abelian group. By Pontryagin duality, there is a one-to-one correspondence between algebraic $\mathbb{Z}^d$-actions and modules over the ring R_d of Laurent polynomials with integer coefficients in d commuting variables.

This correspondence establishes a close connection between algebraic and arithmetical properties of R_d-modules and dynamical properties of algebraic $\mathbb{Z}^d$-actions, which is the subject of this article.

1. Algebraic $\mathbb{Z}^d$-Actions and Their Dual Modules

An *algebraic $\mathbb{Z}^d$-action* is an action $\alpha\colon \mathbf{n} \mapsto \alpha^{\mathbf{n}}$ of $\mathbb{Z}^d$, $d \geq 1$, by continuous automorphisms of a compact abelian group X with Borel field $\mathcal{B}_X$ and normalized Haar measure λ_X. Two algebraic $\mathbb{Z}^d$-actions α and β on compact abelian groups X and Y are *algebraically conjugate* if there exists a continuous group isomorphism $\phi\colon X \longrightarrow Y$ with

$$\phi \cdot \alpha^{\mathbf{n}} = \beta^{\mathbf{n}} \cdot \phi \tag{1}$$

for every $\mathbf{n} \in \mathbb{Z}^d$. If the map ϕ in (1) is a homeomorphism then α and β are *topologically conjugate*. Finally we call α and β *measurably conjugate* if there exists a measure space isomorphism $\phi\colon (X, \mathcal{B}_X, \lambda_X) \to (Y, \mathcal{B}_Y, \lambda_Y)$ satisfying (1) λ_X-a.e. for every $\mathbf{n} \in \mathbb{Z}^d$.

In [4] and [13], Pontryagin duality was shown to imply a one-to-one correspondence between algebraic $\mathbb{Z}^d$-actions (up to algebraic conjugacy) and modules over the ring of Laurent polynomials $R_d = \mathbb{Z}[u_1^{\pm 1}, \ldots, u_d^{\pm 1}]$ with integral coefficients in the commuting variables $u_1, \ldots, u_d$ (up to module isomorphism).

In order to explain this correspondence we write a typical element $f \in R_d$ as

$$f = \sum_{\mathbf{m} \in \mathbb{Z}^d} c_f(\mathbf{m}) u^{\mathbf{m}} \tag{2}$$

with $u^{\mathbf{m}} = u_1^{m_1} \cdots u_d^{m_d}$ and $c_f(\mathbf{m}) \in \mathbb{Z}$ for every $\mathbf{m} = (m_1, \ldots, m_d) \in \mathbb{Z}^d$, where $c_f(\mathbf{m}) = 0$ for all but finitely many $\mathbf{m}$. If α is an algebraic $\mathbb{Z}^d$-action on a compact abelian group X, then the additively-written dual group $M = \widehat{X}$ is a module over the ring R_d with operation

$$f \cdot a = \sum_{\mathbf{m} \in \mathbb{Z}^d} c_f(\mathbf{m}) \widehat{\alpha^{\mathbf{m}}}(a) \tag{3}$$

for $f \in R_d$ and $a \in M$, where $\widehat{\alpha^{\mathbf{m}}}$ is the automorphism of $M = \widehat{X}$ dual to $\alpha^{\mathbf{m}}$. In particular,

$$u^{\mathbf{m}} \cdot a = \widehat{\alpha^{\mathbf{m}}}(a) \tag{4}$$

for $\mathbf{m} \in \mathbb{Z}^d$ and $a \in M$. Conversely, any R_d-module M determines an algebraic $\mathbb{Z}^d$-action α_M on the compact abelian group $X_M = \widehat{M}$ with $\alpha_M^{\mathbf{m}}$ dual to multiplication by $u^{\mathbf{m}}$ on M for every $\mathbf{m} \in \mathbb{Z}^d$ (cf. (4)). Note that X_M is metrizable if and only if its *dual module* M is countable.

Examples 1.1.

(1) *Let $M = R_d$. Since R_d is isomorphic to the direct sum $\sum_{\mathbb{Z}^d} \mathbb{Z}$ of copies of $\mathbb{Z}$, indexed by $\mathbb{Z}^d$, the dual group $X = \widehat{R_d}$ is isomorphic to the Cartesian product $\mathbb{T}^{\mathbb{Z}^d}$ of copies of $\mathbb{T} = \mathbb{R}/\mathbb{Z}$. We write a typical element $x \in \mathbb{T}^{\mathbb{Z}^d}$ as $x = (x_{\mathbf{n}})$ with $x_{\mathbf{n}} \in \mathbb{T}$ for every $\mathbf{n} \in \mathbb{Z}^d$ and choose the following identification of $X_{R_d} = \widehat{R_d}$ and $\mathbb{T}^{\mathbb{Z}^d}$: for every $x \in \mathbb{T}^{\mathbb{Z}^d}$ and $f \in R_d$,*

$$\langle x, f \rangle = e^{2\pi i \sum_{\mathbf{n} \in \mathbb{Z}^d} c_f(\mathbf{n}) x_{\mathbf{n}}} , \tag{5}$$

where f is given by (2). Under this identification the $\mathbb{Z}^d$-action α_{R_d} on $X_{R_d} = \mathbb{T}^{\mathbb{Z}^d}$ becomes the shift-action

$$(\alpha_{R_d}^{\mathbf{m}} x)_{\mathbf{n}} = x_{\mathbf{m}+\mathbf{n}} . \tag{6}$$

(2) *Let $I \subset R_d$ be an ideal and $M = R_d/I$. Since M is a quotient of the additive group R_d by an $\widehat{\alpha_{R_d}}$-invariant subgroup (i.e. by a submodule), the dual group $X_M = \widehat{M}$ is the closed α_{R_d}-invariant subgroup*

$$X_{R_d/I} = \{x \in X_{R_d} = \mathbb{T}^{\mathbb{Z}^d} : \langle x, f \rangle = 1 \text{ for every } f \in I\}$$

$$= \left\{ x \in \mathbb{T}^{\mathbb{Z}^d} : \sum_{\mathbf{n} \in \mathbb{Z}^d} c_f(\mathbf{n}) x_{\mathbf{m}+\mathbf{n}} = 0 \pmod 1 \atop \text{for every } f \in I \text{ and } \mathbf{m} \in \mathbb{Z}^d \right\}, \tag{7}$$

and $\alpha_{R_d/I}$ is the restriction of the shift-action α_{R_d} in (6) to the shift-invariant subgroup $X_{R_d/I} \subset \mathbb{T}^{\mathbb{Z}^d}$.

Conversely, let $X \subset \mathbb{T}^{\mathbb{Z}^d} = \widehat{R_d}$ be a closed subgroup, and let

$$X^{\perp} = \{f \in R_d : \langle x, f \rangle = 1 \text{ for every } x \in X\}$$

be the annihilator of X in $\widehat{R_d}$. Then X is shift-invariant if and only if $X^{\perp}$ is an ideal in R_d.

The correspondence between algebraic $\mathbb{Z}^d$-actions $\alpha = \alpha_M$ and R_d-modules M yields a correspondence (or 'dictionary') between dynamical properties of α_M and algebraic properties of the module M (cf. [16]). It turns out that some of the principal dynamical properties of α_M can be expressed entirely in terms of the prime ideals associated with the module M, where a prime ideal $\mathfrak{p} \subset R_d$ is *associated with* M if

$$\mathfrak{p} = \{f \in R_d : f \cdot a = 0_M\}$$

for some $a \in M$. The set of all prime ideals associated with M is denoted by $\mathrm{asc}(M)$; if M is Noetherian, then $\mathrm{asc}(M)$ is finite.

Figure 1 provides a small illustration of this correspondence; all the relevant results can be found in [16]. In the third column we assume that the R_d-module $M = \widehat{X}$ defining α is of the form $R_d/\mathfrak{p}$, where $\mathfrak{p} \subset R_d$ is a prime ideal, and describe the algebraic condition on $\mathfrak{p}$ equivalent to the dynamical condition on $\alpha = \alpha_{R_d/\mathfrak{p}}$ appearing in the second column. In the fourth column we consider a countable R_d-module M and state the algebraic property of M corresponding to the property of $\alpha = \alpha_M$ in the second column.

	Property of α	$\alpha = \alpha_{R_d/\mathfrak{p}}$	$\alpha = \alpha_M$
(1)	α is expansive	$V_{\mathbb{C}}(\mathfrak{p}) \cap \mathbb{S}^d = \varnothing$	M is Noetherian and $\alpha_{R_d/\mathfrak{p}}$ is expansive for every $\mathfrak{p} \in \mathrm{asc}(M)$
(2)	$\alpha^{\mathbf{n}}$ is ergodic for some $\mathbf{n} \in \mathbb{Z}^d$	$u^{k\mathbf{n}} - 1 \notin \subset \mathfrak{p}$ for every $k \geq 1$	$\alpha^{\mathbf{n}}_{R_d/\mathfrak{p}}$ is ergodic for every $\mathfrak{p} \in \mathrm{asc}(M)$
(3)	α is ergodic	$\{u^{k\mathbf{n}} - 1 : \mathbf{n} \in \mathbb{Z}^d\} \not\subset \mathfrak{p}$ for every $k \geq 1$	$\alpha_{R_d/\mathfrak{p}}$ is ergodic for every $\mathfrak{p} \in \mathrm{asc}(M)$
(4)	α is mixing	$u^{\mathbf{n}} - 1 \notin \mathfrak{p}$ for every non-zero $\mathbf{n} \in \mathbb{Z}^d$	$\alpha_{R_d/\mathfrak{p}}$ is mixing for every $\mathfrak{p} \in \mathrm{asc}(M)$
(5)	α is mixing of every order	Either $\mathfrak{p}$ is equal to pR_d for some rational prime p, or $\mathfrak{p} \cap \mathbb{Z} = \{0\}$ and $\alpha_{R_d/\mathfrak{p}}$ is mixing	For every $\mathfrak{p} \in \mathrm{asc}(M)$, $\alpha_{R_d/\mathfrak{p}}$ is mixing of every order
(6)	$h(\alpha) > 0$	$\mathfrak{p}$ is principal and $\alpha_{R_d/\mathfrak{p}}$ is mixing	$h(\alpha_{R_d/\mathfrak{p}}) > 0$ for at least one $\mathfrak{p} \in \mathrm{asc}(M)$
(7)	$h(\alpha) < \infty$	$\mathfrak{p} \neq \{0\}$	If M is Noetherian: $\mathfrak{p} \neq \{0\}$ for every $\mathfrak{p} \in \mathrm{asc}(M)$
(8)	α has completely positive entropy (or is Bernoulli)	$h(\alpha^{R_d/\mathfrak{p}}) > 0$	$h(\alpha_{R_d/\mathfrak{p}}) > 0$ for every $\mathfrak{p} \in \mathrm{asc}(M)$

FIGURE 1: A POCKET DICTIONARY

The notation in Figure 1 is as follows. In (1),

$$V_{\mathbb{C}}(\mathfrak{p}) = \{c \in (\mathbb{C} \smallsetminus \{0\})^d : f(c) = 0 \text{ for every } f \in \mathfrak{p}\}$$

is the *variety* of $\mathfrak{p}$, and $\mathbb{S} = \{c \in \mathbb{C} : |c| = 1\}$. From (2)–(4) it is clear that α is ergodic if and only if $\alpha^{\mathbf{n}}$ is ergodic for *some* $\mathbf{n} \in \mathbb{Z}^d$, and that α is mixing if and only if $\alpha^{\mathbf{n}}$ is ergodic for *every nonzero* $\mathbf{n} \in \mathbb{Z}^d$. In (5), α is mixing of order $r \geq 2$ if

$$\lim_{\substack{\mathbf{n}_1,\dots,\mathbf{n}_r \in \mathbb{Z}^d \\ \|\mathbf{n}_i - \mathbf{n}_j\| \to \infty \text{ for } 1 \leq i < j \leq d}} \lambda_X \left(\bigcap_{i=1}^{r} \alpha^{-\mathbf{n}_i} B_i \right) = \prod_{i=1}^{r} \lambda_X(B_i)$$

for all Borel sets $B_i \subset X$, $i = 1, \dots, r$. In (6)–(8), $h(\alpha)$ stands for the topological entropy of α (which coincides with the metric entropy $h_{\lambda_X}(\alpha)$). In [8] and [16] there is an explicit entropy formula for algebraic $\mathbb{Z}^d$-actions. In the special case where $\alpha = \alpha_{R_d/\mathfrak{p}}$ for some prime ideal $\mathfrak{p} \subset R_d$ this formula reduces to

$$h(\alpha) = \begin{cases} |\log \mathsf{M}(f)| & \text{if } \mathfrak{p} = (f) = fR_d \text{ is principal}, \\ 0 & \text{otherwise}, \end{cases}$$

where

$$\mathsf{M}(f) = \begin{cases} \exp\left(\int_{\mathbb{S}^d} \log|f(\mathbf{s})|\,d\mathbf{s}\right) & \text{if } f \neq 0, \\ 0 & \text{if } f = 0, \end{cases}$$

is the *Mahler measure* of the polynomial f. Here $d\mathbf{s}$ denotes integration with respect to the normalized Haar measure on the multiplicative subgroup $\mathbb{S}^d \subset \mathbb{C}^d$.

For background, details and proofs of these and further results we refer to [16] and the original articles cited there. The remainder of this note is devoted to two particular problems: the higher order mixing behaviour and the conjugacy problem for algebraic $\mathbb{Z}^d$-actions.

2. Higher Order Mixing Properties of Algebraic $\mathbb{Z}^d$-Actions

In this section we describe the connection between higher order mixing properties of algebraic $\mathbb{Z}^d$-actions and certain diophantine results on additive relations in fields due to Mahler ([9]), Masser ([10, 5]) and Schlickewei, W. Schmidt and van der Poorten ([1, 17]). In the discussion below we shall use the following elementary consequence of Pontryagin duality:

Lemma 2.1. *Let α be an algebraic $\mathbb{Z}^d$-action on a compact abelian group X with dual module M. Then X is connected if and only if no prime ideal $\mathfrak{p} \in \mathrm{asc}(M)$ contains a nonzero constant, and X is zero-dimensional if and only if every $\mathfrak{p} \in \mathrm{asc}(M)$ contains a nonzero constant.*

Let $\mathfrak{p} \subset R_d$ be a prime ideal, and let $\alpha = \alpha_{R_d/\mathfrak{p}}$ be the algebraic $\mathbb{Z}^d$-action with dual module $M = R_d/\mathfrak{p} = \widehat{X}$. If α is not mixing, then there exist Borel sets $B_1, B_2 \subset X$ and a sequence $(\mathbf{n}_k,\, k \geq 1)$ in $\mathbb{Z}^d$ with $\lim_{k \to \infty} \mathbf{n}_k = \infty$ and

$$\lim_{k \to \infty} \lambda_X(B_1 \cap \alpha^{-\mathbf{n}_k} B_2) = c$$

for some $c \neq \lambda_X(B_1)\lambda_X(B_2)$. Fourier expansion implies that the latter condition is equivalent to the existence of nonzero elements $a_1, a_2 \in M$ such that

$$a_1 + u^{\mathbf{n}_k} \cdot a_2 = 0$$

for infinitely many $k \geq 1$. In particular,

$$(u^{\mathbf{m}} - 1) \cdot a_2 = 0 \tag{8}$$

for some nonzero $\mathbf{m} \in \mathbb{Z}^d$ (cf. figure 1(4)). A very similar argument shows that α is not mixing of order $r \geq 2$ if and only if there exist elements $a_1, \ldots, a_r$ in M, not all equal to zero, and a sequence $((\mathbf{n}_k^{(1)}, \ldots, \mathbf{n}_k^{(r)}), k \geq 1)$ in $(\mathbb{Z}^d)^r$ such that $\lim_{k \to \infty} \|\mathbf{n}_k^{(i)} - \mathbf{n}_k^{(j)}\| = \infty$ for all i, j with $1 \leq i < j \leq r$, and with

$$u^{\mathbf{n}_k^{(1)}} \cdot a_1 + \cdots + u^{\mathbf{n}_k^{(r)}} \cdot a_r = 0 \tag{9}$$

for every $k \geq 1$.

Below we shall see that higher order mixing of an algebraic $\mathbb{Z}^d$-action α on a compact abelian group X can break down in a particularly regular way (cf. examples 2.7 and 2.10). We call a nonempty finite subset $S \subset \mathbb{Z}^d$ *mixing* under α if

$$\lim_{k \to \infty} \lambda_X \left(\bigcap_{\mathbf{n} \in S} \alpha^{-k\mathbf{n}} B_{\mathbf{n}} \right) = \prod_{\mathbf{n} \in S} \lambda_X(B_{\mathbf{n}}) \tag{10}$$

for all Borel sets $B_{\mathbf{n}} \subset X$, $\mathbf{n} \in S$, and *nonmixing* otherwise. If α is r-mixing, then every set $S \subset \mathbb{Z}^d$ with cardinality $|S| = r$ is obviously mixing. The validity of the reverse implication for algebraic $\mathbb{Z}^d$-actions is an open problem (cf. Problem 2.11 and Conjecture 2.12).

As in (10) one sees that a nonempty finite set $S \subset \mathbb{Z}^d$ is nonmixing if and only if there exist elements $a_{\mathbf{n}} \in M$, $\mathbf{n} \in S$, not all equal to zero, such that

$$\sum_{\mathbf{n} \in S} u^{k\mathbf{n}} \cdot a_{\mathbf{n}} = 0 \tag{11}$$

for infinitely many $k \geq 1$.

The higher order mixing behaviour of an algebraic $\mathbb{Z}^d$-action α with dual module M is again completely determined by that of the actions $\alpha_{R_d/\mathfrak{p}}$ with $\mathfrak{p} \in \mathrm{asc}(M)$.

Theorem 2.2. *Let α be an algebraic $\mathbb{Z}^d$-action on a compact abelian group X with dual module $M = \widehat{X}$.*

(1) *For every $r \geq 2$, the following conditions are equivalent:*
    a) *α is r-mixing,*
    b) *$\alpha_{R_d/\mathfrak{p}}$ is r-mixing for every $\mathfrak{p} \in \mathrm{asc}(M)$.*

(2) *For every nonempty finite set $S \subset \mathbb{Z}^d$, the following conditions are equivalent:*
    a) *S is α-mixing,*
    b) *S is $\alpha_{R_d/\mathfrak{p}}$-mixing for every $\mathfrak{p} \in \mathrm{asc}(M)$.*

In order to exhibit the connection between mixing properties and additive relations in fields we begin with a theorem by Mahler.

Theorem 2.3. ([9]) *Let K be a field of characteristic 0, $r \geq 2$, and let $x_1, \ldots, x_r$ be nonzero elements of K. If we can find nonzero elements $c_1, \ldots, c_r$ such that the equation*

$$\sum_{i=1}^{r} c_i x_i^k = 0$$

has infinitely many solutions $k \geq 0$, then there exist integers $s \geq 1$ and i, j with $1 \leq i < j \leq r$ such that $x_i^s = x_j^s$.

We denote by K the field of fractions of the integral domain $R_d/\mathfrak{p}$, choose a finite set $S = \{\mathbf{n}_1, \ldots, \mathbf{n}_r\} \subset \mathbb{Z}^d$ with $r \geq 2$, and set $x_i = u^{\mathbf{n}_i}$ for $i = 1, \ldots, r$. In view of Figure 1(4)–(5), Lemma 2.1, (8), (11) and Theorem 2.2, Theorem 2.3 implies (and is, in fact, equivalent to) the following statement:

Theorem 2.4. ([14]) *Let α be a mixing algebraic $\mathbb{Z}^d$-action on a compact connected abelian group X. Then every nonempty finite subset $S \subset \mathbb{Z}^d$ is mixing.*

If an algebraic $\mathbb{Z}^d$-action α is not mixing of every order, then there exists a smallest integer $r \geq 2$ such that α is not r-mixing. As a consequence of Lemma 2.1 and (9) one obtains the equivalence of Theorems 2.5 and 2.6 below.

Theorem 2.5. ([1, 17]) *Let K be a field of characteristic 0 and G a finitely generated multiplicative subgroup of $K^\times = K \smallsetminus \{0\}$. If $r \geq 2$ and $(c_1, \ldots, c_r) \in (K^\times)^r$, then the equation*

$$\sum_{i=1}^{r} c_i x_i = 0 \tag{12}$$

has only finitely many solutions $(x_1, \ldots, x_r) \in G^r$ such that no sub-sum of (12) vanishes.

Theorem 2.6. ([15]) *Let α be a mixing algebraic $\mathbb{Z}^d$-action on a compact connected abelian group X. Then α is mixing of every order.*

The 'absolute' version of the S-unit theorem in [1] contains a bound on the number of solutions of (12) without vanishing subsums which is expressed purely in terms of the integer r and the rank of the group G (in our setting: the order of mixing and the rank of the group $\mathbb{Z}^d$). This bound could be used, for example, to obtain quite remarkable uniform statements on the speed of multiple mixing for all irreducible and mixing algebraic $\mathbb{Z}^d$-actions (cf. Definition 3.1).

For algebraic $\mathbb{Z}^d$-actions on disconnected groups the situation is considerably more complicated due to the possible presence of nonmixing sets (cf. (10)).

Example 2.7. ([7]) *Let $\mathfrak{p} = (2, 1 + u_1 + u_2) = 2R_2 + (1 + u_1 + u_2)R_2$, $M = R_2/\mathfrak{p}$, and let $\alpha = \alpha_M$ be the algebraic $\mathbb{Z}^2$-action on $X = X_M = \widehat{M}$ defined in Example 1.1(2). Then α is mixing by Figure 1(4), but not three-mixing.*

Indeed, $(1 + u_1 + u_2)^{2^n} \cdot a = 0$ for every $n \geq 0$ and $a \in M$. For $a = 1 + (2, 1 + u_1 + u_2) \in M$ our identification of M with $\widehat{X}$ in Example 1.1(2) implies that $x_{(0,0)} + x_{(2^n,0)} + x_{(0,2^n)} = 0 \pmod 1$ *for every* $x \in X$ *and* $n \geq 0$. *For* $B = \{x \in X : x_{(0,0)} = 0\}$ *it follows that*

$$B \cap \alpha^{-(2^n,0)}(B) \cap \alpha^{-(0,2^n)}(B) = B \cap \alpha^{-(2^n,0)}(B),$$

and hence that

$$\lambda_X(B \cap \alpha^{-(2^n,0)}(B) \cap \alpha^{-(0,2^n)}(B)) = \lambda_X(B \cap \alpha^{-(2^n,0)}(B)) = 1/4$$

for every $n \geq 0$. *If* α *were three-mixing, we would have that*

$$\lim_{n \to \infty} \lambda_X(B \cap \alpha^{-(2^n,0)}(B) \cap \alpha^{-(0,2^n)}(B)) = \lambda_X(B)^3 = 1/8.$$

By comparing this with (10) *we see that the set* $S = \{(0,0),(1,0),(0,1)\} \subset \mathbb{Z}^2$ *is nonmixing.*

A mixing algebraic $\mathbb{Z}^d$-action α on a disconnected compact abelian group X has nonmixing sets if and only if it is not Bernoulli (cf. Figure 1(8), [5] and [16, Section 27]). In particular, if α is an ergodic algebraic $\mathbb{Z}^d$-action on a compact zero-dimensional abelian group X with zero entropy, then α has nonmixing sets. The description of the nonmixing sets of such an action α is facilitated by a Theorem of Masser ([5, 10]), which should be seen as an analogue of Theorem 2.3 in positive characteristic.

Theorem 2.8. *Let* K *be an algebraically closed field of characteristic* $p > 0$, $r \geq 2$, *and let* $(x_1, \ldots, x_r) \in (K^\times)^r$. *The following conditions are equivalent:*
(1) *There exists an element* $(c_1, \ldots, c_r) \in (K^\times)^r$ *such that*

$$\sum_{i=1}^r c_i x_i^k = 0$$

for infinitely many $k \geq 0$;
(2) *There exists a rational number* $s > 0$ *such that the set* $\{x_1^s, \ldots, x_r^s\}$ *is linearly dependent over the algebraic closure* $\bar{F}_p \subset K$ *of the prime field* $F_p = \mathbb{Z}/p\mathbb{Z}$.

Corollary 2.9. *Let* $\mathfrak{p} \subset R_d$ *be a prime ideal containing a rational prime* $p > 1$, *and let* $\alpha = \alpha_{R_d/\mathfrak{p}}$ *be the algebraic* $\mathbb{Z}^d$-action on $X = X_{R_d/\mathfrak{p}}$ *defined in Example 1.1(2). We denote by* $K = Q(R_2/\mathfrak{p}) \supset R_2/\mathfrak{p}$ *the quotient field of* $R_d/\mathfrak{p}$, *write* $\bar{K}$ *for its algebraic closure, and set* $x_{\mathbf{n}} = u^{\mathbf{n}} + \mathfrak{p} \in R_d/\mathfrak{p} \subset K \subset \bar{K}$ *for every* $\mathbf{n} \in \mathbb{Z}^d$. *If* $S \subset \mathbb{Z}^d$ *is a nonempty finite set, then the following conditions are equivalent:*
(1) S *is not* α-mixing;
(2) *There exists a rational number* $s > 0$ *such that the set* $\{x_1^s, \ldots, x_r^s\} \subset \bar{K}$ *is linearly dependent over* $\bar{F}_p \subset K$.

Examples 2.10. ([5])
(1) *In the notation of Examples 2.7 and 1.1(2) we set* $f = 1 + u_1 + u_2 + u_1^2 + u_1 u_2 + u_2^2 \in R_2$ *and put* $\mathfrak{p} = (2, f) \subset R_2$, $M = R_2/\mathfrak{p}$, $\alpha = \alpha_M$ *and* $X = X_M = \widehat{M}$. *We claim that the set* $S = \{(0,0),(1,0),(0,1)\}$ *is nonmixing.*

In order to verify this we define $\{x_{\mathbf{n}} : \mathbf{n} \in \mathbb{Z}^2\} \subset K = Q(R_2/\mathfrak{p})$ as in Corollary 2.9 and choose $\omega \in \bar{F}_2 \subset \bar{K}$ with $1 + \omega + \omega^2 = 0$. Since

$$f = (1 + \omega u_1 + \omega^2 u_2)(1 + \omega^2 u_1 + \omega u_2),$$

we obtain that $x_{(0,0)} + \omega x_{(1,0)} + \omega^2 x_{(0,1)} = 0$, so that S is nonmixing by Corollary 2.9.

Since the element $\omega' = \frac{1+u_1}{u_1+u_2} + \mathfrak{p} \in K$ satisfies that $1 + \omega' + \omega'^2 = 0$, we can recover (11) from the fact that

$$(u_1 + u_2) + (1 + u_2)u_1^{3k} + (1 + u_1)u_2^{3k} \in \mathfrak{p}$$

for every $k \geq 0$.

(2) *Let $g = 1 + u_1 + u_2 + u_1^2 + u_1 u_2 + u_2^2 + u_1^3 + u_1^2 u_2 + u_1 u_2^2 + u_2^3$ and $\mathfrak{q} = (2, g) \subset R_2$, $M = R_2/\mathfrak{q}$, $\alpha = \alpha_M$ and $X = X_M = \widehat{M}$. We claim that the set $S = \{(0,0), (1,0), (0,1)\}$ is again nonmixing.*

In Example (1) above we used the fact that f is irreducible over F_2, but not over $\bar{F}_2$. Here the polynomial g is irreducible over $\bar{F}_2$; however, the polynomial $g(u_1^3, u_2^3)$ turns out to be divisible by $1 + u_1 + u_2$, which can be translated into the statement that the set $\{x_{(0,0)}^{1/3}, x_{(1,0)}^{1/3}, x_{(0,1)}^{1/3}\}$ is linearly dependent over $\bar{F}_2$.

The main open question concerning higher order mixing is the following:

Problem 2.11. *Let α be an algebraic $\mathbb{Z}^d$-action on a compact abelian group X, and let $r \geq 2$. If every subset $S \subset \mathbb{Z}^d$ of cardinality r is mixing, is α r-mixing?*

A positive answer to Problem 2.11 would be equivalent to the following analogue of Theorem 2.5 in characteristic $p > 0$:

Conjecture 2.12. *Let K be an algebraically closed field of characteristic $p > 0$, $G \subset K^\times = K \smallsetminus \{0\}$ a finitely generated multiplicative group, $r \geq 2$, and $(c_1, \ldots, c_r) \in (K^\times)^r$. Let us call a solution $(x_1, \ldots, x_r) \in G^r$ of the equation*

$$\sum_{i=1}^{r} c_i x_i = 0 \tag{13}$$

regular *if there exists a rational number $s > 0$ such that $\{x_1^s, \ldots, x_r^s\}$ is linearly dependent over $\bar{F}_p \subset K$, and* irregular *otherwise.*

Then Equation (13) has only finitely many irregular solutions.

3. Conjugacy of Algebraic $\mathbb{Z}^d$-Actions

Every algebraic $\mathbb{Z}^d$-action α with completely positive entropy is measurably conjugate to a Bernoulli shift (cf. Figure 1(8)). Since entropy is a complete invariant for measurable conjugacy of Bernoulli shifts by [11], α is measurably conjugate to the $\mathbb{Z}^d$-action

$$\alpha^A : \mathbf{n} \mapsto \alpha^{A\mathbf{n}}$$

for every $A \in \mathrm{GL}(d, \mathbb{Z})$, since the entropies of all these actions coincide. In general, however, α and α^A are not topologically conjugate.

Every algebraic $\mathbb{Z}^d$-action α with positive entropy has Bernoulli factors by [8] and [12], and two such actions may again be measurably conjugate without being algebraically or topologically conjugate. For zero entropy actions, however, there is some evidence for a very strong form of isomorphism rigidity. Let us begin with a special case.

Definition 3.1. *An algebraic $\mathbb{Z}^d$-action α on a compact abelian group X is irreducible if every closed, α-invariant subgroup $Y \subsetneq X$ is finite.*

Irreducibility is an extremely strong hypothesis: if α is mixing it implies that $\alpha^{\mathbf{n}}$ is Bernoulli with finite entropy for every nonzero $\mathbf{n} \in \mathbb{Z}^d$. If β is a second irreducible and mixing algebraic $\mathbb{Z}^d$-action on a compact abelian group Y such that $h(\alpha^{\mathbf{n}}) = h(\beta^{\mathbf{n}})$ for every $\mathbf{n} \in \mathbb{Z}^d$, then $\alpha^{\mathbf{n}}$ is measurably conjugate to $\beta^{\mathbf{n}}$ for every $\mathbf{n} \in \mathbb{Z}^d$. However, if $d > 1$, then the actions α and β are generally nonconjugate.

Theorem 3.2. *([2, 6]) Let $d > 1$, and let α and β be irreducible and mixing algebraic $\mathbb{Z}^d$-actions on compact abelian groups X and Y, respectively. If $\phi \colon X \longrightarrow Y$ is a measurable conjugacy of α and β, then ϕ is λ_X-a.e. equal to an affine map (a map $\phi \colon X \longrightarrow Y$ is affine if it is of the form $\phi(x) = \psi(x) + y$ for every $x \in X$, where $\psi \colon X \longrightarrow Y$ is a continuous group isomorphism and $y \in Y$). In particular, measurable conjugacy implies algebraic conjugacy.*

If the irreducible actions α and β in Theorem 3.2 are of the form $\alpha = \alpha_{R_d/\mathfrak{p}}$ and $\beta = \alpha_{R_d/\mathfrak{q}}$ for some prime ideals $\mathfrak{p}, \mathfrak{q} \subset R_d$, then measurable conjugacy implies that $\mathfrak{p} = \mathfrak{q}$. This allows the construction of algebraic $\mathbb{Z}^d$-actions with very similar properties which are nevertheless measurably nonconjugate.

Example 3.3. *Consider the algebraic $\mathbb{Z}^2$-actions α, α', α'' on $X = \mathbb{T}^3$ generated by the matrices*

$$A = \begin{pmatrix} 0 & 1 & 0 \\ 0 & 0 & 1 \\ -1 & 8 & 2 \end{pmatrix} \quad and \quad B = \begin{pmatrix} 2 & 1 & 0 \\ 0 & 2 & 1 \\ -1 & 8 & 4 \end{pmatrix},$$

$$A' = \begin{pmatrix} -1 & 2 & 0 \\ -1 & 1 & 1 \\ -6 & 9 & 2 \end{pmatrix} \quad and \quad B' = \begin{pmatrix} 1 & 2 & 0 \\ -1 & 3 & 1 \\ -6 & 9 & 4 \end{pmatrix},$$

$$A'' = \begin{pmatrix} -3 & 4 & 0 \\ -3 & 3 & 1 \\ -10 & 11 & 2 \end{pmatrix} \quad and \quad B'' = \begin{pmatrix} -1 & 4 & 0 \\ -3 & 5 & 1 \\ -10 & 11 & 4 \end{pmatrix},$$

respectively. In [2] it was shown that these actions are not measurably conjugate, although it appears difficult to distinguish them with the usual invariants of measurable conjugacy.

Example 3.4. (Nonconjugacy of $\mathbb{Z}^2$-actions with positive entropy) *Let*

$$f_1 = 1 + u_1 + u_1^2 + u_1 u_2 + u_2^2\,,$$

$$f_2 = 1 + u_1^2 + u_2 + u_1 u_2 + u_2^2\,,$$

$$f_3 = 1 + u_1 + u_1^2 + u_2 + u_2^2\,,$$

$$f_4 = 1 + u_1 + u_1^2 + u_2 + u_1 u_2 + u_2^2\,,$$

in R_2, put $\mathfrak{p}_i = (2, f_i) \subset R_2$, $J_i = (4, f_i) \subset R_2$, $M_i = R_2/J_i$, and define the algebraic $\mathbb{Z}^2$-actions $\alpha_i = \alpha_{R_2/J_i}$ on $X_i = X_{R_2/J_i}$ as in Example 1.1(2). Then $h(\alpha_{R_2/\mathfrak{q}}) = \log 2$ and $h(\alpha_{R_2/\mathfrak{p}_i}) = 0$, and [8, Theorem 6.5] implies that the Pinsker algebra $\pi(\alpha_i)$ of α_i is the sigma-algebra $\mathcal{B}_{X_i/Y_i}$ of Y_i-invariant Borel sets in X_i, where $Y_i = N_i^\perp$ and

$$N_i = \{a \in M_i : \mathfrak{p}_i \cdot a = 0\} = 2M_i \cong R_2/\mathfrak{p}_i\,.$$

In other words, the $\mathbb{Z}^2$-action β_i induced by α_i on the Pinsker algebra $\pi(\alpha_i)$ is measurably conjugate to $\alpha_{R_2/\mathfrak{p}_i}$.

Since any measurable conjugacy of α_i and α_j would map $\pi(\alpha_i)$ to $\pi(\alpha_j)$ and induce a conjugacy of β_i and β_j, Theorem 3.2 implies that α_i and α_j are measurably nonconjugate for $1 \le i < j \le 4$.

The basic idea of the proof of Theorem 3.2 in [2] and [5] was suggested by Thouvenot: if $\phi \colon X \longrightarrow Y$ is a measurable conjugacy of α and β, then there exists a unique probability measure ν on the graph $\Gamma(\phi) = \{(x, \phi(x)) : x \in X\} \subset X \times Y$ which projects to λ_X and λ_Y, respectively, and which is invariant under the product-action $\alpha \times \beta \colon \mathbf{n} \mapsto \alpha^{\mathbf{n}} \times \beta^{\mathbf{n}}$ of $\mathbb{Z}^d$ on $X \times Y$. Since $\alpha \times \beta$, acting on $(X \times Y, \nu)$, is measurably conjugate both to α and to β, the measure ν is mixing and has positive entropy under $\alpha^{\mathbf{n}} \times \beta^{\mathbf{n}}$ for every nonzero $\mathbf{n} \in \mathbb{Z}^d$. The proof of Theorem 3.2 consists of showing that ν is a translate of the Haar measure of some closed $(\alpha \times \beta)$-invariant subgroup of $X \times Y$ (this obviously implies that ϕ is affine). If X and Y are connected, the relevant property of ν follows from [3], and if X and Y are zero-dimensional, the nonmixing sets of ν provide the necessary tool in [6].

Since there are considerable difficulties in extending either of these techniques to general algebraic $\mathbb{Z}^d$-actions with zero entropy, the following conjecture may seem a little premature, but I would still like to risk stating it:

Conjecture 3.5. *Let $d > 1$, and let α and β be mixing algebraic $\mathbb{Z}^d$-actions on compact abelian groups X and Y, respectively. If $h(\alpha) = 0$, and if $\phi \colon X \longrightarrow Y$ is a measurable conjugacy of α and β, then ϕ is λ_X-a.e. equal to an affine map. In particular, measurable conjugacy implies algebraic conjugacy.*

References

[1] J.-H. Evertse, H.-P. Schlickewei and W. Schmidt, *Linear equations in variables which lie in a multiplicative group* (Preprint).

[2] A. Katok, S. Katok and K. Schmidt, *Rigidity of measurable structure for algebraic actions of higher-rank abelian groups*, ESI-Preprint: ftp://ftp.esi.ac.at/pub/Preprints/esi850.ps.

[3] A. Katok and R. J. Spatzier, *Invariant measures for higher-rank hyperbolic abelian actions*, Ergod. Th. & Dynam. Sys. **16** (1996), 751–778; *Corrections*, **18** (1998), 507–507.

[4] B. Kitchens and K. Schmidt, *Automorphisms of compact groups*, Ergod. Th. & Dynam. Sys. **9** (1989), 691–735.

[5] B. Kitchens and K. Schmidt, *Mixing sets and relative entropies for higher dimensional Markov shifts*, Ergod. Th. & Dynam. Sys. **13** (1993), 705–735.

[6] B. Kitchens and K. Schmidt, *Isomorphism rigidity of irreducible algebraic $\mathbf{Z}^d$-actions*, ESI-Preprint: ftp://ftp.esi.ac.at/pub/Preprints/esi761.ps.

[7] F. Ledrappier, *Un champ markovien peut être d'entropie nulle et mélangeant*, C. R. Acad. Sci. Paris Sér. I Math. **287** (1978), 561–562.

[8] D. Lind, K. Schmidt and T. Ward, *Mahler measure and entropy for commuting automorphisms of compact groups*, Invent. Math. **101** (1990), 593–629.

[9] K. Mahler, *Eine arithmetische Eigenschaft der Taylor-Koeffizienten rationaler Funktionen*, Nederl. Akad. Wetensch. Proc. Ser. A **38** (1935), 50–60.

[10] D. Masser, *Two letters to D. Berend*, dated 12th and 19th September, 1985.

[11] D. S. Ornstein and B. Weiss, *Entropy and isomorphism theorems for actions of amenable groups*, J. Analyse Math. **48** (1987), 1–141.

[12] D. J. Rudolph and K. Schmidt, *Almost block independence and Bernoullicity of $\mathbb{Z}^d$-actions by automorphisms of compact groups*, Invent. Math. **120** (1995), 455–488.

[13] K. Schmidt, *Automorphisms of compact abelian groups and affine varieties*, Proc. London Math. Soc. **61** (1990), 480–496.

[14] K. Schmidt, *Mixing automorphisms of compact groups and a theorem by Kurt Mahler*, Pacific J. Math. **137** (1989), 371–384.

[15] K. Schmidt and T. Ward, *Mixing automorphisms of compact groups and a theorem of Schlickewei*, Invent. Math. **111** (1993), 69–76.

[16] K. Schmidt, *Dynamical Systems of Algebraic Origin*, Birkhäuser, Basel-Berlin-Boston, 1995.

[17] A. J. van der Poorten and H. P. Schlickewei, *Additive relations in fields*, J. Austral. Math. Soc. Ser. A **51** (1991), 154–170.

Mathematics Institute, University of Vienna
Strudlhofgasse 4, A-1090 Vienna, Austria

Erwin Schrödinger Institute for Mathematical Physics
Boltzmanngasse 9, A-1090 Vienna, Austria
E-mail address: klaus.schmidt@univie.ac.at

Self-Interacting Random Motions

Bálint Tóth

Abstract. We present a brief survey of results concerning self-interacting random walks and self-repelling continuous random motions. A self-interacting random walk (SIRW) is a nearest neighbour walk on the one-dimensional integer lattice $\mathbb{Z}$ which starts from the origin and at each step jumps to a neighbouring site, the probability of jumping along a bond being proportional to w(number of previous jumps along that lattice bond), where $w\colon \mathbb{N} \to \mathbb{R}_+$ is a monotone weight function. We consider various weight functions, the most natural and most interesting one being $w(n) = \exp\{-\beta n\}$, where $\beta > 0$ is a fixed constant parameter. This weight function defines the so-called 'myopic self-repelling' random walk. Other weight functions are also considered. We present functional limit theorems for the local time processes of these random walks and limit theorems for the position of the random walker at late times, under anomalous scaling rate. A generalization of the Ray-Knight theory of local time is in the background of these results.

In the second part of this note we present results concerning the construction and primary properties of a continuous, locally self-repelling process X_t. The process is a.s. continuous and recurrent, it has a regular occupation time density (local time) denoted $L_t(x)$, and the self-repellence of its trajectories is achieved by the dynamical driving mechanism formally expressed as $dX_t = -\operatorname{grad} L_t(X_t)dt$. This means that the process X_t is instantaneously pushed in the direction of the decrease of its local time. The constructed process is self-similar with scale-exponent $\nu = 2/3$ and has non-trivial local variation of order $3/2$ (in contrast with the finite quadratic variation of semi-martingales).

This note is an abridged version of the survey paper [16]. Full proofs of the cited results can be found in [12]–[15] and [17].

1. Self-Interacting Random Walks on $\mathbb{Z}$

The need for investigation of late-time asymptotics of random walks with long memory arose naturally in the probability and statistical physics literature. See e.g. [1, 2, 3, 6, 9, 10] etc. and the survey chapters of [7] and [8] for the historical origins of the problems. We present a survey of results concerning the long time asymptotics of *self-interacting random walks* on $\mathbb{Z}$ defined as follows: the walk X_i starts from the origin of the lattice and at time $i + 1$ it jumps to one of the two neighbouring sites of X_i, so that the probability of jumping along a bond of the lattice is proportional to

$$w(\text{number of previous jumps along that bond})$$

where

$$w \colon \mathbb{N} \to \mathbb{R}_+$$

is a weight function to be specified later. Complicated long-time memory effects are built up by the self-interaction mechanism driving these random walk. In papers [12]–[15] we developed a method which allowed us to prove limit theorems for the distributions of the late time position of a wide class of SIRWs.

1.1. Setup and main examples

Let $\mathbb{N} \ni i \mapsto X_i \in \mathbb{Z}$ be a nearest neighbour walk, starting from the origin of the lattice. I.e. $X_0 = 0$ and $|X_{i+1} - X_i| = 1$ for any $i \in \mathbb{N}$. For such a walk we define the (edge) local time process in the most natural usual way:

$$L_i(x) := \#\{j \in [0, i) : (X_j, X_{j+1}) = (x, x+1) \text{ or } (x+1, x)\}.$$

That is: $L_i(x)$ is the number of jumps across the edge $\langle x, x+1 \rangle$ in either direction, performed before time i.

Let $w \colon \mathbb{N} \to \mathbb{R}_+$ be a weight function, either monotone non-decreasing or monotone non-increasing. The SIRW defined by the weight function $w(\cdot)$ is a nearest neighbour walk X_i on $\mathbb{Z}$ with $X_0 = 0$ and governed by the law

$$\mathbf{P}\left(X_{i+1} = X_i + 1 \,\middle|\, \underline{X}_0^i\right) = \frac{w\big(L_i(X_i)\big)}{w\big(L_i(X_i)\big) + w\big(L_i(X_i - 1)\big)}$$

$$= 1 - \mathbf{P}\left(X_{i+1} = X_i - 1 \,\middle|\, \underline{X}_0^i\right), \tag{1}$$

where we used the shorthand notation $\underline{X}_0^i := (X_0, X_1, \ldots, X_i)$. In plain words: the random walker jumps to one of the two nearest neighbour sites so that the probability of jumping across an edge of the lattice is always proportional to the weight associated to the number of previous jumps across that edge. It is intuitively clear that monotone non-increasing weight functions define self-repelling walks while monotone non-decreasing weight functions define self-attracting walks. It is also clear that in the conditional jump probabilities (1) the full past history of the walk plays an important role, so these walks have extremely long memory, they are strongly non-Markovian. So, one should expect interesting, non-trivial long time asymptotic behaviour.

It turns out that the asymptotic behaviour of the SIRW is very sensitive to the choice of the weight function $w(\cdot)$. The following classes of weight functions have been considered:

(1) In the *exponentially self-repelling* case

$$w(n) = \exp(-\beta n), \qquad \beta > 0,$$

the random walk scales like $n^{-2/3} X_n$ (see [12]).

(2) The *subexponentially self-repelling* random walk governed by

$$w(n) = \exp(-\beta n^\kappa), \qquad \beta > 0, \ \kappa \in (0,1)$$

scales like $n^{-(\kappa+1)/(\kappa+2)} X_n$ (see [13]).

(3) In case of *power law self-interaction*

$$w(n) = \frac{1}{1-\alpha} \left(\frac{n}{2}\right)^\alpha - \frac{B}{(1-\alpha)^2} \left(\frac{n}{2}\right)^{\alpha-1} + \mathcal{O}(n^{\alpha-2}), \qquad B \in \mathbb{R}, \ \alpha \in (-\infty, 1).$$

We found three very different sub-cases (see [14]): If $\alpha \in (-\infty, 0)$ (polynomial self-repellence), then the correct scaling is $n^{-1/2} X_n$ but the limit law is not gaussian and does not depend essentially on the parameters. If $\alpha = 0$ (asymptotic freedom), then the correct scaling is again $n^{-1/2} X_n$ but the limit law will depend essentially on the value of

$$\delta := 2w(0)^{-1} + 2 \sum_{j=1}^\infty \left(w(2j)^{-1} - w(2j-1)^{-1}\right).$$

For related results on the so-called random walk and Brownian motion perturbed at extrema see also [4, 5, 11, 18] etc.

(4) Finally, if $\alpha \in (0, 1)$ (weak reinforcement) then the walk scales like $n^{-(1-\alpha)/(2-\alpha)} X_n$ (see [15]).

1.2. Limit theorems for the exponentially self-repelling random walk

Pars pro toto we present in some detail the main results (limit theorems) concerning the 'physically' most interesting case, the so-called 'true self-avoiding' or exponentially self-repelling walk.

For a nearest neighbour random walk $\mathbb{N} \ni i \mapsto X_i \in \mathbb{Z}$ we define the upcrossing, downcrossing processes:

$$U_i(x) := \#\{j \in [0,i) : (X_j, X_{j+1}) = (x, x+1)\},$$
$$D_i(x) := \#\{j \in [0,i) : (X_j, X_{j+1}) = (x+1, x)\}, \qquad i \in \mathbb{N}, \ x \in \mathbb{Z}.$$

Clearly $L_i(x) := U_i(x) + D_i(x)$. It is straightforward that for any $i \geq 0$ we have

$$i = \sum_{y \in \mathbb{Z}} L_i(y). \tag{2}$$

The inverse local time processes are

$$T_{x,m}^{\mathrm{U}} := \inf\{i \in \mathbb{N} : U_i(x) \geq m\}, \qquad T_{x,m}^{\mathrm{D}} := \inf\{i \in \mathbb{N} : D_i(x) \geq m\}.$$

Finally, the local time processes stopped at inverse local times are

$$\Lambda_{x,m}^*(y) := L_{T_{x,m}^*}(y).$$

Hereafter the superscript $*$ stands for either U or D. In $\Lambda_{x,m}^*(y)$ one should think about $x \in \mathbb{Z}$ and $m \in \mathbb{Z}_+$ as fixed parameters and $y \in \mathbb{Z}$ variable. From (2) it follows that for any $x \in \mathbb{Z}$ and $m \in \mathbb{N}$

$$T^*_{x,m} = \sum_{y \in \mathbb{Z}} \Lambda^*_{x,m}(y).$$ (3)

The quantities defined above make perfectly good sense for any nearest neighbour walk on $\mathbb{Z}$.

We also need the *reflected-absorbed Brownian motion* process. Fix the parameters $x \in \mathbb{R}$ and $h \in \mathbb{R}_+$ and let $\mathbb{R} \ni y \mapsto W(y) \in \mathbb{R}$ be a two-sided standard Brownian motion with $W(0) = 0$. Then

$$\Lambda_{x,h}(y) := \left|W(y - x) + h\right| \mathbb{1}_{\{\omega^-_{x,h} \leq y \leq \omega^+_{x,h}\}}$$

where

$$\omega^-_{x,h} := \sup\left\{y < \min\{x, 0\} : \left|W(y - x) + h\right| = 0\right\},$$
$$\omega^+_{x,h} := \inf\left\{y > \max\{x, 0\} : \left|W(y - x) + h\right| = 0\right\}.$$

In plain words: $y \mapsto \Lambda_{x,h}(y)$ is a two-sided Brownian motion starting at 'time' $y = x$ from level $\Lambda_{x,h}(x) = h$ which in the 'time' interval $[\min\{x, 0\}, \max\{x, 0\}]$ is reflected at level 0 and outside this interval is absorbed at first hitting of level 0. The total area under the curve $y \mapsto \Lambda_{x,h}(y)$ is

$$T_{x,h} := \int_{-\infty}^{\infty} \Lambda_{x,h}(y)dy = \int_{\omega^-_{x,h}}^{\omega^+_{x,h}} \Lambda_{x,h}(y)dy.$$

If $(x, h) \in \mathbb{R} \times \mathbb{R}_+$ and $(x, h) \neq (0, 0)$ the random variable $T_{x,h}$ has an absolutely continuous distribution. Its density is $\varrho(t, x, h)$ with Laplace transform (in the t variable) $\widehat{\varrho}(s, x, h)$:

$$\varrho(t, x, h) := \frac{\partial}{\partial t}\mathbf{P}\left(T_{x,h} < t\right), \qquad \widehat{\varrho}(s, x, h) := s\int_0^{\infty} \varrho(t; x, h)e^{-st}dt.$$

We shall also use

$$\pi(t, x) := \int_0^{\infty} \varrho(t, x, h)dh, \qquad \widehat{\pi}(s, x) := \int_0^{\infty} \widehat{\varrho}(s, x, h)dh.$$

Scale invariance of Brownian motion implies:

$$\alpha\varrho(\alpha t, \alpha^{2/3}x, \alpha^{1/3}h) = \varrho(t, x, h), \qquad \alpha^{2/3}\pi(\alpha t, \alpha^{2/3}x) = \pi(t, x),$$
$$\alpha\widehat{\varrho}(\alpha^{-1}s, \alpha^{2/3}x, \alpha^{1/3}h) = \widehat{\varrho}(s, x, h), \qquad \alpha^{2/3}\widehat{\pi}(\alpha^{-1}s, \alpha^{2/3}x) = \widehat{\pi}(s, x).$$

Theorem 1.1. *Given $t \in (0, \infty)$ (respectively, $s \in (0, \infty)$) fixed, $(x, h) \mapsto \varrho(t, x, h)$ (respectively, $(x, h) \mapsto \widehat{\varrho}(s, x, h)$) is a probability density on $\mathbb{R} \times \mathbb{R}_+$. That is:*

$$\int_{-\infty}^{\infty} \pi(t, x)dx = 1 = \int_{-\infty}^{\infty} \widehat{\pi}(s, x)dx.$$

The following Ray-Knight-type invariance principle for the local time process (stopped at inverse local times) and its corollary are the clue to the asymptotic description of the self-interacting walk:

Theorem 1.2. *Let $x \in \mathbb{R}$, $h \in \mathbb{R}_+$ be fixed and the superscript $*$ stand for either U (upcrossing) or D (downcrossing). Then, as $n \to \infty$,*

$$(\sqrt{n}\sigma)^{-1}\Lambda^*_{[nx],[\sqrt{n}\sigma h]}([ny]) \Rightarrow \Lambda_{x,h}(y)\,,$$

in the function space $D(-\infty, \infty)$ endowed with the Skorohod topology. The value of σ is given in (1.23) of [12].

Using the identity (3) we get the following:

Corollary 1.3. *Under the same conditions*

$$\left(n^{3/2}\sigma\right)^{-1}T^*_{[nx],[\sqrt{n}\sigma h]} \Rightarrow T_{x,h}\,.$$

Let $s > 0$ be fixed and θ_n a geometrically distributed stopping time which is independent of the walk X_i,

$$\mathbf{P}\Big(\theta_N = k\Big) = (1 - e^{-s/n})e^{-sk/n}\,. \tag{4}$$

The next statement is a limit theorem for the distribution of the location of the walk X_i, stopped at the random stopping time θ_n, with $n \to \infty$. It follows from the conversion of the full information contained in the previous corollary.

Theorem 1.4. *Let $s > 0$ and $x \in \mathbb{R}$ be fixed and θ_n a geometric stopping time, independent of the walk X_i, distributed according to (4). Then, as $n \to \infty$,*

$$\mathbf{P}\Big(n^{-2/3}X_{\theta_n} < x\Big) \to \int_{-\infty}^x \widehat{\pi}(\sigma s, y)dy\,. \tag{5}$$

Remarks 1.5.
(1) *In the original paper a slightly stronger statement was proved: the local version of this limit theorem, i.e. pointwise convergence of the properly defined density functions, rather than convergence of the distribution functions.*
(2) *The statement in Theorem 1.4 is a little bit short of stating the limit theorem for deterministic time:*

$$\mathbf{P}\Big(n^{-2/3}X_{[nt]} < x\Big) \to \int_{-\infty}^x \pi(\sigma^{-1}t, y)dy\,. \tag{6}$$

In order to convert (5) to (6) some refined Tauberian argument would be needed, which we were not able to push through. But, of course, we can conclude that, if $X_{[nt]}$ obeys any limit law as $n \to \infty$, then (6) also must hold.

2. The True Self-Repelling Motion

2.1. Setup

In the previous section a limit theorem was stated for the *one-dimensional marginal distributions* of $X_t^{(n)} := n^{-2/3}X_{[nt]}$, where X_j was the exponentially self-repelling (or myopic self-avoiding) random walk on $\mathbb{Z}$, defined by the weight function $w(n) = \exp(-\beta n)$. Now we consider the problem of *invariance principle*, i.e. that of the weak convergence of the *process* $X_t^{(n)}$, as $n \to \infty$. The results presented in this section are quoted from [17], where we constructed the presumed limit-process: a robust, self-similar stochastic process $\mathbb{R}_+ \ni t \mapsto X_t \in \mathbb{R}$, with all the natural properties requested from a locally self-repelling continuous motion.

The following are the fundamental properties of the true self-repelling motion X_t constructed in [17]:

Continuity, recurrence: Almost surely, $X_0 = 0$, the process $t \mapsto X_t$ is continuous on $[0, \infty)$ and for any $x \in \mathbb{R}$, the set of times $\{t \geq 0 : X_t = x\}$ is unbounded.

Scaling: For all $\alpha > 0$, $(X_{\alpha t}, t \geq 0)$ and $(\alpha^{2/3} X_t, t \geq 0)$ are identical in law.

Local variation: For all $\varepsilon > 0$, define by induction $\theta_0^\varepsilon := 0$ and for all $n \geq 1$,

$$\theta_n^\varepsilon := \inf\{t > \theta_{n-1}^\varepsilon \ : \ |X_t - X_{\theta_{n-1}^\varepsilon}| = \varepsilon\} \,.$$

Then, for all $t \geq 0$,

$$\text{P-}\lim_{\varepsilon \downarrow 0} \varepsilon^{3/2} \sup\{n \geq 0 \ : \ \theta_n^\varepsilon \leq t\} = \frac{2}{\sqrt{\pi}} t \,.$$

(Here and in the sequel P-lim stands for limit in probability.)

Occupation-time density: Almost surely, for all $t \geq 0$, the occupation-time measure of X_s on the time-interval $[0, t]$ has a bounded density with respect to the Lebesgue measure and this density has a continuous version that we denote by $L_t(\cdot)$. We call $L_t(x)$ the local time of X at time t and position x.

Markov property of (X_t, μ_t): The process $(X_t, L_t(\cdot))_{t \geq 0}$, or equivalently the process $(X_t, \mu_t)_{t \geq 0}$, is a Markov process.

Locality: The self-interaction is *local* in the following sense: For all $t \geq 0$, the law of X just after t depends only on L_t restricted to the immediate neighbourhood of the point X_t. In other words, the process X_t is 'feeling' only the self-interaction due to *the germ* of its own past occupation-time measure at the points it is currently visiting.

The following property is of *crucial importance:* it describes in proper mathematical terms the phenomenon of local self-repellence.

Dynamical driving mechanism: There exists a random set $I \subset \mathbb{R}_+$, which is a.s. of full Lebesgue measure, such that for any $T \in I$

$$\text{P-}\lim_{\varepsilon \downarrow 0} \int_0^T \frac{L_s(X_s + \varepsilon) - L_s(X_s - \varepsilon)}{2\varepsilon} ds$$

$$= -X_T + \frac{1}{4}\left(\sup_{0 \leq s \leq T} X_s + \inf_{0 \leq s \leq T} X_s\right) . \qquad (7)$$

Unfortunately, we could prove (7) only for the random set of (stopping) times $I \subset \mathbb{R}_+$. Actually, this property should hold for all $T \in \mathbb{R}_+$. Phenomenologically, this equation states that the motion is driven by the negative gradient of the local time at the actual position, as long as the moving point is in the interior of the range swept in the past. This behaviour entitles us to call this process 'truly self-repelling'. In addition, at the edges of this range an instantaneous partial reflection (moving boundary condition) is felt. Indeed: writing (7) *formally* in differential form we find:

$$dX_t = -\frac{\partial L_t(X_t)}{\partial x} dt + \left(\text{ boundary effects at } \sup_{0 \le s \le t} X_s \text{ and } \inf_{0 \le s \le t} X_s \right). \qquad (8)$$

Strictly speaking, (8) does not make sense mathematically: the local time process is so singular that a 'differential equation' involving its gradient cannot be rigorously defined ($L_t(\cdot)$ has the same regularity properties as Brownian motion.) Nevertheless, this formal way of writing may help the intuition about the dynamics of the process. Note, that there is no 'external noice', or 'external source of randomness' in the driving mechanism. One could think naively that such a mechanism would give rise to a deterministic motion. This is not the case: due to the extremely high singularity of this "differential equation", (8) has only truly stochastic solutions.

One of the main novelties of the process X_t is exactly the fact that it is in striking contrast with our traditional intuition about a random motion being driven by local drift and external noice.

The full proof of the results can be found in [17]. The construction and derivation of key features of the process rely essentially on the construction and analysis of systems of independent coalescing Brownian paths, emerging from every point of a two-dimensional space-time.

In the next subsection we present a not fully rigorous, phenomenological derivation of the driving mechanism. For the technical parts of the construction and analysis see the original publication.

2.2. Phenomenological derivation of the dynamical driving mechanism

In [12] a limit theorem was proved, essentially for the distribution of $n^{-2/3}X_n$ as $n \uparrow \infty$, but the natural question of the asymptotics of the *process*

$$X_t^{(N)} := N^{-2/3} X_{[Nt]}, \qquad t \in \mathbb{R}_+ \qquad (9)$$

in the limit $N \uparrow \infty$ remained open. In the following paragraphs we argue that, if the sequence of processes $t \mapsto X_t^{(N)}$ converges in distribution to a process $t \mapsto X_t^{(\infty)}$, as $N \uparrow \infty$, then the limit process is driven by the gradient of its local time, as claimed in (8). The forthcoming argument is based on a somewhat *formal computation* and it is by no means mathematically rigorous, but it sheds light on the essential phenomenon of local self-repellence.

Beside the scaled position process $t \mapsto X_t^{(N)}$ defined in (9) we define the properly scaled local time process of the exponentially self-repelling walk

$$L_t^{(N)}(x) := N^{-1/3} L_{[Nt]}([N^{2/3}x]), \qquad t \in \mathbb{R}_+, \quad x \in \mathbb{R} \qquad (10)$$

and we assume that the sequences of processes $X_t^{(N)}$ and $L_t^{(N)}(x)$ converge jointly weakly (in some vague topological space):

$$\left(X_t^{(N)}, L_t^{(N)}(x)\right) \Rightarrow \left(X_t^{(\infty)}, L_t^{(\infty)}(x)\right)$$

where $(t, x) \mapsto L_t^{(\infty)}(x)$ is assumed to be the local time of the process $t \mapsto X_t^{(\infty)}$. Let $\mathcal{F}_n$ be the σ-algebra generated by $(X_0, \ldots, X_n)$, then

$$\mathbf{E}\left(X_{n+1} - X_n \,\middle|\, \mathcal{F}_n\right) = -\tanh\left(\beta(L_n(X_n) - L_n(X_n - 1))\right)$$

$$\mathbf{Var}\left(X_{n+1} - X_n \,\middle|\, \mathcal{F}_n\right) = \cosh^{-2}\left(\beta(L_n(X_n) - L_n(X_n - 1))\right).$$

So:

$$X_n + \sum_{k=0}^{n-1} \tanh\left(\beta(L_k(X_k) - L_k(X_k - 1))\right) =: M_n \tag{11}$$

is a martingale with quadratic variation process

$$\langle M \rangle_n = \sum_{k=0}^{n-1} \cosh^{-2}\left(\beta(L_k(X_k) - L_k(X_k - 1))\right) < n. \tag{12}$$

Our object of study is the scaled form of (11):

$$N^{-2/3} X_{[Nt]} + N^{-2/3} \sum_{k=0}^{[Nt]-1} \tanh\left(\beta(L_k(X_k) - L_k(X_k - 1))\right) = N^{-2/3} M_{[Nt]}. \tag{13}$$

The first term on the left-hand side of (13) is just $X_t^{(N)}$. From (12) in particular it follows that for any $T < \infty$

$$\text{P-} \lim_{N \uparrow \infty} \left(\sup_{0 \le t \le T} \left|N^{-2/3} M_{[Nt]}\right|\right) = 0 \tag{14}$$

so that the right-hand side of (13) is asymptotically negligible. A formal computation of the second term on the left-hand side of (13) follows: the first two steps are straightforward transformations using the definitions (9) and (10) of the scaled process and scaled local time:

$$N^{-2/3} \sum_{k=0}^{[Nt]-1} \tanh\left(\beta(L_k(X_k) - L_k(X_k - 1))\right)$$

$$= N^{-1} \sum_{k=0}^{[Nt]-1} N^{1/3} \tanh\left(\beta(L_{Nk/N}(N^{2/3} X_{k/N}^{(N)}) - L_{Nk/N}(N^{2/3} X_{k/N}^{(N)} - 1))\right)$$

$$= N^{-1} \sum_{k=0}^{[Nt]-1} N^{1/3} \tanh\left(\beta N^{1/3}(L_{k/N}^{(N)}(X_{k/N}^{(N)}) - L_{k/N}^{(N)}(X_{k/N}^{(N)} - N^{-2/3}))\right).$$

The next step is the formal, non-rigorous one: we treat formally $L_t^{(N)}(x)$ as a smooth function and replace

$$L_t^{(N)}(x) - L_t^{(N)}(x - \delta x) \qquad \text{by} \qquad \frac{\partial L_t^{(N)}(x)}{\partial x}\delta x$$

to get

$$N^{-2/3} \sum_{k=0}^{[Nt]-1} \tanh\big(\beta(L_k(X_k) - L_k(X_k - 1))\big)$$

$$\text{``}=\text{''}\ N^{-1} \sum_{k=0}^{[Nt]-1} N^{1/3} \tanh\Big(\beta N^{1/3} N^{-2/3}\frac{\partial L_{k/N}^{(N)}(X_{k/N}^{(N)})}{\partial x}\Big)$$

$$\text{``}=\text{''}\ \beta N^{-1} \sum_{k=0}^{[Nt]-1} \frac{\partial L_{k/N}^{(N)}(X_{k/N}^{(N)})}{\partial x} + \mathcal{O}(N^{-1/3})$$

$$\text{``}\Rightarrow\text{''}\ \beta \int_0^t \frac{\partial L_s^{(\infty)}(X_s^{(\infty)})}{\partial x}ds. \tag{15}$$

With the quotation marks "$\cdots$" we intend to emphasize that these last equalities and convergence need more careful consideration. From (13), (14) and (15) we get

$$X_t^{(\infty)} + \text{const.}\int_0^t \frac{\partial L_s^{(\infty)}(X_s^{(\infty)})}{\partial x}ds = 0$$

which is indeed somewhat reminiscent of (7). The effect of 'pushing the boundaries of the range' and the right constant in front of the gradient term can not be recovered on this level of formal computations. We repeat again: this computation is nothing like rigorous, but on the phenomenological level it is convincing.

The same reasoning (on the same level of 'rigour') can be applied to the 'polymer model' proposed by Durrett and Rogers in [6]:

$$X_t = B_t + \int_0^t \left\{ \int_0^s f(X_s - X_u)du \right\} ds$$

where $f\colon \mathbb{R} \to \mathbb{R}$ is a smooth function of compact support and satisfies $f(-x) = -f(x)$ and $\text{sgn}(f(x)) = \text{sgn}(x)$. Defining $X_t^{(N)} = N^{-2/3}X_{Nt}$, in the limit $N \to \infty$ f transforms into δ' and the same dynamical driving mechanism is found.

Acknowledgements

This work was partially supported by the following grants: OTKA T 26176 (National Fund for Scientific Research), FKFP 0638/1999 (Ministry of Culture and Education), TKI 'Stochastics@TUB' (Hungarian Academy of Sciences).

References

[1] D. Amit, G. Parisi and L. Peliti, *Asymptotic behaviour of the 'true' self-avoiding walk,* Phys. Rev. B, **27** (1983), 1635–1645.

[2] D. Coppersmith and P. Diaconis, *Random walks with reinforcements,* Stanford Univ. Preprint(1987)

[3] B. Davis, *Reinforced random walk,* Probab. Theory Relat. Fields, **84** (1990), 203–229.

[4] B. Davis, *Weak limits of preturbed random walks and the equation $Y_t = B_t + \alpha \sup_{s \leq t} Y_s + \beta \inf_{s \leq t} Y_s$,* Ann. Probab., **24** (1997), 2007–2023.

[5] B. Davis, *Brownian motion and random walk perturbed at extrema,* Probab. Theory Relat. Fields, **113** (1999), 501–518.

[6] R. T. Durrett and L. C. G. Rogers, *Asymptotic behaviour of a Brownian polymer,* Probab. Theory Relat. Fields,**92** (1992), 337–349.

[7] G. Lawler, *Intersections of Random Walks,* Birkhäuser, Boston–Basel–Berlin, 1991.

[8] N. Madras and G. Slade, *The Self-Avoiding Walk,* Birkhäuser, Boston–Basel–Berlin, 1993.

[9] S. P. Obukhov and L. Peliti, *Renormalisation of the "true" self-avoiding walk,* J. Phys. A, **16** (1983), L147–L151.

[10] L. Peliti and L. Pietronero, *Random walks with memory,* Riv. Nuovo Cimento, **10** (1987), 1–33.

[11] M. Perman and W. Werner, *Perturbed Brownian motions,* Probab. Theory Relat. Fields, **108** (1997), 357–383.

[12] B. Tóth, *The 'true' self-avoiding walk with bond repulsion on $\mathbb{Z}$: limit theorems,* Ann. Probab., **23** (1995), 1523–1556.

[13] B. Tóth, *'True' self-avoiding walk with generalized bond repulsion on $\mathbb{Z}$,* J. Statist. Phys., **77** (1994), 17–33.

[14] B. Tóth, *Generalized Ray-Knight theory and limit theorems for self-interacting random walks on $\mathbb{Z}$,* Ann. Probab., **24** (1996), 1324–1367.

[15] B. Tóth, *Limit theorems for weakly reinforced random walks,* Studia Sci. Math. Hungar., **33** (1997), 321–337.

[16] B. Tóth, *Self-interacting random motions – a survey,* In: P. Révész, B. Tóth (eds): Random Walks, Bolyai Society Mathematical Studies, **9** (1999), 349–384.

[17] B. Tóth and W. Werner, *The true self-repelling motion,* Probab. Theory Relat. Fields, **111** (1998), 375–452.

[18] W. Werner, *Some remarks on perturbed reflecting Brownian motion,* Séminaire de Probabilités XXIX., Lecture Notes in Mathematics 1613, 37–43, Springer-Verlag, 1995.

Institute of Mathematics
Technical University Budapest
H-1111 Budapest, Hungary
E-mail address: `balint@math.bme.hu`

Harmonic Analysis
on Reductive Symmetric Spaces

Erik van den Ban and Henrik Schlichtkrull

Abstract. We give a relatively non-technical survey of some recent advances
in the Fourier theory for semisimple symmetric spaces. There are three major results: An inversion formula for the Fourier transform, a Paley–Wiener
theorem, which describes the Fourier image of the space of compactly supported smooth functions, and the Plancherel theorem, which describes the
decomposition into irreducibles of the regular representation.

1. Introduction

The beautiful theory of Fourier series has many generalizations that match its
beauty. The majority of these generalizations concern the decomposition of functions on homogeneous spaces of a Lie group, such as, for example, the n-sphere S^n,
which is a homogeneous space for the rotation group $O(n+1)$. The harmonic analysis on S^n is the theory of expansions in spherical harmonics Y_l^m, and the Plancherel
theorem for S^n is the statement that these functions form an orthonormal basis
for $L^2(S^n)$ (with respect to the rotation invariant surface measure). This theory of
harmonic analysis on S^n was generalized to the compact Riemannian symmetric
spaces by Cartan, [25]. These are homogeneous spaces G/K, where G is a connected semisimple compact Lie group, and K is the subgroup consisting of all
points fixed by a given involution σ of G.

The key to Cartan's result is representation theory. The irreducible repre-
sentations $(\pi_\lambda, \mathcal{H}_\lambda)$ of G are determined by a discrete parameter λ, the highest
weight. For some (explicitly known, see [39, p. 535]) values of λ there exist non-
trivial K-fixed vectors in $\mathcal{H}_\lambda$. If that is the case, the space $\mathcal{H}_\lambda^K$ of such vectors is
one-dimensional. The spherical harmonics Y_l^m generalize as the functions (matrix
elements) $x \mapsto \langle \pi_\lambda(x)u, v \rangle$, where $u \in \mathcal{H}_\lambda^K$ is fixed and v runs through an orthonormal basis for $\mathcal{H}_\lambda$. The matrix element is a function on G/K, since u is K-fixed.
The Plancherel theorem for G/K asserts that these functions, appropriately normalized, form an orthonormal basis for $L^2(G/K)$ (with respect to an invariant
measure), and it gives the explicit decomposition of the regular representation of
G on this space, as the direct sum of those $\mathcal{H}_\lambda$ for which $\mathcal{H}_\lambda^K \neq 0$.

The non-compact Riemannian symmetric spaces also carry a beautiful theory of harmonic analysis. These spaces are realized similarly as G/K, with G a connected semisimple Lie group of the non-compact type, and K the subgroup of fixed points of an involution, such that K is compact (in fact, K is then a maximal compact subgroup of G). In this theory, which is due to Harish-Chandra, [32], and Helgason, the Fourier transform of a function on G/K is given by integration against (generalized) spherical functions with respect to invariant measure. The parameter set for the spherical functions is basically a Euclidean space, and the Fourier expansion of a function on G/K is an integral, as in the ordinary Fourier theory on $\mathbb{R}$. The measure, with respect to which the integral is taken, is determined through Harish-Chandra's c-function, which is explicitly known through the formula of Gindikin–Karpelevic (see [39] for details).

Again, the harmonic analysis just described is best understood through representation theory, but since G is non-compact, the representations will necessarily be infinite dimensional, as there exists no finite dimensional non-trivial unitary representation of G. As before, the space $\mathcal{H}^K$ of K-fixed vectors in an irreducible representation $(\pi, \mathcal{H})$ of G is one-dimensional if it is not trivial, and the spherical functions are obtained as matrix elements $\langle \pi(x)u, v \rangle$ with $u \in \mathcal{H}^K$, $v \in \mathcal{H}$.

In a different, but nevertheless closely related, direction, a Fourier theory exists for functions on G, where G is a semisimple Lie group. This theory is the formidable achievement of Harish-Chandra, [33], [34]-[36]. For a general semisimple Lie group, the Fourier decomposition of a function is a mixture of sums and integrals. The generalized spherical functions that serve as the building blocks in the expansion of a function on G are matrix elements of irreducible representations of G. From the point of view of representation theory, the harmonic analysis is the decomposition into irreducibles of the (left or right) regular representation of G on $L^2(G)$. See [42, 49] or [50] for details.

In the present survey, we discuss a theory of harmonic analysis, which generalizes both the theory for G/K and that for G. This is the theory of harmonic analysis on semisimple (or reductive) symmetric spaces. Again, the analysis takes place on a homogeneous space G/H with G a connected semisimple Lie group and H the subgroup of fixed points for an involution, but neither G nor H are assumed to be compact. That this is a generalization of the theory for G/K is obvious. To see that it also generalizes the theory for G, we note that G is a homogeneous space for $G \times G$ (which is semisimple, when G is) through the left times right action. In this fashion, G is identified with the quotient space of $G \times G$ by the diagonal subgroup, which is the set of fixed point for the involution $(x, y) \mapsto (y, x)$. Examples of semisimple symmetric spaces that are not of the form G/K, and not of the form G, are the hyperbolic spaces

$$\{x \in \mathbb{R}^{p+q} \mid x_1^2 + \cdots + x_p^2 - x_{p+1}^2 - \cdots - x_{p+q}^2 = 1\}$$

which are homogeneous spaces for the pseudo-orthogonal groups $O(p, q)$; only the special cases where $q = 0$, $p = 1$ or $p = q = 2$ are of the previously discussed types. There exists a classification of the semisimple symmetric spaces (up to local isomorphism), see [20]. All semisimple symmetric spaces carry an invariant measure, which is unique up to normalization (both G and H are unimodular groups).

The study of harmonic analysis on semisimple symmetric spaces in general was initiated by M. Flensted-Jensen, [30], who constructed a general family of representations in the discrete series. By definition this series consists of the irreducible representations of G that enter as discrete summands in the decomposition of the regular representation on $L^2(G/H)$, which means simply that they are subrepresentations of the regular representation. The study was continued by T. Oshima and T. Matsuki, [45], who gave a quite complete description of the discrete series, which generalizes Harish-Chandra's description for the group case. In particular, a necessary and sufficient condition on G/H for the existence of discrete series representations is given in [30] and [45]. For further details we refer to the survey paper [10] and the references given there.

From a representation theoretic point of view, the representations of the discrete series are probably the most interesting components in $L^2(G/H)$, since they are in some sense most singular. The Langlands parameters of these representations are known, see [47, 31]. However, for the purpose of harmonic analysis, the decomposition of the remainder of $L^2(G/H)$, which by definition is not discrete, is of equal importance. This will be our main focus in the present survey.

The complete decomposition of the non-discrete part of $L^2(G/H)$, the Plancherel theorem for G/H, has been accomplished more recently, through work of P. Delorme, [29], and of the authors, [19]. The various representation components in the decomposition of $L^2(G/H)$ are naturally grouped into a finite number of 'series', according to their spectral properties for the algebra $\mathbb{D}(G/H)$ of invariant differential operators on G/H. One of these series is the discrete series, when it exists. Another comprise the so-called most continuous part. The decomposition of the corresponding subspace $L^2_{\mathrm{mc}}(G/H) \subset L^2(G/H)$ was determined in [15]. Some details will be presented in Section 2 below. Each of the series carry both a discrete and a continuous parameter (in this sense, they resemble a mixture of the compact and non-compact Riemannian cases G/K discussed at the outset). The continuous parameter runs in a Euclidean space, and the most continuous series is distinguished by having that parameter space of maximal dimension. The series that are not discrete and not most continuous are called intermediate series. The representations that enter in the most continuous series, as well as in the intermediate series, are induced representations. The subgroup from which induction takes place is a parabolic subgroup, which is characteristic for the series.

We would like to mention also a closely related problem, which is the determination of the Paley–Wiener space for G/H. Recall that the Paley–Wiener theorem for $\mathbb{R}$ describes the Fourier image of the space of compactly supported L^2 functions. A variant of the theorem is the Paley–Wiener–Schwartz theorem, which describes the Fourier image of the space $C_c^\infty(\mathbb{R})$ of compactly supported smooth functions as the space of entire functions of exponential type (see for example [41,

Thm. 7.3.1]). This theorem was generalized to non-compact Riemannian symmetric spaces G/K by Helgason and Gangolli, see [41, Sect. IV.7], and to semisimple Lie groups G by Arthur, [2], and it has been further generalized to all reductive symmetric spaces G/H by the authors. The Fourier transform is the one associated with the most continuous series. It is a remarkable fact, established in [15], that the orthogonal projection of $C_c^\infty(G/H)$ onto $L_{\mathrm{mc}}^2(G/H)$ is injective, hence the Fourier transform associated with the most continuous series is injective on $C_c^\infty(G/H)$. The Paley–Wiener theorem for G/H describes the image under this Fourier transform of $C_c^\infty(G/H)$ (in fact, more precisely of its dense subspace of K-finite functions).

Both the Plancherel theorem and the Paley–Wiener theorem for G/H were announced in the seminar at the Mittag-Leffler Institute, Stockholm, in the fall of 1995. The Plancherel theorem was announced by Delorme; the proof has appeared in [28]-[29] (previously, Oshima (see [44, p. 604]) had announced a Plancherel formula, but the details have not appeared). The Paley–Wiener theorem was announced by the authors. Together with that the authors also announced that their proof implies the Plancherel formula under the hypothesis that certain identities, the so-called Maass–Selberg relations, are valid for G/H. The validity of these relations, which also play a main role in Delorme's work, as well as in Harish-Chandra's work for the group case, has been established for $L_{\mathrm{mc}}^2(G/H)$ in [7], and for the general case by Carmona and Delorme in [24]. Some details of the work of the authors have appeared in [16] and [17], the rest will appear in [18] and [19]. In the latter paper we will also include an independent proof, found later, of the Maass–Selberg relations. The methods of the two approaches to the Plancherel decomposition are in many respects different, though they both rely on the Maass–Selberg relations. For example, an important ingredient in Delorme's work, which is not used in the other approach, is an a priori characterization of the support of the Plancherel measure (cf. [23, Appendix C]), which in turn is derived from a result of Bernstein [21]. On the other hand, in the work of the authors, an inversion formula for the Fourier transform on $C_c^\infty(G/H)$ plays a crucial role (see [17]). The proof of this formula is based on a theory of multivariable residue calculus, developed for the occasion, [16], using ideas from Langlands [43] and Heckman–Opdam [37].

The present paper is meant to be a relatively non-technical survey. For more elaborate expositions, we refer to [38, Part II], and to [10].

2. The Fourier Transform

In this section we will briefly describe the Fourier transform associated with the most continuous part of $L^2(G/H)$. It is constructed out of the so-called *Eisenstein integrals* on G/H. We will also discuss the Plancherel theorem for $L_{\mathrm{mc}}^2(G/H)$. The main references are [6, 7, 13] and [15].

For the sake of completeness, let us first give the precise definition of the concept of a semisimple symmetric space, since it is actually slightly more general than what was described above. We assume that G is a connected semisimple Lie group with finite center, and that σ is an involution of G. The set G^σ of σ-fixed points in G is in general not connected, but its connected components are finite in number. The natural generality of spaces G/H is obtained by requiring that H is a subgroup of G^σ and that it contains the identity component of G^σ. For the purpose of the proofs of our main results, it would actually be more convenient to work at the outset with the more general class of reductive symmetric spaces, where the group G is allowed to be reductive (of so-called Harish-Chandra class). This is the generality of spaces used for example in [29] and [19], but in the present survey we will not go into this.

In order to describe the Fourier transform, the central issue is to determine the appropriate spherical functions. The point of departure is the matrix element formula $\langle \pi(x)u, v \rangle$ of the Riemannian case G/K. Recall that $(\pi, \mathcal{H})$ is an irreducible unitary representation of G, and that $u \in \mathcal{H}$ is K-fixed. Imitating this formula, we must look for the irreducible representations with a non-trivial H-fixed vector. Unfortunately, in general no such representations exist, unless the notion of representation vectors is extended. This complication appears already in the case of harmonic analysis on the group G, viewed as the symmetric space $G \times G / \operatorname{diag}(G)$. The irreducible unitary representations of $G \times G$ are of the form $(\pi_1 \otimes \pi_2, \mathcal{H}_1 \otimes \mathcal{H}_2)$, where the tensor product is taken in the category of Hilbert spaces. It can be seen (from Schur's lemma) that this representation can carry a $\operatorname{diag}(G)$-fixed vector only if π_1 is equivalent with the contragredient representation π_2^* of π_2. Then $\mathcal{H}_1 \otimes \mathcal{H}_2$ is identified with the space of Hilbert–Schmidt operators on $\mathcal{H}_1$. Again by Schur's lemma, the only $\operatorname{diag}(G)$-fixed operators on $\mathcal{H}_1$ are the constant multiples of the identity, which are not Hilbert–Schmidt if the representation is not finite-dimensional. Thus, in general there exists no non-trivial $\operatorname{diag}(G)$-fixed vector in $\mathcal{H}_1 \otimes \mathcal{H}_2$. Returning to the general case, we must therefore allow representations that carry H-fixed vectors in a suitable generalized sense. For Lie groups there exists a notion of generalized representation vectors, based on the notion of smooth representation vectors. Rather than explaining these notions in general, we will describe explicitly what they amount to in the present case.

We first need some notation, in order to describe the representations that occur. We will, however, try to keep the notation at a minimum, which means that some ambiguities may evolve. The group G has an Iwasawa decomposition, $G = KA_0N_0$. The subgroup K is a maximal compact subgroup, by conjugation we may arrange that $\sigma(K) = K$. Since K is the set of fixed points for the associated Cartan involution θ on G, the assumption $\sigma(K) = K$ amounts to assuming that σ and θ commute. Thus we have at our disposal three commuting involutions, σ, θ and their product $\sigma\theta$. We may also arrange that the abelian group A_0 is preserved by these involutions (whereas the nilpotent part N_0 will not be preserved, in general, by any of the involutions) and that $A_0 \cap H$ is of minimal dimension. Let M_0 be the centralizer of A_0 in K. The product $P_0 = M_0A_0N_0$ is a minimal parabolic subgroup of G. The principal series of representations of G is a series of representations induced from P_0. This is the series of representations that contribute to the decomposition of $L^2(G/K)$ as well as $L^2_{\mathrm{mc}}(G)$. However, for the general case of $L^2_{\mathrm{mc}}(G/H)$ we need to induce from a (possibly) different parabolic subgroup of G, in order to ensure the existence of H-fixed (generalized) vectors.

Let $\mathfrak{g}$ be the Lie algebra of G. By differentiation, an involution of G induces an involution of $\mathfrak{g}$, which we will denote by the same symbol. The involutions θ and σ then give rise to the decompositions

$$\mathfrak{g} = \mathfrak{k} \oplus \mathfrak{p} = \mathfrak{h} \oplus \mathfrak{q},$$

where $\mathfrak{k}$, $\mathfrak{p}$, $\mathfrak{h}$ and $\mathfrak{q}$ are the $+1$ and -1 eigenspaces. In particular, $\mathfrak{k}$ and $\mathfrak{h}$ are the Lie algebras of K and H. The Lie algebra $\mathfrak{a}_0$ of A_0 is a maximal abelian subspace of $\mathfrak{p}$. The intersection $\mathfrak{a}_{\mathrm{q}} := \mathfrak{a}_0 \cap \mathfrak{q}$ is a maximal abelian subspace of $\mathfrak{p} \cap \mathfrak{q}$. It determines a parabolic subgroup $P \supset P_0$ whose Levy part is the centralizer of $\mathfrak{a}_{\mathrm{q}}$ in G. As all parabolic subgroups, P admits a so-called Langlands decomposition, which is written as $P = MAN$. The Levy part of P is the product of M and A, and $N \subset N_0$ is the nilpotent part. The subgroup A is central in MA, its Lie algebra $\mathfrak{a}$ is contained in $\mathfrak{a}_0$ and satisfies $\mathfrak{a} \cap \mathfrak{q} = \mathfrak{a}_{\mathrm{q}}$. The parabolic subgroup P is characterized among the parabolic subgroups containing P_0 by being minimal among those which are stable for the involution $\sigma\theta$.

It is by induction from P that we construct the principal series of representations for G/H. More precisely, let $(\xi, \mathcal{H}_\xi)$ be a finite dimensional unitary representation of M and $\lambda \in i\mathfrak{a}^*$ an imaginary linear functional on $\mathfrak{a}$. The induced representation

$$\pi_{\xi,\lambda} = \mathrm{Ind}_P^G(\xi \otimes e^\lambda \otimes 1)$$

(where 1 stands for the trivial representation of N) is defined as follows. The representation space consists of functions $f \colon G \to \mathcal{H}_\xi$ satisfying

$$f(manx) = a^{\lambda+\rho}\xi(m)f(x), \quad (m \in M, a \in A, n \in N, x \in G) \tag{1}$$

and the action of G is the right action. Here $\rho \in \mathfrak{a}^*$ denotes half the trace of the adjoint action of $\mathfrak{a}$ on $\mathfrak{n} = \mathrm{Lie}(N)$ (the appearance of this is standard), and $a^{\lambda+\rho} = e^{(\lambda+\rho)(\log a)}$ by definition. It follows from (1) that f is uniquely determined by its restriction to K, since $G = PK$. The Hilbert space $\mathcal{H}_{\xi,\lambda}$ is the space of functions satisfying (1), for which $k \mapsto \|f(k)\|_\xi$ is in $L^2(K)$, and its inner product is given by

$$\langle f, g \rangle = \int_K \langle f(k), g(k) \rangle_\xi \, dk,$$

where dk is Haar measure on K. The space of smooth functions satisfying (1) is denoted $C^\infty_{\xi,\lambda}$; it is dense in $\mathcal{H}_{\xi,\lambda}$. The space of generalized functions (in the sense of distribution theory) satisfying the identity is denoted $C^{-\infty}_{\xi,\lambda}$. Then

$$C^\infty_{\xi,\lambda} \subset \mathcal{H}_{\xi,\lambda} \subset C^{-\infty}_{\xi,\lambda},$$

and $\langle\,\cdot\,,\,\cdot\,\rangle$ makes sense as a sesquilinear pairing between the spaces $C_{\xi,\lambda}^{-\infty}$ and $C_{\xi,\lambda}^{\infty}$. In particular, it makes sense to form the matrix element $\langle\pi_{\xi,\lambda}(x)u,v\rangle$ with $u \in C_{\xi,\lambda}^{-\infty}$ and $v \in C_{\xi,\lambda}^{\infty}$. The elements of $C_{\xi,\lambda}^{\infty}$ and $C_{\xi,\lambda}^{-\infty}$ are exactly the smooth and the generalized vectors for the representation $\pi_{\xi,\lambda}$.

We have thus described the representations that occur in the most continuous part of $L^2(G/H)$. The next issue is the H-fixed (generalized) vector u that will be used to form the matrix element by which the appropriate spherical functions are defined. It is an element of $C_{\xi,\lambda}^{-\infty}$. We will not give the details of the construction, but only mention that it requires the extension of the definition of $C_{\xi,\lambda}^{-\infty}$ to $\lambda \in \mathfrak{a}_{\mathrm{q}\mathbb{C}}^*$, the space of complex linear functionals on $\mathfrak{a}_{\mathrm{q}}$. The construction of the H-fixed vector then involves analytic continuation with respect to λ. See [6] and [8] for details. We note that in contrast to the cases of G/K and G, the space of H-fixed (generalized) vectors for a given irreducible representation can have dimension > 1. This means that in the decomposition of $L^2(G/H)$ there can (and sometimes will) occur multiplicities. However, the multiplicities are finite (see [4, 12]).

In order to proceed with the definition of the generalized spherical functions that will be used in the harmonic analysis, one further observation is needed. A vector v in a representation space $(\pi, \mathcal{H})$ for G is called K-finite if the vectors $\pi(k)v$, where $k \in K$, span a finite dimensional subspace of $\mathcal{H}$. The space of K-finite vectors in $L^2(G/H)$ is dense, and hence, in order to describe the Plancherel decomposition it suffices to consider K-finite functions on G/H. In fact, we shall fix a finite dimensional representation of K, and consider only such functions on G/H, whose left translates span a space on which the action of K is equivalent with the fixed representation. Let (τ, V_τ) be a fixed finite dimensional representation of K. Then for convenience, we shall actually move our focus from scalar valued functions on G/H to V_τ-valued functions f that satisfy $f(kx) = \tau(k)f(x)$ for $k \in K$, $x \in G/H$. Such functions are called τ-spherical. It is not difficult to see that there are natural maps allowing this change (see [13, Lemma 5]). Thus, our generalized spherical functions will be τ-spherical functions, and the Fourier transform will be defined for τ-spherical functions. The space of τ-spherical square integrable functions on G/H is denoted $L^2(G/H : \tau)$.

Let (τ, V_τ) be as in the previous paragraph. The generalized spherical function to be defined will depend linearly on an element $\psi \in V_\tau^{K \cap H \cap M}$, the space of $K \cap H \cap M$-fixed vectors in V_τ, and it will depend meromorphically on a parameter $\lambda \in \mathfrak{a}_{\mathrm{q}\mathbb{C}}^*$. Given ψ and λ we define the following V_τ-valued function $\tilde{\psi}_\lambda$ on G. The value of $\tilde{\psi}_\lambda$ at x is to be $a^{\lambda+\rho}\tau(m)\psi$ if $x = manh$ belongs to the product $PH = MANH$, and 0 otherwise. It can be shown that PH is an open subset of G. We then define

$$E(\psi : \lambda)(x) = \int_K \tau(k)\tilde{\psi}_\lambda(k^{-1}x)\,dk \in V_\tau. \qquad (2)$$

Then $E(\psi : \lambda)$ is a τ-spherical function on G/H (by the invariance of the Haar measure on K) which is called an *Eisenstein integral*. In fact, the integral in (2) only converges for λ in a certain region of $\mathfrak{a}_{q\mathbb{C}}^*$, and to give a proper definition one must allow analytic continuation to $\mathfrak{a}_{q\mathbb{C}}^*$. The Eisenstein integrals are our basic generalized spherical functions, and they are the corner stones of the harmonic analysis on G/H. For details, see [7] and [13]. In particular, the components of the vector valued Eisenstein integrals are indeed (linear combinations of) matrix coefficients of the $\pi_{\xi,\lambda}$, with vectors $u \in (C_{\xi,\lambda}^{-\infty})^H$ and $v \in C_{\xi,\lambda}^\infty$ cf. [13, eq. (25)]. Thus, the integral over K in (2) is essentially the same as that in the definition of the pairing between $C_{\xi,\lambda}^{-\infty}$ and $C_{\xi,\lambda}^\infty$. In fact, a somewhat more general construction is possible, in which one considers all the open subsets of G of the form PxH, where $x \in G$. These open subsets are finite in number, and can be parametrized by taking x in a suitable subset $\mathcal{W}$ of K. Then the parameter ψ will be taken in a finite dimensional Hilbert space $^\circ\mathcal{C}$ which contains $V_\tau^{K \cap H \cap M}$ as a subspace. We will not give its definition here, but refer to [13, Sect. 2]. The notation $^\circ\mathcal{C}$ will nevertheless be used freely in the sequel. The reader may think of it as the space $V_\tau^{K \cap H \cap M}$, to which it is equal under the simplifying assumption on G/H that PH is dense. The Eisenstein integral $E(\psi : \lambda)(x)$ depends linearly on $\psi \in {}^\circ\mathcal{C}$. We write $E(\lambda)$ for the corresponding $\mathrm{Hom}(^\circ\mathcal{C}, V_\tau)$-valued function on G/H, and $E(\lambda : x)$ for its value at x.

When G/H is a Riemannian symmetric space, the Eisenstein integrals are equal to the generalized spherical functions of [40, Ch. III, 2], and when G/H is a semisimple Lie group viewed as a symmetric space, they are multiples of the Eisenstein integrals introduced by Harish-Chandra.

There is, however, one serious problem associated with the application of these Eisenstein integrals to the harmonic analysis. This problem does not exist seriously in any of the cases G/K and G. The Eisenstein integrals are constructed as meromorphic functions in the parameter $\lambda \in \mathfrak{a}_{q\mathbb{C}}^*$, but they will be used eventually for purely imaginary functionals λ. These are the values of λ for which $\pi_{\xi,\lambda}$ is unitary, and it is only these representations that contribute in the decomposition of $L_{\mathrm{mc}}^2(G/H)$. However, the meromorphic function $\lambda \mapsto E(\lambda : x)$ may have (and in general, has) singularities on $i\mathfrak{a}_q^*$. This problem is handled in [13] where a suitable *normalized Eisenstein integral* $E^\circ(\lambda : x) = E(\lambda : x) \circ C^{-1}(\lambda)$ is introduced. The normalization factor $C(\lambda)$, which is a linear operator on $^\circ\mathcal{C}$ depending meromorphically on λ, is determined from the asymptotic behaviour of $E(\lambda)$ in a particular direction. The normalized Eisenstein integrals are regular on $i\mathfrak{a}_q^*$, see [13, Thm. 2]. See also [9] for a more general result, with a different proof and valid for the intermediate series as well.

The Fourier transform $\mathcal{F}f$ of a τ-spherical function f on G/H, say continuous with compact support, is now defined as follows. For an element $A \in \mathrm{Hom}(^\circ\mathcal{C}, V_\tau)$ we denote by A^* the adjoint operator in $\mathrm{Hom}(V_\tau, {}^\circ\mathcal{C})$, defined with respect to the inner products of these finite dimensional spaces. Then

$$\mathcal{F}f(\lambda) = \int_{G/H} E^\circ(-\bar\lambda : x)^* f(x)\, dx \in {}^\circ\mathcal{C}, \tag{3}$$

where dx is the invariant measure on G/H. Notice that $\mathcal{F}f(\lambda)$ is a meromorphic function of $\lambda \in \mathfrak{a}_{\mathrm{q}\mathbb{C}}^*$.

As in the ordinary Fourier theory on $\mathbb{R}$, the transpose of the Fourier transform plays an important role in harmonic analysis. The transpose is the map that forms *wave packets* out of functions in λ. More precisely, let $\mathcal{S}(i\mathfrak{a}_{\mathrm{q}}^*)$ denote the space of Schwartz functions on the Euclidean space $i\mathfrak{a}_{\mathrm{q}}^*$. For $\varphi \in \mathcal{S}(i\mathfrak{a}_{\mathrm{q}}^*) \otimes {}^\circ\mathcal{C}$ we define

$$\mathcal{J}\varphi(x) = \int_{i\mathfrak{a}_{\mathrm{q}}^*} E^\circ(\lambda : x)\varphi(\lambda)\, d\lambda \in V_\tau \tag{4}$$

where $d\lambda$ is Lebesgue measure on $i\mathfrak{a}_{\mathrm{q}}^*$, suitably normalized. The function $\mathcal{J}\varphi$ on G/H is obviously τ-spherical. That $\mathcal{J}$ is the transpose of $\mathcal{F}$, in the sense that $\langle \mathcal{F}f, \varphi\rangle = \langle f, \mathcal{J}\varphi\rangle$ for f and φ as above, where the inner products are those of L^2 on $i\mathfrak{a}_{\mathrm{q}}^*$ and G/H, respectively, is essentially an application of Fubini's theorem. However, it requires estimates to see that the integral in (4) converges and produces an element of $L^2(G/H)$. See [15, p. 301].

We can now state what is essentially the Plancherel theorem for $L^2_{\mathrm{mc}}(G/H)$. We define the space $L^2_{\mathrm{mc}}(G/H : \tau)$ as the closure in $L^2(G/H : \tau)$ of the space of all wave packets $\mathcal{J}\varphi$, with φ as above. In order to avoid a constant in the last statement of the following theorem, we assume all measures to be suitably normalized. (The precise normalizations are given in [15, p. 278].)

Theorem 2.1. ([15, Cor. 17.5]) *The Fourier transform $f \mapsto \mathcal{F}f$ allows a unique extension to a continuous linear map $\mathcal{F}\colon L^2(G/H : \tau) \to L^2(i\mathfrak{a}_{\mathrm{q}}^*) \otimes {}^\circ\mathcal{C}$, whose kernel is the orthocomplement of $L^2_{\mathrm{mc}}(G/H : \tau)$, and whose restriction to the latter space is an isometry. Likewise, the wave packet transform $\varphi \mapsto \mathcal{J}\varphi$ allows a unique extension to a continuous linear map $\mathcal{J}\colon L^2(i\mathfrak{a}_{\mathrm{q}}^*) \otimes {}^\circ\mathcal{C} \to L^2_{\mathrm{mc}}(G/H : \tau)$. The map $\mathcal{J}\mathcal{F}$ is the orthogonal projection of $L^2(G/H : \tau)$ onto $L^2_{\mathrm{mc}}(G/H : \tau)$.*

The proof of Theorem 2.1 involves a detailed analysis of the algebra $\mathbb{D}(G/H)$ of invariant differential operators. Results of [3] and [11] are used. A central step consists of showing the existence of an element $D \in \mathbb{D}(G/H)$, depending on τ, which acts injectively on $C_c^\infty(G/H : \tau)$ and which satisfies

$$D\mathcal{J}\mathcal{F}f = Df \tag{5}$$

for all $f \in C_c^\infty(G/H : \tau)$. The orthocomplement of $L^2_{\mathrm{mc}}(G/H : \tau)$ in $L^2(G/H : \tau)$ is then identified as the kernel of D (acting on the L^2-space in the distribution sense).

3. Fourier Inversion and Paley–Wiener Theorem

In addition to Theorem 2.1 it is also proved in [15] (see Thm. 15.1) that the Fourier transform $\mathcal{F}$ is injective, when restricted to the space $C_c^\infty(G/H : \tau)$ of smooth, compactly supported τ-spherical functions. It is therefore an obvious problem to retrieve f from $\mathcal{F}f$. Moreover, it is natural to seek a description of the image $\mathcal{F}(C_c^\infty(G/H : \tau))$. We will now discuss the solution to these problems, the details are to be found in [17] and [19].

The inversion formula for the Fourier transform on $\mathbb{R}$ is the formula $f(x) = \mathcal{J}\mathcal{F}f(x)$. It is valid for all x when $f \in C_c^\infty(\mathbb{R})$ (of course, weaker hypotheses are sufficient). The same holds true for the non-compact Riemannian symmetric spaces G/K, and also for certain other semisimple symmetric spaces, for which $L^2_{\mathrm{mc}}(G/H) = L^2(G/H)$ (see [17, Sect. 11]). However, it is false when $L^2_{\mathrm{mc}}(G/H)$ is a proper subspace of $L^2(G/H)$, since a contradiction would arise from the fact that $\mathcal{J}\mathcal{F}f \in L^2_{\mathrm{mc}}(G/H)$ combined with the density of $C_c^\infty(G/H)$ in $L^2(G/H)$. The function $\mathcal{J}\mathcal{F}f \in L^2_{\mathrm{mc}}(G/H)$ is smooth, but it need not have compact support.

In order to explain our inversion formula, let us consider again a non-compact Riemannian symmetric space G/K, with the trivial K-type $\tau = 1$. In this case $C_c^\infty(G/K : 1) = C_c^\infty(K\backslash G/K)$ and $E^\circ(\lambda : x) = c(\lambda)^{-1}\varphi_\lambda(x)$, where $\varphi_\lambda(x)$ is the spherical function and $c(\lambda)$ is Harish-Chandra's c-function. The inversion formula $f(x) = \mathcal{J}\mathcal{F}f(x)$ was established by Harish-Chandra. It takes the form

$$f(x) = \int_{i\mathfrak{a}_0^*} \varphi_\lambda(x)\mathcal{F}f(\lambda)c(\lambda)^{-1}\,d\lambda. \tag{6}$$

Note that the usual Plancherel measure $|c(\lambda)|^{-2}\,d\lambda$ is hidden by the fact that $\mathcal{F}f(\lambda)$ equals $1/c(-\bar\lambda)^*$ times the usual spherical Fourier transform $\tilde{f}(\lambda)$; the star denotes complex conjugation. A simpler proof of this formula was later given by J. Rosenberg, [46], based on a part of Helgason's proof of the Paley–Wiener theorem (see [39, Ch. IV, 7]). We will now discuss a part of this proof, since its generalization to G/H is used in the statement of the inversion formula.

In Helgason's Paley–Wiener argument, one exploits the Harish-Chandra expansion formula

$$\varphi_\lambda(a) = \sum_{w\in W} c(w\lambda)\Phi_{w\lambda}(a)\,, \tag{7}$$

which is valid for a in the positive chamber A^+ associated with P_0. The function Φ_λ on A^+ is given by a series expansion so that $\Phi_\lambda(a) \sim a^{\lambda-\rho}$ as $a \to \infty$ in A^+. Furthermore, one uses the relations

$$c(\lambda)c(-\bar\lambda)^* = c(w\lambda)c(-w\bar\lambda)^*\,. \tag{8}$$

Insertion of (7) in the right-hand side of (6), and use of (8) together with the W-invariance of $\tilde{f}$ yields the expression

$$\mathcal{J}\mathcal{F}f(a) = |W| \int_{i\mathfrak{a}_0^*} \Phi_\lambda(a)\mathcal{F}f(\lambda)\,d\lambda. \tag{9}$$

(It requires some estimates to see that the latter integral converges.) By Cauchy's theorem one can shift the domain of integration in (9) from $i\mathfrak{a}_0^*$ to $\eta + i\mathfrak{a}_0^*$, where $\eta \in \mathfrak{a}_0^*$ is antidominant (i.e., $\eta(Y) \leq 0$ for all $Y \in \mathfrak{a}_0^+$). The idea is to let η pass to infinity, which will allow one to obtain an estimate showing that $\mathcal{J}\mathcal{F}f$ is compactly supported, as the first step towards the equality with f. Here we will, however, leave Rosenberg's proof and return to the general case of G/H.

In the general case of G/H, there exists for the Eisenstein integral an expansion formula as (7), see [14, Thm. 11.1]. The c-functions are generalized, and there are identities generalizing (8). In fact, these are the Maass–Selberg relations, mentioned in the introduction. For the most continuous series, these identities are established in [7, Thm. 16.3] (see also [8]). By arguments similar to the above, one can thus rewrite the wave packet (4) in a form similar to (9), provided x belongs to the open dense subset $(G/H)_+ = \cup_{w \in W} K A_{\mathrm{q}}^+ w H$ of G/H (for details, see [10, p. 211], or [17, Sect. 4]). However, if one wants to perform the shift by Cauchy's theorem as before, one must take into account that the integrand is no longer holomorphic in λ. This is due to the fact that the Eisenstein integrals are defined by an analytic continuation that only produces a meromorphic function. Thus, the shifted integral differs from the wave packet by some residual terms. When η is moved sufficiently far in the antidominant direction, the integrand becomes holomorphic, and no more residual terms are produced. From then on, the integral is independent of η. We call the sufficiently far shifted integral a *pseudo wave packet*, and denote it by $\mathcal{T}\mathcal{F}f$. The inversion formula for the Fourier transform on $C_c^\infty(G/H : \tau)$ can now be stated as follows.

Theorem 3.1. ([17, Thm. 4.7]) *Let $f \in C_c^\infty(G/H : \tau)$ and let $x \in (G/H)_+$. Then*
$$\mathcal{T}\mathcal{F}f(x) = f(x).$$

Since by Theorem 2.1 the wave packet $\mathcal{J}\mathcal{F}f$ is the projection of f onto the space $L^2_{\mathrm{mc}}(G/H : \tau)$, it follows from Theorem 3.1 that the difference between $\mathcal{T}\mathcal{F}f$ and $\mathcal{J}\mathcal{F}f$ is the projection of f onto $L^2_{\mathrm{mc}}(G/H : \tau)^\perp$. In the process described above, this projection is thus exhibited as the sum of the residual terms.

The main result of [16] describes a suitable grouping of the residual terms leading to an expression of the form
$$\mathcal{T}\varphi = \sum_{F \subset \Delta} \mathcal{T}_F \varphi, \tag{10}$$

with $\varphi = \mathcal{F}f$ and with $\mathcal{T}_\emptyset \varphi = \mathcal{J}\varphi$. Here $\Delta \subset \mathfrak{a}_{\mathrm{q}}^*$ is the set of simple roots for the root system of $\mathfrak{a}_{\mathrm{q}}$ in $\mathfrak{g}$, determined by the parabolic subgroup P, and $\mathcal{T}_F \varphi \in C^\infty((G/H)_+ : \tau)$ for each $F \subset \Delta$. A crucial part of the proof of Theorem 3.1 consists of showing that the functions $\mathcal{T}_F \mathcal{F}f$ extend smoothly to G/H for $F \subsetneq \Delta$. This involves the analogue of the theorem for a lower dimensional symmetric space determined by F. It is thus possible to proceed by induction. The proof of the theorem is then completed by using (5). Notice that as a consequence of the theorem $\mathcal{T}\mathcal{F}f$ extends smoothly to G/H. Hence by (10) also $\mathcal{T}_\Delta \mathcal{F}f$ extends.

Closely related to Theorem 3.1 is the Paley–Wiener theorem, which was conjectured in [15, Rem. 21.8], and proved there for the case that $\dim \mathfrak{a}_{\mathrm{q}} = 1$. Its formulation requires the introduction of a Paley–Wiener space $\mathrm{PW}(G/H : \tau)$, whose precise definition is given in [15, Def. 21.6], see also the forthcoming paper [19]. Here we restrict ourselves to mentioning that $\mathrm{PW}(G/H : \tau)$ consists of meromorphic $°\mathcal{C}$-valued functions on $\mathfrak{a}_{q\mathbb{C}}^*$, satisfying (a) suitable estimates of Paley–Wiener–Schwartz type and (b) the so-called Arthur–Campoli relations. The latter are defined as all the relations of a particular type which hold for all the functions $\lambda \mapsto E°(-\bar{\lambda} : x)^* v$, for $x \in G/H$, $v \in V_\tau$. They generalize the relations used in [2].

Theorem 3.2. (Paley–Wiener theorem, [19]) *The Fourier transform $\mathcal{F}$ is a linear isomorphism from $C_c^\infty(G/H : \tau)$ onto* $\mathrm{PW}(G/H : \tau)$.

The Fourier transform maps injectively into the Paley–Wiener space, so the problem is to establish surjectivity. The proof of this is based on the inversion formula in Theorem 3.1. If $\varphi \in \mathrm{PW}(G/H : \tau)$, then the pseudo-wave packet $\mathcal{T}\varphi$ is a well-defined smooth function on $(G/H)_+$. According to Theorem 3.1, this function is the natural candidate for the inverse image of φ. By the Paley–Wiener shift argument it is seen that the support of $\mathcal{T}\varphi$ is contained in a compact subset of G/H. The main difficulty of the proof is to show that $\mathcal{T}\varphi$ admits a smooth extension to G/H. This is established by observing first that (10) is valid by residue calculus. It is then shown that $\mathcal{T}_F\varphi$ extends smoothly, for each $F \subset \Delta$. For $F = \Delta$ this is a relatively simple consequence of the Arthur–Campoli relations, since the smooth extension is known to hold when φ is already of the form $\mathcal{F}f$. For $F \subsetneq \Delta$ it again involves induction from a lower dimensional symmetric space determined by F. An important ingredient of the proof is that the Arthur–Campoli relations for that space can be induced up to relations for G/H. This principle, which we call induction of relations, is based on a detailed study of asymptotic expansions of meromorphic families of eigenfunctions in [18]. Once it has been established that $\mathcal{T}\varphi \in C_c(G/H : \tau)$, the identity $\mathcal{F}\mathcal{T}\varphi = \varphi$ follows by means of (5).

Our Paley–Wiener theorem generalizes that of Arthur [2] for the group case. However, our approach differs from Arthur's in a number of ways. Firstly, Arthur's result involves unnormalized Eisenstein integrals whereas ours involves normalized ones. Nevertheless, the associated Paley–Wiener theorems are equivalent. Next, Arthur uses residue calculus in the spirit of Langlands' work [43], but at the same time invokes Harish-Chandra's Plancherel theorem for the group. We use the residue calculus to also establish the Plancherel theorem, which is analogous to Langlands' approach. Finally, [2] relies on a lifting principle of Casselman, the proof of which has not been published. In [18] we establish a normalized version of Casselman's principle in close connection with the above mentioned induction of relations.

4. Plancherel Decomposition

Theorem 3.1 is also the starting point for our derivation of the Plancherel formula. In order to simplify the presentation, we will from now on assume that PH is the only open P-orbit on G/H so that the set $\mathcal{W}$ discussed in Section 2 has only one element. We refer to [19] for the general case.

In the residue calculus leading to (10) an important role is played by the concept of a residue weight, as introduced in [16, Sect. 3.2]. It is a certain function on the set of all $\sigma\theta$-stable parabolic subgroups containing A_q, and its purpose is to assign weights according to which residual contributions are counted.

Let W denote the Weyl group of the root system on $\mathfrak{a}_q$ generated by Δ. If $F \subset \Delta$, let $\mathfrak{a}_{Fq}$ be the intersection in $\mathfrak{a}_q$ of the root hyperplanes for the roots from F, and let W_F denote the centralizer of $\mathfrak{a}_{Fq}$ in W. The set F determines a unique $\sigma\theta$-stable parabolic subgroup P_F containing P.

If t is any choice of a residue weight, then from the mentioned procedure of taking residues it follows that, for $f \in C_c^\infty(G/H : \tau)$, $F \subset \Delta$,

$$\mathcal{T}_F \mathcal{F} f(x) = t(P_F)\,|W| \int_{\epsilon_F + i\mathfrak{a}_{Fq}^*} \int_{G/H} K_F(\nu : x : y) f(y)\, dy\, d\nu\,, \qquad (11)$$

with $K_F(\nu : \cdot : \cdot)$ a smooth $\mathrm{End}(V_\tau)$-valued function on $G/H \times G/H$, depending meromorphically on $\nu \in \mathfrak{a}_{Fq\mathbb{C}}^*$. This function turns out to be independent of the particular choice of the residue weight. In the above formula, $d\nu$ is a translate of suitably normalized Lebesgue measure. Finally, ϵ_F is a sufficiently close approximation of the origin in the $\Delta \setminus F$-positive chamber of $\mathfrak{a}_{Fq}^*$; it enters here in order to avoid singularities of the function $\nu \mapsto K_F(\nu : \cdot : \cdot)$ that may possibly lie in the space $i\mathfrak{a}_{Fq}^*$.

From its asymptotic behaviour it is deduced in [19] that the *kernel function* K_F can be expressed as

$$K_F(\nu : x : y) = \frac{1}{|W_F|}\, E_F^\circ(\nu : x) E_F^\circ(-\bar\nu : y)^*\,, \qquad (12)$$

where $E_F^\circ(\nu : x) \in \mathrm{Hom}(\mathcal{A}_{2,F}, V_\tau)$ depends smoothly on $x \in G/H$ and meromorphically on ν. Here $\mathcal{A}_{2,F}$ is a finite dimensional Hilbert space, built from the discrete series representations of the lower dimensional symmetric space determined by F. More precisely, let $P_F = M_F A_F N_F$ be the Langlands decomposition of the parabolic subgroup determined by F. Then $\mathcal{A}_{2,F} = L_d^2(M_F/M_F \cap H : \tau)$, where L_d^2 denotes the discrete part of L^2. It follows in the course of the proof that the space $L_d^2(M_F/M_F \cap H : \tau)$ is finite dimensional, i.e., there are only finitely many discrete series representations that admit the existence of τ-spherical functions (for the given K-type τ). For $F = \emptyset$ we have, in particular, that $\mathcal{A}_{2,\emptyset} \simeq {}^\circ\mathcal{C}$. On the other hand, for $F = \Delta$ and if $\mathfrak{a}_{\Delta q} = \{0\}$, we have $\mathcal{A}_{2,\Delta} = L_d^2(G/H : \tau)$, and $|W| K_\Delta(0)$ turns out to be the integral kernel of the orthogonal projection from $L^2(G/H : \tau)$ onto $L_d^2(G/H : \tau)$.

The functions $E_F^\circ(\nu : x)$ are called normalized Eisenstein integrals determined by the (standard) parabolic subgroup P_F. At a later stage they will appear to be (linear combinations of) matrix coefficients for the generalized principal series determined by P_F. The Eisenstein integrals allow converging series expansions which describe their asymptotic behaviour towards infinity. The highest order asymptotic behaviour is determined by the following asymptotic expression along sets of the form mA_{Fq}^+, as $m \in M_F/M_F \cap H$. Let $\psi \in \mathcal{A}_{2,F}$, then

$$a^{\rho_F} E_F^\circ(\nu : ma)\psi \sim \sum_{s \in W(\mathfrak{a}_{Fq})} a^{s\nu} C_{P_F|P_F}^\circ(s : \nu)\psi(m). \qquad (13)$$

Here $W(\mathfrak{a}_{Fq}) = W_F^*/W_F$, with W_F^* the normalizer of $\mathfrak{a}_{Fq}$ in the Weyl group W. Moreover, $C_{P_F|P_F}^\circ(s : \nu) \in \text{End}(\mathcal{A}_{2,F})$ depends meromorphically on $\nu \in \mathfrak{a}_{Fq\mathbb{C}}^*$. We require that $C_{P_F|P_F}^\circ(1 : \nu)$ is the identity operator for all ν; this requirement of normalization, which can be accomplished, together with (12) determines the Eisenstein integral completely. The expression on the right-hand side of (13) is called the *constant term* of the Eisenstein integral (along P_F).

As in Harish-Chandra's theory for the group, a key role is played by the following Maass–Selberg relations for the (matrix-valued) C-functions

$$C_{P_F|P_F}^\circ(s : \nu)C_{P_F|P_F}^\circ(s : -\bar{\nu})^* = I_{\mathcal{A}_{2,F}}.$$

We derive these relations from $W(\mathfrak{a}_{Fq})$-invariance of the function $\nu \mapsto K_F(\nu)$, which in turn is inherited from a similar invariance of the kernel $K_\emptyset$, in view of W-equivariance of our residue calculus. In this fashion, Maass–Selberg relations for the generalized principal series are deduced from the Maass–Selberg relations in the most continuous case (see Section 3) by means of the residue calculus. The Maass–Selberg relations imply that the constant term on the right-hand side of (13) is regular on the set $i\mathfrak{a}_{Fq}^*$, as a meromorphic function of ν. From the theory of the constant term in [22] and from the results of [9], both of which papers generalize similar results of Harish-Chandra for the group case, it then follows that also the Eisenstein integral on the left-hand side of (13) is regular on $i\mathfrak{a}_{Fq}^*$. It should be mentioned that in the application of the theory of the constant term, it is of crucial importance that all the discrete series representations have an infinitesimal $\mathbb{D}(G/H)$-character which is real and regular. This result has been established in [45]. Beyond this, no further information on the discrete series is needed.

The regularity of the Eisenstein integral, together with suitable estimates, allow us to replace ϵ_F in (11) by the limit value zero. Thus we arrive at wave packets over the sets $i\mathfrak{a}_{Fq}^*$, where the Plancherel measure will ultimately turn out to be supported. The estimates mentioned above are of a uniform nature in both ν and x. They are obtained via a detailed analysis of the differential equations satisfied by the Eisenstein integral in [7] and [19]. To conclude that the Eisenstein integrals have tempered behaviour in the variable x we use information on the location of the residues involved in the definition of K_F, see [16] and [17].

The estimates for the Eisenstein integrals also allow us to define a Fourier transform $\mathcal{F}_F \colon C_c^\infty(G/H : \tau) \to \mathcal{S}(i\mathfrak{a}_{Fq}^*) \otimes \mathcal{A}_{2,F}$ and an adjoint wave packet transform $\mathcal{J}_F \colon \mathcal{S}(i\mathfrak{a}_{Fq}^*) \otimes \mathcal{A}_{2,F} \to L^2(G/H : \tau)$ by the formulas (3) and (4) with $\mathfrak{a}_{Fq}$ in place of $\mathfrak{a}_q$ and with E_F° in place of E°.

Formulas (10), (11) with $\epsilon_F = 0$ for all $F \subset \Delta$, and (12) now lead to the formula

$$I = \sideset{}{'}\sum_{F \subset \Delta} t(P_F)\,[W : W_F]\,\mathcal{J}_F \circ \mathcal{F}_F \qquad \text{on} \quad C_c^\infty(G/H : \tau). \tag{14}$$

Define the equivalence relation $\sim$ on the collection $\mathcal{P}(\Delta)$ of subsets of Δ by $F \sim F'$ if and only if $\mathfrak{a}_{F\mathrm{q}}$ and $\mathfrak{a}_{F'\mathrm{q}}$ are conjugate under the Weyl group. The following result is essentially the Plancherel formula for $L^2(G/H)$.

Theorem 4.1. *If $F \subset \Delta$, the Fourier transform $\mathcal{F}_F$ has a unique extension to a continuous linear operator $L^2(G/H : \tau) \to L^2(i\mathfrak{a}_{F\mathrm{q}}^*) \otimes \mathcal{A}_{2,F}$. Moreover, the wave packet transform $\mathcal{J}_F$ has a unique extension to a continuous linear operator $L^2(i\mathfrak{a}_{F\mathrm{q}}^*) \otimes \mathcal{A}_{2,F} \to L^2(G/H : \tau)$. The operator $\mathcal{J}_F \circ \mathcal{F}_F$ only depends on the class of F in $\mathcal{P}(\Delta)/\sim$; its extension to $L^2(G/H : \tau)$ is $[W : W_F^*]^{-1}$ times the orthogonal projection onto a closed subspace $L^2_{[F]}(G/H : \tau)$. Finally,*

$$L^2(G/H : \tau) = \bigoplus_{[F] \in \mathcal{P}(\Delta)/\sim} L^2_{[F]}(G/H : \tau), \tag{15}$$

with summands that are mutually orthogonal.

We sketch a few important ingredients of the proof. From the Maass–Selberg relations combined with a spectral analysis it is deduced that $\mathcal{J}_F \circ \mathcal{F}_F$ only depends on F through its class in $\mathcal{P}(\Delta)/\sim$; moreover, if $F, F' \subset \Delta$ are not equivalent, then the images of $\mathcal{J}_F \circ \mathcal{F}_F$ and $\mathcal{J}_{F'} \circ \mathcal{F}_{F'}$ are orthogonal. Thus, the identity (14) gives rise to a decomposition of the form (15), where the projection onto $L^2_{[F]}(G/H : \tau)$ is given by the part of the sum in (14) ranging over the elements of the class $[F]$. This part equals $\mathcal{J}_F \circ \mathcal{F}_F$ multiplied by the constant

$$[W : W_F] \sum_{F' \in [F]} t(P_{F'}) = [W : W_F^*],$$

where the equality follows from a straightforward counting argument.

This final argument clarifies the role of the residue weight in the harmonic analysis. It determines the distribution of the projection onto $L^2_{[F]}(G/H : \tau)$ over the elements in the class $[F]$.

In the above discussion, the details of which will appear in [19], Theorem 4.1 has been derived from Theorem 2.1 by a residual calculus, without reference to representation theory. In particular, the Eisenstein integrals $E_F^\circ(\nu)$, for $F \neq \emptyset$, are introduced as residues and not as matrix coefficients of generalized principal series representations. In [19] it is shown that the Eisenstein integral $E_F^\circ(\nu : \cdot)$ indeed is a (generalized) matrix coefficient for the generalized principal series obtained by induction from P_F. This is achieved by an asymptotic analysis in the spirit of the proof of the subrepresentation theorem, see [26]. This then allows us to identify the decomposition described in Theorem 4.1 as the τ-spherical part of the Plancherel decomposition of $L^2(G/H)$ in the sense of representation theory. The fact that the summation in (15) ranges over classes then corresponds to the fact that standard intertwining operators connect the generalized principal series obtained by induction from $P_{F'}$, for $F' \in [F]$.

Delorme's proof in [29] follows a completely different road towards the Plancherel formula. In [27] the normalized Eisenstein integrals are *introduced* as matrix coefficients of generalized principal series representations. Moreover, in the same paper their temperedness is established by an argument involving the use of translation functors. The Maass–Selberg relations for the Eisenstein integrals are established in [24] by the so-called truncation method, see [28], which is inspired by a method used in [1]. In [24] it is then shown that the sum on the right-hand side of (15) is orthogonal. Finally, in [29] it is proved that the orthocomplement of the mentioned sum in $L^2(G/H : \tau)$ is trivial by combining the method of truncation with ideas of [21] that lead to an a priori proof that the Plancherel measure is supported on the representations that are tempered relative to G/H.

References

[1] J. Arthur, A local trace formula. Inst. Hautes Etudes Sci. Publ. Math. **73** (1991), 5–96.

[2] J. Arthur, A Paley–Wiener theorem for real reductive groups. Acta Math. **150** (1983), 1–89.

[3] E. P. van den Ban, A convexity theorem for semisimple symmetric spaces. Pac. J. Math. **124** (1986), 21–55.

[4] E. P. van den Ban, Invariant differential operators on a semisimple symmetric space and finite multiplicities in a Plancherel formula. Arkiv för Mat. **25** (1987), 175–187.

[5] E. P. van den Ban, Asymptotic behaviour of matrix coefficients related to reductive symmetric spaces. Indag. Math. **49** (1987), 225–249.

[6] E. P. van den Ban, The principal series for a reductive symmetric space I. H-fixed distribution vectors. Ann. Scient. Éc. Norm. Sup. **21** (1988), 359–412.

[7] E. P. van den Ban, The principal series for a reductive symmetric space II. Eisenstein integrals. J. Funct. Anal. **109** (1992), 331–441.

[8] E. P. van den Ban, The action of intertwining operators on H-fixed generalized vectors in the minimal principal series of a reductive symmetric space. Indag. Math. **8** (1997), 317–347.

[9] E. P. van den Ban, J. Carmona and P. Delorme, Paquets d'ondes dans l'espace de Schwartz d'un espace symétrique réductif. J. Funct. Anal. **139** (1996), 225–243.

[10] E. P. van den Ban, M. Flensted-Jensen and H. Schlichtkrull, Harmonic analysis on semisimple symmetric spaces: A survey of some general results. Proc. Symp. Pure Math. **61** (1997), 191–217.

[11] E. P. van den Ban and H. Schlichtkrull, Convexity for invariant differential operators on semisimple symmetric spaces. Compositio Math. **89** (1993), 301–313.

[12] E. P. van den Ban and H. Schlichtkrull, Multiplicities in the Plancherel decomposition for a semisimple symmetric space. In: Representation Theory of Groups and Algebras, 163–180. Contemp. Math. **145**, Amer. Math. Soc., 1993.

[13] E. P. van den Ban and H. Schlichtkrull, Fourier transforms on a semisimple symmetric space. Invent. Math. **130** (1997), 517–574.

[14] E. P. van den Ban and H. Schlichtkrull, Expansions for Eisenstein integrals on semisimple symmetric spaces. Arkiv för Mat. **35** (1997), 59–86.

[15] E. P. van den Ban and H. Schlichtkrull, The most continuous part of the Plancherel decomposition for a reductive symmetric space. Annals Math. **145** (1997), 267–364.

[16] E. P. van den Ban and H. Schlichtkrull, A residue calculus for root systems. Compositio Math. **123** (2000), 27–72.

[17] E. P. van den Ban and H. Schlichtkrull, Fourier inversion on a reductive symmetric space. Acta Math. **182** (1999), 25–85.

[18] E. P. van den Ban and H. Schlichtkrull, Analytic families of eigenfunctions on a reductive symmetric space. arXiv: math.RT/0104050.

[19] E. P. van den Ban and H. Schlichtkrull, The Paley-Wiener and the Plancherel theorem for a reductive symmetric space. In preparation.

[20] M. Berger, Les espaces symétriques non compactes. Ann. Sci. Ecole Norm. Sup. (3) **74** (1957), 85–177.

[21] J. N. Bernstein, On the support of the Plancherel measure. J. Geom. Phys. **5** (1988), 663–710.

[22] J. Carmona, Terme constant des fonctions tempérées sur un espace symétrique réductif. J. Reine Angew. Math. **491** (1997), 17–63. Erratum **507** (1999), 233.

[23] J. Carmona and P. Delorme, Base méromorphe de vecteurs distributions H-invariants pour les séries principales géneralisées d'espaces symétriques réductifs. Equation fonctionelle. J. Funct. Anal. **122** (1994), 152–221.

[24] J. Carmona and P. Delorme, Transformation Fourier pour les espaces symétriques réductifs. Invent. Math. **134** (1998), 59–99.

[25] É. Cartan, Sur la détermination d'un système orthogonal complet dans un espace de Riemann symétrique clos. Rend. Circ. Mat. Palermo **53** (1929), 217–252.

[26] W. Casselman and D. Milicić, Asymptotic behaviour of matrix coefficients of admissible representations. Duke Math. J. **49** (1982), 869–930.

[27] P. Delorme, Intégrales d'Eisenstein pour les espaces symétriques réductifs. Tempérance. Majorations. Petite matrice B. J. Funct. Anal. **136** (1996), 422–509.

[28] P. Delorme, Troncature pour les espaces symétriques réductifs. Acta Math. **179** (1997), 41–77.

[29] P. Delorme, Formule de Plancherel pour les espaces symétriques réductifs. Annals of Math. **147** (1998), 417–452.

[30] M. Flensted-Jensen, Discrete series for semisimple symmetric spaces. Annals of Math. **111** (1980), 253–311.

[31] P. Friedman, Langlands parameters of derived functor modules and Vogan diagrams. To appear.

[32] Harish-Chandra, Spherical functions on a semisimple Lie group, I, II. Amer. J. Math. **80** (1958), 241–310, 553–613.

[33] Harish-Chandra, Discrete series for semisimple Lie groups, I, II. Acta. Math. **113** (1965), 241–318, **116** (1966), 1–111.

[34] Harish-Chandra, Harmonic analysis on real reductive groups I. The theory of the constant term. J. Funct. Anal. **19** (1975), 104–204.

[35] Harish-Chandra, Harmonic analysis on real reductive groups II. Wave packets in the Schwartz space. Invent. Math. **36** (1976), 1–55.

[36] Harish-Chandra, Harmonic analysis on real reductive groups III. The Maass–Selberg relations and the Plancherel formula. Ann. Math. **104** (1976), 117–201.

[37] G. Heckman and E. Opdam, Yang's system of particles and Hecke algebras. Annals of Math. **145** (1997), 139–173.

[38] G. Heckman and H. Schlichtkrull, Harmonic analysis and special functions on symmetric spaces. Academic Press, 1994.

[39] S. Helgason, Groups and geometric analysis. Academic Press, 1984.

[40] S. Helgason, Geometric analysis on symmetric spaces. Amer. Math. Soc., 1994.

[41] L. Hörmander, The analysis of linear partial differential operators I. Springer Verlag, 1983.

[42] A. W. Knapp, Representation theory of semisimple groups. Princeton Univ. Press, 1986.

[43] R. P. Langlands, On the functional equations satisfied by Eisenstein series. Lecture Notes in Mathematics **544**, Springer-Verlag, 1976.

[44] T. Oshima, A realization of semisimple symmetric spaces and construction of boundary value maps. Adv. Studies in Pure Math. **14** (1988), 603–650.

[45] T. Oshima and T. Matsuki, A description of discrete series for semisimple symmetric spaces. Adv. Studies in Pure Math. **4** (1984), 331–390.

[46] J. Rosenberg, A quick proof of Harish-Chandra's Plancherel theorem for spherical functions on a semisimple Lie group. Proc. Amer. Math. Soc. **63** (1977), 143–149.

[47] H. Schlichtkrull, The Langlands parameters of Flensted–Jensen's discrete series for semisimple symmetric spaces. J. Funct. Anal. **50** (1983), 133–150.

[48] H. Schlichtkrull, Hyperfunctions and harmonic analysis on symmetric spaces. Birkhäuser, 1984.

[49] V. S. Varadarajan, An introduction to harmonic analysis on semisimple Lie groups. Cambridge Univ. Press, 1989.

[50] N. R. Wallach, Real reductive groups, I-II. Academic Press, 1988, 1992.

Erik van den Ban
Department of Mathematics
University of Utrecht
P.O. Box 80010
NL-3508 TA Utrecht, The Netherlands
E-mail address: ban@math.uu.nl

Henrik Schlichtkrull
Department of Mathematics
University of Copenhagen
Universitetsparken 5
DK-2100 Copenhagen Ø, Denmark
E-mail address: schlicht@math.ku.dk